Microeconometrics Using Stata

Revised Edition

Microeconometrics Using Stata

Revised Edition

A. COLIN CAMERON
Department of Economics
University of California
Davis, CA

PRAVIN K. TRIVEDI
Department of Economics
Indiana University
Bloomington, IN

A Stata Press Publication
StataCorp LP
College Station, Texas

Published by Stata Press, 4905 Lakeway Drive, College Station, Texas 77845
Typeset in LaTeX 2_ε
Printed in the United States of America

10 9 8 7 6 5 4 3 2

ISBN-10: 1-59718-073-4
ISBN-13: 978-1-59718-073-3

Contents

Tables

Figures

Preface to the Revised Edition

Microeconometrics Using Stata, published in December 2008, was written for Stata 10.1. The book incorporated version 10.1 additions to Stata 10.0, most notably, the new random-number generators.

In this revised edition, we present other additions to Stata 10 that appear for the first time in Stata 11. With few exceptions, we present these additions in a way that reproduces the results given in the first edition.

First, we introduce the new construct of factor variables. These provide a simple way to specify models with sets of indicator variables formed from a categorical variable and to specify models with interactions. Factor variables replace the `xi` prefix command. See especially section 1.3.4 and the end of section 2.4.7.

Second, we describe the new `margins` command for prediction and for computation of marginal effects in regression models. The `margins` command with options including the `dydx()` option replaces the Stata `mfx` command and the user-written `margeff` command. Additionally, the `margins` command when used in conjunction with factor variables can simplify computation of marginal effects in models with interactions. See sections 10.5 and 10.6, especially subsections 10.5.7 and 10.6.5. Throughout this revised edition, notably, in chapters 14–17, we replace `mfx` and `margeff` with the `margins` command.

In the first edition, we most often calculated the marginal effect at the mean (MEM), rather than the average marginal effect (AME), because the `mfx` command did not compute the AME. The new `margins` command can compute both the MEM and the AME. In this revised edition, we have endeavored to replicate the results given in the first edition. For that reason, we continue to most frequently calculate the MEM, though in practice, the AME is usually preferred.

Third, we describe the new `gmm` command for generalized method of moments and nonlinear instrumental-variables estimation. See sections 10.3.8 and 17.5.2.

Fourth, we present some minor changes that need to be made to the existing `ml` command when the `d1` and `d2` methods are used. These changes arise because the `ml` command is now a front-end to the new Mata `moptimize()` function. We also present the new `lf0`, `lf1`, and `lf2` methods. See section 11.6. The Mata `optimize()` v evaluator has been renamed to gf evaluator; see section 11.7.

We thank the Stata staff, especially Patricia Branton, David Drukker, Lisa Gilmore, Deirdre Patterson, and Brian Poi, for their assistance in preparing this revised edition.

Davis, CA   A. Colin Cameron
Bloomington, IN   Pravin K. Trivedi
January 2010

Preface to the First Edition

This book explains how an econometrics computer package, Stata, can be used to perform regression analysis of cross-section and panel data. The term microeconometrics is used in the book title because the applications are to economics-related data and because the coverage includes methods such as instrumental-variables regression that are emphasized more in economics than in some other areas of applied statistics. However, many issues, models, and methodologies discussed in this book are also relevant to other social sciences.

The main audience is graduate students and researchers. For them, this book can be used as an adjunct to our own *Microeconometrics: Methods and Applications* (Cameron and Trivedi 2005), as well as to other graduate-level texts such as Greene (2008) and Wooldridge (2002). By comparison to these books, we present little theory and instead emphasize practical aspects of implementation using Stata. More advanced topics we cover include quantile regression, weak instruments, nonlinear optimization, bootstrap methods, nonlinear panel-data methods, and Stata's matrix programming language, Mata.

At the same time, the book provides introductions to topics such as ordinary least-squares regression, instrumental-variables estimation, and logit and probit models so that it is suitable for use in an undergraduate econometrics class, as a complement to an appropriate undergraduate-level text. The following table suggests sections of the book for an introductory class, with the caveat that in places formulas are provided using matrix algebra.

Stata basics	Chapter 1.1–1.4
Data management	Chapter 2.1–2.4, 2.6
OLS	Chapter 3.1–3.6
Simulation	Chapter 4.6–4.7
GLS (heteroskedasticity)	Chapter 5.3
Instrumental variables	Chapter 6.2–6.3
Linear panel data	Chapter 8
Logit and probit models	Chapter 14.1–14.4
Tobit model	Chapter 16.1–16.3

Although we provide considerable detail on Stata, the treatment is by no means complete. In particular, we introduce various Stata commands but avoid detailed listing and description of commands as they are already well documented in the Stata manuals and online help. Typically, we provide a pointer and a brief discussion and often an example.

As much as possible, we provide template code that can be adapted to other problems. Keep in mind that to shorten output for this book, our examples use many fewer regressors than necessary for serious research. Our code often suppresses intermediate output that is important in actual research, because of extensive use of command `quietly` and options `nolog`, `nodots`, and `noheader`. And we minimize the use of graphs compared with typical use in exploratory data analysis.

We have used Stata 10, including Stata updates.[1] Instructions on how to obtain the datasets and the do-files used in this book are available on the Stata Press web site at http://www.stata-press.com/data/mus.html. Any corrections to the book will be documented at http://www.stata-press.com/books/mus.html.

We have learned a lot of econometrics, in addition to learning Stata, during this project. Indeed, we feel strongly that an effective learning tool for econometrics is hands-on learning by opening a Stata dataset and seeing the effect of using different methods and variations on the methods, such as using robust standard errors rather than default standard errors. This method is beneficial at all levels of ability in econometrics. Indeed, an efficient way of familiarizing yourself with Stata's leading features might be to execute the commands in a relevant chapter on your own dataset.

We thank the many people who have assisted us in preparing this book. The project grew out of our 2005 book, and we thank Scott Parris for his expert handling of that book. Juan Du, Qian Li, and Abhijit Ramalingam carefully read many of the book chapters. Discussions with John Daniels, Oscar Jorda, Guido Kuersteiner, and Doug Miller were particularly helpful. We thank Deirdre Patterson for her excellent editing and Lisa Gilmore for managing the LATEX formatting and production of this book. Most especially, we thank David Drukker for his extensive input and encouragement at all stages of this project, including a thorough reading and critique of the final draft, which led to many improvements in both the econometrics and Stata components of this book. Finally, we thank our respective families for making the inevitable sacrifices as we worked to bring this multiyear project to completion.

Davis, CA   A. Colin Cameron
Bloomington, IN   Pravin K. Trivedi
October 2008

---

1. To see whether you have the latest update, type `update query`. For those with earlier versions of Stata, some key changes are the following: Stata 9 introduced the matrix programming language, Mata. The syntax for Stata 10 uses the `vce(robust)` option rather than the `robust` option to obtain robust standard errors. A mid-2008 update of version 10 introduced new random-number functions, such as `runiform()` and `rnormal()`.

1 Stata basics

This chapter provides some of the basic information about issuing commands in Stata. Sections 1.1–1.3 enable a first-time user to begin using Stata interactively. In this book, we instead emphasize storing these commands in a text file, called a Stata do-file, that is then executed. This is presented in section 1.4. Sections 1.5–1.7 present more-advanced Stata material that might be skipped on a first reading.

The chapter concludes with a summary of some commonly used Stata commands and with a template do-file that demonstrates many of the tools introduced in this chapter. Chapters 2 and 3 then demonstrate many of the Stata commands and tools used in applied microeconometrics. Additional features of Stata are introduced throughout the book and in appendices A and B.

1.1 Interactive use

Interactive use means that Stata commands are initiated from within Stata.

A graphical user interface (GUI) for Stata is available. This enables almost all Stata commands to be selected from drop-down menus. Interactive use is then especially easy, as there is no need to know in advance the Stata command.

All implementations of Stata allow commands to be directly typed in; for example, entering `summarize` yields summary statistics for the current dataset. This is the primary way that Stata is used, as it is considerably faster than working through drop-down menus. Furthermore, for most analyses, the standard procedure is to aggregate the various commands needed into one file called a do-file (see section 1.4) that can be run with or without interactive use. We therefore provide little detail on the Stata GUI.

For new Stata users, we suggest entering Stata, usually by clicking on the Stata icon, opening one of the Stata example datasets, and doing some basic statistical analysis. To obtain example data, select **File > Example Datasets...**, meaning from the **File** menu, select the entry **Example Datasets...**. Then click on the link to **Example datasets installed with Stata**. Work with the dataset `auto.dta`; this is used in many of the introductory examples presented in the Stata documentation. First, select **describe** to obtain descriptions of the variables in the dataset. Second, select **use** to read the dataset into Stata. You can then obtain summary statistics either by typing `summarize` in the Command window or by selecting **Statistics > Summaries, tables, and tests > Summary and descriptive statistics > Summary statistics**. You can run a simple regression by typing `regress mpg weight` or by selecting **Statistics**

> **Linear models and related** > **Linear regression** and then using the drop-down
lists in the **Model** tab to choose `mpg` as the dependent variable and `weight` as the
independent variable.

The Stata manual [GS] *Getting Started with Stata* is very helpful, especially [GS] **1 Introducing Stata—sample session**, which uses typed-in commands, and [GS] **2 The
Stata user interface**.

The extent to which you use Stata in interactive mode is really a personal preference.
There are several reasons for at least occasionally using interactive mode. First, it can
be useful for learning how to use Stata. Second, it can be useful for exploratory analysis
of datasets because you can see in real time the effect of, for example, adding or dropping
regressors. If you do this, however, be sure to first start a session log file (see section 1.4)
that saves the commands and resulting output. Third, you can use `help` and related
commands to obtain online information about Stata commands. Fourth, one way to
implement the preferred method of running do-files is to use the Stata Do-file Editor in
interactive mode.

Finally, components of a given version of Stata, such as version 10, are periodically
updated. Entering `update query` determines the current update level and provides the
option to install official updates to Stata. You can also install user-written commands
in interactive mode once the relevant software is located using, for example, the `findit`
command.

1.2   Documentation

Stata documentation is extensive; you can find it in hard copy, in Stata (online), or on
the web.

1.2.1   Stata manuals

For first-time users, see [GS] *Getting Started with Stata*. The most useful manual
is [U] *User's Guide*. Entries within manuals are referred to using shorthand such as
[U] **11.1.4 in range**, which denotes section 11.1.4 of [U] *User's Guide* on the topic **in
range**.

Many commands are described in [R] *Base Reference Manual*, which spans three
volumes. For version 11, these are A–H, I–P, and Q–Z. Not all Stata commands appear here, however, because some appear instead in the appropriate topical manual.
These topical manuals are [D] *Data Management Reference Manual*, [G] *Graphics Reference Manual*, [M] *Mata Reference Manual* (two volumes), [MI] *Multiple-Imputation
Reference Manual*, [MV] *Multivariate Statistics Reference Manual*, [P] *Programming
Reference Manual*, [ST] *Survival Analysis and Epidemiological Tables Reference Manual*, [SVY] *Survey Data Reference Manual*, [TS] *Time-Series Reference Manual*, and [XT]
Longitudinal/Panel-Data Reference Manual. For example, the `generate` command appears in [D] **generate** rather than in [R].

For a complete list of documentation, see [U] **1 Read this—it will help** and also [I] *Quick Reference and Index.*

1.2.2   Additional Stata resources

The *Stata Journal* (SJ) and its predecessor, the *Stata Technical Bulletin* (STB), present examples and code that go beyond the current installation of Stata. SJ articles over three years old and all STB articles are available online from the Stata web site at no charge. You can find this material by using various Stata help commands given later in this section, and you can often install code as a free user-written command.

The Stata web site has a lot of information. This includes a summary of what Stata does. A good place to begin is http://www.stata.com/support/. In particular, see the answers to frequently asked questions (FAQs).

The University of California–Los Angeles web site http://www.ats.ucla.edu/STAT/stata/ provides many Stata tutorials.

1.2.3   The help command

Stata has extensive help available once you are in the program.

The `help` command is most useful if you already know the name of the command for which you need help. For example, for help on the `regress` command, type

```
. help regress
  (output omitted )
```

Note that here and elsewhere the dot (.) is not typed in but is provided to enable distinction between Stata commands (preceded by a dot) and subsequent Stata output, which appears with no dot.

The `help` command is also useful if you know the class of commands for which you need help. For example, for help on functions, type

```
. help function
  (output omitted )
```

(Continued on next page)

Often, however, you need to start with the basic `help` command, which will open the Viewer window shown in figure 1.1.

```
. help
```

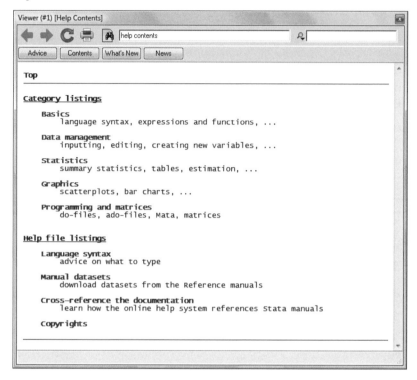

Figure 1.1. Basic `help` contents

For further details, click on a category and subsequent subcategories.

For help with the Stata matrix programming language, Mata, add the term `mata` after `help`. Often, for Mata, it is necessary to start with the very broad command

```
. help mata
  (output omitted)
```

and then narrow the results by selecting the appropriate categories and subcategories.

1.2.4   The search, findit, and hsearch commands

There are several search-related commands that do not require knowledge of command names.

For example, the `search` command does a keyword search. It is especially useful if you do not know the Stata command name or if you want to find the many places that a command or method might be used. The default for `search` is to obtain information from official help files, FAQs, examples, the SJ, and the STB but not from Internet sources. For example, for ordinary least squares (OLS) the command

>   . search ols
>   (*output omitted*)

finds references in the manuals [R], [MV], [SVY], and [XT]; in FAQs; in examples; and in the SJ and the STB. It also gives `help` commands that you can click on to get further information without the need to consult the manuals. The `net search` command searches the Internet for installable packages, including code from the SJ and the STB.

The `findit` command provides the broadest possible keyword search for Stata-related information. You can obtain details on this command by typing `help findit`. To find information on weak instruments, for example, type

>   . findit weak instr
>   (*output omitted*)

This finds joint occurrences of keywords beginning with the letters "weak" and the letters "instr".

The `search` and `findit` commands lead to keyword searches only. A more detailed search is not restricted to keywords. For example, the `hsearch` command searches all words in the help files (extension `.sthlp` or `.hlp`) on your computer, for both official Stata commands and user-written commands. Unlike the `findit` command, `hsearch` uses a whole word search. For example,

>   . hsearch weak instrument
>   (*output omitted*)

actually leads to more results than `hsearch weak instr`.

The `hsearch` command is especially useful if you are unsure whether Stata can perform a particular task. In that case, use `hsearch` first, and if the task is not found, then use `findit` to see if someone else has developed Stata code for the task.

1.3   Command syntax and operators

Stata command syntax describes the rules of the Stata programming language.

1.3.1   Basic command syntax

The basic command syntax is almost always some subset of

$$\left[\textit{prefix:}\right]\ \textit{command}\ \left[\textit{varlist}\right]\ \left[=\ \textit{exp}\right]\ \left[\textit{if}\right]\ \left[\textit{in}\right]\ \left[\textit{weight}\right]$$
$$\left[\texttt{using}\ \textit{filename}\right]\ \left[,\ \textit{options}\right]$$

The square brackets denote qualifiers that in most instances are optional. Words in the typewriter font are to be typed into Stata like they appear on the page. Italicized words are to be substituted by the user, where

- *prefix* denotes a command that repeats execution of *command* or modifies the input or output of *command*,
- *command* denotes a Stata command,
- *varlist* denotes a list of variable names,
- *exp* is a mathematical expression,
- *weight* denotes a weighting expression,
- *filename* is a filename, and
- *options* denotes one or more options that apply to *command*.

The greatest variation across commands is in the available options. Commands can have many options, and these options can also have options, which are given in parentheses.

Stata is case sensitive. We generally use lowercase throughout, though occasionally we use uppercase for model names.

Commands and output are displayed following the style for Stata manuals. For Stata commands given in the text, the typewriter font is used. For example, for OLS, we use the `regress` command. For displayed commands and output, the commands have the prefix . (a period followed by a space), whereas output has no prefix. For Mata commands, the prefix is a colon (:) rather than a period. Output from commands that span more than one line has the continuation prefix > (greater-than sign). For a Stata or Mata `program`, the lines within the program do not have a prefix.

1.3.2   Example: The summarize command

The `summarize` command provides descriptive statistics (e.g., mean, standard deviation) for one or more variables.

You can obtain the syntax of `summarize` by typing `help summarize`. This yields output including

summarize [*varlist*] [*if*] [*in*] [*weight*] [, *options*]

It follows that, at the minimum, we can give the command without any qualifiers. Unlike some commands, `summarize` does not use [= *exp*] or [`using` *filename*].

As an example, we use a commonly used, illustrative dataset installed with Stata called `auto.dta`, which has information on various attributes of 74 new automobiles. You can read this dataset into memory by using the `sysuse` command, which accesses Stata-installed datasets. To read in the data and obtain descriptive statistics, we type

```
. sysuse auto
(1978 Automobile Data)

. summarize
     Variable |        Obs        Mean    Std. Dev.       Min        Max
--------------+--------------------------------------------------------
         make |          0
        price |         74    6165.257    2949.496       3291      15906
          mpg |         74     21.2973    5.785503         12         41
        rep78 |         69    3.405797    .9899323          1          5
     headroom |         74    2.993243    .8459948        1.5          5
--------------+--------------------------------------------------------
        trunk |         74    13.75676    4.277404          5         23
       weight |         74    3019.459    777.1936       1760       4840
       length |         74    187.9324    22.26634        142        233
         turn |         74    39.64865    4.399354         31         51
 displacement |         74    197.2973    91.83722         79        425
--------------+--------------------------------------------------------
   gear_ratio |         74    3.014865    .4562871       2.19       3.89
      foreign |         74    .2972973    .4601885          0          1
```

The dataset comprises 12 variables for 74 automobiles. The average price of the automobiles is $6,165, and the standard deviation is $2,949. The column Obs gives the number of observations for which data are available for each variable. The make variable has zero observations because it is a string (or text) variable giving the make of the automobile, and summary statistics are not applicable to a nonnumeric variable. The rep78 variable is available for only 69 of the 74 observations.

A more focused use of **summarize** restricts attention to selected variables and uses one or more of the available options. For example,

```
. summarize mpg price weight, separator(1)
     Variable |        Obs        Mean    Std. Dev.       Min        Max
--------------+--------------------------------------------------------
          mpg |         74     21.2973    5.785503         12         41
--------------+--------------------------------------------------------
        price |         74    6165.257    2949.496       3291      15906
--------------+--------------------------------------------------------
       weight |         74    3019.459    777.1936       1760       4840
```

provides descriptive statistics for the mpg, price, and weight variables. The option separator(1) inserts a line between the output for each variable.

1.3.3 Example: The regress command

The regress command implements OLS regression.

You can obtain the syntax of regress by typing help regress. This yields output including

regress *depvar* [*indepvars*] [*if*] [*in*] [*weight*] [, *options*]

It follows that, at the minimum, we need to include the variable name for the dependent variable (in that case, the regression is on an intercept only). Although not explicitly stated, prefixes can be used. Many estimation commands have similar syntax.

Suppose that we want to run an OLS regression of the `mpg` variable (fuel economy in miles per gallon) on `price` (auto price in dollars) and `weight` (weight in pounds). The basic command is simply

```
. regress mpg price weight

      Source |       SS       df       MS              Number of obs =      74
-------------+------------------------------           F(  2,    71) =   66.85
       Model | 1595.93249      2  797.966246           Prob > F      =  0.0000
    Residual | 847.526967     71  11.9369995           R-squared     =  0.6531
-------------+------------------------------           Adj R-squared =  0.6434
       Total | 2443.45946     73  33.4720474           Root MSE      =   3.455

------------------------------------------------------------------------------
         mpg |      Coef.   Std. Err.      t    P>|t|     [95% Conf. Interval]
-------------+----------------------------------------------------------------
       price |  -.0000935   .0001627    -0.57   0.567    -.000418     .0002309
      weight |  -.0058175   .0006175    -9.42   0.000    -.0070489    -.0045862
       _cons |   39.43966   1.621563    24.32   0.000     36.20635     42.67296
------------------------------------------------------------------------------
```

The coefficient of $-.0058175$ for `weight` implies that fuel economy falls by 5.8 miles per gallon when the car's weight increases by 1,000 pounds.

A more complicated version of `regress` that demonstrates much of the command syntax is the following:

```
. by foreign: regress mpg price weight if weight < 4000, vce(robust)
  (output omitted )
```

For each value of the `foreign` variable, here either 0 or 1, this command fits distinct OLS regressions of `mpg` on `price` and `weight`. The `if` qualifier limits the sample to cars with `weight` less than 4,000 pounds. The `vce(robust)` option leads to heteroskedasticity-robust standard errors being used.

Output from commands is not always desired. We can suppress output by using the `quietly` prefix. For example,

```
. quietly regress mpg price weight
```

The `quietly` prefix does not require a colon, for historical reasons, even though it is a command prefix. In this book, we use this prefix extensively to suppress extraneous output.

The preceding examples used one of the available options for `regress`. From `help regress`, we find that the `regress` command has the following options: `noconstant`, `hascons`, `tsscons`, `vce(`*vcetype*`)`, `level(`#`)`, `beta`, `eform(`*string*`)`, `depname(`*varname*`)`, *display_options*, `noheader`, `notable`, `plus`, `mse1`, and `coeflegend`.

1.3.4 Factor variables

Factor variables, introduced in Stata 11, enable reference to a set of indicator variables based on a (nonnegative and integer-valued) categorical variable by inserting the i. operator in front of the name of the categorical variable. Factor variables can be used in the variable list of most Stata commands.

As an example, consider the variable rep78. This takes five distinct values that are 1, 2, 3, 4, and 5, though any other nonnegative integer values will do. Additionally, variable rep78 is missing for five observations. We have

```
. * Factor variable for rep78 - base category is omitted
. summarize i.rep78
    Variable |        Obs        Mean    Std. Dev.       Min        Max
-------------+--------------------------------------------------------
       rep78 |
          2  |         69     .115942    .3225009          0          1
          3  |         69   .4347826    .4993602          0          1
          4  |         69   .2608696    .4423259          0          1
          5  |         69   .1594203    .3687494          0          1
```

The default is to omit one category, that for the lowest value taken by the categorical variable. For variable rep78, this is the value 1. To see what category is the base (or omitted) category, add the allbaselevels option after the command (here summarize). To change the base category, use the ib. operator instead of the i. operator. For example, the summarize ib2.rep78 command will omit the second category (here rep78 = 2), and the summarize ib(last).rep78 command will omit the highest-valued category (here rep78 = 5).

A complete set of indicators, with no category omitted, is obtained using the ibn. operator. For example,

```
. * Factor variable for rep78 - no category is omitted
. summarize ibn.rep78
    Variable |        Obs        Mean    Std. Dev.       Min        Max
-------------+--------------------------------------------------------
       rep78 |
          1  |         69   .0289855    .1689948          0          1
          2  |         69     .115942    .3225009          0          1
          3  |         69   .4347826    .4993602          0          1
          4  |         69   .2608696    .4423259          0          1
          5  |         69   .1594203    .3687494          0          1
```

A complete set of interactions between two (or more) categorical variables can be created using the # operator. For example, consider an interaction between categorical variable rep78 and categorical variable foreign (a binary indicator). We have

```
. * Factor variables for interaction between two categorical variables
. summarize i.rep78#i.foreign, allbaselevels
```

Variable	Obs	Mean	Std. Dev.	Min	Max
rep78#					
foreign					
1 0	69	(base)			
1 1	69	(empty)			
2 0	69	.115942	.3225009	0	1
2 1	69	(empty)			
3 0	69	.3913043	.4916177	0	1
3 1	69	.0434783	.2054251	0	1
4 0	69	.1304348	.3392485	0	1
4 1	69	.1304348	.3392485	0	1
5 0	69	.0289855	.1689948	0	1
5 1	69	.1304348	.3392485	0	1

Here the base (omitted) category is rep78 = 1 and foreign = 0 (the lowest-valued joint category). Additionally, there are zero observations falling into two of the categories: rep78 = 1 and foreign = 1, and rep78 = 2 and foreign = 1.

The ## operator creates a factorial interaction that includes sets of indicator variables for each of the two categorical variables, in addition to the interactions given by the # operator. For example, the command summarize i.rep78##i.foreign is equivalent to the command summarize i.rep78 i.foreign i.rep78#i.foreign.

Factor variables can also be used to create interactions between indicator variables and continuous regressors. In that case, the prefix c. needs to be used to signal that the interaction is with a continuous variable. For example,

```
. * Factor variables for interaction between categorical and continuous variables
. summarize i.rep78#c.weight
```

Variable	Obs	Mean	Std. Dev.	Min	Max
rep78#					
c.weight					
1	69	89.85507	527.7129	0	3470
2	69	388.8406	1091.012	0	3900
3	69	1434.348	1719.108	0	4840
4	69	748.6957	1348.094	0	4130
5	69	370.2899	870.8548	0	3170

In this continuous interaction example, there is no omitted category—all five possible values of rep78 are interacted with the continuous variable weight.

Factor variables also permit interaction of continuous variables with continuous variables. For example, the following performs OLS regression of mpg on price and a quadratic in weight.

```
. * Factor variables for interaction between two continuous variables
. regress mpg price c.weight c.weight#c.weight, noheader
```

mpg	Coef.	Std. Err.	t	P>|t|	[95% Conf. Interval]	
price	-.0002597	.0001696	-1.53	0.130	-.000598	.0000786
weight	-.016047	.0040403	-3.97	0.000	-.024105	-.0079889
c.weight#						
c.weight	1.72e-06	6.71e-07	2.56	0.013	3.79e-07	3.06e-06
_cons	54.66807	6.150716	8.89	0.000	42.40086	66.93529

For more on factor variables, type `help factor variables` or see [U] **11.4.3 Factor variables** and [U] **25 Working with categorical data and factor variables**. To check whether the `regress` command, for example, supports factor variables, type the command `help regress` and the output below the syntax summary includes a note that "*indepvars* may contain factor variables; see fvvarlist." Some multinomial model estimation commands do not support factor variables; see section 15.2.5.

1.3.5   Abbreviations, case sensitivity, and wildcards

Commands and parts of commands can be abbreviated to the shortest string of characters that uniquely identify them, often just two or three characters. For example, we can shorten `summarize` to `su`. For expositional clarity, we do not use such abbreviations in this book; a notable exception is that we may use abbreviations in the options to graphics commands because these commands can get very lengthy. Not using abbreviations makes it much easier to read your do-files.

Variable names can be up to 32 characters long, where the characters can be A–Z, a–z, 0–9, and _ (underscore). Some names, such as `in`, are reserved. Stata is case sensitive, and the norm is to use lowercase.

We can use the wildcard * (asterisk) for variable names in commands, provided there is no ambiguity such as two potential variables for a one-variable command. For example,

```
. summarize t*
```

Variable	Obs	Mean	Std. Dev.	Min	Max
trunk	74	13.75676	4.277404	5	23
turn	74	39.64865	4.399354	31	51

provides summary statistics for all variables with names beginning with the letter `t`. Where ambiguity may arise, wildcards are not permitted.

(*Continued on next page*)

1.3.6   Arithmetic, relational, and logical operators

The arithmetic operators in Stata are + (addition), - (subtraction), * (multiplication), / (division), ^ (raised to a power), and the prefix - (negation). For example, to compute and display $-2 \times \{9/(8+2-7)\}^2$, which simplifies to -2×3^2, we type

```
. display -2*(9/(8+2-7))^2
-18
```

If the arithmetic operation is not possible, or data are not available to perform the operation, then a missing value denote by . is displayed. For example,

```
. display 2/0
.
```

The relational operators are > (greater than), < (less than), >= (greater than or equal), <= (less than or equal), == (equal), and != (not equal). These are the obvious symbols, except that a pair of equal-signs is used for equality, and != denotes not equal. Relational operators are often used in `if` qualifiers that define the sample for analysis.

Logical operators return 1 for true and 0 for false. The logical operators are & (and), | (or), and ! (not). The operator ~ can be used in place of !. Logical operators are also used to define the sample for analysis. For example, to restrict regression analysis to smaller less expensive cars, type

```
. regress mpg price weight if weight <= 4000 & price <= 10000
  (output omitted )
```

The string operator + is used to concatenate two strings into a single, longer string.

The order of evaluation of all operators is ! (or ~), ^, - (negation), /, *, - (subtraction), +, != (or ~=), >, <, <=, >=, ==, &, and |.

1.3.7   Error messages

Stata produces error messages when a command fails. These messages are brief, but a fuller explanation can be obtained from the manual or directly from Stata.

For example, if we regress `mpg` on `notthere` but the `notthere` variable does not exist, we get

```
. regress mpg notthere
variable notthere not found
r(111);
```

Here r(111) denotes return code 111. You can obtain further details by clicking on r(111); if in interactive mode or by typing

```
. search rc 111
  (output omitted )
```

1.4   Do-files and log files

For Stata analysis requiring many commands, or requiring lengthy commands, it is best to collect all the commands into a program (or script) that is stored in a text file called a do-file.

In this book, we perform data analysis using a do-file. We assume that the do-file and, if relevant, any input and output files are in a common directory and that Stata is executed from that directory. Then we only need to provide the filename rather than the complete directory structure. For example, we can refer to a file as mus02data.dta rather than C:\mus\chapter2\mus02data.dta.

1.4.1   Writing a do-file

A do-file is a text file with extension .do that contains a series of Stata commands.

As an example, we write a two-line program that reads in the Stata example dataset auto.dta and then presents summary statistics for the mpg variable that we already know is in the dataset. The commands are sysuse auto.dta, clear, where the clear option is added to remove the current dataset from memory, and summarize mpg. The two commands are to be collected into a command file called a do-file. The filename should include no spaces, and the file extension is .do. In this example, we suppose this file is given the name example.do and is stored in the current working directory.

To see the current directory, type cd without any arguments. To change to another directory, cd is used with an argument. For example, in Windows, to change to the directory C:\Program Files\Stata11\, we type

```
. cd "c:\Program Files\Stata11"
c:\Program Files\Stata11
```

The directory name is given in double quotes because it includes spaces. Otherwise, the double quotes are unnecessary.

One way to create the do-file is to start Stata and use the Do-file Editor. Within Stata, we select **Window > Do-file Editor > New Do-file**, type in the commands, and save the do-file.

Alternatively, type in the commands outside Stata by using a preferred text editor. Ideally, this text editor supports multiple windows, reads large files (datasets or output), and gives line numbers and column numbers.

The type command lists the contents of the file. We have

```
. type example.do
sysuse auto.dta, clear
summarize mpg
```

1.4.2   Running do-files

You can run (or execute) an already-written do-file by using the Command window. Start Stata and, in the Command window, change directory (`cd`) to the directory that has the do-file, and then issue the `do` command. We obtain

```
. do example.do
. sysuse auto.dta, clear
(1978 Automobile Data)
. summarize mpg
    Variable |       Obs        Mean    Std. Dev.       Min        Max
-------------+--------------------------------------------------------
         mpg |        74     21.2973    5.785503         12         41
.
end of do-file
```

where we assume that `example.do` is in directory `C:\Program Files\Stata11\`.

An alternative method is to run the do-file from the Do-file Editor. Select **Window > Do-file Editor > New Do-file**, and then select **File > Open...** and the appropriate file, and finally select **Tools > Do**. An advantage to using the Do-file Editor is that you can highlight or select just part of the do-file and then execute this part by selecting **Tools > Do Selection**.

You can also run do-files noninteractively, using batch mode. This initiates Stata, executes the commands in the do-file, and (optionally) exits Stata. The term batch mode is a throwback to earlier times when each line of a program was entered on a separate computer card, so that a program was a collection or "batch" of computer cards. For example, to run `example.do` in batch mode, double-click on `example.do` in Windows Explorer. This initiates Stata and executes the file's Stata commands. You can also use the `do` command. (In Unix, you would use the `stata -b example.do` command.)

It can be useful to include the `set more off` command at the start of a do-file so that output scrolls continuously rather than pausing after each page of output.

1.4.3   Log files

By default, Stata output is sent to the screen. For reproducibility, you should save this output in a separate file. Another advantage to saving output is that lengthy output can be difficult to read on the screen; it can be easier to review results by viewing an output file using a text editor.

A Stata output file is called a log file. It stores the commands in addition to the output from these commands. The default Stata extension for the file is `.log`, but you can choose an alternative extension, such as `.txt`. An extension name change may be worthwhile because several other programs, such as LaTeX compilers, also create files with the `.log` extension. Log files can be read as either standard text or in a special

Stata code called `smcl` (Stata Markup and Control Language). We use text throughout this book, because it is easier to read in a text editor. A useful convention can be to give the log the same filename as that for the do-file. For example, for `example.do`, we save the output as `example.txt`.

A log file is created by using the `log` command. In a typical analysis, the do-file will change over time, in which case the output file will also change. The Stata default is to protect against an existing log being accidentally overwritten. To create a log file in text form named `example.txt`, the usual command is

```
. log using example.txt, text replace
```

The `replace` option permits the existing version of `example.txt`, if there is one, to be overwritten. Without `replace`, Stata will refuse to open the log file if there is already a file called `example.txt`.

In some cases, we may not want to overwrite the existing log, in which case we would not specify the `replace` option. The most likely reason for preserving a log is that it contains important results, such as those from final analysis. Then it can be good practice to rename the log after analysis is complete. Thus `example.txt` might be renamed `example07052008.txt`.

When a program is finished, you should close the log file by typing `log close`.

The log can be very lengthy. If you need a hard copy, you can edit the log to include only essential results. The text editor you use should use a monospace font such as Courier New, where each character takes up the same space, so that output table columns will be properly aligned.

The log file includes the Stata commands, with a dot (.) prefix, and the output. You can use a log file to create a do-file, if a do-file does not already exist, by deleting the dot and all lines that are command results (no dot). By this means, you can do initial work using the Stata GUI and generate a do-file from the session, provided that you created a log file at the beginning of the session.

1.4.4   A three-step process

Data analysis using Stata can repeatedly use the following three-step process:

1. Create or change the do-file.
2. Execute the do-file in Stata.
3. Read the resulting log with a text editor.

The initial do-file can be written by editing a previously written do-file that is a useful template or starting point, especially if it uses the same dataset or the same commands as the current analysis. The resulting log may include Stata errors or estimation results that lead to changes in the original do-file and so on.

Suppose we have fit several models and now want to fit an additional model. In interactive mode, we would type in the new command, execute it, and see the results. Using the three-step process, we add the new command to the do-file, execute the do-file, and read the new output. Because many Stata programs execute in seconds, this adds little extra time compared with using interactive mode, and it has the benefit of having a do-file that can be modified for later use.

1.4.5   Comments and long lines

Stata do-files can include comments. This can greatly increase understanding of a program, which is especially useful if you return to a program and its output a year or two later. Lengthy single-line comments can be allowed to span several lines, ensuring readability. There are several ways to include comments:

- For single-line comments, begin the line with an asterisk (*); Stata ignores such lines.

- For a comment on the same line as a Stata command, use two slashes (//) after the Stata command.

- For multiple-line comments, place the commented text between slash-star (/*) and star-slash (*/).

The Stata default is to view each line as a separate Stata command, where a line continues until a carriage return (end-of-line or *Enter* key) is encountered. Some commands, such as those for nicely formatted graphs, can be very long. For readability, these commands need to span more than one line. The easiest way to break a line at, say, the 70th column is by using three slashes (///) and then continuing the command on the next line.

The following do-file code includes several comments to explain the program and demonstrates how to allow a command to span more than one line.

```
* Demonstrate use of comments
* This program reads in system file auto.dta and gets summary statistics
clear // Remove data from memory
* The next code shows how to allow a single command to span two lines
sysuse ///
auto.dta
summarize
```

For long commands, you can alternatively use the command **#delimit** command. This changes the delimiter from the Stata default, which is a carriage return (i.e., end-of-line), to a semicolon. This also permits more than one command on a single line. The following code changes the delimiter from the default to a semicolon and back to the default:

```
* Change delimiter from cr to semicolon and back to cr
#delimit ;
* More than one command per line and command spans more than one line;
clear; sysuse
auto.dta; summarize;
#delimit cr
```

We recommend using /// instead of changing the delimiter because the comment method produces more readable code.

1.4.6 Different implementations of Stata

The different platforms for Stata share the same command syntax; however, commands can change across versions of Stata. For this book, we use Stata 11. To ensure that later versions of Stata will continue to work with our code, we include the **version 11** command near the beginning of the do-file.

Different implementations of Stata have different limits. A common limit encountered is the memory allocated to Stata, which restricts the size of dataset that can be handled by Stata. The default is small, e.g., 1 megabyte, so that Stata does not occupy too much memory, permitting other tasks to run while Stata is used. Another common limit is the size of matrix, which limits the number of variables in the dataset.

You can increase or decrease the limits with the **set** command. For example,

```
. set matsize 300
```

sets the maximum number of variables in an estimation command to 300.

The maximum possible values vary with the version of Stata: Small Stata, Stata/IC, Stata/SE, or Stata/MP. The **help limits** command provides details on the limits for the current implementation of Stata. The **query** and **creturn list** commands detail the current settings.

1.5 Scalars and matrices

Scalars can store a single number or a single string, and matrices can store several numbers or strings as an array. We provide a very brief introduction here, sufficient for use of the scalars and matrices in section 1.6.

1.5.1 Scalars

A scalar can store a single number or string. You can display the contents of a scalar by using the **display** command.

For example, to store the number 2×3 as the scalar **a** and then display the scalar, we type

```
. * Scalars: Example
. scalar a = 2*3
. scalar b = "2 times 3 = "
. display b a
2 times 3 = 6
```

One common use of scalars, detailed in section 1.6, is to store the scalar results of estimation commands that can then be accessed for use in subsequent analysis. In section 1.7, we discuss the relative merits of using a scalar or a macro to store a scalar quantity.

1.5.2   Matrices

Stata provides two distinct ways to use matrices, both of which store several numbers or strings as an array. One way is through Stata commands that have the `matrix` prefix. More recently, beginning with version 9, Stata includes a matrix programming language, Mata. These two methods are presented in, respectively, appendices A and B.

The following Stata code illustrates the definition of a specific 2×3 matrix, the listing of the matrix, and the extraction and display of a specific element of the matrix.

```
. * Matrix commands: Example
. matrix define A = (1,2,3 \ 4,5,6)
. matrix list A
A[2,3]
     c1  c2  c3
r1    1   2   3
r2    4   5   6
. scalar c = A[2,3]
. display c
6
```

1.6   Using results from Stata commands

One goal of this book is to enable analysis that uses more than just Stata built-in commands and printed output. Much of this additional analysis entails further computations after using Stata commands.

1.6.1   Using results from the r-class command summarize

The Stata commands that analyze the data but do not estimate parameters are r-class commands. All r-class commands save their results in `r()`. The contents of `r()` vary with the command and are listed by typing `return list`.

As an example, we list the results stored after using `summarize`:

```
. * Illustrate use of return list for r-class command summarize
. summarize mpg
    Variable |        Obs        Mean    Std. Dev.       Min        Max
-------------+--------------------------------------------------------
         mpg |         74     21.2973    5.785503         12         41
. return list
scalars:
                  r(N) =  74
              r(sum_w) =  74
               r(mean) =  21.2972972972973
                r(Var) =  33.47204738985561
                 r(sd) =  5.785503209735141
                r(min) =  12
                r(max) =  41
                r(sum) =  1576
```

There are eight separate results stored as Stata scalars with the names `r(N)`, `r(sum_w)`, ..., `r(sum)`. These are fairly obvious aside from `r(sum_w)`, which gives the sum of the weights. Several additional results are returned if the `detail` option to `summarize` is used; see [R] **summarize**.

The following code calculates and displays the range of the data:

```
. * Illustrate use of r()
. quietly summarize mpg
. scalar range = r(max) - r(min)
. display "Sample range = " range
Sample range = 29
```

The results in `r()` disappear when a subsequent r-class or e-class command is executed. We can always save the value as a scalar. It can be particularly useful to save the sample mean.

```
. * Save a result in r() as a scalar
. scalar mpgmean = r(mean)
```

1.6.2   Using results from the e-class command regress

Estimation commands are e-class commands (or estimation-class commands), such as `regress`. The results are stored in `e()`, the contents of which you can view by typing `ereturn list`.

(Continued on next page)

A leading example is **regress** for OLS regression. For example, after typing

```
. regress mpg price weight
      Source |       SS       df       MS              Number of obs =      74
-------------+------------------------------           F(  2,    71) =   66.85
       Model | 1595.93249      2  797.966246           Prob > F      =  0.0000
    Residual | 847.526967     71  11.9369995           R-squared     =  0.6531
-------------+------------------------------           Adj R-squared =  0.6434
       Total | 2443.45946     73  33.4720474           Root MSE      =   3.455

------------------------------------------------------------------------------
         mpg |      Coef.   Std. Err.      t    P>|t|     [95% Conf. Interval]
-------------+----------------------------------------------------------------
       price |  -.0000935   .0001627    -0.57   0.567    -.000418     .0002309
      weight |  -.0058175   .0006175    -9.42   0.000    -.0070489   -.0045862
       _cons |   39.43966   1.621563    24.32   0.000     36.20635    42.67296
------------------------------------------------------------------------------
```

ereturn list yields

```
. * ereturn list after e-class command regress
. ereturn list
scalars:
                  e(N) =  74
               e(df_m) =  2
               e(df_r) =  71
                  e(F) =  66.84814256414501
                 e(r2) =  .6531446579233134
               e(rmse) =  3.454996314099513
                e(mss) =  1595.932492798133
                e(rss) =  847.5269666613265
               e(r2_a) =  .6433740849070687
                 e(ll) =  -195.2169813478502
               e(ll_0) =  -234.3943376482347
               e(rank) =  3

macros:
             e(cmdline) : "regress mpg price weight"
               e(title) : "Linear regression"
           e(marginsok) : "XB default"
                 e(vce) : "ols"
              e(depvar) : "mpg"
                 e(cmd) : "regress"
          e(properties) : "b V"
             e(predict) : "regres_p"
               e(model) : "ols"
           e(estat_cmd) : "regress_estat"

matrices:
                 e(b) :  1 x 3
                 e(V) :  3 x 3

functions:
             e(sample)
```

The key numeric output in the analysis-of-variance table is stored as scalars. As an example of using scalar results, consider the calculation of R^2. The model sum of squares is stored in e(mss), and the residual sum of squares is stored in e(rss), so that

```
. * Use of e() where scalar
. scalar r2 = e(mss)/(e(mss)+e(rss))
. display "r-squared = " r2
r-squared = .65314466
```

The result is the same as the 0.6531 given in the original regression output.

The remaining numeric output is stored as matrices. Here we present methods to extract scalars from these matrices and manipulate them. Specifically, we obtain the OLS coefficient of **price** from the 1×3 matrix **e(b)**, the estimated variance of this estimate from the 3×3 matrix **e(V)**, and then we form the t statistic for testing whether the coefficient of **price** is zero:

```
. * Use of e() where matrix
. matrix best = e(b)
. scalar bprice = best[1,1]
. matrix Vest = e(V)
. scalar Vprice = Vest[1,1]
. scalar tprice = bprice/sqrt(Vprice)
. display "t statistic for H0: b_price = 0 is " tprice
t statistic for H0: b_price = 0 is -.57468079
```

The result is the same as the -0.57 given in the original regression output.

The results in **e()** disappear when a subsequent e-class command is executed. However, you can save the results by using **estimates store**, detailed in section 3.4.4.

1.7   Global and local macros

A macro is a string of characters that stands for another string of characters. For example, you can use the macro **xlist** in place of **"price weight"**. This substitution can lead to code that is shorter, easier to read, and that can be easily adapted to similar problems.

Macros can be global or local. A global macro is accessible across Stata do-files or throughout a Stata session. A local macro can be accessed only within a given do-file or in the interactive session.

1.7.1   Global macros

Global macros are the simplest macro and are adequate for many purposes. We use global macros extensively throughout this book.

Global macros are defined with the **global** command. To access what was stored in a global macro, put the character $ immediately before the macro name. For example, consider a regression of the dependent variable **mpg** on several regressors, where the global macro **xlist** is used to store the regressor list.

```
. * Global macro definition and use
. global xlist price weight
. regress mpg $xlist, noheader        // $ prefix is necessary
```

mpg	Coef.	Std. Err.	t	P>|t|	[95% Conf. Interval]	
price	-.0000935	.0001627	-0.57	0.567	-.000418	.0002309
weight	-.0058175	.0006175	-9.42	0.000	-.0070489	-.0045862
_cons	39.43966	1.621563	24.32	0.000	36.20635	42.67296

Global macros are frequently used when fitting several different models with the same regressor list because they ensure that the regressor list is the same in all instances and they make it easy to change the regressor list. A single change to the global macro changes the regressor list in all instances.

A second example might be where several different models are fit, but we want to hold a key parameter constant throughout. For example, suppose we obtain standard errors by using the bootstrap. Then we might define the global macro `nbreps` for the number of bootstrap replications. Exploratory data analysis might set `nbreps` to a small value such as 50 to save computational time, whereas final results set `nbreps` to an appropriately higher value such as 400.

A third example is to highlight key program parameters, such as the variable used to define the cluster if cluster–robust standard errors are obtained. By gathering all such global macros at the start of the program, it can be clear what the settings are for key program parameters.

1.7.2   Local macros

Local macros are defined with the `local` command. To access what was stored in the local macro, enclose the macro name in single quotes. These quotes differ from how they appear on this printed page. On most keyboards, the left quote is located at the upper left, under the tilde, and the right quote is located at the middle right, under the double quote.

As an example of a local macro, consider a regression of the `mpg` variable on several regressors. We define the local macro `xlist` and subsequently access its contents by enclosing the name in single quotes as `` `xlist' ``.

```
. * Local macro definition and use
. local xlist "price weight"
. regress mpg `xlist', noheader      // single quotes are necessary
```

mpg	Coef.	Std. Err.	t	P>|t|	[95% Conf. Interval]	
price	-.0000935	.0001627	-0.57	0.567	-.000418	.0002309
weight	-.0058175	.0006175	-9.42	0.000	-.0070489	-.0045862
_cons	39.43966	1.621563	24.32	0.000	36.20635	42.67296

The double quotes used in defining the local macro as a string are unnecessary, which is why we did not use them in the earlier global macro example. Using the double quotes does emphasize that a text substitution has been made. The single quotes in subsequent references to xlist are necessary.

We could also use a macro to define the dependent variable. For example,

```
. * Local macro definition without double quotes
. local y mpg
. regress `y' `xlist', noheader
```

mpg	Coef.	Std. Err.	t	P>|t|	[95% Conf. Interval]	
price	-.0000935	.0001627	-0.57	0.567	-.000418	.0002309
weight	-.0058175	.0006175	-9.42	0.000	-.0070489	-.0045862
_cons	39.43966	1.621563	24.32	0.000	36.20635	42.67296

Note that here `y' is not a variable with $N$ observations. Instead, it is the string mpg. The **regress** command simply replaces `y' with the text mpg, which in turn denotes a variable that has N observations.

We can also define a local macro through evaluation of a function. For example,

```
. * Local macro definition through function evaluation
. local z = 2+2
. display `z'
4
```

leads to 'z' being the string 4. Using the equality sign when defining a macro causes the macro to be evaluated as an expression. For numerical expressions, using the equality sign stores the result of the expression and not the characters in the expression itself in the macro. For string assignments, it is best not to use the equality sign. This is especially true when storing lists of variables in macros. Strings in Stata expressions can contain only 244 characters, fewer characters than many variable lists. Macros assigned without an equality sign can hold 165,200 characters in Stata/IC and 1,081,511 characters in Stata/MP and Stata/SE.

Local macros are especially useful for programming in Stata; see appendix A. Then, for example, you can use `y' and `x' as generic notation for the dependent variable and regressors, making the code easier to read.

Local macros apply only to the current program and have the advantage of no potential conflict with other programs. They are preferred to global macros, unless there is a compelling reason to use global macros.

1.7.3 Scalar or macro?

A macro can be used in place of a scalar, but a scalar is simpler. Furthermore, [P] **scalar** points out that using a scalar will usually be faster than using a macro, because a macro requires conversion into and out of internal binary representation. This reference also gives an example where macros lead to a loss of accuracy because of these conversions.

One drawback of a scalar, however, is that the scalar is dropped whenever `clear all` is used. By contrast, a macro is still retained. Consider the following example:

```
. * Scalars disappear after clear all but macro does not
. global b 3
. local c 4
. scalar d = 5
. clear
. display $b _skip(3) `c´   // display macros
3    4
. display d               // display the scalar
5
. clear all
. display $b _skip(3) `c´   // display macros
3    4
. display d               // display the scalar
d not found
r(111);
```

Here the scalar `d` has been dropped after `clear all`, though not after `clear`.

We use global macros in this text because there are cases in which we want the contents of our macros to be accessible across do-files. A second reason for using global macros is that the required $ prefix makes it clear that a global parameter is being used.

1.8 Looping commands

Loops provide a way to repeat the same command many times. We use loops in a variety of contexts throughout the book.

Stata has three looping constructs: `foreach`, `forvalues`, and `while`. The `foreach` construct loops over items in a list, where the list can be a list of variable names (possibly given in a macro) or a list of numbers. The `forvalues` construct loops over consecutive values of numbers. A `while` loop continues until a user-specified condition is not met.

We illustrate how to use these three looping constructs in creating the sum of four variables, where each variable is created from the uniform distribution. There are many variations in the way you can use these loop commands; see [P] **foreach**, [P] **forvalues**, and [P] **while**.

The `generate` command is used to create a new variable. The `runiform()` function provides a draw from the uniform distribution. Whenever random numbers are generated, we set the seed to a specific value with the `set seed` command so that subsequent runs of the same program lead to the same random numbers being drawn. We have, for example,

```
. * Make artificial dataset of 100 observations on 4 uniform variables
. clear
. set obs 100
obs was 0, now 100
. set seed 10101
. generate x1var = runiform()
. generate x2var = runiform()
. generate x3var = runiform()
. generate x4var = runiform()
```

We want to sum the four variables. The obvious way to do this is

```
. * Manually obtain the sum of four variables
. generate sum = x1var + x2var + x3var + x4var
. summarize sum
```

Variable	Obs	Mean	Std. Dev.	Min	Max
sum	100	2.093172	.594672	.5337163	3.204005

We now present several ways to use loops to progressively sum these variables. Although only four variables are considered here, the same methods can potentially be applied to hundreds of variables.

1.8.1 The foreach loop

We begin by using `foreach` to loop over items in a list of variable names. Here the list is x1var, x2var, x3var, and x4var.

The variable ultimately created will be called **sum**. Because **sum** already exists, we need to first drop **sum** and then generate **sum=0**. The `replace sum=0` command collapses these two steps into one step, and the `quietly` prefix suppresses output stating that 100 observations have been replaced. Following this initial line, we use a `foreach` loop and additionally use `quietly` within the loop to suppress output following `replace`. The program is

```
. * foreach loop with a variable list
. quietly replace sum = 0
. foreach var of varlist x1var x2var x3var x4var {
  2.      quietly replace sum = sum + `var´
  3. }
. summarize sum
```

Variable	Obs	Mean	Std. Dev.	Min	Max
sum	100	2.093172	.594672	.5337163	3.204005

The result is the same as that obtained manually.

The preceding code is an example of a program (see appendix A) with the { brace appearing at the end of the first line and the } brace appearing on its own at the last line of the program. The numbers 2. and 3. do not actually appear in the program but are produced as output. In the `foreach` loop, we refer to each variable in the variable list `varlist` by the local macro named `var`, so that `` `var` `` with single quotes is needed in subsequent uses of `var`. The choice of `var` as the local macro name is arbitrary and other names can be used. The word `varlist` is necessary, though types of lists other than variable lists are possible, in which case we use `numlist`, `newlist`, `global`, or `local`; see [P] **foreach**.

An attraction of using a variable list is that the method can be applied when variable names are not sequential. For example, the variable names could have been `incomehusband`, `incomewife`, `incomechild1`, and `incomechild2`.

1.8.2 The forvalues loop

A `forvalues` loop iterates over consecutive values. In the following code, we let the index be the local macro `i`, and `` `i` `` with single quotes is needed in subsequent uses of `i`. The program

```
. * forvalues loop to create a sum of variables
. quietly replace sum = 0
. forvalues i = 1/4 {
  2.     quietly replace sum = sum + x`i'var
  3. }
. summarize sum
```

Variable	Obs	Mean	Std. Dev.	Min	Max
sum	100	2.093172	.594672	.5337163	3.204005

produces the same result.

The choice of the name `i` for the local macro was arbitrary. In this example, the increment is one, but you can use other increments. For example, if we use `forvalues i = 1(2)11`, then the index goes from 1 to 11 in increments of 2.

1.8.3 The while loop

A `while` loop continues until a condition is no longer met. This method is used when `foreach` and `forvalues` cannot be used. For completeness, we apply it to the summing example.

In the following code, the local macro `i` is initialized to 1 and then incremented by 1 in each loop; looping continues, provided that $i \leq 4$.

```
. * While loop and local macros to create a sum of variables
. quietly replace sum = 0
. local i 1
. while `i' <= 4 {
  2.     quietly replace sum = sum + x`i'var
  3.     local i = `i' + 1
  4. }
```

```
. summarize sum

    Variable |      Obs        Mean    Std. Dev.       Min        Max
-------------+--------------------------------------------------------
         sum |      100    2.093172     .594672    .5337163   3.204005
```

1.8.4   The continue command

The `continue` command provides a way to prematurely cease execution of the current loop iteration. This may be useful if, for example, the loop includes taking the log of a number and we want to skip this iteration if the number is negative. Execution then resumes at the start of the next loop iteration, unless the `break` option is used. For details, see `help continue`.

1.9   Some useful commands

We have mentioned only a few Stata commands. See [U] **27.1 43 commands** for a list of 43 commands that everyone will find useful.

1.10   Template do-file

The following do-file provides a template. It captures most of the features of Stata presented in this chapter, aside from looping commands.

```
* 1. Program name
* mus01p2template.do written 2/15/2008 is a template do-file
* 2. Write output to a log file
log using mus01p2template.txt, text replace
* 3. Stata version
version 11              // so will still run in a later version of Stata
* 4. Program explanation
* This illustrative program creates 100 uniform variates
* 5. Change Stata default settings - two examples are given
set more off           // scroll screen output by at full speed
set mem 20m            // set aside 20 mb for memory space
* 6. Set program parameters using global macros
global numobs 100
local seed 10101
local xlist xvar
* 7. Generate data and summarize
set obs $numobs
set seed `seed'
generate xvar = runiform()
generate yvar = xvar^2
summarize
* 8. Demonstrate use of results stored in r()
summarize xvar
display "Sample range = " r(max)-r(min)
regress yvar `xlist'
scalar r2 = e(mss)/(e(mss)+e(rss))
display "r-squared = " r2
```

```
* 9. Close output file and exit Stata
log close
exit, clear
```

1.11   User-written commands

We make extensive use of user-written commands. These are freely available ado-files
(see section A.2.8) that are easy to install, provided you are connected to the Internet
and, for computer lab users, that the computer lab places no restriction on adding
components to Stata. They are then executed in the same way as Stata commands.

As an example, consider instrumental-variables (IV) estimation. In some cases, we
know which user-written commands we want. For example, a leading user-written
command for IV is ivreg2, and we type findit ivreg2 to get it. More generally, we
can type the broader command

```
. findit instrumental variables
  (output omitted)
```

This gives information on IV commands available both within Stata and packages available on the web, provided you are connected to the Internet.

Many entries are provided, often with several potential user-written commands and
several versions of a given user-written command. The best place to begin can be a
recent *Stata Journal* article because this code is more likely to have been closely vetted
for accuracy and written in a way suited to a range of applications. The listing from
the findit command includes

```
SJ-7-4  st0030_3  . . . .  Enhanced routines for IV/GMM estimation and testing
. . . . . . . . . . . . .  C. F. Baum, M. E. Schaffer, and S. Stillman
(help ivactest, ivendog, ivhettest, ivreg2, ivreset,
overid, ranktest if installed)
Q4/07   SJ 7(4):465--506
extension of IV and GMM estimation addressing hetero-
skedasticity- and autocorrelation-consistent standard
errors, weak instruments, LIML and k-class estimation,
tests for endogeneity and Ramsey's regression
specification-error test, and autocorrelation tests
for IV estimates and panel-data IV estimates
```

The entry means that it is the third revision of the package (st0030_3), and the package
is discussed in detail in *Stata Journal*, volume 7, number 4 (SJ-7-4).

By left-clicking on the highlighted text st0030_3 on the first line of the entry, you will
see a new window with title, description/author(s), and installation files for the package.
By left-clicking on the help files, you can obtain information on the commands. By left-
clicking on the (click here to install), you will install the files into an ado-directory.

1.12   Stata resources

For first-time users, [GS] *Getting Started with Stata* is very helpful, along with analyzing an example dataset such as `auto.dta` interactively in Stata. The next source is [U] *Users Guide*, especially the early chapters.

1.13   Exercises

1. Find information on the estimation method `clogit` using `help`, `search`, `findit`, and `hsearch`. Comment on the relative usefulness of these search commands.

2. Download the Stata example dataset `auto.dta`. Obtain summary statistics for `mpg` and `weight` according to whether the car type is foreign (use the `by foreign:` prefix). Comment on any differences between foreign and domestic cars. Then `regress mpg` on `weight` and `foreign`. Comment on any difference for foreign cars.

3. Write a do-file to repeat the previous question. This do-file should include a log file. Run the do-file and then use a text editor to view the log file.

4. Using `auto.dta`, obtain summary statistics for the `price` variable. Then use the results stored in `r()` to compute a scalar, `cv`, equal to the coefficient of variation (the standard deviation divided by the mean) of `price`.

5. Using `auto.dta`, `regress mpg` on `price` and `weight`. Then use the results stored in `e()` to compute a scalar, `r2adj`, equal to $\overline{R}^2$. The adjusted R^2 equals $R^2 - (1 - R^2)(k-1)/(N-k)$, where N is the number of observations and k is the number of regressors including the intercept. Also use the results stored in `e()` to calculate a scalar, `tweight`, equal to the t statistic to test that the coefficient of `weight` is zero.

6. Using `auto.dta`, define a global macro named `varlist` for a variable list with `mpg`, `price`, and `weight`, and then obtain summary statistics for `varlist`. Repeat this exercise for a local macro named `varlist`.

7. Using `auto.dta`, use a `foreach` loop to create a variable, `total`, equal to the sum of `headroom` and `length`. Confirm by using `summarize` that `total` has a mean equal to the sum of the means of `headroom` and `length`.

8. Create a simulated dataset with 100 observations on two random variables that are each drawn from the uniform distribution. Use a seed of 12345. In theory, these random variables have a mean of 0.5 and a variance of 1/12. Does this appear to be the case here?

2 Data management and graphics

2.1 Introduction

The starting point of an empirical investigation based on microeconomic data is the collection and preparation of a relevant dataset. The primary sources are often government surveys and administrative data. We assume the researcher has such a primary dataset and do not address issues of survey design and data collection. Even given primary data, it is rare that it will be in a form that is exactly what is required for ultimate analysis.

The process of transforming original data to a form that is suitable for econometric analysis is referred to as data management. This is typically a time-intensive task that has important implications for the quality and reliability of modeling carried out at the next stage.

This process usually begins with a data file or files containing basic information extracted from a census or a survey. They are often organized by data record for a sampled entity such as an individual, a household, or a firm. Each record or observation is a vector of data on the qualitative and quantitative attributes of each individual. Typically, the data need to be cleaned up and recoded, and data from multiple sources may need to be combined. The focus of the investigation might be a particular group or subpopulation, e.g., employed women, so that a series of criteria need to be used to determine whether a particular observation in the dataset is to be included in the analysis sample.

In this chapter, we present the tasks involved in data preparation and management. These include reading in and modifying data, transforming data, merging data, checking data, and selecting an analysis sample. The rest of the book focuses on analyzing a given sample, though special features of handling panel data and multinomial data are given in the relevant chapters.

2.2 Types of data

All data are ultimately stored in a computer as a sequence of 0s and 1s because computers operate on binary digits, or bits, that are either 0 or 1. There are several different ways to do this, with potential to cause confusion.

2.2.1   Text or ASCII data

A standard text format is ASCII, an acronym for American Standard Code for Information Interchange. Regular ASCII represents $2^7 = 128$ and extended ASCII represents $2^8 = 256$ different digits, letters (uppercase and lowercase), and common symbols and punctuation marks. In either case, eight bits (called a byte) are used. As examples, 1 is stored as 00110001, 2 is stored as 00110010, 3 is stored as 00110011, A is stored as 01010001, and a is stored as 00110001. A text file that is readable on a computer screen is stored in ASCII.

A leading text-file example is a spreadsheet file that has been stored as a "comma-separated values" file, usually a file with the `.csv` extension. Here a comma is used to separate each data value; however, more generally, other separators can be used.

Text-file data can also be stored as fixed-width data. Then no separator is needed provided we use the knowledge that, say, columns 1–7 have the first data entry, columns 8–9 have the second data entry, and so on.

Text data can be numeric or nonnumeric. The letter a is clearly nonnumeric, but depending on the context, the number 3 might be numeric or nonnumeric. For example, the number 3 might represent the number of doctor visits (numeric) or be part of a street address, such as 3 Main Street (nonnumeric).

2.2.2   Internal numeric data

When data are numeric, the computer stores them internally using a format different from text to enable application of arithmetic operations and to reduce storage. The two main types of numeric data are integer and floating point. Because computers work with 0s and 1s (a binary digit or bit), data are stored in base-2 approximations to their base-10 counterparts.

For integer data, the exact integer can be stored. The size of the integer stored depends on the number of bytes used, where a byte is eight bits. For example, if one byte is used, then in theory $2^8 = 256$ different integers could be stored, such as -127, -126, ..., 127, 128.

Noninteger data, or often even integer data, are stored as floating-point data. Standard floating-point data are stored in four bytes, where the first bit may represent the sign, the next 8 bits may represent the exponent, and the remaining 23 bits may represent the digits. Although all integers have an exact base-2 representation, not all base-10 numbers do. For example, the base-10 number 0.1 is $0.0\overline{0011}$ in base 2. For this reason, the more bytes in the base-2 approximation, the more precisely it approximates the base-10 number. Double-precision floating-point data use eight bytes, have about 16 digits precision (in base 10), and are sufficiently accurate for statistical calculations.

Stata has the numeric storage types listed in table 2.1: three are integer and two are floating point.

Table 2.1. Stata's numeric storage types

Storage type	Bytes	Minimum	Maximum
byte	1	-127	100
int	2	$-32,767$	32,740
long	4	$-2,147,483,647$	2,147,483,620
float	4	$-1.70141173319 \times 10^{38}$	$1.70141173319 \times 10^{38}$
double	8	$-8.9984656743 \times 10^{307}$	$8.9984656743 \times 10^{307}$

These internal data types have the advantage of taking fewer bytes to store the same amount of data. For example, the integer 123456789 takes up 9 bytes if stored as text but only 4 bytes if stored as an integer (long) or floating point (float). For large or long numbers, the savings can clearly be much greater. The Stata default is for floating-point data to be stored as float and for computations to be stored as double.

Data read into Stata are stored using these various formats, and Stata data files (.dta) use these formats. One disadvantage is that numbers in internal-storage form cannot be read in the same way that text can; we need to first reconvert them to a text format. A second disadvantage is that it is not easy to transfer data in internal format across packages, such as transferring Excel's .xls to Stata's .dta, though commercial software is available that transfers data across leading packages.

It is much easier to transfer data that is stored as text data. Downsides, however, are an increase in the size of the dataset compared with the same dataset stored in internal numeric form, and possible loss of precision in converting floating-point data to text format.

2.2.3 String data

Nonnumeric data in Stata are recorded as strings, typically enclosed in double quotes, such as "3 Main Street". The format command str20, for example, states that the data should be stored as a string of length 20 characters.

In this book, we focus on numeric data and seldom use strings. Stata has many commands for working with strings. Two useful commands are destring, which converts string data to integer data, and tostring, which does the reverse.

2.2.4 Formats for displaying numeric data

Stata output and text files written by Stata format data for readability. The format is automatically chosen by Stata but can be overridden.

The most commonly used format is the f format, or the fixed format. An example is %7.2f, which means the number will be right-justified and fill 7 columns with 2 digits after the decimal point. For example, 123.321 is represented as 123.32.

The format type always begins with %. The default of right-justification is replaced by left-justification if an optional – follows. Then follows an integer for the width (number of columns), a period (.), an integer for the number of digits following the decimal point, and an e or f or g for the format used. An optional c at the end leads to comma format.

The usual format is the f format, or fixed format, e.g., 123.32. The e, or exponential, format (scientific notation) is used for very large or small numbers, e.g., 1.23321e+02. The g, or general format, leads to e or f being chosen by Stata in a way that will work well regardless of whether the data are very large or very small. In particular, the format %#.(#-1)g will vary the number of columns after the decimal point optimally. For example, %8.7g will present a space followed by the first six digits of the number and the appropriately placed decimal point.

2.3   Inputting data

The starting point is the computer-readable file that contains the raw data. Where large datasets are involved, this is typically either a text file or the output of another computer program, such as Excel, SAS, or even Stata.

2.3.1   General principles

For a discussion of initial use of Stata, see chapter 1. We generally assume that Stata is used in batch mode.

To replace any existing dataset in memory, you need to first clear the current dataset.

```
. * Remove current dataset from memory
. clear
```

This removes data and any associated value labels from memory. If you are reading in data from a Stata dataset, you can instead use the clear option with the use command. Various arguments of clear lead to additional removal of Mata functions, saved results, and programs. The clear all command removes all these.

Some datasets are large. In that case, we need to assign more memory than the Stata default by using the set memory command. For example, if 100 megabytes are needed, then we type

```
. * Set memory to 100 mb
. set memory 100m
```

Various commands are used to read in data, depending on the format of the file being read. These commands, discussed in detail in the rest of this section, include the following:

- `use` to read a Stata dataset (with extension `.dta`)

- `edit` and `input` to enter data from the keyboard or the Data Editor

- `insheet` to read comma-separated or tab-separated text data created by a spreadsheet

- `infile` to read unformatted or fixed-format text data

- `infix` to read formatted data

As soon as data are inputted into Stata, you should save the data as a Stata dataset. For example,

```
. * Save data as a Stata dataset
. save mydata.dta, replace
  (output omitted )
```

The `replace` option will replace any existing dataset with the same name. If you do not want this to happen, then do not use the option.

To check that data are read in correctly, list the first few observations, use `describe`, and obtain the summary statistics.

```
. * Quick check that data are read in correctly
. list in 1/5      // list the first five observations
  (output omitted )
. describe         // describe the variables
  (output omitted )
. summarize        // descriptive statistics for the variables
  (output omitted )
```

Examples illustrating the output from `describe` and `summarize` are given in sections 2.4.1 and 3.2.

2.3.2 Inputting data already in Stata format

Data in the Stata format are stored with the `.dta` extension, e.g., `mydata.dta`. Then the data can be read in with the `use` command. For example,

```
. * Read in existing Stata dataset
. use c:\research\mydata.dta, clear
```

The `clear` option removes any data currently in memory, even if the current data have not been saved, enabling the new file to be read in to memory.

If Stata is initiated from the current directory, then we can more simply type

```
. * Read in dataset in current directory
. use mydata.dta, clear
```

The `use` command also works over the Internet, provided that your computer is connected. For example, you can obtain an extract from the 1980 U.S. Census by typing

```
. * Read in dataset from an Internet web site
. use http://www.stata-press.com/data/r11/census.dta, clear
(1980 Census data by state)
. clear
```

2.3.3   Inputting data from the keyboard

The `input` command enables data to be typed in from the keyboard. It assumes that
data are numeric. If instead data are character, then `input` should additionally define
the data as a string and give the string length. For example,

```
. * Data input from keyboard
. input str20 name age female income
                   name      age     female    income
  1.    "Barry" 25 0 40.990
  2.    "Carrie" 30 1 37.000
  3.    "Gary" 31 0 48.000
  4. end
```

The quotes here are not necessary; we could use `Barry` rather than `"Barry"`. If the
name includes a space, such as `"Barry Jr"`, then double quotes are needed; otherwise,
`Barry` would be read as a string, and then `Jr` would be read as a number, leading to a
program error.

To check that the data are read in correctly, we use the `list` command. Here we
add the `clean` option, which lists the data without divider and separator lines.

```
. list, clean
         name    age   female    income
  1.    Barry     25        0     40.99
  2.   Carrie     30        1        37
  3.     Gary     31        0        48
```

In interactive mode, you can instead use the Data Editor to type in data (and to
edit existing data).

2.3.4   Inputting nontext data

By nontext data, we mean data that are stored in the internal code of a software package
other than Stata. It is easy to establish whether a file is a nontext file by viewing the
file using a text editor. If strange characters appear, then the file is a nontext file. An
example is an Excel `.xls` file.

Stata supports several special formats.   The `fdause` command reads SAS XPORT Transport format files; the `haver` command reads Haver Analytics database files; the `odbc` command reads Open Database Connectivity (ODBC) data files; and the `xmluse` command reads XML files.

Other formats such as an Excel `.xls` file cannot be read by Stata. One solution is to use the software that created the data to write the data out into one of the readable text format files discussed below, such as a comma-separated values text file. For example, just save an Excel worksheet as a `.csv` file. A second solution is to purchase software such as Stat/Transfer that will change data from one format to another. For conversion programs, see http://www.ats.ucla.edu/stat/Stata/faq/convert_pkg.htm.

2.3.5   Inputting text data from a spreadsheet

The `insheet` command reads data that are saved by a spreadsheet or database program as comma-separated or tab-separated text data. For example, `mus02file1.csv`, a file with comma-separated values, has the following data:

```
name,age,female,income
Barry,25,0,40.990
Carrie,30,1,37.000
Gary,31,0,48.000
```

To read these data, we use `insheet`. Thus

```
. * Read data from a csv file that includes variable names using insheet
. clear
. insheet using mus02file1.csv
(4 vars, 3 obs)
. list, clean
           name    age   female    income
  1.      Barry     25        0     40.99
  2.     Carrie     30        1        37
  3.       Gary     31        0        48
```

Stata automatically recognized the **name** variable to be a string variable, the **age** and **female** variables to be integer, and the **income** variable to be floating point.

A major advantage of `insheet` is that it can read in a text file that includes variable names as well as data, making mistakes less likely. There are some limitations, however. The `insheet` command is restricted to files with a single observation per line. And the data must be comma-separated or tab-separated, but not both. It cannot be space-separated, but other delimiters can be specified by using the `delimiter` option.

The first line with variable names is optional. Let `mus02file2.csv` be the same as the original file, except without the header line:

```
Barry,25,0,40.990
Carrie,30,1,37.000
Gary,31,0,48.000
```

The `insheet` command still works. By default, the variables read in are given the names v1, v2, v3, and v4. Alternatively, you can assign more meaningful names in `insheet`. For example,

```
. * Read data from a csv file without variable names and assign names
. clear

. insheet name age female income using mus02file2.csv
(4 vars, 3 obs)
```

2.3.6   Inputting text data in free format

The `infile` command reads free-format text data that are space-separated, tab-separated, or comma-separated.

We again consider `mus02file2.csv`, which has no header line. Then

```
. * Read data from free-format text file using infile
. clear

. infile str20 name age female income using mus02file2.csv
(3 observations read)

. list, clean
          name   age   female   income
   1.    Barry    25        0    40.99
   2.   Carrie    30        1       37
   3.     Gary    31        0       48
```

By default, `infile` reads in all data as numbers that are stored as floating point. This causes obvious problems if the original data are string. By inserting `str20` before `name`, the first variable is instead a string that is stored as a string of at most 20 characters.

For `infile`, a single observation is allowed to span more than one line, or there can be more than one observation per line. Essentially every fourth entry after `Barry` will be read as a string entry for `name`, every fourth entry after `25` will be read as a numeric entry for `age`, and so on.

The `infile` command is the most flexible command to read in data and will also read in fixed-format data.

2.3.7   Inputting text data in fixed format

The `infix` command reads fixed-format text data that are in fixed-column format. For example, suppose `mus02file3.txt` contains the same data as before, except without the header line and with the following fixed format:

```
Barry      250 40.990
Carrie     301 37.000
Gary       310 48.000
```

Here columns 1–10 store the `name` variable, columns 11–12 store the `age` variable, column 13 stores the `female` variable, and columns 14–20 store the `income` variable.

Note that a special feature of fixed-format data is that there need be no separator between data entries. For example, for the first observation, the sequence 250 is not `age` of 250 but is instead two variables: `age` = 25 and `female` = 0. It is easy to make errors when reading fixed-format data.

To use `infix`, we need to define the columns in which each entry appears. There are a number of ways to do this. For example,

```
. * Read data from fixed-format text file using infix
. clear
. infix str20 name 1-10 age 11-12 female 13 income 14-20 using mus02file3.txt
(3 observations read)
. list, clean

          name   age   female    income
  1.     Barry    25        0     40.99
  2.    Carrie    30        1        37
  3.      Gary    31        0        48
```

Similarly to `infile`, we include `str20` to indicate that `name` is a string rather than a number.

A single observation can appear on more than one line. Then we use the symbol / to skip a line or use the entry 2:, for example, to switch to line 2. For example, suppose `mus02file4.txt` is the same as `mus02file3.txt`, except that `income` appears on a separate second line for each observation in columns 1–7. Then

```
. * Read data using infix where an observation spans more than one line
. clear
. infix str20 name 1-10 age 11-12 female 13 2: income 1-7 using mus02file4.txt
(3 observations read)
```

2.3.8   Dictionary files

For more complicated text datasets, the format for the data being read in can be stored in a dictionary file, a text file created by a word processor, or editor. Details are provided in [D] **infile (fixed format)**. Suppose this file is called `mus02dict.dct`. Then we simply type

```
. * Read in data with dictionary file
. infile using mus02dict
```

where the dictionary file `mus02dict.dct` provides variable names and formats as well as the name of the file containing the data.

2.3.9   Common pitfalls

It can be surprisingly difficult to read in data. With fixed-format data, wrong column alignment leads to errors. Data can unexpectedly include string data, perhaps with embedded blanks. Missing values might be coded as NA, causing problems if a numeric value is expected. An observation can span several lines when a single line was erroneously assumed.

It is possible to read a dataset into Stata without Stata issuing an error message; no error message does not mean that the dataset has been successfully read in. For example, transferring data from one computer type to another, such as a file transfer using File Transfer Protocol (FTP), can lead to an additional carriage return, or *Enter*, being typed at the end of each line. Then `infix` reads the dataset as containing one line of data, followed by a blank line, then another line of data, and so on. The blank lines generate extraneous observations with missing values.

You should always perform checks, such as using `list` and `summarize`. Always view the data before beginning analysis.

2.4 Data management

Once the data are read in, there can be considerable work in cleaning up the data, transforming variables, and selecting the final sample. All data-management tasks should be recorded, dated, and saved. The existence of such a record makes it easier to track changes in definitions and eases the task of replication. By far, the easiest way to do this is to have the data-management manipulations stored in a do-file rather than to use commands interactively. We assume that a do-file is used.

2.4.1 PSID example

Data management is best illustrated using a real-data example. Typically, one needs to download the entire original dataset and an accompanying document describing the dataset. For some major commonly used datasets, however, there may be cleaned-up versions of the dataset, simple data extraction tools, or both.

Here we obtain a very small extract from the 1992 Individual-Level data from the Panel Study of Income Dynamics (PSID), a U.S. longitudinal survey conducted by the University of Michigan. The extract was downloaded from the Data Center at the web site http://psidonline.isr.umich.edu/, using interactive tools to select just a few variables. The extracted sample was restricted to men aged 30–50 years. The output conveniently included a Stata do-file in addition to the text data file. Additionally, a codebook describing the variables selected was provided. The data download included several additional variables that enable unique identifiers and provide sample weights. These should also be included in the final dataset but, for brevity, have been omitted below.

Reading the text dataset `mus02psid92m.txt` using a text editor reveals that the first two observations are

```
4^ 3^ 1^ 2^ 1^ 2482^ 1^ 10^ 40^ 9^ 22000^ 2340
4^ 170^ 1^ 2^ 1^ 6974^ 1^ 10^ 37^ 12^ 31468^ 2008
```

The data are text data delimited by the symbol ^.

Several methods could be used to read the data, but the simplest is to use `insheet`. This is especially simple here given the provided do-file. The `mus02psid92m.do` file contains the following information:

```
. * Commands to read in data from PSID extract
. type mus02psid92m.do
* mus02psid92m.do
clear
#delimit ;
*  PSID DATA CENTER ******************************************************
    JOBID             : 10654
    DATA_DOMAIN       : PSID
    USER_WHERE        : ER32000=1 and ER30736 ge 30 and ER
    FILE_TYPE         : All Individuals Data
    OUTPUT_DATA_TYPE  : ASCII Data File
    STATEMENTS        : STATA Statements
    CODEBOOK_TYPE     : PDF
    N_OF_VARIABLES    : 12
    N_OF_OBSERVATIONS : 4290
    MAX_REC_LENGTH    : 56
    DATE & TIME       : November 3, 2003 @ 0:28:35
********************************************************************
;
insheet
    ER30001 ER30002 ER32000 ER32022 ER32049 ER30733 ER30734 ER30735 ER30736
     ER30748 ER30750 ER30754
using mus02psid92m.txt, delim("^") clear
;
destring, replace ;
label variable er30001  "1968 INTERVIEW NUMBER"  ;
label variable er30002  "PERSON NUMBER                   68"  ;
label variable er32000  "SEX OF INDIVIDUAL"  ;
label variable er32022  "# LIVE BIRTHS TO THIS INDIVIDUAL"  ;
label variable er32049  "LAST KNOWN MARITAL STATUS"  ;
label variable er30733  "1992 INTERVIEW NUMBER"  ;
label variable er30734  "SEQUENCE NUMBER              92"  ;
label variable er30735  "RELATION TO HEAD            92"  ;
label variable er30736  "AGE OF INDIVIDUAL           92"  ;
label variable er30748  "COMPLETED EDUCATION         92"  ;
label variable er30750  "TOT LABOR INCOME            92"  ;
label variable er30754  "ANN WORK HRS                92"  ;
#delimit cr;    // Change delimiter to default cr
```

To read the data, only `insheet` is essential. The code separates commands using the delimiter ; rather than the default `cr` (the *Enter* key or carriage return) to enable comments and commands that span several lines. The `destring` command, unnecessary here, converts any string data into numeric data. For example, $1,234 would become 1234. The `label variable` command provides a longer description of the data that will be reproduced by using `describe`.

Executing this code yields output that includes the following:

```
(12 vars, 4290 obs)
. destring, replace ;
er30001 already numeric; no replace
  (output omitted)
er30754 already numeric; no replace
```

The statement `already numeric` is output for all variables because all the data in `mus02psid92m.txt` are numeric.

The `describe` command provides a description of the data:

```
. * Data description
. describe
Contains data
  obs:         4,290
  vars:           12
  size:       98,670 (99.1% of memory free)
```

variable name	storage type	display format	value label	variable label
er30001	int	%8.0g		1968 INTERVIEW NUMBER
er30002	int	%8.0g		PERSON NUMBER 68
er32000	byte	%8.0g		SEX OF INDIVIDUAL
er32022	byte	%8.0g		# LIVE BIRTHS TO THIS INDIVIDUAL
er32049	byte	%8.0g		LAST KNOWN MARITAL STATUS
er30733	int	%8.0g		1992 INTERVIEW NUMBER
er30734	byte	%8.0g		SEQUENCE NUMBER 92
er30735	byte	%8.0g		RELATION TO HEAD 92
er30736	byte	%8.0g		AGE OF INDIVIDUAL 92
er30748	byte	%8.0g		COMPLETED EDUCATION 92
er30750	long	%12.0g		TOT LABOR INCOME 92
er30754	int	%8.0g		ANN WORK HRS 92

```
Sorted by:
    Note:  dataset has changed since last saved
```

The `summarize` command provides descriptive statistics:

```
. * Data summary
. summarize
```

Variable	Obs	Mean	Std. Dev.	Min	Max
er30001	4290	4559.2	2850.509	4	9308
er30002	4290	60.66247	79.93979	1	227
er32000	4290	1	0	1	1
er32022	4290	21.35385	38.20765	1	99
er32049	4290	1.699534	1.391921	1	9
er30733	4290	4911.015	2804.8	1	9829
er30734	4290	3.179487	11.4933	1	81
er30735	4290	13.33147	12.44482	10	98
er30736	4290	38.37995	5.650311	30	50
er30748	4290	14.87249	15.07546	0	99
er30750	4290	27832.68	31927.35	0	999999
er30754	4290	1929.477	899.5496	0	5840

Satisfied that the original data have been read in carefully, we proceed with cleaning the data.

2.4.2   Naming and labeling variables

Just as the Data Editor can be used to input and manage data, the Variables Manager can be used to manage the properties of variables, such as their names and labels. We use Stata commands below to rename and label variables, but we could also have used the Variables Manager.

The first step is to give more meaningful names to variables by using the `rename` command. We do so just for the variables used in subsequent analysis.

```
. * Rename variables
. rename er32000 sex
. rename er30736 age
. rename er30748 education
. rename er30750 earnings
. rename er30754 hours
```

The renamed variables retain the descriptions that they were originally given. Some of these descriptions are unnecessarily long, so we use `label variable` to shorten output from commands, such as `describe`, that give the variable labels.

```
. * Relabel some of the variables
. label variable age "AGE OF INDIVIDUAL"
. label variable education "COMPLETED EDUCATION"
. label variable earnings "TOT LABOR INCOME"
. label variable hours "ANN WORK HRS"
```

For categorical variables, it can be useful to explain the meanings of the variables. For example, from the codebook discussed in section 2.4.4, the `er32000` variable takes on the value 1 if male and 2 if female. We may prefer that the output of variable values uses a label in place of the number. These labels are provided by using `label define` together with `label values`.

```
. * Define the label gender for the values taken by variable sex
. label define gender 1 male 2 female
. label values sex gender
. list sex in 1/2, clean
        sex
  1.    male
  2.    male
```

After renaming, we obtain

```
. * Data summary of key variables after renaming
. summarize sex age education earnings hours
```

Variable	Obs	Mean	Std. Dev.	Min	Max
sex	4290	1	0	1	1
age	4290	38.37995	5.650311	30	50
education	4290	14.87249	15.07546	0	99
earnings	4290	27832.68	31927.35	0	999999
hours	4290	1929.477	899.5496	0	5840

Data exist for these variables for all 4,290 sample observations. The data have $30 \leq$ age ≤ 50 and sex $= 1$ (male) for all observations, as expected. The maximum value for earnings is \$999,999, an unusual value that most likely indicates top-coding. The maximum value of hours is quite high and may also indicate top-coding ($365 \times 16 = 5840$). The maximum value of 99 for education is clearly erroneous; the most likely explanation is that this is a missing-value code, because numbers such as 99 or -99 are often used to denote a missing value.

2.4.3 Viewing data

The standard commands for viewing data are summarize, list, and tabulate.

We have already illustrated the summarize command. Additional statistics, including key percentiles and the five largest and smallest observations, can be obtained by using the detail option; see section 3.2.4.

The list command can list every observation, too many in practice. But you could list just a few observations:

```
. * List first 2 observations of two of the variables
. list age hours in 1/2, clean

        age    hours
  1.     40     2340
  2.     37     2008
```

The list command with no variable list provided will list all the variables. The clean option eliminates dividers and separators.

The tabulate command lists each distinct value of the data and the number of times it occurs. It is useful for data that do not have too many distinctive values. For education, we have

```
. * Tabulate all values taken by a single variable
. tabulate education
```

COMPLETED EDUCATION	Freq.	Percent	Cum.
0	82	1.91	1.91
1	7	0.16	2.07
2	20	0.47	2.54
3	32	0.75	3.29
4	26	0.61	3.89
5	30	0.70	4.59
6	123	2.87	7.46
7	35	0.82	8.28
8	78	1.82	10.09
9	117	2.73	12.82
10	167	3.89	16.71
11	217	5.06	21.77
12	1,510	35.20	56.97
13	263	6.13	63.10
14	432	10.07	73.17
15	172	4.01	77.18
16	535	12.47	89.65
17	317	7.39	97.04
99	127	2.96	100.00
Total	4,290	100.00	

Note that the variable label rather than the variable name is used as a header. The values are generally plausible, with 35% of the sample having a highest grade completed of exactly 12 years (high school graduate). The 7% of observations with 17 years most likely indicates a postgraduate degree (a college degree is only 16 years). The value 99 for 3% of the sample most likely is a missing-data code. Surprisingly, 2% appear to have completed no years of schooling. As we explain next, these are also observations with missing data.

2.4.4 Using original documentation

At this stage, it is really necessary to go to the original documentation.

The `mus02psid92mcb.pdf` file, generated as part of the data extraction from the PSID web site, states that for the `er30748` variable a value of 0 means "inappropriate" for various reasons given in the codebook; the values 1–16 are the highest grade or year of school completed; 17 is at least some graduate work; and 99 denotes not applicable (NA) or did not know (DK).

Clearly, the `education` values of both 0 and 99 denote missing values. Without using the codebook, we may have misinterpreted the value of 0 as meaning zero years of schooling.

2.4.5 Missing values

It is best at this stage to flag missing values and to keep all observations rather than to immediately drop observations with missing data. In later analysis, only those observations with data missing on variables essential to the analysis need to be dropped. The characteristics of individuals with missing data can be compared with those having complete data. Data with a missing value are recoded with a missing-value code.

For education, the missing-data values 0 or 99 are replaced by . (a period), which is the default Stata missing-value code. Rather than create a new variable, we modify the current variable by using replace, as follows:

```
. * Replace missing values with missing-data code
. replace education = . if education == 0 | education == 99
(209 real changes made, 209 to missing)
```

Using the double equality and the symbol | for the logical operator or is detailed in section 1.3.6. As an example of the results, we list observations 46–48:

```
. * Listing of variable including missing value
. list education in 46/48, clean

        educat~n
46.        12
47.         .
48.        16
```

Evidently, the original data on education for the 47th observation equaled 0 or 99. This has been changed to missing.

Subsequent commands using the education variable will drop observations with missing values. For example,

```
. * Example of data analysis with some missing values
. summarize education age

    Variable |      Obs       Mean    Std. Dev.      Min       Max
-------------+-------------------------------------------------------
   education |     4081    12.5533    2.963696        1        17
         age |     4290   38.37995    5.650311       30        50
```

For education, only the 4,081 nonmissing values are used, whereas for age, all 4,290 of the original observations are available.

If desired, you can use more than one missing-value code. This can be useful if you want to keep track of reasons why a variable is missing. The extended missing codes are .a, .b, ..., .z. For example, we could instead have typed

```
. * Assign more than one missing code
. replace education = .a if education == 0
. replace education = .b if education == 99
```

When we want to apply multiple missing codes to a variable, it is more convenient to use the mvdecode command, which is similar to the recode command (discussed in section 2.4.7), which changes variable values or ranges of values into missing-value codes. The reverse command, mvencode, changes missing values to numeric values.

Care is needed once missing values are used. In particular, missing values are treated as large numbers, higher than any other number. The ordering is that all numbers are less than ., which is less than .a, and so on. The command

```
. * This command will include missing values
. list education in 40/60 if education > 16, clean

        educat~n
45.         17
47.          .
60.         17
```

lists the missing value for observation 47 in addition to the two values of 17. If this is not desired, we should instead use

```
. * This command will not include missing values
. list education in 40/60 if education > 16 & education < . , clean

        educat~n
45.         17
60.         17
```

Now observation 47 with the missing observation has been excluded.

The issue of missing values also arises for `earnings` and `hours`. From the codebook, we see that a zero value may mean missing for various reasons, or it may be a true zero if the person did not work. True zeros are indicated by `er30749=0` or 2, but we did not extract this variable. For such reasons, it is not unusual to have to extract data several times. Rather than extract this additional variable, as a shortcut we note that `earnings` and `hours` are missing for the same reasons that `education` is missing. Thus

```
. * Replace missing values with missing-data code
. replace earnings = . if education >= .
(209 real changes made, 209 to missing)
. replace hours = . if education >= .
(209 real changes made, 209 to missing)
```

2.4.6   Imputing missing data

The standard approach in microeconometrics is to drop observations with missing values, called listwise deletion. The loss of observations generally leads to less precise estimation and inference. More importantly, it may lead to sample-selection bias in regression if the retained observations have unrepresentative values of the dependent variable conditional on regressors.

An alternative to dropping observations is to impute missing values. The `impute` command uses predictions from regression to impute. The `ipolate` command uses interpolation methods. We do not cover these commands because these imputation methods have limitations, and the norm in microeconometrics studies is to use only the original data.

A more promising approach, though one more advanced, is multiple imputation. This produces M different imputed datasets (e.g., $M = 20$), fits the model M times, and performs inference that allows for the uncertainty in both estimation and data imputation. For implementation, see the `mi` command, introduced in Stata 11, and see the user-written `ice` and `hotdeck` commands. You can find more information in Cameron and Trivedi (2005) and from `findit multiple imputation`.

2.4.7 Transforming data (generate, replace, egen, recode)

After handling missing values, we have the following for the key variables:

```
. * Summarize cleaned up data
. summarize sex age education earnings

    Variable |       Obs        Mean    Std. Dev.        Min         Max
-------------+--------------------------------------------------------
         sex |      4290           1            0          1           1
         age |      4290    38.37995     5.650311         30          50
   education |      4081     12.5533     2.963696          1          17
    earnings |      4081    28706.65     32279.12          0      999999
```

We now turn to recoding existing variables and creating new variables. The basic commands are `generate` and `replace`. It can be more convenient, however, to use the additional commands `recode`, `egen`, and `tabulate`. These are often used in conjunction with the `if` qualifier and the `by:` prefix. We present many examples throughout the book.

The generate and replace commands

The `generate` command is used to create new variables, often using standard mathematical functions. The syntax of the command is

generate $\big[$ *type* $\big]$ *newvar* = *exp* $\big[$ *if* $\big]$ $\big[$ *in* $\big]$

where for numeric data the default type is `float`, but this can be changed, for example, to `double`.

It is good practice to assign a unique identifier to each observation if one does not already exist. A natural choice is to use the current observation number stored as the system variable _n.

```
. * Create identifier using generate command
. generate id = _n
```

We use this identifier for simplicity, though for these data the `er30001` and `er30002` variables when combined provide a unique PSID identifier.

The following command creates a new variable for the natural logarithm of earnings:

```
. * Create new variable using generate command
. generate lnearns = ln(earnings)
(498 missing values generated)
```

Missing values for `ln(earnings)` are generated whenever `earnings` data are missing. Additionally, missing values arise when `earnings` ≤ 0 because it is then not possible to take on the logarithm.

The `replace` command is used to replace some or all values of an existing variable. We already illustrated this when we created missing-values codes.

The egen command

The `egen` command is an extension to `generate` that enables creation of variables that would be difficult to create using `generate`. For example, suppose we want to create a variable that for each observation equals sample average earnings provided that sample earnings are nonmissing. The command

```
. * Create new variable using egen command
. egen aveearnings = mean(earnings) if earnings < .
(209 missing values generated)
```

creates a variable equal to the average of earnings for those observations not missing data on `earnings`.

The recode command

The `recode` command is an extension to `replace` that recodes categorical variables and generates a new variable if the `generate()` option is used. The command

```
. * Replace existing data using the recode command
. recode education (1/11=1) (12=2) (13/15=3) (16/17=4), generate(edcat)
(4074 differences between education and edcat)
```

creates a new variable, `edcat`, that takes on a value of 1, 2, 3, or 4 corresponding to, respectively, less than high school graduate, high school graduate, some college, and college graduate or higher. The `edcat` variable is set to missing if `education` does not lie in any of the ranges given in the `recode` command.

The by prefix

The `by` *varlist*: prefix repeats a command for each group of observations for which the variables in *varlist* are the same. The data must first be sorted by *varlist*. This can be done by using the `sort` command, which orders the observations in ascending order according to the variable(s) given in the command.

The `sort` command and the `by` prefix are more compactly combined into the `bysort` prefix. For example, suppose we want to create for each individual a variable that equals the sample average earnings for all persons with that individual's years of education. Then we type

```
. * Create new variable using bysort: prefix
. bysort education: egen aveearnsbyed = mean(earnings)
(209 missing values generated)

. sort id
```

The final command, one that returns the ordering of the observation to the original ordering, is not required. But it could make a difference in subsequent analysis if, for example, we were to work with a subsample of the first 1,000 observations.

Indicator variables

Consider creating a variable indicating whether earnings are positive. While there are several ways to proceed, we only describe our recommended method.

The most direct way is to use `generate` with logical operators:

```
. * Create indicator variable using generate command with logical operators
. generate d1 = earnings > 0 if earnings < .
(209 missing values generated)
```

The expression `d1 = earnings > 0` creates an indicator variable equal to 1 if the condition holds and 0 otherwise. Because missing values are treated as large numbers, we add the condition `if earnings < .` so that in those cases `d1` is set equal to missing.

Using `summarize`,

```
. summarize d1
```

Variable	Obs	Mean	Std. Dev.	Min	Max
d1	4081	.929184	.2565486	0	1

we can see that about 93% of the individuals in this sample had some earnings in 1992. We can also see that we have $0.929184 \times 4081 = 3792$ observations with a value of 1, 289 observations with a value of 0, and 209 missing observations.

Set of indicator variables

A complete set of mutually exclusive categorical indicator dummy variables can be created in several ways.

For example, suppose we want to create mutually exclusive indicator variables for less than high school graduate, high school graduate, some college, and college graduate or more. The starting point is the `edcat` variable, created earlier, which takes on the values 1–4.

We can use `tabulate` with the `generate()` option.

```
. * Create a set of indicator variables using tabulate with generate() option
. quietly tabulate edcat, generate(eddummy)

. summarize eddummy*

    Variable |        Obs        Mean    Std. Dev.       Min        Max
-------------+--------------------------------------------------------
    eddummy1 |       4081    .2087724    .4064812          0          1
    eddummy2 |       4081    .3700074    .4828655          0          1
    eddummy3 |       4081    .2124479    .4090902          0          1
    eddummy4 |       4081    .2087724    .4064812          0          1
```

The four means sum to one, as expected for four mutually exclusive categories. Note that if `edcat` had taken on values 4, 5, 7, and 9, rather than 1–4, it would still generate variables numbered `eddummy1`–`eddummy4`.

It is usually not necessary to actually create a set of indicator variables. Instead, we can include factor variables in the variable list; see section 1.3.4. For example,

```
. * Set of indicator variables using factor variables - no category is omitted
. summarize ibn.edcat

    Variable |        Obs        Mean    Std. Dev.       Min        Max
-------------+--------------------------------------------------------
       edcat |
          1  |       4081    .2087724    .4064812          0          1
          2  |       4081    .3700074    .4828655          0          1
          3  |       4081    .2124479    .4090902          0          1
          4  |       4081    .2087724    .4064812          0          1
```

No category is omitted because we used the `ibn.` operator. If instead we used the simpler `i.` operator, then the lowest category (here `edcat = 1`) would be omitted.

Almost all commands with a variable list permit use of factor variables in the variable list. Exceptions include a few estimation commands such as the `asmprobit` and `exlogistic` commands. In such cases, the older `xi` prefix command can be used instead. For details, type `help xi`.

Interactions

Interactive variables can be created in the obvious manner. For example, to create an interaction between the binary earnings indicator `d1` and the continuous variable `education`, type

```
. * Create interactive variable using generate commands
. generate d1education = d1*education
(209 missing values generated)
```

(Continued on next page)

Rather than create the interactive variables, we can use factor variables. For example,

```
. * Set of interactions using factor variables
. summarize i.edcat#c.earnings
    Variable |       Obs        Mean    Std. Dev.        Min         Max
-------------+--------------------------------------------------------
      edcat#|
   c.earnings |
           1 |      4081    3146.368    8286.325          0       80000
           2 |      4081    8757.823    15710.76          0      215000
           3 |      4081    6419.347    16453.14          0      270000
           4 |      4081    10383.11    32316.32          0      999999
```

Here the `#` operator is used to create interactions, the `i.` operator is applied to a categorical variable, and the `c.` operator is for a continuous variable.

We can also create interactions between categorical variables and interactions between continuous variables; see section 1.3.4.

Demeaning

Suppose we want to include a quadratic in age as a regressor. The marginal effect of age is much easier to interpret if we use the demeaned variables $(\mathtt{age} - \overline{\mathtt{age}})$ and $(\mathtt{age} - \overline{\mathtt{age}})^2$ as regressors.

```
. * Create demeaned variables
. egen double aveage = mean(age)
. generate double agedemean = age - aveage
. generate double agesqdemean = agedemean^2
. summarize agedemean agesqdemean
    Variable |       Obs        Mean    Std. Dev.         Min         Max
-------------+---------------------------------------------------------
   agedemean |      4290     2.32e-15    5.650311   -8.379953    11.62005
 agesqdemean |      4290     31.91857    32.53392    .1443646    135.0255
```

We expect the `agedemean` variable to have an average of zero. We specified `double` to obtain additional precision in the floating-point calculations. In the case at hand, the mean of `agedemean` is on the order of 10^{-15} instead of 10^{-6}, which is what single-precision calculations would yield.

2.4.8 Saving data

At this stage, the dataset may be ready for saving. The `save` command creates a Stata data file. For example,

```
. * Save as Stata data file
. save mus02psid92m.dta, replace
file mus02psid92m.dta saved
```

The `replace` option means that an existing dataset with the same name, if it exists, will be overwritten. The `.dta` extension is unnecessary because it is the default extension.

The related command `saveold` saves a data file that can be read by versions 8 and 9 of Stata.

The data can also be saved in another format that can be read by programs other than Stata. The `outsheet` command allows saving as a text file in a spreadsheet format. For example,

```
. * Save as comma-separated values spreadsheet
. outsheet age education eddummy* earnings d1 hours using mus02psid92m.csv,
> comma replace
```

Note the use of the wildcard `*` in `eddummy`. The `outsheet` command expands this to `eddummy1`–`eddummy4` per the rules for wildcards, given in section 1.3.5. The `comma` option leads to a `.csv` file with comma-separated variable names in the first line. The first two lines in `mus02psid92m.csv` are then

```
age,education,eddummy1,eddummy2,eddummy3,eddummy4,earnings,d1,hours
40,9,1,0,0,0,22000,1,2340
```

A space-delimited formatted text file can also be created by using the `outfile` command:

```
. * Save as formatted text (ascii) file
. outfile age education eddummy* earnings d1 hours using mus02psid92m.asc,
> replace
```

The first line in `mus02psid92m.asc` is then

```
   40        9        1        0        0        0        22000
    1     2340
```

This file will take up a lot of space; less space is taken if the `comma` option is used. The format of the file can be specified using Stata's dictionary format.

2.4.9 Selecting the sample

Most commands will automatically drop missing values in implementing a given command. We may want to drop additional observations, for example, to restrict analysis to a particular age group.

This can be done by adding an appropriate `if` qualifier after the command. For example, if we want to summarize data for only those individuals 35–44 years old, then

```
. * Select the sample used in a single command using the if qualifier
. summarize earnings lnearns if age >= 35 & age <= 44
```

Variable	Obs	Mean	Std. Dev.	Min	Max
earnings	2114	30131.05	37660.11	0	999999
lnearns	1983	10.04658	.9001594	4.787492	13.81551

Different samples are being used here for the two variables, because for the 131 observations with zero earnings, we have data on `earnings` but not on `lnearns`. The `if` qualifier uses logical operators, defined in section 1.3.6.

However, for most purposes, we would want to use a consistent sample. For example, if separate earnings regressions were run in levels and in logs, we would usually want to use the same sample in the two regressions.

The `drop` and `keep` commands allow sample selection for the rest of the analysis. The `keep` command explicitly selects the subsample to be retained. Alternatively, we can use the `drop` command, in which case the subsample retained is the portion not dropped. The sample dropped or kept can be determined by using an `if` qualifier, a variable list, or by defining a range of observations.

For the current example, we use

```
. * Select the sample using command keep
. keep if (lnearns != .) & (age >= 35 & age <= 44)
(2307 observations deleted)
. summarize earnings lnearns
    Variable |       Obs        Mean    Std. Dev.        Min         Max
-------------+----------------------------------------------------------
    earnings |      1983    32121.55     38053.31        120      999999
     lnearns |      1983    10.04658     .9001594   4.787492    13.81551
```

This command keeps the data provided: `lnearns` is nonmissing and $35 \leq \mathtt{age} \leq 44$. Note that now `earnings` and `lnearns` are summarized for the same 1,983 observations.

As a second example, the commands

```
. * Select the sample using keep and drop commands
. use mus02psid92m.dta, clear
. keep lnearns age
. drop in 1/1000
(1000 observations deleted)
```

will lead to a sample that contains data on all but the first one thousand observations for just the two variables `lnearns` and `age`. The `use mus02psid92m.dta` command is added because the previous example had already dropped some of the data.

2.5 Manipulating datasets

Useful manipulations of datasets include reordering observations or variables, temporarily changing the dataset but then returning to the original dataset, breaking one observation into several observations (and vice versa), and combining more than one dataset.

2.5.1   Ordering observations and variables

Some commands, such as those using the **by** prefix, require sorted observations. The **sort** command orders observations in ascending order according to the variable(s) in the command. The **gsort** command allows ordering to be in descending order.

You can also reorder the variables by using the **order** command. This can be useful if, for example, you want to distribute a dataset to others with the most important variables appearing as the first variables in the dataset.

2.5.2   Preserving and restoring a dataset

In some cases, it is desirable to temporarily change the dataset, perform some calculation, and then return the dataset to its original form. An example involving the computation of marginal effects is presented in section 10.5.4. The **preserve** command preserves the data, and the **restore** command restores the data to the form it had immediately before **preserve**.

```
. * Commands preserve and restore illustrated
. use mus02psid92m.dta, clear
. list age in 1/1, noheader clean
     1.    40
. preserve
. replace age = age + 1000
age was byte now int
(4290 real changes made)
. list age in 1/1, noheader clean
     1.    1040
. restore
. list age in 1/1, noheader clean
     1.    40
```

As desired, the data have been returned to original values.

2.5.3   Wide and long forms for a dataset

Some datasets may combine several observations into a single observation. For example, a single household observation may contain data for several household members, or a single individual observation may have data for each of several years. This format for data is called wide form. If instead these data are broken out so that an observation is for a distinct household member, or for a distinct individual–year pair, the data are said to be in long form.

The **reshape** command is detailed in section 8.11. It converts data from wide form to long form and vice versa. This is necessary if an estimation command requires data to be in long form, say, but the original dataset is in wide form. The distinction is important especially for analysis of panel data and multinomial data.

2.5.4   Merging datasets

The `merge` command combines two datasets to create a wider dataset, i.e., new variables from the second dataset are added to existing variables of the first dataset. Common examples are data on the same individuals obtained from two separate sources that then need to be combined, and data on supplementary variables or additional years of data.

Merging two datasets involves adding information from a dataset on disk to a dataset in memory. The dataset in memory is known as the master dataset.

Merging two datasets is straightforward if the datasets have the same number of observations and the merge is a line-to-line merge. Then line 10, for example, of one dataset is combined with line 10 of the other dataset to create a longer line 10. We consider instead a match-merge, where observations in the two datasets are combined if they have the same values for one or more identifying variables that are used to determine the match. In either case, when a match is made if a variable appears in both datasets, then the master dataset value is retained unless it is missing, in which case it is replaced by the value in the second dataset. If a variable exists only in the second dataset, then it is added as a variable to the master dataset.

To demonstrate a match-merge, we create two datasets from the dataset used in this chapter. The first dataset comprises every third observation with data on `id`, `education`, and `earnings`:

```
. * Create first dataset with every third observation
. use mus02psid92m.dta, clear
. keep if mod(_n,3) == 0
(2860 observations deleted)
. keep id education earnings
. list in 1/4, clean
      educat~n    earnings    id
  1.        16       38708     3
  2.        12        3265     6
  3.        11       19426     9
  4.        11       30000    12
. quietly save merge1.dta, replace
```

The `keep if mod(_n,3) == 0` command keeps an observation if the observation number (`_n`) is exactly divisible by 3, so every third observation is kept. Because `id=_n` for these data, by saving every third observation we are saving observations with `id` equal to 3, 6, 9, ...

The second dataset comprises every second observation with data on `id`, `education`, and `hours`:

```
. * Create second dataset with every second observation
. use mus02psid92m.dta, clear
. keep if mod(_n,2) == 0
(2145 observations deleted)
. keep id education hours
```

```
. list in 1/4, clean
      educat~n   hours   id
  1.        12    2008    2
  2.        12    2200    4
  3.        12     552    6
  4.        17    3750    8
. quietly save merge2.dta, replace
```

Now we are saving observations with id equal to 2, 4, 6, ...

Now we merge the two datasets by using the merge command.

In our case, the datasets differ in both the observations included and the variables included, though there is considerable overlap. We perform a match-merge on id to obtain

```
. * Merge two datasets with some observations and variables different
. clear
. use merge1.dta
. sort id
. merge 1:1 id using merge2.dta

    Result                           # of obs.

    not matched                          2,145
        from master                        715  (_merge==1)
        from using                       1,430  (_merge==2)

    matched                                715  (_merge==3)

. sort id
. list in 1/4, clean
      educat~n   earnings   id   hours           _merge
  1.        12          .    2    2008    using only (2)
  2.        16      38708    3       .   master only (1)
  3.        12          .    4    2200    using only (2)
  4.        12       3265    6     552       matched (3)
```

Recall that observations from the master dataset have id equal to 3, 6, 9, ..., and observations from the second dataset have id equal to 2, 4, 6, ... Data for education and earnings are always available because they are in the master dataset. But observations for hours come from the second dataset; they are available when id is 2, 4, 6, ... and are missing otherwise.

merge creates a variable, _merge, that takes on a value of 1 if the variables for an observation all come from the master dataset, a value of 2 if they all come from only the second dataset, and a value of 3 if for an observation some variables come from the master and some from the second dataset. After using merge, you should check that the number of observations for each value of _merge matches your expectations.

There are several options when using merge. The update option varies the action merge takes when an observation is matched. By default, the master dataset is held inviolate—if update is specified, values from the master dataset are retained if the same variables are found in both datasets. However, the values from the merging dataset are used in cases where the variable is missing in the master dataset. The `replace` option, allowed only with the `update` option, specifies that even if the master dataset contains nonmissing values, they are to be replaced with corresponding values from the merging dataset when corresponding values are not equal. A nonmissing value, however, will never be replaced with a missing value.

2.5.5   Appending datasets

The `append` command creates a longer dataset, with the observations from the second dataset appended after all the observations from the first dataset. If the same variable has different names in the two datasets, the variable name in one of the datasets should be changed by using the `rename` command so that the names match.

```
. * Append two datasets with some observations and variables different
. clear
. use merge1.dta
. append using merge2.dta
. sort id
. list in 1/5, clean
        educat~n    earnings    id    hours
   1.        12           .      2     2008
   2.        16       38708      3        .
   3.        12           .      4     2200
   4.        12        3265      6        .
   5.        12           .      6      552
```

Now `merge2.dta` is appended to the end of `merge1.dta`. The combined dataset has observations 3, 6, 9, ..., 4290 followed by observations 2, 4, 6, ..., 4290. We then sort on `id`. Now both every second and every third observation is included, so after sorting we have observations 2, 3, 4, 6, 8, 9, ... Note, however, that no attempt has been made to merge the datasets. In particular, for the observation with `id` = 6, the `hours` variable is missing in observation 4 and the `earnings` variable is missing in observation 5. This is because the `hours` variable is missing from the master dataset and the `earnings` variable is missing from the using dataset. There was no attempt to merge the data.

In this example, to take full advantage of the data, we would need to merge the two datasets using the first dataset as the master, merge the two datasets using the second dataset as the master, and then append the two datasets.

2.6   Graphical display of data

Graphs visually demonstrate important features of the data. Different types of data require distinct graph formats to bring out these features. We emphasize methods for numerical data taking many values, particularly, nonparametric methods.

2.6.1   Stata graph commands

The Stata graph commands begin with the word `graph` (in some cases, this is optional) followed by the graph plottype, usually `twoway`. We cover several leading examples but ignore the plottypes `bar` and `pie` for categorical data.

Example graph commands

The basic graph commands are very short and simple to use. For example,

```
. use mus02psid92m.dta, clear
. twoway scatter lnearns hours
```

produces a scatterplot of `lnearns` on `hours`, shown in figure 2.1. Most graph commands support the `if` and `in` qualifiers, and some support weights.

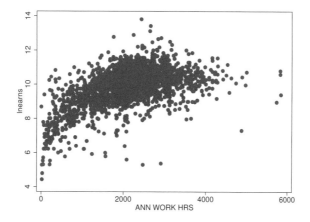

Figure 2.1. A basic scatterplot of log earnings on hours

In practice, however, customizing is often desirable. For example, we may want to display the relationship between `lnearns` and `hours` by showing both the data scatterplot and the ordinary least-squares (OLS) fitted line on the same graph. Additionally, we may want to change the size of the scatterplot data points, change the width of the regression line, and provide a title for the graph. We type

```
. * More advanced graphics command with two plots and with several options
. graph twoway (scatter lnearns hours, msize(small))
> (lfit lnearns hours, lwidth(medthick)),
> title("Scatterplot and OLS fitted line")
```

The two separate components `scatter` and `lfit` are specified separately within parentheses. Each of these commands is given with one option, after the comma but within the relevant parentheses. The `msize(small)` option makes the scatterplot dots smaller than the default, and the `lwidth(medthick)` option makes the OLS fitted line thicker than the default. The `title()` option for `twoway` appears after the last comma. The graph produced is shown in figure 2.2.

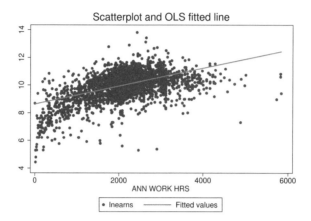

Figure 2.2. A more elaborate scatterplot of log earnings on hours

We often use lengthy graph commands that span multiple lines to produce template graphs that are better looking than those produced with default settings. In particular, these commands add titles and rescale the points, lines, and axes to a suitable size because the graphs printed in this book are printed in a much smaller space than a full-page graph in landscape mode. These templates can be modified for other applications by changing variable names and title text.

Saving and exporting graphs

Once a graph is created, it can be saved. Stata uses the term `save` to mean saving the graph in Stata's internal graph format, as a file with the `.gph` extension. This can be done by using the `saving()` option in a `graph` command or by typing `graph save` after the graph is created. When saved in this way, the graphs can be reaccessed and further manipulated at a later date.

Two or more Stata graphs can be combined into a single figure by using the `graph combine` command. For example, we save the first graph as `graph1.gph`, save the second graph as `graph2.gph`, and type the command

```
. * Combine graphs saved as graph1.gph and graph2.gph
. graph combine graph1 graph2
  (output omitted )
```

Section 3.2.7 provides an example.

The Stata internal graph format (`.gph`) is not recognized by other programs, such as word processors. To save a graph in an external format, you would use the `graph export` command. For example,

```
 . * Save graph as a Windows meta-file
 . graph export mygraph.wmf
```
   (*output omitted*)

Various formats are available, including PostScript (.ps), Encapsulated PostScript (.eps), Windows Metafile (.wmf), PDF (.pdf), and Portable Network Graphics (.png). The best format to select depends in part on what word processor is used; some trial and error may be needed.

Learning how to use graph commands

The Stata graph commands are extremely rich and provide an exceptional range of user control through a multitude of options.

A good way to learn the possibilities is to create a graph interactively in Stata. For example, from the menus, select **Graphics > Twoway graph (scatter, line, etc.)**. In the **Plots** tab of the resulting dialog box, select **Create...**, choose **Scatter**, provide a *Y variable* and an *X variable*, and then click on **Marker properties**. From the **Symbol** drop-down list, change the default to, say, **Triangle**. Similarly, cycle through the other options and change the default settings to something else.

Once an initial graph is created, the point-and-click Stata Graph Editor allows further customizing of the graph, such as adding text and arrows wherever desired. This is an exceptionally powerful tool that we do not pursue here; for a summary, see [G] **graph editor**. The Graph Recorder can even save sequences of changes to apply to similar graphs created from different samples.

Even given familiarity with Stata's graph commands, you may need to tweak a graph considerably to make it useful. For example, any graph that analyzes the earnings variable using all observations will run into problems because one observation has a large outlying value of $999,999. Possibilities in that case are to drop outliers, plot with the yscale(log) option, or use log earnings instead.

2.6.2 Box-and-whisker plot

The graph box command produces a box-and-whisker plot that is a graphical way to display data on a single series. The boxes cover the interquartile range, from the lower quartile to the upper quartile. The whiskers, denoted by horizontal lines, extend to cover most or all the range of the data. Stata places the upper whisker at the upper quartile plus 1.5 times the interquartile range, or at the maximum of the data if this is smaller. Similarly, the lower whisker is the lower quartile minus 1.5 times the interquartile range, or the minimum should this be larger. Any data values outside the whiskers are represented with dots. Box-and-whisker plots can be especially useful for identifying outliers.

The essential command for a box-and-whisker plot of the `hours` variable is

```
. * Simple box-and-whisker plot
. graph box hours
  (output omitted)
```

We want to present separate box plots of `hours` for each of four education groups by using the `over()` option. To make the plot more intelligible, we first provide labels for the four education categories as follows:

```
. use mus02psid92m.dta, clear
. label define edtype 1 "< High School" 2 "High School" 3 "Some College"
> 4 "College Degree"
. label values edcat edtype
```

The `scale(1.2)` graph option is added for readability; it increases the size of text, markers, and line widths (by a multiple 1.2). The `marker()` option is added to reduce the size of quantities within the box; the `ytitle()` option is used to present the title; and the `yscale(titlegap(*5))` option is added to increase the gap between the *y*-axis title and the tick labels. We have

```
. * Box and whisker plot of single variable over several categories
. graph box hours, over(edcat) scale(1.2) marker(1,msize(vsmall))
> ytitle("Annual hours worked by education") yscale(titlegap(*5))
```

The result is given in figure 2.3. The labels for `edcat`, rather than the values, are automatically given, making the graph much more readable. The filled-in boxes present the interquartile range, the intermediate line denotes the median, and data outside the whiskers appear as dots. For these data, annual hours are clearly lower for the lowest schooling group, and there are quite a few outliers. About 30 individuals appear to work in excess of 4,000 hours per year.

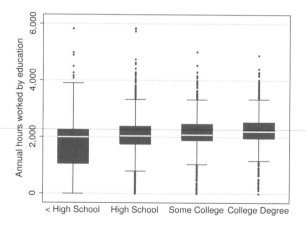

Figure 2.3. Box-and-whisker plots of annual hours for four categories of educational attainment

2.6.3   Histogram

The probability mass function or density function can be estimated using a histogram produced by the `histogram` command. The command can be used with `if` and `in` qualifiers and with weights. The key options are `width(#)` to set the bin width, `bin(#)` to set the number of bins, `start(#)` to set the lower limit of the first bin, and `discrete` to indicate that the data are discrete. The default number of bins is $\min(\sqrt{N}, 10 \ln N / \ln 10)$. Other options overlay a fitted normal density (the `normal` option) or a kernel density estimate (the `kdensity` option).

For discrete data taking relatively few values, there is usually no need to use the options.

For continuous data or for discrete data taking many values, it can be necessary to use options because the Stata defaults set bin widths that are not nicely rounded numbers and the number of bins might also not be desirable. For example, the output from `histogram lnearns` states that there are 35 bins, a bin width of 0.268, and a start value of 4.43. A better choice may be

```
. * Histogram with bin width and start value set
. histogram lnearns, width(0.25) start(4.0)
(bin=40, start=4, width=.25)
```

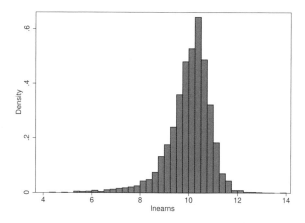

Figure 2.4. A histogram for log earnings

2.6.4   Kernel density plot

For continuous data taking many values, a better alternative to the histogram is a kernel density plot. This provides a smoother version of the histogram in two ways: First, it directly connects the midpoints of the histogram rather than forming the histogram step function. Second, rather than giving each entry in a bin equal weight, it gives more weight to data that are closest to the point of evaluation.

Let $f(x)$ denote the density. The kernel density estimate of $f(x)$ at $x = x_0$ is

$$\widehat{f}(x_0) = \frac{1}{Nh} \sum_{i=1}^{N} K\left(\frac{x_i - x_0}{h}\right) \tag{2.1}$$

where $K(\cdot)$ is a kernel function that places greater weight on points x_i close to x_0. More precisely, $K(z)$ is symmetric around zero, integrates to one, and either $K(z) = 0$ if $|z| \geq z_0$ (for some z_0) or $K(z) \to 0$ as $z \to \infty$. A histogram with a bin width of $2h$ evaluated at x_0 can be shown to be the special case $K(z) = 1/2$ if $|z| < 1$, and $K(z) = 0$ otherwise.

A kernel density plot is obtained by choosing a kernel function, $K(\cdot)$; choosing a width, h; evaluating $\widehat{f}(x_0)$ at a range of values of x_0; and plotting $\widehat{f}(x_0)$ against these x_0 values.

The **kdensity** command produces a kernel density estimate. The command can be used with **if** and **in** qualifiers and with weights. The default kernel function is the Epanechnikov, which sets $K(z) = (3/4)(1 - z^2/5)/\sqrt{5}$ if $|z| < \sqrt{5}$, and $K(z) = 0$ otherwise. The **kernel()** option allows other kernels to be chosen, but unless the width is relatively small, the choice of kernel makes little difference. The default window width or bandwidth is $h = 0.9m/n^{1/5}$, where $m = \min(s_x, iqr_x/1.349)$ and iqr_x is the interquartile range of x. The **bwidth(#)** option allows a different width (h) to be specified, with larger choices of h leading to smoother density plots. The **n(#)** option changes the number of evaluation points, x_0, from the default of $\min(N, 50)$. Other options overlay a fitted normal density (the **normal** option) or a fitted t density (the **student(#)** option).

The output from **kdensity lnearns** states that the Epanechnikov kernel is used and the bandwidth equals 0.1227. If we desire a smoother density estimate with a bandwidth of 0.2, one overlaid by a fitted normal density, we type the command

```
. * Kernel density plot with bandwidth set and fitted normal density overlaid
. kdensity lnearns, bwidth(0.20) normal n(4000)
```

which produces the graph in figure 2.5. This graph shows that the kernel density is more peaked than the normal and is somewhat skewed.

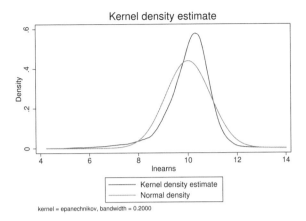

Figure 2.5. The estimated density of log earnings

The following code instead presents a histogram overlaid by a kernel density estimate. The histogram bin width is set to 0.25, the kernel density bandwidth is set to 0.2 using the `kdenopts()` option, and the kernel density plot line thickness is increased using the `lwidth(medthick)` option. Other options used here were explained in section 2.6.2. We have

```
. * Histogram and nonparametric kernel density estimate
. histogram lnearns if lnearns > 0, width(0.25) kdensity
> kdenopts(bwidth(0.2) lwidth(medthick))
> plotregion(style(none)) scale(1.2)
> title("Histogram and density for log earnings")
> xtitle("Log annual earnings", size(medlarge)) xscale(titlegap(*5))
> ytitle("Histogram and density", size(medlarge)) yscale(titlegap(*5))
(bin=38, start=4.4308167, width=.25)
```

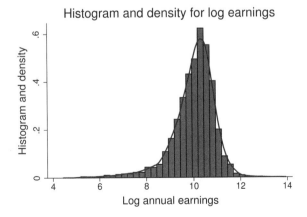

Figure 2.6. Histogram and kernel density plot for natural logarithm of earnings

The result is given in figure 2.6. Both the histogram and the kernel density estimate indicate that the natural logarithm of earnings has a density that is mildly left-skewed. A similar figure for the level of earnings is very right-skewed.

2.6.5 Twoway scatterplots and fitted lines

As we saw in figure 2.1, scatterplots provide a quick look at the relationship between two variables.

For scatterplots with discrete data that take on few values, it can be necessary to use the `jitter()` option. This option adds random noise so that points are not plotted on top of one another; see section 14.6.4 for an example.

It can be useful to additionally provide a fitted curve. Stata provides several possibilities for estimating a global relationship between y against x, where by global we mean that a single relationship is estimated for all observations, and then for plotting the fitted values of y against x.

The `twoway lfit` command does so for a fitted OLS regression line, the `twoway qfit` command does so for a fitted quadratic regression curve, and the `twoway fpfit` command does so for a curve fit by fractional polynomial regression. The related `twoway` commands `lfitci`, `qfitci`, and `fpfitci` additionally provide confidence bands for predicting the conditional mean $E(y|x)$ (by using the `stdp` option) or for forecasting of the actual value of $y|x$ (by using the `stdf` option).

For example, we may want to provide a scatterplot and fitted quadratic with confidence bands for the forecast value of $y|x$ (the result is shown in figure 2.7):

```
. * Two-way scatterplot and quadratic regression curve with 95% ci for y|x
. twoway (qfitci lnearns hours, stdf)  (scatter lnearns hours, msize(small))
```

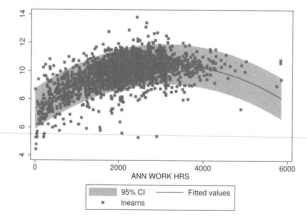

Figure 2.7. Twoway scatterplot and fitted quadratic with confidence bands

2.6.6   Lowess, kernel, local linear, and nearest-neighbor regression

An alternative curve-fitting approach is to use nonparametric methods that fit a local relationship between y and x, where by local we mean that separate fitted relationships are obtained at different values of x. There are several methods. All depend on a bandwidth parameter or smoothing parameter. There are well-established methods to automatically select the bandwidth parameter, but these choices in practice can undersmooth or oversmooth the data so that the bandwidth then needs to be set by using the `bwidth()` option.

An easily understood example is a median-band plot. The range of x is broken into, say, 20 intervals; the medians of y and x in each interval are obtained; and the 20 medians of y are plotted against the 20 medians of x, with connecting lines between the points. The `twoway mband` command does this, and the related `twoway mspline` command uses a cubic spline to obtain a smoother version of the median-band plot.

Most nonparametric methods instead use variants of local regression. Consider the regression model $y = m(x) + u$, where x is a scalar and the conditional mean function $m(\cdot)$ is not specified. A local regression estimate of $m(x)$ at $x = x_0$ is a local weighted average of y_i, $i = 1, \ldots, N$, that places great weight on observations for which x_i is close to x_0 and little or no weight on observations for which x_i is far from x_0. Formally,

$$\widehat{m}(x_0) = \sum_{i=1}^{N} w(x_i, x_0, h) y_i$$

where the weights $w(x_i, x_0, h)$ sum over i to one and decrease as the distance between x_i and x_0 increases. As the bandwidth parameter h increases, less weight is placed on observations for which x_i is close to x_0.

A plot is obtained by choosing a weighting function, $w(x_i, x_0, h)$; choosing a bandwidth, h; evaluating $\widehat{m}(x_0)$ at a range of values of x_0; and plotting $\widehat{m}(x_0)$ against these x_0 values.

The kth-nearest-neighbor estimator uses just the k observations for which x_i is closest to x_0 and equally weights these k closest values. This estimator can be obtained by using the user-written `knnreg` command (Salgado-Ugarte, Shimizu, and Taniuchi 1996).

Kernel regression uses the weight $w(x_i, x_0, h) = K\{(x_i - x_0)/h\}/\sum_{i=1}^{N} K\{(x_i - x_0)/h\}$, where $K(\cdot)$ is a kernel function defined after (2.1). This estimator can be obtained by using the user-written `kernreg` command (Salgado-Ugarte, Shimizu, and Taniuchi 1996). It can also be obtained by using the `lpoly` command, which we present next.

The kernel regression estimate at $x = x_0$ can equivalently be obtained by minimizing $\sum_i K\{(x_i - x_0)/h\}(y_i - \alpha_0)^2$, which is weighted regression on a constant where the kernel weights are largest for observations with x_i close to x_0. The local linear estimator additionally includes a slope coefficient and at $x = x_0$ minimizes

$$\sum_{i=1}^{N} K\left(\frac{x_i - x_0}{h}\right) \{y_i - \alpha_0 - \beta_0(x_i - x_0)\}^2 \tag{2.2}$$

The local polynomial estimator of degree p more generally uses a polynomial of degree p in $(x_i - x_0)$ in (2.2). This estimator is obtained by using `lpoly`. The `degree(#)` option specifies the degree p, the `kernel()` option specifies the kernel, the `bwidth(#)` option specifies the kernel bandwidth h, and the `generate()` option saves the evaluation points x_0 and the estimates $\widehat{m}(x_0)$. The local linear estimator with $p \geq 1$ does much better than the preceding methods at estimating $m(x_0)$ at values of x_0 near the endpoints of the range of x, as it allows for any trends near the endpoints.

The locally weighted scatterplot smoothing estimator (lowess) is a variation of the local linear estimator that uses a variable bandwidth, a tricubic kernel, and downweights observations with large residuals (using a method that greatly increases the computational burden). This estimator is obtained by using the `lowess` command. The bandwidth gives the fraction of the observations used to calculate $\widehat{m}(x_0)$ in the middle of the data, with a smaller fraction used towards the endpoints. The default value of 0.8 can be changed by using the `bwidth(#)` option, so unlike the other methods, a smoother plot is obtained by increasing the bandwidth.

The following example illustrates the relationship between log earnings and hours worked. The one graph includes a scatterplot (`scatter`), a fitted lowess curve (`lowess`), and a local linear curve (`lpoly`). The command is lengthy because of the detailed formatting commands used to produce a nicely labeled and formatted graph. The `msize(tiny)` option is used to decrease the size of the dots in the scatterplot. The `lwidth(medthick)` option is used to increase the thickness of lines, and the `clstyle(p1)` option changes the style of the line for `lowess`. The `title()` option provides the overall title for the graph. The `xtitle()` and `ytitle()` options provide titles for the x axis and y axis, and the `size(medlarge)` option defines the size of the text for these titles. The `legend()` options place the graph legend at four o'clock (`pos(4)`) with text size `small` and provide the legend labels. We have

```
. * Scatterplot with lowess and local linear nonparametric regression
. graph twoway (scatter lnearns hours, msize(tiny))
> (lowess lnearns hours, clstyle(p1) lwidth(medthick))
> (lpoly lnearns hours, kernel(epan2) degree(1) lwidth(medthick)
> bwidth(500)), plotregion(style(none))
> title("Scatterplot, lowess, and local linear regression")
> xtitle("Annual hours", size(medlarge))
> ytitle("Natural logarithm of annual earnings", size(medlarge))
> legend(pos(4) ring(0) col(1)) legend(size(small))
> legend(label(1 "Actual Data") label(2 "Lowess") label(3 "Local linear"))
```

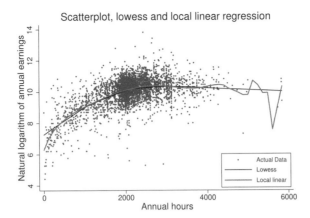

Figure 2.8. Scatterplot, lowess, and local linear curves for natural logarithm of earnings plotted against hours

From figure 2.8, the scatterplot, fitted OLS line, and nonparametric regression all indicate that log earnings increase with hours until about 2,500 hours and that a quadratic relationship may be appropriate. The graph uses the default bandwidth setting for `lowess` and greatly increases the `lpoly` bandwidth from its automatically selected value of 84.17 to 500. Even so, the local linear curve is too variable at high hours where the data are sparse. At low hours, however, the lowess estimator overpredicts while the local linear estimator does not.

2.6.7 Multiple scatterplots

The `graph matrix` command provides separate bivariate scatterplots between several variables. Here we produce bivariate scatterplots (shown in figure 2.9) of `lnearns`, `hours`, and `age` for each of the four education categories:

```
. * Multiple scatterplots
. label variable age "Age"
. label variable lnearns "Log earnings"
. label variable hours "Annual hours"
. graph matrix lnearns hours age, by(edcat) msize(small)
```

(*Continued on next page*)

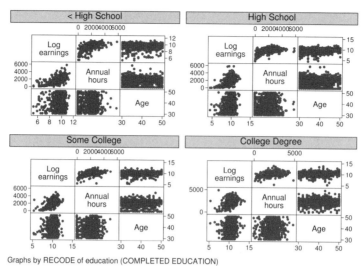

Graphs by RECODE of education (COMPLETED EDUCATION)

Figure 2.9. Multiple scatterplots for each level of education

Stata does not provide three-dimensional graphs, such as that for a nonparametric bivariate density estimate or for nonparametric regression of one variable on two other variables.

2.7 Stata resources

The key data-management references are [U] *Users Guide* and [D] *Data Management Reference Manual.* Useful online `help` categories include 1) `double`, `string`, and `format` for data types; 2) `clear`, `use`, `insheet`, `infile`, and `outsheet` for data input; 3) `summarize`, `list`, `label`, `tabulate`, `generate`, `egen`, `keep`, `drop`, `recode`, `by`, `sort`, `merge`, `append`, and `collapse` for data management; and 4) `graph`, `graph box`, `histogram`, `kdensity`, `twoway`, `lowess`, and `graph matrix` for graphical analysis.

The Stata graphics commands were greatly enhanced in version 8 and are still relatively underutilized. The Stata Graph Editor was introduced in version 10; see [G] **graph editor**. *A Visual Guide to Stata Graphics* by Mitchell (2008) provides many hundreds of template graphs with the underlying Stata code and an explanation for each.

2.8 Exercises

1. Type the command `display %10.5f 123.321`. Compare the results with those you obtain when you change the format `%10.5f` to, respectively, `%10.5e`, `%10.5g`, `%-10.5f`, `%10,5f`, and when you do not specify a format.

2. Consider the example of section 2.3 except with the variables reordered. Specifically, the variables are in the order `age`, `name`, `income`, and `female`. The three observations are `29 "Barry" 40.990 0`; `30 "Carrie" 37.000 1`; and `31 "Gary" 48.000 0`. Use `input` to read these data, along with names, into Stata and list the results. Use a text editor to create a comma-separated values file that includes variable names in the first line, read this file into Stata by using `insheet`, and list the results. Then drop the first line in the text file, read in the data by using `insheet` with variable names assigned, and list the results. Finally, replace the commas in the text file with blanks, read the data in by using `infix`, and list the results.

3. Consider the dataset in section 2.4. The `er32049` variable is the last known marital status. Rename this variable as `marstatus`, give the variable the label "marital status", and tabulate `marstatus`. From the codebook, marital status is married (1), never married (2), widowed (3), divorced or annulment (4), separated (5), not answered or do not know (8), and no marital history collected (9). Set `marstatus` to missing where appropriate. Use `label define` and `label values` to provide descriptions for the remaining categories, and tabulate `marstatus`. Create a binary indicator variable equal to 1 if the last known marital status is married, and equal to 0 otherwise, with appropriate handling of any missing data. Provide a summary of earnings by marital status. Create a set of indicator variables for marital status based on `marstatus`. Create a set of variables that interact these marital status indicators with earnings.

4. Consider the dataset in section 2.6. Create a box-and-whisker plot of `earnings` (in levels) for all the data and for each year of educational attainment (use variable `education`). Create a histogram of `earnings` (in levels) using 100 bins and a kernel density estimate. Do earnings in levels appear to be right-skewed? Create a scatterplot of `earnings` against `education`. Provide a single figure that uses `scatterplot`, `lfit`, and `lowess` of `earnings` against `education`. Add titles for the axes and graph heading.

5. Consider the dataset in section 2.6. Create kernel density plots for `lnearns` using the `kernel(epan2)` option with kernel $K(z) = (3/4)(1 - z^2/5)$ for $|z| < 1$, and using the `kernel(epan2)` option with kernel $K(z) = 1/2$ for $|z| < 1$. Repeat with the bandwidth increased from the default to 0.3. What makes a bigger difference, choice of kernel or choice of bandwidth? The comparison is easier if the four graphs are saved using the `saving()` option and then combined using the `graph combine` command.

6. Consider the dataset in section 2.6. Perform lowess regression of `lnearns` on `hours` using the default bandwidth and using bandwidth of 0.01. Does the bandwidth make a difference? A moving average of y after data are sorted by x is a simple case of nonparametric regression of y on x. Sort the data by `hours`. Create a centered 15-period moving average of `lnearns` with ith observation $yma_i = 1/25 \sum_{j=-12}^{j=12} y_{i+j}$. This is easiest using `forvalues`. Plot this moving average against `hours` using the `twoway connected` graph command. Compare to the lowess plot.

3 Linear regression basics

3.1 Introduction

Linear regression analysis is often the starting point of an empirical investigation. Because of its relative simplicity, it is useful for illustrating the different steps of a typical modeling cycle that involves an initial specification of the model followed by estimation, diagnostic checks, and model respecification. The purpose of such a linear regression analysis may be to summarize the data, generate conditional predictions, or test and evaluate the role of specific regressors. We will illustrate these aspects using a specific data example.

This chapter is limited to basic regression analysis on cross-section data of a continuous dependent variable. The setup is for a single equation and exogenous regressors. Some standard complications of linear regression, such as misspecification of the conditional mean and model errors that are heteroskedastic, will be considered. In particular, we model the natural logarithm of medical expenditures instead of the level. We will ignore other various aspects of the data that can lead to more sophisticated nonlinear models presented in later chapters.

3.2 Data and data summary

The first step is to decide what dataset will be used. In turn, this decision depends on the population of interest and the research question itself. We discussed how to convert a raw dataset to a form amenable to regression analysis in chapter 2. In this section, we present ways to summarize and gain some understanding of the data, a necessary step before any regression analysis.

3.2.1 Data description

We analyze medical expenditures of individuals 65 years and older who qualify for health care under the U.S. Medicare program. The original data source is the Medical Expenditure Panel Survey (MEPS).

Medicare does not cover all medical expenses. For example, copayments for medical services and expenses of prescribed pharmaceutical drugs were not covered for the time period studied here. About half of eligible individuals therefore purchase supplementary insurance in the private market that provides insurance coverage against various out-of-pocket expenses.

In this chapter, we consider the impact of this supplementary insurance on total annual medical expenditures of an individual, measured in dollars. A formal investigation must control for the influence of other factors that also determine individual medical expenditure, notably, sociodemographic factors such as age, gender, education and income, geographical location, and health-status measures such as self-assessed health and presence of chronic or limiting conditions. In this chapter, as in other chapters, we instead deliberately use a short list of regressors. This permits shorter output and simpler discussion of the results, an advantage because our intention is to simply explain the methods and tools available in Stata.

3.2.2   Variable description

Given the Stata dataset for analysis, we begin by using the `describe` command to list various features of the variables to be used in the linear regression. The command without a variable list describes all the variables in the dataset. Here we restrict attention to the variables used in this chapter.

```
. * Variable description for medical expenditure dataset
. use mus03data.dta

. describe totexp ltotexp posexp suppins phylim actlim totchr age female income

              storage  display    value
variable name  type    format     label        variable label
-------------------------------------------------------------------
totexp         double  %12.0g                   Total medical expenditure
ltotexp        float   %9.0g                    ln(totexp) if totexp > 0
posexp         float   %9.0g                    =1 if total expenditure > 0
suppins        float   %9.0g                    =1 if has supp priv insurance
phylim         double  %12.0g                   =1 if has functional limitation
actlim         double  %12.0g                   =1 if has activity limitation
totchr         double  %12.0g                   # of chronic problems
age            double  %12.0g                   Age
female         double  %12.0g                   =1 if female
income         double  %12.0g                   annual household income/1000
```

The variable types and format columns indicate that all the data are numeric. In this case, some variables are stored in single precision (`float`) and some in double precision (`double`). From the variable labels, we expect `totexp` to be nonnegative; `ltotexp` to be missing if `totexp` equals zero; `posexp`, `suppins`, `phylim`, `actlim`, and `female` to be 0 or 1; `totchr` to be a nonnegative integer; `age` to be positive; and `income` to be negative or positive. Note that the integer variables could have been stored much more compactly as `integer` or `byte`. The variable labels provide a short description that is helpful but may not fully describe the variable. For example, the key regressor `suppins` was created by aggregating across several types of private supplementary insurance. No labels for the values taken by the categorical variables have been provided.

3.2.3 Summary statistics

It is essential in any data analysis to first check the data by using the `summarize` command.

```
. * Summary statistics for medical expenditure dataset
. summarize totexp ltotexp posexp suppins phylim actlim totchr age female income

    Variable |      Obs        Mean    Std. Dev.       Min        Max
-------------+--------------------------------------------------------
      totexp |     3064    7030.889    11852.75         0      125610
     ltotexp |     2955    8.059866    1.367592   1.098612   11.74094
      posexp |     3064    .9644256    .1852568         0          1
     suppins |     3064    .5812663    .4934321         0          1
      phylim |     3064    .4255875    .4945125         0          1
-------------+--------------------------------------------------------
      actlim |     3064    .2836162    .4508263         0          1
      totchr |     3064    1.754243    1.307197         0          7
         age |     3064    74.17167    6.372938        65         90
      female |     3064    .5796345    .4936982         0          1
      income |     3064    22.47472    22.53491        -1     312.46
```

On average, 96% of individuals incur medical expenditures during a year; 58% have supplementary insurance; 43% have functional limitations; 28% have activity limitations; and 58% are female, as the elderly population is disproportionately female because of the greater longevity of women. The only variable to have missing data is `ltotexp`, the natural logarithm of `totexp`, which is missing for the $(3064 - 2955) = 109$ observations with `totexp` = 0.

All variables have the expected range, except that income is negative. To see how many observations on `income` are negative, we use the `tabulate` command, restricting attention to nonpositive observations to limit output.

```
. * Tabulate variable
. tabulate income if income <= 0

       annual |
    household |
income/1000 |      Freq.     Percent        Cum.
-------------+-----------------------------------
          -1 |          1        1.14        1.14
           0 |         87       98.86      100.00
-------------+-----------------------------------
       Total |         88      100.00
```

Only one observation is negative, and negative income is possible for income from self-employment or investment. We include the observation in the analysis here, though checking the original data source may be warranted.

Much of the subsequent regression analysis will drop the 109 observations with zero medical expenditures, so in a research paper, it would be best to report summary statistics without these observations.

3.2.4   More-detailed summary statistics

Additional descriptive analysis of key variables, especially the dependent variable, is useful. For `totexp`, the level of medical expenditures, `summarize, detail` yields

```
. * Detailed summary statistics of a single variable
. summarize totexp, detail
                        Total medical expenditure

            Percentiles      Smallest
    1%            0               0
    5%          112               0
   10%          393               0        Obs                  3064
   25%         1271               0        Sum of Wgt.          3064

   50%       3134.5                        Mean             7030.889
                              Largest      Std. Dev.        11852.75
   75%         7151          104823
   90%        17050          108256        Variance         1.40e+08
   95%        27367          123611        Skewness         4.165058
   99%        62346          125610        Kurtosis         26.26796
```

Medical expenditures vary greatly across individuals, with a standard deviation of 11,853, which is almost twice the mean. The median of 3,134 is much smaller than the mean of 7,031, reflecting the skewness of the data. For variable x, the skewness statistic is a scale-free measure of skewness that estimates $E\{(x - \mu)^3\}/\sigma^{3/2}$, the third central moment standardized by the second central moment. The skewness is zero for symmetrically distributed data. The value here of 4.16 indicates considerable right skewness. The kurtosis statistic is an estimate of $E\{(x - \mu)^4\}/\sigma^4$, the fourth central moment standardized by the second central moment. The reference value is 3, the value for normally distributed data. The much higher value here of 26.26 indicates that the tails are much thicker than those of a normal distribution. You can obtain additional summary statistics by using the `centile` command to obtain other percentiles and by using the `table` command, which is explained in section 3.2.5.

We conclude that the distribution of the dependent variable is considerably skewed and has thick tails. These complications often arise for commonly studied individual-level economic variables such as expenditures, income, earnings, wages, and house prices. It is possible that including regressors will eliminate the skewness, but in practice, much of the variation in the data will be left unexplained ($R^2 < 0.3$ is common for individual-level data) and skewness and excess kurtosis will remain.

Such skewed, thick-tailed data suggest a model with multiplicative errors instead of additive errors. A standard solution is to transform the dependent variable by taking the natural logarithm. Here this is complicated by the presence of 109 zero-valued observations. We take the expedient approach of dropping the zero observations from analysis in either logs or levels. This should make little difference here because only 3.6% of the sample is then dropped. A better approach, using two-part or selection models, is covered in chapter 16.

The output for `tabstat` in section 3.2.5 reveals that taking the natural logarithm for these data essentially eliminates the skewness and excess kurtosis.

The user-written `fsum` command (Wolfe 2002) is an enhancement of `summarize` that enables formatting the output and including additional information such as percentiles and variable labels. The user-written `outsum` command (Papps 2006) produces a text file of means and standard deviations for one or more subsets of the data, e.g., one column for the full sample, one for a male subsample, and one for a female subsample.

3.2.5 Tables for data

One-way tables can be created by using the `table` command, which produces just frequencies, or the `tabulate` command, which additionally produces percentages and cumulative percentages; an example was given in section 3.2.3.

Two-way tables can also be created by using these commands. For frequencies, only `table` produces clean output. For example,

```
. * Two-way table of frequencies
. table female totchr
```

=1 if female	# of chronic problems							
	0	1	2	3	4	5	6	7
0	239	415	323	201	82	23	4	1
1	313	466	493	305	140	46	11	2

provides frequencies for a two-way tabulation of gender against the number of chronic conditions. The `tabulate` command is much richer. For example,

```
. * Two-way table with row and column percentages and Pearson chi-squared
. tabulate female suppins, row col chi2
```

Key
frequency
row percentage
column percentage

=1 if female	=1 if has supp priv insurance		Total
	0	1	
0	488	800	1,288
	37.89	62.11	100.00
	38.04	44.92	42.04
1	795	981	1,776
	44.76	55.24	100.00
	61.96	55.08	57.96
Total	1,283	1,781	3,064
	41.87	58.13	100.00
	100.00	100.00	100.00

```
          Pearson chi2(1) =  14.4991   Pr = 0.000
```

Comparing the row percentages for this sample, we see that while a woman is more likely to have supplemental insurance than not, the probability that a woman in this sample has purchased supplemental insurance is lower than the probability that a man in this sample has purchased supplemental insurance. Although we do not have the information to draw these inferences for the population, the results for Pearson's chi-squared test soundly reject the null hypothesis that these variables are independent. Other tests of association are available. The related command `tab2` will produce all possible two-way tables that can be obtained from a list of several variables.

For multiway tables, it is best to use `table`. For the example at hand, we have

```
. * Three-way table of frequencies
. table female totchr suppins
```

| | =1 if has supp priv insurance and # of chronic problems | | | | | | | |
| | | | | 0 | | | | |
=1 if female	0	1	2	3	4	5	6	7
0	102	165	121	68	25	6	1	
1	135	212	233	134	56	22	1	2

| | =1 if has supp priv insurance and # of chronic problems | | | | | | | |
| | | | | 1 | | | | |
=1 if female	0	1	2	3	4	5	6	7
0	137	250	202	133	57	17	3	1
1	178	254	260	171	84	24	10	

An alternative is to use `tabulate` with the `by` prefix, but the results are not as neat as those from `table`.

The preceding tabulations will produce voluminous output if one of the variables being tabulated takes on many values. Then it is much better to use `table` with the `contents()` option to present tables that give key summary statistics for that variable, such as the mean and standard deviation. Such tabulations can be useful even when variables take on few values. For example, when summarizing the number of chronic problems by gender, `table` yields

```
. * One-way table of summary statistics
. table female, contents(N totchr mean totchr sd totchr p50 totchr)
```

=1 if female	N(totchr)	mean(totchr)	sd(totchr)	med(totchr)
0	1,288	1.659937888	1.261175	1
1	1,776	1.822635135	1.335776	2

Women on average have more chronic problems (1.82 versus 1.66 for men). The option contents() can produce many other statistics, including the minimum, maximum, and key percentiles.

The table command with the contents() option can additionally produce two-way and multiway tables of summary statistics. As an example,

```
. * Two-way table of summary statistics
. table female suppins, contents(N totchr mean totchr)
```

=1 if female	=1 if has supp priv insurance 0	1
0	488 1.530737705	800 1.73875
1	795 1.803773585	981 1.837920489

shows that those with supplementary insurance on average have more chronic problems. This is especially so for males (1.74 versus 1.53).

The tabulate, summarize() command can be used to produce one-way and two-way tables with means, standard deviations, and frequencies. This is a small subset of the statistics that can be produced using table, so we might as well use table.

The tabstat command provides a table of summary statistics that permits more flexibility than summarize. The following output presents summary statistics on medical expenditures and the natural logarithm of expenditures that are useful in determining skewness and kurtosis.

```
. * Summary statistics obtained using command tabstat
. tabstat totexp ltotexp, stat (count mean p50 sd skew kurt) col(stat)
```

variable	N	mean	p50	sd	skewness	kurtosis
totexp	3064	7030.889	3134.5	11852.75	4.165058	26.26796
ltotexp	2955	8.059866	8.111928	1.367592	-.3857887	3.842263

This reproduces information given in section 3.2.4 and shows that taking the natural logarithm eliminates most skewness and kurtosis. The col(stat) option presents the results with summary statistics given in the columns and each variable being given in a separate row. Without this option, we would have summary statistics in rows and variables in the columns. A two-way table of summary statistics can be obtained by using the by() option.

(Continued on next page)

3.2.6   Statistical tests

The `ttest` command can be used to test hypotheses about the population mean of a single variable (H_0: $\mu = \mu^*$ for specified value μ^*) and to test the equality of means ($H_0 : \mu_1 = \mu_2$). For more general analysis of variance and analysis of covariance, the `oneway` and `anova` commands can be used, and several other tests exist for more specialized examples such as testing the equality of proportions. These commands are rarely used in microeconometrics because they can be recast as a special case of regression with an intercept and appropriate indicator variables. Furthermore, regression has the advantage of reliance on less restrictive distributional assumptions, provided samples are large enough for asymptotic theory to provide a good approximation.

For example, consider testing the equality of mean medical expenditures for those with and without supplementary health insurance. The `ttest totexp, by(suppins)` `unequal` command performs the test but makes the restrictive assumption of a common variance for all those with `suppins=0` and a (possibly different) common variance for all those with `suppins=1`. An alternative method is to perform ordinary least-squares (OLS) regression of `totexp` on an intercept and `suppins` and then test whether `suppins` has coefficient zero. Using this latter method, we can permit all observations to have a different variance by using the `vce(robust)` option for `regress` to obtain heteroskedastic-consistent standard errors; see section 3.3.4.

3.2.7   Data plots

It is useful to plot a histogram or a density estimate of the dependent variable. Here we use the `kdensity` command, which provides a kernel estimate of the density.

The data are highly skewed, with a 97th percentile of approximately $40,000 and a maximum of $1,000,000. The `kdensity totexp` command will therefore bunch 97% of the density in the first 4% of the x axis. One possibility is to type `kdensity totexp` `if totexp < 40000`, but this produces a kernel density estimate assuming the data are truncated at $40,000. Instead, we use command `kdensity totexp`, we save the evaluation points in `kx1` and the kernel density estimates in `kd1`, and then we line-plot `kd1` against `kx1`.

We do this for both the level and the natural logarithm of medical expenditures, and we use `graph combine` to produce a figure that includes both density graphs (shown in figure 3.1). We have

```
. * Kernel density plots with adjustment for highly skewed data
. kdensity totexp if posexp==1, generate (kx1 kd1) n(500)
. graph twoway (line kd1 kx1) if kx1 < 40000, name(levels)
. kdensity ltotexp if posexp==1, generate (kx2 kd2) n(500)
. graph twoway (line kd2 kx2) if kx2 < ln(40000), name(logs)
. graph combine levels logs, iscale(1.0)
```

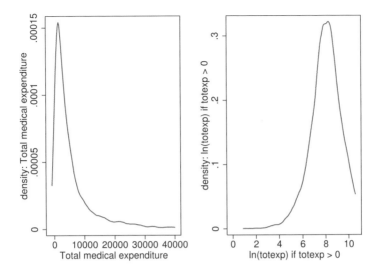

Figure 3.1. Comparison of densities of level and natural logarithm of medical expenditures

Only positive expenditures are considered, and for graph readability, the very long right tail of `totexp` has been truncated at $40,000. In figure 3.1, the distribution of `totexp` is very right-skewed, whereas that of `ltotexp` is fairly symmetric.

3.3   Regression in levels and logs

We present the linear regression model, first in levels and then for a transformed dependent variable, here in logs.

3.3.1   Basic regression theory

We begin by introducing terminology used throughout the rest of this book. Let $\boldsymbol{\theta}$ denote the vector of parameters to be estimated, and let $\widehat{\boldsymbol{\theta}}$ denote an estimator of $\boldsymbol{\theta}$. Ideally, the distribution of $\widehat{\boldsymbol{\theta}}$ is centered on $\boldsymbol{\theta}$ with small variance, for precision, and a known distribution, to permit statistical inference. We restrict analysis to estimators that are consistent for $\boldsymbol{\theta}$, meaning that in infinitely large samples, $\widehat{\boldsymbol{\theta}}$ equals $\boldsymbol{\theta}$ aside from negligible random variation. This is denoted by $\widehat{\boldsymbol{\theta}} \xrightarrow{p} \boldsymbol{\theta}$ or more formally by $\widehat{\boldsymbol{\theta}} \xrightarrow{p} \boldsymbol{\theta}_0$, where $\boldsymbol{\theta}_0$ denotes the unknown "true" parameter value. A necessary condition for consistency is correct model specification or, in some leading cases, correct specification of key components of the model, most notably the conditional mean.

Under additional assumptions, the estimators considered in this book are asymptotically normally distributed, meaning that their distribution is well approximated by the multivariate normal in large samples. This is denoted by

$$\widehat{\boldsymbol{\theta}} \overset{a}{\sim} N\{\boldsymbol{\theta},\ \text{Var}(\widehat{\boldsymbol{\theta}})\}$$

where $\text{Var}(\widehat{\boldsymbol{\theta}})$ denotes the (asymptotic) variance–covariance matrix of the estimator (VCE). More efficient estimators have smaller VCEs. The VCE depends on unknown parameters, so we use an estimate of the VCE, denoted by $\widehat{V}(\widehat{\boldsymbol{\theta}})$. Standard errors of the parameter estimates are obtained as the square root of diagonal entries in $\widehat{V}(\widehat{\boldsymbol{\theta}})$. Different assumptions about the data-generating process (DGP), such as heteroskedasticity, can lead to different estimates of the VCE.

Test statistics based on asymptotic normal results lead to the use of the standard normal distribution and chi-squared distribution to compute critical values and p-values. For some estimators, notably, the OLS estimator, tests are instead based on the t distribution and the F distribution. This makes essentially no difference in large samples with, say, degrees of freedom greater than 100, but it may provide a better approximation in smaller samples.

3.3.2   OLS regression and matrix algebra

The goal of linear regression is to estimate the parameters of the linear conditional mean

$$E(y|\mathbf{x}) = \mathbf{x}'\boldsymbol{\beta} = \beta_1 x_1 + \beta_2 x_2 + \cdots + \beta_K x_K \tag{3.1}$$

where usually an intercept is included so that $x_1 = 1$. Here $\mathbf{x}$ is a $K \times 1$ column vector with the jth entry—the jth regressor x_j—and $\boldsymbol{\beta}$ is a $K \times 1$ column vector with the jth entry β_j.

Sometimes $E(y|\mathbf{x})$ is of direct interest for prediction. More often, however, econometrics studies are interested in one or more of the associated marginal effects (MEs),

$$\frac{\partial E(y|\mathbf{x})}{\partial x_j} = \beta_j$$

for the jth regressor. For example, we are interested in the marginal effect of supplementary private health insurance on medical expenditures. An attraction of the linear model is that estimated MEs are given directly by estimates of the slope coefficients.

The linear regression model specifies an additive error so that, for the typical ith observation,

$$y_i = \mathbf{x}_i'\boldsymbol{\beta} + u_i, \quad i = 1, \ldots, N$$

The OLS estimator minimizes the sum of squared errors, $\sum_{i=1}^{N}(y_i - \mathbf{x}_i'\boldsymbol{\beta})^2$.

Matrix notation provides a compact way to represent the estimator and variance matrix formulas that involve sums of products and cross products. We define the $N \times 1$

column vector $\mathbf{y}$ to have the ith entry y_i, and we define the $N \times K$ regressor matrix $\mathbf{X}$ to have the ith row $\mathbf{x}_i'$. Then the OLS estimator can be written in several ways, with

$$\widehat{\boldsymbol{\beta}} = (\mathbf{X}'\mathbf{X})^{-1}\mathbf{X}'\mathbf{y}$$

$$= \left(\sum_{i=1}^{N} \mathbf{x}_i\mathbf{x}_i'\right)^{-1} \sum_{i=1}^{N} \mathbf{x}_iy_i$$

$$= \begin{bmatrix} \sum_{i=1}^{N} x_{1i}^2 & \sum_{i=1}^{N} x_{1i}x_{2i} & \cdots & \sum_{i=1}^{N} x_{1i}x_{Ki} \\ \sum_{i=1}^{N} x_{2i}x_{1i} & \sum_{i=1}^{N} x_{2i}^2 & & \vdots \\ & & \ddots & \\ \sum_{i=1}^{N} x_{Ki}x_{1i} & & \cdots & \sum_{i=1}^{N} x_{Ki}^2 \end{bmatrix}^{-1} \begin{bmatrix} \sum_{i=1}^{N} x_{1i}y_i \\ \sum_{i=1}^{N} x_{2i}y_i \\ \vdots \\ \sum_{i=1}^{N} x_{Ki}y_i \end{bmatrix}$$

We define all vectors as column vectors, with a transpose if row vectors are desired. By contrast, Stata commands and Mata commands define vectors as row vectors, so in parts of Stata and Mata code, we need to take a transpose to conform to the notation in the book.

3.3.3   Properties of the OLS estimator

The properties of any estimator vary with the assumptions made about the DGP. For the linear regression model, this reduces to assumptions about the regression error u_i.

The starting point for analysis is to assume that u_i satisfies the following classical conditions:

1. $E(u_i|\mathbf{x}_i) = \mathbf{0}$ (exogeneity of regressors)
2. $E(u_i^2|\mathbf{x}_i) = \sigma^2$ (conditional homoskedasticity)
3. $E(u_iu_j|\mathbf{x}_i, \mathbf{x}_j) = \mathbf{0}$, $i \neq j$, (conditionally uncorrelated observations)

Assumption 1 is essential for consistent estimation of $\boldsymbol{\beta}$ and implies that the conditional mean given in (3.1) is correctly specified. This means that the conditional mean is linear and that all relevant variables have been included in the regression. Assumption 1 is relaxed in chapter 6.

Assumptions 2 and 3 determine the form of the VCE of $\widehat{\boldsymbol{\beta}}$. Assumptions 1–3 lead to $\widehat{\boldsymbol{\beta}}$ being asymptotically normally distributed with the default estimator of the VCE

$$\widehat{V}_{\text{default}}(\widehat{\boldsymbol{\beta}}) = s^2(\mathbf{X}'\mathbf{X})^{-1}$$

where

$$s^2 = (N - k)^{-1} \sum_i \widehat{u}_i^2 \tag{3.2}$$

and $\widehat{u}_i = y_i - \mathbf{x}_i'\widehat{\boldsymbol{\beta}}$. Under assumptions 1–3, the OLS estimator is fully efficient. If, additionally, u_i is normally distributed, then "t statistics" are exactly t distributed. This fourth assumption is not made, but it is common to continue to use the t distribution in the hope that it provides a better approximation than the standard normal in finite samples.

When assumptions 2 and 3 are relaxed, OLS is no longer fully efficient. In chapter 5, we present examples of more-efficient feasible generalized least-squares (FGLS) estimation. In the current chapter, we continue to use the OLS estimator, as is often done in practice, but we use alternative estimates of the VCE that are valid when assumption 2, assumption 3, or both are relaxed.

3.3.4 Heteroskedasticity-robust standard errors

Given assumptions 1 and 3, but not 2, we have heteroskedastic uncorrelated errors. Then a robust estimator, or more precisely a heteroskedasticity-robust estimator, of the VCE of the OLS estimator is

$$\widehat{V}_{\text{robust}}(\widehat{\boldsymbol{\beta}}) = (\mathbf{X}'\mathbf{X})^{-1} \left(\frac{N}{N-k} \sum_i \widehat{u}_i^2 \mathbf{x}_i \mathbf{x}_i' \right) (\mathbf{X}'\mathbf{X})^{-1} \tag{3.3}$$

For cross-section data that are independent, this estimator, introduced by White (1980), has supplanted the default variance matrix estimate in most applied work because heteroskedasticity is the norm, and in that case, the default estimate of the VCE is incorrect.

In Stata, a robust estimate of the VCE is obtained by using the `vce(robust)` option of the `regress` command, as illustrated in section 3.4.2. Related options are `vce(hc2)` and `vce(hc3)`, which may provide better heteroskedasticity-robust estimates of the VCE when the sample size is small; see [R] **regress**. The robust estimator of the VCE has been extended to other estimators and models, and a feature of Stata is the `vce(robust)` option, which is applicable for many estimation commands. Some user-written commands use `robust` in place of `vce(robust)`.

3.3.5 Cluster–robust standard errors

When errors for different observations are correlated, assumption 3 is violated. Then both default and robust estimates of the VCE are invalid. For time-series data, this is the case if errors are serially correlated, and the `newey` command should be used. For cross-section data, this can arise when errors are clustered.

Clustered or grouped errors are errors that are correlated within a cluster or group and are uncorrelated across clusters. A simple example of clustering arises when sampling is of independent units but errors for individuals within the unit are correlated. For example, 100 independent villages may be sampled, with several people from each village surveyed. Then, if a regression model overpredicts y for one village member, it is likely to overpredict for other members of the same village, indicating positive correlation. Similar comments apply when sampling is of households with several individuals in each household. Another leading example is panel data with independence over individuals but with correlation over time for a given individual.

Given assumption 1, but not 2 or 3, a cluster–robust estimator of the VCE of the OLS estimator is

$$\widehat{V}_{\text{cluster}}(\widehat{\boldsymbol{\beta}}) = (\mathbf{X}'\mathbf{X})^{-1} \left(\frac{G}{G-1} \frac{N-1}{N-k} \sum_g \mathbf{X}_g \widehat{\mathbf{u}}_g \widehat{\mathbf{u}}_g' \mathbf{X}_g' \right) (\mathbf{X}'\mathbf{X})^{-1}$$

where $g = 1, \ldots, G$ denotes the cluster (such as village), $\widehat{\mathbf{u}}_g$ is the vector of residuals for the observations in the gth cluster, and $\mathbf{X}_g$ is a matrix of the regressors for the observations in the gth cluster. The key assumptions made are error independence across clusters and that the number of clusters $G \to \infty$.

Cluster–robust standard errors can be computed by using the vce(cluster *clustvar*) option in Stata, where clusters are defined by the different values taken by the *clustvar* variable. The estimate of the VCE is in fact heteroskedasticity-robust and cluster–robust, because there is no restriction on $\text{Cov}(u_{gi}, u_{gj})$. The cluster VCE estimate can be applied to many estimators and models; see section 9.6.

Cluster–robust standard errors must be used when data are clustered. For a scalar regressor x, a rule of thumb is that cluster–robust standard errors are $\sqrt{1 + \rho_x \rho_u (M-1)}$ times the incorrect default standard errors, where ρ_x is the within-cluster correlation coefficient of the regressor, ρ_u is the within-cluster correlation coefficient of the error, and M is the average cluster size.

It can be necessary to use cluster–robust standard errors even where it is not immediately obvious. This is particularly the case when a regressor is an aggregated or macro variable, because then $\rho_x = 1$. For example, suppose we use data from the U.S. Current Population Survey and regress individual earnings on individual characteristics and a state-level regressor that does not vary within a state. Then, if there are many individuals in each state so M is large, even slight error correlation for individuals in the same state can lead to great downward bias in default standard errors and in heteroskedasticity-robust standard errors. Clustering can also be induced by the design of sample surveys. This topic is pursued in section 5.5.

3.3.6 Regression in logs

The medical expenditure data are very right-skewed. Then a linear model in levels can provide very poor predictions because it restricts the effects of regressors to be additive. For example, aging 10 years is assumed to increase medical expenditures by the same amount regardless of observed health status. Instead, it is more reasonable to assume that aging 10 years has a multiplicative effect. For example, it may increase medical expenditures by 20%.

We begin with an exponential mean model for positive expenditures, with error that is also multiplicative, so $y_i = \exp(\mathbf{x}_i'\boldsymbol{\beta})\varepsilon_i$. Defining $\varepsilon_i = \exp(u_i)$, we have $y_i = \exp(\mathbf{x}_i'\boldsymbol{\beta} + u_i)$, and taking the natural logarithm, we fit the log-linear model

$$\ln y_i = \mathbf{x}_i'\boldsymbol{\beta} + u_i$$

by OLS regression of $\ln y$ on $\mathbf{x}$. The conditional mean of $\ln y$ is being modeled, rather than the conditional mean of y. In particular,

$$E(\ln y|\mathbf{x}) = \mathbf{x}'\boldsymbol{\beta}$$

assuming u_i is independent with conditional mean zero.

Parameter interpretation requires care. For regression of $\ln y$ on $\mathbf{x}$, the coefficient β_j measures the effect of a change in regressor x_j on $E(\ln y|\mathbf{x})$, but ultimate interest lies instead on the effect on $E(y|\mathbf{x})$. Some algebra shows that β_j measures the proportionate change in $E(y|\mathbf{x})$ as x_j changes, called a semielasticity, rather than the level of change in $E(y|\mathbf{x})$. For example, if $\beta_j = 0.02$, then a one-unit change in x_j is associated with a proportionate increase of 0.02, or 2%, in $E(y|\mathbf{x})$.

Prediction of $E(y|\mathbf{x})$ is substantially more difficult because it can be shown that $E(\ln y|\mathbf{x}) \neq \exp(\mathbf{x}'\boldsymbol{\beta})$. This is pursued in section 3.6.3.

3.4 Basic regression analysis

We use `regress` to run an OLS regression of the natural logarithm of medical expenditures, `ltotexp`, on `suppins` and several demographic and health-status measures. Using $\ln y$ rather than y as the dependent variable leads to no change in the implementation of OLS but, as already noted, will change the interpretation of coefficients and predictions.

Many of the details we provide in this section are applicable to all Stata estimation commands, not just to `regress`.

3.4.1 Correlations

Before regression, it can be useful to investigate pairwise correlations of the dependent variables and key regressor variables by using `correlate`. We have

```
. * Pairwise correlations for dependent variable and regressor variables
. correlate ltotexp suppins phylim actlim totchr age female income
(obs=2955)
```

	ltotexp	suppins	phylim	actlim	totchr	age
ltotexp	1.0000					
suppins	0.0941	1.0000				
phylim	0.2924	-0.0243	1.0000			
actlim	0.2888	-0.0675	0.5904	1.0000		
totchr	0.4283	0.0124	0.3334	0.3260	1.0000	
age	0.0858	-0.1226	0.2538	0.2394	0.0904	1.0000
female	-0.0058	-0.0796	0.0943	0.0499	0.0557	0.0774
income	0.0023	0.1943	-0.1142	-0.1483	-0.0816	-0.1542

	female	income
female	1.0000	
income	-0.1312	1.0000

Medical expenditures are most highly correlated with the health-status measures `phylim`, `actlim`, and `totchr`. The regressors are only weakly correlated with each other, aside from the health-status measures. Note that **correlate** restricts analysis to the 2,955 observations where data are available for all variables in the variable list. The related command `pwcorr`, not demonstrated, with the `sig` option gives the statistical significance of the correlations.

3.4.2 The regress command

The `regress` command performs OLS regression and yields an analysis-of-variance table, goodness-of-fit statistics, coefficient estimates, standard errors, t statistics, p-values, and confidence intervals. The syntax of the command is

regress *depvar* [*indepvars*] [*if*] [*in*] [*weight*] [*, options*]

Other Stata estimation commands have similar syntaxes. The output from `regress` is similar to that from many linear regression packages.

For independent cross-section data, the standard approach is to use the `vce(robust)` option, which gives standard errors that are valid even if model errors are heteroskedastic; see section 3.3.4. In that case, the analysis-of-variance table, based on the assumption of homoskedasticity, is dropped from the output. We obtain

```
. * OLS regression with heteroskedasticity-robust standard errors
. regress ltotexp suppins phylim actlim totchr age female income, vce(robust)
Linear regression                               Number of obs =     2955
                                                F(  7,  2947) =   126.97
                                                Prob > F      =   0.0000
                                                R-squared     =   0.2289
                                                Root MSE      =   1.2023
```

ltotexp	Coef.	Robust Std. Err.	t	P>|t|	[95% Conf.	Interval]
suppins	.2556428	.0465982	5.49	0.000	.1642744	.3470112
phylim	.3020598	.057705	5.23	0.000	.1889136	.415206
actlim	.3560054	.0634066	5.61	0.000	.2316797	.4803311
totchr	.3758201	.0187185	20.08	0.000	.3391175	.4125228
age	.0038016	.0037028	1.03	0.305	-.0034587	.011062
female	-.0843275	.045654	-1.85	0.065	-.1738444	.0051894
income	.0025498	.0010468	2.44	0.015	.0004973	.0046023
_cons	6.703737	.2825751	23.72	0.000	6.149673	7.257802

The regressors are jointly statistically significant, because the overall F statistic of 126.97 has a p-value of 0.000. At the same time, much of the variation is unexplained with $R^2 = 0.2289$. The root MSE statistic reports s, the standard error of the regression, defined in (3.2). By using a two-sided test at level 0.05, all regressors are individually statistically significant because $p < 0.05$, aside from `age` and `female`. The strong statistical insignificance of age may be due to sample restriction to elderly people and the inclusion of several health-status measures that capture well the health effect of age.

Statistical significance of coefficients is easily established. More important is the economic significance of coefficients, meaning the measured impact of regressors on medical expenditures. This is straightforward for regression in levels, because we can directly use the estimated coefficients. But here the regression is in logs. From section 3.3.6, in the log-linear model, parameters need to be interpreted as semielasticities. For example, the coefficient on `suppins` is 0.256. This means that private supplementary insurance is associated with a 0.256 proportionate rise, or a 25.6% rise, in medical expenditures. Similarly, large effects are obtained for the health-status measures, whereas health expenditures for women are 8.4% lower than those for men after controlling for other characteristics. The `income` coefficient of 0.0025 suggests a very small effect, but this is misleading. The standard deviation of `income` is 22, so a 1–standard deviation in `income` leads to a 0.055 proportionate rise, or 5.5% rise, in medical expenditures.

MEs in nonlinear models are discussed in more detail in section 10.6. The preceding interpretations are based on calculus methods that consider very small changes in the regressor. For larger changes in the regressor, the finite-difference method is more appropriate. Then the interpretation in the log-linear model is similar to that for the exponential conditional mean model; see section 10.6.4. For example, the estimated effect of going from no supplementary insurance (`suppins=0`) to having supplementary insurance (`suppins=1`) is more precisely a $100 \times (e^{0.256} - 1)$, or 29.2%, rise.

The `regress` command provides additional results that are not listed. In particular, the estimate of the VCE is stored in the matrix `e(V)`. Ways to access this and other stored results from regression have been given in section 1.6. Various postestimation commands enable prediction, computation of residuals, hypothesis testing, and model specification tests. Many of these are illustrated in subsequent sections. Two useful commands are

```
. * Display stored results and list available postestimation commands
. ereturn list
```
  (*output omitted*)
```
. help regress postestimation
```
  (*output omitted*)

3.4.3   Hypothesis tests

The `test` command performs hypothesis tests using the Wald test procedure that uses the estimated model coefficients and VCE. We present some leading examples here, with a more extensive discussion deferred to section 12.3. The F statistic version of the Wald test is used after `regress`, whereas for many other estimators the chi-squared version is instead used.

A common test is one of equality of coefficients. For example, consider testing that having a functional limitation has the same impact on medical expenditures as having an activity limitation. The test of $H_0: \beta_{\text{phylim}} = \beta_{\text{actlim}}$ against $H_a: \beta_{\text{phylim}} \neq \beta_{\text{actlim}}$ is implemented as

```
. * Wald test of equality of coefficients
. quietly regress ltotexp suppins phylim actlim totchr age female
> income, vce(robust)
. test phylim = actlim
( 1)  phylim - actlim = 0
      F(  1,  2947) =    0.27
            Prob > F =    0.6054
```

Because $p = 0.61 > 0.05$, we do not reject the null hypothesis at the 5% significance level. There is no statistically significant difference between the coefficients of the two variables.

The model can also be fit subject to constraints. For example, to obtain the least-squares estimates subject to $\beta_{\text{phylim}} = \beta_{\text{actlim}}$, we define the constraint using constraint define and then fit the model using cnsreg for constrained regression with the constraints() option. See exercise 2 at the end of this chapter for an example.

Another common test is one of the joint statistical significance of a subset of the regressors. A test of the joint significance of the health-status measures is one of H_0: $\beta_{\text{phylim}} = 0$, $\beta_{\text{actlim}} = 0$, $\beta_{\text{totchr}} = 0$ against H_a: at least one is nonzero. This is implemented as

```
. *  Joint test of statistical significance of several variables
. test phylim actlim totchr
( 1)  phylim = 0
( 2)  actlim = 0
( 3)  totchr = 0
      F(  3,  2947) =  272.36
            Prob > F =    0.0000
```

These three variables are jointly statistically significant at the 0.05 level because $p = 0.000 < 0.05$.

3.4.4   Tables of output from several regressions

It is very useful to be able to tabulate key results from multiple regressions for both one's own analysis and final report writing.

The estimates store command after regression leads to results in e() being associated with a user-provided model name and preserved even if subsequent models are fit. Given one or more such sets of stored estimates, estimates table presents a table of regression coefficients (the default) and, optionally, additional results. The estimates stats command lists the sample size and several likelihood-based statistics.

We compare the original regression model with a variant that replaces income with educyr. The example uses several of the available options for estimates table.

```
. * Store and then tabulate results from multiple regressions
. quietly regress ltotexp suppins phylim actlim totchr age female income,
> vce(robust)

. estimates store REG1

. quietly regress ltotexp suppins phylim actlim totchr age female educyr,
> vce(robust)

. estimates store REG2

. estimates table REG1 REG2, b(%9.4f) se stats(N r2 F ll)
> keep(suppins income educyr)
```

Variable	REG1	REG2
suppins	0.2556	0.2063
	0.0466	0.0471
income	0.0025	
	0.0010	
educyr		0.0480
		0.0070
N	2955	2955
r2	0.2289	0.2406
F	126.9723	132.5337
ll	-4.73e+03	-4.71e+03

legend: b/se

This table presents coefficients (b) and standard errors (se), with other available options including t statistics (t) and p-values (p). The statistics given are the sample size, the R^2, the overall F statistic (based on the robust estimate of the VCE), and the log likelihood (based on the strong assumption of normal homoskedastic errors). The keep() option, like the drop() option, provides a way to tabulate results for just the key regressors of interest. Here educyr is a much stronger predictor than income, because it is more highly statistically significant and R^2 is higher, and there is considerable change in the coefficient of suppins.

3.4.5   Even better tables of regression output

The preceding table is very useful for model comparison but has several limitations. It would be more readable if the standard errors appeared in parentheses. It would be beneficial to be able to report a p-value for the overall F statistic. Also some work may be needed to import the table into a table format in external software such as Excel, Word, or LaTeX.

The user-written esttab command (Jann 2007) provides a way to do this, following the estimates store command. A cleaner version of the previous table is given by

```
. * Tabulate results using user-written command esttab to produce cleaner output
. esttab REG1 REG2, b(%10.4f) se scalars(N r2 F ll) mtitles
> keep(suppins income educyr) title("Model comparison of REG1-REG2")
Model comparison of REG1-REG2
```

	(1) REG1	(2) REG2
suppins	0.2556***	0.2063***
	(0.0466)	(0.0471)
income	0.0025*	
	(0.0010)	
educyr		0.0480***
		(0.0070)
N	2955	2955
r2	0.2289	0.2406
F	126.9723	132.5337
ll	-4733.4476	-4710.9578

```
Standard errors in parentheses
* p<0.05, ** p<0.01, *** p<0.001
```

Now standard errors are in parentheses, the strength of statistical significance is given using stars that can be suppressed by using the **nostar** option, and a title is added.

The table can be written to a file that, for example, creates a table in LaTeX.

```
. * Write tabulated results to a file in latex table format
. quietly esttab REG1 REG2 using mus03table.tex, replace b(%10.4f) se
> scalars(N r2 F ll) mtitles keep(suppins age income educyr _cons)
> title("Model comparison of REG1-REG2")
```

Other formats include .rtf for rich text format (Word), .csv for comma-separated values, and .txt for fixed and tab-delimited text.

As mentioned earlier, this table would be better if the p-value for the overall F statistic were provided. This is not stored in e(). However, it is possible to calculate the p-value given other variables in e(). The user-written **estadd** command (Jann 2005) allows adding this computed p-value to stored results that can then be tabulated with **esttab**. We demonstrate this for a smaller table to minimize output.

```
. * Add a user-calculated statistic to the table
. estimates drop REG1 REG2
. quietly regress ltotexp suppins phylim actlim totchr age female income,
> vce(robust)
. estadd scalar pvalue = Ftail(e(df_r),e(df_m),e(F))
  (output omitted)
. estimates store REG1
. quietly regress ltotexp suppins phylim actlim totchr age female educyr,
> vce(robust)
```

```
. estadd scalar pvalue = Ftail(e(df_r),e(df_m),e(F))
  (output omitted)
. estimates store REG2
. esttab REG1 REG2, b(%10.4f) se scalars(F pvalue) mtitles keep(suppins)
```

	(1)	(2)
	REG1	REG2
suppins	0.2556***	0.2063***
	(0.0466)	(0.0471)
N	2955	2955
F	126.9723	132.5337
pvalue	0.0000	0.0000

```
Standard errors in parentheses
* p<0.05, ** p<0.01, *** p<0.001
```

The `estimates drop` command saves memory by dropping stored estimates that are no longer needed. In particular, for large samples the sample inclusion indicator `e(sample)` can take up much memory.

Related user-written commands by Jann (2005, 2007) are `estout`, a richer but more complicated version of `esttab`, and `eststo`, which extends `estimates store`. Several earlier user-written commands, notably, `outreg`, also create tables of regression output but are generally no longer being updated by their authors. The user-written `reformat` command (Brady 2002) allows formatting of the usual table of output from a single estimation command.

3.4.6   Factor variables for categorical variables and interactions

Suppose we wish to add as regressors to the regression model a set of indicator variables for family size and this set of indicators interacted with income. From sections 1.3.4 and 2.4.7, the factor variables `i.famsze` form a set of indicator variables based on the nonnegative, integer-valued categorical variable `famsze`, and the factor variables `c.income#i.famsze` denote the continuous variable `income` interacted with the set of indicators.

```
. * Factor variables for sets of indicator variables and interactions
. regress ltotexp suppins phylim actlim totchr age female c.income
> i.famsze c.income#i.famsze, vce(robust) noheader allbaselevels
note: 8.famsze#c.income omitted because of collinearity
note: 13.famsze#c.income omitted because of collinearity
```

ltotexp	Coef.	Robust Std. Err.	t	P>|t|	[95% Conf. Interval]	
suppins	.2393808	.0466804	5.13	0.000	.1478511	.3309104
phylim	.3053458	.0575971	5.30	0.000	.192411	.4182807
actlim	.3464812	.0631655	5.49	0.000	.2226279	.4703345
totchr	.3743755	.0187983	19.92	0.000	.3375162	.4112347
age	.00313	.0037607	0.83	0.405	-.0042438	.0105039
female	-.0725641	.0475022	-1.53	0.127	-.1657051	.0205769
income	.0028057	.0015684	1.79	0.074	-.0002695	.0058809
famsze						
1	(base)					
2	.0759158	.0722829	1.05	0.294	-.0658145	.2176462
3	-.2180488	.1310662	-1.66	0.096	-.4750399	.0389423
4	-.2928383	.1983967	-1.48	0.140	-.6818493	.0961727
5	.393989	.4501513	0.88	0.382	-.4886557	1.276634
6	-.3438142	.4524585	-0.76	0.447	-1.230983	.5433545
7	-1.101773	.5046005	-2.18	0.029	-2.09118	-.1123653
8	.216274	.0625337	3.46	0.001	.0936596	.3388884
10	1.482976	.2976336	4.98	0.000	.8993834	2.066568
13	-1.874285	.0712566	-26.30	0.000	-2.014003	-1.734567
famsze# c.income						
1	(base)					
2	-.0012899	.0020704	-0.62	0.533	-.0053495	.0027697
3	.004134	.0039464	1.05	0.295	-.003604	.0118719
4	.0160613	.0083284	1.93	0.054	-.0002688	.0323915
5	-.0251491	.017609	-1.43	0.153	-.0596764	.0093781
6	.0280329	.0227835	1.23	0.219	-.0166403	.0727062
7	-.0324118	.0279151	-1.16	0.246	-.087147	.0223234
8	(omitted)					
10	-.1759027	.0169673	-10.37	0.000	-.2091717	-.1426337
13	(omitted)					
_cons	6.748094	.3005551	22.45	0.000	6.158773	7.337414

Here there are 10 possible indicator variables for family size (1–8, 10, and 13), and the indicator for the lowest-valued of these (famsze = 1) is the base category that is omitted from the regression. In principle, there should be as many interactions with income included in the regression, but those corresponding to famsze equal to 8 and 13 are omitted because they are not identified for these data where only one observation has famsze equal to 8 and only one has famsze equal to 13.

We can test for joint significance of the sets of indicator variables, including their interaction with income, with the following command:

```
. * Test joint significance of sets of indicator variables and interactions
. testparm i.famsz c.income#i.famsze
 ( 1)   2.famsze = 0
 ( 2)   3.famsze = 0
 ( 3)   4.famsze = 0
 ( 4)   5.famsze = 0
 ( 5)   6.famsze = 0
 ( 6)   7.famsze = 0
 ( 7)   8.famsze = 0
 ( 8)  10.famsze = 0
 ( 9)  13.famsze = 0
 (10)   2.famsze#c.income = 0
 (11)   3.famsze#c.income = 0
 (12)   4.famsze#c.income = 0
 (13)   5.famsze#c.income = 0
 (14)   6.famsze#c.income = 0
 (15)   7.famsze#c.income = 0
 (16)  10.famsze#c.income = 0
        F( 16,  2931) =    83.76
            Prob > F =    0.0000
```

The sets of indicator variables for `famsze` are jointly statistically significant at level 0.05 because $p = 0.00 < 0.05$. The number of degrees of freedom is 16, so the additional two omitted variables (interaction with `income` when `famsze` equals 8 or 13) were correctly accounted for.

Calculation of the MEs with respect to income or family size will be complicated in this model. We calculate MEs in section 3.6.2.

3.5   Specification analysis

The fitted model has $R^2 = 0.23$, which is reasonable for cross-section data, and most regressors are highly statistically significant with the expected coefficient signs. Therefore, it is tempting to begin interpreting the results.

However, before doing so, it is useful to subject this regression to some additional scrutiny because a badly misspecified model may lead to erroneous inferences. We consider several specification tests, with the notable exception of testing for regressor exogeneity, which is deferred to chapter 6.

3.5.1   Specification tests and model diagnostics

In microeconometrics, the most common approach to deciding on the adequacy of a model is a Wald-test approach that fits a richer model and determines whether the data support the need for a richer model. For example, we may add additional regressors to the model and test whether they have a zero coefficient.

Stata also presents the user with an impressive and bewildering menu of choices of diagnostic checks for the currently fitted regression; see [R] **regress postestimation**. Some are specific to OLS regression, whereas others apply to most regression models. Some are visual aids such as plots of residuals against fitted values. Some are diagnostic statistics such as influence statistics that indicate the relative importance of individual observations. And some are formal tests that test for the failure of one or more assumptions of the model. We briefly present plots and diagnostic statistics, before giving a lengthier treatment of specification tests.

3.5.2   Residual diagnostic plots

Diagnostic plots are used less in microeconometrics than in some other branches of statistics, for several reasons. First, economic theory and previous research provide a lot of guidance as to the likely key regressors and functional form for a model. Studies rely on this and shy away from excessive data mining. Secondly, microeconometric studies typically use large datasets and regressions with many variables. Many variables potentially lead to many diagnostic plots, and many observations make it less likely that any single observation will be very influential, unless data for that observation are seriously miscoded.

We consider various residual plots that can aid in outlier detection, where an outlier is an observation poorly predicted by the model. One way to do this is to plot actual values against fitted values of the dependent variable. The postestimation command **rvfplot** gives a transformation of this, plotting the residuals $\widehat{u}_i = y_i - \widehat{y}_i$ against the fitted values $\widehat{y}_i = \mathbf{x}_i'\widehat{\boldsymbol{\beta}}$. We have

```
. * Plot of residuals against fitted values
. quietly regress ltotexp suppins phylim actlim totchr age female income,
> vce(robust)

. rvfplot
```

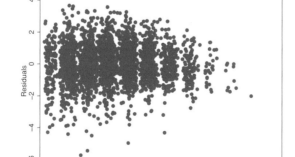

Figure 3.2. Residuals plotted against fitted values after OLS regression

Figure 3.2 does not indicate any extreme outliers, though the three observations with a residual less than -5 may be worth investigating. To do so, we need to generate $\hat{u}$ by using the `predict` command, detailed in section 3.6, and we need to list some details on those observations with $\hat{u} < -5$. We have

```
. * Details on the outlier residuals
. predict uhat, residual
(109 missing values generated)

. predict yhat, xb

. list totexp ltotexp yhat uhat if uhat < -5, clean

          totexp     ltotexp        yhat        uhat
     1.         3    1.098612    7.254341   -6.155728
     2.         6    1.791759    7.513358   -5.721598
     3.         9    2.197225    7.631211   -5.433987
```

The three outlying residuals are for three observations with the very smallest annual medical expenditures of, respectively, \$3, \$6, and \$9. The model evidently greatly overpredicts for these observations, with the predicted logarithm of total expenditures (`yhat`) much greater than `ltotexp`.

Stata provides several other residual plots. The `rvpplot` postestimation command plots residuals against an individual regressor. The `avplot` command provides an added-variable plot, or partial regression plot, that is a useful visual aid to outlier detection. Other commands give component-plus-residual plots that aid detection of nonlinearities and leverage plots. For details and additional references, see [R] **regress postestimation**.

3.5.3 Influential observations

Some observations may have unusual influence in determining parameter estimates and resulting model predictions.

Influential observations can be detected using one of several measures that are large if the residual is large, the leverage measure is large, or both. The leverage measure of the ith observation, denoted by h_i, equals the ith diagonal entry in the so-called hat matrix $\mathbf{H} = \mathbf{X}(\mathbf{X'X})^{-1}\mathbf{X}$. If h_i is large, then y_i has a big influence on its OLS prediction $\hat{y}_i$ because $\hat{\mathbf{y}} = \mathbf{Hy}$. Different measures, including h_i, can be obtained by using different options of `predict`.

A commonly used measure is dfits$_i$, which can be shown to equal the (scaled) difference between predictions of y_i with and without the ith observation in the OLS regression (so dfits means difference in fits). Large absolute values of dfits indicate an influential data point. One can plot dfits and investigate further observations with outlying values of dfits. A rule of thumb is that observations with $|\text{dfits}| > 2\sqrt{k/N}$ may be worthy of further investigation, though for large datasets this rule can suggest that many observations are influential.

The `dfits` option of `predict` can be used after `regress` provided that regression is with default standard errors because the underlying theory presumes homoskedastic errors. We have

```
. * Compute dfits that combines outliers and leverage
. quietly regress ltotexp suppins phylim actlim totchr age female income
. predict dfits, dfits
(109 missing values generated)
. scalar threshold = 2*sqrt((e(df_m)+1)/e(N))
. display "dfits threshold = "  %6.3f threshold
dfits threshold =   0.104
. tabstat dfits, stat (min p1 p5 p95 p99 max) format(%9.3f) col(stat)
```

variable	min	p1	p5	p95	p99	max
dfits	-0.421	-0.147	-0.083	0.085	0.127	0.221

```
. list dfits totexp ltotexp yhat uhat if abs(dfits) > 2*threshold & e(sample),
> clean
```

	dfits	totexp	ltotexp	yhat	uhat
1.	-.2319179	3	1.098612	7.254341	-6.155728
2.	-.3002994	6	1.791759	7.513358	-5.721598
3.	-.2765266	9	2.197225	7.631211	-5.433987
10.	-.2170063	30	3.401197	8.348724	-4.947527
42.	-.2612321	103	4.634729	7.57982	-2.945091
44.	-.4212185	110	4.70048	8.993904	-4.293423
108.	-.2326284	228	5.429346	7.971406	-2.54206
114.	-.2447627	239	5.476463	7.946239	-2.469776
137.	-.2177336	283	5.645447	7.929719	-2.284273
211.	-.211344	415	6.028278	8.028338	-2.00006
2925.	.2207284	62346	11.04045	8.660131	2.380323

Here over 2% of the sample has |dfits| greater than the suggested threshold of 0.104. But only 11 observations have |dfits| greater than two times the threshold. These correspond to observations with relatively low expenditures, or in one case, relatively high expenditures. We conclude that no observation has unusual influence.

3.5.4 Specification tests

Formal model-specification tests have two limitations. First, a test for the failure of a specific model assumption may not be robust with respect to the failure of another assumption that is not under test. For example, the rejection of the null hypothesis of homoskedasticity may be due to a misspecified functional form for the conditional mean. An example is given in section 3.5.5. Second, with a very large sample, even trivial deviations from the null hypothesis of correct specification will cause the test to reject the null hypothesis. For example, if a previously omitted regressor has a very small coefficient, say, 0.000001, then with an infinitely large sample the estimate will be sufficiently precise that we will always reject the null of zero coefficient.

Test of omitted variables

The most common specification test is to include additional regressors and test whether they are statistically significant by using a Wald test of the null hypothesis that the coefficient is zero. The additional regressor may be a variable not already included, a transformation of a variable(s) already included such as a quadratic in age, or a quadratic with interaction terms in age and education. If groups of regressors are included, such as a set of region dummies, `test` can be used after `regress` to perform a joint test of statistical significance.

In some branches of biostatistics, it is common to include only regressors with $p < 0.05$. In microeconometrics, it is common instead to additionally include regressors that are statistically insignificant if economic theory or conventional practice includes the variable as a control. This reduces the likelihood of inconsistent parameter estimation due to omitted-variables bias at the expense of reduced precision in estimation.

Test of the Box–Cox model

A common specification-testing approach is to fit a richer model that tests the current model as a special case and perform a Wald test of the parameter restrictions that lead to the simpler model. The preceding omitted-variable test is an example.

Here we consider a test specific to the current example. We want to decide whether a regression model for medical expenditures is better in logs than in levels. There is no obvious way to compare the two models because they have different dependent variables. However, the Box–Cox transform leads to a richer model that includes the linear and log-linear models as special cases. Specifically, we fit the model with the transformed dependent variable

$$g(y_i, \theta) \equiv \frac{y_i^\theta - 1}{\theta} = \mathbf{x}_i'\boldsymbol{\beta} + u_i$$

where θ and $\boldsymbol{\beta}$ are estimated under the assumption that $u_i \sim N(0, \sigma^2)$. Three leading cases are 1) $g(y, \theta) = y - 1$ if $\theta = 1$; 2) $g(y, \theta) = \ln y$ if $\theta = 0$; and 3) $g(y, \theta) = 1 - 1/y$ if $\theta = -1$. The log-linear model is supported if $\widehat{\theta}$ is close to 0, and the linear model is supported if $\widehat{\theta} = 1$.

The Box–Cox transformation introduces a nonlinearity and an additional unknown parameter θ into the model. This moves the modeling exercise into the domain of nonlinear models. The model is straightforward to fit, however, because Stata provides the `boxcox` command to fit the model. We obtain

```
. * Boxcox model with lhs variable transformed
. boxcox totexp suppins phylim actlim totchr age female income if totexp>0, nolog
Fitting comparison model

Fitting full model
```

			Number of obs	=	2955
			LR chi2(7)	=	773.02
Log likelihood = -28518.267			Prob > chi2	=	0.000

totexp	Coef.	Std. Err.	z	P>|z|	[95% Conf. Interval]
/theta	.0758956	.0096386	7.87	0.000	.0570042 .0947869

```
Estimates of scale-variant parameters
```

	Coef.
Notrans	
suppins	.4459618
phylim	.577317
actlim	.6905939
totchr	.6754338
age	.0051321
female	-.1767976
income	.0044039
_cons	8.930566
/sigma	2.189679

Test H0:	Restricted log likelihood	LR statistic chi2	P-value Prob > chi2
theta = -1	-37454.643	17872.75	0.000
theta = 0	-28550.353	64.17	0.000
theta = 1	-31762.809	6489.08	0.000

The null hypothesis of $\theta = 0$ is strongly rejected, so the log-linear model is rejected. However, the Box–Cox model with general θ is difficult to interpret and use, and the estimate of $\widehat{\theta} = 0.0759$ gives much greater support for a log-linear model ($\theta = 0$) than the linear model ($\theta = 1$). Thus we prefer to use the log-linear model.

Test of the functional form of the conditional mean

The linear regression model specifies that the conditional mean of the dependent variable (whether measured in levels or in logs) equals $\mathbf{x}_i'\boldsymbol{\beta}$. A standard test that this is the correct specification is a variable augmentation test. A common approach is to add powers of $\widehat{y}_i = \mathbf{x}_i'\widehat{\boldsymbol{\beta}}$, the fitted value of the dependent variable, as regressors and a test for the statistical significance of the powers.

The `estat ovtest` postestimation command provides a RESET test that regresses y on $\mathbf{x}$ and $\widehat{y}^2$, $\widehat{y}^3$, and $\widehat{y}^4$, and jointly tests that the coefficients of $\widehat{y}^2$, $\widehat{y}^3$, and $\widehat{y}^4$ are zero. We have

```
. * Variable augmentation test of conditional mean using estat ovtest
. quietly regress ltotexp suppins phylim actlim totchr age female income,
> vce(robust)

. estat ovtest

Ramsey RESET test using powers of the fitted values of ltotexp
      Ho:  model has no omitted variables
                  F(3, 2944) =          9.04
                      Prob > F =        0.0000
```

The model is strongly rejected because $p = 0.000$.

An alternative, simpler test is provided by the `linktest` command. This regresses y on $\widehat{y}$ and $\widehat{y}^2$, where now the original model regressors $\mathbf{x}$ are omitted, and it tests whether the coefficient of $\widehat{y}^2$ is zero. We have

```
. * Link test of functional form of conditional mean
. quietly regress ltotexp suppins phylim actlim totchr age female income,
> vce(robust)

. linktest
```

Source	SS	df	MS		
Model	1301.41696	2	650.708481		
Residual	4223.47242	2952	1.43071559		
Total	5524.88938	2954	1.87030785		

	Number of obs =	2955
	F(2, 2952) =	454.81
	Prob > F =	0.0000
	R-squared =	0.2356
	Adj R-squared =	0.2350
	Root MSE =	1.1961

ltotexp	Coef.	Std. Err.	t	P>|t|	[95% Conf. Interval]	
_hat	4.429216	.6779517	6.53	0.000	3.09991	5.758522
_hatsq	-.2084091	.0411515	-5.06	0.000	-.2890976	-.1277206
_cons	-14.01127	2.779936	-5.04	0.000	-19.46208	-8.56046

Again the null hypothesis that the conditional mean is correctly specified is rejected. A likely reason is that so few regressors were included in the model, for pedagogical reasons.

The two preceding commands had different formats. The first test used the `estat ovtest` command, where `estat` produces various statistics following estimation and the particular statistics available vary with the previous estimation command. The second test used `linktest`, which is available for a wider range of models.

Heteroskedasticity test

One consequence of heteroskedasticity is that default OLS standard errors are incorrect. This can be readily corrected and guarded against by routinely using heteroskedasticity-robust standard errors.

Nonetheless, there may be interest in formally testing whether heteroskedasticity is present. For example, the retransformation methods for the log-linear model used in section 3.6.3 assume homoskedastic errors. In section 5.3, we present diagnostic plots for heteroskedasticity. Here we instead present a formal test.

A quite general model of heteroskedasticity is

$$\text{Var}(y|\mathbf{x}) = h(\alpha_1 + \mathbf{z}'\boldsymbol{\alpha}_2)$$

where $h(\cdot)$ is a positive monotonic function such as $\exp(\cdot)$ and the variables in $\mathbf{z}$ are functions of the variables in $\mathbf{x}$. Tests for heteroskedasticity are tests of

$$H_0 : \boldsymbol{\alpha}_2 = \mathbf{0}$$

and can be shown to be independent of the choice of function $h(\cdot)$. We reject H_0 at the α level if the test statistic exceeds the α critical value of a chi-squared distribution with degrees of freedom equal to the number of components of $\mathbf{z}$. The test is performed by using the `estat hettest` postestimation command. The simplest version is the Breusch–Pagan Lagrange multiplier test, which is equal to N times the uncentered explained sum of squares from the regression of the squared residuals on an intercept and $\mathbf{z}$. We use the `iid` option to obtain a different version of the test that relaxes the default assumption that the errors are normally distributed.

Several choices of the components of $\mathbf{z}$ are possible. By far, the best choice is to use variables that are a priori likely determinants of heteroskedasticity. For example, in regressing the level of earnings on several regressors including years of schooling, it is likely that those with many years of schooling have the greatest variability in earnings. Such candidates rarely exist. Instead, standard choices are to use the OLS fitted value $\widehat{y}$, the default for `estat hettest`, or to use all the regressors so $\mathbf{z} = \mathbf{x}$. White's test for heteroskedasticity is equivalent to letting $\mathbf{z}$ equal unique terms in the products and cross products of the terms in $\mathbf{x}$.

We consider $\mathbf{z} = \widehat{y}$ and $\mathbf{z} = \mathbf{x}$. Then we have

```
. * Heteroskedasticity tests using estat hettest and option iid
. quietly regress ltotexp suppins phylim actlim totchr age female income
. estat hettest, iid
Breusch-Pagan / Cook-Weisberg test for heteroskedasticity
        Ho: Constant variance
        Variables: fitted values of ltotexp
        chi2(1)      =      32.87
        Prob > chi2  =     0.0000
. estat hettest suppins phylim actlim totchr age female income, iid
Breusch-Pagan / Cook-Weisberg test for heteroskedasticity
        Ho: Constant variance
        Variables: suppins phylim actlim totchr age female income
        chi2(7)      =      93.13
        Prob > chi2  =     0.0000
```

Both versions of the test, with $\mathbf{z} = \widehat{y}$ and with $\mathbf{z} = \mathbf{x}$, have $p = 0.0000$ and strongly reject homoskedasticity.

Omnibus test

An alternative to separate tests of misspecification is an omnibus test, which is a joint test of misspecification in several directions. A leading example is the information matrix (IM) test (see section 12.7), which is a test for correct specification of a fully parametric model based on whether the IM equality holds. For linear regression with normal homoskedastic errors, the IM test can be shown to be a joint test of heteroskedasticity, skewness, and nonnormal kurtosis compared with the null hypothesis of homoskedasticity, symmetry, and kurtosis coefficient of 3; see Hall (1987).

The `estat imtest` postestimation command computes the joint IM test and also splits it into its three components. We obtain

```
. * Information matrix test
. quietly regress ltotexp suppins phylim actlim totchr age female income
. estat imtest
Cameron & Trivedi's decomposition of IM-test
```

Source	chi2	df	p
Heteroskedasticity	139.90	31	0.0000
Skewness	35.11	7	0.0000
Kurtosis	11.96	1	0.0005
Total	186.97	39	0.0000

The overall joint IM test rejects the model assumption that $y \sim N(\mathbf{x}'\boldsymbol{\beta}, \sigma^2 \mathbf{I})$, because $p = 0.0000$ in the `Total` row. The decomposition indicates that all three assumptions of homoskedasticity, symmetry, and normal kurtosis are rejected. Note, however, that the decomposition assumes correct specification of the conditional mean. If instead the mean is misspecified, then that could be the cause of rejection of the model by the IM test.

3.5.5 Tests have power in more than one direction

Tests can have power in more than one direction, so that if a test targeted to a particular type of model misspecification rejects a model, it is not necessarily the case that this particular type of model misspecification is the underlying problem. For example, a test of heteroskedasticity may reject homoskedasticity, even though the underlying cause of rejection is that the conditional mean is misspecified rather than that errors are heteroskedastic.

To illustrate this example, we use the following simulation exercise. The DGP is one with homoskedastic normal errors

$$y_i = \exp(1 + 0.25 \times x_i + 4 \times x_i^2) + u_i,$$
$$x_i \sim U(0,1), \quad u_i \sim N(0,1)$$

We instead fit a model with a misspecified conditional mean function:

$$y = \beta_0 + \beta_1 x + \beta_2 x^2 + v$$

We consider a simulation with a sample size of 50. We generate the regressors and the dependent variable by using commands detailed in section 4.2. We obtain

```
. * Simulation to show tests have power in more than one direction
. clear all
. set obs 50
obs was 0, now 50
. set seed 10101
. generate x = runiform()               // x ~ uniform(0,1)
. generate u = rnormal()                // u ~ N(0,1)
. generate y = exp(1 + 0.25*x + 4*x^2) + u
. generate xsq = x^2
. regress y x xsq
```

Source	SS	df	MS		Number of obs =	50
					F(2, 47) =	168.27
Model	76293.9057	2	38146.9528		Prob > F =	0.0000
Residual	10654.8492	47	226.698919		R-squared =	0.8775
					Adj R-squared =	0.8722
Total	86948.7549	49	1774.46438		Root MSE =	15.057

y	Coef.	Std. Err.	t	P>|t|	[95% Conf. Interval]	
x	-228.8379	29.3865	-7.79	0.000	-287.9559	-169.7199
xsq	342.7992	28.71815	11.94	0.000	285.0258	400.5727
_cons	28.68793	6.605434	4.34	0.000	15.39951	41.97635

The misspecified model seems to fit the data very well with highly statistically significant regressors and an R^2 of 0.88.

Now consider a test for heteroskedasticity:

```
. * Test for heteroskedasticity
. estat hettest
Breusch-Pagan / Cook-Weisberg test for heteroskedasticity
         Ho: Constant variance
         Variables: fitted values of y
         chi2(1)     =     22.70
         Prob > chi2 =    0.0000
```

This test strongly suggests that the errors are heteroskedastic because $p = 0.0000$, even though the DGP had homoskedastic errors.

(Continued on next page)

The problem is that the regression function itself was misspecified. A RESET test yields

```
. * Test for misspecified conditional mean
. estat ovtest

Ramsey RESET test using powers of the fitted values of y
      Ho:  model has no omitted variables
                  F(3, 44) =    2702.16
                  Prob > F =     0.0000
```

This strongly rejects correct specification of the conditional mean because $p = 0.0000$.

Going the other way, could misspecification of other features of the model lead to rejection of the conditional mean, even though the conditional mean itself was correctly specified? This is an econometrically subtle question. The answer, in general, is yes. However, for the linear regression model, this is not the case essentially because consistency of the OLS estimator requires only that the conditional mean be correctly specified.

3.6 Prediction

For the linear regression model, the estimator of the conditional mean of y given $\mathbf{x} = \mathbf{x}_p$, $E(y|\mathbf{x}_p) = \mathbf{x}'_p\boldsymbol{\beta}$, is the conditional predictor $\widehat{y} = \mathbf{x}'_p\widehat{\boldsymbol{\beta}}$. We focus here on prediction for each observation in the sample. We begin with prediction from a linear model for medical expenditures, because this is straightforward, before turning to the log-linear model.

Further details on prediction are presented in section 3.7, where weighted average prediction is discussed, and in sections 10.5 and 10.6, where many methods are presented.

3.6.1 In-sample prediction

The most common type of prediction is in-sample, where evaluation is at the observed regressor values for each observation. Then $\widehat{y}_i = \mathbf{x}'_i\widehat{\boldsymbol{\beta}}$ predicts $E(y_i|\mathbf{x}_i)$ for $i = 1, \ldots, N$.

To do this, we use **predict** after **regress**. The syntax for **predict** is

predict [*type*] *newvar* [*if*] [*in*] [, *options*]

The user always provides a name for the created variable, *newvar*. The default option is the prediction $\widehat{y}_i$. Other options yield residuals (usual, standardized, and studentized), several leverage and influential observation measures, predicted values, and associated standard errors of prediction. We have already used some of these options in section 3.5. The **predict** command can also be used for out-of-sample prediction. When used for in-sample prediction, it is good practice to add the **if e(sample)** qualifier, because this ensures that prediction is for the same sample as that used in estimation.

We consider prediction based on a linear regression model in levels rather than logs. We begin by reporting the regression results with `totexp` as the dependent variable.

```
. * Change dependent variable to level of positive medical expenditures
. use mus03data.dta, clear

. keep if totexp > 0
(109 observations deleted)

. regress totexp suppins phylim actlim totchr age female income, vce(robust)
```

Linear regression

```
Number of obs =     2955
F(  7,  2947) =    40.58
Prob > F      =   0.0000
R-squared     =   0.1163
Root MSE      =    11285
```

totexp	Coef.	Robust Std. Err.	t	P>|t|	[95% Conf. Interval]	
suppins	724.8632	427.3045	1.70	0.090	-112.9824	1562.709
phylim	2389.019	544.3493	4.39	0.000	1321.675	3456.362
actlim	3900.491	705.2244	5.53	0.000	2517.708	5283.273
totchr	1844.377	186.8938	9.87	0.000	1477.921	2210.832
age	-85.36264	37.81868	-2.26	0.024	-159.5163	-11.20892
female	-1383.29	432.4759	-3.20	0.001	-2231.275	-535.3044
income	6.46894	8.570658	0.75	0.450	-10.33614	23.27402
_cons	8358.954	2847.802	2.94	0.003	2775.07	13942.84

We then predict the level of medical expenditures:

```
. * Prediction in model linear in levels
. predict yhatlevels
(option xb assumed; fitted values)

. summarize totexp yhatlevels
```

Variable	Obs	Mean	Std. Dev.	Min	Max
totexp	2955	7290.235	11990.84	3	125610
yhatlevels	2955	7290.235	4089.624	-236.3781	22559

The summary statistics show that on average the predicted value `yhatlevels` equals the dependent variable. This suggests that the predictor does a good job. But this is misleading because this is always the case after OLS regression in a model with an intercept, since then residuals sum to zero implying $\sum y_i = \sum \widehat{y}_i$. The standard deviation of `yhatlevels` is \$4,090, so there is some variation in the predicted values.

For this example, a more discriminating test is to compare the median predicted and actual values. We have

```
. * Compare median prediction and median actual value
. tabstat totexp yhatlevels, stat (count p50) col(stat)
```

variable	N	p50
totexp	2955	3334
yhatlevels	2955	6464.692

There is considerable difference between the two, a consequence of the right-skewness of the original data, which the linear regression model does not capture.

The `stdp` option provides the standard error of the prediction, and the `stdf` option provides the standard error of the prediction for each sample observation, provided the original estimation command used the default VCE. We therefore reestimate without `vce(robust)` and use `predict` to obtain

```
. * Compute standard errors of prediction and forecast with default VCE
. quietly regress totexp suppins phylim actlim totchr age female income
. predict yhatstdp, stdp
. predict yhatstdf, stdf
. summarize yhatstdp yhatstdf
    Variable |      Obs        Mean    Std. Dev.       Min        Max
-------------+--------------------------------------------------------
   yhatstdp |     2955       572.7    129.6575   393.5964   2813.983
   yhatstdf |     2955    11300.52    10.50946   11292.12    11630.8
```

The first quantity views $\mathbf{x}_i'\widehat{\boldsymbol{\beta}}$ as an estimate of the conditional mean $\mathbf{x}_i'\boldsymbol{\beta}$ and is quite precisely estimated because the average standard deviation is \$573 compared with an average prediction of \$7,290. The second quantity views $\mathbf{x}_i'\widehat{\boldsymbol{\beta}}$ as an estimate of the actual value y_i and is very imprecisely estimated because $y_i = \mathbf{x}_i'\boldsymbol{\beta} + u_i$, and the error u_i here has relatively large variance because the levels equation has $s = 11285$.

More generally, microeconometric models predict poorly for a given individual, as evidenced by the typically low values of R^2 obtained from regression on cross-section data. These same models may nonetheless predict the conditional mean well, and it is this latter quantity that is needed for policy analysis that focuses on average behavior.

3.6.2   MEs and elasticities

The computation of MEs and elasticities using the postestimation `margins` command, introduced in Stata 11, is detailed in section 10.6 in the context of nonlinear models. Here we provide a brief summary for the linear model after OLS regression.

The `margins` command calculates predictions (with no option), marginal effects (with the `dydx()` option), and elasticities (with the `eyex()` option). These can be evaluated at the sample average values and then averaged (the default option), or evaluated at the sample means of the regressors (with the `atmean` option), or evaluated at specified values of the regressors (with the `at()` option). The `margins` command also produces associated standard errors and confidence intervals.

The default for the `margins` command is to obtain predictions, MEs, and elasticities for the quantity that is the default for the postestimation `predict` command. For many estimation commands, including `regress`, this is the conditional mean. Then the `margins, dydx()` command computes for each regressor the derivative $\partial E(y|\mathbf{x})/\partial x$. For binary indicator variables that explicitly enter the regression as factor variables, `margins` instead computes the finite difference $\Delta E(y|\mathbf{x})/\Delta x$.

For the linear model, the estimated ME of the jth regressor is $\widehat{\beta}_j$. As a result, the command `margins, dydx(income)` reproduces the slope coefficients, associated standard errors, and confidence intervals for the regressor `income`; the command `margins, dydx(*)` does so for all regressors. Therefore, there is often no need to use `margins` to compute the ME.

Once interactions between variables are introduced, however, computation of the ME becomes more complicated. For example, the effect of a change in `income` on the predicted conditional mean of `ltotexp` will be quite burdensome in the model of section 3.4.6, where `income` is interacted with sets of indicator variables for family size. However, `margins` takes care of this automatically, provided that we use factor variables to define the key variables in the original regression. Continuing the example of section 3.4.6, but with the dependent variable now `totexp` rather than `ltotexp` and using the full sample, we have

```
. * Compute the average marginal effect in model with interactions
. quietly regress totexp suppins phylim actlim totchr age female c.income
> i.famsze c.income#i.famsze, vce(robust) noheader allbaselevels

. margins, dydx(income)
Average marginal effects                        Number of obs    =       3064
Model VCE     : Robust

Expression    : Linear prediction, predict()
dy/dx w.r.t.  : income
```

		Delta-method				
	dy/dx	Std. Err.	z	P>|z|	[95% Conf.	Interval]
income	3.893248	8.387865	0.46	0.643	-12.54667	20.33316

By comparison, for the simpler model of section 3.6.1, running `margins, dydx(income)` gives an ME of 6.469 with a standard error of 8.571.

In this example, the average ME is obtained; that is, the ME for each individual observation is calculated and then they are averaged. Alternative points at which to evaluate the ME are detailed in section 10.6.

Another use of the `margins` command is to compute elasticities (and semielasticities). The elasticity of y with respect to x is $\partial y/\partial x \times (x/y)$. Because the elasticity can be rewritten as $(\partial y/y)/(\partial x/x)$, it is interpreted as the proportionate change in y divided by the proportionate change in x.

To compute the elasticity, we use the `eyex()` option of the `margins` command. The default is to compute the sample average elasticity, but usually there is no intrinsic interest in this quantity because it is a nonlinear transformation of the ME. Instead, it is more useful to evaluate the elasticity at a specific value of the regressors; most simply, at the sample mean of the regressors by using the `atmean` option. For example, to obtain the elasticity of `totexp` with respect to variable `totchr` in the model above, with evaluation at the sample means of `totchr` and the regressors, we type the following commands:

```
. * Compute elasticity for a specified regressor
. quietly regress totexp suppins phylim actlim totchr age female income,
> vce(robust)

. margins, eyex(totchr) atmean
Conditional marginal effects                   Number of obs    =    3064
Model VCE    : Robust

Expression   : Linear prediction, predict()
ey/ex w.r.t. : totchr
at           : suppins          =    .5812663  (mean)
               phylim           =    .4255875  (mean)
               actlim           =    .2836162  (mean)
               totchr           =    1.754243  (mean)
               age              =    74.17167  (mean)
               female           =    .5796345  (mean)
               income           =    22.47472  (mean)
```

	ey/ex	Delta-method Std. Err.	z	P>|z|	[95% Conf. Interval]
totchr	.4839724	.0433653	11.16	0.000	.398978 .5689668

A 1% increase in chronic problems is associated with a 0.48% increase in medical expenditures. The eyex(*) option computes elasticities for all regressors.

3.6.3   Prediction in logs: The retransformation problem

Transforming the dependent variable by taking the natural logarithm complicates prediction. It is easy to predict $E(\ln y|\mathbf{x})$, but we are instead interested in $E(y|\mathbf{x})$ because we want to predict the level of medical expenditures rather than the natural logarithm. The obvious procedure of predicting $\ln y$ and taking the exponential is wrong because $\exp\{E(\ln y)\} \neq E(y)$, just as, for example, $\sqrt{E(y^2)} \neq E(y)$.

The log-linear model $\ln y = \mathbf{x}'\boldsymbol{\beta} + u$ implies that $y = \exp(\mathbf{x}'\boldsymbol{\beta})\exp(u)$. It follows that

$$E(y_i|\mathbf{x}_i) = \exp(\mathbf{x}_i'\boldsymbol{\beta})E\{\exp(u_i)\}$$

The simplest prediction is $\exp(\mathbf{x}_i'\widehat{\boldsymbol{\beta}})$, but this is wrong because it ignores the multiple $E\{\exp(u_i)\}$. If it is assumed that $u_i \sim N(0, \sigma^2)$, then it can be shown that $E\{\exp(u_i)\} = \exp(0.5\sigma^2)$, which can be estimated by $\exp(0.5\widehat{\sigma}^2)$, where $\widehat{\sigma}^2$ is an unbiased estimator of the log-linear regression model error. A weaker assumption is to assume that u_i is independent and identically distributed, in which case we can consistently estimate $E\{\exp(u_i)\}$ by the sample average $N^{-1}\sum_{j=1}^{N}\exp(\widehat{u}_j)$; see Duan (1983).

Applying these methods to the medical expenditure data yields

```
. * Prediction in levels from a logarithmic model
. quietly regress ltotexp suppins phylim actlim totchr age female income

. quietly predict lyhat

. generate yhatwrong = exp(lyhat)

. generate yhatnormal = exp(lyhat)*exp(0.5*e(rmse)^2)
```

```
. quietly predict uhat, residual
. generate expuhat = exp(uhat)
. quietly summarize expuhat
. generate yhatduan = r(mean)*exp(lyhat)
. summarize totexp yhatwrong yhatnormal yhatduan yhatlevels
```

Variable	Obs	Mean	Std. Dev.	Min	Max
totexp	2955	7290.235	11990.84	3	125610
yhatwrong	2955	4004.453	3303.555	959.5991	37726.22
yhatnormal	2955	8249.927	6805.945	1976.955	77723.13
yhatduan	2955	8005.522	6604.318	1918.387	75420.57
yhatlevels	2955	7290.235	4089.624	-236.3781	22559

Ignoring the retransformation bias leads to a very poor prediction, because yhatwrong has a mean of \$4,004 compared with the sample mean of \$7,290. The two alternative methods yield much closer average values of \$8,250 and \$8,006. Furthermore, the predictions from log regression, compared with those in levels, have the desirable feature of always being positive and have greater variability. The standard deviation of yhatnormal, for example, is \$6,806 compared with \$4,090 from the levels model.

3.6.4 Prediction exercise

There are several ways that predictions can be used to simulate the effects of a policy experiment. We consider the effect of a binary treatment, whether a person has supplementary insurance, on medical expenditure. Here we base our predictions on estimates that assume supplementary insurance is exogenous. A more thorough analysis could instead use methods that more realistically permit insurance to be endogenous. As we discuss in section 6.2.1, a variable is endogenous if it is related to the error term. Our analysis here assumes that supplementary insurance is not related to the error term.

An obvious comparison is to compare the difference in sample means $(\overline{y}_1 - \overline{y}_0)$, where the subscript 1 denotes those with supplementary insurance and the subscript 0 denotes those without supplementary insurance. This measure does not control for individual characteristics. A measure that does control for individual characteristics is the difference in mean predictions $(\overline{\hat{y}}_1 - \overline{\hat{y}}_0)$, where, for example, $\overline{\hat{y}}_1$ denotes the average prediction for those with health insurance.

We implement the first two approaches for the complete sample based on OLS regression in levels and in logs. We obtain

(Continued on next page)

```
. * Predicted effect of supplementary insurance: methods 1 and 2
. bysort suppins: summarize totexp yhatlevels yhatduan
```

```
-> suppins = 0
    Variable |     Obs        Mean    Std. Dev.       Min        Max
-------------+--------------------------------------------------------
      totexp |    1207    6824.303    11425.94          9     104823
  yhatlevels |    1207    6824.303    4077.064    -236.3781   20131.43
    yhatduan |    1207    6745.959    5365.255    1918.387    54981.73
```

```
-> suppins = 1
    Variable |     Obs        Mean    Std. Dev.       Min        Max
-------------+--------------------------------------------------------
      totexp |    1748    7611.963    12358.83          3     125610
  yhatlevels |    1748    7611.963    4068.397    502.9237     22559
    yhatduan |    1748    8875.255    7212.993    2518.538    75420.57
```

The average difference is \$788 (from $7612 - 6824$) using either the difference in sample means or the difference in fitted values from the linear model. Equality of the two is a consequence of OLS regression and prediction using the estimation sample. The log-linear model, using the prediction based on Duan's method, gives a larger average difference of \$2,129 (from $8875 - 6746$).

A third measure is the difference between the mean predictions, one with suppins set to 1 for all observations and one with suppins = 0. For the linear model, this is simply the estimated coefficient of suppins, which is \$725.

For the log-linear model, we need to make separate predictions for each individual with suppins set to 1 and with suppins set to 0. For simplicity, we make predictions in levels from the log-linear model assuming normally distributed errors. To make these changes and after the analysis have suppins returned to its original sample values, we use preserve and restore (see section 2.5.2). We obtain

```
. * Predicted effect of supplementary insurance: method 3 for log-linear model
. quietly regress ltotexp suppins phylim actlim totchr age female income
. preserve
. quietly replace suppins = 1
. quietly predict lyhat1
. generate yhatnormal1 = exp(lyhat1)*exp(0.5*e(rmse)^2)
. quietly replace suppins = 0
. quietly predict lyhat0
. generate yhatnormal0 = exp(lyhat0)*exp(0.5*e(rmse)^2)
. generate treateffect = yhatnormal1 - yhatnormal0
. summarize yhatnormal1 yhatnormal0 treateffect
    Variable |     Obs        Mean    Std. Dev.       Min        Max
-------------+--------------------------------------------------------
 yhatnormal1 |    2955    9077.072    7313.963    2552.825    77723.13
 yhatnormal0 |    2955    7029.453    5664.069    1976.955    60190.23
 treateffect |    2955    2047.619    1649.894    575.8701    17532.91
. restore
```

While the average treatment effect of $2,048 is considerably larger than that obtained by using the difference in sample means of the linear model, it is comparable to the estimate produced by Duan's method.

3.7 Sampling weights

The analysis to date has presumed simple random sampling, where sample observations have been drawn from the population with equal probability. In practice, however, many microeconometric studies use data from surveys that are not representative of the population. Instead, groups of key interest to policy makers that would have too few observations in a purely random sample are oversampled, with other groups then undersampled. Examples are individuals from racial minorities or those with low income or living in sparsely populated states.

As explained below, weights should be used for estimation of population means and for postregression prediction and computation of MEs. However, in most cases, the regression itself can be fit without weights, as is the norm in microeconometrics. If weighted analysis is desired, it can be done using standard commands with a weighting option, which is the approach of this section and the standard approach in microeconometrics. Alternatively, one can use survey commands as detailed in section 5.5.

3.7.1 Weights

Sampling weights are provided by most survey datasets. These are called probability weights or `pweights` in Stata, though some others call them inverse-probability weights because they are inversely proportional to the probability of inclusion of the sample. A `pweight` of 1,400 in a survey of the U.S. population, for example, means that the observation is representative of 1,400 U.S. residents and the probability of this observation being included in the sample is 1/1400.

Most estimation commands allow probability weighted estimators that are obtained by adding `[pweight=weight]`, where *weight* is the name of the weighting variable.

To illustrate the use of sampling weights, we create an artificial weighting variable (sampling weights are available for the MEPS data but were not included in the data extract used in this chapter). We manufacture weights that increase the weight given to those with more chronic problems. In practice, such weights might arise if the original sampling framework oversampled people with few chronic problems and undersampled people with many chronic problems. In this section, we analyze levels of expenditures, including expenditures of zero. Specifically,

(Continued on next page)

```
. * Create artificial sampling weights
. use mus03data.dta, clear
. generate swght = totchr^2 + 0.5
. summarize swght
    Variable │       Obs        Mean    Std. Dev.        Min         Max
─────────────┼─────────────────────────────────────────────────────────
       swght │      3064    5.285574    6.029423          .5        49.5
```

What matters in subsequent analysis is the relative values of the sampling weights rather than the absolute values. The sampling weight variable swght takes on values from 0.5 to 49.5, so weighted analysis will give some observations as much as $49.5/0.5 = 99$ times the weight given to others.

Stata offers three other types of weights that for most analyses can be ignored. Analytical weights, called aweights, are used for the quite different purpose of compensating for different observations having different variances that are known up to scale; see section 5.3.4. For duplicated observations, fweights provide the number of duplicated observations. So-called importance weights, or iweights, are sometimes used in more advanced programming.

3.7.2 Weighted mean

If an estimate of a population mean is desired, then we should clearly weight. In this example, by oversampling those with few chronic problems, we will have oversampled people who on average have low medical expenditures, so that the unweighted sample mean will understate population mean medical expenditures.

Let w_i be the population weight for individual i. Then, by defining $W = \sum_{i=1}^{N} w_i$ to be the sum of the weights, the weighted mean $\bar{y}_W$ is

$$\bar{y}_W = \frac{1}{W} \sum_{i=1}^{N} w_i y_i$$

with variance estimator (assuming independent observations) $\widehat{V}(\bar{y}_W) = \{1/W(W-1)\} \sum_{i=1}^{N} w_i (y_i - \bar{y}_W)^2$. These formulas reduce to those for the unweighted mean if equal weights are used.

The weighted mean downweights oversampled observations because they will have a value of pweights (and hence w_i) that is smaller than that for most observations. We have

```
. * Calculate the weighted mean
. mean totexp [pweight=swght]
Mean estimation                         Number of obs    =      3064

             │       Mean    Std. Err.     [95% Conf. Interval]
─────────────┼────────────────────────────────────────────────
      totexp │   10670.83    428.5148       9830.62     11511.03
```

The weighted mean of \$10,671 is much larger than the unweighted mean of \$7,031 (see section 3.2.4) because the unweighted mean does not adjust for the oversampling of individuals with few chronic problems.

3.7.3   Weighted regression

The weighted least-squares estimator for the regression of y_i on $\mathbf{x}_i$ with the weights w_i is given by

$$\widehat{\boldsymbol{\beta}}_W = \left(\sum_{i=1}^{N} w_i \mathbf{x}_i \mathbf{x}_i'\right)^{-1} \sum_{i=1}^{N} w_i \mathbf{x}_i y_i$$

The OLS estimator is the special case of equal weights with $w_i = w_j$ for all i and j. The default estimator of the VCE is a weighted version of the heteroskedasticity-robust version in (3.3), which assumes independent observations. If observations are clustered, then the option vce(cluster *clustvar*) should be used.

Although the weighted estimator is easily obtained, for legitimate reasons many microeconometric analyses do not use weighted regression even where sampling weights are available. We provide a brief explanation of this conceptually difficult issue. For a more complete discussion, see Cameron and Trivedi (2005, 818–821).

Weighted regression should be used if a census parameter estimate is desired. For example, suppose we want to obtain an estimate for the U.S. population of the average change in earnings associated with one more year of schooling. Then, if disadvantaged minorities are oversampled, we most likely will understate the earnings increase, because disadvantaged minorities are likely to have earnings that are lower than average for their given level of schooling. A second example is when aggregate state-level data are used in a natural experiment setting, where the goal is to measure the effect of an exogenous policy change that affects some states and not other states. Intuitively, the impact on more populous states should be given more weight. Note that these estimates are being given a correlative rather than a causal interpretation.

Weighted regression is not needed if we make the stronger assumptions that the DGP is the specified model $y_i = \mathbf{x}_i'\boldsymbol{\beta} + u_i$ and sufficient controls are assumed to be added so that the error $E(u_i|\mathbf{x}_i) = 0$. This approach, called a control-function approach or a model approach, is the approach usually taken in microeconometric studies that emphasize a causal interpretation of regression. Under the assumption that $E(u_i|\mathbf{x}_i) = 0$, the weighted least-squares estimator will be consistent for $\boldsymbol{\beta}$ for any choice of weights including equal weights, and if u_i is homoskedastic, the most efficient estimator is the OLS estimator, which uses equal weights. For the assumption that $E(u_i|\mathbf{x}_i) = 0$ to be reasonable, the determinants of the sampling frame should be included in the controls $\mathbf{x}$ and should not be directly determined by the dependent variable y.

These points carry over directly to nonlinear regression models. In most cases, microeconometric analyses take on a model approach. In that case, unweighted estimation is appropriate, with any weighting based on efficiency grounds. If a census-parameter approach is being taken, however, then it is necessary to weight.

For our data example, we obtain

```
. * Perform weighted regression
. regress totexp suppins phylim actlim totchr age female income [pweight=swght]
(sum of wgt is   1.6195e+04)
```

Linear regression						
				Number of obs	=	3064
				F(7, 3056)	=	14.08
				Prob > F	=	0.0000
				R-squared	=	0.0977
				Root MSE	=	13824

totexp	Coef.	Robust Std. Err.	t	P>|t|	[95% Conf.	Interval]
suppins	278.1578	825.6959	0.34	0.736	-1340.818	1897.133
phylim	2484.52	933.7116	2.66	0.008	653.7541	4315.286
actlim	4271.154	1024.686	4.17	0.000	2262.011	6280.296
totchr	1819.929	349.2234	5.21	0.000	1135.193	2504.666
age	-59.3125	68.01237	-0.87	0.383	-192.6671	74.04212
female	-2654.432	911.6422	-2.91	0.004	-4441.926	-866.9381
income	5.042348	16.6509	0.30	0.762	-27.60575	37.69045
_cons	7336.758	5263.377	1.39	0.163	-2983.359	17656.87

The estimated coefficients of all statistically significant variables aside from `female` are within 10% of those from unweighted regression (not given for brevity). Big differences between weighted and unweighted regression would indicate that $E(u_i|\mathbf{x}_i) \neq 0$ because of model misspecification. Note that robust standard errors are reported by default.

3.7.4 Weighted prediction and MEs

After regression, unweighted prediction will provide an estimate of the sample-average value of the dependent variable. We may instead want to estimate the population-mean value of the dependent variable. Then sampling weights should be used in forming an average prediction.

This point is particularly easy to see for OLS regression. Because $1/N \sum_i (y_i - \hat{y}_i) = 0$, because in-sample residuals sum to zero if an intercept is included, the average prediction $1/N \sum_i \hat{y}_i$ equals the sample mean $\bar{y}$. But given an unrepresentative sample, the unweighted sample mean $\bar{y}$ may be a poor estimate of the population mean. Instead, we should use the weighted average prediction $1/N \sum_i w_i \hat{y}_i$, even if $\hat{y}_i$ is obtained by using unweighted regression.

For this to be useful, however, the prediction should be based on a model that includes as regressors variables that control for the unrepresentative sampling.

For our example, we obtain the weighted prediction by typing

```
. * Weighted prediction
. quietly predict yhatwols
. mean yhatwols [pweight=swght], noheader
```

	Mean	Std. Err.	[95% Conf.	Interval]
yhatwols	10670.83	138.0828	10400.08	10941.57

```
. mean yhatwols, noheader        // unweighted prediction
```

	Mean	Std. Err.	[95% Conf.	Interval]
yhatwols	7135.206	78.57376	6981.144	7289.269

The population mean for medical expenditures is predicted to be \$10,671 using weighted prediction, whereas the unweighted prediction gives a much lower value of \$7,135.

Weights similarly should be used in computing average MEs. For the linear model, the standard ME $\partial E(y_i|\mathbf{x}_i)/\partial x_{ij}$ equals β_j for all observations, so weighting will make no difference in computing the marginal effect. Weighting will make a difference for averages of other marginal effects, such as elasticities, and for MEs in nonlinear models.

3.8   OLS using Mata

Stata offers two different ways to perform computations using matrices: Stata `matrix` commands and Mata functions (which are discussed, respectively, in appendices A and B).

Mata, introduced in Stata 9, is much richer. We illustrate the use of Mata by using the same OLS regression as that in section 3.4.2.

The program is written for the dependent variable provided in the local macro `y` and the regressors in the local macro `xlist`. We begin by reading in the data and defining the local macros.

```
. * OLS with White robust standard errors using Mata
. use mus03data.dta, clear
. keep if totexp > 0    // Analysis for positive medical expenditures only
(109 observations deleted)
. generate cons = 1
. local y ltotexp
. local xlist suppins phylim actlim totchr age female income cons
```

We then move into Mata. The `st_view()` Mata function is used to transfer the Stata data variables to Mata matrices `y` and `X`, with `tokens("")` added to convert `` `xlist' `` to a comma-separated list with each entry in double quotes, necessary for `st_view()`.

The key part of the program forms $\widehat{\boldsymbol{\beta}} = (\mathbf{X}'\mathbf{X})^{-1}\mathbf{X}'\mathbf{y}$ and $\widehat{V}(\widehat{\boldsymbol{\beta}}) = (N/N - K)$ $(\mathbf{X}'\mathbf{X})^{-1}(\sum_i \widehat{u}_i^2 \mathbf{x}_i \mathbf{x}_i')(\mathbf{X}'\mathbf{X})^{-1}$. The cross-product function `cross(X,X)` is used to form $\mathbf{X}'\mathbf{X}$ because this handles missing values and is more efficient than the more obvious `X'X`. The matrix inverse is formed by using `cholinv()` because this is the fastest method in the special case that the matrix is symmetric positive definite. We calculate the $K \times K$ matrix $\sum_i \widehat{u}_i^2 \mathbf{x}_i \mathbf{x}_i'$ as $\sum_i (\widehat{u}_i \mathbf{x}_i')'(\widehat{u}_i \mathbf{x}_i') = \mathbf{A}'\mathbf{A}$, where the $N \times K$ matrix $\mathbf{A}$ has an ith row equal to $\widehat{u}_i \mathbf{x}_i'$. Now $\widehat{u}_i \mathbf{x}_i'$ equals the ith row of the $N \times 1$ residual vector $\widehat{\mathbf{u}}$ times the ith row of the $N \times K$ regressor matrix $\mathbf{X}$, so $\mathbf{A}$ can be computed by element-by-element multiplication of $\widehat{\mathbf{u}}$ by $\mathbf{X}$, or `(e:*X)`, where `e` is $\widehat{\mathbf{u}}$. Alternatively, $\sum_i \widehat{u}_i^2 \mathbf{x}_i \mathbf{x}_i' = \mathbf{X}'\mathbf{D}\mathbf{X}$, where $\mathbf{D}$ is an $N \times N$ diagonal matrix with entries $\widehat{u}_i^2$, but the matrix $\mathbf{D}$ becomes exceptionally large, unnecessarily so, for a large N.

The Mata program concludes by using `st_matrix()` to pass the estimated $\widehat{\boldsymbol{\beta}}$ and $\widehat{V}(\widehat{\boldsymbol{\beta}})$ back to Stata.

```
. mata
                                            ──────── mata (type end to exit) ────────
:   // Create y vector and X matrix from Stata dataset
:   st_view(y=., ., "`y'")                 // y is nx1
:   st_view(X=., ., tokens("`xlist'"))     // X is nxk
:   XXinv = cholinv(cross(X,X))            // XXinv is inverse of X´X
:   b = XXinv*cross(X,y)                   // b = [(X´X)^-1]*X´y
:   e = y - X*b
:   n = rows(X)
:   k = cols(X)
:   s2 = (e´e)/(n-k)
:   vdef = s2*XXinv                        // default VCE not used here
:   vwhite = XXinv*((e:*X)´(e:*X)*n/(n-k))*XXinv  // robust VCE
:   st_matrix("b",b´)                      // pass results from Mata to Stata
:   st_matrix("V",vwhite)                  // pass results from Mata to Stata
: end
```

Once back in Stata, we use `ereturn` to display the results in a format similar to that for built-in commands, first assigning names to the columns and rows of `b` and `V`.

```
. * Use Stata ereturn display to present nicely formatted results
. matrix colnames b = `xlist'
. matrix colnames V = `xlist'
. matrix rownames V = `xlist'
. ereturn post b V
```

. ereturn display

	Coef.	Std. Err.	z	P>|z|	[95% Conf.	Interval]
suppins	.2556428	.0465982	5.49	0.000	.1643119	.3469736
phylim	.3020598	.057705	5.23	0.000	.18896	.4151595
actlim	.3560054	.0634066	5.61	0.000	.2317308	.48028
totchr	.3758201	.0187185	20.08	0.000	.3391326	.4125077
age	.0038016	.0037028	1.03	0.305	-.0034558	.011059
female	-.0843275	.045654	-1.85	0.065	-.1738076	.0051526
income	.0025498	.0010468	2.44	0.015	.0004981	.0046015
cons	6.703737	.2825751	23.72	0.000	6.1499	7.257575

The results are exactly the same as those given in section 3.4.2, when we used `regress` with the `vce(robust)` option.

3.9 Stata resources

The key Stata references are [U] *User's Guide* and [R] **regress**, [R] **regress postestimation**, [R] **estimates**, [R] **predict**, and [R] **test**. A useful user-written command is `estout`. The material in this chapter appears in many econometrics texts, such as Greene (2008).

3.10 Exercises

1. Fit the model in section 3.4 using only the first 100 observations. Compute standard errors in three ways: default, heteroskedastic, and cluster–robust where clustering is on the number of chronic problems. Use `estimates` to produce a table with three sets of coefficients and standard errors, and comment on any appreciable differences in the standard errors. Construct a similar table for three alternative sets of heteroskedasticity-robust standard errors, obtained by using the `vce(robust)`, `vce(hc2)`, and `vce(hc3)` options, and comment on any differences between the different estimates of the standard errors.

2. Fit the model in section 3.4 with robust standard errors reported. Test at 5% the joint significance of the demographic variables `age`, `female`, and `income`. Test the hypothesis that being male (rather than female) has the same impact on medical expenditures as aging 10 years. Fit the model under the constraint that $\beta_{\text{phylim}} = \beta_{\text{actlim}}$ by first typing `constraint 1 phylim = actlim` and then by using `cnsreg` with the `constraints(1)` option.

3. Fit the model in section 3.5, and implement the RESET test manually by regressing y on $\mathbf{x}$ and $\hat{y}^2$, $\hat{y}^3$, and $\hat{y}^4$ and jointly testing that the coefficients of $\hat{y}^2$, $\hat{y}^3$, and $\hat{y}^4$ are zero. To get the same results as `estat ovtest`, do you need to use default or robust estimates of the VCE in this regression? Comment. Similarly, implement `linktest` by regressing y on $\hat{y}$ and $\hat{y}^2$ and testing that the coefficient of $\hat{y}^2$ is zero. To get the same results as `linktest`, do you need to use default or robust estimates of the VCE in this regression? Comment.

4. Fit the model in section 3.5, and perform the standard Lagrange multiplier test for heteroskedasticity by using `estat hettest` with $\mathbf{z} = \mathbf{x}$. Then implement the test manually as 0.5 times the explained sum of squares from the regression of y_i^* on an intercept and $\mathbf{z}_i$, where $y_i^* = \{\widehat{u}_i^2/(1/N)\sum_j \widehat{u}_j^2\} - 1$ and $\widehat{u}_i$ is the residual from the original OLS regression. Next use `estat hettest` with the `iid` option and show that this test is obtained as $N \times R^2$, where R^2 is obtained from the regression of $\widehat{u}_i^2$ on an intercept and $\mathbf{z}_i$.

5. Fit the model in section 3.6 on levels, except use all observations rather than those with just positive expenditures, and report robust standard errors. Predict medical expenditures. Use `correlate` to obtain the correlation coefficient between the actual and fitted value and show that, upon squaring, this equals R^2. Show that for the linear model `margins` with the `dydx(*)` option reproduces the OLS coefficients. Now use `margins` with an appropriate option to obtain the average income elasticity of medical expenditures.

6. Fit the model in section 3.6 on levels, using the first 2,000 observations. Use these estimates to predict medical expenditures for the remaining 1,064 observations, and compare these with the actual values. Note that the model predicts very poorly in part because the data were ordered by `totexp`.

4 Simulation

4.1 Introduction

Simulation by Monte Carlo experimentation is a useful and powerful methodology for investigating the properties of econometric estimators and tests. The power of the methodology derives from being able to define and control the statistical environment in which the investigator specifies the data-generating process (DGP) and generates data used in subsequent experiments.

Monte Carlo experiments can be used to verify that valid methods of statistical inference are being used. An obvious example is checking a new computer program or algorithm. Another example is investigating the robustness of an established estimation or test procedure to deviations from settings where the properties of the procedure are known.

Even when valid methods are used, they often rely on asymptotic results. We may want to check whether these provide a good approximation in samples of the size typically available to the investigators. Also asymptotically equivalent procedures may have different properties in finite samples. Monte Carlo experiments enable finite-sample comparisons.

This chapter deals with the basic elements common to Monte Carlo experiments: computer generation of random numbers that mimic the theoretical properties of realizations of random variables; commands for repeated execution of a set of instructions; and machinery for saving, storing, and processing the simulation output, generated in an experiment, to obtain the summary measures that are used to evaluate the properties of the procedures under study. We provide a series of examples to illustrate various aspects of Monte Carlo analyses.

The chapter appears early in the book. Simulation is a powerful pedagogic tool for exposition and illustration of statistical concepts. At the simplest level, we can use pseudorandom samples to illustrate distributional features of artificial data. The goal of this chapter is to use simulation to study the distributional and moment properties of statistics in certain idealized statistical environments. Another possible use of the Monte Carlo methodology is to check the correctness of computer code. Many applied studies use methods complex enough that it is easy to make mistakes. Often these mistakes could be detected by an appropriate simulation exercise. We believe that simulation is greatly underutilized, even though Monte Carlo experimentation is relatively straightforward in Stata.

4.2   Pseudorandom-number generators: Introduction

Suppose we want to use simulation to study the properties of the ordinary least-squares estimator (OLS) estimator in the linear regression model with normal errors. Then, at the minimum, we need to make draws from a specified normal distribution. The literature on (pseudo) random-number generation contains many methods of generating such sequences of numbers. When we use packaged functions, we usually do not need to know the details of the method. Yet the match between the theoretical and the sample properties of the draws does depend upon such details.

Stata introduced a new suite of fast and easy-to-use random-number functions (generators) in mid-2008. These functions begin with the letter r (from random) and can be readily installed via an update to version 10. The suite includes the uniform, normal, binomial, gamma, and Poisson functions that we will use in this chapter, as well as several others that we do not use. The functions for generating pseudorandom numbers are summarized in `help functions`.

To a large extent, these new functions obviate the previous methods of using one's own generators or user-written commands to generate pseudorandom numbers other than the uniform. Nonetheless, there can sometimes be a need to make draws from distributions that are not included in the suite. For these draws, the uniform distribution is often the starting point. The new `runiform()` function generates exactly the same uniform draws as `uniform()`, which it replaces.

4.2.1   Uniform random-number generation

The term random-number generation is an oxymoron. It is more accurate to use the term pseudorandom numbers. Pseudorandom-number generators use deterministic devices to produce long chains of numbers that mimic the realizations from some target distribution. For uniform random numbers, the target distribution is the uniform distribution from 0 to 1, for which any value between 0 and 1 is equally likely. Given such a sequence, methods exist for mapping these into sequences of nonuniform draws from desired distributions such as the normal.

A standard simple generator for uniform draws uses the deterministic rule $X_j = (kX_{j-1} + c) \bmod m$, $j = 1, \ldots, J$, where the modulus operator $a \bmod b$ forms the remainder when a is divided by b, to produce a sequence of J integers between 0 and $(m-1)$. Then $R_j = X_j/m$ is a sequence of J numbers between 0 and 1. If computation is done using 32-bit integer arithmetic, then $m = 2^{31} - 1$ and the maximum periodicity is $2^{31} - 1 \simeq 2.1 \times 10^9$, but it is easy to select poor values of k, c, and X_0 so that the cycle repeats much more often than that.

This generator is implemented using Stata function `runiform()`, a 32-bit KISS generator that uses good values of k and c. The initial value for the cycle, X_0, is called the seed. The default is to have this set by Stata, based on the computer clock. For reproducibility of results, however, it is best to actually set the initial seed by using `set seed`. Then, if the program is rerun at a later time or by a different researcher, the same results will be obtained.

To obtain and display one draw from the uniform, type

```
. * Single draw of a uniform number
. set seed 10101
. scalar u = runiform()
. display u
.16796649
```

This number is internally stored at much greater precision than the eight displayed digits.

The following code obtains 1,000 draws from the uniform distribution and then provides some details on these draws:

```
. * 1000 draws of uniform numbers
. quietly set obs 1000
. set seed 10101
. generate x = runiform()
. list x in 1/5, clean

           x
  1.    .1679665
  2.    .3197621
  3.    .7911349
  4.    .7193382
  5.    .5408687
. summarize x
```

Variable	Obs	Mean	Std. Dev.	Min	Max
x	1000	.5150332	.2934123	.0002845	.9993234

The 1,000 draws have a mean of 0.515 and a standard deviation of 0.293, close to the theoretical values of 0.5 and $\sqrt{1/12} = 0.289$. A histogram, not given, has ten equal-width bins with heights that range from 0.8 to 1.2, close to the theory of equal heights of 1.0.

The draws should be serially uncorrelated, despite a deterministic rule being used to generate the draws. To verify this, we create a time-identifier variable, t, equal to the observation number (_n), and we use tsset to declare the data to be time series with time-identifier t. We could then use the corrgram, ac, and pac commands to test whether autocorrelations and partial autocorrelations are zero. We more simply use pwcorr to produce the first three autocorrelations, where L2.x is the x variable lagged twice and the star(0.05) option puts a star on correlations that are statistically significantly different from zero at level 0.05.

```
. * First three autocorrelations for the uniform draws
. generate t = _n
. tsset t
        time variable:  t, 1 to 1000
                delta:  1 unit
```

```
. pwcorr x L.x L2.x L3.x, star(0.05)
```

	x	L.x	L2.x	L3.x
x	1.0000			
L.x	-0.0185	1.0000		
L2.x	-0.0047	-0.0199	1.0000	
L3.x	0.0116	-0.0059	-0.0207	1.0000

The autocorrelations are low, and none are statistically different from zero at the 0.05 level. Uniform random-number generators used by packages such as Stata are, of course, subjected to much more stringent tests than these.

4.2.2 Draws from normal

For simulations of standard estimators such as OLS, nonlinear least squares (NLS), and instrumental variables (IV), all that is needed are draws from the uniform and normal distributions, because normal errors are a natural starting point and the most common choices of distribution for generated regressors are normal and uniform.

The command for making draws from the standard normal has the following simple syntax:

```
generate varname = rnormal()
```

To make draws from $N(m, s^2)$, the corresponding command is

```
generate varname = rnormal(m,s)
```

Note that $s > 0$ is the standard deviation. The arguments m and s can be numbers or variables.

Draws from the standard normal distribution also can be obtained as a transformation of draws from the uniform by using the inverse probability transformation method explained in section 4.4.1; that is, by using

```
generate varname = invnormal(runiform())
```

where the new function `runiform()` replaces `uniform()` in the older versions.

The following code generates and summarizes three pseudorandom variables with 1,000 observations each. The pseudorandom variables have distributions uniform$(0, 1)$, standard normal, and normal with a mean of 5 and a standard deviation of 2.

```
. * normal and uniform
. clear
. quietly set obs 1000
. set seed 10101                          // set the seed
. generate uniform = runiform()           // uniform(0,1)
```

```
. generate stnormal = rnormal()              // N(0,1)
. generate norm5and2 = rnormal(5,2)
. tabstat uniform stnormal norm5and2, stat(mean sd skew kurt min max) col(stat)
    variable |      mean        sd  skewness  kurtosis       min       max
-------------+------------------------------------------------------------------
     uniform |  .5150332  .2934123 -.0899003  1.818878  .0002845  .9993234
    stnormal |  .0109413  1.010856  .0680232  3.130058 -2.978147  3.730844
   norm5and2 |  4.995458  1.970729 -.0282467  3.050581 -3.027987  10.80905
```

The sample mean and other sample statistics are random variables; therefore, their values will, in general, differ from the true population values. As the number of observations grows, each sample statistic will converge to the population parameter because each sample statistic is a consistent estimator for the population parameter.

For `norm5and2`, the sample mean and standard deviation are very close to the theoretical values of 5 and 2. Output from `tabstat` gives a skewness statistic of -0.028 and a kurtosis statistic of 3.051, close to 0 and 3, respectively.

For draws from the truncated normal, see section 4.4.4, and for draws from the multivariate normal, see section 4.4.5.

4.2.3 Draws from t, chi-squared, F, gamma, and beta

Stata's library of functions contains a number of generators that allow the user to draw directly from a number of common continuous distributions. The function formats are similar to that of the `rnormal()` function, and the argument(s) can be a number or a variable.

Let $t(n)$ denote Students' t distribution with n degrees of freedom, $\chi^2(m)$ denote the chi-squared distribution with m degrees of freedom, and $F(h, n)$ denote the F distribution with h and n degrees of freedom. Draws from $t(n)$ and $\chi^2(h)$ can be made directly by using the `rt`(df) and `rchi2`(df) functions. We then generate $F(h, n)$ draws by transformation because a function for drawing directly from the F distribution is not available.

The following example generates draws from $t(10)$, $\chi^2(10)$, and $F(10, 5)$.

```
. * t, chi-squared, and F with constant degrees of freedom
. clear
. quietly set obs 2000
. set seed 10101
. generate xt = rt(10)              // result xt ~ t(10)
. generate xc = rchi2(10)           // result xc ~ chisquared(10)
. generate xfn = rchi2(10)/10       // result numerator of F(10,5)
. generate xfd = rchi2(5)/5         // result denominator of F(10,5)
. generate xf = xfn/xfd             // result xf ~ F(10,5)
```

```
. summarize xt xc xf

    Variable |        Obs        Mean    Std. Dev.         Min         Max
-------------+--------------------------------------------------------------
          xt |       2000    .0295636    1.118426   -5.390713    4.290518
          xc |       2000    9.967206    4.530771    .7512587    35.23849
          xf |       2000    1.637549    2.134448    .0511289    34.40774
```

The $t(10)$ draws have a sample mean and a standard deviation close to the theoretical values of 0 and $\sqrt{10/(10-2)} = 1.118$; the $\chi^2(10)$ draws have a sample mean and a standard deviation close to the theoretical values of 10 and $\sqrt{20} = 4.472$; and the $F(10,5)$ draws have a sample mean close to the theoretical value of $5/(5-2) = 1.7$. The sample standard deviation of 2.134 differs from the theoretical standard deviation of $\sqrt{2 \times 5^2 \times 13/(10 \times 3^2 \times 1)} = 2.687$. This is because of randomness, and a much larger number of draws eliminates this divergence.

Using `rbeta(`a,b`)`, we can draw from a two-parameter beta with the shape parameters $a, b > 0$, mean $a/(a+b)$, and variance $ab/(a+b)^2(a+b+1)$. Using `rgamma(`a,b`)`, we can draw from a two-parameter gamma with the shape parameter $a > 0$, scale parameter $b > 0$, mean ab, and variance ab^2.

4.2.4   Draws from binomial, Poisson, and negative binomial

Stata functions also generate draws from some leading discrete distributions. Again the argument(s) can be a number or a variable.

Let $\text{Bin}(n, p)$ denote the binomial distribution with positive integer n trials (**n**) and success probability p, $0 < p < 1$, and let $\text{Poisson}(m)$ denote the Poisson distribution with the mean or rate parameter m. The `rbinomial(`n,p`)` function generates random draws from the binomial distribution, and the `rpoisson(`m`)` function makes draws from the Poisson distribution.

We demonstrate these functions with an argument that is a variable so that the parameters differ across draws.

Independent (but not identically distributed) draws from binomial

As illustration, we consider draws from the binomial distribution, when both the probability p and the number of trials n may vary over i.

```
. * Discrete rv´s: binomial
. set seed 10101
. generate p1 = runiform()              // here p1~uniform(0,1)
. generate trials = ceil(10*runiform()) // here # trials varies btwn 1 & 10
. generate xbin = rbinomial(trials,p1)  // draws from binomial(n,p1)
```

```
. summarize p1 trials xbin
    Variable |        Obs        Mean    Std. Dev.        Min        Max
-------------+--------------------------------------------------------
          p1 |       2000    .5155468    .2874989    .0002845    .9995974
       trials |       2000       5.438    2.887616           1         10
        xbin |       2000       2.753    2.434328           0         10
```

The DGP setup implies that the number of trials n is a random variable with an expected value of 5.5 and that the probability p is a random variable with an expected value of 0.5. Thus we expect that `xbin` has a mean of $5.5 \times 0.5 = 2.75$, and this is approximately the case here.

Independent (but not identically distributed) draws from Poisson

For simulating a Poisson regression DGP, denoted $y \sim \text{Poisson}(\mu)$, we need to make draws that are independent but not identically distributed, with the mean μ varying across draws because of regressors.

We do so in two ways. First, let μ_i equal `xb=4+2*x` with `x=runiform()`. Then $4 < \mu_i < 6$. Second, let μ_i equal `xb` times `xg` where `xg=rgamma(1,1)`, which yields a draw from the gamma distribution with a mean of $1 \times 1 = 1$ and a variance of $1 \times 1^2 = 1$. Then $\mu_i > 0$. In both cases, the setup can be shown to be such that the ultimate draw has a mean of 5, but the variance differs from 5 for the independent and identically distributed (i.i.d.) Poisson because in neither case are the draws from an identical distribution. We obtain

```
. * Discrete rv's: independent poisson and negbin draws
. set seed 10101
. generate xb= 4 + 2*runiform()
. generate xg = rgamma(1,1)              // draw from gamma;E(v)=1
. generate xbh = xb*xg                   // apply multiplicative heterogeneity
. generate xp = rpoisson(5)              // result xp ~ Poisson(5)
. generate xp1 = rpoisson(xb)            // result xp1 ~ Poisson(xb)
. generate xp2 = rpoisson(xbh)           // result xp2 ~ NB(xb)
. summarize xg xb xp xp1 xp2
    Variable |        Obs        Mean    Std. Dev.        Min        Max
-------------+--------------------------------------------------------
          xg |       2000    1.032808    1.044434     .000112    8.00521
          xb |       2000    5.031094    .5749978    4.000569   5.999195
          xp |       2000       5.024    2.300232           0         14
         xp1 |       2000       4.976    2.239851           0         14
         xp2 |       2000      5.1375    5.676945           0         44
```

The `xb` variable lies between 4 and 6, as expected, and the `xg` gamma variable has a mean and variance close to 1, as expected. For a benchmark comparison, we make draws of `xp` from Poisson(5), which has a sample mean close to 5 and a sample standard deviation close to $\sqrt{5} = 2.236$. Both `xp1` and `xp2` have means close to 5. In the case of `xp2`, the model has the multiplicative unobserved heterogeneity term `xg` that is itself drawn from a gamma distribution with shape and scale parameter both set to 1. Introducing this type of heterogeneity means that `xp2` is drawn from a distribution with the same mean as that of `xp1`, but the variance of the distribution is larger. More specifically, Var(`xp2`|`xb`) = `xb*(1+xb)`, using results in section 17.2.2, leading to the much larger standard deviation for `xp2`.

The second example makes a draw from the Poisson–gamma mixture, yielding the negative binomial distribution. The `rnbinomial()` function draws from a different parameterization of the negative binomial distribution. For this reason, we draw from the Poisson–gamma mixture here and in chapter 17.

Histograms and density plots

For a visual depiction, it is often useful to plot a histogram or kernel density estimate of the generated random numbers. Here we do this for the draws `xc` from $\chi^2(10)$ and `xp` from Poisson(5). The results are shown in figure 4.1.

```
. * Example of histogram and kernel density plus graph combine
. quietly twoway (histogram xc, width(1)) (kdensity xc, lwidth(thick)),
> title("Draws from chisquared(10)")

. quietly graph save mus04cdistr.gph, replace

. quietly twoway (histogram xp, discrete) (kdensity xp, lwidth(thick) w(1)),
> title("Draws from Poisson(mu) for 5<mu<6")

. quietly graph save mus04poissdistr.gph, replace

. graph combine mus04cdistr.gph mus04poissdistr.gph,
> title("Random-number generation examples", margin(b=2) size(vlarge))
```

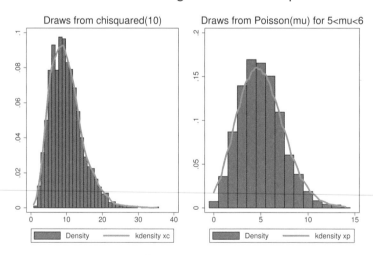

Figure 4.1. $\chi^2(10)$ and Poisson(5) draws

4.3 Distribution of the sample mean

As an introductory example of simulation, we demonstrate the central limit theorem result, $(\bar{x}_N - \mu)/(\sigma/\sqrt{N}) \to N(0,1)$; i.e., the sample mean is approximately normally distributed as $N \to \infty$. We consider a random variable that has the uniform distribution, and a sample size of 30.

We begin by drawing a single sample of size 30 of the random variable X that is uniformly distributed on $(0,1)$, using the `runiform()` random-number function. To ensure the same results are obtained in future runs of the same code or on other machines, we use `set seed`. We have

```
. * Draw 1 sample of size 30 from uniform distribution
. quietly set obs 30
. set seed 10101
. generate x = runiform()
```

To see the results, we use `summarize` and `histogram`. We have

```
. * Summarize x and produce a histogram
. summarize x
```

Variable	Obs	Mean	Std. Dev.	Min	Max
x	30	.5459987	.2803788	.0524637	.9983786

```
. quietly histogram x, width(0.1) xtitle("x from one sample")
```

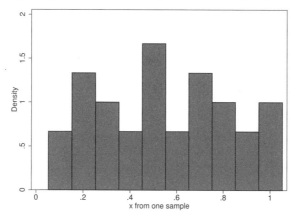

Figure 4.2. Histogram for one sample of size 30

The summary statistics show that 30 observations were generated and that for this sample $\bar{x} = 0.546$. The histogram for this single sample of 30 uniform draws, given in figure 4.2, looks nothing like the bell-shaped curve of a normal, because we are sampling from the uniform distribution. For very large samples, this histogram approaches a horizontal line with a density value of 1.

To obtain the distribution of the sample mean by simulation, we redo the preceding 10,000 times, obtaining 10,000 samples of size 30 and 10,000 sample means $\bar{x}$. These 10,000 sample means are draws from the distribution of the sample-mean estimator. By the central limit theorem, the distribution of the sample-mean estimator has approximately a normal distribution. Because the mean of a uniform$(0, 1)$ distribution is 0.5, the mean of the distribution of the sample-mean estimator is 0.5. Because the standard deviation of a uniform$(0, 1)$ distribution is $\sqrt{1/12}$ and each of the 10,000 samples is of size 30, the standard deviation of the distribution of the sample-mean estimator is $\sqrt{(1/12)/30} = 0.0527$.

4.3.1   Stata program

A mechanism for repeating the same statistical procedure 10,000 times is to write a program (see appendix A.2 for more details) that does the procedure once and use the `simulate` command to run the program 10,000 times.

We name the program `onesample` and define it to be r-class, meaning that the ultimate result, the sample mean for one sample, is returned in `r()`. Because we name this result `meanforonesample`, it will be returned in `r(meanforonesample)`. The program has no inputs, so there is no need for program arguments. The program drops any existing data on variable `x`, sets the sample size to 30, draws 30 uniform variates, and obtains the sample mean with `summarize`. The `summarize` command is itself an r-class command that stores the sample mean in `r(mean)`; see section 1.6.1. The last line of the program returns `r(mean)` as the result `meanforonesample`.

The program is

```
. * Program to draw 1 sample of size 30 from uniform and return sample mean
. program onesample, rclass
  1.      drop _all
  2.      quietly set obs 30
  3.      generate x = runiform()
  4.      summarize x
  5.      return scalar meanforonesample = r(mean)
  6. end
```

To check the program, we run it once, using the same seed as earlier. We obtain

```
. * Run program onesample once as a check
. set seed 10101
. onesample
    Variable |        Obs        Mean    Std. Dev.        Min        Max
-------------+--------------------------------------------------------
           x |         30    .5459987    .2803788    .0524637    .9983786
. return list

scalars:
    r(meanforonesample) =  .5459987225631873
```

The results for one sample are exactly the same as those given earlier.

4.3.2 The simulate command

The `simulate` command runs a specified command # times, where the user specifies #. The basic syntax is

simulate [*exp_list*], reps(#) [*options*]: *command*

where *command* is the name of the command, often a user-written program, and # is the number of simulations or replications. The quantities to be calculated and stored from *command* are given in *exp_list*. We provide additional details on `simulate` in section 4.6.1.

After `simulate` is run, the Stata dataset currently in memory is replaced by a dataset that has # observations, with a separate variable for each of the quantities given in *exp_list*.

4.3.3 Central limit theorem simulation

The `simulate` command can be used to run the `onesample` program 10,000 times, yielding 10,000 sample means from samples of size 30 of uniform variates. We additionally used options that set the seed and suppress the output of a dot for each of the 10,000 simulations. We have

```
. * Run program onesample 10,000 times to get 10,000 sample means
. simulate xbar = r(meanforonesample), seed(10101) reps(10000) nodots:
> onesample
      command:  onesample
         xbar:  r(meanforonesample)
```

The result from each sample, `r(meanforonesample)`, is stored as the variable `xbar`.

The `simulate` command overwrites any existing data with a dataset of 10,000 "observations" on $\overline{x}$. We summarize these values, expecting them to have a mean of 0.5 and a standard deviation of 0.0527. We also present a histogram overlaid by a normal density curve with a mean and standard deviation, which are those of the 10,000 values of $\overline{x}$. We have

```
. * Summarize the 10,000 sample means and draw histogram
. summarize xbar
```

Variable	Obs	Mean	Std. Dev.	Min	Max
xbar	10000	.4995835	.0533809	.3008736	.6990562

```
. quietly histogram xbar, normal xtitle("xbar from many samples")
```

(*Continued on next page*)

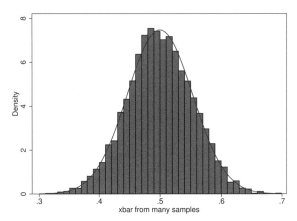

Figure 4.3. Histogram of the 10,000 sample means, each from a sample of size 30

The histogram given in figure 4.3 is very close to the bell-shaped curve of the normal.

There are several possible variations on this example. Different distributions for x can be used with different random-number functions in the `generate` command for x. As sample size (`set obs`) and number of simulations (`reps`) increases, the results become closer to a normal distribution.

4.3.4   The postfile command

In this book, we generally use `simulate` to perform simulations. An alternative method is to use a looping command, such as `forvalues`, and within each iteration of the loop use `post` to write (or post) key results to a file that is declared in the `postfile` command. After the loop ends, we then analyze the data in the posted file.

The `postfile` command has the following basic syntax:

`postfile` *postname newvarlist* `using` *filename* [`, every(`#`)` `replace`]

where *postname* is an internal filename, *newvarlist* contains the names of the variables to be put in the dataset, and *filename* is the external filename.

The `post` *postname* (*exp1*)(*exp2*)... command is used to write *exp1*, *exp2*, ... to the file. Each expression needs to be enclosed in parentheses.

The `postclose` *postname* command ends the posting of observations.

The `postfile` command offers more flexibility than `simulate` and, unlike `simulate`, does not lead to the dataset in memory being overwritten. For the examples in this book, `simulate` is adequate.

4.3.5   Alternative central limit theorem simulation

We illustrate the use of `postfile` for the central limit theorem example. We have

```
. * Simulation using postfile
. set seed 10101
. postfile sim_mem xmean using simresults, replace
. forvalues i = 1/10000 {
  2.       drop _all
  3.       quietly set obs 30
  4.       tempvar x
  5.       generate `x´ = runiform()
  6.       quietly  summarize `x´
  7.       post sim_mem (r(mean))
  8. }
. postclose sim_mem
```

The `postfile` command declares the memory object in which the results are stored, the names of variables in the results dataset, and the name of the results dataset file. In this example, the memory object is named `sim_mem`, `xmean` will be the only variable in the results dataset file, and `simresults.dta` will be the results dataset file. (The `replace` option causes any existing `simresults.dta` to be replaced.) The `forvalues` loop (see section 1.8) is used to perform 10,000 repetitions. At each repetition, the sample mean, result `r(mean)`, is posted and will be included as an observation in the new `xmean` variable in `simresults.dta`.

To see the results, we need to open `simresults.dta` and `summarize`.

```
. * See the results stored in simresults
. use simresults, clear
. summarize
    Variable |        Obs        Mean    Std. Dev.        Min        Max
-------------+--------------------------------------------------------
       xmean |      10000    .4995835    .0533809    .3008736    .6990562
```

The results are identical to those in section 4.3.3 with `simulate` due to using the same seed and same sequence of evaluation of random-number functions.

The `simulate` command suppresses all output within the simulations. This is not the case for the `forvalues` loop, so the `quietly` prefix was used in two places in the code above. It can be more convenient to instead apply the `quietly` prefix to all commands in the entire `forvalues` loop.

4.4   Pseudorandom-number generators: Further details

In this section, we present further details on random-number generation that explain the methods used in section 4.2 and are useful for making draws from additional distributions.

Commonly used methods for generating pseudorandom samples include inverse-probability transforms, direct transformations, accept–reject methods, mixing and compounding, and Markov chains. In what follows, we emphasize application and refer the interested reader to Cameron and Trivedi (2005, chap. 12) or numerous other texts for additional details.

4.4.1   Inverse-probability transformation

Let $F(x) = \Pr(X \leq x)$ denote the cumulative distribution function of a random variable x. Given a draw of a uniform variate r, $0 \leq r \leq 1$, the inverse transformation $x = F^{-1}(r)$ gives a unique value of x because $F(x)$ is nondecreasing in x. If r approximates well a random draw from the uniform, then $x = F^{-1}(r)$ will approximate well a random draw from $F(x)$.

A leading application is to the standard normal distribution. Then the inverse of the cumulative distribution function (c.d.f.),

$$F(x) = \Phi(x) = \int_{-\infty}^{x} \frac{1}{\sqrt{2\pi}} e^{-z^2/2} dz$$

has no closed-form solution, and there is consequently no analytical expression for $\Phi^{-1}(x)$. Nonetheless, the inverse-transformation method is easy to implement because numerical analysis provides functions that calculate a very good approximation to $\Phi^{-1}(x)$. In Stata, the function is `invnormal()`. Combining the two steps of drawing a random uniform variate and evaluating the inverse c.d.f., we have

```
. * Inverse probability transformation example: standard normal
. quietly set obs 2000

. set seed 10101

. generate xstn = invnormal(runiform())
```

This method was presented in section 4.2.2 but is now superseded by the `rnormal()` function.

As another application, consider drawing from the unit exponential, with c.d.f. $F(x) = 1 - e^{-x}$. Solving $r = 1 - e^{-x}$ yields $x = -\ln(1 - r)$. If the uniform draw is, say, 0.640, then $x = -\ln(1 - 0.640) = 1.022$. With continuous monotonically increasing c.d.f., the inverse transformation yields a unique value of x, given r. The Stata code for generating a draw from the unit exponential illustrates the method:

```
. * Inverse probability transformation example: unit exponential
. generate xue = -ln(1-runiform())
```

For discrete random variables, the c.d.f. is a step function. Then the inverse is not unique, but it can be uniquely determined by a convention for choosing a value on the flat portion of the c.d.f., e.g., the left limit of the segment.

In the simplest case, we consider a Bernoulli random variable taking a value of 1 with a probability of p and a value of 0 with a probability of $1 - p$. Then we take a uniform draw, u, and set $y = 1$ if $u \leq p$ and $y = 0$ if $u > p$. Thus, if $p = 0.6$, we obtain the following:

```
. * Inverse probability transformation example: Bernoulli (p = 0.6)
. generate xbernoulli = runiform() > 0.6   // Bernoulli(0.6)

. summarize xstn xue xbernoulli
    Variable |       Obs        Mean    Std. Dev.       Min        Max
-------------+--------------------------------------------------------
        xstn |      2000     .0481581    1.001728   -3.445941   3.350993
         xue |      2000     .9829519    1.000921    .0003338   9.096659
  xbernoulli |      2000        .4055    .4911113           0          1
```

This code uses a logical operator that sets $y = 1$ if the condition is met and $y = 0$ otherwise; see section 2.4.7.

A more complicated discrete example is the Poisson distribution because then the random variable can potentially take an infinite number of values. The method is to sequentially calculate the c.d.f. $\Pr(Y \leq k)$ for $k = 0, 1, 2, \ldots$. Then stop when the first $\Pr(Y \leq k) > u$, where u is the uniform draw, and set $y = k$. For example, consider the Poisson with a mean of 2 and a uniform draw of 0.701. We first calculate $\Pr(y \leq 0) = 0.135 < u$, then calculate $\Pr(y \leq 1) = 0.406 < u$, then calculate $\Pr(y \leq 2) = 0.677 < u$, and finally calculate $\Pr(y \leq 3) = 0.857$. This last calculation exceeds the uniform draw of 0.701, so stop and set $y = 3$. $\Pr(Y \leq k)$ is computed by using the recursion $\Pr(Y \leq k) = \Pr(Y \leq k - 1) + \Pr(Y = k)$.

4.4.2   Direct transformation

Suppose we want to make draws from the random variable Y, and from probability theory, it is known that Y is a transformation of the random variable X, say, $Y = g(X)$.

In this situation, the direct transformation method obtains draws of Y by drawing X and then applying the transformation $g(\cdot)$. The method is clearly attractive when it is easy to draw X and evaluate $g(\cdot)$.

Direct transformation is particularly easy to illustrate for well-known transforms of a standard normally distributed random variable. A $\chi^2(1)$ draw can be obtained as the square of a draw from the standard normal; a $\chi^2(m)$ draw is the sum of m independent draws from $\chi^2(1)$; an $F(m_1, m_2)$ draw is $(v_1/m_1)/(v_2/m_2)$, where v_1 and v_2 are independent draws from $\chi^2(m_1)$ and $\chi^2(m_2)$; and a $t(m)$ draw is $u/\sqrt{v/m}$ where u and v are independent draws from $N(0, 1)$ and $\chi^2(m)$.

4.4.3   Other methods

In some cases, a distribution can be obtained as a mixture of distributions. A leading example is the negative binomial, which can be obtained as a Poisson–gamma mixture (see section 4.2.4). Specifically, if $y|\lambda$ is Poisson(λ) and $\lambda|\mu, \alpha$ is gamma with a mean of μ and a variance of $\alpha\mu$, then $y|\mu, \alpha$ is a negative binomial distributed with a mean of μ and a variance of $\mu + \alpha\mu^2$. This implies that we can draw from the negative binomial by using a two-step method in which we first draw (say, ν) from the gamma distribution with a mean equal to 1 and then, conditional on ν, draw from Poisson$(\mu\nu)$. This example, using mixing, is used again in chapter 17.

More-advanced methods include accept–reject algorithms and importance sampling. Many of Stata's pseudorandom-number generators use accept–reject algorithms. Type `help random number functions` for more information on the methods used by Stata.

4.4.4 Draws from truncated normal

In simulation-based estimation for latent normal models with censoring or selection, it is often necessary to generate draws from a truncated normal distribution. The inverse-probability transformation can be extended to obtain draws in this case.

Consider making draws from a truncated normal. Then $X \sim TN_{(a,b)}(\mu, \sigma^2)$, where without truncation $X \sim N(\mu, \sigma^2)$. With truncation, realizations of X are restricted to lie between left truncation point a and right truncation point b.

For simplicity, first consider the standard normal case ($\mu = 0$, $\sigma = 1$) and let $Z \sim N(0, 1)$. Given the draw u from the uniform distribution, x is defined by the solution of the inverse-probability transformation equation

$$u = F(x) = \frac{\Pr(a \le Z \le x)}{\Pr(a \le Z \le b)} = \frac{\Phi(x) - \Phi(a)}{\Phi(b) - \Phi(a)}$$

Rearranging, $\Phi(x) = \Phi(a) + \{\Phi(b) - \Phi(a)\}u$ so that solving for x we obtain

$$x = \Phi^{-1}[\Phi(a) + \{\Phi(b) - \Phi(a)\}\, u]$$

To extend this to the general case, note that if $Z \sim N(\mu, \sigma^2)$ then $Z^* = (Z - \mu)/\sigma \sim N(0, 1)$, and the truncation points for Z^*, rather than Z, are $a^* = (a - \mu)/\sigma$ and $b^* = (b - \mu)/\sigma$. Then

$$x = \mu + \sigma\Phi^{-1}[\Phi(a^*) + \{\Phi(b^*) - \Phi(a^*)\}\, u]$$

As an example, we consider draws from $N(5, 4^2)$ for a random variable truncated to the range $[0, 12]$.

```
. * Draws from truncated normal x ~ N(mu,sigma^2) in [a,b]
. quietly set obs 2000
. set seed 10101
. scalar a = 0                          // lower truncation point
. scalar b = 12                         // upper truncation point
. scalar mu = 5                         // mean
. scalar sigma = 4                      // standard deviation
. generate u = runiform()
```

```
. generate w=normal((a-mu)/sigma)+u*(normal((b-mu)/sigma)-normal((a-mu)/sigma))
. generate xtrunc = mu + sigma*invnormal(w)
. summarize xtrunc
```

Variable	Obs	Mean	Std. Dev.	Min	Max
xtrunc	2000	5.605522	2.944887	.005319	11.98411

Here there is more truncation from below, because a is 1.25σ from μ whereas b is 1.75σ from μ, so we expect the truncated mean to exceed the untruncated mean. Accordingly, the sample mean is 5.606 compared with the untruncated mean of 5. Truncation reduces the range and, for most but not all distributions, will reduce the variability. The sample standard deviation of 2.945 is less than the untruncated standard deviation of 4.

An alternative way to draw $X \sim TN_{(a,b)}(\mu, \sigma^2)$ is to keep drawing from untruncated $N(\mu, \sigma^2)$ until the realization lies in (a, b). This method will be very inefficient if, for example, $(a, b) = (-0.01, 0.01)$. A Poisson example is given in section 17.3.5.

4.4.5 Draws from multivariate normal

Making draws from multivariate distributions is generally more complicated. The method depends on the specific case under consideration, and inverse-transformation methods and transformation methods that work in the univariate case may no longer apply.

However, making draws from the multivariate normal is relatively straightforward because, unlike most other distributions, linear combinations of normals are also normal.

Direct draws from multivariate normal

The `drawnorm` command generates draws from $N(\boldsymbol{\mu}, \boldsymbol{\Sigma})$ for the user-specified vector $\boldsymbol{\mu}$ and matrix $\boldsymbol{\Sigma}$. For example, consider making 200 draws from a standard bivariate normal distribution with means of 10 and 20, variances of 4 and 9, and a correlation of 0.5 (so the covariance is 3).

```
. * Bivariate normal example:
. * means 10, 20; variances 4, 9; and correlation 0.5
. clear
. quietly set obs 1000
. set seed 10101
. matrix MU = (10,20)                  // MU is 2 x 1
. scalar sig12 = 0.5*sqrt(4*9)
. matrix SIGMA = (4, sig12 \ sig12, 9) // SIGMA is 2 x 2
. drawnorm y1 y2, means(MU) cov(SIGMA)
```

```
. summarize y1 y2
    Variable |        Obs        Mean    Std. Dev.        Min        Max
-------------+--------------------------------------------------------
          y1 |       1000     10.0093    1.985447    3.176768    16.38992
          y2 |       1000    20.07195    2.941012    9.202241    29.94806
. correlate y1 y2
(obs=1000)
             |      y1        y2
-------------+------------------
          y1 |  1.0000
          y2 |  0.4724    1.0000
```

The sample means are close to 10 and 20, and the standard deviations are close to $\sqrt{4} = 2$ and $\sqrt{9} = 3$. The sample correlation of 0.4724 differs somewhat from 0.50, though this difference disappears for much larger sample sizes.

Transformation using Cholesky decomposition

The method uses the result that if $\mathbf{z} \sim N(\mathbf{0}, \mathbf{I})$ then $\mathbf{x} = \boldsymbol{\mu} + \mathbf{Lz} \sim N(\boldsymbol{\mu}, \mathbf{LL}')$. It is easy to draw $\mathbf{z} \sim N(\mathbf{0}, \mathbf{I})$ because $\mathbf{z}$ is just a column vector of univariate normal draws. The transformation method to make draws of $\mathbf{x} \sim N(\boldsymbol{\mu}, \boldsymbol{\Sigma})$ evaluates $\mathbf{x} = \boldsymbol{\mu} + \mathbf{Lz}$, where the matrix $\mathbf{L}$ satisfies $\mathbf{LL}' = \boldsymbol{\Sigma}$. More than one matrix $\mathbf{L}$ satisfies $\mathbf{LL}' = \boldsymbol{\Sigma}$, the matrix analog of the square root of $\boldsymbol{\Sigma}$. Standard practice is to use the Cholesky decomposition that restricts $\mathbf{L}$ to be a lower triangular matrix. Specifically, for the trivariate normal distribution, let $E(\mathbf{zz}') = \boldsymbol{\Sigma} = \mathbf{Lzz'L}'$, where $\mathbf{z} \sim N(\mathbf{0}, \mathbf{I}_3)$ and

$$\mathbf{L} = \begin{bmatrix} l_{11} & 0 & 0 \\ l_{21} & l_{22} & 0 \\ l_{31} & l_{32} & l_{33} \end{bmatrix}$$

Then the following transformations of $\mathbf{z}' = (z_1 \ z_2 \ z_3)$ yield the desired multivariate normal vector $\mathbf{x} \sim N(\boldsymbol{\mu}, \boldsymbol{\Sigma})$:

$$x_1 = \mu_1 + l_{11}z_1$$
$$x_2 = \mu_2 + l_{21}z_1 + l_{22}z_2$$
$$x_3 = \mu_3 + l_{31}z_1 + l_{32}z_2 + l_{33}z_3$$

4.4.6   Draws using Markov chain Monte Carlo method

In some cases, making direct draws from a target joint (multivariate) distribution is difficult, so the objective must be achieved in a different way. However, if it is also possible to make draws from the distribution of a subset, conditional on the rest, then one can create a Markov chain of draws. If one recursively makes draws from the conditional distribution and if a sufficiently long chain is constructed, then the distribution of the draws will, under some conditions, converge to the distribution of independent draws from the stationary joint distribution. This so-called Markov chain Monte Carlo method is now standard in modern Bayesian inference.

To be concrete, let $\mathbf{Y} = (Y_1, Y_2)$ have a bivariate density of $f(\mathbf{Y}) = f(Y_1, Y_2)$, and suppose the two conditional densities $f(Y_1|Y_2)$ and $f(Y_2|Y_1)$ are known and that it is possible to make draws from these. Then it can be shown that alternating sequential draws from $f(Y_1|Y_2)$ and $f(Y_2|Y_1)$ converge in the limit to draws from $f(Y_1, Y_2)$, even though in general $f(Y_1, Y_2) \neq f(Y_1|Y_2)f(Y_2|Y_1)$ (recall that $f(Y_1, Y_2) = f(Y_1|Y_2)f(Y_2)$). The repeated recursive sampling from $f(Y_1|Y_2)$ and $f(Y_2|Y_1)$ is called the Gibbs sampler.

We illustrate the Markov chain Monte Carlo approach by making draws from a bivariate normal distribution, $f(Y_1, Y_2)$. Of course, using the **drawnorm** command, it is quite straightforward to draw samples from the bivariate normal. So the application presented is illustrative rather than practical. The relative simplicity of this method comes from the fact that the conditional distributions $f(Y_1|Y_2)$ and $f(Y_2|Y_1)$ derived from a bivariate normal are also normal.

We draw bivariate normal data with means of 0, variances of 1, and a correlation of $\rho = 0.9$. Then $Y_1|Y_2 \sim N\left\{0, (1 - \rho^2)\right\}$ and $Y_2|Y_1 \sim N\left\{0, (1 - \rho^2)\right\}$. Implementation requires looping that is much easier using matrix programming language commands. The following Mata code implements this algorithm by using commands explained in appendix B.2.

```
. * MCMC example: Gibbs for bivariate normal mu´s=0 v´s=1 corr=rho=0.9
. set seed 10101
. clear all
. set obs 1000
obs was 0, now 1000
. generate double y1 =.
(1000 missing values generated)
. generate double y2 =.
(1000 missing values generated)
. mata:
————————————————————————————————————————— mata (type end to exit) ———————
:   s0 = 10000              // Burn-in for the Gibbs sampler (to be discarded)
:   s1 = 1000               // Actual draws used from the Gibbs sampler
:   y1 = J(s0+s1,1,0)       // Initialize y1
:   y2 = J(s0+s1,1,0)       // Initialize y2
:   rho = 0.90              // Correlation parameter
:   for(i=2; i<=s0+s1; i++) {
>       y1[i,1] = ((1-rho^2)^0.5)*(rnormal(1, 1, 0, 1)) + rho*y2[i-1,1]
>       y2[i,1] = ((1-rho^2)^0.5)*(rnormal(1, 1, 0, 1)) + rho*y1[i,1]
>   }
:   y = y1,y2
:   y = y[|(s0+1),1 \ (s0+s1),.|]   // Drop the burn-ins
:   mean(y)                         // Means of y1, y2
                 1                2
    1 |   .0831308345    .0647158328
```

```
:    variance(y)                        // Variance matrix of y1, y2
[symmetric]
                     1              2

     1 |  1.104291499
     2 |  1.005053494    1.108773741


:    correlation(y)                     // Correlation matrix of y1, y2
[symmetric]
                     1              2

     1 |          1
     2 |  .9082927488                1


: end
```

Many draws may be needed before the chain converges. Here we assume that 11,000 draws are sufficient, and we discard the first 10,000 draws; the remaining 1,000 draws are kept. In a real application, one should run careful checks to ensure that the chain has indeed converged to the desired bivariate normal. For the example here, the sample means of Y_1 and Y_2 are 0.08 and 0.06, differing quite a bit from 0. Similarly, the sample variances of 1.10 and 1.11 differ from 1 and the sample covariance of 1.01 differs from 0.9, while the implied correlation is 0.91 as desired. A longer Markov chain or longer burn-in may be needed to generate numbers with desired properties for this example with relatively high ρ.

Even given convergence of the Markov chain, the sequential draws of any random variable will be correlated. The output below shows that for the example here, the first-order correlation of sequential draws of y_2 is 0.823.

```
. mata:
                                        ——— mata (type end to exit) ———
:    y2 = y[|2,2 \ s1,2|]
:    y2lag1 = y[|1,2 \ (s1-1),2|]
:    y2andlag1 = y2,y2lag1
:    correlation(y2andlag1,1)          // Correlation between y2 and y2 lag 1
[symmetric]
                     1              2

     1 |          1
     2 |  .822692407                1


: end
```

4.5   Computing integrals

Some estimation problems may involve definite or indefinite integrals. In such cases, the integral may be numerically calculated.

4.5.1   Quadrature

For one-dimensional integrals of the form $\int_a^b f(y)dy$, where possibly $a = -\infty$, $b = \infty$, or both, Gaussian quadrature is the standard method. This approximates the integral by a weighted sum of m terms, where a larger m gives a better approximation and often even $m = 20$ can give a good approximation. The formulas for the weights are quite complicated but are given in standard numerical analysis books.

One-dimensional integrals often appear in regression models with a random intercept or random effect. In many nonlinear models, this random effect does not integrate out analytically. Most often, the random effect is normal so that integration is over $(-\infty, \infty)$ and Gauss–Hermite quadrature is used. A leading example is the random-effects estimator for nonlinear panel models fit using various xt commands. For Stata code, see, for example, the user-written command rfprobit.do for a random-effects probit package or file gllamm.ado for generalized linear additive models.

4.5.2   Monte Carlo integration

Suppose the integral is of the form

$$E\left\{h(Y)\right\} = \int_a^b h(y)g(y)dy$$

where $g(y)$ is a density function. This can be estimated by the direct Monte Carlo integral estimate

$$\widehat{E}\left\{h(Y)\right\} = S^{-1} \sum_{s=1}^{S} h(y^s)$$

where $y^1, \ldots, y^S$ are S independent pseudorandom numbers from the density $g(y)$, obtained by using methods described earlier. This method works if $E\left\{h(Y)\right\}$ exists and $S \to \infty$.

This method can be applied to both definite and indefinite integrals. It has the added advantage of being immediately applicable to multidimensional integrals, provided we can draw from the appropriate multivariate distribution. It has the disadvantage that it will always provide an estimate, even if the integral does not exist. For example, to obtain $E(Y)$ for the Cauchy distribution, we could average S draws from the Cauchy. But this would be wrong because the mean of the Cauchy does not exist.

As an example, we consider the computation of $E[\exp\{-\exp(Y)\}]$ when $y \sim N(0,1)$. This is the integral:

$$E\left[\exp\left\{-\exp(Y)\right\}\right] = \int_{-\infty}^{\infty} \frac{1}{\sqrt{2\pi}} \exp\left\{-\exp(y)\right\} \exp\left(-y^2/2\right) dy$$

It has no closed-form solution but can be proved to exist. We use the estimate

$$\widehat{E}\left[\exp\left\{-\exp(Y)\right\}\right] = \frac{1}{S} \sum_{s=1}^{S} \exp\left\{-\exp(y^s)\right\}$$

where y^s is the sth draw of S draws from the $N(0,1)$ distribution.

This approximation task can be accomplished for a specified value of S, say, 100, by using the following code.

```
. * Integral evaluation by Monte Carlo simulation with S=100
. clear all
. quietly set obs 100
. set seed 10101
. generate double y = invnormal(runiform())
. generate double gy = exp(-exp(y))
. quietly summarize gy, meanonly
. scalar Egy = r(mean)
. display "After 100 draws the MC estimate of E[exp(-exp(x))] is " Egy
After 100 draws the MC estimate of E[exp(-exp(x))] is .3524417
```

The Monte Carlo estimate of the integral is 0.352, based on 100 draws.

4.5.3 Monte Carlo integration using different S

It is not known in advance what value of S will yield a good Monte Carlo approximation to the integral. We can compare the outcome for several different values of S (including $S = 100$), stopping when the estimates stabilize.

To investigate this, we replace the preceding code by a Stata program that has as an argument S, the number of simulations. The program can then be called and run several times with different values of S.

The program is named `mcintegration`. The number of simulations is passed to the program as a named positional argument, `numsims`. This variable is a local variable within the program that needs to be referenced using quotes. The call to the program needs to include a value for `numsims`. Appendix A.2 provides the details on writing a Stata program. The program is r-class and returns results for a single scalar, $E\{g(y)\}$, where $g(y) = \exp\{-\exp(y)\}$.

```
. * Program mcintegration to compute E{g(y)} numsims times
. program mcintegration, rclass
  1.     version 11
  2.     args numsims        // Call to program will include value for numsims
  3.     drop _all
  4.     quietly set obs `numsims´
  5.     set seed 10101
  6.     generate double y = rnormal(0)
  7.     generate double gy = exp(-exp(y))
  8.     quietly summarize gy, meanonly
  9.     scalar Egy = r(mean)
 10.     display "#simulations: " %9.0g `numsims´  ///
  >        "  MC estimate of  E[exp(-exp(x))] is " Egy
 11. end
```

The program is then run several times, for $S = 10, 100, 1000, 10000$, and 100000.

```
. * Run program mcintegration S = 10, 100, ...., 100000 times
. mcintegration 10
#simulations:         10  MC estimate of  E[exp-exp(x)] is .30979214

. mcintegration 100
#simulations:        100  MC estimate of  E[exp-exp(x)] is .3714466

. mcintegration 1000
#simulations:       1000  MC estimate of  E[exp-exp(x)] is .38146534

. mcintegration 10000
#simulations:      10000  MC estimate of  E[exp-exp(x)] is .38081373

. mcintegration 100000
#simulations:     100000  MC estimate of  E[exp-exp(x)] is .38231031
```

The estimates of $E\{g(y)\}$ stabilize as $S \to \infty$, but even with $S = 10^5$, the estimate changes in the third decimal place.

4.6 Simulation for regression: Introduction

The simplest use of simulation methods is to generate a single dataset and estimate the DGP parameter $\boldsymbol{\theta}$. Under some assumptions, if the estimated parameter $\widehat{\boldsymbol{\theta}}$ differs from $\boldsymbol{\theta}$ for a large sample size, the estimator is probably inconsistent. We defer an example of this simpler simulation to section 4.6.4.

More often, $\boldsymbol{\theta}$ is estimated from each of S generated datasets, and the estimates are stored and summarized to learn about the distribution of $\widehat{\boldsymbol{\theta}}$ for a given DGP. For example, this approach is necessary if one wants to check the validity of a standard error estimator or the finite-sample size of a test. This approach requires the ability to perform the same analysis S times and to store the results from each simulation. The simplest approach is to write a Stata program for the analysis of one simulation and then use **simulate** to run this program many times.

4.6.1 Simulation example: OLS with χ^2 errors

In this section, we use simulation methods to investigate the finite-sample properties of the OLS estimator with random regressors and skewed errors. If the errors are i.i.d., the fact that they are skewed has no effect on the large-sample properties of the OLS estimator. However, when the errors are skewed, we will need a larger sample size for the asymptotic distribution to better approximate the finite-sample distribution of the OLS estimator than when the errors are normal. This example also highlights an important modeling decision: when y is skewed, we sometimes choose to model $E(\ln y|\mathbf{x})$ instead of $E(y|\mathbf{x})$ because we believe the disturbances enter multiplicatively instead of additively. This choice is driven by the multiplicative way the error affects the outcome and is independent of the functional form of its distribution. As illustrated in this simulation, the asymptotic theory for the OLS estimator works well when the errors are i.i.d. from a skewed distribution.

We consider the following DGP,

$$y = \beta_1 + \beta_2 x + u; \quad u \sim \chi^2(1) - 1; \quad x \sim \chi^2(1)$$

where $\beta_1 = 1$, $\beta_2 = 2$, and the sample size $N = 150$. For this DGP, the error u is independent of the regressor x (ensuring consistency of OLS) and has a mean of 0, variance of 2, skewness of $\sqrt{8}$, and kurtosis of 15. By contrast, a normal error has a skewness of 0 and a kurtosis of 3.

We wish to perform 1,000 simulations, where in each simulation we obtain parameter estimates, standard errors, t-values for the t test of $H_0 : \beta_2 = 2$, and the outcome of a two-sided test of H_0 at level 0.05.

Two of the most frequently changed parameters in a simulation study are the sample size and the number of simulations. For this reason, these two parameters are almost always stored in something that can easily be changed. We use global macros. In the output below, we store the number of observations in the global macro **numobs** and the number of repetitions in the global macro **numsims**. We use these global macros in the examples in this section.

```
. * defining global macros for sample size and number of simulations
. global numobs 150              // sample size N
. global numsims "1000"          // number of simulations
```

We first write the **chi2data** program, which generates data on y, performs OLS, and returns $\widehat{\beta}_2$, $s_{\widehat{\beta}_2}$, $t_2 = (\widehat{\beta}_2 - 2)/s_{\widehat{\beta}_2}$, a rejection indicator $r_2 = 1$ if $|t_2| > t_{0.025}(148)$, and the p-value for the two-sided t test. The **chi2data** program is an r-class program, so these results are returned in **r()** using the **return** command.

```
. * Program for finite-sample properties of OLS
. program chi2data, rclass
  1.      version 11
  2.      drop _all
  3.      set obs $numobs
  4.      generate double x = rchi2(1)
  5.      generate y = 1 + 2*x + rchi2(1)-1      // demeaned chi^2 error
  6.      regress y x
  7.      return scalar b2 =_b[x]
  8.      return scalar se2 = _se[x]
  9.      return scalar t2 = (_b[x]-2)/_se[x]
 10.      return scalar r2 = abs(return(t2))>invttail($numobs-2,.025)
 11.      return scalar p2 = 2*ttail($numobs-2,abs(return(t2)))
 12. end
```

Instead of computing the t statistic and p-value by hand, we could have used **test**, which would have computed an F statistic with the same p-value. We perform the computations manually for pedagogical purposes. The following output illustrates that **test** and the manual calculations yield the same p-value.

```
. set seed 10101
. quietly chi2data
```

```
. return list

scalars:
                r(p2) =  .0419507319188168
                r(r2) =  1
                r(t2) =  2.051809742705669
               r(se2) =  .0774765767688598
                r(b2) =  2.15896719504583

. quietly test x=2

. return list

scalars:
              r(drop) =  0
              r(df_r) =  148
                 r(F) =  4.209923220261905
                r(df) =  1
                 r(p) =  .0419507319188168
```

Below we use **simulate** to call **chi2data** $numsims times and to store the results; here $numsims = 1000. The current dataset is replaced by one with the results from each simulation. These results can be displayed by using **summarize**, where **obs** in the output refers to the number of simulations and not the sample size in each simulation. The **summarize** output indicates that 1) the mean of the point estimates is very close to the true value of 2, 2) the standard deviation of the point estimates is close to the mean of the standard errors, and 3) the rejection rate of 0.046 is very close to the size of 0.05.

```
. * Simulation for finite-sample properties of OLS
. simulate b2f=r(b2) se2f=r(se2) t2f=r(t2) reject2f=r(r2) p2f=r(p2),
> reps($numsims) saving(chi2datares, replace) nolegend nodots: chi2data

. summarize b2f se2f reject2f
```

Variable	Obs	Mean	Std. Dev.	Min	Max
b2f	1000	2.000502	.0842622	1.719513	2.40565
se2f	1000	.0839736	.0172607	.0415919	.145264
reject2f	1000	.046	.2095899	0	1

Below we use **mean** to obtain 95% confidence intervals for the simulation averages. The results for **b2f** and the rejection rate indicate that there is no significant bias and that the asymptotic distribution approximated the finite-sample distribution well for this DGP with samples of size 150. The confidence interval for the standard errors includes the sample standard deviation for **b2f**, which is another indication that the large-sample theory provides a good approximation to the finite-sample distribution.

```
. mean b2f se2f reject2f
```

Mean estimation   Number of obs   =   1000

	Mean	Std. Err.	[95% Conf. Interval]	
b2f	2.000502	.0026646	1.995273	2.005731
se2f	.0839736	.0005458	.0829025	.0850448
reject2f	.046	.0066278	.032994	.059006

Further information on the distribution of the results can be obtained by using the `summarize, detail` and `kdensity` commands.

4.6.2   Interpreting simulation output

We consider in turn unbiasedness of $\widehat{\beta}_2$, correctness of the standard-error formula for $s_{\widehat{\beta}_2}$, distribution of the t statistic, and test size.

Unbiasedness of estimator

The average of $\widehat{\beta}_2$ over the 1,000 estimates, $\overline{\widehat{\beta}_2} = (1/1000)\sum_{s=1}^{1000} \widehat{\beta}_s$, is the simulation estimate of $E(\widehat{\beta}_2)$. Here $\overline{\widehat{\beta}_2} = 2.001$ (see the mean of `b2f`) is very close to the DGP value $\beta_2 = 2.0$, suggesting that the estimator is unbiased. However, this comparison should account for simulation error. From the `mean` command, the simulation yields a 95% confidence interval for $E(\widehat{\beta}_2)$ of $[1.995, 2.006]$. This interval is quite narrow and includes 2.0, so we conclude that $E(\widehat{\beta}_2)$ is unbiased.

Many estimators, particularly nonlinear estimators, are biased in finite samples. Then exercises such as this can be used to estimate the magnitude of the bias in typical sample sizes. If the estimator is consistent, then any bias should disappear as the sample size N goes to infinity.

Standard errors

The variance of $\widehat{\beta}_2$ over the 1,000 estimates, $s_{\widehat{\beta}_2}^2 = (1/999)\sum_{s=1}^{1000}(\widehat{\beta}_s - \overline{\widehat{\beta}_2})^2$, is the simulation estimate of $\sigma_{\widehat{\beta}_2}^2 = \mathrm{Var}(\widehat{\beta}_2)$, the variance of $\widehat{\beta}_2$. Similarly, $s_{\widehat{\beta}_2} = 0.084$ (see the standard deviation of `b2f`) is the simulation estimate of $\sigma_{\widehat{\beta}_2}$. Here $\widehat{\mathrm{se}(\widehat{\beta}_2)} = 0.084$ (see the mean of `se2f`) and the 95% confidence interval for $\mathrm{se}(\widehat{\beta}_2)$ is $[0.083, 0.085]$. Because this interval includes $s_{\widehat{\beta}_2} = 0.084$, there is no evidence that $\mathrm{se}(\widehat{\beta}_2)$ is biased for $\sigma_{\widehat{\beta}_2}$, which means that the asymptotic distribution is approximating the finite-sample distribution well.

In general, that $\{\mathrm{se}(\widehat{\beta}_2)\}^2$ is unbiased for $\sigma_{\widehat{\beta}_2}^2$ does not imply that upon taking the square root $\mathrm{se}(\widehat{\beta}_2)$ is unbiased for $\sigma_{\widehat{\beta}_2}$.

t statistic

Because we impose looser restrictions on the DGP, t statistics are not exactly t distributed and z statistics are not exactly z distributed. However, the extent to which they diverge from the reference distribution disappears as the sample size increases. The output below generates the graph in figure 4.4, which compares the density of the t statistics with the $t(148)$ distribution.

```
. kdensity t2f,  n(1000) gen(t2_x t2_d) nograph
. generate double t2_d2 = tden(148, t2_x)
. graph twoway (line t2_d t2_x) (line t2_d2 t2_x)
```

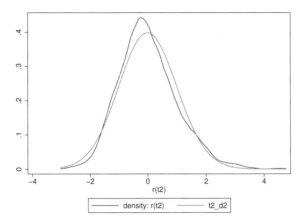

Figure 4.4. t statistic density against asymptotic distribution

Although the graph highlights some differences between the finite-sample and the asymptotic distributions, the divergence between the two does not appear to be great. Rather than focus on the distribution of the t statistics, we instead focus on the size of tests or coverage of confidence intervals based on these statistics.

Test size

The size of the test is the probability of rejecting H_0 when H_0 is true. Because the DGP sets $\beta_2 = 2$, we consider a two-sided test of H_0: $\beta_2 = 2$ against H_a: $\beta_2 \neq 2$. The level or nominal size of the test is set to 0.05, and the t test is used. The proportion of simulations that lead to a rejection of H_0 is known as the rejection rate, and this proportion is the simulation estimate of the true test size. Here the estimated rejection rate is 0.046 (see the mean of `reject2f`). The associated 95% confidence interval (from `mean reject2f`) is $[0.033, 0.059]$, which is quite wide but includes 0.05. The width of this confidence interval is partially a result of having run only 1,000 repetitions, and partially an indication that, with 150 observations, the true size of the test can differ from the nominal size. When this simulation is rerun with 10,000 repetitions, the estimated rejection rate is 0.049 and the confidence interval is $[0.044, 0.052]$.

The simulation results also include the variable `p2f`, which stores the p-values of each test. If the $t(148)$ distribution is the correct distribution for the t test, then `p2f` should be uniformly distributed on (0,1). A histogram, not shown, reveals this to be the case.

More simulations are needed to accurately measure test size (and power) than are needed for bias and standard-error calculations. For a test with estimated size a based on S simulations, a 95% confidence interval for the true size is $a \pm 1.96 \times \sqrt{a(1-a)/S}$. For example, if $a = 0.06$ and $S = 10,000$ then the 95% confidence interval is $[0.055, 0.065]$. A more detailed Monte Carlo experiment for test size and power is given in section 12.6.

Number of simulations

Ideally, 10,000 simulations or more would be run in reported results, but this can be computationally expensive. With only 1,000 simulations, there can be considerable simulation noise, especially for estimates of test size (and power).

4.6.3   Variations

The preceding code is easily adapted to other problems of interest.

Different sample size and number of simulations

Sample size can be changed by changing the global macro `numobs`. Many simulation studies focus on finite-sample deviations from asymptotic theory. For some estimators, most notably IV with weak instruments, such deviations can occur even with samples of many thousands of observations.

Changing the global macro `numsims` can increase the number of simulations to yield more-precise simulation results.

Test power

The power of a test is the probability that it rejects a false null hypothesis. To simulate the power of a test, we estimate the rejection rate for a test against a false null hypothesis. The larger the difference between the tested value and the true value, the greater the power and the rejection rate. The example below modifies `chi2data` to estimate the power of a test against the false null hypothesis that $\beta_2 = 2.1$.

```
. * Program for finite-sample properties of OLS: power
. program chi2datab, rclass
  1.      version 11
  2.      drop _all
  3.      set obs $numobs
  4.      generate double x = rchi2(1)
  5.      generate y = 1 + 2*x + rchi2(1)-1      // demeaned chi^2 error
  6.      regress y x
  7.      return scalar b2  =_b[x]
  8.      return scalar se2 =_se[x]
  9.      test x=2.1
 10.      return scalar r2 = (r(p)<.05)
 11. end
```

Below we use `simulate` to run the simulation 1,000 times, and then we summarize the results.

```
. * Power simulation for finite-sample properties of OLS
. simulate b2f=r(b2) se2f=r(se2) reject2f=r(r2), reps($numsims)
> saving(chi2databres, replace) nolegend nodots: chi2datab
. mean b2f se2f reject2f
Mean estimation                       Number of obs    =    1000
```

	Mean	Std. Err.	[95% Conf.	Interval]
b2f	2.001816	.0026958	1.996526	2.007106
se2f	.0836454	.0005591	.0825483	.0847426
reject2f	.241	.0135315	.2144465	.2675535

The sample mean of `reject2f` provides an estimate of the power. The estimated power is 0.241, which is not high. Increasing the sample size or the distance between the tested value and the true value will increase the power of the test.

A useful way to incorporate power estimation is to define the hypothesized value of β_2 to be an argument of the program `chi2datab`. This is demonstrated in the more detailed Monte Carlo experiment in section 12.6.

Different error distributions

We can investigate the effect of using other error distributions by changing the distribution used in `chi2data`. For linear regression, the t statistic becomes closer to t distributed as the error distribution becomes closer to i.i.d. normal. For nonlinear models, the exact finite-sample distribution of estimators and test statistics is unknown even if the errors are i.i.d. normal.

The example in section 4.6.2 used different draws of both regressors and errors in each simulation. This corresponds to simple random sampling where we jointly sample the pair (y, x), especially relevant to survey data where individuals are sampled, and we use data (y, x) for the sampled individuals. An alternative approach is that of fixed regressors in repeated trials, especially relevant to designed experiments. Then we draw a sample of x only once, and we use the same sample of x in each simulation while redrawing only the error u (and hence y). In that case, we create `fixedx.dta`, which has 150 observations on a variable, x, that is drawn from the $\chi^2(1)$ distribution, and we replace lines 2–4 of `chi2data` by typing `use fixedx, clear`.

4.6.4   Estimator inconsistency

Establishing estimator inconsistency requires less coding because we need to generate data and obtain estimates only once, with a large N, and then compare the estimates with the DGP values.

We do so for a classical errors-in-variables model of measurement error. Not only is it known that the OLS estimator is inconsistent, but in this case, the magnitude of the inconsistency is also known, so we have a benchmark for comparison.

The DGP considered is

$$y = \beta x^* + u; \quad x^* \sim N(0,9); \quad u \sim N(0,1)$$
$$x = x^* + v; \quad v \sim N(0,1)$$

OLS regression of y on x^* consistently estimates β. However, only data on x rather than x^* are available, so we instead obtain $\widehat{\beta}$ from an OLS regression of y on x. It is a well-known result that then $\widehat{\beta}$ is inconsistent, with a downward bias, $s\beta$, where $s = \sigma_v^2/(\sigma_v^2 + \sigma_{x^*}^2)$ is the noise–signal ratio. For the DGP under consideration, this ratio is $1/(1+9) = 0.1$, so plim $\widehat{\beta} = \beta - s\beta = 1 - 0.1 \times 1 = 0.9$.

The following simulation checks this theoretical prediction, with sample size set to $10,000$. We use `drawnorm` to jointly draw (x^*, u, v), though we could have more simply made three separate standard normal draws. We set $\beta = 1$.

```
. * Inconsistency of OLS in errors-in-variables model (measurement error)
. clear

. quietly set obs 10000

. set seed 10101

. matrix mu = (0,0,0)

. matrix sigmasq = (9,0,0\0,1,0\0,0,1)

. drawnorm xstar u v, means(mu) cov(sigmasq)

. generate y = 1*xstar + u    // DGP for y depends on xstar

. generate x = xstar + v      // x is mismeasured xstar

. regress y x, noconstant
```

Source	SS	df	MS		
Model	80231.7664	1	80231.7664	Number of obs =	10000
Residual	19114.8283	9999	1.91167399	F(1, 9999) =41969.38	
				Prob > F =	0.0000
				R-squared =	0.8076
				Adj R-squared =	0.8076
Total	99346.5946	10000	9.93465946	Root MSE =	1.3826

y	Coef.	Std. Err.	t	P>|t|	[95% Conf. Interval]	
x	.8997697	.004392	204.86	0.000	.8911604	.9083789

The OLS estimate is very precisely estimated, given the large sample size. The estimate of 0.8998 clearly differs from the DGP value of 1.0, so OLS is inconsistent. Furthermore, the simulation estimate essentially equals the theoretical value of 0.9.

4.6.5 Simulation with endogenous regressors

Endogeneity is one of the most frequent causes of estimator inconsistency. A simple method to generate an endogenous regressor is to first generate the error u and then generate the regressor x to be the sum of a multiple of u and an independent component.

We adapt the previous DGP as follows:

$$y = \beta_1 + \beta_2 x + u; \quad u \sim N(0, 1);$$
$$x = z + 0.5u; \quad z \sim N(0, 1)$$

We set $\beta_1 = 10$ and $\beta_2 = 2$. For this DGP, the correlation between x and u equals 0.5. We let $N = 150$.

The following program generates the data:

```
. * Endogenous regressor
. clear

. set seed 10101

. program endogreg, rclass
  1.        version 11
  2.        drop _all
  3.        set obs $numobs
  4.        generate u = rnormal(0)
  5.        generate x = 0.5*u + rnormal(0)          // endogenous regressors
  6.        generate y = 10 + 2*x + u
  7.        regress y x
  8.        return scalar b2 =_b[x]
  9.        return scalar se2 = _se[x]
 10.        return scalar t2 = (_b[x]-2)/_se[x]
 11.        return scalar r2 = abs(return(t2))>invttail($numobs-2,.025)
 12.        return scalar p2 = 2*ttail($numobs-2,abs(return(t2)))
 13. end
```

Below we run the simulations and summarize the results.

```
. simulate b2r=r(b2) se2r=r(se2) t2r=r(t2) reject2r=r(r2) p2r=r(p2),
> reps($numsims) nolegend nodots: endogreg

. mean b2r se2r reject2r
Mean estimation                     Number of obs    =     1000
```

	Mean	Std. Err.	[95% Conf.	Interval]
b2r	2.399301	.0020709	2.395237	2.403365
se2r	.0658053	.0001684	.0654747	.0661358
reject2r	1	0	.	.

The results from these 1,000 repetitions indicate that for $N = 150$, the OLS estimator is biased by about 20%, the standard error is about 32 times too small, and we always reject the true null hypothesis that $\beta_2 = 2$.

By setting N large, we could also show that the OLS estimator is inconsistent with a single repetition. As a variation, we could instead estimate by IV, with z an instrument for x, and verify that the IV estimator is consistent.

4.7 Stata resources

The key reference for random-number functions is `help functions`. This covers most of the generators illustrated in this chapter and several other standard ones that have not been used. Note, however, that the `rnbinomial(k,p)` function for making draws from the negative binomial distribution has a different parameterization from that used in this book. The key Stata commands for simulation are [R] **simulate** and [P] **postfile**. The `simulate` command requires first collecting commands into a program; see [P] **program**.

A standard book that presents algorithms for random-number generation is Press et al. (1992). Cameron and Trivedi (2005) discuss random-number generation and present a Monte Carlo study; see also chapter 12.7.

4.8 Exercises

1. Using the normal generator, generate a random draw from a 50–50 scale mixture of $N(1,1)$ and $N(1,3^2)$ distributions. Repeat the exercise with the $N(1,3^2)$ component replaced by $N(3,1)$. For both cases, display the features of the generated data by using a kernel density plot.

2. Generate 1,000 observations from the $F(5,10)$ distribution. Use `rchi2()` to obtain draws from the $\chi^2(5)$ and the $\chi^2(10)$ distributions. Compare the sample moments with their theoretical counterparts.

3. Make 1,000 draws from the $N(6,2^2)$ distribution by making a transformation of draws from $N(0,1)$ and then making the transformation $Y = \mu + \sigma Z$.

4. Generate 1,000 draws from the $t(6)$ distribution, which has a mean of 0 and a variance of 4. Compare your results with those from exercise 3.

5. Generate a large sample from the $N(\mu = 1, \sigma^2 = 1)$ distribution and estimate σ/μ, the coefficient of variation. Verify that the sample estimate is a consistent estimate.

6. Generate a draw from a multivariate normal distribution, $N(\boldsymbol{\mu}, \boldsymbol{\Sigma} = \mathbf{LL'})$, with $\boldsymbol{\mu'} = [0\ 0\ 0]$ and

$$\mathbf{L} = \begin{bmatrix} 1 & 0 & 0 \\ 1 & \sqrt{3} & 0 \\ 0 & \sqrt{3} & \sqrt{6} \end{bmatrix}, \text{ or } \boldsymbol{\Sigma} = \begin{bmatrix} 1 & 1 & 0 \\ 1 & 4 & 3 \\ 0 & 3 & 9 \end{bmatrix}$$

using transformations based on this Cholesky decomposition. Compare your results with those based on using the `drawnorm` command.

7. Let s denote the sample estimate of σ and $\bar{x}$ denote the sample estimate of μ. The coefficient of variation (CV) σ/μ, which is the ratio of the standard deviation to the mean, is a dimensionless measure of dispersion. The asymptotic distribution of the sample CV $s/\bar{x}$ is $N[\sigma/\mu, (N-2)^{-1/2}(\sigma/\mu)^2 \{0.5 + (\sigma/\mu)^2\}]$; see Miller (1991). For $N = 25$, using either `simulate` or `postfile`, compare the Monte

Carlo and asymptotic variance of the sample CV with the following specification of the DGP: $x \sim N(\mu, \sigma^2)$ with three different values of CV = 0.1, 0.33, and 0.67.

8. It is suspected that making draws from the truncated normal using the method given in section 4.4.4 may not work well when sampling from the extreme tails of the normal. Using different truncation points, check this suggestion.

9. Repeat the example of section 4.6.1 (OLS with χ^2 errors), now using the `postfile` command. Use `postfile` to save the estimated slope coefficient, standard error, the t statistic for H_0: $\beta = 2$, and an indicator for whether H_0 is rejected at 0.05 level in a Stata file named `simresults`. The template program is as follows:

```
* Postfile and post example: repeat OLS with chi-squared errors example
clear
set seed 10101
program simbypost
    version 11
    tempname simfile
    postfile `simfile´ b2 se2 t2 reject2 p2 using simresults, replace
    quietly {
        forvalues i = 1/$numsims {
            drop _all
            set obs $numobs
            generate x = rchi2(1)
            generate y = 1 + 2*x + rchi2(1) - 1    // demeaned chi^2 error
            regress y x
            scalar b2 =_b[x]
            scalar se2 = _se[x]
            scalar t2 = (_b[x]-2)/_se[x]
            scalar reject2 = abs(t2) > invttail($numobs-2,.025)
            scalar p2 = 2*ttail($numobs-2,abs(t2))
            post `simfile´ (b2) (se2) (t2) (reject2) (p2)
        }
    }
    postclose `simfile´
end
simbypost
use simresults, clear
summarize
```

5   GLS regression

5.1   Introduction

This chapter presents generalized least-squares (GLS) estimation in the linear regression model.

GLS estimators are appropriate when one or more of the assumptions of homoskedasticity and noncorrelation of regression errors fails. We presented in chapter 3 ordinary least-squares (OLS) estimation with inference based on, respectively, heteroskedasticity-robust or cluster–robust standard errors. Now we go further and present GLS estimation based on a richer correctly specified model for the error. This is more efficient than OLS estimation, leading to smaller standard errors, narrower confidence intervals, and larger t statistics.

Here we detail GLS for single-equation regression on cross-section data with heteroskedastic errors, and for multiequation seemingly unrelated regressions (SUR), an example of correlated errors. Other examples of GLS include the three-stage least-squares estimator for simultaneous-equations systems (section 6.6), the random-effects estimator for panel data (section 8.7), and systems of nonlinear equations (section 15.10.2).

This chapter concludes with a stand-alone presentation of a quite distinct topic: survey estimation methods that explicitly control for the three complications of data from complex surveys—sampling that is weighted, clustered, and stratified.

5.2   GLS and FGLS regression

We provide an overview of theory for GLS and feasible GLS (FGLS) estimation.

5.2.1   GLS for heteroskedastic errors

A simple example is the single-equation linear regression model with heteroskedastic independent errors, where a specific model for heteroskedasticity is given. Specifically,

$$y_i = \mathbf{x}_i'\boldsymbol{\beta} + u_i, \quad i = 1, \dots, N \qquad (5.1)$$
$$u_i = \sigma(\mathbf{z}_i)\varepsilon_i$$

where ε_i satisfies $E(\varepsilon_i|\mathbf{x}_i, \mathbf{z}_i) = 0$, $E(\varepsilon_i\varepsilon_j|\mathbf{x}_i, \mathbf{z}_i, \mathbf{x}_j, \mathbf{z}_j) = 0$, $i \neq j$, and $E(\varepsilon_i^2|\mathbf{x}_i, \mathbf{z}_i) = 1$. The function $\sigma(\mathbf{z}_i)$, called a skedasticity function, is a specified scalar-valued function of the observable variables $\mathbf{z}_i$. The special case of homoskedastic regression arises if $\sigma(\mathbf{z}_i) = \sigma$, a constant. The elements of the vectors $\mathbf{z}$ and $\mathbf{x}$ may or may not overlap.

Under these assumptions, the errors u_i in (5.1) have zero mean and are uncorrelated but are heteroskedastic with variance $\sigma^2(\mathbf{z}_i)$. Then OLS estimation of (5.1) yields consistent estimates, but more-efficient estimation is possible if we instead estimate OLS by a transformed model that has homoskedastic errors. Transforming the model by multiplying by $w_i = 1/\sigma(\mathbf{z}_i)$ yields the homoskedastic regression

$$\left\{ \frac{y_i}{\sigma(\mathbf{z}_i)} \right\} = \left\{ \frac{\mathbf{x}_i}{\sigma(\mathbf{z}_i)} \right\}' \boldsymbol{\beta} + \varepsilon_i \tag{5.2}$$

because $u_i/\sigma(\mathbf{z}_i) = \{\sigma(\mathbf{z}_i)\varepsilon_i\}/\sigma(\mathbf{z}_i) = \varepsilon_i$ and ε_i is homoskedastic. The GLS estimator is the OLS estimator of this transformed model. This regression can also be interpreted as a weighted linear regression of y_i on $\mathbf{x}_i$ with the weight $w_i = 1/\sigma(\mathbf{z}_i)$ assigned to the ith observation. In practice, $\sigma(\mathbf{z}_i)$ may depend on unknown parameters, leading to the feasible GLS estimator that uses the estimated weights $\widehat{\sigma}(\mathbf{z}_i)$ as explained below.

5.2.2 GLS and FGLS

More generally, we begin with the linear model in matrix notation:

$$\mathbf{y} = \mathbf{X}\boldsymbol{\beta} + \mathbf{u} \tag{5.3}$$

By the Gauss–Markov theorem, the OLS estimator is efficient among linear unbiased estimators if the linear regression model errors are zero-mean independent and homoskedastic.

We suppose instead that $E(\mathbf{u}\mathbf{u}'|\mathbf{X}) = \boldsymbol{\Omega}$, where $\boldsymbol{\Omega} \neq \sigma^2\mathbf{I}$ for a variety of reasons that may include heteroskedasticity or clustering. Then the efficient GLS estimator is obtained by OLS estimation of the transformed model

$$\boldsymbol{\Omega}^{-1/2}\mathbf{y} = \boldsymbol{\Omega}^{-1/2}\mathbf{X}\boldsymbol{\beta} + \boldsymbol{\varepsilon}$$

where $\boldsymbol{\Omega}^{-1/2}\boldsymbol{\Omega}\boldsymbol{\Omega}^{-1/2\prime} = \mathbf{I}$ so that the transformed error $\boldsymbol{\varepsilon} = \boldsymbol{\Omega}^{-1/2}\mathbf{u} \sim [\mathbf{0}, \mathbf{I}]$ is homoskedastic. In the heteroskedastic case, $\boldsymbol{\Omega} = \text{Diag}\{\sigma^2(\mathbf{z}_i)\}$, so $\boldsymbol{\Omega}^{-1/2} = \text{Diag}\{1/\sigma(\mathbf{z}_i)\}$.

In practice, $\boldsymbol{\Omega}$ is not known. Instead, we specify an error variance matrix model, $\boldsymbol{\Omega} = \boldsymbol{\Omega}(\boldsymbol{\gamma})$, that depends on a finite-dimensional parameter vector $\boldsymbol{\gamma}$ and, possibly, data. Given a consistent estimate $\widehat{\boldsymbol{\gamma}}$ of $\boldsymbol{\gamma}$, we form $\widehat{\boldsymbol{\Omega}} = \boldsymbol{\Omega}(\widehat{\boldsymbol{\gamma}})$. Different situations correspond to different models for $\boldsymbol{\Omega}(\boldsymbol{\gamma})$ and estimates of $\widehat{\boldsymbol{\Omega}}$. The FGLS estimator is the OLS estimator from the regression of $\widehat{\boldsymbol{\Omega}}^{-1/2}\mathbf{y}$ on $\widehat{\boldsymbol{\Omega}}^{-1/2}\mathbf{X}$ and equals

$$\widehat{\boldsymbol{\beta}}_{\text{FGLS}} = (\mathbf{X}'\widehat{\boldsymbol{\Omega}}^{-1}\mathbf{X})^{-1}\mathbf{X}'\widehat{\boldsymbol{\Omega}}^{-1}\mathbf{y}$$

Under the assumption that $\boldsymbol{\Omega}(\boldsymbol{\gamma})$ is correctly specified, the variance–covariance matrix of the estimator (VCE) of $\widehat{\boldsymbol{\beta}}_{\text{FGLS}}$ is $(\mathbf{X}'\widehat{\boldsymbol{\Omega}}^{-1}\mathbf{X})^{-1}$ because it can be shown that estimating $\boldsymbol{\Omega}$ by $\widehat{\boldsymbol{\Omega}}$ makes no difference asymptotically.

5.2.3 Weighted least squares and robust standard errors

The FGLS estimator requires specification of a model, $\boldsymbol{\Omega}(\boldsymbol{\gamma})$, for the error variance matrix. Usually, it is clear what general complication is likely to be present. For example, heteroskedastic errors are likely with cross-section data, but it is not clear what specific model for that complication is appropriate. If the model for $\boldsymbol{\Omega}(\boldsymbol{\gamma})$ is misspecified, then FGLS is still consistent, though it is no longer efficient. More importantly, the usual VCE of $\widehat{\boldsymbol{\beta}}_{\mathrm{FGLS}}$ will be incorrect. Instead, a robust estimator of the VCE should be used.

We therefore distinguish between the true error variance matrix, $\boldsymbol{\Omega} = E(\mathbf{u}\mathbf{u}'|\mathbf{X})$, and the specified model for the error variance, denoted by $\boldsymbol{\Sigma} = \boldsymbol{\Sigma}(\boldsymbol{\gamma})$. In the statistics literature, especially that for generalized linear models, $\boldsymbol{\Sigma}(\boldsymbol{\gamma})$ is called a working variance matrix. Form the estimate $\widehat{\boldsymbol{\Sigma}} = \boldsymbol{\Sigma}(\widehat{\boldsymbol{\gamma}})$, where $\widehat{\boldsymbol{\gamma}}$ is an estimate of $\boldsymbol{\gamma}$. Then do FGLS with the weighting matrix $\widehat{\boldsymbol{\Sigma}}^{-1}$, but obtain a robust estimate of the VCE. This estimator is called a weighted least-squares (WLS) estimator to indicate that we no longer maintain that $\boldsymbol{\Sigma}(\boldsymbol{\gamma}) = \boldsymbol{\Omega}$.

Table 5.1 presents the lengthy formula for the estimated VCE of the WLS estimator, along with corresponding formulas for OLS and FGLS. Heteroskedasticity-robust standard errors can be obtained after OLS and after FGLS; see section 5.3.5, which uses the `vce(robust)` option. The cluster–robust case is presented for panel data in chapter 8.

Table 5.1. Least-squares estimators and their asymptotic variance

Estimator	Definition	Estimated asymptotic variance
OLS	$\widehat{\boldsymbol{\beta}} = (\mathbf{X}'\mathbf{X})^{-1}\mathbf{X}'\mathbf{y}$	$(\mathbf{X}'\mathbf{X})^{-1}\mathbf{X}'\widehat{\boldsymbol{\Omega}}\mathbf{X}(\mathbf{X}'\mathbf{X})^{-1}$
FGLS	$\widehat{\boldsymbol{\beta}} = (\mathbf{X}'\widehat{\boldsymbol{\Omega}}^{-1}\mathbf{X})^{-1}\mathbf{X}'\widehat{\boldsymbol{\Omega}}^{-1}\mathbf{y}$	$(\mathbf{X}'\widehat{\boldsymbol{\Omega}}^{-1}\mathbf{X})^{-1}$
WLS	$\widehat{\boldsymbol{\beta}} = (\mathbf{X}'\widehat{\boldsymbol{\Sigma}}^{-1}\mathbf{X})^{-1}\mathbf{X}'\widehat{\boldsymbol{\Sigma}}^{-1}\mathbf{y}$	$(\mathbf{X}'\widehat{\boldsymbol{\Sigma}}^{-1}\mathbf{X})^{-1}\mathbf{X}'\widehat{\boldsymbol{\Sigma}}^{-1}\widehat{\boldsymbol{\Omega}}\widehat{\boldsymbol{\Sigma}}^{-1}\mathbf{X}(\mathbf{X}'\widehat{\boldsymbol{\Sigma}}^{-1}\mathbf{X})^{-1}$

Note: All results are for a linear regression model whose errors have a conditional variance matrix $\boldsymbol{\Omega}$. For FGLS, it is assumed that $\widehat{\boldsymbol{\Omega}}$ is consistent for $\boldsymbol{\Omega}$. For OLS and WLS, the heteroskedasticity-robust VCE of $\widehat{\boldsymbol{\beta}}$ uses $\widehat{\boldsymbol{\Omega}}$ equal to a diagonal matrix with squared residuals on the diagonals. A cluster–robust VCE can also be used.

5.2.4 Leading examples

The GLS framework is relevant whenever $\boldsymbol{\Omega} \neq \sigma^2\mathbf{I}$. We summarize several leading cases.

Heteroskedastic errors have already been discussed at some length, and can arise in many different ways. In particular, they may reflect specification errors associated with the functional form of the model. Examples include neglected random or systematic parameter variation; incorrect functional form of the conditional mean; incorrect scaling of the variables in the regression; and incorrect distributional assumptions regarding the dependent variables. A proper treatment of the problem of heteroskedasticity may therefore require analysis of the functional form of the regression. For example, in chapter 3, a log-linear model was found to be more appropriate than a linear model.

For multivariate linear regression, such as the estimation of systems of equations, errors can be correlated across the equations for a specific individual. In this case, the model consists of m linear regression equations $y_{ij} = \mathbf{x}'_{ij}\boldsymbol{\beta}_j + u_{ij}$, where the errors u_{ij} are correlated over j for a given i, but are uncorrelated over i. Then GLS estimation refers to efficient joint estimation of all m regressions. The three-stage least-squares estimator is an extension to the case of simultaneous-equations systems.

Another common example is that of clustered (or grouped) errors, with errors being correlated within clusters but uncorrelated between clusters. A cluster consists of a group of observations that share some social, geographical, or economic trait that induces within-cluster dependence between observations, even after controlling for sources of observable differences. Such dependence can also be induced by other latent factors such as shared social norms, habits, or influence of a common local environment. In this case, $\boldsymbol{\Omega}$ can be partitioned by cluster. If all observations can be partitioned into C mutually exclusive and exhaustive groups, then $\boldsymbol{\Omega}$ can be partitioned into C submatrices, with each submatrix having its own intracluster correlation. A leading example is the random-effects estimator for panel data, where clustering is on the individual with independence across individuals. Then algorithms exist to simplify the necessary inversion of the potentially very large $N \times N$ matrix $\boldsymbol{\Omega}$.

5.3   Modeling heteroskedastic data

Heteroskedastic errors are pervasive in microeconometrics. The failure of homoskedasticity in the standard regression model, introduced in chapter 3, leads to the OLS estimator being inefficient, though it is still a consistent estimator. Given heteroskedastic errors, there are two leading approaches. The first, taken in chapter 3, is to obtain robust estimates of the standard errors of regression coefficients without assumptions about the functional form of heteroskedasticity. Under this option, the form of heteroskedasticity has no interest for the investigator who only wants to report correct standard errors, t statistics, and p-values. This approach is easily implemented in Stata, using the vce(robust) option. The second approach seeks to model the heteroskedasticity and to obtain more-efficient FGLS estimates. This enables more precise estimation of parameters and marginal effects and more precise prediction of the conditional mean.

Unlike some other standard settings for FGLS, there is no direct Stata command for FGLS estimation given heteroskedastic errors. However, it is straightforward to obtain the FGLS estimator manually, as we now demonstrate.

5.3.1   Simulated dataset

We use a simulated dataset, one where the conditional mean of y depends on regressors x_2 and x_3, while the conditional variance depends on only x_2. The specific data-generating process (DGP) is

$$y = 1 + 1 \times x_2 + 1 \times x_3 + u; \quad x_2, x_3 \sim N(0, 25)$$
$$u = \sqrt{\exp(-1 + 0.2 \times x_2)} \times \varepsilon; \quad \varepsilon \sim N(0, 25)$$

Then the error u is heteroskedastic with a conditional variance of $25 \times \exp(-1+0.2 \times x_2)$ that varies across observations according to the value taken by x_2.

We generate a sample of size 500 from this DGP:

```
. * Generated data for heteroskedasticity example
. set seed 10101
. quietly set obs 500
. generate double x2 = 5*rnormal(0)
. generate double x3 = 5*rnormal(0)
. generate double e  = 5*rnormal(0)
. generate double u  = sqrt(exp(-1+0.2*x2))*e
. generate double y  = 1 + 1*x2 + 1*x3 + u
. summarize
```

Variable	Obs	Mean	Std. Dev.	Min	Max
x2	500	-.0357347	4.929534	-17.05808	15.1011
x3	500	.08222	5.001709	-14.89073	15.9748
e	500	-.04497	5.130303	-12.57444	18.65422
u	500	-.1564096	3.80155	-17.38211	16.09441
y	500	.8900757	7.709741	-21.65168	28.89449

The generated normal variables x2, x3, and e have, approximately, means of 0 and standard deviations of 5 as expected.

5.3.2   OLS estimation

OLS regression with default standard errors yields

```
. * OLS regression with default standard errors
. regress y x2 x3
```

Source	SS	df	MS		
Model	22566.6872	2	11283.3436		
Residual	7093.92492	497	14.2734908		
Total	29660.6122	499	59.4401046		

	Number of obs =	500
	F(2, 497) =	790.51
	Prob > F =	0.0000
	R-squared =	0.7608
	Adj R-squared =	0.7599
	Root MSE =	3.778

y	Coef.	Std. Err.	t	P>|t|	[95% Conf. Interval]	
x2	.9271964	.0343585	26.99	0.000	.8596905	.9947023
x3	.9384295	.0338627	27.71	0.000	.8718977	1.004961
_cons	.8460511	.168987	5.01	0.000	.5140341	1.178068

The coefficient estimates are close to their true values and just within or outside the upper limit of the 95% confidence intervals. The estimates are quite precise because there are 500 observations, and for this generated dataset, the $R^2 = 0.76$ is very high.

The standard procedure is to obtain heteroskedasticity-robust standard errors for the same OLS estimators. We have

```
. * OLS regression with heteroskedasticity-robust standard errors
. regress y x2 x3, vce(robust)
Linear regression                               Number of obs =      500
                                                F(  2,   497) =   652.33
                                                Prob > F      =   0.0000
                                                R-squared     =   0.7608
                                                Root MSE      =    3.778
```

y	Coef.	Robust Std. Err.	t	P>|t|	[95% Conf. Interval]
x2	.9271964	.0452823	20.48	0.000	.8382281 1.016165
x3	.9384295	.0398793	23.53	0.000	.8600767 1.016782
_cons	.8460511	.170438	4.96	0.000	.5111833 1.180919

In general, failure to control for heteroskedasticity leads to default standard errors being wrong, though a priori it is not known whether they will be too large or too small. In our example, we expect the standard errors for the coefficient of x2 to be most effected because the heteroskedasticity depends on x2. This is indeed the case. For x2, the robust standard error is 30% higher than the incorrect default (0.045 versus 0.034). The original failure to control for heteroskedasticity led to wrong standard errors, in this case, considerable understatement of the standard error of x2. For x3, there is less change in the standard error.

5.3.3 Detecting heteroskedasticity

A simple informal diagnostic procedure is to plot the absolute value of the fitted regression residual, $|\hat{u}_i|$, against a variable assumed to be in the skedasticity function. The regressors in the model are natural candidates.

The following code produces separate plots of $|\hat{u}_i|$ against x_{2i} and $|\hat{u}_i|$ against x_{3i}, and then combines these into one graph (shown in figure 5.1) by using the **graph combine** command; see section 2.6. Several options for the **twoway** command are used to improve the legibility of the graph.

```
. * Heteroskedasticity diagnostic scatterplot
. quietly regress y x2 x3
. predict double uhat, resid
. generate double absu = abs(uhat)
. quietly twoway (scatter absu x2) (lowess absu x2, bw(0.4) lw(thick)),
> scale(1.2) xscale(titleg(*5)) yscale(titleg(*5))
> plotr(style(none)) name(gls1)
```

```
. quietly twoway (scatter absu x3) (lowess absu x3, bw(0.4) lw(thick)),
> scale(1.2) xscale(titleg(*5)) yscale(titleg(*5))
> plotr(style(none)) name(gls2)

. graph combine gls1 gls2
```

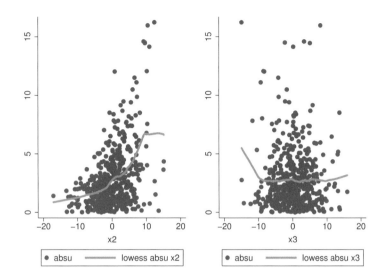

Figure 5.1. Absolute residuals graphed against x_2 and x_3

It is easy to see that the range of the scatterplot becomes wider as x_2 increases, with a nonlinear relationship, and is unchanging as x_3 increases. These observations are to be expected given the DGP.

We can go beyond a visual representation of heteroskedasticity by formally testing the null hypothesis of homoskedasticity against the alternative that residual variances depend upon a) x_2 only, b) x_3 only, and c) x_2 and x_3 jointly. Given the previous plot (and our knowledge of the DGP), we expect the first test and the third test to reject homoskedasticity, while the second test should not reject homoskedasticity.

These tests can be implemented using Stata's postestimation command `estat hettest`, introduced in section 3.5.4. The simplest test is to use the `mtest` option, which performs multiple tests that separately test each component and then test all components. We have

(Continued on next page)

```
. * Test heteroskedasticity depending on x2, x3, and x2 and x3
. estat hettest x2 x3, mtest

Breusch-Pagan / Cook-Weisberg test for heteroskedasticity
        Ho: Constant variance
```

Variable	chi2	df	p
x2	180.80	1	0.0000 #
x3	2.16	1	0.1413 #
simultaneous	185.62	2	0.0000

```
                        # unadjusted p-values
```

The p-value for x2 is 0.000, causing us to reject the null hypothesis that the skedasticity function does not depend on x2. We conclude that there is heteroskedasticity due to x2 alone. In contrast, the p-value for x3 is 0.1413, so we cannot reject the null hypothesis that the skedasticity function does not depend on x3. We conclude that there is no heteroskedasticity due to x3 alone. Similarly, the p-value of 0.000 for the joint (simultaneous) hypothesis leads us to conclude that the skedasticity function depends on x2 and x3.

The mtest option is especially convenient if there are many regressors and, hence, many candidates for causing heteroskedasticity. It does, however, use the version of hettest that assumes that errors are normally distributed. To relax this assumption to one of independent and identically distributed errors, we need to use the iid option (see section 3.5.4) and conduct separate tests. Doing this leads to test statistics (not reported) with values lower than those obtained above without iid, but leads to the same conclusion: the heteroskedasticity is due to x2.

5.3.4   FGLS estimation

For potential gains in efficiency, we can estimate the parameters of the model by using the two-step FGLS estimation method presented in section 5.2.2. For heteroskedasticity, this is easy: from (5.2), we need to 1) estimate $\widehat{\sigma}_i^2$ and 2) OLS regress $y_i/\widehat{\sigma}_i$ on $\mathbf{x}_i/\widehat{\sigma}_i$.

At the first step, we estimate the linear regression by OLS, save the residuals $\widehat{u}_i = y - \mathbf{x}'\widehat{\boldsymbol{\beta}}_{\mathrm{OLS}}$, estimate the skedasticity function $\sigma^2(\mathbf{z}_i, \boldsymbol{\gamma})$ by regressing $\widehat{u}_i^2$ on $\sigma^2(\mathbf{z}_i, \boldsymbol{\gamma})$, and get the predicted values $\widehat{\sigma}^2(\mathbf{z}_i, \widehat{\boldsymbol{\gamma}})$. Here our tests suggest that the skedasticity function should include only x2. We specify the skedasticity function $\sigma^2(\mathbf{z}) = \exp(\gamma_1 + \gamma_2 x_2)$, because taking the exponential ensures a positive variance. This is a nonlinear model that needs to be estimated by nonlinear least squares. We use the nl command, which is explained in section 10.3.5.

The first step of FGLS yields

```
. * FGLS: First step get estimate of skedasticity function
. quietly regress y x2 x3                      // get bols
. predict double uhat, resid
```

```
. generate double uhatsq = uhat^2          // get squared residual

. generate double one = 1

. nl (uhatsq = exp({xb: x2 one})), nolog    // NLS of uhatsq on exp(z´a)
(obs = 500)
```

Source	SS	df	MS
Model	188726.865	2	94363.4324
Residual	384195.497	498	771.476902
Total	572922.362	500	1145.84472

```
                                    Number of obs =        500
                                    R-squared      =     0.3294
                                    Adj R-squared =     0.3267
                                    Root MSE       =   27.77547
                                    Res. dev.      =   4741.088
```

uhatsq	Coef.	Std. Err.	t	P>|t|	[95% Conf. Interval]
/xb_x2	.1427541	.0128147	11.14	0.000	.1175766 .1679317
/xb_one	2.462675	.1119496	22.00	0.000	2.242723 2.682626

```
. predict double varu, yhat            // get sigmahat^2
```

Note that x2 explains a good deal of the heteroskedasticity ($R^2 = 0.33$) and is highly statistically significant. For our DGP, $\sigma^2(\mathbf{z}) = 25 \times \exp(-1 + 0.2x_2) = \exp(\ln 25 - 1 + 0.2x_2) = \exp(2.22 + 0.2x_2)$, and the estimates of 2.46 and 0.14 are close to these values.

At the second step, the predictions $\hat{\sigma}^2(\mathbf{z})$ define the weights that are used to obtain the FGLS estimator. Specifically, we regress $y_i/\hat{\sigma}_i$ on $\mathbf{x}_i/\hat{\sigma}_i$ where $\hat{\sigma}_i^2 = \exp(\hat{\gamma}_1 + \hat{\gamma}_2 x_{2i})$. This weighting can be done automatically by using aweight in estimation. If the aweight variable is w_i, then OLS regression is of $\sqrt{w_i}y_i$ on $\sqrt{w_i}\mathbf{x}_i$. Here we want the aweight variable to be $1/\hat{\sigma}_i^2$, or 1/varu. Then

```
. * FGLS: Second step get estimate of skedasticity function
. regress y x2 x3 [aweight=1/varu]
(sum of wgt is    5.4993e+01)
```

Source	SS	df	MS
Model	29055.2584	2	14527.6292
Residual	3818.72635	497	7.68355402
Total	32873.9847	499	65.8797289

```
                                    Number of obs =        500
                                    F( 2,   497) =    1890.74
                                    Prob > F       =     0.0000
                                    R-squared      =     0.8838
                                    Adj R-squared =     0.8834
                                    Root MSE       =     2.7719
```

y	Coef.	Std. Err.	t	P>|t|	[95% Conf. Interval]
x2	.9880644	.0246626	40.06	0.000	.9396087 1.03652
x3	.9783926	.025276	38.71	0.000	.9287315 1.028054
_cons	.9522962	.1516564	6.28	0.000	.6543296 1.250263

Comparison with previous results for OLS with the correct robust standard errors shows that the estimated confidence intervals are narrower for FGLS. For example, for x2 the improvement is from [0.84, 1.02] to [0.94, 1.04]. As predicted by theory, FGLS with a correctly specified model for heteroskedasticity is more efficient than OLS.

In practice, the form of heteroskedasticity is not known. Then a similar favorable outcome may not occur, and we should create more robust standard errors as we next consider.

5.3.5   WLS estimation

The FGLS standard errors are based on the assumption of a correct model for heteroskedasticity. To guard against misspecification of this model, we use the WLS estimator presented in section 5.2.3, which is equal to the FGLS estimator but uses robust
standard errors that do not rely on a model for heteroskedasticity. We have

```
. * WLS estimator is FGLS with robust estimate of VCE
. regress y x2 x3 [aweight=1/varu], vce(robust)
(sum of wgt is    5.4993e+01)

Linear regression                              Number of obs =      500
                                               F( 2,    497) = 2589.73
                                               Prob > F      =   0.0000
                                               R-squared     =   0.8838
                                               Root MSE      =   2.7719
```

y	Coef.	Robust Std. Err.	t	P>|t|	[95% Conf. Interval]
x2	.9880644	.0218783	45.16	0.000	.9450791	1.03105
x3	.9783926	.0242506	40.35	0.000	.9307462	1.026039
_cons	.9522962	.1546593	6.16	0.000	.6484296	1.256163

The standard errors are quite similar to those for FGLS, as expected because here FGLS
is known to use the DGP model for heteroskedasticity.

5.4   System of linear regressions

In this section, we extend GLS estimation to a system of linear equations with errors
that are correlated across equations for a given individual but are uncorrelated across
individuals. Then cross-equation correlation of the errors can be exploited to improve
estimator efficiency. This multivariate linear regression model is usually referred to
in econometrics as a set of SUR equations. It arises naturally in many contexts in
economics—a system of demand equations is a leading example. The GLS methods
presented here can be extended to systems of simultaneous equations (three-stage least-
squares estimation presented in section 6.6), panel data (chapter 8), and to systems of
nonlinear equations (section 15.10.2).

We also illustrate how to test or impose restrictions on parameters across equations. This additional complication can arise with systems of equations. For example,
consumer demand theory may impose symmetry restrictions.

5.4.1   SUR model

The model consists of m linear regression equations for N individuals. The jth equation
for individual i is $y_{ij} = \mathbf{x}'_{ij}\boldsymbol{\beta}_j + u_{ij}$. With all observations stacked, the model for the
jth equation can be written as $\mathbf{y}_j = \mathbf{X}_j\boldsymbol{\beta}_j + \mathbf{u}_j$. We then stack the m equations to give
the SUR model

$$
\begin{bmatrix} \mathbf{y}_1 \\ \mathbf{y}_2 \\ \vdots \\ \mathbf{y}_m \end{bmatrix} = \begin{bmatrix} \mathbf{X}_1 & \mathbf{0} & \cdots & \mathbf{0} \\ \mathbf{0} & \mathbf{X}_2 & & \vdots \\ \vdots & & \ddots & \mathbf{0} \\ \mathbf{0} & \cdots & \mathbf{0} & \mathbf{X}_m \end{bmatrix} \begin{bmatrix} \boldsymbol{\beta}_1 \\ \boldsymbol{\beta}_2 \\ \vdots \\ \boldsymbol{\beta}_m \end{bmatrix} + \begin{bmatrix} \mathbf{u}_1 \\ \mathbf{u}_2 \\ \vdots \\ \mathbf{u}_m \end{bmatrix} \tag{5.4}
$$

This has a compact representation:

$$
\mathbf{y} = \mathbf{X}\boldsymbol{\beta} + \mathbf{u} \tag{5.5}
$$

The error terms are assumed to have zero mean and to be independent across individuals and homoskedastic. The complication is that for a given individual the errors are correlated across equations, with $E(u_{ij}u_{ij'}|\mathbf{X}) = \sigma_{jj'}$ and $\sigma_{jj'} \neq 0$ when $j \neq j'$. It follows that the $N \times 1$ error vectors $\mathbf{u}_j$, $j = 1, \ldots, m$, satisfy the assumptions 1) $E(\mathbf{u}_j|\mathbf{X}) = \mathbf{0}$; 2) $E(\mathbf{u}_j\mathbf{u}_j'|\mathbf{X}) = \sigma_{jj}\mathbf{I}_N$; and 3) $E(\mathbf{u}_j\mathbf{u}_{j'}'|\mathbf{X}) = \sigma_{jj'}\mathbf{I}_N$, $j \neq j'$. Then for the entire system, $\boldsymbol{\Omega} = E(\mathbf{uu}') = \boldsymbol{\Sigma} \otimes \mathbf{I}_N$, where $\boldsymbol{\Sigma} = (\sigma_{jj'})$ is an $m \times m$ positive-definite matrix and $\otimes$ denotes the Kronecker product of two matrices.

OLS applied to each equation yields a consistent estimator of $\boldsymbol{\beta}$, but the optimal estimator for this model is the GLS estimator. Using $\boldsymbol{\Omega}^{-1} = \boldsymbol{\Sigma}^{-1} \otimes \mathbf{I}_N$, because $\boldsymbol{\Omega} = \boldsymbol{\Sigma} \otimes \mathbf{I}_N$, the GLS is

$$
\widehat{\boldsymbol{\beta}}_{\text{GLS}} = \left\{ \mathbf{X}' \left(\boldsymbol{\Sigma}^{-1} \otimes \mathbf{I}_N \right) \mathbf{X} \right\}^{-1} \left\{ \mathbf{X}' \left(\boldsymbol{\Sigma}^{-1} \otimes \mathbf{I}_N \right) \mathbf{y} \right\} \tag{5.6}
$$

with a VCE given by

$$
\text{Var}(\widehat{\boldsymbol{\beta}}) = \left\{ \mathbf{X}' \left(\boldsymbol{\Sigma}^{-1} \otimes \mathbf{I}_N \right) \mathbf{X} \right\}^{-1}
$$

FGLS estimation is straightforward, and the estimator is called the SUR estimator. We require only estimation and inversion of the $m \times m$ matrix $\boldsymbol{\Sigma}$. Computation is in two steps. First, each equation is estimated by OLS, and the residuals from the m equations are used to estimate $\boldsymbol{\Sigma}$, using $\widehat{\mathbf{u}}_j = \mathbf{y}_j - \mathbf{X}_j\widehat{\boldsymbol{\beta}}_j$, and $\widehat{\sigma}_{jj'} = \widehat{\mathbf{u}}_j'\widehat{\mathbf{u}}_{j'}/N$. Second, $\widehat{\boldsymbol{\Sigma}}$ is substituted for $\boldsymbol{\Sigma}$ in (5.6) to obtain the FGLS estimator $\widehat{\boldsymbol{\beta}}_{\text{FGLS}}$. An alternative is to further iterate these two steps until the estimation converges, called the iterated FGLS (IFGLS) estimator. Although asymptotically there is no advantage from iterating, in finite samples there may be. Asymptotic theory assumes that m is fixed while $N \to \infty$.

There are two cases where FGLS reduces to equation-by-equation OLS. First is the obvious case of errors uncorrelated across equations, so $\boldsymbol{\Sigma}$ is diagonal. The second case is less obvious but can often arise in practice. Even if $\boldsymbol{\Sigma}$ is nondiagonal, if each equation contains exactly the same set of regressors, so $\mathbf{X}_j = \mathbf{X}_{j'}$ for all j and j', then it can be shown that the FGLS systems estimator reduces to equation-by-equation OLS.

5.4.2 The sureg command

The SUR estimator is performed in Stata by using the command `sureg`. This command requires specification of dependent and regressor variables for each of the m equations. The basic syntax for `sureg` is

```
sureg (depvar1 varlist1) ... (depvarm varlistm) [if] [in] [weight] [, options]
```

where each pair of parentheses contains the model specification for each of the m linear regressions. The default is two-step SUR estimation. Specifying the `isure` option causes `sureg` to produce the iterated estimator.

5.4.3   Application to two categories of expenditures

The application of SUR considered here involves two dependent variables that are the logarithm of expenditure on prescribed drugs (`ldrugexp`) and expenditure on all categories of medical services other than drugs (`ltotothr`).

This data extract from the Medical Expenditure Panel Survey (MEPS) is similar to that studied in chapter 3 and covers the Medicare-eligible population of those aged 65 years and more. The regressors are socioeconomic variables (`educyr` and a quadratic in `age`), health-status variables (`actlim` and `totchr`), and supplemental insurance indicators (`private` and `medicaid`). We have

```
. * Summary statistics for seemingly unrelated regressions example
. clear all
. use mus05surdata.dta
. summarize ldrugexp ltotothr age age2 educyr actlim totchr medicaid private
```

Variable	Obs	Mean	Std. Dev.	Min	Max
ldrugexp	3285	6.936533	1.300312	1.386294	10.33773
ltotothr	3350	7.537196	1.61298	1.098612	11.71892
age	3384	74.38475	6.388984	65	90
age2	3384	5573.898	961.357	4225	8100
educyr	3384	11.29108	3.7758	0	17
actlim	3384	.3454492	.4755848	0	1
totchr	3384	1.954492	1.326529	0	8
medicaid	3384	.161643	.3681774	0	1
private	3384	.5156619	.4998285	0	1

The parameters of the SUR model are estimated by using the `sureg` command. Because SUR estimation reduces to OLS if exactly the same set of regressors appears in each equation, we omit `educyr` from the model for `ldrugexp`, and we omit `medicaid` from the model for `ltotothr`. We use the `corr` option because this yields the correlation matrix for the fitted residuals that is used to form a test of the independence of the errors in the two equations. We have

```
. * SUR estimation of a seemingly unrelated regressions model
. sureg (ldrugexp age age2 actlim totchr medicaid private)
> (ltotothr age age2 educyr actlim totchr private), corr
Seemingly unrelated regression
```

Equation	Obs	Parms	RMSE	"R-sq"	chi2	P
ldrugexp	3251	6	1.133657	0.2284	962.07	0.0000
ltotothr	3251	6	1.491159	0.1491	567.91	0.0000

	Coef.	Std. Err.	z	P>|z|	[95% Conf. Interval]	
ldrugexp						
age	.2630418	.0795316	3.31	0.001	.1071627	.4189209
age2	-.0017428	.0005287	-3.30	0.001	-.002779	-.0007066
actlim	.3546589	.046617	7.61	0.000	.2632912	.4460266
totchr	.4005159	.0161432	24.81	0.000	.3688757	.432156
medicaid	.1067772	.0592275	1.80	0.071	-.0093065	.2228608
private	.0810116	.0435596	1.86	0.063	-.0043636	.1663867
_cons	-3.891259	2.975898	-1.31	0.191	-9.723911	1.941394
ltotothr						
age	.2927827	.1046145	2.80	0.005	.087742	.4978234
age2	-.0019247	.0006955	-2.77	0.006	-.0032878	-.0005617
educyr	.0652702	.00732	8.92	0.000	.0509233	.0796172
actlim	.7386912	.0608764	12.13	0.000	.6193756	.8580068
totchr	.2873668	.0211713	13.57	0.000	.2458719	.3288618
private	.2689068	.055683	4.83	0.000	.1597701	.3780434
_cons	-5.198327	3.914053	-1.33	0.184	-12.86973	2.473077

```
Correlation matrix of residuals:
          ldrugexp  ltotothr
ldrugexp   1.0000
ltotothr   0.1741    1.0000
Breusch-Pagan test of independence: chi2(1) =    98.590, Pr = 0.0000
```

There are only 3,251 observations in this regression because of missing values for ldrugexp and ltotothr. The lengthy output from sureg has three components.

The first set of results summarizes the goodness-of-fit for each equation. For the dependent variable ldrugexp, we have $R^2 = 0.23$. A test for joint significance of all regressors in the equation (aside from the intercept) has a value of 962.07 with a p-value of $p = 0.000$ obtained from the $\chi^2(6)$ distribution because there are six regressors. The regressors are jointly significant in each equation.

The middle set of results presents the estimated coefficients. Most regressors are statistically significant at the 5% level, and the regressors generally have a bigger impact on other expenditures than they do on drug expenditures. As you will see in exercise 6 at the end of this chapter, the coefficient estimates are similar to those from OLS, and the efficiency gains of SUR compared with OLS are relatively modest, with standard errors reduced by roughly 1%.

The final set of results are generated by the `corr` option. The errors in the two equations are positively correlated, with $r_{12} = \hat{\sigma}_{12}/\sqrt{\hat{\sigma}_{11}\hat{\sigma}_{22}} = 0.1741$. The Breusch–Pagan Lagrange multiplier test for error independence, computed as $Nr_{12}^2 = 3251 \times 0.1741^2 = 98.54$, has $p = 0.000$, computed by using the $\chi^2(1)$ distribution. Because r_{12} is not exactly equal to 0.1741, the hand calculation yields 98.54, which is not exactly equal to the 98.590 in the output. This indicates statistically significant correlation between the errors in the two equations, as should be expected because the two categories of expenditures may have similar underlying determinants. At the same time, the correlation is not particularly strong, so the efficiency gains to SUR estimation are not great in this example.

5.4.4   Robust standard errors

The standard errors reported from `sureg` impose homoskedasticity. This is a reasonable assumption in this example, because taking the natural logarithm of expenditures greatly reduces heteroskedasticity. But in other applications, such as using the levels of expenditures, this would not be reasonable.

There is no option available with `sureg` to allow the errors to be heteroskedastic. However, the `bootstrap` prefix, explained in chapter 13, can be used. It resamples over individuals and provides standard errors that are valid under the weaker assumption that $E(u_{ij}u_{ij'}|\mathbf{X}) = \sigma_{i,jj'}$, while maintaining the assumption of independence over individuals. As you will learn in section 13.3.4, it is good practice to use more bootstraps than the Stata default and to set a seed. We have

```
. * Bootstrap to get heteroskedasticity-robust SEs for SUR estimator
. bootstrap, reps(400) seed(10101) nodots: sureg
> (ldrugexp age age2 actlim totchr medicaid private)
> (ltotothr age age2 educyr actlim totchr private)
Seemingly unrelated regression
```

Equation	Obs	Parms	RMSE	"R-sq"	chi2	P
ldrugexp	3251	6	1.133657	0.2284	962.07	0.0000
ltotothr	3251	6	1.491159	0.1491	567.91	0.0000

	Coef.	Bootstrap Std. Err.	z	P>|z|	Normal-based [95% Conf. Interval]	
ldrugexp						
age	.2630418	.0799786	3.29	0.001	.1062866	.4197969
age2	-.0017428	.0005319	-3.28	0.001	-.0027853	-.0007003
actlim	.3546589	.0460193	7.71	0.000	.2644627	.4448551
totchr	.4005159	.0160369	24.97	0.000	.3690841	.4319477
medicaid	.1067772	.0578864	1.84	0.065	-.0066781	.2202324
private	.0810116	.042024	1.93	0.054	-.0013539	.163377
_cons	-3.891259	2.993037	-1.30	0.194	-9.757504	1.974986
ltotothr						
age	.2927827	.1040127	2.81	0.005	.0889216	.4966438
age2	-.0019247	.0006946	-2.77	0.006	-.0032861	-.0005633
educyr	.0652702	.0082043	7.96	0.000	.0491902	.0813503
actlim	.7386912	.0655458	11.27	0.000	.6102238	.8671586
totchr	.2873668	.0212155	13.55	0.000	.2457853	.3289483
private	.2689068	.057441	4.68	0.000	.1563244	.3814891
_cons	-5.198327	3.872773	-1.34	0.180	-12.78882	2.392168

The output shows that the bootstrap standard errors differ little from the default standard errors. So, as expected for this example for expenditures in logs, heteroskedasticity makes little difference to the standard errors.

5.4.5   Testing cross-equation constraints

Testing and imposing cross-equation constraints is not possible using equation-by-equation OLS but is possible using SUR estimation. We begin with testing.

To test the joint significance of the age regressors, we type

```
. * Test of variables in both equations
. quietly sureg (ldrugexp age age2 actlim totchr medicaid private)
> (ltotothr age age2 educyr actlim totchr private)

. test age age2
 ( 1)  [ldrugexp]age = 0
 ( 2)  [ltotothr]age = 0
 ( 3)  [ldrugexp]age2 = 0
 ( 4)  [ltotothr]age2 = 0

          chi2( 4) =    16.55
        Prob > chi2 =    0.0024
```

This command automatically conducted the test for both equations.

The format used to refer to coefficient estimates is [*depname*] *varname*, where *depname* is the name of the dependent variable in the equation of interest, and *varname* is the name of the regressor of interest.

A test for significance of regressors in just the first equation is therefore

```
. * Test of variables in just the first equation
. test [ldrugexp]age [ldrugexp]age2
 ( 1)  [ldrugexp]age = 0
 ( 2)  [ldrugexp]age2 = 0
           chi2(  2) =    10.98
         Prob > chi2 =    0.0041
```

The quadratic in age in the first equation is jointly statistically significant at the 5% level.

Now consider a test of a cross-equation restriction. Suppose we want to test the null hypothesis that having private insurance has the same impact on both dependent variables. We can set up the test as follows:

```
. * Test of a restriction across the two equations
. test [ldrugexp]private = [ltotothr]private
 ( 1)  [ldrugexp]private - [ltotothr]private = 0
           chi2(  1) =     8.35
         Prob > chi2 =    0.0038
```

The null hypothesis is rejected at the 5% significance level. The coefficients in the two equations differ.

In the more general case involving cross-equation restrictions in models with three or more equations, then the `accumulate` option of the `test` command should be used.

5.4.6   Imposing cross-equation constraints

We now obtain estimates that impose restrictions on parameters across equations. Usually, such constraints are based on economic theory. As an illustrative example, we impose the constraint that having private insurance has the same impact on both dependent variables.

We first use the `constraint` command to define the constraint.

```
. * Specify a restriction across the two equations
. constraint 1 [ldrugexp]private = [ltotothr]private
```

Subsequent commands imposing the constraint will refer to it by the number 1 (any integer between 1 and 1,999 can be used).

We then impose the constraint using the `constraints()` option. We have

```
. * Estimate subject to the cross-equation constraint
. sureg (ldrugexp age age2 actlim totchr medicaid private)
> (ltotothr age age2 educyr actlim totchr private), constraints(1)
Seemingly unrelated regression
```

Equation	Obs	Parms	RMSE	"R-sq"	chi2	P
ldrugexp	3251	6	1.134035	0.2279	974.09	0.0000
ltotothr	3251	6	1.492163	0.1479	559.71	0.0000

 (1)   [ldrugexp]private - [ltotothr]private = 0

	Coef.	Std. Err.	z	P>|z|	[95% Conf. Interval]	
ldrugexp						
age	.2707053	.0795434	3.40	0.001	.1148031	.4266076
age2	-.0017907	.0005288	-3.39	0.001	-.0028271	-.0007543
actlim	.3575386	.0466396	7.67	0.000	.2661268	.4489505
totchr	.3997819	.0161527	24.75	0.000	.3681233	.4314405
medicaid	.1473961	.0575962	2.56	0.010	.0345096	.2602827
private	.1482936	.0368364	4.03	0.000	.0760955	.2204917
_cons	-4.235088	2.975613	-1.42	0.155	-10.06718	1.597006
ltotothr						
age	.2780287	.1045298	2.66	0.008	.073154	.4829034
age2	-.0018298	.0006949	-2.63	0.008	-.0031919	-.0004677
educyr	.0703523	.0071112	9.89	0.000	.0564147	.0842899
actlim	.7276336	.0607791	11.97	0.000	.6085088	.8467584
totchr	.2874639	.0211794	13.57	0.000	.245953	.3289747
private	.1482936	.0368364	4.03	0.000	.0760955	.2204917
_cons	-4.62162	3.910453	-1.18	0.237	-12.28597	3.042727

As desired, the `private` variable has the same coefficient in the two equations: 0.148.

More generally, separate `constraint` commands can be typed to specify many constraints, and the `constraints()` option will then have as an argument a list of the constraint numbers.

5.5 Survey data: Weighting, clustering, and stratification

We now turn to a quite different topic: adjustments to standard estimation methods when the data are not from a simple random sample, as we have implicitly assumed, but instead come from complex survey data. The issues raised apply to all estimation methods, including single-equation least-squares estimation of the linear model, on which we focus here.

Complex survey data lead to a sample that can be weighted, clustered, and stratified. From section 3.7, weighted estimation, if desired, can be performed by using the estimation command modifier [pweight=*weight*]. (This is a quite different reason for weighting than is that leading to the use of `aweight`s in section 5.3.4.) Valid standard errors that control for clustering can be obtained by using the vce(cluster *clustvar*) option. This is the usual approach in microeconometric analysis—standard errors should always control for any clustering of errors, and weighted analysis may or may not be appropriate depending on whether a census coefficient approach or a model approach is taken; see section 3.7.3.

The drawback to this approach is that while it yields valid estimates it ignores the improvement in precision of these estimates that arises because of stratification. This leads to conservative inference that uses overestimates of the standard errors, though for regression analysis this overestimation need not be too large. The attraction of survey commands, performed in Stata by using the svy prefix, is that they simultaneously do all three adjustments, including that for stratification.

5.5.1 Survey design

As an example of complex survey data, we use nhanes2.dta provided at the Stata web site. These data come from the second National Health and Nutrition Examination Survey (NHANES II), a U.S. survey conducted in 1976–1980.

We consider models for the hemoglobin count, a measure of the amount of the oxygen-transporting protein hemoglobin present in one's blood. Low values are associated with anemia. We estimate both the mean and the relationship with age and gender, restricting analysis to nonelderly adults. The question being asked is a purely descriptive one of how does hemoglobin vary with age and gender in the population. To answer the question, we should use sampling weights because the sample design is such that different types of individuals appear in the survey with different probabilities.

Here is a brief explanation of the survey design for the data analyzed: The country is split into 32 geographical strata. Each stratum contains a number of primary sampling units (PSUs), where a PSU represents a county or several contiguous counties with an average population of several hundred thousand people. Exactly two PSUs were chosen from each of the 32 strata, and then several hundred individuals were sampled from each PSU. The sampling of PSUs and individuals within the PSU was not purely random, so sampling weights are provided to enable correct estimation of population means at the national level. Observations on individuals may be correlated within a given PSU but are uncorrelated across PSUs, so there is clustering on the PSU. And the strata are defined so that PSUs are more similar within strata than they are across strata. This stratification improves estimator efficiency.

We can see descriptions and summary statistics for the key survey design variables and key analysis variables by typing

```
. * Survey data example: NHANES II data
. clear all
. use mus05nhanes2.dta
. quietly keep if age >= 21 & age <= 65
```

```
. describe sampl finalwgt strata psu

              storage  display    value
variable name   type   format     label      variable label

sampl           long   %9.0g                 unique case identifier
finalwgt        long   %9.0g                 sampling weight (except lead)
strata          byte   %9.0g                 stratum identifier, 1-32
psu             byte   %9.0g                 primary sampling unit, 1 or 2
. summarize sampl finalwgt strata psu

    Variable |        Obs        Mean    Std. Dev.        Min         Max

       sampl |       8136    33518.94     18447.04       1400       64702
    finalwgt |       8136    12654.81     7400.205       2079       79634
      strata |       8136    16.67146     9.431087          1          32
         psu |       8136    1.487955     .4998856          1           2
```

There are three key survey design variables. The sample weights are given in `finalwgt` and take on a wide range of values, so weighting may be important. The strata are given in `strata` and are numbered 1 to 32. The `psu` variable defines each PSU within strata and takes on only values of 1 and 2 because there are two PSUs per strata.

Before survey commands can be used, the survey design must be declared by using the `svyset` command. For a single-stage survey, the command syntax is

svyset [*psu*] [*weight*] [, *design_options options*]

For our data, we are able to provide all three of these quantities, as follows:

```
. * Declare survey design
. svyset psu [pweight=finalwgt], strata(strata)
      pweight: finalwgt
          VCE: linearized
  Single unit: missing
     Strata 1: strata
         SU 1: psu
        FPC 1: <zero>
```

For our dataset, the PSU variable was named `psu` and the strata variable was named `strata`, but other names could have been used. The output `VCE: linearized` means that the VCE will be estimated using Taylor linearization, which is analogous to cluster–robust methods in the nonsurvey case. An alternative that we do not consider is balanced repeated replication, which can be an improvement on linearization and requires provision of replicate-weight variables that ensure respondent confidentiality, whereas provision of variables for the strata and PSU may not. The output `FPC 1: <zero>` means that no finite-population correction (FPC) is provided. The FPC corrects for the complication that sampling is without replacement rather than with replacement, but this correction is only necessary if a considerable portion of the PSU is actually sampled. The FPC is generally unnecessary for a national survey of individuals, unless the PSU is very small.

The design information is given for a single-stage survey. In fact, the NHANES II is a multistage survey with sample segments (usually city blocks) chosen from within each PSU, households chosen from within each segment, and individuals chosen from within each household. This additional information can also be provided in svyset but is often not available for confidentiality reasons, and by far the most important information is declaring the first-stage sampling units.

The svydescribe command gives details on the survey design:

```
. * Describe the survey design
. svydescribe
Survey: Describing stage 1 sampling units
        pweight: finalwgt
            VCE: linearized
    Single unit: missing
       Strata 1: strata
           SU 1: psu
          FPC 1: <zero>
```

			#Obs per Unit		
Stratum	#Units	#Obs	min	mean	max
1	2	286	132	143.0	154
2	2	138	57	69.0	81
3	2	255	103	127.5	152
4	2	369	179	184.5	190
5	2	215	93	107.5	122
6	2	245	112	122.5	133
7	2	349	145	174.5	204
8	2	250	114	125.0	136
9	2	203	88	101.5	115
10	2	205	97	102.5	108
11	2	226	105	113.0	121
12	2	253	123	126.5	130
13	2	276	121	138.0	155
14	2	327	163	163.5	164
15	2	295	145	147.5	150
16	2	268	128	134.0	140
17	2	321	142	160.5	179
18	2	287	117	143.5	170
20	2	221	95	110.5	126
21	2	170	84	85.0	86
22	2	242	98	121.0	144
23	2	277	136	138.5	141
24	2	339	162	169.5	177
25	2	210	94	105.0	116
26	2	210	103	105.0	107
27	2	230	110	115.0	120
28	2	229	106	114.5	123
29	2	351	165	175.5	186
30	2	291	134	145.5	157
31	2	251	115	125.5	136
32	2	347	166	173.5	181
31	62	8136	57	131.2	204

For this data extract, only 31 of the 32 strata are included (stratum 19 is excluded) and each stratum has exactly two PSUs, so there are 62 distinct PSUs in all.

5.5.2 Survey mean estimation

We consider estimation of the population mean of hgb, the hemoglobin count with a normal range of approximately 12–15 for women and 13.5–16.5 for men. To estimate the population mean, we should definitely use the sampling weights.

To additionally control for clustering and stratification, we give the svy: prefix before mean. We have

```
. * Estimate the population mean using svy:
. svy: mean hgb
(running mean on estimation sample)

Survey: Mean estimation

Number of strata =      31      Number of obs   =       8136
Number of PSUs   =      62      Population size = 102959526
                                Design df       =         31
```

	Mean	Linearized Std. Err.	[95% Conf.	Interval]
hgb	14.29713	.0345366	14.22669	14.36757

The population mean is quite precisely estimated with a 95% confidence interval [14.23, 14.37].

What if we completely ignored the survey design? We have

```
. * Estimate the population mean using no weights and no cluster
. mean hgb

Mean estimation                        Number of obs   =      8136
```

	Mean	Std. Err.	[95% Conf.	Interval]
hgb	14.28575	.0153361	14.25569	14.31582

In this example, the estimate of the population mean is essentially unchanged. There is a big difference in the standard errors. The default standard error estimate of 0.015 is wrong for two reasons: it is underestimated because of failure to control for clustering, and it is overestimated because of failure to control for stratification. Here $0.015 < 0.035$, so, as in many cases, the failure to control for clustering dominates and leads to great overstatement of the precision of the estimator.

5.5.3 Survey linear regression

The svy prefix before regress simultaneously controls for weighting, clustering, and stratification declared in the preceding svyset command. We type

```
. * Regression using svy:
. svy: regress hgb age female
(running regress on estimation sample)

Survey: Linear regression

Number of strata   =        31          Number of obs    =       8136
Number of PSUs     =        62          Population size  =  102959526
                                        Design df        =         31
                                        F(  2,    30)    =    2071.57
                                        Prob > F         =     0.0000
                                        R-squared        =     0.3739
```

hgb	Coef.	Linearized Std. Err.	t	P>|t|	[95% Conf. Interval]	
age	.0021623	.0010488	2.06	0.048	.0000232	.0043014
female	-1.696847	.0261232	-64.96	0.000	-1.750125	-1.643568
_cons	15.0851	.0651976	231.38	0.000	14.95213	15.21807

The hemoglobin count increases slightly with age and is considerably lower for women when compared with the sample mean of 14.3.

The same weighted estimates, with standard errors that control for clustering but not stratification, can be obtained without using survey commands. To do so, we first need to define a single variable that uniquely identifies each PSU, whereas survey commands can use two separate variables, here **strata** and **psu**, to uniquely identify each PSU. Specifically, **strata** took 31 different integer values while **psu** took only the values 1 and 2. To make 62 unique PSU identifiers, we multiply **strata** by two and add **psu**. Then we have

```
. * Regression using weights and cluster on PSU
. generate uniqpsu = 2*strata + psu  // make unique identifier for each psu

. regress hgb age female [pweight=finalwgt], vce(cluster uniqpsu)
(sum of wgt is   1.0296e+08)

Linear regression                       Number of obs =       8136
                                        F(  2,    61) =    1450.50
                                        Prob > F      =     0.0000
                                        R-squared     =     0.3739
                                        Root MSE      =     1.0977

                         (Std. Err. adjusted for 62 clusters in uniqpsu)
```

hgb	Coef.	Robust Std. Err.	t	P>|t|	[95% Conf. Interval]	
age	.0021623	.0011106	1.95	0.056	-.0000585	.0043831
female	-1.696847	.0317958	-53.37	0.000	-1.760426	-1.633267
_cons	15.0851	.0654031	230.65	0.000	14.95432	15.21588

The regression coefficients are the same as before. The standard errors for the slope coefficients are roughly 5% and 20% larger than those obtained when the **svy** prefix is used, so using survey methods to additionally control for stratification improves estimator efficiency.

Finally, consider a naïve OLS regression without weighting or obtaining cluster–robust VCE:

```
. * Regression using no weights and no cluster
. regress hgb age female
      Source │      SS       df       MS              Number of obs =     8136
─────────────┼────────────────────────────           F(  2,  8133) = 2135.79
       Model │ 5360.48245        2  2680.24123        Prob > F      =  0.0000
    Residual │ 10206.2566     8133  1.25491905        R-squared     =  0.3444
─────────────┼────────────────────────────           Adj R-squared =  0.3442
       Total │ 15566.7391     8135  1.91355121        Root MSE      =  1.1202

─────────────┬──────────────────────────────────────────────────────────────
         hgb │      Coef.   Std. Err.      t    P>|t|     [95% Conf. Interval]
─────────────┼──────────────────────────────────────────────────────────────
         age │   .0013372   .0008469     1.58   0.114    -.0003231    .0029974
      female │  -1.624161    .024857   -65.34   0.000    -1.672887   -1.575435
       _cons │   15.07118   .0406259   370.97   0.000     14.99154    15.15081
─────────────┴──────────────────────────────────────────────────────────────
```

Now the coefficient of **age** has changed considerably and standard errors are, erroneously, considerably smaller because of failure to control for clustering on the PSU.

For most microeconometric analyses, one should always obtain standard errors that control for clustering, if clustering is present. Many data extracts from complex survey datasets do not include data on the PSU, for confidentiality reasons or because the researcher did not extract the variable. Then a conservative approach is to use nonsurvey methods and obtain standard errors that cluster on a variable that subsumes the PSUs, for example, a geographic region such as a state.

As emphasized in section 3.7, the issue of whether to weight in regression analysis (rather than mean estimation) with complex survey data is a subtle one. For the many microeconometrics applications that take a control-function approach, it is unnecessary.

5.6   Stata resources

The **sureg** command introduces multiequation regression. Related multiequation commands are [R] **mvreg**, [R] **nlsur**, and [R] **reg3**. The multivariate regression command **mvreg** is essentially the same as **sureg**. The **nlsur** command generalizes **sureg** to nonlinear equations; see section 15.10. The **reg3** command generalizes the SUR model to handle endogenous regressors; see section 6.6.

Econometrics texts give little coverage of survey methods, and the survey literature is a stand-alone literature that is relatively inaccessible to econometricians. The Stata [SVY] *Survey Data Reference Manual* is quite helpful. Econometrics references include Bhattacharya (2005), Cameron and Trivedi (2005), and Kreuter and Valliant (2007).

5.7   Exercises

1. Generate data by using the same DGP as that in section 5.3, and implement the first step of FGLS estimation to get the predicted variance **varu**. Now com-

pare several different methods to implement the second step of weighted estimation. First, use `regress` with the modifier [`aweight=1/varu`], as in the text. Second, manually implement this regression by generating the transformed variable `try=y/sqrt(varu)` and regressing `try` on the similarly constructed variables `trx2`, `trx3`, and `trone`, using `regress` with the `noconstant` option. Third, use `regress` with [`pweight=1/varu`], and show that the default standard errors using `pweight`s differ from those using `aweight`s because the `pweight`s default is to compute robust standard errors.

2. Consider the same DGP as that in section 5.3. Given this specification of the model, the rescaled equation $y/w = \beta_1(1/w) + \beta_2(x_2/w) + \beta_3(x_3/w) + e$, where $w = \sqrt{\exp(-1 + 0.2 * x_2)}$, will have the error e, which is normally distributed and homoskedastic. Treat w as known and estimate this rescaled regression in Stata by using `regress` with the `noconstant` option. Compare the results with those given in section 5.3, where the weight w was estimated. Is there a big difference here between the GLS and FGLS estimators?

3. Consider the same DGP as that in section 5.3. Suppose that we incorrectly assume that $u \sim N(0, \sigma^2 x_2^2)$. Then FGLS estimates can be obtained by using `regress` with [`pweight=1/x2sq`], where `x2sq=x2`^{2}. How different are the estimates of $(\beta_1, \beta_2, \beta_3)$ from the OLS results? Can you explain what has happened in terms of the consequences of using the wrong skedasticity function? Do the standard errors change much if robust standard errors are computed? Use the `estat hettest` command to check whether the regression errors in the transformed model are homoskedastic.

4. Consider the same dataset as in section 5.4. Repeat the analysis of section 5.4 using the dependent variables `drugexp` and `totothr`, which are in levels rather than logs (so heteroskedasticity is more likely to be a problem). First, estimate the two equations using OLS with default standard errors and robust standard errors, and compare the standard errors. Second, estimate the two equations jointly using `sureg`, and compare the estimates with those from OLS. Third, use the `bootstrap` prefix to obtain robust standard errors from `sureg`, and compare the efficiency of joint estimation with that of OLS. Hint: It is much easier to compare estimates across methods if the `estimates` command is used; see section 3.4.4.

5. Consider the same dataset as in section 5.5. Repeat the analysis of section 5.5 using all observations rather than restricting the sample to ages between 21 and 65 years. First, declare the survey design. Second, compare the unweighted mean of `hgb` and its standard error, ignoring survey design, with the weighted mean and standard error allowing for all features of the survey design. Third, do a similar comparison for least-squares regression of `hgb` on `age` and `female`. Fourth, estimate this same regression using `regress` with `pweight`s and cluster–robust standard errors, and compare with the survey results.

6. Reconsider the dataset from section 5.4.3. Estimate the parameters of each equation by OLS. Compare these OLS results with the SUR results reported in section 5.4.3.

6 Linear instrumental-variables regression

6.1 Introduction

The fundamental assumption for consistency of least-squares estimators is that the model error term is unrelated to the regressors, i.e., $E(u|\mathbf{x}) = 0$.

If this assumption fails, the ordinary least-squares (OLS) estimator is inconsistent and the OLS estimator can no longer be given a causal interpretation. Specifically, the OLS estimate $\widehat{\beta}_j$ can no longer be interpreted as estimating the marginal effect on the dependent variable y of an exogenous change in the jth regressor variable x_j. This is a fundamental problem because such marginal effects are a key input to economic policy.

The instrumental-variables (IV) estimator provides a consistent estimator under the very strong assumption that valid instruments exist, where the instruments $\mathbf{z}$ are variables that are correlated with the regressors $\mathbf{x}$ that satisfy $E(u|\mathbf{z}) = 0$.

The IV approach is the original and leading approach for estimating the parameters of the models with endogenous regressors and errors-in-variables models. Mechanically, the IV method is no more difficult to implement than OLS regression. Conceptually, the IV method is more difficult than other regression methods. Practically, it can be very difficult to obtain valid instruments, so $E(u|\mathbf{z}) = 0$. Even where such instruments exist, they may be so weakly correlated with endogenous regressors that standard asymptotic theory provides a poor guide in finite samples.

6.2 IV estimation

IV methods are more widely used in econometrics than in other applied areas of statistics. This section provides some intuition for IV methods and details the methods.

6.2.1 Basic IV theory

We introduce IV methods in the simplest regression model, one where the dependent variable y is regressed on a single regressor x:

$$y = \beta x + u \tag{6.1}$$

The model is written without the intercept. This leads to no loss of generality if both y and x are measured as deviations from their respective means.

For concreteness, suppose y measures earnings, x measures years of schooling, and u is the error term. The simple regression model assumes that x is uncorrelated with the errors in (6.1). Then the only effect of x on y is a direct effect via the term βx. Schematically, we have the following path diagram:

$$x \longrightarrow y$$
$$\nearrow$$
$$u$$

The absence of a directional arrow from u to x means that there is no association between x and u. Then the OLS estimator $\widehat{\beta} = \sum_i x_i y_i / \sum_i x_i^2$ is consistent for β.

The error u embodies all factors other than schooling that determine earnings. One such factor in u is ability. However, high ability will induce correlation between x and u because high (low) ability will on average be associated with high (low) years of schooling. Then a more appropriate schematic diagram is

$$x \longrightarrow y$$
$$\uparrow \quad \nearrow$$
$$u$$

where now there is an association between x and u.

The OLS estimator $\widehat{\beta}$ is then inconsistent for β, because $\widehat{\beta}$ combines the desired direct effect of schooling on earnings (β) with the indirect effect that people with high schooling are likely to have high ability, high u, and hence high y. For example, if one more year of schooling is found to be associated on average with a \$1,000 increase in annual earnings, we are not sure how much of this increase is due to schooling per se (β) and how much is due to people with higher schooling having on average higher ability (so higher u).

The regressor x is said to be endogenous, meaning it arises within a system that influences u. By contrast, an exogenous regressor arises outside the system and is unrelated to u. The inconsistency of $\widehat{\beta}$ is referred to as endogeneity bias, because the bias does not disappear asymptotically.

The obvious solution to the endogeneity problem is to include as regressors controls for ability, a solution called the control-function approach. But such regressors may not be available. Few earnings–schooling datasets additionally have measures of ability such as IQ tests; even if they do, there are questions about the extent to which they measure inherent ability.

The IV approach provides an alternative solution. We introduce a (new) instrumental variable, z, which has the property that changes in z are associated with changes in x but do not lead to changes in y (except indirectly via x). This leads to the following path diagram:

$$z \longrightarrow x \longrightarrow y$$
$$\uparrow \quad \nearrow$$
$$u$$

For example, proximity to college (z) may determine college attendance (x) but not directly determine earnings (y).

The IV estimator in this simple example is $\widehat{\beta}_{\mathrm{IV}} = \sum_i z_i y_i / \sum_i z_i x_i$. This can be interpreted as the ratio of the correlation of y with z to the correlation of x with z or, after some algebra, as the ratio of dy/dz to dx/dz. For example, if a one-unit increase in z is associated with 0.2 more years of education and with \$500 higher earnings, then $\widehat{\beta}_{\mathrm{IV}} = \$500/0.2 = \$2,500$, so one more year of schooling increases earnings by \$2,500.

The IV estimator $\widehat{\beta}_{\mathrm{IV}}$ is consistent for β provided that the instrument z is uncorrelated with the error u and correlated with the regressor x.

6.2.2 Model setup

We now consider the more general regression model with the scalar dependent variable y_1, which depends on m endogenous regressors, denoted by $\mathbf{y}_2$, and K_1 exogenous regressors (including an intercept), denoted by $\mathbf{x}_1$. This model is called a structural equation, with

$$y_{1i} = \mathbf{y}_{2i}'\boldsymbol{\beta}_1 + \mathbf{x}_{1i}'\boldsymbol{\beta}_2 + u_i, \quad i = 1, \ldots, N \tag{6.2}$$

The regression errors u_i are assumed to be uncorrelated with $\mathbf{x}_{1i}$ but are correlated with $\mathbf{y}_{2i}$. This correlation leads to the OLS estimator being inconsistent for $\boldsymbol{\beta}$.

To obtain a consistent estimator, we assume the existence of at least m IV $\mathbf{x}_2$ for $\mathbf{y}_2$ that satisfy the assumption that $E(u_i | \mathbf{x}_{2i}) = 0$. The instruments $\mathbf{x}_2$ need to be correlated with $\mathbf{y}_2$ so that they provide some information on the variables being instrumented. One way to motivate this is to assume that each component y_{2j} of $\mathbf{y}_2$ satisfies the first-stage equation (also called a reduced-form model)

$$y_{2ji} = \mathbf{x}_{1i}'\boldsymbol{\pi}_{1j} + \mathbf{x}_{2i}'\boldsymbol{\pi}_{2j} + v_{ji}, \quad j = 1, \ldots, m \tag{6.3}$$

The first-stage equations have only exogenous variables on the right-hand side. The exogenous regressors $\mathbf{x}_1$ in (6.2) can be used as instruments for themselves. The challenge is to come up with at least one additional instrument $\mathbf{x}_2$. Often $\mathbf{y}_2$ is scalar, $m = 1$, and we need to find one additional instrument $\mathbf{x}_2$. More generally, with m endogenous regressors, we need at least m additional instruments $\mathbf{x}_2$. This can be difficult because $\mathbf{x}_2$ needs to be a variable that can be legitimately excluded from the structural model (6.2) for y_1.

The model (6.2) can be more simply written as

$$y_i = \mathbf{x}_i'\boldsymbol{\beta} + u_i \tag{6.4}$$

where the regressor vector $\mathbf{x}_i' = [\mathbf{y}_{2i}' \ \mathbf{x}_{1i}']$ combines endogenous and exogenous variables, and the dependent variable is denoted by y rather than y_1. We similarly combine the instruments for these variables. Then the vector of IV (or, more simply, instruments) is $\mathbf{z}_i' = [\mathbf{x}_{1i}'\ \mathbf{x}_{2i}']$, where $\mathbf{x}_1$ serves as the (ideal) instrument for itself and $\mathbf{x}_2$ is the instrument for $\mathbf{y}_2$, and the instruments $\mathbf{z}$ satisfy the conditional moment restriction

$$E(u_i|\mathbf{z}_i) = 0 \tag{6.5}$$

In summary, we regress y on $\mathbf{x}$ using instruments $\mathbf{z}$.

6.2.3 IV estimators: IV, 2SLS, and GMM

The key (and in many cases, heroic) assumption is (6.5). This implies that $E(\mathbf{z}_i u_i) = \mathbf{0}$, and hence the moment condition, or population zero-correlation condition,

$$E\{\mathbf{z}_i'(y_i - \mathbf{x}_i'\boldsymbol{\beta})\} = \mathbf{0} \tag{6.6}$$

IV estimators are solutions to the sample analog of (6.6).

We begin with the case where $\dim(\mathbf{z}) = \dim(\mathbf{x})$, called the just-identified case, where the number of instruments exactly equals the number of regressors. Then the sample analog of (6.6) is $\sum_{i=1}^N \mathbf{z}_i'(y_i - \mathbf{x}_i'\boldsymbol{\beta}) = \mathbf{0}$. As usual, stack the vectors $\mathbf{x}_i'$ into the matrix $\mathbf{X}$, the scalars y_i into the vector $\mathbf{y}$, and the vectors $\mathbf{z}_i'$ into the matrix $\mathbf{Z}$. Then we have $\mathbf{Z}'(\mathbf{y} - \mathbf{X}\boldsymbol{\beta}) = \mathbf{0}$. Solving for $\boldsymbol{\beta}$ leads to the IV estimator

$$\widehat{\boldsymbol{\beta}}_{\text{IV}} = (\mathbf{Z}'\mathbf{X})^{-1}\mathbf{Z}'\mathbf{y}$$

A second case is where $\dim(\mathbf{z}) < \dim(\mathbf{x})$, called the not-identified or underidentified case, where there are fewer instruments than regressors. Then no consistent IV estimator exists. This situation often arises in practice. Obtaining enough instruments, even just one in applications with a single endogenous regressor, can require considerable ingenuity or access to unusually rich data.

A third case is where $\dim(\mathbf{z}) > \dim(\mathbf{x})$, called the overidentified case, where there are more instruments than regressors. This can happen especially when economic theory leads to clear exclusion of variables from the equation of interest, freeing up these variables to be used as instruments if they are correlated with the included endogenous regressors. Then $\mathbf{Z}'(\mathbf{y} - \mathbf{X}\boldsymbol{\beta}) = \mathbf{0}$ has no solution for $\boldsymbol{\beta}$ because it is a system of $\dim(\mathbf{z})$ equations in only $\dim(\mathbf{x})$ unknowns. One possibility is to arbitrarily drop instruments to get to the just-identified case. But there are more-efficient estimators. One estimator is the two-stage least-squares (2SLS) estimator,

$$\widehat{\boldsymbol{\beta}}_{\text{2SLS}} = \left\{\mathbf{X}'\mathbf{Z}(\mathbf{Z}'\mathbf{Z})^{-1}\mathbf{Z}'\mathbf{X}\right\}^{-1}\mathbf{X}'\mathbf{Z}(\mathbf{Z}'\mathbf{Z})^{-1}\mathbf{Z}'\mathbf{y}$$

This estimator is the most efficient estimator if the errors u_i are independent and homoskedastic. And it equals $\widehat{\boldsymbol{\beta}}_{\text{IV}}$ in the just-identified case. The term 2SLS arises because the estimator can be computed in two steps. First, estimate by OLS the first-stage regressions given in (6.3), and second, estimate by OLS the structural regression (6.2) with endogenous regressors replaced by predictions from the first step.

A quite general estimator is the generalized method of moments (GMM) estimator

$$\widehat{\boldsymbol{\beta}}_{\mathrm{GMM}} = \left(\mathbf{X'ZWZ'X}\right)^{-1}\mathbf{X'ZWZ'y} \tag{6.7}$$

where $\mathbf{W}$ is any full-rank symmetric-weighting matrix. In general, the weights in $\mathbf{W}$ may depend both on data and on unknown parameters. For just-identified models, all choices of $\mathbf{W}$ lead to the same estimator. This estimator minimizes the objective function

$$Q(\boldsymbol{\beta}) = \left\{\frac{1}{N}(\mathbf{y}-\mathbf{X}\boldsymbol{\beta})'\mathbf{Z}\right\}\mathbf{W}\left\{\frac{1}{N}\mathbf{Z}'(\mathbf{y}-\mathbf{X}\boldsymbol{\beta})\right\} \tag{6.8}$$

which is a matrix-weighted quadratic form in $\mathbf{Z}'(\mathbf{y}-\mathbf{X}\boldsymbol{\beta})$.

For GMM, some choices of $\mathbf{W}$ are better than others. The 2SLS estimator is obtained with weighting matrix $\mathbf{W}=(\mathbf{Z'Z})^{-1}$. The optimal GMM estimator uses $\mathbf{W}=\widehat{\mathbf{S}}^{-1}$, so

$$\widehat{\boldsymbol{\beta}}_{\mathrm{OGMM}} = \left(\mathbf{X'Z}\widehat{\mathbf{S}}^{-1}\mathbf{Z'X}\right)^{-1}\mathbf{X'Z}\widehat{\mathbf{S}}^{-1}\mathbf{Z'y}$$

where $\widehat{\mathbf{S}}$ is an estimate of $\mathrm{Var}(N^{-1/2}\mathbf{Z'u})$. If the errors u_i are independent and heteroskedastic, then $\widehat{\mathbf{S}} = 1/N\sum_{i=1}^{N}\widehat{u}_i^2\mathbf{z}_i\mathbf{z}_i'$, where $\widehat{u}_i = y_i - \mathbf{x}_i'\widehat{\boldsymbol{\beta}}$ and $\widehat{\boldsymbol{\beta}}$ is a consistent estimator, usually $\widehat{\boldsymbol{\beta}}_{\mathrm{2SLS}}$. The estimator reduces to $\widehat{\boldsymbol{\beta}}_{\mathrm{IV}}$ in the just-identified case.

In later sections, we consider additional estimators. In particular, the limited-information maximum-likelihood (LIML) estimator, while asymptotically equivalent to 2SLS, has been found in recent research to outperform both 2SLS and GMM in finite samples.

6.2.4 Instrument validity and relevance

All the preceding estimators have the same starting point. The instruments must satisfy condition (6.5). This condition is impossible to test in the just-identified case. And even in the overidentified case, where a test is possible (see section 6.3.7), instrument validity relies more on persuasive argument, economic theory, and norms established in prior related empirical studies.

Additionally, the instruments must be relevant. For the model of section 6.2.2, this means that after controlling for the remaining exogenous regressors $\mathbf{x}_1$, the instruments $\mathbf{x}_2$ must account for significant variation in $\mathbf{y}_2$. Intuitively, the stronger the association between the instruments $\mathbf{z}$ and $\mathbf{x}$, the stronger will be the identification of the model. Conversely, instruments that are only marginally relevant are referred to as weak instruments.

The first consequence of an instrument being weak is that estimation becomes much less precise, so standard errors can become many times larger, and t statistics many times smaller, compared with those from (inconsistent) OLS. Then a promising t statistic of 5 from OLS estimation may become a t statistic of 1 from IV estimation. If this loss of precision is critical, then one needs to obtain better instruments or more data.

The second consequence is that even though IV estimators are consistent, the standard asymptotic theory may provide a poor approximation to the actual sampling distribution of the IV estimator in typical finite-sample sizes. For example, the asymptotic critical values of standard Wald tests will lead to tests whose actual size differs considerably from the nominal size, and hence the tests are potentially misleading. This problem arises because in finite samples the IV estimator is not centered on $\boldsymbol{\beta}$ even though in infinite samples it is consistent for $\boldsymbol{\beta}$. The problem is referred to as finite-sample bias of IV, even in situations where formally the mean of the estimator does not exist. The question "How large of a sample size does one need before these biases become unimportant?" does not have a simple answer. This issue is considered further in sections 6.4 and 6.5.

6.2.5 Robust standard-error estimates

Table 6.1 provides a summary of three leading variants of the IV family of estimators. For just-identified models, we use the IV estimator because the other models collapse to the IV estimator in that case. For overidentified models, the standard estimators are 2SLS and optimal GMM.

The formulas given for estimates of the VCEs are robust estimates, where $\widehat{\mathbf{S}}$ is an estimate of the asymptotic variance of $\sqrt{N}\mathbf{Z}'\mathbf{u}$. For heteroskedastic errors, $\widehat{\mathbf{S}} = N^{-1}\sum_{i=1}^{N}\widehat{u}_i^2\mathbf{z}_i\mathbf{z}_i'$, and we use the vce(robust) option. For clustered errors, we use the vce(cluster *clustvar*) option.

Table 6.1. IV estimators and their asymptotic variances

Estimator	Definition and estimate of the VCE
IV (just-identified)	$\widehat{\boldsymbol{\beta}}_{\mathrm{IV}} = (\mathbf{Z}'\mathbf{X})^{-1}\mathbf{Z}'\mathbf{y}$ $\widehat{V}(\widehat{\boldsymbol{\beta}}) = (\mathbf{Z}'\mathbf{X})^{-1}\widehat{\mathbf{S}}(\mathbf{Z}'\mathbf{X})^{-1}\mathbf{Z}$
2SLS	$\widehat{\boldsymbol{\beta}}_{\mathrm{2SLS}} = \left\{\mathbf{X}'\mathbf{Z}(\mathbf{Z}'\mathbf{Z})^{-1}\mathbf{Z}'\mathbf{X}\right\}^{-1}\mathbf{X}'\mathbf{Z}(\mathbf{Z}'\mathbf{Z})^{-1}\mathbf{Z}'\mathbf{y}$ $\widehat{V}(\widehat{\boldsymbol{\beta}}) = \left(\mathbf{X}'\mathbf{Z}(\mathbf{Z}'\mathbf{Z})^{-1}\mathbf{Z}'\mathbf{X}\right)^{-1}\mathbf{X}'\mathbf{Z}(\mathbf{Z}'\mathbf{Z})^{-1}\widehat{\mathbf{S}}(\mathbf{Z}'\mathbf{Z})^{-1}\mathbf{Z}'\mathbf{X}$ $\times\left\{\mathbf{X}'\mathbf{Z}(\mathbf{Z}'\mathbf{Z})^{-1}\mathbf{Z}'\mathbf{X}\right\}^{-1}$
Optimal GMM	$\widehat{\boldsymbol{\beta}}_{\mathrm{OGMM}} = \left(\mathbf{X}'\mathbf{Z}\widehat{\mathbf{S}}^{-1}\mathbf{Z}'\mathbf{X}\right)^{-1}\mathbf{X}'\mathbf{Z}\widehat{\mathbf{S}}^{-1}\mathbf{Z}'\mathbf{y}$ $\widehat{V}(\widehat{\boldsymbol{\beta}}_{\mathrm{OGMM}}) = \left(\mathbf{X}'\mathbf{Z}\widehat{\mathbf{S}}^{-1}\mathbf{Z}'\mathbf{X}\right)^{-1}$

6.3   IV example

All estimators in this chapter use the `ivregress` command. This command, introduced in Stata 10, is a significant enhancement of the earlier `ivreg` command and incorporates several of the features of the user-written `ivreg2` command. The rest of this section provides application to an example with a single endogenous regressor.

6.3.1   The ivregress command

The `ivregress` command performs IV regression and yields goodness-of-fit statistics, coefficient estimates, standard errors, t statistics, p-values, and confidence intervals. The syntax of the command is

`ivregress` *estimator depvar* $\big[$ *varlist1* $\big]$ (*varlist2* = *varlist_iv*) $\big[$ *if* $\big]$ $\big[$ *in* $\big]$
   $\big[$ *weight* $\big]$ $\big[$, *options* $\big]$

Here *estimator* is one of `2sls` (2SLS), `gmm` (optimal GMM), or `liml` (limited-information maximum likelihood); *depvar* is the scalar dependent variable; *varlist1* is the list of exogenous regressors; *varlist2* is the list of endogenous regressors; and *varlist_iv* is the list of instruments for the endogenous regressors. Note that endogenous regressors and their instruments appear inside parentheses. If the model has several endogenous variables, they are all listed on the left-hand side of the equality. Because there is no `iv` option for *estimator*, in the just-identified case we use the `2sls` option, because 2SLS is equivalent to IV in that case.

An example of the command is `ivregress 2sls y x1 x2 (y2 y3 = x3 x4 x5)`. This performs 2SLS estimation of a structural-equation model with the dependent variable, `y`; two exogenous regressors, `x1` and `x2`; two endogenous regressors, `y2` and `y3`; and three instruments, `x3`, `x4`, and `x5`. The model is overidentified because there is one more instrument than there are endogenous regressors.

In terms of the model of section 6.2.2, y_1 is *depvar*, $\mathbf{x}_1$ is *varlist1*, $\mathbf{y}_2$ is *varlist2*, and $\mathbf{x}_2$ is *varlist_iv*. In the just-identified case, *varlist2* and *varlist_iv* have the same number of variables, and we use the `2sls` option to obtain the IV estimator. In the overidentified case, *varlist_iv* has more variables than does *varlist2*.

The `first` option yields considerable output from the first-stage regression. Several useful tests regarding the instruments and the goodness of fit of the first-stage regression are displayed; therefore, this option is more convenient than the user running the first-stage regression and conducting tests.

The `vce(`*vcetype*`)` option specifies the type of standard errors reported by Stata. The options for *vcetype* are `robust`, which yields heteroskedasticity-robust standard errors; `unadjusted`, which yields nonrobust standard errors; `cluster` *clustvar*; `bootstrap`; `jackknife`; and `hac` *kernel*. Various specification test statistics that are automatically produced by Stata are made more robust if the `vce(robust)` option is used.

For the overidentified models fit by GMM, the `wmatrix(`*wmtype*`)` option specifies the type of weighting matrix used in the objective function [see **W** in (6.7)] to obtain optimal GMM. Different choices of *wmtype* lead to different estimators. For heteroskedastic errors, set *wmtype* to `robust`. For correlation between elements of a cluster, set *wmtype* to `cluster` *clustvar*, where *clustvar* specifies the variable that identifies the cluster. For time-series data with heteroskedasticity- and autocorrelation-consistent (HAC) errors, set *wmtype* to `hac` *kernel* or `hac` *kernel #* or `hac` *kernel* `opt`; see [R] **ivregress** for additional details. If `vce()` is not specified when `wmatrix()` is specified, then *vcetype* is set to *wmtype*. The `igmm` option yields an iterated version of the GMM estimator.

6.3.2   Medical expenditures with one endogenous regressor

We consider a model with one endogenous regressor, several exogenous regressors, and one or more excluded exogenous variables that serve as the identifying instruments.

The dataset is an extract from the Medical Expenditure Panel Survey (MEPS) of individuals over the age of 65 years, similar to the dataset described in section 3.2.1. The equation to be estimated has the dependent variable `ldrugexp`, the log of total out-of-pocket expenditures on prescribed medications. The regressors are an indicator for whether the individual holds either employer or union-sponsored health insurance (`hi_empunion`), number of chronic conditions (`totchr`), and four sociodemographic variables: age in years (`age`), indicators for whether female (`female`) and whether black or Hispanic (`blhisp`), and the natural logarithm of annual household income in thousands of dollars (`linc`).

We treat the health insurance variable `hi_empunion` as endogenous. The intuitive justification is that having such supplementary insurance on top of the near universal Medicare insurance for the elderly may be a choice variable. Even though most individuals in the sample are no longer working, those who expected high future medical expenses might have been more likely to choose a job when they were working that would provide supplementary health insurance upon retirement. Note that Medicare did not cover drug expenses for the time period we study.

We use the global macro `x2list` to store the names of the variables that are treated as exogenous regressors. We have

```
. * Read data, define global x2list, and summarize data
. use mus06data.dta

. global x2list totchr age female blhisp linc

. summarize ldrugexp hi_empunion $x2list
```

Variable	Obs	Mean	Std. Dev.	Min	Max
ldrugexp	10391	6.479668	1.363395	0	10.18017
hi_empunion	10391	.3796555	.4853245	0	1
totchr	10391	1.860745	1.290131	0	9
age	10391	75.04639	6.69368	65	91
female	10391	.5797325	.4936256	0	1
blhisp	10391	.1703397	.3759491	0	1
linc	10089	2.743275	.9131433	-6.907755	5.744476

Around 38% of the sample has either employer or union-sponsored health insurance in addition to Medicare insurance. Subsequent analysis drops those observations with missing data on `linc`.

6.3.3 Available instruments

We consider four potential instruments for `hi_empunion`. Two reflect the income status of the individual and two are based on employer characteristics.

The `ssiratio` instrument is the ratio of an individual's social security income to the individual's income from all sources, with high values indicating a significant income constraint. The `lowincome` instrument is a qualitative indicator of low-income status. Both these instruments are likely to be relevant, because they are expected to be negatively correlated with having supplementary insurance. To be valid instruments, we need to assume they can be omitted from the equation for `ldrugexp`, arguing that the direct role of income is adequately captured by the regressor `linc`.

The `firmsz` instrument measures the size of the firm's employed labor force, and the `multlc` instrument indicates whether the firm is a large operator with multiple locations. These variables are intended to capture whether the individual has access to supplementary insurance through the employer. These two variables are irrelevant for those who are retired, self-employed, or purchase insurance privately. In that sense, these two instruments could potentially be weak.

```
. * Summarize available instruments
. summarize ssiratio lowincome multlc firmsz if linc!=.

    Variable |       Obs        Mean    Std. Dev.       Min        Max
-------------+--------------------------------------------------------
    ssiratio |     10089    .5365438    .3678175          0    9.25062
   lowincome |     10089    .1874319    .3902771          0          1
      multlc |     10089    .0620478    .2412543          0          1
      firmsz |     10089    .1405293    2.170389          0         50
```

We have four available instruments for one endogenous regressor. The obvious approach is to use all available instruments, because in theory this leads to the most efficient estimator. In practice, it may lead to larger small-sample bias because the small-sample biases of IV estimators increase with the number of instruments (Hahn and Hausman 2002).

At a minimum, it is informative to use `correlate` to view the gross correlation between endogenous variables and instruments and between instruments. When multiple instruments are available, as in the case of overidentified models, then it is actually the partial correlation after controlling for other available instruments that matters. This important step is deferred to sections 6.4.2 and 6.4.3.

6.3.4 IV estimation of an exactly identified model

We begin with IV regression of ldrugexp on the endogenous regressor hi_empunion, instrumented by the single instrument ssiratio, and several exogenous regressors.

We use ivregress with the 2sls estimator and the options vce(robust) to control for heteroskedastic errors and first to provide output that additionally reports results from the first-stage regression. The output is in two parts:

```
. * IV estimation of a just-identified model with single endog regressor
. ivregress 2sls ldrugexp (hi_empunion = ssiratio) $x2list, vce(robust) first
First-stage regressions
```

```
                                   Number of obs   =      10089
                                   F(   6,  10082) =     119.18
                                   Prob > F        =     0.0000
                                   R-squared       =     0.0761
                                   Adj R-squared   =     0.0755
                                   Root MSE        =     0.4672
```

hi_empunion	Coef.	Robust Std. Err.	t	P>|t|	[95% Conf. Interval]	
totchr	.0127865	.0036655	3.49	0.000	.0056015	.0199716
age	-.0086323	.0007087	-12.18	0.000	-.0100216	-.0072431
female	-.07345	.0096392	-7.62	0.000	-.0923448	-.0545552
blhisp	-.06268	.0122742	-5.11	0.000	-.08674	-.0386201
linc	.0483937	.0066075	7.32	0.000	.0354417	.0613456
ssiratio	-.1916432	.0236326	-8.11	0.000	-.2379678	-.1453186
_cons	1.028981	.0581387	17.70	0.000	.9150172	1.142944

```
Instrumental variables (2SLS) regression    Number of obs =      10089
                                             Wald chi2(6)  =    2000.86
                                             Prob > chi2   =     0.0000
                                             R-squared     =     0.0640
                                             Root MSE      =     1.3177
```

ldrugexp	Coef.	Robust Std. Err.	z	P>|z|	[95% Conf. Interval]	
hi_empunion	-.8975913	.2211268	-4.06	0.000	-1.330992	-.4641908
totchr	.4502655	.0101969	44.16	0.000	.43028	.470251
age	-.0132176	.0029977	-4.41	0.000	-.0190931	-.0073421
female	-.020406	.0326114	-0.63	0.531	-.0843232	.0435113
blhisp	-.2174244	.0394944	-5.51	0.000	-.294832	-.1400167
linc	.0870018	.0226356	3.84	0.000	.0426368	.1313668
_cons	6.78717	.2688453	25.25	0.000	6.260243	7.314097

```
Instrumented:  hi_empunion
Instruments:   totchr age female blhisp linc ssiratio
```

The first part, added because of the `first` option, reports results from the first-stage regression of the endogenous variable `hi_empunion` on all the exogenous variables, here `ssiratio` and all the exogenous regressors in the structural equation. The first-stage regression has reasonable explanatory power, and the coefficient of `ssiratio` is negative, as expected, and highly statistically significant. In models with more than one endogenous regressor, more than one first-stage regression is reported if the `first` option is used.

The second part reports the results of intrinsic interest, those from the IV regression of `ldrugexp` on `hi_empunion` and several exogenous regressors. Supplementary insurance has a big effect. The estimated coefficient of `hi_empunion` is -0.898, indicating that supplementary-insured individuals have out-of-pocket drug expenses that are 90% lower than those for people without employment or union-related supplementary insurance.

6.3.5   IV estimation of an overidentified model

We next consider estimation of an overidentified model. Then different estimates are obtained by 2SLS estimation and by different variants of GMM.

We use two instruments, `ssiratio` and `multlc`, for `hi_empunion`, the endogenous regressor. The first estimator is 2SLS; obtained by using `2sls` with standard errors that correct for heteroskedasticity with the `vce(robust)` option. The second estimator is optimal GMM given heteroskedastic errors; obtained by using `gmm` with the `wmatrix(robust)` option. These are the two leading estimators for overidentified IV with cross-section data and no clustering of the errors. The third estimator adds `igmm` to iterate to convergence. The fourth estimator is one that illustrates optimal GMM with clustered errors by clustering on `age`. The final estimator is the same as the first but reports default standard errors that do not adjust for heteroskedasticity.

```
. * Compare 5 estimators and variance estimates for overidentified models
. global ivmodel "ldrugexp (hi_empunion = ssiratio multlc) $x2list"
. quietly ivregress 2sls $ivmodel, vce(robust)
. estimates store TwoSLS
. quietly ivregress gmm  $ivmodel, wmatrix(robust)
. estimates store GMM_het
. quietly ivregress gmm  $ivmodel, wmatrix(robust) igmm
. estimates store GMM_igmm
. quietly ivregress gmm  $ivmodel, wmatrix(cluster age)
. estimates store GMM_clu
. quietly ivregress 2sls  $ivmodel
. estimates store TwoSLS_def
```

```
. estimates table TwoSLS GMM_het GMM_igmm GMM_clu TwoSLS_def, b(%9.5f) se
```

Variable	TwoSLS	GMM_het	GMM_igmm	GMM_clu	TwoSLS_~f
hi_empunion	-0.98993	-0.99328	-0.99329	-1.03587	-0.98993
	0.20459	0.20467	0.20467	0.20438	0.19221
totchr	0.45121	0.45095	0.45095	0.44822	0.45121
	0.01031	0.01031	0.01031	0.01325	0.01051
age	-0.01414	-0.01415	-0.01415	-0.01185	-0.01414
	0.00290	0.00290	0.00290	0.00626	0.00278
female	-0.02784	-0.02817	-0.02817	-0.02451	-0.02784
	0.03217	0.03219	0.03219	0.02919	0.03117
blhisp	-0.22371	-0.22310	-0.22311	-0.20907	-0.22371
	0.03958	0.03960	0.03960	0.05018	0.03870
linc	0.09427	0.09446	0.09446	0.09573	0.09427
	0.02188	0.02190	0.02190	0.01474	0.02123
_cons	6.87519	6.87782	6.87783	6.72769	6.87519
	0.25789	0.25800	0.25800	0.50588	0.24528

legend: b/se

Compared with the just-identified IV estimates of section 6.3.4, the parameter estimates have changed by less than 10% (aside from those for the statistically insignificant regressor `female`). The standard errors are little changed except for that for `hi_empunion`, which has fallen by about 7%, reflecting efficiency gain due to additional instruments.

Next we compare the five different overidentified estimators. The differences between 2SLS, optimal GMM given heteroskedasticity, and iterated optimal GMM are negligible. Optimal GMM with clustering differs more. And the final column shows that the default standard errors for 2SLS differ little from the robust standard errors in the first column, reflecting the success of the log transformation in eliminating heteroskedasticity.

6.3.6 Testing for regressor endogeneity

The preceding analysis treats the insurance variable, `hi_empunion`, as endogenous. If instead the variable is exogenous, then the IV estimators (IV, 2SLS, or GMM) are still consistent, but they can be much less efficient than the OLS estimator.

The Hausman test principle provides a way to test whether a regressor is endogenous. If there is little difference between OLS and IV estimators, then there is no need to instrument, and we conclude that the regressor was exogenous. If instead there is considerable difference, then we needed to instrument and the regressor is endogenous. The test usually compares just the coefficients of the endogenous variables. In the case of just one potentially endogenous regressor with a coefficient denoted by β, the Hausman test statistic

$$T_H = \frac{(\widehat{\beta}_{\mathrm{IV}} - \widehat{\beta}_{\mathrm{OLS}})^2}{\widehat{V}(\widehat{\beta}_{\mathrm{IV}} - \widehat{\beta}_{\mathrm{OLS}})}$$

is $\chi^2(1)$ distributed under the null hypothesis that the regressor is exogenous.

Before considering implementation of the test, we first obtain the OLS estimates to compare them with the earlier IV estimates. We have

```
. * Obtain OLS estimates to compare with preceding IV estimates
. regress ldrugexp hi_empunion $x2list, vce(robust)
Linear regression                               Number of obs =    10089
                                                F(  6, 10082) =   376.85
                                                Prob > F      =   0.0000
                                                R-squared     =   0.1770
                                                Root MSE      =    1.236
```

		Robust				
ldrugexp	Coef.	Std. Err.	t	P>|t|	[95% Conf.	Interval]
hi_empunion	.0738788	.0259848	2.84	0.004	.0229435	.1248141
totchr	.4403807	.0093633	47.03	0.000	.4220268	.4587346
age	-.0035295	.001937	-1.82	0.068	-.0073264	.0002675
female	.0578055	.0253651	2.28	0.023	.0080848	.1075262
blhisp	-.1513068	.0341264	-4.43	0.000	-.2182013	-.0844122
linc	.0104815	.0137126	0.76	0.445	-.0163979	.037361
_cons	5.861131	.1571037	37.31	0.000	5.553176	6.169085

The OLS estimates differ substantially from the just-identified IV estimates given in section 6.3.4. The coefficient of hi_empunion has an OLS estimate of 0.074, greatly different from the IV estimate of -0.898. This is strong evidence that hi_empunion is endogenous. Some coefficients of exogenous variables also change, notably, those for age and female. Note also the loss in precision in using IV. Most notably, the standard error of the instrumented regressor increases from 0.026 for OLS to 0.221 for IV, an eightfold increase, indicating the potential loss in efficiency due to IV estimation.

The hausman command can be used to compute T_H under the assumption that $\widehat{V}(\widehat{\beta}_{IV} - \widehat{\beta}_{OLS}) = \widehat{V}(\widehat{\beta}_{IV}) - \widehat{V}(\widehat{\beta}_{OLS})$; see section 12.7.5. This greatly simplifies analysis because then all that is needed are coefficient estimates and standard errors from separate IV estimation (IV, 2SLS, or GMM) and OLS estimation. But this assumption is too strong. It is correct only if $\widehat{\beta}_{OLS}$ is the fully efficient estimator under the null hypothesis of exogeneity, an assumption that is valid only under the very strong assumption that model errors are independent and homoskedastic. One possible variation is to perform an appropriate bootstrap; see section 13.4.6.

The postestimation estat endogenous command implements the related Durbin–Wu–Hausman (DWH) test. Because the DWH test uses the device of augmented regressors, it produces a robust test statistic (Davidson 2000). The essential idea is the following. Consider the model as specified in section 6.2.1. Rewrite the structural equation (6.2) with an additional variable, v_1, that is the error from the first-stage equation (6.3) for y_2. Then

$$y_{1i} = \beta_1 y_{2i} + \mathbf{x}'_{1i}\boldsymbol{\beta}_2 + \rho v_{1i} + u_i$$

Under the null hypothesis that y_{2i} is exogenous, $E(v_{1i}u_i|y_{2i}, \mathbf{x}_{1i}) = 0$. If v_1 could be observed, then the test of exogeneity would be the test of $H_0: \rho = 0$ in the OLS regression of y_1 on y_2, $\mathbf{x}_1$, and v_1. Because v_1 is not directly observed, the fitted residual vector

$\widehat{v}_1$ from the first-stage OLS regression (6.3) is instead substituted. For independent homoskedastic errors, this test is asymptotically equivalent to the earlier Hausman test. In the more realistic case of heteroskedastic errors, the test of $H_0 : \rho = 0$ can still be implemented provided that we use robust variance estimates. This test can be extended to the multiple endogenous regressors case by including multiple residual vectors and testing separately for correlation of each with the error on the structural equation.

We apply the test to our example with one potentially endogenous regressor, hi_empunion, instrumented by ssiratio. Then

```
. * Robust Durbin-Wu-Hausman test of endogeneity implemented by estat endogenous
. ivregress 2sls ldrugexp (hi_empunion = ssiratio) $x2list, vce(robust)
Instrumental variables (2SLS) regression           Number of obs =    10089
                                                   Wald chi2(6)  = 2000.86
                                                   Prob > chi2   =  0.0000
                                                   R-squared     =  0.0640
                                                   Root MSE      =  1.3177
```

ldrugexp	Coef.	Robust Std. Err.	z	P>|z|	[95% Conf. Interval]	
hi_empunion	-.8975913	.2211268	-4.06	0.000	-1.330992	-.4641908
totchr	.4502655	.0101969	44.16	0.000	.43028	.470251
age	-.0132176	.0029977	-4.41	0.000	-.0190931	-.0073421
female	-.020406	.0326114	-0.63	0.531	-.0843232	.0435113
blhisp	-.2174244	.0394944	-5.51	0.000	-.294832	-.1400167
linc	.0870018	.0226356	3.84	0.000	.0426368	.1313668
_cons	6.78717	.2688453	25.25	0.000	6.260243	7.314097

```
Instrumented:  hi_empunion
Instruments:   totchr age female blhisp linc ssiratio

. estat endogenous

  Tests of endogeneity
  Ho: variables are exogenous

  Robust score chi2(1)          =   24.935   (p = 0.0000)
  Robust regression F(1,10081)  =   26.4333  (p = 0.0000)
```

The last line of output is the robustified DWH test and leads to strong rejection of the null hypothesis that hi_empunion is exogenous. We conclude that it is endogenous.

We obtain exactly the same test statistic when we manually perform the robustified DWH test. We have

```
. * Robust Durbin-Wu-Hausman test of endogeneity implemented manually
. quietly regress hi_empunion ssiratio $x2list
. quietly predict v1hat, resid
. quietly regress ldrugexp hi_empunion v1hat $x2list, vce(robust)
. test v1hat

 ( 1)  v1hat = 0

       F(  1, 10081) =   26.43
            Prob > F =    0.0000
```

6.3.7   Tests of overidentifying restrictions

The validity of an instrument cannot be tested in a just-identified model. But it is possible to test the validity of overidentifying instruments in an overidentified model provided that the parameters of the model are estimated using optimal GMM. The same test has several names, including overidentifying restrictions (OIR) test, overidentified (OID) test, Hansen's test, Sargan's test, and Hansen–Sargan test.

The starting point is the fitted value of the criterion function (6.8) after optimal GMM, i.e., $Q(\widehat{\boldsymbol{\beta}}) = \{(1/N)(\mathbf{y} - \mathbf{X}\widehat{\boldsymbol{\beta}})'\mathbf{Z}\}\widehat{\mathbf{S}}^{-1}\{(1/N)\mathbf{Z}'(\mathbf{y} - \mathbf{X}\widehat{\boldsymbol{\beta}})\}$. If the population moment conditions $E\{\mathbf{Z}'(\mathbf{y} - \mathbf{X}\boldsymbol{\beta})\} = \mathbf{0}$ are correct, then $\mathbf{Z}'(\mathbf{y} - \mathbf{X}\widehat{\boldsymbol{\beta}}) \simeq \mathbf{0}$, so $Q(\widehat{\boldsymbol{\beta}})$ should be close to zero. Under the null hypothesis that all instruments are valid, it can be shown that $Q(\widehat{\boldsymbol{\beta}})$ has an asymptotic chi-squared distribution with degrees of freedom equal to the number of overidentifying restrictions.

Large values of $Q(\widehat{\boldsymbol{\beta}})$ lead to rejection of $H_0 \colon E\{\mathbf{Z}'(\mathbf{y} - \mathbf{X}\boldsymbol{\beta})\} = \mathbf{0}$. Rejection is interpreted as indicating that at least one of the instruments is not valid. Tests can have power in other directions, however, as emphasized in section 3.5.5. It is possible that rejection of H_0 indicates that the model $\mathbf{X}\boldsymbol{\beta}$ for the conditional mean is misspecified. Going the other way, the test is only one of validity of the overidentifying instruments, so failure to reject H_0 does not guarantee that all the instruments are valid.

The test is implemented with the postestimation `estat overid` command following the `ivregress gmm` command for an overidentified model. We do so for the optimal GMM estimator with heteroskedastic errors and instruments, `ssiratio` and `multc`. The example below implements `estat overid` under the overidentifying restriction.

```
. * Test of overidentifying restrictions following ivregress gmm
. quietly ivregress gmm ldrugexp (hi_empunion = ssiratio multlc)
> $x2list, wmatrix(robust)

. estat overid

  Test of overidentifying restriction:

  Hansen´s J chi2(1) = 1.04754 (p = 0.3061)
```

The test statistic is $\chi^2(1)$ distributed because the number of overidentifying restrictions equals $2 - 1 = 1$. Because $p > 0.05$, we do not reject the null hypothesis and conclude that the overidentifying restriction is valid.

(Continued on next page)

A similar test using all four available instruments yields

```
. * Test of overidentifying restrictions following ivregress gmm
. ivregress gmm ldrugexp (hi_empunion = ssiratio lowincome multlc firmsz)
> $x2list, wmatrix(robust)
Instrumental variables (GMM) regression          Number of obs =     10089
                                                  Wald chi2(6)  =   2042.12
                                                  Prob > chi2   =    0.0000
                                                  R-squared     =    0.0829
GMM weight matrix: Robust                         Root MSE      =    1.3043
```

ldrugexp	Coef.	Robust Std. Err.	z	P>|z|	[95% Conf. Interval]	
hi_empunion	-.8124043	.1846433	-4.40	0.000	-1.174299	-.45051
totchr	.449488	.010047	44.74	0.000	.4297962	.4691799
age	-.0124598	.0027466	-4.54	0.000	-.0178432	-.0070765
female	-.0104528	.0306889	-0.34	0.733	-.0706019	.0496963
blhisp	-.2061018	.0382891	-5.38	0.000	-.2811471	-.1310566
linc	.0796532	.0203397	3.92	0.000	.0397882	.1195183
_cons	6.7126	.2425973	27.67	0.000	6.237118	7.188081

```
Instrumented:  hi_empunion
Instruments:   totchr age female blhisp linc ssiratio lowincome multlc
               firmsz

. estat overid

  Test of overidentifying restriction:

  Hansen's J chi2(3) = 11.5903 (p = 0.0089)
```

Now we reject the null hypothesis at level 0.05 and, barely, at level 0.01. Despite this rejection, the coefficient of the endogenous regressor `hi_empunion` is -0.812, not all that different from the estimate when `ssiratio` is the only instrument.

6.3.8   IV estimation with a binary endogenous regressor

In our example, the endogenous regressor `hi_empunion` is a binary variable. The IV methods we have used are valid under the assumption that $E(u_i|\mathbf{z}_i) = 0$, which in our example means that the error in the structural equation for `ldrugexp` has a mean of zero conditional on the exogenous regressors [$\mathbf{x}_1$ in (6.2)] and any instruments such as `ssiratio`. The reasonableness of this assumption does not change when the endogenous regressor `hi_empunion` is binary.

An alternative approach adds more structure to explicitly account for the binary nature of the endogenous regressor by changing the first-stage model to be a latent-variable model similar to the probit model presented in chapter 14. Let y_1 depend in part on y_2, a binary endogenous regressor. We introduce an unobserved latent variable, y_2^*, that determines whether $y_2 = 1$ or 0. The models (6.2) and (6.3) become

$$y_{1i} = \beta_1 y_{2i} + \mathbf{x}'_{1i}\boldsymbol{\beta}_2 + u_i \qquad (6.9)$$
$$y_{2i}^* = \mathbf{x}'_{1i}\boldsymbol{\pi}_{1j} + \mathbf{x}'_{2i}\boldsymbol{\pi}_{2j} + v_i$$
$$y_{2i} = \begin{cases} 1 & \text{if } y_{2i}^* > 0 \\ 0 & \text{otherwise} \end{cases}$$

The errors (u_i, v_i) are assumed to be correlated bivariate normal with $\text{Var}(u_i) = \sigma^2$, $\text{Var}(v_i) = 1$, and $\text{Cov}(u_i, v_i) = \rho\sigma^2$.

The binary endogenous regressor y_2 can be viewed as a treatment indicator. If $y_2 = 1$, we receive treatment (here access to employer- or union-provided insurance), and if $y_2 = 0$, we do not receive treatment. The Stata documentation refers to (6.9) as the treatment-effects model, though the treatment-effects literature is vast and encompasses many models and methods.

The `treatreg` command fits (6.9) by maximum likelihood (ML), the default, or two-step methods. The basic syntax is

`treatreg` *depvar* $\big[$ *indepvars* $\big]$ `, treat(`*depvar_t* `=` *indepvars_t*`)` $\big[$ `twostep` $\big]$

where *depvar* is y_1, *indepvars* is $\mathbf{x}_1$, *depvar_t* is y_2^*, and *indepvars_t* is $\mathbf{x}_1$ and $\mathbf{x}_2$.

We apply this estimator to the exactly identified setup of section 6.3.4, with the single instrument `ssiratio`. We obtain

```
. * Regression with a dummy variable regressor
. treatreg ldrugexp $x2list, treat(hi_empunion = ssiratio $x2list)
  (output omitted )
Treatment-effects model -- MLE              Number of obs   =       10089
                                            Wald chi2(6)    =     1931.55
Log likelihood = -22721.082                 Prob > chi2     =      0.0000
```

	Coef.	Std. Err.	z	P>|z|	[95% Conf.	Interval]
ldrugexp						
totchr	.4555085	.0110291	41.30	0.000	.4338919	.4771252
age	-.0183563	.0022975	-7.99	0.000	-.0228594	-.0138531
female	-.0618901	.0295655	-2.09	0.036	-.1198374	-.0039427
blhisp	-.2524937	.0391998	-6.44	0.000	-.3293239	-.1756635
linc	.1275888	.0171264	7.45	0.000	.0940217	.1611559
hi_empunion	-1.412868	.0821001	-17.21	0.000	-1.573781	-1.251954
_cons	7.27835	.1905198	38.20	0.000	6.904938	7.651762
hi_empunion						
ssiratio	-.4718775	.0344656	-13.69	0.000	-.5394288	-.4043262
totchr	.0385586	.0099715	3.87	0.000	.0190148	.0581023
age	-.0243318	.0019918	-12.22	0.000	-.0282355	-.020428
female	-.1942343	.0260033	-7.47	0.000	-.2451998	-.1432688
blhisp	-.1950778	.0359513	-5.43	0.000	-.265541	-.1246146
linc	.1346908	.0150101	8.97	0.000	.1052715	.16411
_cons	1.462713	.1597052	9.16	0.000	1.149696	1.775729
/athrho	.7781623	.044122	17.64	0.000	.6916848	.8646399
/lnsigma	.3509918	.0151708	23.14	0.000	.3212577	.380726
rho	.6516507	.0253856			.5990633	.6986405
sigma	1.420476	.0215497			1.378861	1.463347
lambda	.925654	.048921			.8297705	1.021537

```
LR test of indep. eqns. (rho = 0):   chi2(1) =    86.80   Prob > chi2 = 0.0000
```

The key output is the first set of regression coefficients. Compared with IV estimates in section 6.3.4, the coefficient of `hi_empunion` has increased in absolute value from -0.898 to -1.413, and the standard error has fallen greatly from 0.221 to 0.082. The coefficients and standard errors of the exogenous regressors change much less.

The quantities `rho`, `sigma`, and `lambda` denote, respectively, ρ, σ, and $\rho\sigma$. To ensure that $\hat{\sigma} > 0$ and $|\hat{\rho}| < 1$, `treatreg` estimates the transformed parameters $0.5 \times \ln\{(1 + \rho)/(1-\rho)\}$, reported as `/athrho`, and $\ln\sigma$, reported as `/lnsigma`. If the error correlation $\rho = 0$, then the errors u and v are independent and there is no endogeneity problem. The last line of output clearly rejects $H_0 : \rho = 0$, so `hi_empunion` is indeed an endogenous regressor.

Which method is better: regular IV or (6.9)? Intuitively, (6.9) imposes more structure. The benefit may be increased precision of estimation, as in this example. The cost is a greater chance of misspecification error. If the errors are heteroskedastic, as is likely, the IV estimator remains consistent but the treatment-effects estimator given here becomes inconsistent.

More generally, when regressors in nonlinear models, such as binary-data models and count-data models, include endogenous regressors, there is more than one approach to model estimation; see also section 17.5.

6.4   Weak instruments

In this section, we assume that the chosen instrument is valid, so IV estimators are consistent.

Instead, our concern is with whether the instrument is weak, because then asymptotic theory can provide a poor guide to actual finite-sample distributions.

Several diagnostics and tests are provided by the `estat firststage` command following `ivregress`. These are not exhaustive; other tests have been proposed and are currently being developed. The user-written `ivreg2` command (Baum, Schaffer, and Stillman 2007) provides similar information via a one-line command and stores many of the resulting statistics in `e()`. We focus on `ivregress` because it is fully supported by Stata.

6.4.1   Finite-sample properties of IV estimators

Even when IV estimators are consistent, they are biased in finite samples. This result has been formally established in overidentified models. In just-identified models, the first moment of the IV estimator does not exist, but for simplicity, we follow the literature and continue to use the term "bias" in this case.

The finite-sample properties of IV estimators are complicated. However, there are three cases in which it is possible to say something about the finite-sample bias; see Davidson and MacKinnon (2004, chap. 8.4).

First, when the number of instruments is very large relative to the sample size and the first-stage regression fits very well, the IV estimator may approach the OLS estimator and hence will be similarly biased. This case of many instruments is not very relevant for cross-section microeconometric data, though it can be relevant for panel-data IV estimators such as Arellano–Bond. Second, when the correlation between the structural-equation error u and some components of the vector $\mathbf{v}$ of first-stage–equation errors is high, then asymptotic theory may be a poor guide to the finite-sample distribution. Third, if we have weak instruments in the sense that one or more of the first-stage regressions have a poor fit, then asymptotic theory may provide a poor guide to the finite-sample distribution of the IV estimator, even if the sample has thousands of observations.

In what follows, our main focus will be on the third case, that of weak instruments. More precise definitions of weak instruments are considered in the next section.

6.4.2 Weak instruments

There are several approaches for investigating the weak IV problem, based on analysis of the first-stage reduced-form equation(s) and, particularly, the F statistic for the joint significance of the key instruments.

Diagnostics for weak instruments

The simplest method is to use the pairwise correlations between any endogenous regressor and instruments. For our example, we have

```
. * Correlations of endogenous regressor with instruments
. correlate hi_empunion ssiratio lowincome multlc firmsz if linc!=.
(obs=10089)
```

	hi_emp~n	ssiratio	lowinc~e	multlc	firmsz
hi_empunion	1.0000				
ssiratio	-0.2124	1.0000			
lowincome	-0.1164	0.2539	1.0000		
multlc	0.1198	-0.1904	-0.0625	1.0000	
firmsz	0.0374	-0.0446	-0.0082	0.1873	1.0000

The gross correlations of instruments with the endogenous regressor hi_empunion are low. This will lead to considerable efficiency loss using IV compared to OLS. But the correlations are not so low as to immediately flag a problem of weak instruments.

For IV estimation that uses more than one instrument, we can consider the joint correlation of the endogenous regressor with the several instruments. Possible measures of this correlation are R^2 from regression of the endogenous regressor y_2 on the several instruments $\mathbf{x}_2$, and the F statistic for test of overall fit in this regression. Low values of R^2 or F are indicative of weak instruments. However, this neglects the presence of the structural-model exogenous regressors $\mathbf{x}_1$ in the first-stage regression (6.3) of y_2 on both $\mathbf{x}_2$ and $\mathbf{x}_1$. If the instruments $\mathbf{x}_2$ add little extra to explaining y_1 after controlling for $\mathbf{x}_1$, then the instruments are weak.

One commonly used diagnostic is, therefore, the F statistic for joint significance of the instruments $\mathbf{x}_2$ in first-stage regression of the endogenous regressor y_2 on $\mathbf{x}_2$ and $\mathbf{x}_1$. This is a test that $\boldsymbol{\pi}_2 = \mathbf{0}$ in (6.3). A widely used rule of thumb suggested by Staiger and Stock (1997) views an F statistic of less than 10 as indicating weak instruments. This rule of thumb is ad hoc and may not be sufficiently conservative when there are many overidentifying restrictions. There is no clear critical value for the F statistic because it depends on the criteria used, the number of endogenous variables, and the number of overidentifying restrictions (excess instruments). Stock and Yogo (2005) proposed two test approaches, under the assumption of homoskedastic errors, that lead to critical values for the F statistic, which are provided in the output from `estat firststage`, discussed next. The first approach, applicable only if there are at least two overidentifying restrictions, suggests that the rule of thumb is reasonable. The second approach can lead to F statistic critical values that are much greater than 10 in models that are overidentified.

A second diagnostic is the partial R^2 between y_2 and $\mathbf{x}_2$ after controlling for $\mathbf{x}_1$. This is the R^2 from OLS regression of 1) the residuals from OLS of y_2 on $\mathbf{x}_1$ on 2) the residuals from OLS of $\mathbf{x}_2$ on $\mathbf{x}_1$. There is no consensus on how low of a value indicates a problem. For structural equations with more than one endogenous regressor and hence more than one first-stage regression, a generalization called Shea's partial R^2 is used.

Formal tests for weak instruments

Stock and Yogo (2005) proposed two tests of weak instruments. Both use the same test statistic, but they use different critical values based on different criteria. The test statistic is the aforementioned F statistic for joint significance of instruments in the first-stage regression, in the common special case of just one endogenous regressor in the original structural model. With more than one endogenous regressor in the structural model, however, there will be more than one first-stage regression and more than one F statistic. Then the test statistic used is the minimum eigenvalue of a matrix analog of the F statistic that is defined in Stock and Yogo (2005, 84) or in [R] **ivregress postestimation**. This statistic was originally proposed by Cragg and Donald (1993) to test nonidentification. Stock and Yogo presume identification and interpret a low minimum eigenvalue (equals the F statistic if there is just one endogenous regressor) to mean the instruments are weak. So the null hypothesis is that the instruments are weak against the alternative that they are strong. Critical values are obtained by using two criteria we now elaborate.

The first criterion addresses the concern that the estimation bias of the IV estimator resulting from the use of weak instruments can be large, sometimes even exceeding the bias of OLS. To apply the test, one first chooses b, the largest relative bias of the 2SLS estimator relative to OLS, that is acceptable. Stock and Yogo's tables provide the test critical value that varies with b and with the number of endogenous regressors (m) and the number of exclusion restrictions (K_2). For example, if $b = 0.05$ (only a 5% relative bias toleration), $m = 1$, and $K_2 = 3$, then they compute the critical value of the test to be 13.91, so we reject the null hypothesis of weak instruments if the F statistic (which equals the minimum eigenvalue when $m = 1$) exceeds 13.91. For a larger 10% relative-bias toleration, the critical value decreases to 9.08. Unfortunately, critical values are only available when the model has at least two overidentifying restrictions. So with one endogenous regressor, we need at least three instruments.

The second test, which can be applied to both just-identified and overidentified models, addresses the concern that weak instruments can lead to size distortion of Wald tests on the parameters in finite samples. The Wald test is a joint statistical significance of the endogenous regressors in the structural model [$\boldsymbol{\beta}_1 = \mathbf{0}$ in (6.2)] at a level of 0.05. The practitioner chooses a tolerance for the size distortion of this test. For example, if we will not tolerate an actual test size greater than $r = 0.10$, then with $m = 1$ and $K_2 = 3$, the critical value of the F test from the Stock–Yogo tables is 22.30. If, instead, $r = 0.15$, then the critical value is 12.83.

The test statistic and critical values are printed following the `ivregress` postestimation `estat firststage` command. The critical values are considerably larger than the values used for a standard F test of the joint significance of a set of regressors. We focus on the 2SLS estimator, though critical values for the LIML estimator are also given.

6.4.3 The estat firststage command

Following `ivregress`, various diagnostics and tests for weak instruments are provided by `estat firststage`. The syntax for this command is

```
estat firststage [ , forcenonrobust all]
```

The `forcenonrobust` option is used to allow use of `estat firststage` even when the preceding `ivregress` command used the `vce(robust)` option. The reason is that the underlying theory for the tests in `estat firststage` assumes that regression errors are Gaussian and independent and identically distributed (i.i.d.). By using the `forcenonrobust` option, we are acknowledging that we know this, but we are nonetheless willing to use the tests even if, say, heteroskedasticity is present.

If there is more than one endogenous regressor, the `all` option provides separate sets of results for each endogenous regressor in addition to the key joint statistics for the endogenous regressors that are automatically provided. It also produces Shea's partial R^2.

6.4.4 Just-identified model

We consider a just-identified model with one endogenous regressor, with `hi_empunion` instrumented by one variable, `ssiratio`. Because we use `vce(robust)` in `ivregress`, we need to add the `forcenonrobust` option. We add the `all` option to print Shea's partial R^2, which is unnecessary here because we have only one endogenous regressor. The output is in three parts.

```
. * Weak instrument tests - just-identified model
. quietly ivregress 2sls ldrugexp (hi_empunion = ssiratio) $x2list, vce(robust)

. estat firststage, forcenonrobust all

First-stage regression summary statistics
```

Variable	R-sq.	Adjusted R-sq.	Partial R-sq.	Robust F(1,10082)	Prob > F
hi_empunion	0.0761	0.0755	0.0179	65.7602	0.0000

```
Shea's partial R-squared
```

Variable	Shea's Partial R-sq.	Shea's Adj. Partial R-sq.
hi_empunion	0.0179	0.0174

```
Minimum eigenvalue statistic = 183.98

Critical Values                  # of endogenous regressors:    1
Ho: Instruments are weak         # of excluded instruments:     1
```

	5%	10%	20%	30%
2SLS relative bias		(not available)		

	10%	15%	20%	25%
2SLS Size of nominal 5% Wald test	16.38	8.96	6.66	5.53
LIML Size of nominal 5% Wald test	16.38	8.96	6.66	5.53

The first part is a summary table of key diagnostic statistics that are useful in suggesting weak instruments. The first two statistics are the R^2 and adjusted-R^2 from the first-stage regression. These are around 0.08, so there will be considerable loss of precision because of IV estimation. They are not low enough to flag a weak-instruments problem, although, as already noted, there may still be a problem because ssiratio may be contributing very little to this fit. To isolate the explanatory power of ssiratio in explaining hi_empunion, two statistics are given. The partial R^2 is that between hi_empunion and ssiratio after controlling for totchr, age, female, blhisp, and linc. This is quite low at 0.0179, suggesting some need for caution. The final statistic is an F statistic for the joint significance of the instruments excluded from the structural model. Here this is a test on just ssiratio, and $F = 65.76$ is simply the square of the t statistic from the first-stage regression ($8.11^2 = 65.76$). This F statistic of 65.76 is considerably larger than the rule of thumb value of 10 that is sometimes suggested, so ssiratio does not seem to be a weak instrument.

The second part gives Shea's partial R^2, which equals the previously discussed partial R^2 because there is just one endogenous regressor.

The third part implements the tests of Stock and Yogo. The first test is not available because the model is just-identified rather than overidentified by two or more restrictions. The second test gives critical values for both the 2SLS estimator and the LIML estimator. We are considering the 2SLS estimator. If we are willing to tolerate distortion for a 5% Wald test based on the 2SLS estimator, so that the true size can be at most 10%, then we reject the null hypothesis if the test statistic exceeds 16.38. It is not exactly clear what test statistic to use here since theory does not apply exactly because of heteroskedasticity. The reported minimum eigenvalue statistic of 183.98 equals the F statistic that `ssiratio = 0` if default standard errors are used in the first-stage regression. We instead used robust standard errors (`vce(robust)`), which led to $F = 65.76$. The theory presumes homoskedastic errors, which is clearly not appropriate here. But both F statistics greatly exceed the critical value of 16.38, so we feel comfortable in rejecting the null hypothesis of weak instruments.

6.4.5 Overidentified model

For a model with a single endogenous regressor that is overidentified, the output is of the same format as the previous example. The F statistic will now be a joint test for the several instruments. If there are three or more instruments, so that there are two or more overidentifying restrictions, then the relative-bias criterion can be used.

We consider an example with three overidentifying restrictions:

```
. * Weak instrument tests - two or more overidentifying restrictions
. quietly ivregress gmm ldrugexp (hi_empunion = ssiratio lowincome multlc firmsz)
> $x2list, vce(robust)

. estat firststage, forcenonrobust

First-stage regression summary statistics
```

Variable	R-sq.	Adjusted R-sq.	Partial R-sq.	Robust F(4,10079)	Prob > F
hi_empunion	0.0821	0.0812	0.0243	44.823	0.0000

```
Minimum eigenvalue statistic = 62.749
```

Critical Values Ho: Instruments are weak	.	# of endogenous regressors: 1 # of excluded instruments: 4		
	5%	10%	20%	30%
2SLS relative bias	16.85	10.27	6.71	5.34
	10%	15%	20%	25%
2SLS Size of nominal 5% Wald test	24.58	13.96	10.26	8.31
LIML Size of nominal 5% Wald test	5.44	3.87	3.30	2.98

Using either $F = 44.82$ or the minimum eigenvalue of 62.749, we firmly reject the null hypothesis of weak instruments. The endogenous regressor `hi_empunion`, from structural-model estimates not given, has a coefficient of -0.812 and a standard error of 0.185 compared with -0.898 and 0.221 when `ssiratio` is the only instrument.

6.4.6   More than one endogenous regressor

With more than one endogenous regressor, `estat firststage` reports weak-instruments diagnostics that include Shea's partial R^2 and the tests based on the minimum eigenvalue statistic. The `all` option additionally leads to reporting for each endogenous regressor the first-stage regression and associated F statistic and partial R^2.

6.4.7   Sensitivity to choice of instruments

In the main equation, `hi_empunion` has a strong negative impact on `ldrugexp`. This contrasts with a small positive effect that is observed in the OLS results when `hi_empunion` is treated as exogenous; see section 6.3.6. If our instrument `ssiratio` is valid, then this would suggest a substantial bias in the OLS result. But is this result sensitive to the choice of the instrument?

To address this question, we compare results for four just-identified specifications, each estimated using just one of the four available instruments. We present a table with the structural-equation estimates for OLS and for the four IV estimations, followed by the minimum eigenvalue statistic (which equals the F statistic because we use just one endogenous regressor) for each of the four IV estimations. We have

```
. * Compare 4 just-identified model estimates with different instruments
. quietly regress ldrugexp hi_empunion $x2list, vce(robust)
. estimates store OLS0
. quietly ivregress 2sls ldrugexp (hi_empunion=ssiratio) $x2list, vce(robust)
. estimates store IV_INST1
. quietly estat firststage, forcenonrobust
. scalar me1 = r(mineig)
. quietly ivregress 2sls ldrugexp (hi_empunion=lowincome) $x2list, vce(robust)
. estimates store IV_INST2
. quietly estat firststage, forcenonrobust
. scalar me2 = r(mineig)
. quietly ivregress 2sls ldrugexp (hi_empunion=multlc) $x2list, vce(robust)
. estimates store IV_INST3
. quietly estat firststage, forcenonrobust
. scalar me3 = r(mineig)
. quietly ivregress 2sls ldrugexp (hi_empunion=firmsz) $x2list, vce(robust)
. estimates store IV_INST4
. quietly estat firststage, forcenonrobust
. scalar me4 = r(mineig)
```

```
. estimates table OLS0 IV_INST1 IV_INST2 IV_INST3 IV_INST4, b(%8.4f) se
```

Variable	OLS0	IV_INST1	IV_INST2	IV_INST3	IV_INST4
hi_empunion	0.0739	-0.8976	0.1170	-1.3459	-2.9323
	0.0260	0.2211	0.3594	0.4238	1.4025
totchr	0.4404	0.4503	0.4399	0.4548	0.4710
	0.0094	0.0102	0.0100	0.0116	0.0203
age	-0.0035	-0.0132	-0.0031	-0.0177	-0.0335
	0.0019	0.0030	0.0041	0.0048	0.0143
female	0.0578	-0.0204	0.0613	-0.0565	-0.1842
	0.0254	0.0326	0.0381	0.0449	0.1203
blhisp	-0.1513	-0.2174	-0.1484	-0.2479	-0.3559
	0.0341	0.0395	0.0416	0.0489	0.1098
linc	0.0105	0.0870	0.0071	0.1223	0.2473
	0.0137	0.0226	0.0311	0.0371	0.1130
_cons	5.8611	6.7872	5.8201	7.2145	8.7267
	0.1571	0.2688	0.3812	0.4419	1.3594

```
                                                         legend: b/se
. display "Minimum eigenvalues are:      " me1 _s(2) me2 _s(2) me3 _s(2) me4
Minimum eigenvalues are:      183.97973  54.328603  55.157581  9.9595082
```

The different instruments produce very different IV estimates for the coefficient of the endogenous regressor `hi_empunion`, though they are within two standard errors of each other (with the exception of that with `lowincome` as the instrument). All differ greatly from OLS estimates, aside from when `lowincome` is the instrument (IV_INST2). The coefficient of the most highly statistically significant regressor, `totchr`, changes little with the choice of instrument. Coefficients of some of the other exogenous regressors change considerably, though there is no sign reversal aside from `female`.

Because all models are just-identified with one endogenous regressor, the minimum eigenvalue statistics can be compared with the critical value given in section 6.4.4. If we are willing to tolerate distortion for a 5% Wald test based on the 2SLS estimator so that the true size can be at most 10%, then we reject the null hypothesis if the test statistic exceeds 16.38. By this criterion, only the final instrument, `firmsz`, is weak.

When we do a similar sensitivity analysis by progressively adding `lowincome`, `multlc`, and `firmsz` as instruments to an originally just-identified model with `ssiratio` as the instrument, there is little change in the 2SLS estimates; see the exercises at the end of this chapter.

There are several possible explanations for the sensitivity in the just-identified case. The results could reflect the expected high variability of the IV estimator in a just-identified model, variability that also reflects the differing strength of different instruments. Some of the instruments may not be valid instruments. Perhaps it is equally likely to be the case that the results reflect model misspecification—relative to a model one would fit in a serious empirical analysis of `ldrugexp`, the model used in the example is rather simple. While here we concentrate on the statistical tools for exploring the issue, in practice a careful context-specific investigation based on relevant theory is required to satisfactorily resolve the issue.

6.5   Better inference with weak instruments

The preceding section considers diagnostics and tests for weak instruments. If instruments are not weak, then we use standard asymptotic theory. A different approach when weak instruments are present is to apply an alternative asymptotic theory that may be more appropriate when instruments are weak or to use estimators other than 2SLS, for which the usual asymptotic theory may provide a more reasonable approximation when instruments are weak.

6.5.1   Conditional tests and confidence intervals

The conditional approach focuses on inference on the coefficient of the endogenous regressor in the structural model. Critical values, p-values, and confidence intervals are obtained that are of asymptotically correct size, assuming i.i.d. errors, no matter how weak the instruments.

The user-written `condivreg` command, developed by Mikusheva and Poi (2006), is a significant enhancement of an earlier version of `condivreg`. It implements methods that are surveyed and further developed in Andrews, Moreira, and Stock (2007). The critical values obtained are typically larger than the usual asymptotic critical values, as can be seen in the example below.

The `condivreg` command has the same basic syntax as `ivregress`, except that the specific estimator used (`2sls` or `liml`) is passed as an option. The default is to report a corrected p-value for the test of statistical significance and a corrected 95% confidence interval based on the likelihood-ratio (LR) test statistic. Under standard asymptotic theory the Wald, LR, and Lagrange multiplier (LM) tests are asymptotically equivalent under local alternatives. Here, instead, the LR test has better power than the LM test, and the Wald test has very poor power. The `lm` option computes the LM test and associated confidence interval, and there is no option for the Wald. The `ar` option computes the size-corrected p-value for an alternative test statistic, the Anderson–Rubin test statistic. The `level(#)` option is used to give confidence intervals other than 95%, and the `test(#)` option is used to get p-values for tests of values other than zero for the coefficient of the endogenous regressor.

We consider the original example with the single instrument `ssiratio` for the `hi_empunion` variable. We have

```
. * Conditional test and confidence intervals when weak instruments
. condivreg ldrugexp (hi_empunion = ssiratio) $x2list, lm ar 2sls test(0)
```

Instrumental variables (2SLS) regression

```
First-stage results                          Number of obs =     10089
─────────────────────                        F(  6, 10082) =    319.62
F(  1, 10082) =  183.98                       Prob > F      =    0.0000
Prob > F      =  0.0000                        R-squared     =    0.0640
R-squared     =  0.0761                        Adj R-squared =    0.0634
Adj R-squared =  0.0755                        Root MSE      =    1.318
```

ldrugexp	Coef.	Std. Err.	t	P>|t|	[95% Conf. Interval]	
hi_empunion	-.8975913	.2079906	-4.32	0.000	-1.305294	-.4898882
totchr	.4502655	.0104225	43.20	0.000	.4298354	.4706957
age	-.0132176	.0028759	-4.60	0.000	-.0188549	-.0075802
female	-.020406	.0315518	-0.65	0.518	-.0822538	.0414418
blhisp	-.2174244	.0386879	-5.62	0.000	-.2932603	-.1415884
linc	.0870018	.0220221	3.95	0.000	.0438342	.1301694
_cons	6.78717	.2555229	26.56	0.000	6.286294	7.288046

```
Instrumented:  hi_empunion
Instruments:   totchr age female blhisp linc ssiratio
Confidence set and p-value for hi_empunion are based on normal approximation
```

```
           Coverage-corrected confidence sets and p-values
                     for Ho: _b[hi_empunion] = 0
            LIML estimate of _b[hi_empunion] = -.8975913
```

Test	Confidence Set	p-value
Conditional LR	[-1.331227, -.5061496]	0.0000
Anderson-Rubin	[-1.331227, -.5061496]	0.0000
Score (LM)	[-1.331227, -.5061496]	0.0000

The first set of output is the same as that from `ivregress` with default standard errors that assume i.i.d. errors. It includes the first-stage $F = 183.98$, which strongly suggests weak instruments are not a problem. All three size-corrected tests given in the second set of output give the same 95% confidence interval of $[-1.331, -0.506]$ that is a bit wider than the conventional asymptotic interval of $[-1.305, -0.490]$. This again strongly suggests there is no need to correct for weak instruments. The term "confidence set" is used rather than confidence interval because it may comprise the union of two or more disjointed intervals.

The preceding results assume i.i.d. model errors, but errors are heteroskedastic here. This is potentially a problem, though from the output in section 6.3.4, the robust standard error for the coefficient of `hi_empunion` was 0.221, quite similar to the nonrobust standard error of 0.208.

Recall that tests suggested `firmsz` was a borderline weak instrument. When we repeat the previous command with `firmsz` as the single instrument, the corrected confidence intervals are considerably broader than those using conventional asymptotics; see the exercises at the end of this chapter.

6.5.2 LIML estimator

The literature suggests several alternative estimators that are asymptotically equivalent to 2SLS but may have better finite-sample properties than 2SLS.

The leading example is the LIML estimator. This is based on the assumption of joint normality of errors in the structural and first-stage equations. It is an ML estimator for obvious reasons and is a limited-information estimator when compared with a full-information approach that specifies structural equations (rather than first-stage equations) for all endogenous variables in the model.

The LIML estimator preceded 2SLS but has been less widely used because it is known to be asymptotically equivalent to 2SLS. Both are special cases of the k-class estimators. The two estimators differ in finite samples, however, because of differences in the weights placed on instruments. Recent research has found that LIML has some desirable finite-sample properties, especially if the instruments are not strong. For example, several studies have shown that LIML has a smaller bias than either 2SLS or GMM.

The LIML estimator is a special case of the so-called k-class estimator, defined as

$$\widehat{\beta}_{k\text{-class}} = \left\{ \mathbf{X}'(\mathbf{I} - k\mathbf{M}_z)^{-1}\mathbf{X} \right\}^{-1} \mathbf{X}'(\mathbf{I} - k\mathbf{M}_\mathbf{z})^{-1}\mathbf{y}$$

where the structural equation is denoted here as $\mathbf{y} = \mathbf{X}\boldsymbol{\beta} + \mathbf{u}$. The LIML estimator sets k equal to the minimum eigenvalue of $(\mathbf{Y}'\mathbf{M_Z}\mathbf{Y})^{-1/2}\mathbf{Y}'\mathbf{M_{X_1}}\mathbf{Y}(\mathbf{Y}'\mathbf{M_Z}\mathbf{Y})^{-1/2}$, where $\mathbf{M_{X_1}} = \mathbf{I} - \mathbf{X}_1(\mathbf{X}_1'\mathbf{X}_1)^{-1}\mathbf{X}_1$, $\mathbf{M_Z} = \mathbf{I} - \mathbf{Z}(\mathbf{Z}'\mathbf{Z})^{-1}\mathbf{Z}$, and the first-stage equations are $\mathbf{Y} = \mathbf{Z}\boldsymbol{\Pi} + \mathbf{V}$. The estimator has a VCE of

$$\widehat{V}(\widehat{\beta}_{k\text{-class}}) = s^2 \left\{ \mathbf{X}'(\mathbf{I} - k\mathbf{M_Z})^{-1}\mathbf{X} \right\}^{-1}$$

where $s^2 = \widehat{\mathbf{u}}'\widehat{\mathbf{u}}/N$ under the assumption that the errors $\mathbf{u}$ and $\mathbf{V}$ are homoskedastic. A leading k-class estimator is the 2SLS estimator, when $k = 1$.

The LIML estimator is performed by using the `ivregress liml` command rather than `ivregress 2sls`. The `vce(robust)` option provides a robust estimate of the VCE for LIML when errors are heteroskedastic. In that case, the LIML estimator remains asymptotically equivalent to 2SLS. But in finite samples, studies suggest LIML may be better.

6.5.3 Jackknife IV estimator

The jackknife IV estimator (JIVE) eliminates the correlation between the first-stage fitted values and the structural-equation error term that is one source of bias of the traditional 2SLS estimator. The hope is that this may lead to smaller bias in the estimator.

Let the subscript $(-i)$ denote the leave-one-out operation that drops the ith observation. Denote the structural equation by $\mathbf{y}_i = \mathbf{x}_i'\boldsymbol{\beta} + \mathbf{u}_i$, and consider first-stage equations for both endogenous and exogenous regressors, so $\mathbf{x}_i' = \mathbf{z}_i'\boldsymbol{\Pi} + \mathbf{v}_i$. Then, for each $i = 1, \ldots, N$, we estimate the parameters of the first-stage model with the

ith observation deleted, regressing $\mathbf{X}_{(-i)}$ on $\mathbf{Z}_{(-i)}$, and given estimate $\widehat{\mathbf{\Pi}}_i$ construct the instrument for observation i as $\widetilde{\mathbf{x}}'_i = \mathbf{z}'_i \widehat{\mathbf{\Pi}}_i$. Combining for $i = 1, \ldots, N$ yields an instrument matrix denoted by $\widetilde{\mathbf{X}}_{(-i)}$ with the ith row $\widetilde{\mathbf{x}}'_i$, leading to the JIVE

$$\widetilde{\boldsymbol{\beta}}_{\text{JIVE}} = \{\widetilde{\mathbf{X}}'_{(-i)}\mathbf{X}\}^{-1}\widetilde{\mathbf{X}}'_{(-i)}\mathbf{y}$$

The user-written `jive` command (Poi 2006) has similar syntax to `ivregress`, except that the specific estimator is passed as an option. The variants are `ujive1` and `ujive2` (Angrist, Imbens, and Krueger 1999) and `jive1` and `jive2` (Blomquist and Dahlberg 1999). The default is `ujive1`. The `robust` option gives heteroskedasticity-robust standard errors.

There is mixed evidence to date on the benefits of using JIVE; see the articles cited above and Davidson and MacKinnon (2006). Caution should be exercised in its use.

6.5.4   Comparison of 2SLS, LIML, JIVE, and GMM

Before comparing various estimators, we introduce the user-written `ivreg2` command, most recently described in Baum, Schaffer, and Stillman (2007). This overlaps considerably with `ivregress` but also provides additional estimators and statistics, and stores many results conveniently in `e()`.

The format of `ivreg2` is similar to that of `ivregress`, except that the particular estimator used is provided as an option. We use `ivreg2` with the `gmm` and `robust` options. When applied to an overidentified model, this yields the optimal GMM estimator when errors are heteroskedastic. It is equivalent to `ivreg gmm` with the `wmatrix(robust)` option.

We compare estimators for an overidentified model with four instruments for `hi_empunion`. We have

```
. * Variants of IV Estimators: 2SLS, LIML, JIVE, GMM_het, GMM-het using IVREG2
. global ivmodel "ldrugexp (hi_empunion = ssiratio lowincome multlc firmsz)
> $x2list"
. quietly ivregress 2sls $ivmodel, vce(robust)
. estimates store TWOSLS
. quietly ivregress liml $ivmodel, vce(robust)
. estimates store LIML
. quietly jive $ivmodel, robust
. estimates store JIVE
. quietly ivregress gmm $ivmodel, wmatrix(robust)
. estimates store GMM_het
. quietly ivreg2 $ivmodel, gmm robust
. estimates store IVREG2
```

```
. estimates table TWOSLS LIML JIVE GMM_het IVREG2, b(%7.4f) se
```

Variable	TWOSLS	LIML	JIVE	GMM_het	IVREG2
hi_empunion	-0.8623	-0.9156	-0.9129	-0.8124	-0.8124
	0.1868	0.1989	0.1998	0.1846	0.1861
totchr	0.4499	0.4504	0.4504	0.4495	0.4495
	0.0101	0.0102	0.0102	0.0100	0.0101
age	-0.0129	-0.0134	-0.0134	-0.0125	-0.0125
	0.0028	0.0029	0.0029	0.0027	0.0028
female	-0.0176	-0.0219	-0.0216	-0.0105	-0.0105
	0.0310	0.0316	0.0317	0.0307	0.0309
blhisp	-0.2150	-0.2186	-0.2185	-0.2061	-0.2061
	0.0386	0.0391	0.0391	0.0383	0.0385
linc	0.0842	0.0884	0.0882	0.0797	0.0797
	0.0206	0.0214	0.0214	0.0203	0.0205
_cons	6.7536	6.8043	6.8018	6.7126	6.7126
	0.2446	0.2538	0.2544	0.2426	0.2441

```
                                                        legend: b/se
```

Here there is little variation across estimators in estimated coefficients and standard errors. As expected, the last two columns give exactly the same coefficient estimates, though the standard errors differ slightly.

6.6   3SLS systems estimation

The preceding estimators are asymmetric in that they specify a structural equation for only one variable, rather than for all endogenous variables. For example, we specified a structural model for ldrugexp, but not one for hi_empunion. A more complete model specifies structural equations for all endogenous variables.

Consider a multiequation model with m (≥ 2) linear structural equations, each of the form

$$y_{ji} = \mathbf{y}'_{ji}\boldsymbol{\beta}_j + \mathbf{x}_{ji}'\beta_{j2} + u_{ji}, \quad j = 1, \ldots, m; \; i = 1, \ldots, N$$

For each of the m endogenous regressors y_j, we specify a structural equation with the endogenous regressors $\mathbf{y}_j$, the subset of endogenous variables that determine y_j, and the exogenous regressors $\mathbf{x}_j$, the subset of exogenous variables that determine y_j. Model identification is secured by rank and order conditions, given in standard graduate texts, requiring that some of the endogenous or exogenous regressors are excluded from each y_j equation.

The preceding IV estimators remain valid in this system. And specification of the full system can aid in providing instruments, because any exogenous regressors in the system that do not appear in $\mathbf{x}_j$ can be used as instruments for y_j.

Under the strong assumption that errors are i.i.d., more-efficient estimation is possible by exploiting cross-equation correlation of errors, just as for the SUR model discussed in section 5.4. This estimator is called the three-stage least-squares (3SLS) estimator.

We do not pursue it in detail, however, because the 3SLS estimator becomes inconsistent if errors are heteroskedastic, and errors are often heteroskedastic.

For the example below, we need to provide a structural model for `hi_empunion` in addition to the structural model already specified for `ldrugexp`. We suppose that `hi_empunion` depends on the single instrument `ssiratio`, on `ldrugexp`, and on `female` and `blhisp`. This means that we are (arbitrarily) excluding two regressors, `age` and `linc`. This ensures that the `hi_empunion` equation is overidentified. If instead it was just-identified, then the system would be just-identified because the `ldrugexp` is just-identified, and 3SLS would reduce to equation-by-equation 2SLS.

The syntax for the `reg3` command is similar to that for `sureg`, with each equation specified in a separate set of parentheses. The endogenous variables in the system are simply determined, because they are given as the first variable in each set of parentheses. We have

```
. * 3SLS estimation requires errors to be homoskedastic
. reg3 (ldrugexp hi_empunion totchr age female blhisp linc)
> (hi_empunion ldrugexp totchr female blhisp ssiratio)
```

Three-stage least-squares regression

Equation	Obs	Parms	RMSE	"R-sq"	chi2	P
ldrugexp	10089	6	1.314421	0.0686	1920.03	0.0000
hi_empunion	10089	5	1.709026	-11.3697	61.58	0.0000

	Coef.	Std. Err.	z	P>|z|	[95% Conf. Interval]	
ldrugexp						
hi_empunion	-.8771793	.2057101	-4.26	0.000	-1.280364	-.4739949
totchr	.4501818	.0104181	43.21	0.000	.4297626	.470601
age	-.0138551	.0027155	-5.10	0.000	-.0191774	-.0085327
female	-.0190905	.0314806	-0.61	0.544	-.0807914	.0426104
blhisp	-.2191746	.0385875	-5.68	0.000	-.2948048	-.1435444
linc	.0795382	.0190397	4.18	0.000	.0422212	.1168552
_cons	6.847371	.2393768	28.60	0.000	6.378201	7.316541
hi_empunion						
ldrugexp	1.344501	.3278678	4.10	0.000	.7018922	1.98711
totchr	-.5774437	.1437134	-4.02	0.000	-.8591169	-.2957706
female	-.1343657	.0368424	-3.65	0.000	-.2065754	-.0621559
blhisp	.1587661	.0711773	2.23	0.026	.0192612	.2982709
ssiratio	-.4167723	.05924	-7.04	0.000	-.5328805	-.300664
_cons	-6.982224	1.841294	-3.79	0.000	-10.59109	-3.373353

```
Endogenous variables:   ldrugexp hi_empunion
Exogenous variables:    totchr age female blhisp linc ssiratio
```

6.7   Stata resources

The `ivregress` command, introduced in Stata 10, is a major enhancement of the earlier command `ivreg`. The user-written `ivreg2` command (Baum, Schaffer, and Stillman 2007) has additional features including an extension of Ramsey's RESET test, a test of homoskedasticity, and additional tests of endogeneity. The user-written `condivreg` command enables inference with weak instruments assuming i.i.d. errors. The user-written `jive` command performs JIVE estimation. Estimation and testing with weak instruments and with many instruments is an active research area. Current official commands and user-written commands will no doubt be revised and enhanced, and new user-written commands may be developed.

The `ivregress` command is also important for understanding the approach to IV estimation and Stata commands used in IV estimation of several nonlinear models including the commands `ivprobit` and `ivtobit`.

6.8   Exercises

1. Estimate by 2SLS the same regression model as in section 6.3.4, with the instruments `multlc` and `firmsz`. Compare the 2SLS estimates with OLS estimates. Perform a test of endogeneity of `hi_empunion`. Perform a test of overidentification. State what you conclude. Throughout this exercise, perform inference that is robust to heteroskedasticity.

2. Repeat exercise 1 using optimal GMM.

3. Use the model and instruments of exercise 1. Compare the following estimators: 2SLS, LIML, and optimal GMM given heteroskedastic errors. For the last model, estimate the parameters by using the user-written `ivreg2` command in addition to `ivregress`.

4. Use the model of exercise 1. Compare 2SLS estimates as the instruments `ssiratio`, `lowincome`, `multlc`, and `firmsz` are progressively added.

5. Use the model and instruments of exercise 1. Perform appropriate diagnostics and tests for weak instruments using the 2SLS estimator. State what you conclude. Throughout this exercise, perform inference assuming errors are i.i.d.

6. Use the model and instruments of exercise 1. Use the user-written `condivreg` command to perform inference for the 2SLS estimator. Compare the results with those using conventional asymptotics.

7. Use the model and instruments of exercise 1. Use the user-written `jive` command and compare estimates and standard errors from the four different variants of JIVE and from optimal GMM. Throughout this exercise, perform inference that is robust to heteroskedasticity.

8. Estimate the 3SLS model of section 6.6, and compare the 3SLS coefficient estimates and standard errors in the `ldrugexp` equation with those from 2SLS estimation (with default standard errors).

9. This question considers the same earnings–schooling dataset as that analyzed in Cameron and Trivedi (2005, 111). The data are in `mus06klingdata.dta`. The `describe` command provides descriptions of the regressors. There are three endogenous regressors—years of schooling, years of work experience, and experience-squared—and three instruments—a college proximity indicator, age, and age-squared. Interest lies in the coefficient of schooling. Perform appropriate diagnostics and tests for weak instruments for the following model. State what you conclude. The following commands yield the IV estimator:

```
. use mus06klingdata.dta, clear
. global x2list black south76 smsa76 reg2-reg9 smsa66
> sinmom14 nodaded nomomed daded momed famed1-famed8
. ivregress 2sls wage76 (grade76 exp76 expsq76 = col4 age76 agesq76) $x2list,
> vce(robust)
. estat firststage
```

10. Use the same dataset as the previous question. Treat only `grade76` as endogenous, let `exp76` and `expsq76` be exogenous, and use `col4` as the only instrument. Perform appropriate diagnostics and tests for a weak instrument and state what you conclude. Then use the user-written `condivreg` command to perform inference, and compare the results with those using conventional asymptotics.

11. When an endogenous variable enters the regression nonlinearly, the obvious IV estimator is inconsistent and a modification is needed. Specifically, suppose $y_1 = \beta y_2^2 + u$, and the first-stage equation for y_2 is $y_2 = \pi_2 z + v$, where the zero-mean errors u and v are correlated. Here the endogenous regressor appears in the structural equation as y_2^2 rather than y_2. The IV estimator is $\widehat{\beta}_{IV} = \left(\sum_i z_i y_{2i}^2 \right)^{-1} \sum_i z_i y_{1i}$. This can be implemented by a regular IV regression of y on y_2^2 with the instrument z: regress y_2^2 on z and then regress y_1 on the first-stage prediction $\widehat{y_2^2}$. If instead we regress y_2 on z at the first stage, giving $\widehat{y}_2$, and then regress y_1 on $(\widehat{y}_2)^2$, an inconsistent estimate is obtained. Generate a simulation sample to demonstrate these points. Consider whether this example can be generalized to other nonlinear models where the nonlinearity is in regressors only, so that $y_1 = \mathbf{g}(\mathbf{y}_2)'\boldsymbol{\beta} + u$, where $\mathbf{g}(\mathbf{y}_2)$ is a nonlinear function of $\mathbf{y}_2$.

7 Quantile regression

7.1 Introduction

The standard linear regression is a useful tool for summarizing the average relationship between the outcome variable of interest and a set of regressors, based on the conditional mean function $E(y|\mathbf{x})$. This provides only a partial view of the relationship. A more complete picture would provide information about the relationship between the outcome y and the regressors $\mathbf{x}$ at different points in the conditional distribution of y. Quantile regression (QR) is a statistical tool for building just such a picture.

Quantiles and percentiles are synonymous—the 0.99 quantile is the 99th percentile. The median, defined as the middle value of a set of ranked data, is the best-known specific quantile. The sample median is an estimator of the population median. If $F(y) = \Pr(Y \leq y)$ defines the cumulative distribution function (c.d.f.), then $F(y_{\text{med}}) = 1/2$ is the equation whose solution defines the median $y_{\text{med}} = F^{-1}(1/2)$. The quantile q, $q \in (0,1)$, is defined as that value of y that splits the data into the proportions q below and $1 - q$ above, i.e., $F(y_q) = q$ and $y_q = F^{-1}(q)$. For example, if $y_{0.99} = 200$, then $\Pr(Y \leq 200) = 0.99$. These concepts extend to the conditional quantile regression function, denoted as $Q_q(y|\mathbf{x})$, where the conditional quantile will be taken to be linear in $\mathbf{x}$.

QRs have considerable appeal for several reasons. Median regression, also called least absolute-deviations regression, is more robust to outliers than is mean regression. QR, as we shall see, permits us to study the impact of regressors on both the location and scale parameters of the model, thereby allowing a richer understanding of the data. And the approach is semiparametric in the sense that it avoids assumptions about parametric distribution of regression errors. These features make QR especially suitable for heteroskedastic data.

Recently, computation of QR models has become easier. This chapter explores the application of QR using several of Stata's QR commands. We also discuss the presentation and interpretation of QR computer output using three examples, including an extension to discrete count data.

7.2 QR

In this section, we briefly review the theoretical background of QR analysis.

Let e_i denote the model prediction error. Then ordinary least squares (OLS) minimizes $\sum_i e_i^2$, median regression minimizes $\sum_i |e_i|$, and QR minimizes a sum that gives the asymmetric penalties $(1 - q) |e_i|$ for overprediction and $q |e_i|$ for underprediction. Linear programming methods need to be used to obtain the QR estimator, but it is still asymptotically normally distributed and easily obtained using Stata commands.

7.2.1   Conditional quantiles

Many applied econometrics studies model conditional moments, especially the conditional mean function. Suppose that the main objective of modeling is the conditional prediction of y given $\mathbf{x}$. Let $\widehat{y}(\mathbf{x})$ denote the predictor function and $e(\mathbf{x}) \equiv y - \widehat{y}(\mathbf{x})$ denote the prediction error. Then

$$L\{e(\mathbf{x})\} = L\{y - \widehat{y}(\mathbf{x})\}$$

denotes the loss associated with the prediction error e. The optimal loss-minimizing predictor depends upon the function $L(\cdot)$. If $L(e) = e^2$, then the conditional mean function, $E(y|\mathbf{x}) = \mathbf{x}'\boldsymbol{\beta}$ in the linear case, is the optimal predictor. If the loss criterion is absolute error loss, then the optimal predictor is the conditional median, denoted by $\text{med}(y|\mathbf{x})$. If the conditional median function is linear, so that $\text{med}(y|\mathbf{x}) = \mathbf{x}'\boldsymbol{\beta}$, then the optimal predictor is $\widehat{y} = \mathbf{x}'\widehat{\boldsymbol{\beta}}$, where $\widehat{\boldsymbol{\beta}}$ is the least absolute-deviations estimator that minimizes $\sum_i |y_i - \mathbf{x}_i'\boldsymbol{\beta}|$.

Both the squared-error and absolute-error loss functions are symmetric, which implies that the same penalty is imposed for prediction error of a given magnitude regardless of the direction of the prediction error. The asymmetry parameter q is specified. It lies in the interval $(0, 1)$ with symmetry when $q = 0.5$ and increasing asymmetry as q approaches 0 or 1. Then the optimal predictor is the qth conditional quantile, denoted by $Q_q(y|\mathbf{x})$, and the conditional median is a special case when $q = 0.5$. QR involves inference regarding the conditional quantile function.

Standard conditional QR analysis assumes that the conditional QR $Q_q(y|\mathbf{x})$ is linear in $\mathbf{x}$. This model can be analyzed in Stata. Recent theoretical advances cover nonparametric QR; see Koenker (2005).

Quite apart from the considerations of loss function (on which agreement may be difficult to obtain), there are several attractive features of QR. First, unlike the OLS regression that is sensitive to the presence of outliers and can be inefficient when the dependent variable has a highly nonnormal distribution, the QR estimates are more robust. Second, QR also provides a potentially richer characterization of the data. For example, QR allows us to study the impact of a covariate on the full distribution or any particular percentile of the distribution, not just the conditional mean. Third, unlike OLS, QR estimators do not require existence of the conditional mean for consistency. Finally, it is equivariant to monotone transformations. This means that the quantiles of a transformed variable y, denoted by $h(y)$, where $h(\cdot)$ is a monotonic function, equal the transforms of the quantiles of y, so $Q_q\{h(y)\} = h\{Q_q(y)\}$. Hence, if the quantile model is expressed as $h(y)$, e.g., $\ln y$, then one can use the inverse transformation to translate the results back to y. This is not possible for the mean, because $E\{h(y)\} \neq h\{E(y)\}$. The equivariance property for quantiles continues to hold in the regression context, assuming that the conditional quantile model is correctly specified; see section 7.3.4.

7.2.2   Computation of QR estimates and standard errors

Like OLS and maximum likelihood, QR is an extremum estimator. Computational implementation of QR is different, however, because optimization uses linear programming methods.

The qth QR estimator $\widehat{\boldsymbol{\beta}}_q$ minimizes over $\boldsymbol{\beta}_q$ the objective function

$$Q(\boldsymbol{\beta}_q) = \sum_{i:y_i \geq \mathbf{x}_i'\boldsymbol{\beta}}^{N} q|y_i - \mathbf{x}_i'\boldsymbol{\beta}_q| + \sum_{i:y_i < \mathbf{x}_i'\boldsymbol{\beta}}^{N} (1-q)|y_i - \mathbf{x}_i'\boldsymbol{\beta}_q| \tag{7.1}$$

where $0 < q < 1$, and we use $\boldsymbol{\beta}_q$ rather than $\boldsymbol{\beta}$ to make clear that different choices of q estimate different values of $\boldsymbol{\beta}$. If $q = 0.9$, for example, then much more weight is placed on prediction for observations with $y \geq \mathbf{x}'\boldsymbol{\beta}$ than for observations with $y < \mathbf{x}'\boldsymbol{\beta}$. Often, estimation sets $q = 0.5$, giving the least absolute-deviations estimator that minimizes $\sum_i |y_i - \mathbf{x}_i'\boldsymbol{\beta}_{0.5}|$.

The objective function (7.1) is not differentiable, so the usual gradient optimization methods cannot be applied. Instead it is a linear program. The classic solution method is the simplex method that is guaranteed to yield a solution in a finite number of simplex iterations.

The estimator that minimizes $Q(\boldsymbol{\beta}_q)$ is an m estimator with well-established asymptotic properties. The QR estimator is asymptotically normal under general conditions; see Cameron and Trivedi (2005, 88). It can be shown that

$$\widehat{\boldsymbol{\beta}}_q \overset{a}{\sim} N(\boldsymbol{\beta}_q, \mathbf{A}^{-1}\mathbf{B}\mathbf{A}^{-1}) \tag{7.2}$$

where $\mathbf{A} = \sum_i q(1-q)\mathbf{x}_i\mathbf{x}_i'$, $\mathbf{B} = \sum_i f_{u_q}(0|\mathbf{x}_i)\mathbf{x}_i\mathbf{x}_i'$, and $f_{u_q}(0|\mathbf{x})$ is the conditional density of the error term $u_q = y - \mathbf{x}'\boldsymbol{\beta}_q$ evaluated at $u_q = 0$. This analytical expression involves $f_{u_q}(0|\mathbf{x}_i)$, which is awkward to estimate. Estimates of the VCE using the paired bootstrap method (see chapter 13) are often preferred, though this adds to computational intensity.

7.2.3   The qreg, bsqreg, and sqreg commands

The Stata commands for QR estimation are similar to those for ordinary regression. There are three variants—qreg, bsqreg, and sqreg—that are commonly used. The first two are used for estimating a QR for a specified value of q, without or with bootstrap standard errors, respectively. The sqreg command is used when several different values of q are specified simultaneously. A fourth command, used less frequently, is iqreg, for interquartile range regression.

The basic QR command is `qreg`, with the following syntax:

`qreg` *depvar* [*indepvars*] [*if*] [*in*] [*weight*] [, *options*]

A simple example with the `qreg` options set to default is `qreg y x z`. This will estimate the median regression, $y_{\text{med}} = \beta_1 + \beta_2 x + \beta_3 z$, i.e., the default q is 0.5. The reported standard errors are those obtained using the analytical formula (7.2). The `quantile()` option allows one to choose q. For example, `qreg y x z, quantile(.75)` sets $q = 0.75$. The only other options are `level(#)` to set the level for reported confidence intervals and two optimization-related options. There is no `vce()` option for `qreg`.

The `bsqreg` command is instead used to obtain bootstrap standard errors that assume independence over i but, unlike (7.2), do not require an identical distribution. The standard errors from `bsqreg` are robust in the same sense as those from `vce()` for other commands. The command syntax is the same as for `qreg`. A key option is `reps(#)`, which sets the number of bootstrap replications. This option should be used because the default is only 20. And for replicability of results, one should first issue the `set seed` command. For example, give the commands `set seed 10101` and `bsqreg y x z, reps(400) quantile(.75)`.

The `iqreg` command, used for interquartile range regression, has similar syntax and options. If data are clustered, there is no `vce(cluster` *clustvar*`)` option, but a clustered bootstrap could be used; see chapter 13.

When QR estimates are obtained for several values of q, and we want to test whether regression coefficients for different values of q differ, the `sqreg` command is used. This provides coefficient estimates and an estimate of the simultaneous or joint VCE of $\widehat{\beta}_q$ across different specified values of q, using the bootstrap. The command syntax is again the same as `qreg` and `bsqreg`, and several quantiles can now be specified in the `quantile()` option. For example, `sqreg y x z, quantile(.2,.5,.8) reps(400)` produces QR estimates for $q = 0.2$, $q = 0.5$, and $q = 0.8$, together with bootstrap standard errors based on 400 replications.

7.3   QR for medical expenditures data

We present the basic QR commands applied to the log of medical expenditures.

7.3.1   Data summary

The data used in this example come from the Medical Expenditure Panel Survey (MEPS) and are identical to those discussed in section 3.2. Again we consider a regression model of total medical expenditure by the Medicare elderly. The dependent variable is `ltotexp`, so observations with zero expenditures are omitted. The explanatory variables are an indicator for supplementary private insurance (`suppins`), one health-status variable (`totchr`), and three sociodemographic variables (`age`, `female`, `white`).

We first summarize the data:

```
. * Read in log of medical expenditures data and summarize
. use mus03data.dta, clear

. drop if ltotexp == .
(109 observations deleted)

. summarize ltotexp suppins totchr age female white, separator(0)
```

Variable	Obs	Mean	Std. Dev.	Min	Max
ltotexp	2955	8.059866	1.367592	1.098612	11.74094
suppins	2955	.5915398	.4916322	0	1
totchr	2955	1.808799	1.294613	0	7
age	2955	74.24535	6.375975	65	90
female	2955	.5840948	.4929608	0	1
white	2955	.9736041	.1603368	0	1

The major quantiles of `ltotexp` can be obtained by using `summarize, detail`, and specific quantiles can be obtained by using `centile`. We instead illustrate them graphically, using the user-written `qplot` command. We have

```
. * Quantile plot for ltotexp using user-written command qplot
. qplot ltotexp, recast(line) scale(1.5)
```

The plot, shown in figure 7.1, is the same as a plot of the empirical c.d.f. of `ltotexp`, except that the axes are reversed. We have, very approximately, $q_{0.1} = 6$, $q_{0.25} = 7$, $q_{0.5} = 8$, $q_{0.75} = 9$, and $q_{0.9} = 10$. The distribution appears to be reasonably symmetric, at least for $0.05 < q < 0.95$.

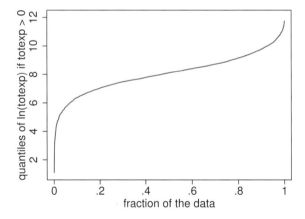

Figure 7.1. Quantiles of the dependent variable

7.3.2 QR estimates

The basic QR output for the median regression, with standard errors computed using the default option, is illustrated with the `qreg` command.

```
. * Basic quantile regression for q = 0.5
. qreg ltotexp suppins totchr age female white
Iteration  1:  WLS sum of weighted deviations = 2801.6338

Iteration  1: sum of abs. weighted deviations = 2801.9971
Iteration  2: sum of abs. weighted deviations = 2799.5941
Iteration  3: sum of abs. weighted deviations = 2799.5058
Iteration  4: sum of abs. weighted deviations = 2799.1722
Iteration  5: sum of abs. weighted deviations = 2797.8184
Iteration  6: sum of abs. weighted deviations = 2797.6548
note:  alternate solutions exist
Iteration  7: sum of abs. weighted deviations = 2797.0458
Iteration  8: sum of abs. weighted deviations = 2797.0439
Iteration  9: sum of abs. weighted deviations = 2797.0309
Iteration 10: sum of abs. weighted deviations =  2797.021
Iteration 11: sum of abs. weighted deviations = 2797.0134
Iteration 12: sum of abs. weighted deviations = 2796.9984
Iteration 13: sum of abs. weighted deviations = 2796.9961
Iteration 14: sum of abs. weighted deviations = 2796.9901
Iteration 15: sum of abs. weighted deviations = 2796.9889
Iteration 16: sum of abs. weighted deviations = 2796.9831

Median regression                                  Number of obs =        2955
  Raw sum of deviations 3110.961 (about 8.111928)
  Min sum of deviations 2796.983                   Pseudo R2    =      0.1009
```

ltotexp	Coef.	Std. Err.	t	P>|t|	[95% Conf. Interval]	
suppins	.2769771	.0471881	5.87	0.000	.1844521	.369502
totchr	.3942664	.0178276	22.12	0.000	.3593106	.4292222
age	.0148666	.003655	4.07	0.000	.0077	.0220331
female	-.0880967	.0468492	-1.88	0.060	-.1799571	.0037637
white	.4987456	.1428856	3.49	0.000	.2185801	.7789112
_cons	5.648891	.3000798	18.82	0.000	5.060504	6.237278

The iterations here refer to simplex iterations rather than the usual Newton–Raphson (or related gradient-method) iterations. All regressors, aside from `female`, are highly statistically significant with expected signs.

7.3.3   Interpretation of conditional quantile coefficients

Consider the standard bivariate regression model, for simplicity, with the conditional mean function $E(y_i|x_i) = \beta_0 + \beta_1 x_i$. This can be written as

$$y_i = \beta_1 + \beta_2 x_i + u_i \qquad (7.3)$$

where the error u_i satisfies $E(y_i|x_i) = 0$.

We denote the qth conditional quantile function of y given x as $Q_q(y|x)$. [The notation $Q_q(y|x)$ is used because it is analogous to $E(y|x)$.] In general, (7.3) implies that

$$Q_q(y_i|x_i) = \beta_1 + \beta_2 x_i + F_{u_i}^{-1}(q)$$

where F_{u_i} is the distribution function of u_i. Conditional on x_i, the quantile depends on the distribution of u_i via the term $F_{u_i}^{-1}(q)$. This will depend on x_i if, for example, errors are heteroskedastic. Then, in general, $Q_q(y|x)$ at different values of q will differ in more than just the intercept and may well even be nonlinear in x.

In the special case that errors are independent and identically distributed (i.i.d.), considerable simplification occurs as $F_{u_i}^{-1}(q) = F_u^{-1}(q)$, which does not vary with i. Then the conditional quantile is

$$Q_q(y_i|x_i) = \{\beta_1 + F_u^{-1}(q)\} + \beta_2 x_i$$

Here the conditional quantile functions have a common slope, differing only in the intercepts $\beta_0 + F_u^{-1}(q)$. In such a simple case, there is no need to use QR to obtain marginal effects (MEs) at different quantiles, because the quantile slope coefficient β_2 does not vary with the quantile.

More generally, errors are not i.i.d., for example, because of heteroskedasticity, and more than just the intercept varies. The standard quantile approach is to specify the conditional quantile function to be linear, though with intercept and slope parameters that may well vary with the quantile. In the general K-regressor, the standard linear conditional quantile function is

$$Q_q(y_i|\mathbf{x}_i) = \mathbf{x}_i'\boldsymbol{\beta}_q$$

The MEs after QR can be obtained in the usual way. For the jth (continuous) regressor, the ME is

$$\frac{\partial Q_q(y|\mathbf{x})}{\partial x_j} = \beta_{qj}$$

As for linear least-squares regression, the ME is given by the slope coefficient and is invariant across individuals, simplifying analysis. The interpretation is somewhat delicate for discrete changes that are more than infinitessimal, however, because the partial derivative measures the impact of a change in x_j under the assumption that the individual remains in the same quantile of the distribution after the change. For larger changes in a regressor, the individual may shift into a different quantile.

7.3.4   Retransformation

For our example, with the dependent variable `ltotexp` $= \ln(\texttt{totexp})$, the results from `qreg` give marginal effects for $\ln(\texttt{totexp})$. We may want instead to compute the ME on `totexp`, not `ltotexp`.

The equivariance property of QR is relevant. Given $Q_q(\ln y|\mathbf{x}) = \mathbf{x}'\boldsymbol{\beta}_q$, we have $Q_q(y|\mathbf{x}) = \exp\{Q_q(\ln y|\mathbf{x})\} = \exp(\mathbf{x}'\boldsymbol{\beta}_q)$. The ME on y in levels, given QR model $\mathbf{x}'\boldsymbol{\beta}_q$ in logs, is then

$$\frac{\partial Q_q(y|\mathbf{x})}{\partial x_j} = \exp(\mathbf{x}'\boldsymbol{\beta}_q)\beta_{qj}$$

This depends on $\mathbf{x}$. The average marginal effect (AME) is $\{N^{-1} \sum_{i=1}^{N} \exp(\mathbf{x}'_i \boldsymbol{\beta}_q)\} \beta_{qj}$ and can be estimated if we use the postestimation `predict` command to obtain $\exp(\mathbf{x}'_i \widehat{\boldsymbol{\beta}}_q)$ and then average. We obtain

```
. * Obtain multiplier to convert QR coeffs in logs to AME in levels.
. quietly predict xb

. generate expxb = exp(xb)

. quietly summarize expxb

. display "Multiplier of QR in logs coeffs to get AME in levels = " r(mean)
Multiplier of QR in logs coeffs to get AME in levels = 3746.7178
```

For example, the AME of `totchr` on ln(`totexp`) is 0.3943 from the output above. The implied AME of `totchr` on the levels variable is therefore $3746.7 \times 0.3943 = 1477$. One more chronic condition increases the conditional median of expenditures by \$1,477.

The equivariance property that $Q_q(y|\mathbf{x}) = \exp\{Q_q(\ln y|\mathbf{x})\}$ is exact only if the conditional quantile function is correctly specified. This is unlikely to be the case because the linear model will inevitably be only an approximation. One case where the linear model is exact is where all regressors are discrete and we specify a fully saturated model with indicator variables as regressors that exhaust all possible interactions between discrete regressors. We pursue this in the second end-of-chapter exercise.

7.3.5 Comparison of estimates at different quantiles

QRs can be performed at different quantiles, specifically the quartiles $q = 0.25, 0.50$, and 0.75. Here we do so and compare the results with one another and with OLS estimates. The QR standard errors use the default formula, (7.2), except that for median regression ($q = 0.50$) we additionally use `bsqreg` to obtain bootstrap standard errors, with the `reps()` option set to 400 and the random-number generator seed set to 10101. We obtain

```
. * Compare (1) OLS; (2-4) coeffs across quantiles; (5) bootstrap SEs
. quietly regress ltotexp suppins totchr age female white

. estimates store OLS

. quietly qreg ltotexp suppins totchr age female white, quantile(.25)

. estimates store QR_25

. quietly qreg ltotexp suppins totchr age female white, quantile(.50)

. estimates store QR_50

. quietly qreg ltotexp suppins totchr age female white, quantile(.75)

. estimates store QR_75

. set seed 10101

. quietly bsqreg ltotexp suppins totchr age female white, quant(.50) reps(400)

. estimates store BSQR_50
```

```
. estimates table OLS QR_25 QR_50 QR_75 BSQR_50, b(%7.3f) se
```

Variable	OLS	QR_25	QR_50	QR_75	BSQR_50
suppins	0.257	0.386	0.277	0.149	0.277
	0.046	0.055	0.047	0.060	0.059
totchr	0.445	0.459	0.394	0.374	0.394
	0.018	0.022	0.018	0.022	0.020
age	0.013	0.016	0.015	0.018	0.015
	0.004	0.004	0.004	0.005	0.004
female	-0.077	-0.016	-0.088	-0.122	-0.088
	0.046	0.054	0.047	0.060	0.052
white	0.318	0.338	0.499	0.193	0.499
	0.141	0.166	0.143	0.182	0.233
_cons	5.898	4.748	5.649	6.600	5.649
	0.296	0.363	0.300	0.381	0.385

```
                                              legend: b/se
```

The coefficients vary across quantiles. Most noticeably, the highly statistically significant regressor suppins (supplementary insurance) has a much greater impact at the lower conditional quantiles of expenditure. The standard errors are smaller for median regression ($q = 0.50$) than for the upper and lower quantiles ($q = 0.25, 0.75$), reflecting more precision at the center of the distribution. OLS coefficients differ considerably from the QR coefficients, even those for median regression. Comparing the third and fifth columns, for median regression the standard errors in this example are 10%–50% higher when estimated using the bootstrap method rather than the default method. We mainly use the default standard errors in this chapter for the simple reason that programs then run considerably faster.

7.3.6 Heteroskedasticity test

One reason for coefficients differing across quantiles is the presence of heteroskedastic errors. From output not shown here, the OLS standard errors are similar whether the default or robust estimates are obtained, suggesting little heteroskedasticity. And the logarithmic transformation of the dependent variable that has been used often reduces heteroskedasticity.

We use estat hettest to test against heteroskedasticity, which depends on the same variables as those in the regression. Then

```
. * Test for heteroskedasticity in linear model using estat hettest
. quietly regress ltotexp suppins totchr age female white
. estat hettest suppins totchr age female white, iid
Breusch-Pagan / Cook-Weisberg test for heteroskedasticity
        Ho: Constant variance
        Variables: suppins totchr age female white

        chi2(5)      =     71.38
        Prob > chi2  =    0.0000
```

The null hypothesis of homoskedasticity is soundly rejected.

7.3.7   Hypothesis tests

It is possible to conduct hypothesis tests of equality of the regression coefficients at different conditional quantiles.

Consider a test of the equality of the coefficient of `suppins` from QR with $q = 0.25$, $q = 0.50$, and $q = 0.75$. We first estimate using `sqreg`, rather than `qreg` or `sqreg`, to obtain the full covariance matrix of coefficients, and then we test. Because this uses the bootstrap, we need to set the seed and number of bootstrap replications.

```
. * Simultaneous QR regression with several values of q
. set seed 10101
. sqreg ltotexp suppins totchr age female white, q(.25 .50 .75) reps(400)
(fitting base model)
(bootstrapping ...............................................................
> ...............................................................................
> ...............................................................................
> ...............................................................................
> ...........................)
Simultaneous quantile regression                Number of obs =      2955
   bootstrap(400) SEs                            .25 Pseudo R2 =    0.1292
                                                 .50 Pseudo R2 =    0.1009
                                                 .75 Pseudo R2 =    0.0873
```

ltotexp	Coef.	Bootstrap Std. Err.	t	P>|t|	[95% Conf. Interval]	
q25						
suppins	.3856797	.059541	6.48	0.000	.2689335	.5024259
totchr	.459022	.0244648	18.76	0.000	.4110522	.5069919
age	.0155106	.0043515	3.56	0.000	.0069783	.0240429
female	-.0160694	.0579008	-0.28	0.781	-.1295996	.0974608
white	.3375935	.1076673	3.14	0.002	.1264829	.5487042
_cons	4.747962	.3454094	13.75	0.000	4.070694	5.42523
q50						
suppins	.2769771	.0535382	5.17	0.000	.1720011	.381953
totchr	.3942664	.0188346	20.93	0.000	.357336	.4311967
age	.0148666	.0044951	3.31	0.001	.0060528	.0236803
female	-.0880967	.0506032	-1.74	0.082	-.1873178	.0111244
white	.4987456	.2135776	2.34	0.020	.0799694	.9175219
_cons	5.648891	.4015098	14.07	0.000	4.861623	6.436159
q75						
suppins	.1488548	.0649661	2.29	0.022	.0214712	.2762383
totchr	.3735364	.022233	16.80	0.000	.3299425	.4171302
age	.0182506	.0049719	3.67	0.000	.0085018	.0279995
female	-.1219365	.0542792	-2.25	0.025	-.2283654	-.0155077
white	.1931923	.192686	1.00	0.316	-.1846205	.5710051
_cons	6.599972	.4187657	15.76	0.000	5.778869	7.421075

`sqreg` estimates a QR function for each specified quantile. Some of the coefficients appear to differ across the quantiles, and we use the `test` command to perform a Wald test on the hypothesis that the coefficients on `suppins` are the same for the three

quantiles. Because we are comparing estimates from different equations, we need a prefix
to indicate the equation. Here the prefix for the model with $q = 0.25$, for example, is
[q25]. To test that coefficients on the same variable have the same value in different
equations, we use the syntax

test [*eqname* = *eqname* ...]: *varlist*

We obtain

```
. * Test of coefficient equality across QR with different q
. test [q25=q50=q75]: suppins
 ( 1)  [q25]suppins - [q50]suppins = 0
 ( 2)  [q25]suppins - [q75]suppins = 0
       F(  2,  2949) =     5.32
            Prob > F =     0.0050
```

The null hypothesis of coefficient equality is rejected at a level of 0.05.

7.3.8   Graphical display of coefficients over quantiles

An attractive way to present QR results is via a graphical display of coefficients of interest
and their respective confidence intervals. This can be done manually by estimating the
parameters of the QR model for a range of values of q, saving the results to file, and
producing separate graphs for each regressor of the estimated coefficient plotted against
the quantile q.

This is done automatically by the user-written grqreg command, which provides
95% confidence intervals in addition to estimated coefficients. One of the qreg, bsqreg,
or sqreg commands must first be executed, and the confidence intervals use the standard errors from whichever command is used. The grqreg command does not have
enormous flexibility. In particular, it plots coefficients for all regressors, not just selected regressors.

We use grqreg with the options cons to include the intercept in the graph, ci to
include a 95% confidence interval, and ols and olsci to include the OLS coefficient and
its 95% confidence interval. The graph option scale(1.1) is added to increase the size
of the axis titles. The command uses variable labels on the y axis of each plot, so we
provide better variable labels for two of the regressors. We have

```
. * Plots of each regressor´s coefficients as quantile q varies
. quietly bsqreg ltotexp suppins totchr age female white, quantile(.50) reps(400)
. label variable suppins "=1 if supp ins"
. label variable totchr "# of chronic condns"
. grqreg, cons ci ols olsci scale(1.1)
```

In figure 7.2, the horizontal lines are the OLS point estimates and confidence intervals
(these do not vary with the quantile). The top middle plot shows that the coefficient
on suppins is positive over most of the range of q, with a much larger effect at lower

quantiles. In the lower quantiles, the point estimates suggest that supplementary insurance is associated with 20–25% higher medical expenditures (recall that because the dependent variable is in logs, coefficients can be interpreted as semielasticities). Notice that confidence intervals widen at both the extreme upper and lower quantiles.

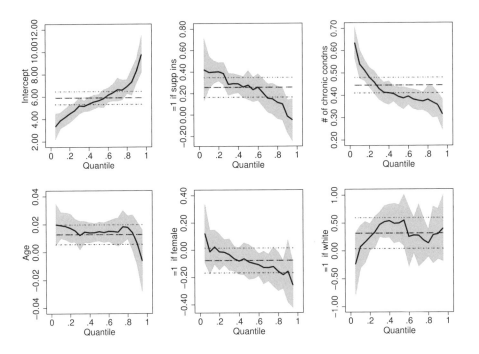

Figure 7.2. QR and OLS coefficients and confidence intervals for each regressor as q varies from 0 to 1

7.4 QR for generated heteroskedastic data

To gain more insight on QR, we consider a simulation example where the quantiles are known to be linear, and we specify a particular form of multiplicative heteroskedasticity.

7.4.1 Simulated dataset

We use a simulated dataset, one where the conditional mean of y depends on the regressors x_2 and x_3, while the conditional variance depends on only x_2.

If $y = \mathbf{x}'\boldsymbol{\beta} + u$ and $u = \mathbf{x}'\boldsymbol{\alpha} \times \varepsilon$, and it is assumed that $\mathbf{x}'\boldsymbol{\alpha} > 0$ and that ε_i is i.i.d., then the quantiles are linear in $\mathbf{x}$ with the qth conditional quantile $Q_q(y|\mathbf{x}) =$

$\mathbf{x}'\{\boldsymbol{\beta}+\boldsymbol{\alpha}\times F_{\varepsilon}^{-1}(q)\}$; see Cameron and Trivedi (2005, 86). So for regressors that appear in the conditional mean but not in the heteroskedasticity function (i.e., $\alpha_j = 0$), the QR coefficients do not change with q, while for other regressors the coefficients change even though the conditional quantile function is linear in $\mathbf{x}$.

If we let $y = \beta_1 + \beta_2 x_2 + \beta_3 x_3 + u$, where $u = (\alpha_1 + \alpha_2 x_2) \times \varepsilon$, then the QR coefficients for x_2 will change with q while those for x_3 will not. This result requires that $\alpha_1 + \alpha_2 x_2 > 0$, so we generate x2 from a $\chi^2(1)$ distribution.

The specific data-generating process (DGP) is

$$y = 1 + 1 \times x_2 + 1 \times x_3 + u; \quad x_2 \sim \chi^2(1), \ x_3 \sim N(0,25)$$
$$u = (0.1 + 0.5 \times x_2) \times \varepsilon; \quad \varepsilon \sim N(0,25)$$

We expect that the QR estimates of the coefficient of x_3 will be relatively unchanged at 1 as the quantiles vary, while the QR estimates of the coefficient of x_2 will increase as q increases (because the heteroskedasticity is increasing in x_2).

We first generate the data as follows:

```
. * Generated dataset with heteroskedastic errors
. set seed 10101
. set obs 10000
obs was 2955, now 10000
. generate x2 = rchi2(1)
. generate x3 = 5*rnormal(0)
. generate e = 5*rnormal(0)
. generate u = (.1+0.5*x2)*e
. generate y = 1 + 1*x2 + 1*x3 + u
. summarize e x2 x3 u y
```

Variable	Obs	Mean	Std. Dev.	Min	Max
e	10000	-.0536158	5.039203	-17.76732	18.3252
x2	10000	1.010537	1.445047	3.20e-08	14.64606
x3	10000	-.0037783	4.975565	-17.89821	18.15374
u	10000	.0013916	4.715262	-51.39212	68.7901
y	10000	2.00815	7.005894	-40.17517	86.42495

The summary statistics confirm that x3 and e have a mean of 0 and a variance of 25 and that x2 has a mean of 1 and a variance of 2, as desired. The output also shows that the heteroskedasticity has induced unusually extreme values of u and y that are more than 10 standard deviations from the mean.

Before we analyze the data, we run a quick check to compare the estimated coefficients with their theoretical values. The output below shows that the estimates are roughly in line with the theory underlying the DGP.

```
. * Quantile regression for q = .25, .50 and .75
. sqreg y x2 x3, quantile(.25 .50 .75)
(fitting base model)
(bootstrapping ....................)
```

Simultaneous quantile regression   Number of obs =   10000
  bootstrap(20) SEs   .25 Pseudo R2 =   0.5186
                                                          .50 Pseudo R2 =   0.5231
                                                          .75 Pseudo R2 =   0.5520

y	Coef.	Bootstrap Std. Err.	t	P>|t|	[95% Conf. Interval]	
q25						
x2	-.6961591	.0675393	-10.31	0.000	-.8285497	-.5637686
x3	.9991559	.0040589	246.16	0.000	.9911996	1.007112
_cons	.6398693	.0225349	28.39	0.000	.5956965	.6840422
q50						
x2	1.070516	.1139481	9.39	0.000	.8471551	1.293877
x3	1.001247	.0036866	271.59	0.000	.9940204	1.008473
_cons	.9688206	.0282632	34.28	0.000	.913419	1.024222
q75						
x2	2.821881	.0787823	35.82	0.000	2.667452	2.97631
x3	1.004919	.0042897	234.26	0.000	.9965106	1.013328
_cons	1.297878	.026478	49.02	0.000	1.245976	1.34978

```
. * Predicted coefficient of x2 for q = .25, .50 and .75
. quietly summarize e, detail

. display "Predicted coefficient of x2 for q = .25, .50, and .75"
> _newline 1+.5*r(p25) _newline 1+.5*r(p50) _newline 1+.5*r(p75)
Predicted coefficient of x2 for q = .25, .50, and .75
-.7404058
.97979342
2.6934063
```

For example, for $q = 0.75$ the estimated coefficient of x2 is 2.822, close to the theoretical 2.693.

We study the distribution of y further by using several plots. We have

```
. * Generate scatterplots and qplot
. quietly kdensity u, scale(1.25) lwidth(medthick) saving(density, replace)

. quietly qplot y, recast(line) scale(1.4) lwidth(medthick)
> saving(quanty, replace)

. quietly scatter y x2, scale(1.25) saving(yversusx2, replace)

. quietly scatter y x3, scale(1.25) saving(yversusx3, replace)

. graph combine density.gph quanty.gph yversusx2.gph yversusx3.gph
```

This leads to figure 7.3. The first panel, with the kernel density of u, shows that the distribution of the error u is essentially symmetric but has very long tails. The second panel shows the quantiles of y and indicates symmetry. The third panel plots y against x_2 and indicates heteroskedasticity and the strongly nonlinear way in which x_2 enters the conditional variance function of y. The fourth panel shows no such relationship between y and x_3.

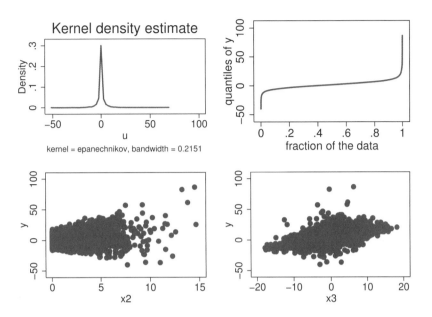

Figure 7.3. Density of u, quantiles of y, and scatterplots of (y, x_2) and (y, x_3)

Here x_2 affects both the conditional mean and variance of y, whereas x_3 enters only the conditional mean function. The regressor x_2 will impact the conditional quantiles differently, whereas x_3 will do so in a constant way. The OLS regression can only display the relationship between average y and (x_2, x_3). QR, however, can show the relationship between the regressors and the distribution of y.

7.4.2   QR estimates

We next estimate the regression using OLS (with heteroskedasticity-robust standard errors) and QR at $q = 0.25$, 0.50, and 0.75, with bootstrap standard errors. The saved results are displayed in a table. The relevant commands and the resulting output are as follows:

```
. * OLS and quantile regression for q = .25, .5, .75
. quietly regress y x2 x3
. estimates store OLS
. quietly regress y x2 x3, vce(robust)
. estimates store OLS_Rob
. quietly bsqreg y x2 x3, quantile(.25) reps(400)
. estimates store QR_25
. quietly bsqreg y x2 x3, quantile(.50) reps(400)
. estimates store QR_50
. quietly bsqreg y x2 x3, quantile(.75) reps(400)
. estimates store QR_75
```

```
. estimates table OLS OLS_Rob QR_25 QR_50 QR_75, b(%7.3f) se
```

Variable	OLS	OLS_Rob	QR_25	QR_50	QR_75
x2	1.079	1.079	-0.696	1.071	2.822
	0.033	0.116	0.070	0.077	0.079
x3	0.996	0.996	0.999	1.001	1.005
	0.009	0.009	0.004	0.003	0.004
_cons	0.922	0.922	0.640	0.969	1.298
	0.058	0.086	0.020	0.020	0.022

```
                                                          legend: b/se
```

The median regression parameter point estimates of $\beta_{0.5,2}$ and $\beta_{0.5,3}$ are close to the true values of 1. Interestingly, the median regression parameter estimates are much more precise than the OLS parameter estimates. This improvement is possible because OLS is no longer fully efficient when there is heteroskedasticity. Because the heteroskedasticity depends on x_2 and not on x_3, the estimates of β_{q2} vary over the quantiles q, while β_{q3} is invariant with respect to q.

We can test whether this is the case by using the **bsqreg** command. A test of $\beta_{0.25,2} = \beta_{0.75,2}$ can be interpreted as a robust test of heteroskedasticity independent of the functional form of the heteroskedasticity. The test is implemented as follows:

```
. * Test equality of coeff of x2 for q=.25 and q=.75
. set seed 10101
. quietly sqreg y x2 x3, q(.25 .75) reps(400)
. test [q25]x2 = [q75]x2
 ( 1)  [q25]x2 - [q75]x2 = 0
        F(  1,  9997) = 1565.58
             Prob > F =    0.0000
. test [q25]x3 = [q75]x3
 ( 1)  [q25]x3 - [q75]x3 = 0
        F(  1,  9997) =    1.94
             Prob > F =    0.1633
```

The test outcome leads to a strong rejection of the hypothesis that x_2 does not affect both the location and scale of y. As expected, the test for x_3 yields a p-value of 0.16, which does not lead to a rejection of the null hypothesis.

7.5   QR for count data

QR is usually applied to continuous-response data because the quantiles of discrete variables are not unique since the c.d.f. is discontinuous with discrete jumps between flat sections. By convention, the lower boundary of the interval defines the quantile in such a case. However, recent theoretical advances have extended QR to a special case of discrete variable model—the count regression.

In this section, we present application of QR to counts, the leading example of ordered discrete data. The method, proposed by Machado and Santos Silva (2005), enables QR methods to be applied by suitably smoothing the count data. We presume no knowledge of count regression.

7.5.1 Quantile count regression

The key step in the quantile count regression (QCR) model of Machado and Santos Silva is to replace the discrete count outcome y with a continuous variable, $z = h(y)$, where $h(\cdot)$ is a smooth continuous transformation. The standard linear QR methods are then applied to z. Point and interval estimates are then retransformed to the original y-scale by using functions that preserve the quantile properties.

The particular transformation used is

$$z = y + u$$

where $u \sim U(0, 1)$ is a pseudorandom draw from the uniform distribution on $(0, 1)$. This step is called "jittering" the count.

Because counts are nonnegative, the conventional count models presented in chapter 17 are based on an exponential model for the conditional mean, $\exp(\mathbf{x}'\boldsymbol{\beta})$, rather than a linear function $\mathbf{x}'\boldsymbol{\beta}$. Let $Q_q(y|\mathbf{x})$ and $Q_q(z|\mathbf{x})$ denote the qth quantiles of the conditional distributions of y and z, respectively. Then, to allow for the exponentiation, the conditional quantile for $Q_q(z|\mathbf{x})$ is specified to be

$$Q_q(z|\mathbf{x}) = q + \exp(\mathbf{x}'\boldsymbol{\beta}_q) \tag{7.4}$$

The additional term q appears in the equation because $Q_q(z|\mathbf{x})$ is bounded from below by q, because of the jittering operation.

To be able to estimate the parameters of a quantile model in the usual linear form $\mathbf{x}'\boldsymbol{\beta}$, a log transformation is applied so that $\ln(z-q)$ is modeled, with the adjustment that if $z - q < 0$ then we use $\ln(\varepsilon)$, where ε is a small positive number. The transformation is justified by the property that quantiles are equivariant to monotonic transformation (see section 7.2.1) and the property that quantiles above the censoring point are not affected by censoring from below. Postestimation transformation of the z quantiles back to y quantiles uses the ceiling function, with

$$Q_q(y|\mathbf{x}) = \lceil Q_q(z|\mathbf{x}) - 1 \rceil \tag{7.5}$$

where the symbol $\lceil r \rceil$ in the right-hand side of (7.5) denotes the smallest integer greater than or equal to r.

To reduce the effect of noise due to jittering, the parameters of the model are estimated multiple times using independent draws from the $U(0, 1)$ distribution, and the multiple estimated coefficients and confidence interval endpoints are averaged. Hence, the estimates of the quantiles of y counts are based on $\widehat{Q}_q(y|\mathbf{x}) = \lceil Q_q(z|\mathbf{x}) - 1 \rceil = \lceil q + \exp(\mathbf{x}'\overline{\overline{\boldsymbol{\beta}}}_q) - 1 \rceil$, where $\overline{\overline{\boldsymbol{\beta}}}$ denotes the average over the jittered replications.

7.5.2 The qcount command

The QCR method of Machado and Santos Silva can be performed by using the user-written qcount command (Miranda 2007). The command syntax is

qcount *depvar* [*indepvars*] [*if*] [*in*], quantile(*number*) [repetition(#)]

where quantile(*number*) specifies the quantile to be estimated and repetition(#) specifies the number of jittered samples to be used to calculate the parameters of the model with the default value being 1,000. The postestimation command qcount_mfx computes MEs for the model, evaluated at the means of the regressors.

For example, qcount y x1 x2, q(0.5) rep(500) estimates a median regression of the count y on x1 and x2 with 500 repetitions. The subsequent command qcount_mfx gives the associated MEs.

7.5.3 Summary of doctor visits data

We illustrate these commands using a dataset on the annual number of doctor visits (docvis) by the Medicare elderly in the year 2003. The regressors are an indicator for having private insurance that supplements Medicare (private), number of chronic conditions (totchr), age in years (age), and indicators for female and white. We have

```
. * Read in doctor visits count data and summarize
. use mus07qrcnt.dta, clear
. summarize docvis private totchr age female white, separator(0)
    Variable |       Obs        Mean    Std. Dev.       Min        Max
      docvis |      3677    6.822682    7.394937          0        144
     private |      3677    .4966005    .5000564          0          1
      totchr |      3677    1.843351    1.350026          0          8
         age |      3677    74.24476    6.376638         65         90
      female |      3677    .6010335    .4897525          0          1
       white |      3677    .9709002    .1681092          0          1
```

The dependent variable, annual number of doctor visits (docvis), is a count. The median number of visits is only 5, but there is a long right tail. The frequency distribution shows that around 0.5% of individuals have over 40 visits, and the maximum value is 144.

To demonstrate the smoothing that occurs with jittering, we create the variable docvisu, which is obtained for each individual by adding a random uniform variate to docvis. We then compare the quantile plot of the smoothed docvisu with that for the discrete count docvis. We have

```
. * Generate jittered values and compare quantile plots
. set seed 10101
. generate docvisu = docvis + runiform()
. quietly qplot docvis if docvis < 40, recast(line) scale(1.25)
> lwidth(medthick) saving(docvisqplot, replace)
```

```
. quietly qplot docvisu if docvis < 40, recast(line) scale(1.25)
> lwidth(medthick) saving(docvisuqplot, replace)
. graph combine docvisqplot.gph docvisuqplot.gph
```

For graph readability, values in excess of 40 were dropped. The graphs are shown in figure 7.4.

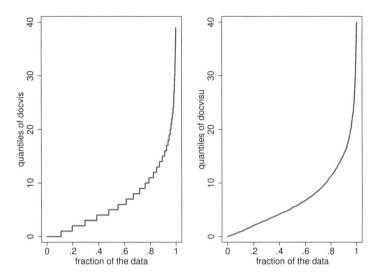

Figure 7.4. Quantile plots of count `docvis` (left) and its jittered transform (right)

The common starting point for regression analysis of counts is Poisson or negative binomial regression. We use the latter and simply print out the MEs of a change in the conditional mean of a change in each regressor, evaluated at sample means of the regressors. These will be compared later with MEs for the median obtained with `qcount`.

```
. * Marginal effects from conventional negative binomial model
. quietly nbreg docvis i.private totchr age i.female i.white, vce(robust)
. margins, dydx(*) atmean noatlegend
```

Conditional marginal effects   Number of obs   =   3677
Model VCE   : Robust

Expression   : Predicted number of events, predict()
dy/dx w.r.t. : 1.private totchr age 1.female 1.white

	dy/dx	Delta-method Std. Err.	z	P>|z|	[95% Conf. Interval]	
1.private	1.082549	.2148132	5.04	0.000	.661523	1.503575
totchr	1.885011	.0771011	24.45	0.000	1.733896	2.036127
age	.0340016	.0176656	1.92	0.054	−.0006224	.0686255
1.female	−.1401461	.2179811	−0.64	0.520	−.5673812	.2870891
1.white	.4905679	.5711726	0.86	0.390	−.62891	1.610046

Note: dy/dx for factor levels is the discrete change from the base level.

In the **nbreg** command, factor variables were used for the indicator variables `private`, `female`, and `white`. This does not affect the parameter estimates but does mean that the subsequent marginal effects for these variables are computed using the finite-difference method rather than calculus methods; see section 10.6.5.

7.5.4 Results from QCR

We estimate the parameters of the QCR model at the median. We obtain

```
. * Quantile count regression
. set seed 10101
. qcount docvis private totchr age female white, q(0.50) rep(500)
....................................................................
> ..................................................................
> ..................................................................
> ..................................................................
> ..................................................................
> ..................................................................
> ..................................
Count Data Quantile Regression
( Quantile 0.50 )
```

| | | | | | Number of obs | = 3677 |
| | | | | | No. jittered samples | = 500 |

docvis	Coef.	Std. Err.	z	P>|z|	[95% Conf.	Interval]
private	.2026897	.0409784	4.95	0.000	.1223735	.283006
totchr	.3464992	.0181838	19.06	0.000	.3108596	.3821387
age	.0084273	.0033869	2.49	0.013	.0017891	.0150655
female	.0025235	.04131	0.06	0.951	-.0784427	.0834896
white	.1200776	.0980302	1.22	0.221	-.072058	.3122132
_cons	.0338338	.2525908	0.13	0.893	-.4612352	.5289027

The statistically significant regressors have the expected signs.

The parameters of the model estimated use an exponential functional form for the conditional quantile. To interpret results, it is easier to use the MEs. The `qcount_mfx` command gives two sets of MEs after conditional QR. The first is for the jittered variable $Q_q(z|\mathbf{x})$, and the second is for the original count $Q_q(y|\mathbf{x})$. We have

```
. * Marginal effects after quantile regression for median
. set linesize 81

. qcount_mfx

Marginal effects after qcount
        y = Qz(0.50|X)
          = 5.05849 (0.0975)
```

	ME	Std. Err.	z	P>|z|	[95% C.I]		X
private	.92617897	.18594172	4.98	0.0000	0.5617	1.2906	0.50
totchr	1.5795119	.07861945	20.1	0.0000	1.4254	1.7336	1.84
age	.03841567	.01533432	2.51	0.0122	0.0084	0.0685	74.24
female	.01150027	.18822481	.0611	0.9513	-0.3574	0.3804	0.60
white	.51759079	.40076951	1.29	0.1965	-0.2679	1.3031	0.97

```
Marginal effects after qcount
        y = Qy(0.50|X)
          = 5
```

	ME	[95% C. Set]		X
private	0	0	1	0.50
totchr	1	1	1	1.84
age	0	0	0	74.24
female	0	-1	0	0.60
white	0	-1	1	0.97

The set linesize 81 command is added to avoid output wraparound, because the output from qcount_mfx takes 81 characters and the Stata default line size is 80 characters.

The estimated MEs for the conditional quantile $Q_q(z|\mathbf{x})$ of the jittered variable defined in (7.4) differ by around 20% from those from the negative binomial model given earlier, aside from a much greater change for the quite statistically insignificant regressor female. Of course, the difference between negative binomial estimates and QCR estimates depends on the quantile q. Comparisons between negative binomial and QCR estimates for some other quantiles will show even larger differences than those given above.

The second set of output gives the estimated MEs for the conditional quantile of the original discrete count variable $Q_q(y|\mathbf{x})$ defined in (7.5). These are discretized, and only that for totchr is positive. We note in passing that if we estimate the model using qreg rather than qcount, then the estimated coefficients are 1 for private, 2 for totchr, 0 for the other three regressors, 1 for the intercept, and all standard errors are zero.

The qcount command allows one to study the impact of a regressor at different points in the distribution. To explore this point, we reestimate with $q = 0.75$. We have

(Continued on next page)

```
. * Quantile count regression for q = 0.75
. set seed 10101

. quietly qcount docvis private totchr age female white, q(0.75) rep(500)

. qcount_mfx

Marginal effects after qcount
      y = Qz(0.75|X)
        =   9.06557 (0.1600)
```

	ME	Std. Err.	z	P>|z|	[95% C.I]		X
private	1.2255773	.33167392	3.7	0.0002	0.5755	1.8757	0.50
totchr	2.3236279	.13394814	17.3	0.0000	2.0611	2.5862	1.84
age	.02647556	.02547965	1.04	0.2988	-0.0235	0.0764	74.24
female	-.00421291	.3283728	-.0128	0.9898	-0.6478	0.6394	0.60
white	1.1880327	.81448878	1.46	0.1447	-0.4084	2.7844	0.97

```
Marginal effects after qcount
      y = Qy(0.75|X)
        = 9
```

	ME	[95% C. Set]	X
private	1	0 1	0.50
totchr	2	2 2	1.84
age	0	0 0	74.24
female	0	-1 0	0.60
white	1	-1 2	0.97

For the highly statistically significant regressors, private and totchr, the MEs are
30–50% higher than those for the conditional median.

7.6   Stata resources

The basic Stata commands related to qreg are bsqreg, iqreg, and sqreg; see [R] qreg
and [R] qreg postestimation. Currently, there are no options for cluster–robust variance estimation. The Stata user-written qplot command is illustrated by its author in
some detail in Cox (2005). The user-written grqreg command was created by Azevedo
(2004). The user-written qcount command was created by Miranda (2007).

7.7   Exercises

1. Consider the medical expenditures data example of section 7.3, except use totexp
   rather than ltotexp as the dependent variable. Use the same sample, so still
   drop if ltotexp==. Estimate the parameters of the model with $q = 0.5$ using
   qreg, and comment on the parameter estimates. Reestimate using bsqreg and
   compare results. Use sqreg to estimate at the quantiles 0.25, 0.50, and 0.75.
   Compare these estimates with each other (and their precision) and also with OLS
   (with robust standard errors). Use the Stata user-written grqreg command after
   bsqreg to further compare estimates as qreg varies.

2. Use the medical expenditures data of section 7.3. Show that the median of ltotexp equals the exponential of the median of totexp. Now add a single regressor, the indicator female. Then any conditional quantile function must be linear in the regressor, with $Q_q(\ln y|\mathbf{x}) = \alpha_{q1} + \alpha_{q2}$female and $Q_q(y|\mathbf{x}) = \beta_{q1} + \beta_{q2}$female. Show that if we estimate qreg ltotexp female, then predict, and finally exponentiate the prediction, we get the same prediction as that directly from qreg totexp female. Now add another regressor, say, totchr. Then the conditional quantile may no longer be linear in female and totchr. Repeat the prediction exercise and show that the invariance under the transformation property no longer holds.

3. Use the medical expenditures data example of section 7.3 with the dependent variable ltotexp. Test the hypothesis that heteroskedasticity is a function of the single variable totchr, which measures the number of chronic conditions. Record the test outcome. Next test the hypothesis that the location and scale of the dependent variable expenditures varies with totchr. What is the connection between the two parts of this question?

4. Use the medical expenditures data of section 7.3, and estimate the parameters of the model for ltotexp using qreg with $q = 0.5$. Then estimate the same parameters using bsqreg with the number of bootstrap replications set at 10, 50, 100, and 400. In each case, use the same seed of 10101. Would you say that a high number of replications produces substantially different standard errors?

5. Consider the heteroskedastic regression example of section 7.4. Change the specification of the variance function so that the variance function is a function of x_3 and not x_2; i.e., reverse the roles of x_2 and x_3. Estimate QRs for the generated data for $q = 0.25, 0.50$, and 0.75. Compare the results you obtain with those given in section 7.3.2. Next vary the coefficient of x_3 in the variance function, and study its impact on the QR estimates.

6. Consider the heteroskedasticity example of section 7.4. There the regression error is symmetrically distributed. Suppose we want to study whether the differences between OLS and QR results are sensitive to the shape of the error distribution. Make suitable changes to the simulation data, and implement an analysis similar to that in section 7.4 with asymmetric errors. For example, first draw u from the uniform distribution and then apply the transformation $-\lambda \log(u)$, where $\lambda > 0$. (This generates draws from the exponential distribution with a mean of λ and a variance of λ^2.)

7. Use the data from section 7.5, except let the dependent count variable be totchr, and drop totchr from the regressors. Using the user-written qcount command, estimate qcount regressions for $q = 0.25, 0.50$, and 0.75. Use qcount_mfx to calculate the MEs. Store and print the results in tabular form. Explain how the mfx command works in the case of qcount regressions and whether there are any differences in the interpretation compared with the standard Poisson regression.

8. When the number of regressors in the QR is very large and one only wants to generate graphs for selected coefficients, it may be necessary to write one's own code to estimate and save the coefficients. This would be followed by a suitable twoway

plot. The following program uses `postfile` to save the output in `bsqrcoef1.dta`,
`forvalues` to loop around values of $q = 0.1(0.1)0.9$, and `bsqreg` to estimate bootstrap standard errors. Run the following program, and use `bsqrcoef1.dta` and
the `twoway` command to generate a plot for the coefficient of `suppins` as q varies.

```
* Save coefficients and generate graph for a range of quantiles
use mus03data.dta, clear
drop if ltotexp == .
capture program drop mus07plot
program mus07plot
postfile myfile percentile b1 upper lower using bsqrcoef1.dta, replace
forvalues tau1=0.10(0.1)0.9 {
    set seed 10101
    quietly bsqreg ltotexp suppins age female white totchr, quantile(`tau1´) ///
        reps(400)
    matrix b = e(b)
    scalar b1=b[1,1]
    matrix V = e(V)
    scalar v1=V[1,1]
    scalar df=e(df_r)
    scalar upper = b1 + invttail(df,.025)*sqrt(v1)
    scalar lower = b1 - invttail(df,.025)*sqrt(v1)
    post myfile (`tau1´) (b1) (upper) (lower)
    matrix drop V b
    scalar drop b1 v1 upper lower df
}
postclose myfile
end
mus07plot
program drop mus07plot
use bsqrcoef1.dta, clear
twoway connected b1 percentile || line upper percentile||line lower percentile, ///
    title("Slope Estimates") subtitle("Coefficient of suppins") ///
    xtitle("Quantile", size(medlarge)) ///
    ytitle("Slope and confidence bands", size(medlarge)) ///
    legend( label(1 "Quantile slope coefficient") ///
    label(2 "Upper 95% bs confidence band") label(3 "Lower 95% bs confidence band"))
graph save bsqrcoef1.gph, replace
```

8 Linear panel-data models: Basics

8.1 Introduction

Panel data or longitudinal data are repeated measurements at different points in time on the same individual unit, such as person, firm, state, or country. Regressions can then capture both variation over units, similar to regression on cross-section data, and variation over time.

Panel-data methods are more complicated than cross-section–data methods. The standard errors of panel-data estimators need to be adjusted because each additional time period of data is not independent of previous periods. Panel data requires the use of much richer models and estimation methods. Also different areas of applied statistics use different methods for essentially the same data. The Stata xt commands, where xt is an acronym for cross-section time series, cover many of these methods.

We focus on methods for a short panel, meaning data on many individual units and few time periods. Examples include longitudinal surveys of many individuals and panel datasets on many firms. And we emphasize microeconometrics methods that attempt to estimate key marginal effects that can be given a causative interpretation.

The essential panel-data methods are given in this chapter, most notably, the important distinction between fixed-effects and random-effects models. Chapter 9 presents many other panel-data methods for the linear model, including those for instrumental-variables (IV) estimation, estimation when lagged dependent variables are regressors, estimation when panels are long rather than short, and estimation of mixed models with slope parameters that vary across individuals. Chapter 9 also shows how methods for short panels are applicable to other forms of clustered data or hierarchical data, such as cross-section individual data from a survey conducted at a number of villages, with clustering at the village level. Nonlinear models are presented in chapter 18.

8.2 Panel-data methods overview

There are many types of panel data and goals of panel-data analysis, leading to different models and estimators for panel data. We provide an overview in this section, with subsequent sections illustrating many of the various models and estimation methods.

8.2.1　Some basic considerations

First, panel data are usually observed at regular time intervals, as is the case for most time-series data. A common exception is growth curve analysis where, for example, children are observed at several irregularly spaced intervals in time, and a measure such as height or IQ is regressed on a polynomial in age.

Second, panel data can be balanced, meaning all individual units are observed in all time periods ($T_i = T$ for all i), or unbalanced ($T_i \neq T$ for some i). Most xt commands can be applied to both balanced and unbalanced data. In either case, however, estimator consistency requires that the sample-selection process does not lead to errors being correlated with regressors. Loosely speaking, the missingness is for random reasons rather than systematic reasons.

Third, the dataset may be a short panel (few time periods and many individuals), a long panel (many time periods and few individuals), or both (many time periods and many individuals). This distinction has consequences for both estimation and inference.

Fourth, model errors are very likely correlated. Microeconometrics methods emphasize correlation (or clustering) over time for a given individual, with independence over individual units. For some panel datasets, such as country panels, there additionally may be correlation across individuals. Regardless of the assumptions made, some correction to default ordinary least-squares (OLS) standard errors is usually necessary and efficiency gains using generalized least squares (GLS) may be possible.

Fifth, regression coefficient identification for some estimators can depend on regressor type. Some regressors, such as gender, may be time invariant with $x_{it} = x_i$ for all t. Some regressors, such as an overall time trend, may be individual invariant with $x_{it} = x_t$ for all i. And some may vary over both time and individuals.

Sixth, some or all model coefficients may vary across individuals or over time.

Seventh, the microeconometrics literature emphasizes the fixed-effects model. This model, explained in the next section, permits regressors to be endogenous provided that they are correlated only with a time-invariant component of the error. Most other branches of applied statistics instead emphasize the random-effects model that assumes that regressors are completely exogenous.

Finally, panel data permit estimation of dynamic models where lagged dependent variables may be regressors. Most panel-data analyses use models without this complication.

In this chapter, we focus on short panels (T fixed and $N \to \infty$) with model errors assumed to be independent over individuals. Long panels are treated separately in section 8.10. We consider linear models with and without fixed effects, and both static and dynamic models. The applications in this chapter use balanced panels. Most commands can also be applied to unbalanced panels, as demonstrated in some of the exercises, though one should also then check for panel-attrition bias.

8.2.2 Some basic panel models

There are several different linear models for panel data.

The fundamental distinction is that between fixed-effects and random-effects models. The term "fixed effects" is misleading because in both types of models individual-level effects are random. Fixed-effects models have the added complication that regressors may be correlated with the individual-level effects so that consistent estimation of regression parameters requires eliminating or controlling for the fixed effects.

Individual-effects model

The individual-specific–effects model for the scalar dependent variable y_{it} specifies that

$$y_{it} = \alpha_i + \mathbf{x}'_{it}\boldsymbol{\beta} + \varepsilon_{it} \tag{8.1}$$

where $\mathbf{x}_{it}$ are regressors, α_i are random individual-specific effects, and ε_{it} is an idiosyncratic error.

Two quite different models for the α_i are the fixed-effects and random-effects models.

Fixed-effects model

In the fixed-effects (FE) model, the α_i in (8.1) are permitted to be correlated with the regressors $\mathbf{x}_{it}$. This allows a limited form of endogeneity. We view the error in (8.1) as $u_{it} = \alpha_i + \varepsilon_{it}$ and permit $\mathbf{x}_{it}$ to be correlated with the time-invariant component of the error (α_i), while continuing to assume that $\mathbf{x}_{it}$ is uncorrelated with the idiosyncratic error ε_{it}. For example, we assume that if regressors in an earnings regression are correlated with unobserved ability, they are correlated only with the time-invariant component of ability, captured by α_i.

One possible estimation method is to jointly estimate $\alpha_1, \ldots, \alpha_N$ and $\boldsymbol{\beta}$. But for a short panel, asymptotic theory relies on $N \to \infty$, and here as $N \to \infty$ so too does the number of fixed effects to estimate. This problem is called the incidental-parameters problem. Interest lies in estimating $\boldsymbol{\beta}$, but first we need to control for the nuisance or incidental parameters, α_i.

Instead, it is still possible to consistently estimate $\boldsymbol{\beta}$, for time-varying regressors, by appropriate differencing transformations applied to (8.1) that eliminate α_i. These estimators are detailed in sections 8.5 and 8.9.

The FE model implies that $E(y_{it}|\alpha_i, \mathbf{x}_{it}) = \alpha_i + \mathbf{x}'_{it}\boldsymbol{\beta}$, assuming $E(\varepsilon_{it}|\alpha_i, \mathbf{x}_{it}) = 0$, so $\beta_j = \partial E(y_{it}|\alpha_i, \mathbf{x}_{it})/\partial x_{j,it}$. The attraction of the FE model is that we can obtain a consistent estimate of the marginal effect of the jth regressor on $E(y_{it}|\alpha_i, \mathbf{x}_{it})$, provided $x_{j,it}$ is time varying, even if the regressors are endogenous (albeit, a limited form of endogeneity).

At the same time, knowledge of $\boldsymbol{\beta}$ does not give complete information on the process generating y_{it}. In particular for prediction, we need an estimate of $E(y_{it}|\mathbf{x}_{it}) = E(\alpha_i|\mathbf{x}_{it}) + \mathbf{x}'_{it}\boldsymbol{\beta}$, and $E(\alpha_i|\mathbf{x}_{it})$ cannot be consistently estimated in short panels.

In nonlinear FE models, these results need to be tempered. It is not always possible to eliminate α_i, which is shown in chapter 18. And even if it is, consistent estimation of $\boldsymbol{\beta}$ may still not lead to a consistent estimate of the marginal effect $\partial E(y_{it}|\alpha_i, \mathbf{x}_{it})/\partial x_{j,it}$.

Random-effects model

In the random-effects (RE) model, it is assumed that α_i in (8.1) is purely random, a stronger assumption implying that α_i is uncorrelated with the regressors.

Estimation is then by a feasible generalized least-squares (FGLS) estimator, given in section 8.6. Advantages of the RE model are that it yields estimates of all coefficients and hence marginal effects, even those of time-invariant regressors, and that $E(y_{it}|\mathbf{x}_{it})$ can be estimated. The big disadvantage is that these estimates are inconsistent if the FE model is appropriate.

Pooled model or population-averaged model

Pooled models assume that regressors are exogenous and simply write the error as u_{it} rather than using the decomposition $\alpha_i + \varepsilon_{it}$. Then

$$y_{it} = \alpha + \mathbf{x}'_{it}\boldsymbol{\beta} + u_{it} \tag{8.2}$$

Note that $\mathbf{x}_{it}$ here does not include a constant, whereas in cross-section chapters, $\mathbf{x}_i$ additionally included a constant term.

OLS estimation of the parameters of this model is straightforward, but inference needs to control for likely correlation of the error u_{it} over time for a given individual (within correlation) and possible correlation over individuals (between correlation). FGLS estimation of (8.2) given an assumed model for the within correlation of u_{it} is presented in section 8.4. In the statistics literature, this is called a population-averaged model. Like RE estimators, consistency of the estimators requires that regressors be uncorrelated with u_{it}.

Two-way–effects model

A standard extension of the individual effects is a two-way–effects model that allows the intercept to vary over individuals and over time:

$$y_{it} = \alpha_i + \gamma_t + \mathbf{x}'_{it}\boldsymbol{\beta} + \varepsilon_{it} \tag{8.3}$$

For short panels, it is common to let the time effects γ_t be fixed effects. Then (8.3) reduces to (8.1), if the regressors in (8.1) include a set of time dummies (with one time dummy dropped to avoid the dummy-variable trap).

Mixed linear models

If the RE model is appropriate, richer models can permit slope parameters to also vary over individuals or time. The mixed linear model is a hierarchical linear model that is quite flexible and permits random parameter variation to depend on observable variables. The random-coefficients model is a special case that specifies

$$y_{it} = \alpha_i + \mathbf{x}'_{it}\boldsymbol{\beta}_i + \varepsilon_{it}$$

where $(\alpha_i\,\boldsymbol{\beta}'_i)' \sim (\boldsymbol{\beta}, \boldsymbol{\Sigma})$. For a long panel with few individuals, α_i and $\boldsymbol{\beta}_i$ can instead be parameters that can be estimated by running separate regressions for each individual.

8.2.3   Cluster–robust inference

Various estimators for the preceding models are given in subsequent sections. These estimators are usually based on the assumption that the idiosyncratic error $\varepsilon_{it} \sim (0, \sigma^2_\varepsilon)$. This assumption is often not satisfied in panel applications. Then many panel estimators still retain consistency, provided that ε_{it} are independent over i, but reported standard errors are incorrect.

For short panels, it is possible to obtain cluster–robust standard errors under the weaker assumptions that errors are independent across individuals and that $N \to \infty$. Specifically, $E(\varepsilon_{it}\varepsilon_{js}) = 0$ for $i \neq j$, $E(\varepsilon_{it}\varepsilon_{is})$ is unrestricted, and ε_{it} may be heteroskedastic. Where applicable, we use cluster–robust standard errors rather than the Stata defaults. For some, but not all, `xt` commands, the `vce(robust)` option is available. This leads to a cluster–robust estimate of the variance–covariance matrix of the estimator (VCE) for some commands and a robust estimate of the VCE for some commands. Otherwise, the `vce(bootstrap)` or `vce(jackknife)` options can be used because, for `xt` commands, these usually resample over clusters.

8.2.4   The xtreg command

The key command for estimation of the parameters of a linear panel-data model is the `xtreg` command. The command syntax is

`xtreg` *depvar* [*indepvars*] [*if*] [*in*] [*weight*] [, *options*]

The individual identifier must first be declared with the `xtset` command.

The key model options are population-averaged model (`pa`), FE model (`fe`), RE model (`re` and `mle`), and between-effects model (`be`). The individual models are discussed in detail in subsequent sections. The *weight* modifier is available only for `fe`, `mle`, and `pa`.

The `vce(robust)` option provides cluster–robust estimates of the standard errors, for all models but `be` and `mle`. Stata 10 labels the estimated VCE as simply "Robust" because the use of `xtreg` implies that we are in a clustered setting.

8.2.5 Stata linear panel-data commands

Table 8.1 summarizes `xt` commands for viewing panel data and estimating the parameters of linear panel-data models.

Table 8.1. Summary of `xt` commands

Data summary	`xtset`; `xtdescribe`; `xtsum`; `xtdata`; `xtline`; `xttab`; `xttrans`
Pooled OLS	`regress`
Pooled FGLS	`xtgee, family(gaussian)`; `xtgls`; `xtpcse`
Random effects	`xtreg, re`; `xtregar, re`
Fixed effects	`xtreg, fe`; `xtregar, fe`
Random slopes	`xtmixed`; `xtrc`
First-differences	`regress` (with differenced data)
Static IV	`xtivreg`; `xthtaylor`
Dynamic IV	`xtabond`; `xtdpdsys`; `xtdpd`

The core methods are presented in this chapter, with more specialized commands presented in chapter 9. Readers with long panels should look at section 8.10 (`xtgls`, `xtpcse`, `xtregar`) and data input may require first reading section 8.11.

8.3 Panel-data summary

In this section, we present various ways to summarize and view panel data and estimate a pooled OLS regression. The dataset used is a panel on log hourly wages and other variables for 595 people over the seven years 1976–1982.

8.3.1 Data description and summary statistics

The data, from Baltagi and Khanti-Akom (1990), were drawn from the Panel Study of Income Dynamics (PSID) and are a corrected version of data originally used by Cornwell and Rupert (1988).

The `mus08psidextract.dta` dataset has the following data:

```
. * Read in dataset and describe
. use mus08psidextract.dta, clear
(PSID wage data 1976-82 from Baltagi and Khanti-Akom (1990))

. describe
Contains data from mus08psidextract.dta
  obs:         4,165                        PSID wage data 1976-82 from Baltagi
                                              and Khanti-Akom (1990)
  vars:           22                        16 Aug 2007 16:29
  size:      295,715 (99.1% of memory free) (_dta has notes)
```

variable name	storage type	display format	value label	variable label
exp	float	%9.0g		years of full-time work experience
wks	float	%9.0g		weeks worked
occ	float	%9.0g		occupation; occ==1 if in a blue-collar occupation
ind	float	%9.0g		industry; ind==1 if working in a manufacturing industry
south	float	%9.0g		residence; south==1 if in the South area
smsa	float	%9.0g		smsa==1 if in the Standard metropolitan statistical area
ms	float	%9.0g		marital status
fem	float	%9.0g		female or male
union	float	%9.0g		if wage set be a union contract
ed	float	%9.0g		years of education
blk	float	%9.0g		black
lwage	float	%9.0g		log wage
id	float	%9.0g		
t	float	%9.0g		
tdum1	byte	%8.0g		t== 1.0000
tdum2	byte	%8.0g		t== 2.0000
tdum3	byte	%8.0g		t== 3.0000
tdum4	byte	%8.0g		t== 4.0000
tdum5	byte	%8.0g		t== 5.0000
tdum6	byte	%8.0g		t== 6.0000
tdum7	byte	%8.0g		t== 7.0000
exp2	float	%9.0g		

```
Sorted by:  id  t
```

There are 4,165 individual–year pair observations. The variable labels describe the variables fairly clearly, though note that `lwage` is the log of hourly wage in cents, the indicator `fem` is 1 if female, `id` is the individual identifier, `t` is the year, and `exp2` is the square of `exp`.

(*Continued on next page*)

Descriptive statistics can be obtained by using the command `summarize`:

```
. * Summary of dataset
. summarize
    Variable |       Obs        Mean    Std. Dev.       Min        Max
-------------+--------------------------------------------------------
         exp |      4165    19.85378    10.96637          1         51
         wks |      4165    46.81152    5.129098          5         52
         occ |      4165    .5111645    .4999354          0          1
         ind |      4165    .3954382    .4890033          0          1
       south |      4165    .2902761    .4539442          0          1
-------------+--------------------------------------------------------
        smsa |      4165    .6537815     .475821          0          1
          ms |      4165    .8144058    .3888256          0          1
         fem |      4165     .112605    .3161473          0          1
       union |      4165    .3639856    .4812023          0          1
          ed |      4165    12.84538    2.787995          4         17
-------------+--------------------------------------------------------
         blk |      4165    .0722689    .2589637          0          1
       lwage |      4165    6.676346    .4615122    4.60517      8.537
          id |      4165         298    171.7821          1        595
           t |      4165           4     2.00024          1          7
       tdum1 |      4165    .1428571    .3499691          0          1
-------------+--------------------------------------------------------
       tdum2 |      4165    .1428571    .3499691          0          1
       tdum3 |      4165    .1428571    .3499691          0          1
       tdum4 |      4165    .1428571    .3499691          0          1
       tdum5 |      4165    .1428571    .3499691          0          1
       tdum6 |      4165    .1428571    .3499691          0          1
-------------+--------------------------------------------------------
       tdum7 |      4165    .1428571    .3499691          0          1
        exp2 |      4165     514.405    496.9962          1       2601
```

The variables take on values that are within the expected ranges, and there are no missing values. Both men and women are included, though from the mean of `fem` only 11% of the sample is female. Wages data are nonmissing in all years, and weeks worked are always positive, so the sample is restricted to individuals who work in all seven years.

8.3.2   Panel-data organization

The `xt` commands require that panel data be organized in so-called long form, with each observation a distinct individual–time pair, here an individual–year pair. Data may instead be organized in wide form, with a single observation combining data from all years for a given individual or combining data on all individuals for a given year. Then the data need to be converted from wide form to long form by using the `reshape` command presented in section 8.11.

Data organization can often be clear from listing the first few observations. For brevity, we list the first three observations for a few variables:

```
. * Organization of dataset
. list id t exp wks occ in 1/3, clean

        id   t   exp   wks   occ
   1.    1   1     3    32     0
   2.    1   2     4    43     0
   3.    1   3     5    40     0
```

The first observation is for individual 1 in year 1, the second observation is for individual 1 in year 2, and so on. These data are thus in long form. From `summarize`, the panel identifier `id` takes on the values 1–595, and the time variable `t` takes on the values 1–7. In general, the panel identifier need just be a unique identifier and the time variable could take on values of, for example, 76–82.

The panel-data `xt` commands require that, at a minimum, the panel identifier be declared. Many `xt` commands also require that the time identifier be declared. This is done by using the `xtset` command. Here we declare both identifiers:

```
. * Declare individual identifier and time identifier
. xtset id t
       panel variable:  id (strongly balanced)
        time variable:  t, 1 to 7
               delta:  1 unit
```

The panel identifier is given first, followed by the optional time identifier. The output indicates that data are available for all individuals in all time periods (strongly balanced) and the time variable increments uniformly by one.

When a Stata dataset is saved, the current settings, if any, from `xtset` are also saved. In this particular case, the original Stata dataset `psidextract.dta` already contained this information, so the preceding `xtset` command was actually unnecessary. The `xtset` command without any arguments reveals the current settings, if any.

8.3.3 Panel-data description

Once the panel data are `xtset`, the `xtdescribe` command provides information about the extent to which the panel is unbalanced.

```
. * Panel description of dataset
. xtdescribe
       id: 1, 2, ..., 595                                     n =      595
        t: 1, 2, ..., 7                                       T =        7
           Delta(t) = 1 unit
           Span(t)  = 7 periods
           (id*t uniquely identifies each observation)
Distribution of T_i:   min      5%     25%     50%     75%     95%     max
                         7       7       7       7       7       7       7

     Freq.  Percent    Cum. |  Pattern
   -------------------------+---------
      595    100.00  100.00 |  1111111
   -------------------------+---------
      595    100.00         |  XXXXXXX
```

In this case, all 595 individuals have exactly 7 years of data. The data are therefore balanced because, additionally, the earlier `summarize` command showed that there are no missing values. Section 18.3 provides an example of `xtdescribe` with unbalanced data.

8.3.4   Within and between variation

Dependent variables and regressors can potentially vary over both time and individuals. Variation over time or a given individual is called within variation, and variation across individuals is called between variation. This distinction is important because estimators differ in their use of within and between variation. In particular, in the FE model the coefficient of a regressor with little within variation will be imprecisely estimated and will be not identified if there is no within variation at all.

The `xtsum`, `xttab`, and `xttrans` commands provide information on the relative importance of within variation and between variation of a variable.

We begin with `xtsum`. The total variation (around grand mean $\bar{x} = 1/NT \sum_i \sum_t x_{it}$) can be decomposed into within variation over time for each individual (around individual mean $\bar{x}_i = 1/T \sum_t x_{it}$) and between variation across individuals (for $\bar{x}$ around $\bar{x}_i$). The corresponding decomposition for the variance is

Within variance: $s_W^2 = \frac{1}{NT-1} \sum_i \sum_t (x_{it} - \bar{x}_i)^2 = \frac{1}{NT-1} \sum_i \sum_t (x_{it} - \bar{x}_i + \bar{x})^2$

Between variance: $s_B^2 = \frac{1}{N-1} \sum_i (\bar{x}_i - \bar{x})^2$

Overall variance: $s_O^2 = \frac{1}{NT-1} \sum_i \sum_t (x_{it} - \bar{x})^2$

The second expression for s_W^2 is equivalent to the first, because adding a constant does not change the variance, and is used at times because $x_{it} - \bar{x}_i + \bar{x}$ is centered on $\bar{x}$, providing a sense of scale, whereas $x_{it} - \bar{x}_i$ is centered on zero. For unbalanced data, replace NT in the formulas with $\sum_i T_i$. It can be shown that $s_O^2 \simeq s_W^2 + s_B^2$.

The `xtsum` command provides this variance decomposition. We do this for selected regressors and obtain

```
. * Panel summary statistics: within and between variation
. xtsum id t lwage ed exp exp2 wks south tdum1
```

Variable		Mean	Std. Dev.	Min	Max	Observations		
id	overall	298	171.7821	1	595	N =		4165
	between		171.906	1	595	n =		595
	within		0	298	298	T =		7
t	overall	4	2.00024	1	7	N =		4165
	between		0	4	4	n =		595
	within		2.00024	1	7	T =		7
lwage	overall	6.676346	.4615122	4.60517	8.537	N =		4165
	between		.3942387	5.3364	7.813596	n =		595
	within		.2404023	4.781808	8.621092	T =		7
ed	overall	12.84538	2.787995	4	17	N =		4165
	between		2.790006	4	17	n =		595
	within		0	12.84538	12.84538	T =		7
exp	overall	19.85378	10.96637	1	51	N =		4165
	between		10.79018	4	48	n =		595
	within		2.00024	16.85378	22.85378	T =		7
exp2	overall	514.405	496.9962	1	2601	N =		4165
	between		489.0495	20	2308	n =		595
	within		90.44581	231.405	807.405	T =		7
wks	overall	46.81152	5.129098	5	52	N =		4165
	between		3.284016	31.57143	51.57143	n =		595
	within		3.941881	12.2401	63.66867	T =		7
south	overall	.2902761	.4539442	0	1	N =		4165
	between		.4489462	0	1	n =		595
	within		.0693042	-.5668667	1.147419	T =		7
tdum1	overall	.1428571	.3499691	0	1	N =		4165
	between		0	.1428571	.1428571	n =		595
	within		.3499691	0	1	T =		7

Time-invariant regressors have zero within variation, so the individual identifier `id` and the variable `ed` are time-invariant. Individual-invariant regressors have zero between variation, so the time identifier `t` and the time dummy `tdum1` are individual-invariant. For all other variables but `wks`, there is more variation across individuals (between variation) than over time (within variation), so within estimation may lead to considerable efficiency loss. What is not clear from the output from `xtsum` is that while variable `exp` has nonzero within variation, it evolves deterministically because for this sample `exp` increments by one with each additional period. The `min` and `max` columns give the minimums and maximums of x_{it} for `overall`, $\bar{x}_i$ for `between`, and $x_{it} - \bar{x}_i + \bar{x}$ for `within`.

In the `xtsum` output, Stata uses lowercase n to denote the number of individuals and uppercase N to denote the total number of individual–time observations. In our notation, these quantities are, respectively, N and $\sum_{i=1}^{N} T_i$.

The `xttab` command tabulates data in a way that provides additional details on the within and between variation of a variable. For example,

```
. * Panel tabulation for a variable
. xttab south
```

south	Overall Freq.	Percent	Between Freq.	Percent	Within Percent
0	2956	70.97	428	71.93	98.66
1	1209	29.03	182	30.59	94.90
Total	4165	100.00	610	102.52	97.54

(n = 595)

The overall summary shows that 71% of the 4,165 individual–year observations had south $= 0$, and 29% had south $= 1$. The between summary indicates that of the 595 people, 72% had south $= 0$ at least once and 31% had south $= 1$ at least once. The between total percentage is 102.52, because 2.52% of the sampled individuals (15 persons) lived some of the time in the south and some not in the south and hence are double counted. The within summary indicates that 95% of people who ever lived in the south always lived in the south during the time period covered by the panel, and 99% who lived outside the south always lived outside the south. The south variable is close to time-invariant.

The `xttab` command is most useful when the variable takes on few values, because then there are few values to tabulate and interpret.

The `xttrans` command provides transition probabilities from one period to the next. For example,

```
. * Transition probabilities for a variable
. xttrans south, freq
```

residence; south==1 if in the South area	residence; south==1 if in the South area 0	1	Total
0	2,527	8	2,535
	99.68	0.32	100.00
1	8	1,027	1,035
	0.77	99.23	100.00
Total	2,535	1,035	3,570
	71.01	28.99	100.00

One time period is lost in calculating transitions, so 3,570 observations are used. For time-invariant data, the diagonal entries will be 100% and the off-diagonal entries will be 0%. For south, 99.2% of the observations ever in the south for one period remain in the south for the next period. And for those who did not live in the south for one period, 99.7% remained outside the south for the next period. The south variable is close to time-invariant.

The `xttrans` command is most useful when the variable takes on few values.

8.3.5   Time-series plots for each individual

It can be useful to provide separate time-series plots for some or all individual units.

Separate time-series plots of a variable for one or more individuals can be obtained by using the `xtline` command. The `overlay` option overlays the plots for each individual on the same graph. For example,

```
. quietly xtline lwage if id<=20, overlay
```

produces overlaid time-series plots of `lwage` for the first 20 individuals in the sample.

We provide time-series plots for the first 20 individuals in the sample. The default is to provide a graph legend that identifies each individual that appears in the graph and takes up much of the graph if the graph uses data from many individuals. This legend can be suppressed by using the `legend(off)` option. Separate plots are obtained for `lwage` and for `wks`, and these are then combined by using the `graph combine` command. We have

```
. * Simple time-series plot for each of 20 individuals
. quietly xtline lwage if id<=20, overlay legend(off) saving(lwage, replace)
. quietly xtline wks if id<=20, overlay legend(off) saving(wks, replace)
. graph combine lwage.gph wks.gph, iscale(1)
```

Figure 8.1 shows that the wage rate increases roughly linearly over time, aside from two individuals with large increases from years 1 to 2, and that weeks worked show no discernible trend over time.

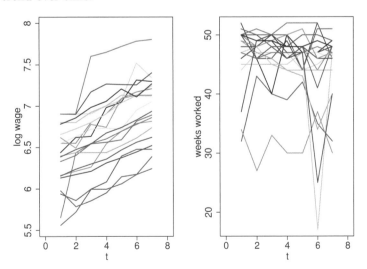

Figure 8.1. Time-series plots of log wage against year and weeks worked against year for each of the first 20 observations

8.3.6 Overall scatterplot

In cases where there is one key regressor, we can begin with a scatterplot of the dependent variable on the key regressor, using data from all panel observations.

The following command adds fitted quadratic regression and lowess regression curves to the scatterplot.

```
. graph twoway (scatter lwage exp) (qfit lwage exp) (lowess lwage exp)
```

This produces a graph that is difficult to read as the scatterplot points are very large, making it hard to then see the regression curves.

The following code presents a better-looking scatterplot of `lnwage` on `exp`, along with the fitted regression lines. It uses the same graph options as those explained in section 2.6.6. We have

```
. * Scatterplot, quadratic fit and nonparametric regression (lowess)
. graph twoway (scatter lwage exp, msize(small) msymbol(o))
> (qfit lwage exp, clstyle(p3) lwidth(medthick))
> (lowess lwage exp, bwidth(0.4) clstyle(p1) lwidth(medthick)),
> plotregion(style(none))
> title("Overall variation: Log wage versus experience")
> xtitle("Years of experience", size(medlarge)) xscale(titlegap(*5))
> ytitle("Log hourly wage", size(medlarge)) yscale(titlegap(*5))
> legend(pos(4) ring(0) col(1)) legend(size(small))
> legend(label(1 "Actual Data") label(2 "Quadratic fit") label(3 "Lowess"))
```

Each point on figure 8.2 represents an individual–year pair. The dashed smooth curve line is fitted by OLS of `lwage` on a quadratic in `exp` (using `qfit`), and the solid line is fitted by nonparametric regression (using `lowess`). Log wage increases until thirty or so years of experience and then declines.

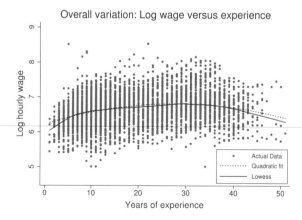

Figure 8.2. Overall scatterplot of log wage against experience using all observations

8.3.7   Within scatterplot

The `xtdata` command can be used to obtain similar plots for within variation, using option `fe`; between variation, using option `be`; and RE variation (the default), using option `re`. The `xtdata` command replaces the data in memory with the specified transform, so you should first `preserve` the data and then `restore` the data when you are finished with the transformed data.

For example, the `fe` option creates deviations from means, so that $(y_{it} - \overline{y}_i + \overline{y})$ is plotted against $(x_{it} - \overline{x}_i + \overline{x})$. For `lwage` plotted against `exp`, we obtain

```
. * Scatterplot for within variation
. preserve
. xtdata, fe
. graph twoway (scatter lwage exp) (qfit lwage exp) (lowess lwage exp),
> plotregion(style(none)) title("Within variation: Log wage versus experience")
. restore
```

The result is given in figure 8.3. At first glance, this figure is puzzling because only seven distinct values of `exp` appear. But the panel is balanced and `exp` (years of work experience) is increasing by exactly one each period for each individual in this sample of people who worked every year. So $(x_{it} - \overline{x}_i)$ increases by one each period, as does $(x_{it} - \overline{x}_i + \overline{x})$. The latter quantity is centered on $\overline{x} = 19.85$ (see section 8.3.1), which is the value in the middle year with $t = 4$. Clearly, it can be very useful to plot a figure such as this.

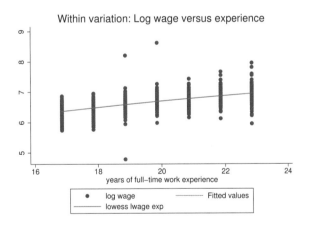

Figure 8.3. Within scatterplot of log-wage deviations from individual means against experience deviations from individual means

8.3.8   Pooled OLS regression with cluster–robust standard errors

A natural starting point is a pooled OLS regression for log wage using data for all individuals in all years.

We include as regressors education, weeks worked, and a quadratic in experience. Education is a time-invariant regressor, taking the same value each year for a given individual. Weeks worked is an example of a time-varying regressor. Experience is also time-varying, though it is so in a deterministic way as the sample comprises people who work full-time in all years, so experience increases by one year as t increments by one.

Regressing y_{it} on $\mathbf{x}_{it}$ yields consistent estimates of $\boldsymbol{\beta}$ if the composite error u_{it} in the pooled model of (8.2) is uncorrelated with $\mathbf{x}_{it}$. As explained in section 8.2, the error u_{it} is likely to be correlated over time for a given individual, so we use cluster–robust standard errors that cluster on the individual. We have

```
. * Pooled OLS with cluster-robust standard errors
. use mus08psidextract.dta, clear
(PSID wage data 1976-82 from Baltagi and Khanti-Akom (1990))

. regress lwage exp exp2 wks ed, vce(cluster id)
```

Linear regression			Number of obs =	4165
			F(4, 594) =	72.58
			Prob > F =	0.0000
			R-squared =	0.2836
			Root MSE =	.39082

(Std. Err. adjusted for 595 clusters in id)

lwage	Coef.	Robust Std. Err.	t	P>|t|	[95% Conf. Interval]
exp	.044675	.0054385	8.21	0.000	.0339941 .055356
exp2	-.0007156	.0001285	-5.57	0.000	-.0009679 -.0004633
wks	.005827	.0019284	3.02	0.003	.0020396 .0096144
ed	.0760407	.0052122	14.59	0.000	.0658042 .0862772
_cons	4.907961	.1399887	35.06	0.000	4.633028 5.182894

The output shows that $R^2 = 0.28$, and the estimates imply that wages increase with experience until a peak at 31 years $[= 0.0447/(2 \times 0.00072)]$ and then decline. Wages increase by 0.6% with each additional week worked. And wages increase by 7.6% with each additional year of education.

For panel data, it is essential that OLS standard errors be corrected for clustering on the individual. In contrast, the default standard errors assume that the regression errors are independent and identically distributed (i.i.d.). Using the default standard errors, we obtain

```
. * Pooled OLS with incorrect default standard errors
. regress lwage exp exp2 wks ed
```

Source	SS	df	MS		
Model	251.491445	4	62.8728613		
Residual	635.413457	4160	.152743619		
Total	886.904902	4164	.212993492		

Number of obs = 4165
F(4, 4160) = 411.62
Prob > F = 0.0000
R-squared = 0.2836
Adj R-squared = 0.2829
Root MSE = .39082

lwage	Coef.	Std. Err.	t	P>|t|	[95% Conf. Interval]	
exp	.044675	.0023929	18.67	0.000	.0399838	.0493663
exp2	-.0007156	.0000528	-13.56	0.000	-.0008191	-.0006121
wks	.005827	.0011827	4.93	0.000	.0035084	.0081456
ed	.0760407	.0022266	34.15	0.000	.0716754	.080406
_cons	4.907961	.0673297	72.89	0.000	4.775959	5.039963

These standard errors are misleadingly small; the cluster–robust standard errors are, respectively, 0.0054, 0.0001, 0.0019, and 0.0052.

It is likely that if log wage is overpredicted in one year for a given person, then it is likely to be overpredicted in other years. Failure to control for this error correlation leads to underestimation of standard errors because, intuitively, each additional observation for a given person actually provides less than an independent piece of new information.

The difference between default and cluster–robust standard errors for pooled OLS can be very large. The difference increases with increasing T, increasing autocorrelation in model errors, and increasing autocorrelation of the regressor of interest. Specifically, the standard-error inflation factor $\tau \simeq \sqrt{1 + \rho_u \rho_x (T-1)}$, where ρ_u is the intraclass correlation of the error, defined below in (8.4), and ρ_x is the intraclass correlation of the regressor. Here $\rho_u \simeq 0.80$, shown below, and for time-invariant regressor ed, $\rho_x = 1$, so $\tau \simeq \sqrt{1 + 0.80 \times 1 \times 6} = \sqrt{5.8} \simeq 2.41$ for ed. Similarly the regressor exp has $\rho_x = 1$ because for this sample experience increases by one year as t increments by one.

Cluster–robust standard errors require that $N \to \infty$ and that errors are independent over i. The assumption of independence over i can be relaxed to independence at a more aggregated level, provided that the number of units is still large and the units nest the individual. For example, the PSID is a household survey and errors for individuals from the same household may be correlated. If, say, houseid is available as a household identifier, then we would use the vce(cluster houseid) option. As a second example, if the regressor of interest is aggregated at the state level, such as a state policy variable, and there are many states, then it may be better to use the vce(cluster state) option.

8.3.9 Time-series autocorrelations for panel data

The Stata time-series operators can be applied to panel data when both panel and time identifiers are set with the xtset command. Examples include L.lwage or L1.lwage for lwage lagged once, L2.lwage for lwage lagged twice, D.lwage for the difference in lwage (equals lwage − L.lwage), LD.lwage for this difference lagged once, and L2D.lwage for this difference lagged twice.

Use of these operators is the best way to create lagged variables because relevant missing values are automatically and correctly created. For example, `regress lwage L2.wage` will use $(7 - 2) \times 595$ observations because forming `L2.wage` leads to a loss of the first two years of data for each of the 595 individuals.

The `corrgram` command for computing autocorrelations of time-series data does not work for panel data. Instead, autocorrelations can be obtained by using the `correlate` command. For example,

```
. * First-order autocorrelation in a variable
. sort id t
. correlate lwage L.lwage
(obs=3570)
```

| | | L. |
	lwage	lwage
lwage		
--.	1.0000	
L1.	0.9189	1.0000

calculates the first-order autocorrelation coefficient for `lwage` to be 0.92.

We now calculate autocorrelations at all lags (here up to six periods). Rather than doing so for `lwage`, we do so for the residuals from the previous pooled OLS regression for `lwage`. We have

```
. * Autocorrelations of residual
. quietly regress lwage exp exp2 wks ed, vce(cluster id)
. predict uhat, residuals
. forvalues j = 1/6 {
  2.      quietly corr uhat L`j´.uhat
  3.      display "Autocorrelation at lag `j´ = " %6.3f r(rho)
  4.      }
Autocorrelation at lag 1 =  0.884
Autocorrelation at lag 2 =  0.838
Autocorrelation at lag 3 =  0.811
Autocorrelation at lag 4 =  0.786
Autocorrelation at lag 5 =  0.750
Autocorrelation at lag 6 =  0.729
```

The `forvalues` loop leads to separate computation of each autocorrelation to maximize the number of observations used. If instead we gave a one-line command to compute the autocorrelations of `uhat` through `L6.uhat`, then only 595 observations would have been used. Here 6×595 observations are used to compute the autocorrelation at lag 1, 5×595 observations are used to compute the autocorrelation at lag 2, and so on. The average of the autocorrelations, 0.80, provides a rough estimate of the intraclass correlation coefficient of the residuals.

Clearly, the errors are serially correlated, and cluster–robust standard errors after pooled OLS are required. The individual-effects model provides an explanation for this correlation. If the error $u_{it} = \alpha_i + \varepsilon_{it}$, then even if ε_{it} is i.i.d. $(0, \sigma_\varepsilon^2)$, we have $\mathrm{Cor}(u_{it}, u_{is}) \neq 0$ for $t \neq s$ if $\alpha_i \neq 0$. The individual effect α_i induces correlation over time for a given individual.

The preceding estimated autocorrelations are constant across years. For example, the correlation of uhat with L.uhat across years 1 and 2 is assumed to be the same as that across years 2 and 3, years 3 and 4, . . . , years 6 and 7. This presumes that the errors are stationary.

In the nonstationary case, the autocorrelations will differ across pairs of years. For example, we consider the autocorrelations one year apart and allow these to differ across the year pairs. We have

```
. * First-order autocorrelation differs in different year pairs
. forvalues s = 2/7 {
  2.        quietly corr uhat L1.uhat if t == `s´
  3.        display "Autocorrelation at lag 1 in year `s´ = " %6.3f r(rho)
  4.        }
Autocorrelation at lag 1 in year 2 =  0.915
Autocorrelation at lag 1 in year 3 =  0.799
Autocorrelation at lag 1 in year 4 =  0.855
Autocorrelation at lag 1 in year 5 =  0.867
Autocorrelation at lag 1 in year 6 =  0.894
Autocorrelation at lag 1 in year 7 =  0.893
```

The lag-1 autocorrelations for individual–year pairs range from 0.80 to 0.92, and their average is 0.87. From the earlier output, the lag-1 autocorrelation equals 0.88 when it is constrained to be equal across all year pairs. It is common to impose equality for simplicity.

8.3.10   Error correlation in the RE model

For the individual-effects model (8.1), the combined error $u_{it} = \alpha_i + \varepsilon_{it}$. The RE model assumes that α_i is i.i.d. with a variance of σ_α^2 and that u_{it} is i.i.d. with a variance of σ_ε^2.

Then u_{it} has a variance of $\mathrm{Var}(u_{it}) = \sigma_\alpha^2 + \sigma_\varepsilon^2$ and a covariance of $\mathrm{Cov}(u_{it}, u_{is}) = \sigma_\alpha^2$, $s \neq t$. It follows that in the RE model,

$$\rho_u = \mathrm{Cor}(u_{it}, u_{is}) = \sigma_\alpha^2/(\sigma_\alpha^2 + \sigma_\varepsilon^2), \text{ for all } s \neq t \tag{8.4}$$

This constant correlation is called the intraclass correlation of the error.

The RE model therefore permits serial correlation in the model error. This correlation can approach 1 if the random effect is large relative to the idiosyncratic error, so that σ_α^2 is large relative to σ_ε^2.

This serial correlation is restricted to be the same at all lags, and the errors u_{it} are then called equicorrelated or exchangeable. From section 8.3.9, the error correlations were, respectively, 0.88, 0.84, 0.81, 0.79, 0.75, and 0.73, so a better model may be one that allows the error correlation to decrease with the lag length.

8.4 Pooled or population-averaged estimators

Pooled estimators simply regress y_{it} on an intercept and $\mathbf{x}_{it}$, using both between (cross-section) and within (time-series) variation in the data. Standard errors need to adjust for any error correlation and, given a model for error correlation, more-efficient FGLS estimation is possible. Pooled estimators, called population-averaged estimators in the statistics literature, are consistent if the RE model is appropriate and are inconsistent if the FE model is appropriate.

8.4.1 Pooled OLS estimator

The pooled OLS estimator can be motivated from the individual-effects model by rewriting (8.1) as the pooled model

$$y_{it} = \alpha + \mathbf{x}'_{it}\boldsymbol{\beta} + (\alpha_i - \alpha + \varepsilon_{it}) \tag{8.5}$$

Any time-specific effects are assumed to be fixed and already included as time dummies in the regressors $\mathbf{x}_{it}$. The model (8.5) explicitly includes a common intercept, and the individual effects $\alpha_i - \alpha$ are now centered on zero.

Consistency of OLS requires that the error term $(\alpha_i - \alpha + \varepsilon_{it})$ be uncorrelated with $\mathbf{x}_{it}$. So pooled OLS is consistent in the RE model but is inconsistent in the FE model because then α_i is correlated with $\mathbf{x}_{it}$.

The pooled OLS estimator for our data example has already been presented in section 8.3.8. As emphasized there, cluster–robust standard errors are necessary in the common case of a short panel with independence across individuals.

8.4.2 Pooled FGLS estimator or population-averaged estimator

Pooled FGLS (PFGLS) estimation can lead to estimators of the parameters of the pooled model (8.5) that are more efficient than OLS estimation. Again we assume that any individual-level effects are uncorrelated with regressors, so PFGLS is consistent.

Different assumptions about the correlation structure for the errors u_{it} lead to different PFGLS estimators. In section 8.10, we present some estimators for long panels, using the `xtgls` and `xtregar` commands.

Here we consider only short panels with errors independent across individuals. We need to model the $T \times T$ matrix of error correlations. An assumed correlation structure, called a working matrix, is specified and the appropriate PFGLS estimator is obtained. To guard against the working matrix being a misspecified model of the error correlation, cluster–robust standard errors are computed. Better models for the error correlation lead to more-efficient estimators, but the use of robust standard errors means that the estimators are not presumed to be fully efficient.

In statistics literature, the pooled approach is called a population-averaged (PA) approach, because any individual effects are assumed to be random and are averaged out. The PFGLS estimator is then called the PA estimator.

8.4.3 The xtreg, pa command

The pooled estimator, or PA estimator, is obtained by using the **xtreg** command (see section 8.2.4) with the **pa** option. The two key additional options are **corr()**, to place different restrictions on the error correlations, and **vce(robust)**, to obtain cluster–robust standard errors that are valid even if **corr()** does not specify the correct correlation model, provided that observations are independent over i and $N \to \infty$.

Let $\rho_{ts} = \text{Cor}(u_{it}u_{is})$, the error correlation over time for individual i, and note the restriction that ρ_{ts} does not vary with i. The **corr()** options all set $\rho_{tt} = 1$ but differ in the model for ρ_{ts} for $t \neq s$. With T time periods, the correlation matrix is $T \times T$, and there are potentially as many as $T(T-1)$ unique off-diagonal entries because it need not necessarily be the case that $\rho_{ts} = \rho_{st}$.

The **corr(independent)** option sets $\rho_{ts} = 0$ for $s \neq t$. Then the PA estimator equals the pooled OLS estimator.

The **corr(exchangeable)** option sets $\rho_{ts} = \rho$ for all $s \neq t$ so that errors are assumed to be equicorrelated. This assumption is imposed by the RE model (see section 8.3.10), and as a result, **xtreg, pa** with this option is asymptotically equivalent to **xtreg, re**.

For panel data, it is often the case that the error correlation ρ_{ts} declines as the time difference $|t-s|$ increases—the application in section 8.3.9 provided an example. The **corr(ar k)** option models this dampening by assuming an autoregressive process of order k, or AR(k) process, for u_{it}. For example, **corr(ar 1)** assumes that $u_{it} = \rho_1 u_{i,t-1} + \varepsilon_{it}$, which implies that $\rho_{ts} = \rho_1^{|t-s|}$. The **corr(stationary g)** option instead uses a moving-average process, or MA(g) process. This sets $\rho_{ts} = \rho_{|t-s|}$ if $|t-s| \leq g$, and $\rho_{ts} = 0$ if $|t-s| > g$.

The **corr(unstructured)** option places no restrictions on ρ_{ts}, aside from equality of $\rho_{i,ts}$ across individuals. Then $\text{Cov}(u_{it}, u_{is}) = 1/N \sum_i (\widehat{u}_{it} - \overline{\widehat{u}_t})(\widehat{u}_{is} - \overline{\widehat{u}_s})$. For small T, this may be the best model, but for larger T, the method can fail numerically because there are $T(T-1)$ unique parameters ρ_{ts} to estimate. The **corr(nonstationary g)** option allows ρ_{ts} to be unrestricted if $|t-s| \leq g$ and sets $\rho_{ts} = 0$ if $|t-s| > g$ so there are fewer correlation parameters to estimate.

The PA estimator is also called the generalized estimating equations estimator in the statistics literature. The **xtreg, pa** command is the special case of **xtgee** with the **family(gaussian)** option. The more general **xtgee** command, presented in section 18.4.4, has other options that permit application to a wide range of nonlinear panel models.

8.4.4 Application of the xtreg, pa command

As an example, we specify an AR(2) error process. We have

```
. * Population-averaged or pooled FGLS estimator with AR(2) error
. xtreg lwage exp exp2 wks ed, pa corr(ar 2) vce(robust) nolog
GEE population-averaged model          Number of obs      =       4165
Group and time vars:              id t  Number of groups   =        595
Link:                         identity  Obs per group: min =          7
Family:                       Gaussian                 avg =        7.0
Correlation:                     AR(2)                  max =          7
                                       Wald chi2(4)       =     873.28
Scale parameter:               .1966639  Prob > chi2       =     0.0000

                                       (Std. Err. adjusted for clustering on id)
```

lwage	Coef.	Semirobust Std. Err.	z	P>|z|	[95% Conf. Interval]	
exp	.0718915	.003999	17.98	0.000	.0640535	.0797294
exp2	-.0008966	.0000933	-9.61	0.000	-.0010794	-.0007137
wks	.0002964	.0010553	0.28	0.779	-.001772	.0023647
ed	.0905069	.0060161	15.04	0.000	.0787156	.1022982
_cons	4.526381	.1056897	42.83	0.000	4.319233	4.733529

The coefficients change considerably compared with those from pooled OLS. The cluster–robust standard errors are smaller than those from pooled OLS for all regressors except ed, illustrating the desired improved efficiency because of better modeling of the error correlations. Note that unlike the pure time-series case, controlling for autocorrelation does not lead to the loss of initial observations.

The estimated correlation matrix is stored in e(R). We have

```
. * Estimated error correlation matrix after xtreg, pa
. matrix list e(R)
symmetric e(R)[7,7]
            c1          c2          c3          c4          c5          c6          c7
r1           1
r2   .89722058           1
r3   .84308581   .89722058           1
r4   .78392846   .84308581   .89722058           1
r5   .73064474   .78392846   .84308581   .89722058           1
r6    .6806209   .73064474   .78392846   .84308581   .89722058           1
r7   .63409777    .6806209   .73064474   .78392846   .84308581   .89722058           1
```

By comparison, from section 8.3.9 the autocorrelations of the errors after pooled OLS estimation were 0.88, 0.84, 0.81, 0.79, 0.75, and 0.73.

In an end-of-chapter exercise, we compare estimates obtained using different error-correlation structures.

8.5 Within estimator

Estimators of the parameters $\boldsymbol{\beta}$ of the FE model (8.1) must remove the fixed effects α_i. The within transform discussed in the next section does so by mean-differencing. The within estimator performs OLS on the mean-differenced data. Because all the observations of the mean-difference of a time-invariant variable are zero, we cannot estimate the coefficient on a time-invariant variable.

Because the within estimator provides a consistent estimate of the FE model, it is often called the FE estimator, though the first-difference estimator given in section 8.9 also provides consistent estimates in the FE model. The within estimator is also consistent under the RE model, but alternative estimators are more efficient in the RE model.

8.5.1 Within estimator

The fixed effects α_i in the model (8.1) can be eliminated by subtraction of the corresponding model for individual means $\overline{y}_i = \overline{\mathbf{x}}_i{}'\boldsymbol{\beta} + \overline{\varepsilon}_i$, leading to the within model or mean-difference model

$$(y_{it} - \overline{y}_i) = (\mathbf{x}_{it} - \overline{\mathbf{x}}_i)'\boldsymbol{\beta} + (\varepsilon_{it} - \overline{\varepsilon}_i) \tag{8.6}$$

where, for example, $\overline{\mathbf{x}}_i = T_i^{-1} \sum_{t=1}^{T_i} \mathbf{x}_{it}$. The within estimator is the OLS estimator of this model.

Because α_i has been eliminated, OLS leads to consistent estimates of $\boldsymbol{\beta}$ even if α_i is correlated with $\mathbf{x}_{it}$, as is the case in the FE model. This result is a great advantage of panel data. Consistent estimation is possible even with endogenous regressors $\mathbf{x}_{it}$, provided that $\mathbf{x}_{it}$ is correlated only with the time-invariant component of the error, α_i, and not with the time-varying component of the error, ε_{it}.

This desirable property of consistent parameter estimation in the FE model is tempered, however, by the inability to estimate the coefficients or a time-invariant regressor. Also the within estimator will be relatively imprecise for time-varying regressors that vary little over time.

Stata actually fits the model

$$(y_{it} - \overline{y}_i + \overline{\overline{y}}) = \alpha + (\mathbf{x}_{it} - \overline{\mathbf{x}}_i + \overline{\overline{\mathbf{x}}})'\boldsymbol{\beta} + (\varepsilon_{it} - \overline{\varepsilon}_i + \overline{\overline{\varepsilon}}) \tag{8.7}$$

where, for example, $\overline{\overline{y}} = (1/N)\overline{y}_i$ is the grand mean of y_{it}. This parameterization has the advantage of providing an intercept estimate, the average of the individual effects α_i, while yielding the same slope estimate $\boldsymbol{\beta}$ as that from the within model.

8.5.2 The xtreg, fe command

The within estimator is computed by using the `xtreg` command (see section 8.2.4) with the `fe` option. The default standard errors assume that after controlling for α_i, the error

ε_{it} is i.i.d. The `vce(robust)` option relaxes this assumption and provides cluster–robust standard errors, provided that observations are independent over i and $N \to \infty$.

8.5.3 Application of the xtreg, fe command

For our data, we obtain

```
. * Within or FE estimator with cluster-robust standard errors
. xtreg lwage exp exp2 wks ed, fe vce(cluster id)
note: ed omitted because of collinearity
Fixed-effects (within) regression         Number of obs      =       4165
Group variable: id                        Number of groups   =        595

R-sq:  within  = 0.6566                    Obs per group: min =          7
       between = 0.0276                                   avg =        7.0
       overall = 0.0476                                   max =          7

                                           F(3,594)           =    1059.72
corr(u_i, Xb)  = -0.9107                    Prob > F           =     0.0000

                               (Std. Err. adjusted for 595 clusters in id)
```

lwage	Coef.	Robust Std. Err.	t	P>|t|	[95% Conf. Interval]	
exp	.1137879	.0040289	28.24	0.000	.1058753	.1217004
exp2	-.0004244	.0000822	-5.16	0.000	-.0005858	-.0002629
wks	.0008359	.0008697	0.96	0.337	-.0008721	.0025439
ed	(omitted)					
_cons	4.596396	.0600887	76.49	0.000	4.478384	4.714408
sigma_u	1.0362039					
sigma_e	.15220316					
rho	.97888036	(fraction of variance due to u_i)				

Compared with pooled OLS, the standard errors have roughly tripled because only within variation of the data is being used. The `sigma_u` and `sigma_e` entries are explained in section 8.8.1, and the R^2 measures are explained in section 8.8.2.

The most striking result is that the coefficient for education is not identified. This is because the data on education is time-invariant. In fact, given that we knew from the `xtsum` output in section 8.3.4 that `ed` had zero within standard deviation, we should not have included it as one of the regressors in the `xtreg, fe` command.

This is unfortunate because how wages depend on education is of great policy interest. It is certainly endogenous, because people with high ability are likely to have on average both high education and high wages. Alternative panel-data methods to control for endogeneity of the `ed` variable are presented in chapter 9. In other panel applications, endogenous regressors may be time-varying and the within estimator will suffice.

8.5.4   Least-squares dummy-variables regression

The within estimator of $\boldsymbol{\beta}$ is also called the FE estimator because it can be shown to equal the estimator obtained from direct OLS estimation of $\alpha_1, \ldots, \alpha_N$ and $\boldsymbol{\beta}$ in the original individual-effects model (8.1). The estimates of the fixed effects are then $\widehat{\alpha}_i = \bar{y}_i - \bar{\mathbf{x}}_i'\widehat{\boldsymbol{\beta}}$. In short panels, $\widehat{\alpha}_i$ is not consistently estimated, because it essentially relies on only T_i observations used to form $\bar{y}_i$ and $\bar{\mathbf{x}}_i$, but $\widehat{\boldsymbol{\beta}}$ is nonetheless consistently estimated.

Another name for the within estimator is the least-squares dummy-variable (LSDV) estimator, because it can be shown to equal the estimator obtained from OLS estimation of y_{it} on $\mathbf{x}_{it}$ and N individual-specific indicator variables $d_{j,it}$, $j = 1, \ldots, N$, where $d_{j,it} = 1$ for the itth observation if $j = 1$, and $d_{j,it} = 0$ otherwise. Thus we fit the model

$$y_{it} = \left(\sum\nolimits_{j=1}^{N} \alpha_i d_{j,it} \right) + \mathbf{x}_{it}'\boldsymbol{\beta} + \varepsilon_{it} \tag{8.8}$$

This equivalence of LSDV and within estimators does not carry over to nonlinear models.

This parameterization provides an alternative way to estimate the parameters of the fixed-effects model, using cross-section OLS commands. The `areg` command, which fits the linear regression (8.8) with one set of mutually exclusive indicators, reports only the estimates of the parameters $\boldsymbol{\beta}$. We have

```
. * LSDV model fit using areg with cluster-robust standard errors
. areg lwage exp exp2 wks ed, absorb(id) vce(cluster id)
note: ed omitted because of collinearity
Linear regression, absorbing indicators          Number of obs =      4165
                                                 F(  3,   594) =    908.44
                                                 Prob > F      =    0.0000
                                                 R-squared     =    0.9068
                                                 Adj R-squared =    0.8912
                                                 Root MSE      =     .1522

                        (Std. Err. adjusted for 595 clusters in id)
```

		Robust				
lwage	Coef.	Std. Err.	t	P>|t|	[95% Conf. Interval]	
exp	.1137879	.0043514	26.15	0.000	.1052418	.1223339
exp2	-.0004244	.0000888	-4.78	0.000	-.0005988	-.00025
wks	.0008359	.0009393	0.89	0.374	-.0010089	.0026806
ed	(omitted)					
_cons	4.596396	.0648993	70.82	0.000	4.468936	4.723856
id	absorbed				(595 categories)	

The coefficient estimates are the same as those from `xtreg, fe`. The cluster–robust standard errors differ because of different small-sample correction, and those from `xtreg, fe` should be used. This difference arises because inference for `areg` is designed for the case where N is fixed and $T \to \infty$, whereas we are considering the short-panel case, where T is fixed and $N \to \infty$.

The model can also be fit using `regress`. One way to include a set of indicator variables for each individual is by inserting the `i.` operator before the categorical variable `id`. To do this, we need to increase the default setting of `matsize` to at least $N + K$, where K is the number of regressors in this model. The output from `regress` is very long because it includes coefficients for all the dummy variables. We instead suppress the output and use `estimates table` to list results for just the coefficients of interest.

```
. * LSDV model fit using factor variables with cluster-robust standard errors
. set matsize 800
. quietly regress lwage exp exp2 wks ed i.id, vce(cluster id)
. estimates table, keep(exp exp2 wks ed _cons) b se b(%12.7f)
```

Variable	active
exp	0.1137879
	0.0043514
exp2	-0.0004244
	0.0000888
wks	0.0008359
	0.0009393
ed	0.1022134
	0.0046744
_cons	4.3476807
	0.0443191

```
                    legend: b/se
```

The coefficient estimates and standard errors are exactly the same as those obtained from `areg`, aside from the constant. For `areg` (and `xtreg, fe`), the intercept is fit so that $\bar{y} - \bar{\bar{\mathbf{x}}}'\widehat{\boldsymbol{\beta}} = 0$, whereas this is not the case using `regress`. The standard errors are the same as those from `areg`, and as already noted, those from `xtreg, fe` should be used.

8.6 Between estimator

The between estimator uses only between or cross-section variation in the data and is the OLS estimator from the regression of $\bar{y}_i$ on $\bar{\mathbf{x}}_i$. Because only cross-section variation in the data is used, the coefficients of any individual-invariant regressors, such as time dummies, cannot be identified. We provide the estimator for completeness, even though it is seldom used because pooled estimators and the RE estimator are more efficient.

8.6.1 Between estimator

The between estimator is inconsistent in the FE model and is consistent in the RE model. To see this, average the individual-effects model (8.1) to obtain the between model

$$\bar{y}_i = \alpha + \bar{\mathbf{x}}_i'\boldsymbol{\beta} + (\alpha_i - \alpha + \bar{\varepsilon}_i)$$

The between estimator is the OLS estimator in this model. Consistency requires that the error term $(\alpha_i - \alpha + \bar{\varepsilon}_i)$ be uncorrelated with $\mathbf{x}_{it}$. This is the case if α_i is a random effect but not if α_i is a fixed effect.

8.6.2 Application of the xtreg, be command

The between estimator is obtained by specifying the **be** option of the **xtreg** command. There is no explicit option to obtain heteroskedasticity-robust standard errors, but these can be obtained by using the **vce(bootstrap)** option.

For our data, the bootstrap standard errors differ from the default by only 10%, because averages are used so that the complication is one of heteroskedastic errors rather than clustered errors. We report the default standard errors that are much more quickly computed. We have

```
. * Between estimator with default standard errors
. xtreg lwage exp exp2 wks ed, be

Between regression (regression on group means)   Number of obs    =     4165
Group variable: id                               Number of groups =      595

R-sq:   within  = 0.1357                          Obs per group: min =        7
        between = 0.3264                                         avg =      7.0
        overall = 0.2723                                         max =        7

                                                  F(4,590)          =    71.48
sd(u_i + avg(e_i.))=   .324656                    Prob > F          =   0.0000
```

lwage	Coef.	Std. Err.	t	P>|t|	[95% Conf. Interval]
exp	.038153	.0056967	6.70	0.000	.0269647	.0493412
exp2	-.0006313	.0001257	-5.02	0.000	-.0008781	-.0003844
wks	.0130903	.0040659	3.22	0.001	.0051048	.0210757
ed	.0737838	.0048985	15.06	0.000	.0641632	.0834044
_cons	4.683039	.2100989	22.29	0.000	4.270407	5.095672

The estimates and standard errors are closer to those obtained from pooled OLS than those obtained from within estimation.

8.7 RE estimator

The RE estimator is the FGLS estimator in the RE model (8.1) under the assumption that the random effect α_i is i.i.d. and the idiosyncratic error ε_{it} is i.i.d. The RE estimator is consistent if the RE model is appropriate and is inconsistent if the FE model is appropriate.

(Continued on next page)

8.7.1 RE estimator

The RE model is the individual-effects model (8.1)

$$y_{it} = \mathbf{x}'_{it}\boldsymbol{\beta} + (\alpha_i + \varepsilon_{it}) \tag{8.9}$$

with $\alpha_i \sim (\alpha, \sigma^2_\alpha)$ and $\varepsilon_{it} \sim (0, \sigma^2_u)$. Then from (8.4), the combined error $u_{it} = \alpha_i + \varepsilon_{it}$ is correlated over t for the given i with

$$\text{Cor}(u_{it}, u_{is}) = \sigma^2_\alpha/(\sigma^2_\alpha + \sigma^2_\varepsilon), \text{ for all } s \neq t \tag{8.10}$$

The RE estimator is the FGLS estimator of $\boldsymbol{\beta}$ in (8.9) given (8.10) for the error correlations.

In several different settings, such as heteroskedastic errors and AR(1) errors, the FGLS estimator can be calculated as the OLS estimator in a model transformed to have homoskedastic uncorrelated errors. This is also possible here. Some considerable algebra shows that the RE estimator can be obtained by OLS estimation in the transformed model

$$(y_{it} - \widehat{\theta}_i\overline{y}_i) = (1 - \widehat{\theta}_i)\alpha + (\mathbf{x}_{it} - \widehat{\theta}_i\overline{\mathbf{x}}_i)'\boldsymbol{\beta} + \{(1 - \widehat{\theta}_i)\alpha_i + (\varepsilon_{it} - \widehat{\theta}_i\overline{\varepsilon}_i)\} \tag{8.11}$$

where $\widehat{\theta}_i$ is a consistent estimate of

$$\theta_i = 1 - \sqrt{\sigma^2_\varepsilon/(T_i\sigma^2_\alpha + \sigma^2_\varepsilon)}$$

The RE estimator is consistent and fully efficient if the RE model is appropriate. It is inconsistent if the FE model is appropriate, because then correlation between $\mathbf{x}_{it}$ and α_i implies correlation between the regressors and the error in (8.11). Also, if there are no fixed effects but the errors exhibit within-panel correlation, then the RE estimator is consistent but inefficient, and cluster–robust standard errors should be obtained.

The RE estimator uses both between and within variation in the data and has special cases of pooled OLS ($\widehat{\theta}_i = 0$) and within estimation ($\widehat{\theta}_i = 1$). The RE estimator approaches the within estimator as T gets large and as σ^2_α gets large relative to σ^2_ε, because in those cases $\widehat{\theta}_i \to 1$.

8.7.2 The xtreg, re command

Three closely related and asymptotically equivalent RE estimators can be obtained by using the xtreg command (see section 8.2.4) with the re, mle, or pa option. These estimators use different estimates of the variance components σ^2_ε and σ^2_α and hence different estimates $\widehat{\theta}_i$ in the RE regression; see [XT] **xtreg** for the formulas.

The RE estimator uses unbiased estimates of the variance components and is obtained by using the re option. The maximum likelihood estimator, under the additional assumption of normally distributed α_i and ε_{it}, is computed by using the mle option. The RE model implies the errors are equicorrelated or exchangeable (see section 8.3.10), so xtreg with the `pa` and `corr(exchangeable)` options yields asymptotically equivalent results.

For panel data, the RE estimator assumption of equicorrelated errors is usually too strong. At the least, one should use the `vce(cluster id)` option to obtain cluster–robust standard errors. And more-efficient estimates can be obtained with `xtreg, pa` with a better error structure than those obtained with the `corr(exchangeable)` option.

8.7.3 Application of the xtreg, re command

For our data, `xtreg, re` yields

```
. * Random-effects estimator with cluster-robust standard errors
. xtreg lwage exp exp2 wks ed, re vce(cluster id) theta
Random-effects GLS regression                 Number of obs     =      4165
Group variable: id                             Number of groups  =       595

R-sq:  within  = 0.6340                         Obs per group: min =         7
       between = 0.1716                                        avg =       7.0
       overall = 0.1830                                        max =         7

Random effects u_i ~ Gaussian                   Wald chi2(4)      =   1598.50
corr(u_i, X)        = 0 (assumed)               Prob > chi2       =    0.0000
theta               = .82280511

                                (Std. Err. adjusted for 595 clusters in id)
```

lwage	Coef.	Robust Std. Err.	z	P>|z|	[95% Conf. Interval]
exp	.0888609	.0039992	22.22	0.000	.0810227	.0966992
exp2	-.0007726	.0000896	-8.62	0.000	-.0009481	-.000597
wks	.0009658	.0009259	1.04	0.297	-.000849	.0027806
ed	.1117099	.0083954	13.31	0.000	.0952552	.1281647
_cons	3.829366	.1333931	28.71	0.000	3.567921	4.090812
sigma_u	.31951859					
sigma_e	.15220316					
rho	.81505521	(fraction of variance due to u_i)				

Unlike the within estimator, the coefficient of the time-invariant regressor `ed` is now estimated. The standard errors are somewhat smaller than those for the within estimator because some between variation is also used. The entries `sigma_u`, `sigma_e`, and `rho`, and the various R^2 measures, are explained in the next section.

The `re`, `mle`, and `pa corr(exchangeable)` options of `xtreg` yield asymptotically equivalent estimators that differ in typical sample sizes. Comparison for these data is left as an exercise.

(Continued on next page)

8.8   Comparison of estimators

Output from `xtreg` includes estimates of the standard deviation of the error components and R^2 measures that measure within, between, and overall fit. Prediction is possible using the postestimation `predict` command. We present these estimates before turning to comparison of OLS, between, RE, and within estimators.

8.8.1   Estimates of variance components

Output from the `fe`, `re`, and `mle` options of `xtreg` includes estimates of the standard deviations of the error components. The combined error in the individual-effects model that we label $\alpha_i + \varepsilon_{it}$ is referred to as $u_i + e_{it}$ in the Stata documentation and output. Thus Stata output `sigma_u` gives the standard deviation of the individual effect α_i, and `sigma_e` gives the standard deviation of the idiosyncratic error ε_{it}.

For the RE model estimates given in the previous section, the estimated standard deviation of α_i is twice that of ε_{it}. So the individual-specific component of the error (the random effect) is much more important than the idiosyncratic error.

The output labeled `rho` equals the intraclass correlation of the error ρ_u defined in (8.4). For the RE model, for example, the estimate of 0.815 is very high. This is expected because, from section 8.3.9, the average autocorrelation of the OLS residuals was computed to be around 0.80.

The `theta` option, available for the `re` option in the case of balanced data, reports the estimate $\widehat{\theta}_i = \widehat{\theta}$. Because $\widehat{\theta} = 0.823$, here the RE estimates will be much closer to the within estimates than to the OLS estimates. More generally, in the unbalanced case the matrix `e(theta)` saves the minimum, 5th percentile, median, 95th percentile, and maximum of $\widehat{\theta}_1, \ldots, \widehat{\theta}_N$.

8.8.2   Within and between R-squared

The table header from `xtreg` provides three R^2 measures, computed using the interpretation of R^2 as the squared correlation between the actual and fitted values of the dependent variable, where the fitted values ignore the contribution of $\widehat{\alpha}_i$.

Let $\widehat{\alpha}$ and $\widehat{\boldsymbol{\beta}}$ be estimates obtained by one of the `xtreg` options (`be`, `fe`, or `re`). Let $\rho^2(x, y)$ denote the squared correlation between x and y. Then

$$
\begin{aligned}
\text{Within } R^2: \quad & \rho^2\{(y_{it} - \overline{y}_i), (\mathbf{x}'_{it}\widehat{\boldsymbol{\beta}} - \overline{\mathbf{x}}'_i\widehat{\boldsymbol{\beta}})\} \\
\text{Between } R^2: \quad & \rho^2(\overline{y}_i, \overline{\mathbf{x}}'_i\widehat{\boldsymbol{\beta}}) \\
\text{Overall } R^2: \quad & \rho^2(y_{it}, \mathbf{x}'_{it}\widehat{\boldsymbol{\beta}})
\end{aligned}
$$

The three R^2 measures are, respectively, 0.66, 0.03, and 0.05 for the within estimator; 0.14, 0.33, and 0.27 for the between estimator; and 0.63, 0.17, and 0.18 for the RE estimator. So the within estimator best explains the within variation ($R^2_{\text{w}} = 0.66$), and the between estimator best explains the between variation ($R_b^2 = 0.33$). The within estimator has a low $R_o^2 = 0.05$ and a much higher $R^2 = 0.91$ in section 8.5.4, because R_o^2 neglects $\widehat{\alpha}_i$.

8.8.3 Estimator comparison

We compare some of the panel estimators and associated standard errors, variance components estimates, and R^2. Pooled OLS is the same as the `xtreg` command with the `corr(independent)` and `pa` options. We have

```
. * Compare OLS, BE, FE, RE estimators, and methods to compute standard errors
. global xlist exp exp2 wks ed
. quietly regress lwage $xlist, vce(cluster id)
. estimates store OLS_rob
. quietly xtreg lwage $xlist, be
. estimates store BE
. quietly xtreg lwage $xlist, fe
. estimates store FE
. quietly xtreg lwage $xlist, fe vce(robust)
. estimates store FE_rob
. quietly xtreg lwage $xlist, re
. estimates store RE
. quietly xtreg lwage $xlist, re vce(robust)
. estimates store RE_rob
. estimates table OLS_rob BE FE FE_rob RE RE_rob,
> b se stats(N r2 r2_o r2_b r2_w sigma_u sigma_e rho) b(%7.4f)
```

Variable	OLS_rob	BE	FE	FE_rob	RE	RE_rob
exp	0.0447	0.0382	0.1138	0.1138	0.0889	0.0889
	0.0054	0.0057	0.0025	0.0040	0.0028	0.0040
exp2	-0.0007	-0.0006	-0.0004	-0.0004	-0.0008	-0.0008
	0.0001	0.0001	0.0001	0.0001	0.0001	0.0001
wks	0.0058	0.0131	0.0008	0.0008	0.0010	0.0010
	0.0019	0.0041	0.0006	0.0009	0.0007	0.0009
ed	0.0760	0.0738	(omitted)	(omitted)	0.1117	0.1117
	0.0052	0.0049			0.0061	0.0084
_cons	4.9080	4.6830	4.5964	4.5964	3.8294	3.8294
	0.1400	0.2101	0.0389	0.0601	0.0936	0.1334
N	4165	4165	4165	4165	4165	4165
r2	0.2836	0.3264	0.6566	0.6566		
r2_o		0.2723	0.0476	0.0476	0.1830	0.1830
r2_b		0.3264	0.0276	0.0276	0.1716	0.1716
r2_w		0.1357	0.6566	0.6566	0.6340	0.6340
sigma_u			1.0362	1.0362	0.3195	0.3195
sigma_e			0.1522	0.1522	0.1522	0.1522
rho			0.9789	0.9789	0.8151	0.8151

legend: b/se

Several features emerge. The estimated coefficients vary considerably across estimators, especially for the time-varying regressors. This reflects quite different results
according to whether within variation or between variation is used. The within estimator did not provide a coefficient estimate for the time-invariant regressor ed (with the
coefficient reported as 0.00). Cluster–robust standard errors for the FE and RE models
exceed the default standard errors by one-third to one-half. The various R^2 measures
and variance-components estimates also vary considerably across models.

8.8.4   Fixed effects versus random effects

The essential distinction in microeconometrics analysis of panel data is that between
FE and RE models. If effects are fixed, then the pooled OLS and RE estimators are
inconsistent, and instead the within (or FE) estimator needs to be used. The within
estimator is otherwise less desirable, because using only within variation leads to less-
efficient estimation and inability to estimate coefficients of time-invariant regressors.

To understand this distinction, consider the scalar regression of y_{it} on x_{it}. Con-
sistency of the pooled OLS estimator requires that $E(u_{it}|x_{it}) = 0$ in the model $y_{it} =
\alpha + \beta x_{it} + u_{it}$. If this assumption fails so that x_{it} is endogenous, IV estimation can yield
consistent estimates. It can be difficult to find an instrument z_{it} for x_{it} that satisfies
$E(u_{it}|z_{it}) = 0$.

Panel data provide an alternative way to obtain consistent estimates. Introduce
the individual-effects model $y_{it} = \alpha_i + \beta x_{it} + \varepsilon_{it}$. Consistency in this model requires
the weaker assumption that $E(\varepsilon_{it}|\alpha_i, x_{it}) = 0$. Essentially, the error has two compo-
nents: the time-invariant component α_i correlated with regressors that we can eliminate
through differencing, and a time-varying component that, given α_i, is uncorrelated with
regressors.

The RE model adds an additional assumption to the individual-effects model: α_i is
distributed independently of x_{it}. This is a much stronger assumption because it implies
that $E(\varepsilon_{it}|\alpha_i, x_{it}) = E(\varepsilon_{it}|x_{it})$, so consistency requires that $E(\varepsilon_{it}|x_{it}) = 0$, as assumed
by the pooled OLS model.

For individual-effects models, the fundamental issue is whether the individual effect
is correlated with regressors.

8.8.5   Hausman test for fixed effects

Under the null hypothesis that individual effects are random, these estimators should
be similar because both are consistent. Under the alternative, these estimators diverge.
This juxtaposition is a natural setting for a Hausman test (see section 12.7), comparing
FE and RE estimators. The test compares the estimable coefficients of time-varying
regressors or can be applied to a key subset of these (often one key regressor).

The hausman command

The `hausman` command implements the standard form of the Hausman test. We have already stored the within estimates as FE and the RE estimates as RE, so we can immediately implement the test.

For these data, the default version of the `hausman FE RE` command leads to a variance estimate $\{\widehat{V}(\widehat{\boldsymbol{\beta}}_{\mathrm{FE}}) - \widehat{V}(\widehat{\boldsymbol{\beta}}_{\mathrm{RE}})\}$ that is negative definite, so estimated standard errors of $(\widehat{\beta}_{j,\mathrm{FE}} - \widehat{\beta}_{j,\mathrm{RE}})$ cannot be obtained. This problem can arise because different estimates of the error variance are used in forming $\widehat{V}(\widehat{\boldsymbol{\beta}}_{\mathrm{FE}})$ and $\widehat{V}(\widehat{\boldsymbol{\beta}}_{\mathrm{RE}})$. Similar issues arise for a Hausman test comparing OLS and two-stage least-squares estimates.

It is better to use the `sigmamore` option, which specifies that both covariance matrices are based on the (same) estimated disturbance variance from the efficient estimator. We obtain

```
. * Hausman test assuming RE estimator is fully efficient under null hypothesis
. hausman FE RE, sigmamore
```

| | ———— Coefficients ———— | | | |
| | (b) | (B) | (b-B) | sqrt(diag(V_b-V_B)) |
	FE	RE	Difference	S.E.
exp	.1137879	.0888609	.0249269	.0012778
exp2	-.0004244	-.0007726	.0003482	.0000285
wks	.0008359	.0009658	-.0001299	.0001108

```
                  b = consistent under Ho and Ha; obtained from xtreg
      B = inconsistent under Ha, efficient under Ho; obtained from xtreg

Test:  Ho:  difference in coefficients not systematic

           chi2(3) = (b-B)´[(V_b-V_B)^(-1)](b-B)
                   =      1513.02
           Prob>chi2 =    0.0000
```

The output from `hausman` provides a nice side-by-side comparison. For the coefficient of regressor `exp`, a test of RE against FE yields $t = 0.0249/0.00128 = 19.5$, a highly statistically significant difference. And the overall statistic, here $\chi^2(3)$, has $p = 0.000$. This leads to strong rejection of the null hypothesis that RE provides consistent estimates.

Robust Hausman test

A serious shortcoming of the standard Hausman test is that it requires the RE estimator to be efficient. This in turn requires that the α_i and ε_{it} are i.i.d., an invalid assumption if cluster–robust standard errors for the RE estimator differ substantially from default standard errors. For our data example, and in many applications, a robust version of the Hausman test is needed. There is no Stata command for this. A panel bootstrap Hausman test can be conducted, using an adaptation of the bootstrap Hausman test example in section 13.4.6.

Simpler is to test $H_0\colon \boldsymbol{\gamma} = 0$ in the auxiliary OLS regression

$$(y_{it} - \widehat{\theta}\overline{y}_i) = (1 - \widehat{\theta})\alpha + (\mathbf{x}_{it} - \widehat{\theta}\overline{\mathbf{x}}_i)'\boldsymbol{\beta} + (\mathbf{x}_{1it} - \overline{\mathbf{x}}_{1i})'\boldsymbol{\gamma} + v_{it}$$

where $\mathbf{x}_1$ denotes only time-varying regressors. A Wald test of $\boldsymbol{\gamma} = 0$ can be shown to be asymptotically equivalent to the standard test when the RE estimator is fully efficient under H_0 and is numerically equivalent to `hausman` with the `sigmaless` option. A summary of related tests for fixed versus random effects is given in Baltagi (2008, 72–78).

In the more likely case that the RE estimator is not fully efficient, Wooldridge (2002) proposes performing the Wald test using cluster–robust standard errors. To implement this test in Stata, we need to generate the RE differences $y_{it} - \widehat{\theta}\overline{y}_i$ and $\mathbf{x}_{it} - \widehat{\theta}\overline{\mathbf{x}}_i$, and the mean-differences $\mathbf{x}_{1it} - \overline{\mathbf{x}}_{1i}$.

```
. * Robust Hausman test using method of Wooldridge (2002)
. quietly xtreg lwage $xlist, re
. scalar theta = e(theta)
. global yandxforhausman lwage exp exp2 wks ed
. sort id
. foreach x of varlist $yandxforhausman {
  2.    by id: egen mean`x' = mean(`x')
  3.    generate md`x' = `x' - mean`x'
  4.    generate red`x' = `x' - theta*mean`x'
  5.  }
. quietly regress redlwage redexp redexp2 redwks reded mdexp mdexp2 mdwks,
> vce(cluster id)
. test mdexp mdexp2 mdwks
 ( 1)  mdexp = 0
 ( 2)  mdexp2 = 0
 ( 3)  mdwks = 0
       F(  3,   594) =   597.47
            Prob > F =    0.0000
```

The test strongly rejects the null hypothesis, and we conclude that the RE model is not appropriate. The code will become more complex in the unbalanced case, because we then need to compute $\widehat{\theta}_i$ for each observation. The user-written command `xtoverid` following command `xtreg, re vce(cluster id)` implements the preceding test in both balanced and unbalanced settings.

8.8.6 Prediction

The postestimation `predict` command after `xtreg` provides estimated residuals and fitted values following estimation of the individual-effects model $y_{it} = \alpha_i + \mathbf{x}_{it}'\boldsymbol{\beta} + \varepsilon_{it}$.

The estimated individual-specific error $\widehat{\alpha}_i = \overline{y}_i - \overline{\mathbf{x}}_{it}'\widehat{\boldsymbol{\beta}}$ is obtained by using the `u` option; the estimated idiosyncratic error $\widehat{\varepsilon}_{it} = y_{it} - \widehat{\alpha}_i - \mathbf{x}_{it}'\widehat{\boldsymbol{\beta}}$ is obtained by using the `e` option; and the `ue` option gives $\widehat{\alpha}_i + \widehat{\varepsilon}_{it}$.

Fitted values of the dependent variable differ according to whether the estimated individual-specific error is used. The fitted value $y_{it} = \widehat{\alpha} + \mathbf{x}'_{it}\widehat{\boldsymbol{\beta}}$, where $\widehat{\alpha} = N^{-1}\sum_i \widehat{\alpha}_i$, is obtained by using the xb option. The fitted value $y_{it} = \widehat{\alpha}_i + \mathbf{x}'_{it}\widehat{\boldsymbol{\beta}}$ is obtained by using the xbu option.

As an example, we contrast OLS and RE in-sample fitted values.

```
. * Prediction after OLS and RE estimation
. quietly regress lwage exp exp2 wks ed, vce(cluster id)
. predict xbols, xb
. quietly xtreg lwage exp exp2 wks ed, re
. predict xbre, xb
. predict xbure, xbu
. summarize lwage xbols xbre xbure
    Variable |        Obs        Mean    Std. Dev.        Min        Max
-------------+--------------------------------------------------------------
       lwage |       4165    6.676346    .4615122     4.60517      8.537
       xbols |       4165    6.676346    .2457572    5.850037   7.200861
        xbre |       4165    6.676346    .6205324    5.028067    8.22958
       xbure |       4165    6.676346    .4082951     5.29993   7.968179
. correlate lwage xbols xbre xbure
(obs=4165)
             |    lwage    xbols     xbre    xbure
-------------+------------------------------------
       lwage |   1.0000
       xbols |   0.5325   1.0000
        xbre |   0.4278   0.8034   1.0000
       xbure |   0.9375   0.6019   0.4836   1.0000
```

The RE prediction $\widehat{\alpha} + \mathbf{x}'_{it}\widehat{\boldsymbol{\beta}}$ is not as highly correlated with lwage as is the OLS prediction (0.43 versus 0.53), which was expected because the OLS estimator maximizes this correlation.

When instead we use $\widehat{\alpha}_i + \mathbf{x}'_{it}\widehat{\boldsymbol{\beta}}$ so the fitted individual effect is included, the correlation of the prediction with lwage increases greatly to 0.94. In a short panel, however, these predictions are not consistent because each individual prediction $\widehat{\alpha}_i = \overline{y}_i - \overline{\mathbf{x}}'_{it}\widehat{\boldsymbol{\beta}}$ is based on only T observations and $T \not\rightarrow \infty$.

8.9   First-difference estimator

Consistent estimation of $\boldsymbol{\beta}$ in the FE model requires eliminating the α_i. One way to do so is to mean-difference, yielding the within estimator. An alternative way is to first-difference, leading to the first-difference estimator. This alternative has the advantage of relying on weaker exogeneity assumptions, explained below, that become important in dynamic models presented in the next chapter. In the current chapter, the within estimator is traditionally favored as it is the more efficient estimator if the ε_{it} are i.i.d.

8.9.1   First-difference estimator

The first-difference (FD) estimator is obtained by performing OLS on the first-differenced variables

$$(y_{it} - y_{i,t-1}) = (\mathbf{x}_{it} - \mathbf{x}_{i,t-1})'\boldsymbol{\beta} + (\varepsilon_{it} - \varepsilon_{i,t-1}) \tag{8.12}$$

First-differencing has eliminated α_i, so OLS estimation of this model leads to consistent estimates of $\boldsymbol{\beta}$ in the FE model. The coefficients of time-invariant regressors are not identified, because then $x_{it} - x_{i,t-1} = 0$, as was the case for the within estimator.

The FD estimator is not provided as an option to `xtreg`. Instead, the estimator can be computed by using `regress` and Stata time-series operators to compute the first-differences. We have

```
. sort id t
. * First-differences estimator with cluster-robust standard errors
. regress D.(lwage exp exp2 wks ed), vce(cluster id) noconstant
note: _delete omitted because of collinearity

Linear regression                               Number of obs =      3570
                                                F(  3,   594) = 1035.19
                                                Prob > F      =    0.0000
                                                R-squared     =    0.2209
                                                Root MSE      =    .18156

                               (Std. Err. adjusted for 595 clusters in id)
```

D.lwage	Coef.	Robust Std. Err.	t	P>|t|	[95% Conf. Interval]	
exp D1.	.1170654	.0040974	28.57	0.000	.1090182	.1251126
exp2 D1.	-.0005321	.0000808	-6.58	0.000	-.0006908	-.0003734
wks D1.	-.0002683	.0011783	-0.23	0.820	-.0025824	.0020459
ed D1.	(omitted)					

Note that the `noconstant` option is used. If instead an intercept is included in (8.12), say, δ, this would imply that the original model had a time trend because $\delta t - \delta(t-1) = \delta$.

As expected, the coefficient for education is not identified because `ed` here is time-invariant. The coefficient for `wks` actually changes sign compared with the other estimators, though it is highly statistically insignificant.

The FD estimator, like the within estimator, provides consistent estimators when the individual effects are fixed. For panels with $T = 2$, the FD and within estimators are equivalent; otherwise, the two differ. For static models, the FE model is used because it is the efficient estimator if the idiosyncratic error ε_{it} is i.i.d.

The FD estimator seemingly uses one less year of data compared with the within estimator, because the FD output lists 3,570 observations rather than 4,165. This, however, is misleading. Using the LSDV interpretation of the within estimator, the within estimator essentially loses 595 observations by estimating the T fixed effects $\alpha_1, \ldots, \alpha_T$.

8.9.2   Strict and weak exogeneity

From (8.6), the within estimator requires that $\varepsilon_{it} - \bar{\varepsilon}_i$ be uncorrelated with $\mathbf{x}_{it} - \bar{\mathbf{x}}_i$. This is the case under the assumption of strict exogeneity or strong exogeneity that

$$E(\varepsilon_{it}|\alpha_i, \mathbf{x}_{i1}, \ldots, \mathbf{x}_{it}, \ldots, \mathbf{x}_{iT}) = 0$$

From (8.12), the FD estimator requires that $\varepsilon_{it} - \varepsilon_{i,t-1}$ be uncorrelated with $\mathbf{x}_{it} - \mathbf{x}_{i,t-1}$. This is the case under the assumption of weak exogeneity that

$$E(\varepsilon_{it}|\alpha_i, \mathbf{x}_{i1}, \ldots, \mathbf{x}_{it}) = 0$$

This is a considerably weaker assumption because it permits future values of the regressors to be correlated with the error, as will be the case if the regressor is a lagged dependent variable.

As long as there is no feedback from the idiosyncratic shock today to a covariate tomorrow, this distinction is unnecessary when estimating static models. It becomes important for dynamic models (see section 9.4), because then strict exogeneity no longer holds and we turn to the FD estimator.

8.10   Long panels

The methods up to this point have focused on short panels. Now we consider long panels with many time periods for relatively few individuals (N is small and $T \rightarrow \infty$). Examples are data on a few regions, firms, or industries followed for many time periods.

Then individual fixed effects, if desired, can be easily handled by including dummy variables for each individual as regressors. Instead, the focus is on more-efficient GLS estimation under richer models of the error process than those specified in the short-panel case. Here we consider only methods for stationary errors, and we only briefly cover the growing area of panel data with unit roots and cointegration.

8.10.1   Long-panel dataset

The dataset used is a U.S. state–year panel from Baltagi, Griffin, and Xiong (2000) on annual cigarette consumption and price for U.S. states over 30 years. The ultimate goal is to measure the responsiveness of per capita cigarette consumption to real cigarette prices. Price varies across states, due in large part to different levels of taxation, as well as over time.

The original data were for $N = 46$ states and $T = 30$, and it is not clear whether we should treat $N \to \infty$, as we have done to date, or $T \to \infty$, or both. This situation is not unusual for a panel that uses aggregated regional data over time. To make explicit that we are considering $T \to \infty$, we use data from only $N = 10$ states, similar to many countries where there may be around 10 major regions (states or provinces).

The mus08cigar.dta dataset has the following data:

```
. * Description of cigarette dataset
. use mus08cigar.dta, clear

. describe
Contains data from mus08cigar.dta
  obs:           300
  vars:            6                              13 Mar 2008 20:45
  size:        8,400  (99.9% of memory free)

              storage   display    value
variable name   type    format     label      variable label

state          float    %9.0g                 U.S. state
year           float    %9.0g                 Year 1963 to 1992
lnp            float    %9.0g                 Log state real price of pack of
                                                cigarettes
lnpmin         float    %9.0g                 Log of min real price in
                                                adjoining states
lnc            float    %9.0g                 Log state cigarette sales in
                                                packs per capita
lny            float    %9.0g                 Log state per capita disposable
                                                income

Sorted by:
```

There are 300 observations, so each state–year pair is a separate observation because $10 \times 30 = 300$. The quantity demanded (lnc) will depend on price (lnp), price of a substitute (lnpmin), and income (lny).

Descriptive statistics can be obtained by using summarize:

```
. * Summary of cigarette dataset
. summarize, separator(6)
    Variable |       Obs        Mean    Std. Dev.       Min        Max

       state |       300         5.5     2.87708          1         10
        year |       300        77.5    8.669903         63         92
         lnp |       300    4.518424    .1406979   4.176332    4.96916
      lnpmin |       300      4.4308    .1379243     4.0428   4.831303
         lnc |       300    4.792591    .2071792   4.212128   5.690022
         lny |       300    8.731014    .6942426   7.300023    10.0385
```

The variables state and year have the expected ranges. The variability in per capita cigarette sales (lnc) is actually greater than the variability in price (lnp), with respective standard deviations of 0.21 and 0.14. All variables are observed for all 300 observations, so the panel is indeed balanced.

8.10.2   Pooled OLS and PFGLS

A natural starting point is the two-way–effects model $y_{it} = \alpha_i + \gamma_t + \mathbf{x}'_{it}\boldsymbol{\beta} + \varepsilon_{it}$. When the panel has few individuals relative to the number of periods, the individual effects α_i (here state effects) can be incorporated into $\mathbf{x}_{it}$ as dummy-variable regressors. Then there are too many time effects γ_t (here year effects). Rather than trying to control for these in ways analogous to the use of xtreg in the short-panel case, it is usually sufficient to take advantage of the natural ordering of time (as opposed to individuals) and simply include a linear or quadratic trend in time.

We therefore focus on the pooled model

$$y_{it} = \mathbf{x}'_{it}\boldsymbol{\beta} + u_{it}, \quad i = 1, \ldots, N, \ t = 1, \ldots, T \tag{8.13}$$

where the regressors $\mathbf{x}_{it}$ include an intercept, often time and possibly time-squared, and possibly a set of individual indicator variables. We assume that N is quite small relative to T.

We consider pooled OLS and PFGLS of this model under a variety of assumptions about the error u_{it}. In the short-panel case, it was possible to obtain standard errors that control for serial correlation in the error without explicitly stating a model for serial correlation. Instead, we could use cluster–robust standard errors, given a small T and $N \to \infty$. Now, however, T is large relative to N, and it is necessary to specify a model for serial correlation in the error. Also given that N is small, it is possible to relax the assumption that u_{it} is independent over i.

8.10.3   The xtpcse and xtgls commands

The xtpcse and xtgls commands are more suited than xtgee for pooled OLS and GLS when data are from a long panel. They allow the error u_{it} in the model to be correlated over i, allow the use of an AR(1) model for u_{it} over t, and allow u_{it} to be heteroskedastic. At the greatest level of generality,

$$u_{it} = \rho_i u_{i,t-1} + \varepsilon_{it} \tag{8.14}$$

where ε_{it} are serially uncorrelated but are correlated over i with $\text{Cor}(\varepsilon_{it}, \varepsilon_{is}) = \sigma_{ts}$.

The xtpcse command yields (long) panel-corrected standard errors for the pooled OLS estimator, as well as for a pooled least-squares estimator with an AR(1) model for u_{it}. The syntax is

xtpcse *depvar* [*indepvars*] [*if*] [*in*] [*weight*] [, *options*]

The correlation() option determines the type of pooled estimator. Pooled OLS is obtained by using correlation(independent). The pooled AR(1) estimator with general ρ_i is obtained by using correlation(psar1). With a balanced panel, $y_{it} - \widehat{\rho}_i y_{it,t-1}$ is regressed on $\mathbf{x}^*_{it} = \mathbf{x}_{it} - \widehat{\rho}\mathbf{x}_{it,t-1}$ for $t > 1$, whereas $\sqrt{(1 - \widehat{\rho}_i)^2} y_{i1}$ is regressed on $\sqrt{(1-\widehat{\rho}_i)^2}\mathbf{x}_{i1}$ for $t = 1$. The pooled estimator with AR(1) error and $\rho_i = \rho$ is obtained by using `correlation(ar1)`. Then $\widehat{\rho}$, calculated as the average of the $\widehat{\rho}_i$, is used.

In all cases, panel-corrected standard errors that allow heteroskedasticity and correlation over i are reported, unless the `hetonly` option is used, in which case independence over i is assumed, or the `independent` option is used, in which case ε_{it} is i.i.d.

The `xtgls` command goes further and obtains PFGLS estimates and associated standard errors assuming the model for the errors is the correct model. The estimators are more efficient asymptotically than those from `xtpcse`, if the model is correctly specified. The command has the usual syntax:

`xtgls` *depvar* [*indepvars*] [*if*] [*in*] [*weight*] [, *options*]

The `panels()` option specifies the error correlation across individuals, where for our data an individual is a state. The `panels(iid)` option specifies u_{it} to be i.i.d., in which case the pooled OLS estimator is obtained. The `panels(heteroskedastic)` option specifies u_{it} to be independent with a variance of $E(u_{it}^2) = \sigma_i^2$ that can be different for each individual. Because there are many observations for each individual, σ_i^2 can be consistently estimated. The `panels(correlated)` option additionally allows correlation across individuals, with independence over time for a given individual, so that $E(u_{it}u_{jt}) = \sigma_{ij}$. This option requires that $T > N$.

The `corr()` option specifies the serial correlation of errors for each individual state. The `corr(independent)` option specifies u_{it} to be serially uncorrelated. The `corr(ar1)` option permits AR(1) autocorrelation of the error with $u_{it} = \rho u_{i,t-1} + \varepsilon_{it}$, where ε_{it} is i.i.d. The `corr(psar1)` option relaxes the assumption of a common AR(1) parameter to allow $u_{it} = \rho_i u_{i,t-1} + \varepsilon_{it}$. The `rhotype()` option provides various methods to compute this AR(1) parameter(s). The default estimator is two-step FGLS, whereas the `igls` option uses iterated FGLS. The `force` option enables estimation even if observations are unequally spaced over time.

Additionally, we illustrate the user-written `xtscc` command (Hoechle 2007). This generalizes `xtpcse` by applying the method of Driscoll and Kraay (1998) to obtain Newey–West-type standard errors that allow autocorrelated errors of general form, rather than restricting errors to be AR(1). Error correlation across panels, often called spatial correlation, is assumed. The error is allowed to be serially correlated for m lags. The default is for the program to determine m. Alternatively, m can be specified using the `lags(m)` option.

8.10.4 Application of the xtgls, xtpcse, and xtscc commands

As an example, we begin with a PFGLS estimator that uses the most flexible model for the error u_{it}, with flexible correlation across states and a distinct AR(1) process for the error in each state. In principle, this is the best estimator to use, but in practice when T is not much larger than N, there can be finite-sample bias in the estimators and standard errors; see Beck and Katz (1995). Then it is best, at the least, to use the more restrictive `corr(ar1)` rather than `corr(psar1)`.

We obtain

```
. * Pooled GLS with error correlated across states and state-specific AR(1)
. xtset state year
       panel variable:  state (strongly balanced)
        time variable:  year, 63 to 92
               delta:  1 unit
. xtgls lnc lnp lny lnpmin year, panels(correlated) corr(psar1)

Cross-sectional time-series FGLS regression

Coefficients:  generalized least squares
Panels:        heteroskedastic with cross-sectional correlation
Correlation:   panel-specific AR(1)

Estimated covariances      =        55          Number of obs      =         300
Estimated autocorrelations =        10          Number of groups   =          10
Estimated coefficients     =         5          Time periods       =          30
                                                Wald chi2(4)       =      342.15
                                                Prob > chi2        =      0.0000
```

lnc	Coef.	Std. Err.	z	P>|z|	[95% Conf. Interval]	
lnp	-.3260683	.0218214	-14.94	0.000	-.3688375	-.2832991
lny	.4646236	.0645149	7.20	0.000	.3381768	.5910704
lnpmin	.0174759	.0274963	0.64	0.525	-.0364159	.0713677
year	-.0397666	.0052431	-7.58	0.000	-.0500429	-.0294902
_cons	5.157994	.2753002	18.74	0.000	4.618416	5.697573

All regressors have the expected effects. The estimated price elasticity of demand for cigarettes is -0.326, the income elasticity is 0.465, demand declines by 4% per year (the coefficient of `year` is a semielasticity because the dependent variable is in logs), and a higher minimum price in adjoining states increases demand in the current state. There are 10 states, so there are $10 \times 11/2 = 55$ unique entries in the 10×10 contemporaneous error covariance matrix, and 10 autocorrelation parameters ρ_i are estimated.

We now use `xtpcse`, `xtgls`, and user-written `xtscc` to obtain the following pooled estimators and associated standard errors: 1) pooled OLS with i.i.d. errors; 2) pooled OLS with standard errors assuming correlation over states; 3) pooled OLS with standard errors assuming general serial correlation in the error (to four lags) and correlation over states; 4) pooled OLS that assumes an AR(1) error and then gets standard errors that additionally permit correlation over states; 5) PFGLS with standard errors assuming an AR(1) error; and 6) PFGLS assuming an AR(1) error and correlation across states. In all cases of AR(1) error, we specialize to $\rho_i = \rho$.

```
. * Comparison of various pooled OLS and GLS estimators
. quietly xtpcse lnc lnp lny lnpmin year, corr(ind) independent nmk
. estimates store OLS_iid
. quietly xtpcse lnc lnp lny lnpmin year, corr(ind)
. estimates store OLS_cor
. quietly xtscc lnc lnp lny lnpmin year, lag(4)
```

```
. estimates store OLS_DK

. quietly xtpcse lnc lnp lny lnpmin year, corr(ar1)

. estimates store AR1_cor

. quietly xtgls lnc lnp lny lnpmin year, corr(ar1) panels(iid)

. estimates store FGLSAR1

. quietly xtgls lnc lnp lny lnpmin year, corr(ar1) panels(correlated)

. estimates store FGLSCAR

. estimates table OLS_iid OLS_cor OLS_DK AR1_cor FGLSAR1 FGLSCAR, b(%7.3f) se
```

Variable	OLS_iid	OLS_cor	OLS_DK	AR1_cor	FGLSAR1	FGLSCAR
lnp	-0.583	-0.583	-0.583	-0.266	-0.264	-0.330
	0.129	0.169	0.273	0.049	0.049	0.026
lny	0.365	0.365	0.365	0.398	0.397	0.407
	0.049	0.080	0.163	0.125	0.094	0.080
lnpmin	-0.027	-0.027	-0.027	0.069	0.070	0.036
	0.128	0.166	0.252	0.064	0.059	0.034
year	-0.033	-0.033	-0.033	-0.038	-0.038	-0.037
	0.004	0.006	0.012	0.010	0.007	0.006
_cons	6.930	6.930	6.930	5.115	5.100	5.393
	0.353	0.330	0.515	0.544	0.414	0.361

```
                                                          legend: b/se
```

For pooled OLS with i.i.d. errors, the **nmk** option normalizes the VCE by $N - k$ rather than N, so that output is exactly the same as that from **regress** with default standard errors. The same results could be obtained by using **xtgls** with the **corr(ind)** **panel(iid) nmk** options. Allowing correlation across states increases OLS standard errors by 30–50%. Additionally, allowing for serial correlation (OLS_DK) leads to another 50–100% increase in the standard errors. The fourth and fifth estimators control for at least an AR(1) error and yield roughly similar coefficients and standard errors. The final column results are similar to those given at the start of this section, where we used the more flexible **corr(psar1)** rather than **corr(ar1)**.

8.10.5 Separate regressions

The pooled regression specifies the same regression model for all individuals in all years. Instead, we could have a separate regression model for each individual unit:

$$y_{it} = \mathbf{x}_{it}'\boldsymbol{\beta}_i + u_{it}$$

This model has NK parameters, so inference is easiest for a long panel with a small N.

For example, suppose for the cigarette example we want to fit separate regressions for each state. Separate OLS regressions for each state can be obtained by using the **statsby** prefix with the **by(state)** option. We have

```
. * Run separate regressions for each state
. statsby, by(state) clear: regress lnc lnp lny lnpmin year
(running regress on estimation sample)

        command:  regress lnc lnp lny lnpmin year
             by:  state

Statsby groups
┼──── 1 ───┼──── 2 ───┼──── 3 ───┼──── 4 ───┼──── 5
.........
```

This leads to a dataset with 10 observations on **state** and the five regression coefficients. We have

```
. * Report regression coefficients for each state
. format _b* %9.2f

. list, clean

        state   _b_lnp   _b_lny   _b_lnp~n   _b_year   _b_cons
  1.        1    -0.36     1.10       0.24     -0.08      2.10
  2.        2     0.12     0.60      -0.45     -0.05      5.14
  3.        3    -0.20     0.76       0.12     -0.05      2.72
  4.        4    -0.52    -0.14      -0.21     -0.00      9.56
  5.        5    -0.55     0.71       0.30     -0.07      4.76
  6.        6    -0.11     0.21      -0.14     -0.02      6.20
  7.        7    -0.43    -0.07       0.18     -0.03      9.14
  8.        8    -0.26     0.89       0.08     -0.07      3.67
  9.        9    -0.03     0.55      -0.36     -0.04      4.69
 10.       10    -1.41     1.12       1.14     -0.08      2.70
```

In all states except one, sales decline as price rises, and in most states, sales increase with income.

One can also test for poolability, meaning to test whether the parameters are the same across states. In this example, there are $5 \times 10 = 50$ parameters in the unrestricted model and 5 in the restricted pooled model, so there are 45 parameters to test.

8.10.6   FE and RE models

As noted earlier, if there are few individuals and many time periods, individual-specific FE models can be fit with the LSDV approach of including a set of dummy variables, here for each time period (rather than for each individual as in the short-panel case).

Alternatively, one can use the **xtregar** command. This model is the individual-effects model $y_{it} = \alpha_i + \mathbf{x}'_{it}\boldsymbol{\beta} + u_{it}$, with AR(1) error $u_{it} = \rho u_{i,t-1} + \varepsilon_{it}$. This is a better model of the error than the i.i.d. error model $u_{it} = \varepsilon_{it}$ assumed in **xtreg**, so **xtregar** potentially will lead to more-efficient parameter estimates.

The syntax of **xtregar** is similar to that for **xtreg**. The two key options are **fe** and **re**. The **fe** option treats α_i as a fixed effect. Given an estimate of $\hat{\rho}$, we first transform to eliminate the effect of the AR(1) error, as described after (8.14), and then transform again (mean-difference) to eliminate the individual effect. The **re** option treats α_i as a random effect.

We compare pooled OLS estimates, RE estimates using xtreg and xtregar, and within estimates using xtreg, xtregar, and xtscc. Recall that xtscc calculates either the OLS or regular within estimator but then estimates the VCE assuming quite general error correlation over time and across states. We have

```
. * Comparison of various RE and FE estimators
. use mus08cigar.dta, clear
. quietly xtscc lnc lnp lny lnpmin, lag(4)
. estimates store OLS_DK
. quietly xtreg lnc lnp lny lnpmin, fe
. estimates store FE_REG
. quietly xtreg lnc lnp lny lnpmin, re
. estimates store RE_REG
. quietly xtregar lnc lnp lny lnpmin, fe
. estimates store FE_REGAR
. quietly xtregar lnc lnp lny lnpmin, re
. estimates store RE_REGAR
. quietly xtscc lnc lnp lny lnpmin, fe lag(4)
. estimates store FE_DK
. estimates table OLS_DK FE_REG RE_REG FE_REGAR RE_REGAR FE_DK, b(%7.3f) se
```

Variable	OLS_DK	FE_REG	RE_REG	FE_RE~R	RE_RE~R	FE_DK
lnp	-0.611	-1.136	-1.110	-0.260	-0.282	-1.136
	0.428	0.101	0.102	0.049	0.052	0.158
lny	-0.027	-0.046	-0.045	-0.066	-0.074	-0.046
	0.026	0.011	0.011	0.064	0.026	0.020
lnpmin	-0.129	0.421	0.394	-0.010	-0.004	0.421
	0.338	0.101	0.102	0.057	0.060	0.168
_cons	8.357	8.462	8.459	6.537	6.708	8.462
	0.633	0.241	0.247	0.036	0.289	0.464

legend: b/se

There are three distinctly different sets of coefficient estimates: those using pooled OLS, those using xtreg to obtain FE and RE estimators, and those using xtregar to obtain FE and RE estimators. The final set of estimates uses the fe option of the user-written xtscc command. This produces the standard within estimator but then finds standard errors that are robust to both spatial (across panels) and serial autocorrelation of the error.

8.10.7 Unit roots and cointegration

Panel methods for unit roots and cointegration are based on methods developed for a single time series and assume that $T \to \infty$. We now consider their application to panel data, a currently active area of research.

If N is small, say $N < 10$, then seemingly unrelated equations methods can be used. When N is large, the panel aspect becomes more important. Complications include the need to control for cross-section unobserved heterogeneity when N is large, asymptotic theory that can vary with exactly how N and T both go to infinity, and the possibility of cross-section dependence. At the same time, statistics that have nonnormal distributions for a single time series can be averaged over cross sections to obtain statistics with a normal distribution.

Unit-root tests can have low power. Panel data may increase the power because of now having time series for several cross sections. The unit-root tests can also be of interest per se, such as testing purchasing power parity, as well as being relevant to consequent considerations of cointegration. A dynamic model with cross-section heterogeneity is

$$y_{it} = \rho_i y_{i,t-1} + \phi_{i1}\Delta y_{i,t-1} + \cdots + \phi_{ip_i}\Delta y_{i,t-p_i} + \mathbf{z}'_{it}\boldsymbol{\gamma}_i + u_{it}$$

where lagged changes are introduced so that u_{it} is i.i.d. Examples of $\mathbf{z}_{it}$ include individual effects $[\mathbf{z}_{it} = (1)]$, individual effects and individual time trends $[\mathbf{z}_{it} = (1\ t)']$, and $\gamma_i = \gamma$ in the case of homogeneity. A unit-root test is a test of $H_0 : \rho_1 = \cdots \rho_N = 1$. Levin, C.-F. Lin, and C.-S. J. Chu (2002) proposed a test against the alternative of homogeneity, $H_a : \rho_1 = \cdots = \rho_N = \rho < 1$, that is based on pooled OLS estimation using specific first-step pooled residuals, where in both steps homogeneity ($\rho_i = \rho$ and $\phi_{ik} = \phi_k$) is imposed. The user-written `levinlin` command (Bornhorst and Baum 2006) performs this test. Im, Pesaran, and Shin (2003) instead test against an alternative of heterogeneity, $H_a : \rho_1 < 1, \ldots, \rho_{N_o} < 1$, for a fraction N_0/N of the ρ_i by averaging separate augmented Dickey–Fuller tests for each cross section. The user-written `ipshin` command (Bornhorst and Baum 2007) implements this test. Both test statistics are asymptotically normal and both assume $N/T \to 0$ so that the time-series dimension dominates the cross-section dimension.

As in the case of a single time series, cointegration tests are used to ensure that statistical relationships between trending variables are not spurious. A quite general cointegrated panel model is

$$y_{it} = \mathbf{x}'_{it}\boldsymbol{\beta}_i + \mathbf{z}'_{it}\boldsymbol{\gamma}_i + u_{it}$$
$$\mathbf{x}_{it} = \mathbf{x}_{i,t-1} + \boldsymbol{\varepsilon}_{it}$$

where $\mathbf{z}_{it}$ is deterministic and can include individual effects and time trends, and $\mathbf{x}_{it}$ are (co)integrated regressors. Most tests of cointegration are based on the OLS residuals $\widehat{u}_{it}$, but the unit-root tests cannot be directly applied if $\text{Cov}(u_{it}, \varepsilon_{it}) \neq 0$, as is likely. Single-equation estimators have been proposed that generalize to panels fully modified OLS and dynamic OLS, and Johanssen's system approach has also been generalized to panels. The user-written `xtpmg` command (Blackburne and Frank 2007) implements the estimators of Pesaran and Smith (1995) and Pesaran, Shin, and Smith (1999) for nonstationary heterogeneous panels with a large N and T. For references, see Baltagi (2008) and Breitung and Pesaran (2005).

8.11   Panel-data management

Stata **xt** commands require panel data to be in long form, meaning that each individual–time pair is a separate observation. Some datasets instead store panel data in wide form, which has the advantage of using less space. Sometimes the observational unit is the individual, and a single observation has all time periods for that individual. And sometimes the observational unit is a time period, and a single observation has all individuals for that time period.

We illustrate how to move from wide form to long form and vice versa by using the **reshape** command. Our example is for panel data, but **reshape** can also be used in other contexts where data are grouped, such as clustered data grouped by village rather than panel data grouped by time.

8.11.1   Wide-form data

We consider a dataset that is originally in wide form, with each observation containing all years of data for an individual. The dataset is a subset of the data from the previous section. Each observation is a state and has all years of data for that state. We have

```
. * Wide form data (observation is a state)
. use mus08cigarwide.dta, clear
. list, clean
         state    lnp63    lnc63    lnp64    lnc64    lnp65    lnc65
    1.       1      4.5      4.5      4.6      4.6      4.5      4.6
    2.       2      4.4      4.8      4.3      4.8      4.3      4.8
    3.       3      4.5      4.6      4.5      4.6      4.5      4.6
    4.       4      4.4      5.0      4.4      4.9      4.4      4.9
    5.       5      4.5      5.1      4.5      5.0      4.5      5.0
    6.       6      4.5      5.1      4.5      5.1      4.5      5.1
    7.       7      4.3      5.5      4.3      5.5      4.3      5.5
    8.       8      4.5      4.9      4.6      4.8      4.5      4.9
    9.       9      4.5      4.7      4.5      4.7      4.6      4.6
   10.      10      4.5      4.6      4.6      4.5      4.5      4.6
```

The data contain a state identifier, **state**; three years of data on log price, **lnp63–lnp65**; and three years of data on log sales, **lnc63–lnc65**. The data are for 10 states.

8.11.2   Convert wide form to long form

The data can be converted from wide form to long form by using **reshape long**. The desired dataset will have an observation as a state–year pair. The variables should be a state identifier, a year identifier, and the current state–year observations on **lnp** and **lnc**.

The simple command **reshape long** actually does this automatically, because it interprets the suffixes 63–65 as denoting the grouping that needs to be expanded to long form. We use a more detailed version of the command that spells out exactly what we want to do and leads to exactly the same result as **reshape long** without arguments. We have

```
. * Convert from wide form to long form (observation is a state-year pair)
. reshape long lnp lnc, i(state) j(year)
(note: j = 63 64 65)
Data                                    wide   ->   long

Number of obs.                            10   ->     30
Number of variables                        7   ->      4
j variable (3 values)                          ->   year
xij variables:
                         lnp63 lnp64 lnp65     ->   lnp
                         lnc63 lnc64 lnc65     ->   lnc
```

The output indicates that we have expanded the dataset from 10 observations (10 states) to 30 observations (30 state–year pairs). A year-identifier variable, year, has been created. The wide-form data lnp63–lnp65 have been collapsed to lnp in long form, and lnc63–lnc65 have been collapse to lnc.

We now list the first six observations of the new long-form data.

```
. * Long-form data (observation is a state)
. list in 1/6, sepby(state)

        state   year   lnp   lnc

  1.        1     63   4.5   4.5
  2.        1     64   4.6   4.6
  3.        1     65   4.5   4.6

  4.        2     63   4.4   4.8
  5.        2     64   4.3   4.8
  6.        2     65   4.3   4.8
```

Any year-invariant variables will also be included in the long-form data. Here the state-identifier variable, state, is the only such variable.

8.11.3   Convert long form to wide form

Going the other way, data can be converted from long form to wide form by using reshape wide. The desired dataset will have an observation as a state. The constructed variables should be a state identifier and observations on lnp and lnc for each of the three years 63–65.

The reshape wide command without arguments actually does this automatically, because it interprets year as the relevant time-identifier and adds suffixes 63–65 to the variables lnp and lnc that are varying with year. We use a more detailed version of the command that spells out exactly what we want to do and leads to exactly the same result. We have

(*Continued on next page*)

```
. * Reconvert from long form to wide form (observation is a state)
. reshape wide lnp lnc, i(state) j(year)
(note: j = 63 64 65)

Data                                    long    ->    wide

Number of obs.                           30     ->      10
Number of variables                       4     ->       7
j variable (3 values)                  year     ->    (dropped)
xij variables:
                                        lnp     ->    lnp63 lnp64 lnp65
                                        lnc     ->    lnc63 lnc64 lnc65
```

The output indicates that we have collapsed the dataset from 30 observations (30 state–year pairs) to 10 observations (10 states). The `year` variable has been dropped. The long-form data `lnp` has been expanded to `lnp63`–`lnp65` in wide form, and `lnc` has been expanded to `lnc63`–`lnc65`.

A complete listing of the wide form dataset is

```
. list, clean

        state   lnp63   lnc63   lnp64   lnc64   lnp65   lnc65
  1.       1     4.5     4.5     4.6     4.6     4.5     4.6
  2.       2     4.4     4.8     4.3     4.8     4.3     4.8
  3.       3     4.5     4.6     4.5     4.6     4.5     4.6
  4.       4     4.4     5.0     4.4     4.9     4.4     4.9
  5.       5     4.5     5.1     4.5     5.0     4.5     5.0
  6.       6     4.5     5.1     4.5     5.1     4.5     5.1
  7.       7     4.3     5.5     4.3     5.5     4.3     5.5
  8.       8     4.5     4.9     4.6     4.8     4.5     4.9
  9.       9     4.5     4.7     4.5     4.7     4.6     4.6
 10.      10     4.5     4.6     4.6     4.5     4.5     4.6
```

This is exactly the same as the original `mus08cigarwide.dta` dataset, listed in section 8.11.1.

8.11.4   An alternative wide-form data

The wide form we considered had each state as the unit of observation. An alternative is that each year is the observation. Then the preceding commands are reversed so that we have `i(year)` `j(state)` rather than `i(state)` `j(year)`.

To demonstrate this case, we first need to create the data in wide form with `year` as the observational unit. We do so by converting the current data, in wide form with `state` as the observational unit, to long form with 30 observations as presented above, and then use `reshape wide` to create wide-form data with `year` as the observational unit.

```
. * Create alternative wide-form data (observation is a year)
. quietly reshape long lnp lnc, i(state) j(year)

. reshape wide lnp lnc, i(year) j(state)
(note: j = 1 2 3 4 5 6 7 8 9 10)

Data                                  long   ->   wide
───────────────────────────────────────────────────────────────
Number of obs.                          30   ->      3
Number of variables                      4   ->     21
j variable (10 values)               state   ->   (dropped)
xij variables:
                                       lnp   ->   lnp1 lnp2 ... lnp10
                                       lnc   ->   lnc1 lnc2 ... lnc10
───────────────────────────────────────────────────────────────

. list year lnp1 lnp2 lnc1 lnc2, clean

        year   lnp1   lnp2   lnc1   lnc2
   1.     63    4.5    4.4    4.5    4.8
   2.     64    4.6    4.3    4.6    4.8
   3.     65    4.5    4.3    4.6    4.8
```

The wide form has 3 observations (one per year) and 21 variables (lnp and lnc for each of 10 states plus year).

We now have data in wide form with year as the observational unit. To use xt commands, we use reshape long to convert to long-form data with an observation for each state–year pair. We have

```
. * Convert from wide form (observation is year) to long form (year-state)
. reshape long lnp lnc, i(year) j(state)
(note: j = 1 2 3 4 5 6 7 8 9 10)

Data                                  wide   ->   long
───────────────────────────────────────────────────────────────
Number of obs.                           3   ->     30
Number of variables                     21   ->      4
j variable (10 values)                       ->   state
xij variables:
                          lnp1 lnp2 ... lnp10   ->   lnp
                          lnc1 lnc2 ... lnc10   ->   lnc
───────────────────────────────────────────────────────────────

. list in 1/6, clean

        year   state   lnp   lnc
   1.     63       1   4.5   4.5
   2.     63       2   4.4   4.8
   3.     63       3   4.5   4.6
   4.     63       4   4.4   5.0
   5.     63       5   4.5   5.1
   6.     63       6   4.5   5.1
```

The data are now in long form, as in section 8.11.2.

8.12   Stata resources

FE and RE estimators appear in many econometrics texts. Panel texts with complete coverage of the basic material are Baltagi (2008) and Hsiao (2003). The key Stata reference is [XT] *Longitudinal/Panel-Data Reference Manual*, especially [XT] **xt** and [XT] **xtreg**. Useful online **help** categories include **xt** and **xtreg**. For estimation with long panels, a useful Stata user-written command is **xtscc**, as well as several others mentioned in section 8.10.

8.13   Exercises

1. For the data of section 8.3, use **xtsum** to describe the variation in **occ**, **smsa**, **ind**, **ms**, **union**, **fem**, and **blk**. Which of these variables are time-invariant? Use **xttab** and **xttrans** to provide interpretations of how **occ** changes for individuals over the seven years. Provide a time-series plot of **exp** for the first ten observations and provide interpretation. Provide a scatterplot of **lwage** against **ed**. Is this plot showing within variation, between variation, or both?

2. For the data of section 8.3, manually obtain the three standard deviations of **lwage** given by the **xtsum** command. For the overall standard deviation, use **summarize**. For the between standard deviation, compute **by id: egen meanwage = mean(lwage)** and apply **summarize** to (meanwage-*grandmean*) for t==1, where *grandmean* is the grand mean over all observations. For the within standard deviation, apply **summarize** to (lwage-meanwage). Compare your standard deviations with those from **xtsum**. Does $s_{\mathrm{O}}^2 \simeq s_{\mathrm{W}}^2 + s_{\mathrm{B}}^2$?

3. For the model and data of section 8.4, compare PFGLS estimators under the following assumptions about the error process: independent, exchangeable, AR(2), and MA(6). Also compare the associated standard-error estimates obtained by using default standard errors and by using cluster–robust standard errors. You will find it easiest if you combine results using **estimates table**. What happens if you try to fit the model with no structure placed on the error correlations?

4. For the model and data of section 8.5, obtain the within estimator by applying **regress** to (8.7). Hint: For example, for variable x, type **by id: egen avex = mean(x)** followed by **summarize** x and then **generate mdx = x - avex + r(mean)**. Verify that you get the same estimated coefficients as you would with **xtreg, fe**.

5. For the model and data of section 8.6, compare the RE estimators obtained by using **xtreg** with the **re**, **mle**, and **pa** options, and **xtgee** with the **corr(exchangeable)** option. Also compare the associated standard-error estimates obtained by using default standard errors and by using cluster–robust standard errors. You will find it easiest if you combine results using **estimates table**.

6. Consider the RE model output given in section 8.7. Verify that, given the estimated values of **e_sigma** and **u_sigma**, application of the formulas in that section leads to the estimated values of **rho** and **theta**.

7. Make an unbalanced panel dataset by using the data of section 8.4 but then typing `set seed 10101` and `drop if runiform() < 0.2`. This will randomly drop 20% of the individual–year observations. Type `xtdescribe`. Do you obtain the expected patterns of missing data? Use `xtsum` to describe the variation in `id`, `t`, `wage`, `ed`, and `south`. How do the results compare with those from the full panel? Use `xttab` and `xttrans` to provide interpretations of how `south` changes for individuals over time. Compare the within estimator with that in section 8.5 using the balanced panel.

9 Linear panel-data models: Extensions

9.1 Introduction

The essential panel methods for linear models, most notably, the important distinction between fixed-effects (FE) and random-effects (RE) models, were presented in chapter 8.

In this chapter, we present other panel methods for the linear model, those for instrumental-variables (IV) estimation, estimation when lagged dependent variables are regressors, and estimation of mixed models with slope parameters that vary across individuals. We also consider estimation methods for clustered data or hierarchical data, such as cross-section individual data from a survey conducted at a number of villages with clustering at the village level, including the use of methods for short panels in this context. Nonlinear panel models are presented in chapter 18.

9.2 Panel IV estimation

IV methods have been extended from cross-section data (see chapter 6 for an explanation) to panel data. Estimation still needs to eliminate the α_i, if the FE model is appropriate, and inference needs to control for the clustering inherent in panel data.

In this section, we detail `xtivreg`, which is a panel extension of the cross-section command `ivregress`. The subsequent two sections present more specialized IV estimators and commands that are applicable in situations where regressors from periods other than the current period are used as instruments.

9.2.1 Panel IV

If a pooled model is appropriate with $y_{it} = \alpha + \mathbf{x}'_{it}\boldsymbol{\beta} + u_{it}$ and instruments $\mathbf{z}_{it}$ exist satisfying $E(u_{it}|\mathbf{z}_{it}) = 0$, then consistent estimation is possible by two-stage least-squares (2SLS) regression of y_{it} on $\mathbf{x}_{it}$ with instruments $\mathbf{z}_{it}$. The `ivregress` command can be used, with subsequent statistical inference based on cluster–robust standard errors.

More often, we use the individual-effects model

$$y_{it} = \mathbf{x}'_{it}\boldsymbol{\beta} + \alpha_i + \varepsilon_{it} \tag{9.1}$$

which has two error components, α_i and ε_{it}. The FE and first-difference (FD) estimators provide consistent estimates of the coefficients of the time-varying regressors under a limited form of endogeneity of the regressors—$\mathbf{x}_{it}$ may be correlated with the fixed effects α_i but not with ε_{it}.

We now consider a richer type of endogeneity, with $\mathbf{x}_{it}$ correlated with ε_{it}. We need to assume the existence of instruments $\mathbf{z}_{it}$ that are correlated with $\mathbf{x}_{it}$ and uncorrelated with ε_{it}. The panel IV procedure is to suitably transform the individual-effects model to control for α_i and then apply IV to the transformed model.

9.2.2   The xtivreg command

The `xtivreg` command implements 2SLS regression with options corresponding to those for the `xtreg` command. The syntax is similar to that for the cross-section `ivregress` command:

`xtivreg` *depvar* [*varlist1*] (*varlist2=varlist_iv*) [*if*] [*in*] [, *options*]

The four main options are `fe`, `fd`, `re`, and `be`. The `fe` option performs within 2SLS regression of $y_{it} - \overline{y}_i$ on an intercept and $\mathbf{x}_{it} - \overline{\mathbf{x}}_i$ with the instruments $\mathbf{z}_{it} - \overline{\mathbf{z}}_i$. The `fd` option, not available in `xtreg`, performs FD 2SLS regression of $y_{it} - y_{i,t-1}$ on an intercept and $\mathbf{x}_{it} - \mathbf{x}_{i,t-1}$ with the instruments $\mathbf{z}_{it} - \mathbf{z}_{i,t-1}$. The `re` option performs RE 2SLS regression of $y_{it} - \widehat{\theta}_i \overline{y}_i$ on an intercept and $\mathbf{x}_{it} - \widehat{\theta}_i \overline{\mathbf{x}}_i$ with the instruments $\mathbf{z}_{it} - \widehat{\theta}_i \overline{\mathbf{z}}_i$, and the additional options `ec2sls` and `nosa` provide variations of this estimator. The `be` option performs between 2SLS regression of $\overline{y}_i$ on $\overline{\mathbf{x}}_i$ with instruments $\overline{\mathbf{z}}_i$. Other options include `first` to report first-stage regression results and `regress` to ignore the instruments and instead estimate the parameters of the transformed model by ordinary least squares (OLS).

The `xtivreg` command has no `vce(robust)` option to produce cluster–robust standard errors. Cluster–robust standard errors can be computed by using the option `vce(bootstrap)`. Alternatively, the user-written `xtivreg2` command (Schaffer 2007) can be used. This command for 2SLS, GMM, or LIML estimation provides cluster–robust standard errors, tests of weak instruments, and overidentifying restrictions.

As usual, exogenous regressors are instrumented by themselves. For endogenous regressors, we can proceed as in the cross-section case, obtaining an additional variable(s) that does not directly determine y_{it} but is correlated with the variable(s) being instrumented. In the simplest case, the instrument is an external instrument, a variable that does not appear directly as a regressor in the model. This is the same IV identification strategy as used with cross-section data.

9.2.3   Application of the xtivreg command

Consider the chapter 8 example of regression of `lwage` on `exp`, `exp2`, `wks`, and `ed`. We assume the experience variables `exp` and `exp2` are exogenous and that `ed` is correlated with the time-invariant component of the error but is uncorrelated with the time-varying component of the error. Given just these assumptions, we need to control for fixed effects. From section 8.6, the within estimator yields consistent estimates of coefficients of `exp`, `exp2`, and `wks`, whereas the coefficient of `ed` is not identified because it is a time-invariant regressor.

Now suppose that the regressor `wks` is correlated with the time-varying component of the error. Then the within estimator becomes inconsistent, and we need to instrument for `wks`. We suppose that `ms` (marital status) is a suitable instrument. This requires an assumption that marital status does not directly determine the wage rate but is correlated with weeks worked. Because the effects here are fixed, the `fe` or `fd` options of `xtivreg` need to be used.

Formally, we have assumed that the instruments—here `exp`, `exp2`, and `ms`—satisfy the strong exogeneity assumption that

$$E(\varepsilon_{it}|\alpha_i, \mathbf{z}_{i1}, \ldots, \mathbf{z}_{it}, \ldots, \mathbf{z}_{iT}) = 0$$

so that instruments and errors are uncorrelated in all periods. One consequence of this strong assumption is that panel IV estimators are consistent even if the ε_{it} are serially correlated, so cluster–robust standard errors could be used. The `xtivreg` command does not provide a direct option for cluster–robust standard errors, so we report the default standard errors.

We use `xtivreg` with the `fe` option to eliminate the fixed effects. We drop the unidentified time-invariant regressor `ed`—the same results are obtained if it is included. We obtain

```
. * Panel IV example: FE with wks instrumented by external instrument ms
. use mus08psidextract.dta, clear
(PSID wage data 1976-82 from Baltagi and Khanti-Akom (1990))

. xtivreg lwage exp exp2 (wks = ms), fe
Fixed-effects (within) IV regression          Number of obs      =        4165
Group variable: id                             Number of groups   =         595

R-sq:  within  =     .                         Obs per group: min =           7
       between = 0.0172                                        avg =         7.0
       overall = 0.0284                                        max =           7

                                               Wald chi2(3)       =   700142.43
corr(u_i, Xb)  = -0.8499                        Prob > chi2        =      0.0000

------------------------------------------------------------------------------
       lwage |      Coef.   Std. Err.      z    P>|z|     [95% Conf. Interval]
-------------+----------------------------------------------------------------
         wks | -.1149742   .2316926    -0.50   0.620    -.5690832    .3391349
         exp |  .1408101   .0547014     2.57   0.010     .0335974    .2480228
        exp2 | -.0011207   .0014052    -0.80   0.425    -.0038748    .0016334
       _cons |   9.83932   10.48955     0.94   0.348    -10.71983    30.39847
-------------+----------------------------------------------------------------
     sigma_u |  1.0980369
     sigma_e |  .51515503
         rho |  .81959748   (fraction of variance due to u_i)
------------------------------------------------------------------------------
F test that all u_i=0:     F(594,3567) =      4.62             Prob > F = 0.0000
------------------------------------------------------------------------------
Instrumented:  wks
Instruments:   exp exp2 ms
------------------------------------------------------------------------------
```

The estimates imply that, surprisingly, wages decrease by 11.5% for each additional week worked, though the coefficient is statistically insignificant. Wages increase with experience until a peak at 64 years $[= 0.1408/(2 \times 0.0011)]$.

Comparing the IV results with those given in section 8.5.3 using `xtreg, fe`, the coefficient of the endogenous variable `wks` has changed sign and is many times larger in absolute value, whereas the coefficients of the exogenous experience regressors are less affected. For these data, the IV standard errors are more than ten times larger. Because the instrument `ms` is not very correlated with `wks`, IV regression leads to a substantial loss in estimator efficiency.

9.2.4　Panel IV extensions

We used the external instrument `ms` as the instrument for `wks`.

An alternative is to use `wks` from a period other than the current period as the instrument. This has the attraction of being reasonably highly correlated with the variable being instrumented, but it is not necessarily a valid instrument. In the simplest panel model $y_{it} = \mathbf{x}'_{it}\boldsymbol{\beta} + \varepsilon_{it}$, if the errors ε_{it} are independent, then any variable in any period that is not in $\mathbf{x}_{it}$ is a valid instrument. Once we introduce an individual effect as in (9.1) and transform the model, more care is needed.

The next two sections present, respectively, the Hausman–Taylor estimator and the Arellano–Bond estimator that use as instruments regressors from periods other than the current period.

9.3　Hausman–Taylor estimator

We consider the FE model. The FE and FD estimators provide consistent estimators but not for the coefficients of time-invariant regressors because these are then not identified. The Hausman–Taylor estimator is an IV estimator that additionally enables the coefficients of time-invariant regressors to be estimated. It does so by making the stronger assumption that some specified regressors are uncorrelated with the fixed effect. Then values of these regressors in periods other than the current period can be used as instruments.

9.3.1　Hausman–Taylor estimator

The key step is to distinguish between regressors uncorrelated with the fixed effect and those potentially correlated with the fixed effect. The method additionally distinguishes between time-varying and time-invariant regressors.

The individual-effects model is then written as

$$y_{it} = \mathbf{x}'_{1it}\boldsymbol{\beta}_1 + \mathbf{x}'_{2it}\boldsymbol{\beta}_2 + \mathbf{w}'_{1i}\boldsymbol{\gamma}_1 + \mathbf{w}'_{2i}\boldsymbol{\gamma}_2 + \alpha_i + \varepsilon_{it} \tag{9.2}$$

where regressors with subscript 1 are specified to be uncorrelated with α_i and regressors with subscript 2 are specified to be correlated with α_i, $\mathbf{w}$ denotes time-invariant regressors, and $\mathbf{x}$ now denotes time-varying regressors. All regressors are assumed to be uncorrelated with ε_{it}, whereas $\texttt{xtivreg}$ explicitly deals with such correlation.

The Hausman–Taylor method is based on the random-effects transformation that leads to the model

$$\widetilde{y}_{it} = \widetilde{\mathbf{x}}'_{1it}\boldsymbol{\beta}_1 + \widetilde{\mathbf{x}}'_{2it}\boldsymbol{\beta}_2 + \widetilde{\mathbf{w}}'_{1i}\boldsymbol{\gamma}_1 + \widetilde{\mathbf{w}}'_{2i}\boldsymbol{\gamma}_2 + \widetilde{\alpha}_i + \widetilde{\varepsilon}_{it}$$

where, for example, $\widetilde{\mathbf{x}}_{1it} = \mathbf{x}_{1it} - \widehat{\theta}_i \overline{\mathbf{x}}_{1i}$, and the formula for $\widehat{\theta}_i$ is given in [XT] $\textbf{xthtaylor}$.

The RE transformation is used because, unlike the within transform, here $\widetilde{\mathbf{w}}_{1i} \neq 0$ and $\widetilde{\mathbf{w}}_{2i} \neq 0$ so $\boldsymbol{\gamma}_1$ and $\boldsymbol{\gamma}_2$ can be estimated. But $\widetilde{\alpha}_i = \alpha_i(1 - \widehat{\theta}_i) \neq 0$, so the fixed effect has not been eliminated, and $\widetilde{\alpha}_i$ is correlated with $\widetilde{\mathbf{x}}_{2it}$ and with $\widetilde{\mathbf{w}}_{2i}$. This correlation is dealt with by IV estimation. For $\widetilde{\mathbf{x}}_{2it}$, the instrument used is $\ddot{\mathbf{x}}_{2it} = \mathbf{x}_{2it} - \overline{\mathbf{x}}_{2i}$, which can be shown to be uncorrelated with α_i. For $\widetilde{\mathbf{w}}_{2i}$, the instrument is $\overline{\mathbf{x}}_{1i}$, so the method requires that the number of time-varying exogenous regressors be at least as large as the number of time-invariant endogenous regressors. The method uses $\ddot{\mathbf{x}}_{1it}$ as an instrument for $\widetilde{\mathbf{x}}_{1it}$ and $\mathbf{w}_{1i}$ as an instrument for $\widetilde{\mathbf{w}}_{1i}$. Essentially, $\mathbf{x}_1$ is used as an instrument twice: as $\ddot{\mathbf{x}}_{1it}$ and as $\overline{\mathbf{x}}_{1i}$. By using the average of $\overline{\mathbf{x}}_{1i}$ in forming instruments, we are using data from other periods to form instruments.

9.3.2   The xthtaylor command

The $\texttt{xthtaylor}$ command performs IV estimation of the parameters of (9.2) using the instruments $\ddot{\mathbf{x}}_{1it}$, $\ddot{\mathbf{x}}_{2it}$, $\mathbf{w}_{1i}$, and $\overline{\mathbf{x}}_{1i}$. The syntax of the command is

$\textbf{xthtaylor}$ *depvar indepvars* $\big[\,if\,\big]$ $\big[\,in\,\big]$ $\big[\,weight\,\big]$, $\texttt{endog}$(*varlist*) $\big[\,options\,\big]$

Here all the regressors are given in *indepvars*, and the subset of these that are potentially correlated with α_i are given in $\texttt{endog}$(*varlist*). The $\texttt{xthtaylor}$ command does not provide an option to compute cluster–robust standard errors.

The options include $\texttt{amacurdy}$, which uses a wider range of instruments. Specifically, the Hausman–Taylor method requires that $\overline{\mathbf{x}}_{1i}$ be uncorrelated with α_i. If each $\mathbf{x}_{1it}$, $t = 1, \ldots, T$, is uncorrelated with α_i, then more instruments are available and we can use as instruments $\ddot{\mathbf{x}}_{1it}$, $\ddot{\mathbf{x}}_{2it}$, $\mathbf{w}_{1i}$, and $\mathbf{x}_{1i1}, \ldots, \mathbf{x}_{1iT}$.

9.3.3   Application of the xthtaylor command

The dataset used in chapter 8 and in this chapter, attributed to Baltagi and Khanti-Akom (1990) and Cornwell and Rupert (1988), was originally applied to the Hausman–Taylor estimator. We reproduce that application here. It uses a wider set of regressors than we have used to this point.

The goal is to obtain a consistent estimate of the coefficient $\texttt{ed}$ because there is great interest in the impact of education on wages. Education is clearly endogenous. It is assumed that education is only correlated with the individual-specific component of the error α_i. In principal, within estimation gives a consistent estimator, but in practice, no estimator is obtained because `ed` is time-invariant so its coefficient cannot be estimated.

The Hausman–Taylor estimator is used instead, assuming that only a subset of the regressors are correlated with α_i. Identification requires that there be at least one time-varying regressor that is uncorrelated with the fixed effect. Cornwell and Ruppert assumed that, for the time-varying regressors, `exp`, `exp2`, `wks`, `ms`, and `union` were endogenous, whereas `occ`, `south`, `smsa`, and `ind` were exogenous. And for the time-invariant regressors, `ed` is endogenous and `fem` and `blk` are exogenous. The `xthtaylor` command requires only distinction between endogenous and exogenous regressors, because it can determine which regressors are time-varying and which are not.

We obtain

```
. * Hausman-Taylor example of Baltagi and Khanti-Akom (1990)
. use mus08psidextract.dta, clear
(PSID wage data 1976-82 from Baltagi and Khanti-Akom (1990))

. xthtaylor lwage occ south smsa ind exp exp2 wks ms union fem blk ed,
> endog(exp exp2 wks ms union ed)
```

```
Hausman-Taylor estimation              Number of obs      =       4165
Group variable: id                     Number of groups   =        595

                                       Obs per group: min =          7
                                                      avg =          7
                                                      max =          7

Random effects u_i ~ i.i.d.            Wald chi2(12)      =    6891.87
                                       Prob > chi2        =     0.0000
```

lwage	Coef.	Std. Err.	z	P>|z|	[95% Conf. Interval]	
TVexogenous						
occ	-.0207047	.0137809	-1.50	0.133	-.0477149	.0063055
south	.0074398	.031955	0.23	0.816	-.0551908	.0700705
smsa	-.0418334	.0189581	-2.21	0.027	-.0789906	-.0046761
ind	.0136039	.0152374	0.89	0.372	-.0162608	.0434686
TVendogenous						
exp	.1131328	.002471	45.79	0.000	.1082898	.1179758
exp2	-.0004189	.0000546	-7.67	0.000	-.0005259	-.0003119
wks	.0008374	.0005997	1.40	0.163	-.0003381	.0020129
ms	-.0298508	.01898	-1.57	0.116	-.0670508	.0073493
union	.0327714	.0149084	2.20	0.028	.0035514	.0619914
TIexogenous						
fem	-.1309236	.126659	-1.03	0.301	-.3791707	.1173234
blk	-.2857479	.1557019	-1.84	0.066	-.5909179	.0194221
TIendogenous						
ed	.137944	.0212485	6.49	0.000	.0962977	.1795902
_cons	2.912726	.2836522	10.27	0.000	2.356778	3.468674
sigma_u	.94180304					
sigma_e	.15180273					
rho	.97467788	(fraction of variance due to u_i)				

Note: TV refers to time varying; TI refers to time invariant.

Compared with the RE estimates given in section 8.7, the coefficient of **ed** has increased from 0.112 to 0.138, and the standard error has increased from 0.0084 to 0.0212.

For the regular IV estimator to be consistent, it is necessary to argue that any instruments are uncorrelated with the error term. Similarly, for the Hausman–Taylor estimator to be consistent, it is necessary to argue that all regressors are uncorrelated with the idiosyncratic error ε_{it} and that a specified subset of the regressors is uncorrelated with the fixed effect α_i. This strong assumption can be tested by the user-written command **xtoverid** following command **xthtaylor**.

9.4   Arellano–Bond estimator

With panel data, the dependent variable is observed over time, opening up the possibility of estimating parameters of dynamic models that specify the dependent variable for an individual to depend in part on its values in previous periods. As in the nonpanel case, however, care is needed because OLS with a lagged dependent variable and serially correlated error leads to inconsistent parameter estimates.

We consider estimation of fixed-effects models for short panels when one or more lags of the dependent variable are included as regressors. Then the fixed effect needs to be eliminated by first-differencing rather than mean-differencing for reasons given at the end of section 9.4.1. Consistent estimators can be obtained by IV estimation of the parameters in the first-difference model, using appropriate lags of regressors as the instruments. This estimator, called the Arellano–Bond estimator, can be performed, with some considerable manipulation, by using the IV commands **ivregress** or **xtivreg**. But it is much easier to use the specialized commands **xtabond**, **xtdpdsys**, and **xtdpd**. These commands also enable more-efficient estimation and provide appropriate model specification tests.

9.4.1   Dynamic model

The general model considered is an autoregressive model of order p in y_{it} [an AR(p) model] with $y_{i,t-1}, \ldots, y_{i,t-p}$ as regressors, as well as the regressors $\mathbf{x}_{it}$. The model is

$$y_{it} = \gamma_1 y_{i,t-1} + \cdots + \gamma_p y_{i,t-p} + \mathbf{x}'_{it}\boldsymbol{\beta} + \alpha_i + \varepsilon_{it}, \quad t = p+1, \ldots, T \qquad (9.3)$$

where α_i is a fixed effect. The regressors $\mathbf{x}_{it}$ are initially assumed to be uncorrelated with ε_{it}, an assumption that is relaxed in section 9.4.8. The goal is to consistently estimate $\gamma_1, \ldots, \gamma_p$ and $\boldsymbol{\beta}$ when α_i is a fixed effect. The estimators are also consistent if α_i is a random effect.

The dynamic model (9.3) provides several quite different reasons for correlation in y over time: 1) directly through y in preceding periods, called true state dependence; 2) directly through observables $\mathbf{x}$, called observed heterogeneity; and 3) indirectly through the time-invariant individual effect α_i, called unobserved heterogeneity.

These reasons have substantively different policy implications. For illustration, consider a pure AR(1) time-series model for earnings $y_{it} = \gamma_1 y_{i,t-1} + \alpha_i + \varepsilon_{it}$ with $\varepsilon_{it} \simeq 0$ for $t > 1$. Suppose in period 1 there is a large positive shock, ε_{i1}, leading to a large value for y_{i1}, moving a low-paid individual to a high-paid job. Then, if $\gamma_1 \simeq 1$, earnings will remain high in future years (because $y_{i,t+1} \simeq y_{it} + \alpha_i$). If instead $\gamma_1 \simeq 0$, earnings will return to α_i in future years (because $y_{i,t+1} \simeq \alpha_i$).

It is important to note that the within estimator is inconsistent once lagged regressors are introduced. This is because the within model will have the first regressor $y_{i,t-1} - \overline{y}_i$ that is correlated with the error $\varepsilon_{it} - \overline{\varepsilon}_i$, because $y_{i,t-1}$ is correlated with $\varepsilon_{i,t-1}$ and hence with $\overline{\varepsilon}_i$. Furthermore, IV estimation using lags is not possible because any lag $y_{i,s}$ will also be correlated with $\overline{\varepsilon}_i$ and hence with $\varepsilon_{it} - \overline{\varepsilon}_i$. By contrast, although the FD estimator is also inconsistent, IV estimators of the FD model that use appropriate lags of y_{it} as instruments do lead to consistent parameter estimates.

9.4.2   IV estimation in the FD model

The FD model is

$$\Delta y_{it} = \gamma_1 \Delta y_{i,t-1} + \cdots + \gamma_p \Delta y_{i,t-p} + \Delta \mathbf{x}'_{it} \boldsymbol{\beta} + \Delta \varepsilon_{it}, \quad t = p+1, \ldots, T \qquad (9.4)$$

We make the crucial assumption that ε_{it} are serially uncorrelated, a departure from most analysis to this point that has permitted ε_{it} to be correlated over time for a given individual. This assumption is testable, is likely to hold if p is sufficiently large, and can be relaxed by using `xtdpd`, presented in section 9.4.8.

In contrast to a static model, OLS on the first-differenced data produces inconsistent parameter estimates because the regressor $\Delta y_{i,t-1}$ is correlated with the error $\Delta \varepsilon_{it}$, even if ε_{it} are serially uncorrelated. For serially uncorrelated ε_{it}, the FD model error $\Delta \varepsilon_{it} = \varepsilon_{it} - \varepsilon_{i,t-1}$ is correlated with $\Delta y_{i,t-1} = y_{i,t-1} - y_{i,t-2}$ because $y_{i,t-1}$ depends on $\varepsilon_{i,t-1}$. At the same time, $\Delta \varepsilon_{it}$ is uncorrelated with $\Delta y_{i,t-k}$ for $k \geq 2$, opening up the possibility of IV estimation using lagged variables as instruments.

Anderson and Hsiao (1981) proposed IV estimation using $y_{i,t-2}$, which is uncorrelated with $\Delta \varepsilon_{it}$, as an instrument for $\Delta y_{i,t-1}$. The other lagged dependent variables can be instruments for themselves. The regressors $\mathbf{x}_{it}$ can be used as instruments for themselves if they are strictly exogenous; otherwise, they can also be instrumented as detailed below.

More-efficient IV estimators can be obtained by using additional lags of the dependent variable as an instrument; see Holtz-Eakin, Newey, and Rosen (1988). The estimator is then called the Arellano–Bond estimator after Arellano and Bond (1991), who detailed implementation of the estimator and proposed tests of the crucial assumption that ε_{it} are serially uncorrelated. Because the instrument set is unbalanced and can be quite complicated, Stata provides the distinct command `xtabond`.

9.4.3   The xtabond command

The xtabond command has the syntax

xtabond *depvar* [*indepvars*] [*if*] [*in*] [, *options*]

The number of lags in the dependent variable, p in (9.4), is defined by using the lags($\#$) option with the default $p = 1$. The regressors are declared in different ways depending on the type of regressor.

First, strictly exogenous regressors are uncorrelated with ε_{it}, require no special treatment, are used as instruments for themselves, and are entered as *indepvars*.

Second, predetermined regressors or weakly exogenous regressors are correlated with past errors but are uncorrelated with future errors: $E(x_{it}\varepsilon_{is}) \neq 0$ for $s < t$, and $E(x_{it}\varepsilon_{is}) = 0$ for $s \geq t$. These regressors can be instrumented in the same way that $y_{i,t-1}$ is instrumented using subsequent lags of $y_{i,t-1}$. Specifically, x_{it} is instrumented by $x_{i,t-1}$, $x_{i,t-2}$, $\ldots$. These regressors are entered by using the pre(*varlist*) option.

Third, a regressor may be contemporaneously endogenous: $E(x_{it}\varepsilon_{is}) \neq 0$ for $s \leq t$, and $E(x_{it}\varepsilon_{is}) = 0$ for $s > t$. Now $E(x_{it}\varepsilon_{it}) \neq 0$, so $x_{i,t-1}$ is no longer a valid instrument in the FD model. The instruments for x_{it} are now $x_{i,t-2}$, $x_{i,t-3}$, $\ldots$. These regressors are entered by using the endogenous(*varlist*) option.

Finally, additional instruments can be included by using the inst(*varlist*) option.

Potentially, many instruments are available, especially if T is large. If too many instruments are used, then asymptotic theory provides a poor finite-sample approximation to the distribution of the estimator. The maxldep($\#$) option sets the maximum number of lags of the dependent variable that can be used as instruments. The maxlags($\#$) option sets the maximum number of lags of the predetermined and endogenous variables that can be used as instruments. Alternatively, the lagstruct(*lags*, *endlags*) suboption can be applied individually to each variable in pre(*varlist*) and endogenous(*varlist*).

Two different IV estimators can be obtained; see section 6.2. The 2SLS estimator, also called the one-step estimator, is the default. Because the model is overidentified, more-efficient estimation is possible using optimal generalized method of moments (GMM), also called the two-step estimator because first-step estimation is needed to obtain the optimal weighting matrix used at the second step. The optimal GMM estimator is obtained by using the twostep option.

The vce(robust) option provides a heteroskedastic-consistent estimate of the variance–covariance matrix of the estimator (VCE). If the ε_{it} are serially correlated, the estimator is no longer consistent, so there is no cluster–robust VCE for this case.

Postestimation commands for xtabond include estat abond, to test the critical assumption of no error correlation, and estat sargan, to perform an overidentifying restrictions test. See section 9.4.6.

9.4.4   Arellano–Bond estimator: Pure time series

For concreteness, consider an AR(2) model for lnwage with no other regressors and seven years of data. Then we have sufficient data to obtain IV estimates in the model

$$\Delta y_{it} = \alpha + \gamma_1 \Delta y_{i,t-1} + \gamma_2 \Delta y_{i,t-2} + \Delta\varepsilon_{it}, \quad t = 4, 5, 6, 7 \qquad (9.5)$$

At $t = 4$, there are two available instruments, y_{i1} and y_{i2}, because these are uncorrelated with $\Delta\varepsilon_{i4}$. At $t = 5$, there are now three instruments, y_{i1}, y_{i2}, and y_{i3}, that are uncorrelated with $\Delta\varepsilon_{i5}$. Continuing in this manner at $t = 6$, there are four instruments, $y_{i1}, \ldots, y_{i4}$; and at $t = 7$, there are five instruments, $y_{i1}, \ldots, y_{i5}$. In all, there are $2+3+4+5 = 14$ available instruments for the two lagged dependent variable regressors. Additionally, the intercept is an instrument for itself. Estimation can be by 2SLS or by the more efficient optimal GMM, which is possible because the model is overidentified. Because the instrument set is unbalanced, it is much easier to use xtabond than it is to manually set up the instruments and use ivregress.

We apply the estimator to an AR(2) model for the wages data, initially without additional regressors.

```
. * 2SLS or one-step GMM for a pure time-series AR(2) panel model
. use mus08psidextract.dta, clear
(PSID wage data 1976-82 from Baltagi and Khanti-Akom (1990))

. xtabond lwage, lags(2) vce(robust)

Arellano-Bond dynamic panel-data estimation   Number of obs     =      2380
Group variable: id                             Number of groups  =       595
Time variable: t
                                               Obs per group:    min =        4
                                                                 avg =        4
                                                                 max =        4

Number of instruments =     15                 Wald chi2(2)      =   1253.03
                                               Prob > chi2       =    0.0000
One-step results

                                  (Std. Err. adjusted for clustering on id)
                          Robust
      lwage     Coef.   Std. Err.      z     P>|z|    [95% Conf. Interval]

      lwage
        L1.   .5707517  .0333941    17.09   0.000    .5053005    .6362029
        L2.   .2675649  .0242641    11.03   0.000    .2200082    .3151216

      _cons  1.203588   .164496     7.32   0.000    .8811814   1.525994

Instruments for differenced equation
        GMM-type: L(2/.).lwage
Instruments for level equation
        Standard: _cons
```

There are $4 \times 595 = 2380$ observations because the first three years of data are lost in order to construct $\Delta y_{i,t-2}$. The results are reported for the original levels model, with the dependent variable y_{it} and the regressors the lagged dependent variables $y_{i,t-1}$ and $y_{i,t-2}$, even though mechanically the FD model is fit. There are 15 instruments, as already explained, with output L(2/.), meaning that $y_{i,t-2}$, $y_{i,t-3}$, ..., $y_{i,1}$ are the instruments used for period t. Wages depend greatly on past wages, with the lag weights summing to $0.57 + 0.27 = 0.84$.

The results given are for the 2SLS or one-step estimator. The standard errors reported are robust standard errors that permit the underlying error ε_{it} to be heteroskedastic but do not allow for any serial correlation in ε_{it}, because then the estimator is inconsistent.

More-efficient estimation is possible using optimal or two-step GMM, because the model is overidentified. Standard errors reported using the standard textbook formulas for the two-step GMM estimator are downward biased in finite samples. A better estimate of the standard errors, proposed by Windmeijer (2005), can be obtained by using the vce(robust) option. As for the one-step estimator, these standard errors permit heteroskedasticity in ε_{it}.

Two-step GMM estimation for our data yields

```
. * Optimal or two-step GMM for a pure time-series AR(2) panel model
. xtabond lwage, lags(2) twostep vce(robust)
Arellano-Bond dynamic panel-data estimation  Number of obs      =       2380
Group variable: id                           Number of groups   =        595
Time variable: t
                                             Obs per group:     min =          4
                                                                avg =          4
                                                                max =          4
Number of instruments =      15              Wald chi2(2)       =    1974.40
                                             Prob > chi2        =     0.0000
Two-step results
                                     (Std. Err. adjusted for clustering on id)
```

lwage	Coef.	WC-Robust Std. Err.	z	P>|z|	[95% Conf. Interval]	
lwage						
L1.	.6095931	.0330542	18.44	0.000	.544808	.6743782
L2.	.2708335	.0279226	9.70	0.000	.2161061	.3255608
_cons	.9182262	.1339978	6.85	0.000	.6555952	1.180857

```
Instruments for differenced equation
        GMM-type: L(2/.).lwage
Instruments for level equation
        Standard: _cons
```

Here the one-step and two-step estimators have similar estimated coefficients, and the standard errors are also similar, so there is little efficiency gain in two-step estimation.

For a large T, the Arellano–Bond method generates many instruments, leading to potential poor performance of asymptotic results. The number of instruments can be restricted by using the maxldep() option. For example, we may use only the first available lag, so that just $y_{i,t-2}$ is the instrument in period t.

```
. * Reduce the number of instruments for a pure time-series AR(2) panel model
. xtabond lwage, lags(2) vce(robust) maxldep(1)
Arellano-Bond dynamic panel-data estimation  Number of obs        =       2380
Group variable: id                           Number of groups     =        595
Time variable: t
                                             Obs per group:  min =          4
                                                             avg =          4
                                                             max =          4
Number of instruments =        5             Wald chi2(2)         =    1372.33
                                             Prob > chi2          =     0.0000
One-step results
                                       (Std. Err. adjusted for clustering on id)
```

		Robust				
lwage	Coef.	Std. Err.	z	P>|z|	[95% Conf. Interval]	
lwage						
L1.	.4863642	.1919353	2.53	0.011	.110178	.8625505
L2.	.3647456	.1661008	2.20	0.028	.039194	.6902973
_cons	1.127609	.2429357	4.64	0.000	.6514633	1.603754

```
Instruments for differenced equation
        GMM-type: L(2/2).lwage
Instruments for level equation
        Standard: _cons
```

Here there are five instruments: y_{i2} when $t = 4$, y_{i3} when $t = 5$, y_{i4} when $t = 6$, y_{i5} when $t = 7$, and the intercept is an instrument for itself.

In this example, there is considerable loss of efficiency because the standard errors are now about six times larger. This inefficiency disappears if we instead use the `maxldep(2)` option, yielding eight instruments rather than the original 15.

9.4.5 Arellano–Bond estimator: Additional regressors

We now introduce regressors that are not lagged dependent variables.

We fit a model for `lwage` similar to the model specified in section 9.3. The time-invariant regressors `fem`, `blk`, and `ed` are dropped because they are eliminated after first-differencing. The regressors `occ`, `south`, `smsa`, and `ind` are treated as strictly exogenous. The regressor `wks` appears both contemporaneously and with one lag, and it is treated as predetermined. The regressors `ms` and `union` are treated as endogenous. The first two lags of the dependent variable `lwage` are also regressors.

The model omits one very important regressor, years of work experience (`exp`). For these data, it is difficult to disentangle the separate effects of previous periods' wages and work experience. When both are included, the estimates become very imprecise. Because here we wish to emphasize the role of lagged wages, we exclude work experience from the model.

We fit the model using optimal or two-step GMM and report robust standard errors. The strictly exogenous variables appear as regular regressors. The predetermined and endogenous variables are instead given as options, with restrictions placed on the number of available instruments that are actually used. The dependent variable appears with two lags, and the `maxldep(3)` option is specified so that at most three lags are used as instruments. For example, when $t = 7$, the instruments are y_{i5}, y_{i4}, and y_{i3}. The `pre(wks,lag(1,2))` option is specified so that `wks` and `L1.wks` are regressors and only two additional lags are to be used as instruments. The `endogenous(ms,lag(0,2))` option is used to indicate that `ms` appears only as a contemporaneous regressor and that at most two additional lags are used as instruments. The `artests(3)` option does not affect the estimation but will affect the postestimation command `estat abond`, as explained in the next section. We have

```
. * Optimal or two-step GMM for a dynamic panel model
. xtabond lwage occ south smsa ind, lags(2) maxldep(3) pre(wks,lag(1,2))
> endogenous(ms,lag(0,2)) endogenous(union,lag(0,2)) twostep vce(robust)
> artests(3)
```

```
Arellano-Bond dynamic panel-data estimation   Number of obs       =       2380
Group variable: id                            Number of groups    =        595
Time variable: t
                                              Obs per group:  min =          4
                                                              avg =          4
                                                              max =          4

Number of instruments =      40               Wald chi2(10)       =    1287.77
                                              Prob > chi2         =     0.0000
Two-step results
                                         (Std. Err. adjusted for clustering on id)
```

lwage	Coef.	WC-Robust Std. Err.	z	P>|z|	[95% Conf. Interval]	
lwage						
L1.	.611753	.0373491	16.38	0.000	.5385501	.6849559
L2.	.2409058	.0319939	7.53	0.000	.1781989	.3036127
wks						
--.	−.0159751	.0082523	−1.94	0.053	−.0321493	.000199
L1.	.0039944	.0027425	1.46	0.145	−.0013807	.0093695
ms	.1859324	.144458	1.29	0.198	−.0972	.4690649
union	−.1531329	.1677842	−0.91	0.361	−.4819839	.1757181
occ	−.0357509	.0347705	−1.03	0.304	−.1038999	.032398
south	−.0250368	.2150806	−0.12	0.907	−.446587	.3965134
smsa	−.0848223	.0525243	−1.61	0.106	−.187768	.0181235
ind	.0227008	.0424207	0.54	0.593	−.0604422	.1058437
_cons	1.639999	.4981019	3.29	0.001	.6637377	2.616261

```
Instruments for differenced equation
        GMM-type: L(2/4).lwage L(1/2).L.wks L(2/3).ms L(2/3).union
        Standard: D.occ D.south D.smsa D.ind
Instruments for level equation
        Standard: _cons
```

With the inclusion of additional regressors, the coefficients of the lagged dependent variables have changed little and the standard errors are about 10–15% higher. The additional regressors are all statistically insignificant at 5%. By contrast, some are statistically significant using the within estimator for a static model that does not include the lagged dependent variables.

The output explains the instruments used. For example, L(2/4).lwage means that $lwage_{i,t-2}$, $lwage_{i,t-3}$, and $lwage_{i,t-4}$ are used as instruments, provided they are available. In the initial period $t = 4$, only the first two of these are available, whereas in $t = 5, 6, 7$, all three are available for a total of $2 + 3 + 3 + 3 = 11$ instruments. By similar analysis, L(1/2).L.wks, L(2/3).ms, and L(2/3).union each provide 8 instruments, and there are five standard instruments. In all, there are $11 + 8 + 8 + 8 + 5 = 40$ instruments, as stated at the top of the output.

9.4.6 Specification tests

For consistent estimation, the xtabond estimators require that the error ε_{it} be serially uncorrelated. This assumption is testable.

Specifically, if ε_{it} are serially uncorrelated, then $\Delta\varepsilon_{it}$ are correlated with $\Delta\varepsilon_{i,t-1}$, because $\text{Cov}(\Delta\varepsilon_{it}, \Delta\varepsilon_{i,t-1}) = \text{Cov}(\varepsilon_{it} - \varepsilon_{i,t-1}, \varepsilon_{i,t-1} - \varepsilon_{i,t-2}) = -\text{Cov}(\varepsilon_{i,t-1}, \varepsilon_{i,t-1}) \neq 0$. But $\Delta\varepsilon_{it}$ will not be correlated with $\Delta\varepsilon_{i,t-k}$ for $k \geq 2$. A test of whether $\Delta\varepsilon_{it}$ are correlated with $\Delta\varepsilon_{i,t-k}$ for $k \geq 2$ can be calculated based on the correlation of the fitted residuals $\Delta\widehat{\varepsilon_{it}}$. This is performed by using the estat abond command.

The default is to test to lag 2, but here we also test the third lag. This can be done in two ways. One way is to use estat abond with the artests(3) option, which leads to recalculation of the estimator defined in the preceding xtabond command. Alternatively, we can include the artests(3) option in xtabond, in which case we simply use estat abond and no recalculation is necessary.

In our case, the artests(3) option was included in the preceding xtabond command. We obtain

```
. * Test whether error is serially correlated
. estat abond

Arellano-Bond test for zero autocorrelation in first-differenced errors
```

Order	z	Prob > z
1	-4.5244	0.0000
2	-1.6041	0.1087
3	.35729	0.7209

```
H0: no autocorrelation
```

The null hypothesis that $\text{Cov}(\Delta\varepsilon_{it}, \Delta\varepsilon_{i,t-k}) = 0$ for $k = 1, 2, 3$ is rejected at a level of 0.05 if $p < 0.05$. As explained above, if ε_{it} are serially uncorrelated, we expect to reject at order 1 but not at higher orders. This is indeed the case. We reject at order 1 because $p = 0.000$. At order 2, $\Delta\varepsilon_{it}$ and $\Delta\varepsilon_{i,t-2}$ are serially uncorrelated because $p = 0.109 > 0.05$. Similarly, at order 3, there is no evidence of serial correlation because $p = 0.721 > 0.05$. There is no serial correlation in the original error ε_{it}, as desired.

A second specification test is a test of overidentifying restrictions; see section 6.3.7. Here 40 instruments were used to estimate 11 parameters, so there are 29 overidentifying restrictions. The `estat sargan` command implements the test. This command is not implemented after `xtabond` if the `vce(robust)` option is used, because the test is then invalid since it requires that the errors ε_{it} be independent and identically distributed (i.i.d.). We therefore need to first run `xtabond` without this option. We have

```
. * Test of overidentifying restrictions (first estimate with no vce(robust))
. quietly xtabond lwage occ south smsa ind, lags(2) maxldep(3) pre(wks,lag(1,2))
> endogenous(ms,lag(0,2)) endogenous(union,lag(0,2)) twostep artests(3)

. estat sargan
Sargan test of overidentifying restrictions
        H0: overidentifying restrictions are valid

        chi2(29)     =   39.87571
        Prob > chi2  =    0.0860
```

The null hypothesis that the population moment conditions are correct is not rejected because $p = 0.086 > 0.05$.

9.4.7   The xtdpdsys command

The Arellano–Bond estimator uses an IV estimator based on the assumption that $E(y_{is}\Delta\varepsilon_{it}) = 0$ for $s \leq t-2$ in (9.3), so that the lags $y_{i,t-2}$, $y_{i,t-3}$, ... can be used as instruments in the first-differenced (9.4). Several papers suggest using additional moment conditions to obtain an estimator with improved precision and better finite-sample properties. In particular, Arellano and Bover (1995) and Blundell and Bond (1998) consider using the additional condition $E(\Delta y_{i,t-1}\varepsilon_{it}) = 0$ so that we also incorporate the levels (9.3) and use as an instrument $\Delta y_{i,t-1}$. Similar additional moment conditions can be added for endogenous and predetermined variables, whose first-differences can be used as instruments.

This estimator is performed by using the `xtdpdsys` command, introduced in Stata 10. It is also performed by using the user-written `xtabond2` command. The syntax is exactly the same as that for `xtabond`.

(Continued on next page)

We refit the model of section 9.4.5 using xtdpdsys rather than xtabond.

```
. * Arellano/Bover or Blundell/Bond for a dynamic panel model
. xtdpdsys lwage occ south smsa ind, lags(2) maxldep(3) pre(wks,lag(1,2))
> endogenous(ms,lag(0,2)) endogenous(union,lag(0,2)) twostep vce(robust)
> artests(3)
System dynamic panel-data estimation          Number of obs       =        2975
Group variable: id                            Number of groups    =         595
Time variable: t

                                              Obs per group:   min =           5
                                                               avg =           5
                                                               max =           5

Number of instruments =        60             Wald chi2(10)       =     2270.88
                                              Prob > chi2         =      0.0000
Two-step results
```

| | | WC-Robust | | | | |
lwage	Coef.	Std. Err.	z	P>|z|	[95% Conf. Interval]	
lwage						
L1.	.6017533	.0291502	20.64	0.000	.5446199	.6588866
L2.	.2880537	.0285319	10.10	0.000	.2321322	.3439752
wks						
--.	-.0014979	.0056143	-0.27	0.790	-.0125017	.009506
L1.	.0006786	.0015694	0.43	0.665	-.0023973	.0037545
ms	.0395337	.0558543	0.71	0.479	-.0699386	.1490061
union	-.0422409	.0719919	-0.59	0.557	-.1833423	.0988606
occ	-.0508803	.0331149	-1.54	0.124	-.1157843	.0140237
south	-.1062817	.083753	-1.27	0.204	-.2704346	.0578713
smsa	-.0483567	.0479016	-1.01	0.313	-.1422422	.0455288
ind	.0144749	.031448	0.46	0.645	-.0471621	.0761118
_cons	.9584113	.3632287	2.64	0.008	.2464961	1.670327

```
Instruments for differenced equation
        GMM-type: L(2/4).lwage L(1/2).L.wks L(2/3).ms L(2/3).union
        Standard: D.occ D.south D.smsa D.ind
Instruments for level equation
        GMM-type: LD.lwage LD.wks LD.ms LD.union
        Standard: _cons
```

There are now 60 instruments rather than 40 instruments because the lagged first-differences in lwage, wks, ms, and union are available for each of the five periods $t = 3, \ldots, 7$. There is some change in estimated coefficients. More noticeable is a reduction in standard errors of 10–60%, reflecting greater precision because of the additional moment conditions.

The procedure assumes the errors ε_{it} are serially uncorrelated. This assumption can be tested by using the postestimation estat abond command, and from output not given, this test confirms that the errors are serially uncorrelated here. If the xtdpdsys command is run with the default standard errors, the estat sargan command can be used to test the overidentifying conditions.

9.4.8   The xtdpd command

The preceding estimators and commands require that the model errors ε_{it} be serially uncorrelated. If this assumption is rejected (it is testable by using the `estat abond` command), then one possibility is to add more lags of the dependent variable as regressors in the hope that this will eliminate any serial correlation in the error.

An alternative is to use the `xtdpd` command, an acronym for dynamic panel data, that allows ε_{it} to follow a moving-average (MA) process of low order. This command also allows predetermined variables to have a more complicated structure.

For `xtdpd`, a very different syntax is used to enter all the variables and instruments in the model; see [XT] **xtdpd**. Essentially, one specifies a variable list with all model regressors (lagged dependent, exogenous, predetermined, and endogenous), followed by options that specify instruments. For exogenous regressors, the `div()` option is used, and for other types of regressors, the `dgmmiv()` option is used with the explicit statement of the lags of each regressor to be used as instruments. Instruments for the levels equation, used in the `xtdpdsys` command, can also be specified with the `lgmmiv()` option.

As an example, we provide without explanation an `xtdpd` command that exactly reproduces the `xtdpdsys` command of the previous section. We have

(*Continued on next page*)

```
. * Use of xtdpd to exactly reproduce the previous xtdpdsys command
. xtdpd L(0/2).lwage L(0/1).wks occ south smsa ind ms union,
> div(occ south smsa ind) dgmmiv(lwage, lagrange(2 4))
> dgmmiv(ms union, lagrange(2 3)) dgmmiv(L.wks, lagrange(1 2))
> lgmmiv(lwage wks ms union) twostep vce(robust) artests(3)
Dynamic panel-data estimation              Number of obs      =        2975
Group variable: id                         Number of groups   =         595
Time variable: t

                                           Obs per group:    min =           5
                                                             avg =           5
                                                             max =           5

Number of instruments =        60          Wald chi2(10)      =     2270.88
                                           Prob > chi2        =      0.0000
Two-step results
                                      (Std. Err. adjusted for clustering on id)
```

lwage	Coef.	WC-Robust Std. Err.	z	P>|z|	[95% Conf. Interval]	
lwage						
L1.	.6017533	.0291502	20.64	0.000	.5446199	.6588866
L2.	.2880537	.0285319	10.10	0.000	.2321322	.3439752
wks						
--.	-.0014979	.0056143	-0.27	0.790	-.0125017	.009506
L1.	.0006786	.0015694	0.43	0.665	-.0023973	.0037545
occ	-.0508803	.0331149	-1.54	0.124	-.1157843	.0140237
south	-.1062817	.083753	-1.27	0.204	-.2704346	.0578713
smsa	-.0483567	.0479016	-1.01	0.313	-.1422422	.0455288
ind	.0144749	.031448	0.46	0.645	-.0471621	.0761118
ms	.0395337	.0558543	0.71	0.479	-.0699386	.1490061
union	-.0422409	.0719919	-0.59	0.557	-.1833423	.0988606
_cons	.9584113	.3632287	2.64	0.008	.2464961	1.670327

```
Instruments for differenced equation
        GMM-type: L(2/4).lwage L(2/3).ms L(2/3).union L(1/2).L.wks
        Standard: D.occ D.south D.smsa D.ind
Instruments for level equation
        GMM-type: LD.lwage LD.wks LD.ms LD.union
        Standard: _cons
```

Now suppose that the error ε_{it} in (9.3) is MA(1), so that $\varepsilon_{it} = \eta_{it} + \delta\eta_{i,t-1}$, where η_{it} is i.i.d. Then $y_{i,t-2}$ is no longer a valid instrument, but $y_{i,t-3}$ and further lags are. Also, for the level equation, $\Delta y_{i,t-1}$ is no longer a valid instrument but $\Delta y_{i,t-2}$ is valid. We need to change the **dgmmiv()** and **lgmmiv()** options for **lwage**. The command becomes

```
. * Previous command if model error is MA(1)
. xtdpd L(0/2).lwage L(0/1).wks occ south smsa ind ms union,
> div(occ south smsa ind) dgmmiv(lwage, lagrange(3 4))
> dgmmiv(ms union, lagrange(2 3)) dgmmiv(L.wks, lagrange(1 2))
> lgmmiv(L.lwage wks ms union) twostep vce(robust) artests(3)
   (output omitted )
```

The output is the same as the results from **xtdpdsys**.

9.5 Mixed linear models

In the RE model, it is assumed that the individual-specific intercept is uncorrelated with the regressors. Richer models can additionally permit slope parameters to vary over individuals or time. We present two models, the mixed linear model and the less flexible random-coefficients model.

These models are more elaborate RE models. They are not often used in microeconometrics panel-data models because attention is more focused on FE models. Even if an RE model is appropriate, then it is simpler and possibly more efficient to use the `xtreg, pa` command with an appropriate working matrix for the time-series correlation of the errors. The mixed linear model is more useful in clustered settings other than panel data; see section 9.6.

9.5.1 Mixed linear model

The mixed linear model specifies a model for the conditional variance and covariances of y_{it} that can depend on observable variables. Maximum likelihood (ML) or feasible generalized least-squares (FGLS) estimation that exploits this model will lead to more-efficient estimation of the parameters of the model for the conditional mean, assuming that the model for the conditional variances and covariances is correctly specified.

The conditional mean of y_{it} is specified to be $\mathbf{x}'_{it}\boldsymbol{\beta}$, where the regressors $\mathbf{x}_{it}$ now include an intercept. The observed value y_{it} equals this conditional mean plus an error term $\mathbf{z}'_{it}\mathbf{u}_i + \varepsilon_{it}$, where $\mathbf{z}_{it}$ are observable variables, and $\mathbf{u}_i$ and ε_{it} are i.i.d. normally distributed random variables with a mean of zero. We have

$$y_{it} = \mathbf{x}'_{it}\boldsymbol{\beta} + \mathbf{z}'_{it}\mathbf{u}_i + \varepsilon_{it} \tag{9.6}$$

where $\mathbf{u}_i \sim N(\mathbf{0}, \boldsymbol{\Sigma}_\mathbf{u})$ and $\varepsilon_{it} \sim N(0, \sigma_\varepsilon^2)$. The variances and covariances in $\boldsymbol{\Sigma}_\mathbf{u}$ are called RE parameters.

The mixed linear models literature refers to the conditional mean parameters $\boldsymbol{\beta}$ as fixed effects, to be contrasted to the error terms $\mathbf{u}_i$ that are called random effects. We minimize use of this terminology because this is a very different use of the term "fixed effects". Indeed, if the fixed effects defined in section 8.2.2 are present, then the estimators of this section are inconsistent.

Specific choices of $\mathbf{z}_{it}$ lead to some standard models. Pooled OLS corresponds to $\mathbf{z}_{it} = \mathbf{0}$. The RE model of section 8.7 corresponds to $\mathbf{z}_{it} = 1$, because only the intercept is random. A model often called the random-coefficients model sets $\mathbf{z}_{it} = \mathbf{x}_{it}$ so that, for the regressor $\mathbf{x}_{it}$, both the intercept and slope coefficients are random. The hierarchical linear models framework (see section 9.6.4) leads to further choices of $\mathbf{z}_{it}$.

Estimation is by ML or by an asymptotically equivalent variation of ML called restricted maximum likelihood (REML), which produces variance estimates that are unbiased in balanced samples. Normality of the errors is assumed. It is not necessary for consistency of $\boldsymbol{\beta}$, because consistency requires essentially that $E(y_{it}|\mathbf{x}_{it}, \mathbf{z}_{it}) = \mathbf{x}'_{it}\boldsymbol{\beta}$.

However, correctness of the reported standard errors does require i.i.d. errors satisfying $\mathbf{u}_i \sim [\mathbf{0}, \boldsymbol{\Sigma_u}]$ and $\varepsilon_{it} \sim [0, \sigma_\varepsilon^2]$.

9.5.2   The xtmixed command

The `xtmixed` command fits a multilevel mixed-effects model. The dependent variable and regressors in (9.6) are defined, followed by two vertical bars || and a definition of the portion of the model for the random effects $\mathbf{u}_i$.

For example, the command `xtmixed y x || id: z, mle` is used if we want to regress y_{it} on an intercept and x_{it}; the variable `id` identifies individual i, the random effects vary with an intercept and variable z_{it}, and estimation is by ML.

The general syntax of the command is

> `xtmixed` *depvar* $\big[$ *fe_equation* $\big]$ $\big[$ || *re_equation* ... $\big]$ $\big[$ || *re_equation* ... $\big]$
> $\big[$, *options* $\big]$

The dependent variable y_{it} is given in *depvar*. The regressors are defined in *fe_equation*, which has the syntax

> *indepvars* $\big[$ *if* $\big]$ $\big[$ *in* $\big]$ $\big[$, *fe_options* $\big]$

where *indepvars* defines the regressors $\mathbf{x}_{it}$, and the *fe_option* `noconstant` is added if no intercept is to be included. The RE model is given in *re_equation*, which has the syntax

> *levelvar* : $\big[$ *varlist* $\big]$ $\big[$, *re_options* $\big]$

where *levelvar* is the individual unit identifier, *varlist* gives the variable $\mathbf{z}_{it}$, and the `noconstant` *re_option* is added if there is to be no random intercept. The *re_option* `covariance(`*vartype*`)` places structure on $\boldsymbol{\Sigma_u}$, where *vartype* includes `independent` (the default) with $\boldsymbol{\Sigma_u}$ diagonal, and `unstructured` with no structure placed on $\boldsymbol{\Sigma_u}$.

Estimation is by ML if the `mle` option is used. Estimation is by asymptotically equivalent REML (the default) if the `reml` option is used. For multilevel models, the additional levels are specified by a series of *re_equations*, each separated by ||. This is pursued in section 9.6.4. There is no option for alternative estimates of the VCE.

9.5.3   Random-intercept model

The random-intercept model restricts $\mathbf{u}_i = 1$ and is identical to the RE model of section 8.7.

The model can be fit by using the `xtmixed` command, with the RE portion of the model defined simply as `id:`. This `id` identifies the individual unit, and the default is to include a random intercept for the individual unit. We use the `mle` option to obtain

```
. use mus08psidextract.dta, clear
(PSID wage data 1976-82 from Baltagi and Khanti-Akom (1990))

. xtmixed lwage exp exp2 wks ed || id:, mle

Performing EM optimization:

Performing gradient-based optimization:

Iteration 0:   log likelihood =  293.69563
Iteration 1:   log likelihood =  293.69563

Computing standard errors:

Mixed-effects ML regression            Number of obs     =      4165
Group variable: id                     Number of groups  =       595

                                       Obs per group: min =         7
                                                      avg =       7.0
                                                      max =         7

                                       Wald chi2(4)      =   6160.60
Log likelihood =  293.69563            Prob > chi2       =    0.0000
```

lwage	Coef.	Std. Err.	z	P>|z|	[95% Conf. Interval]	
exp	.1079955	.0024527	44.03	0.000	.1031883	.1128027
exp2	-.0005202	.0000543	-9.59	0.000	-.0006266	-.0004139
wks	.0008365	.0006042	1.38	0.166	-.0003477	.0020208
ed	.1378558	.0125814	10.96	0.000	.1131968	.1625149
_cons	2.989858	.17118	17.47	0.000	2.654352	3.325365

Random-effects Parameters	Estimate	Std. Err.	[95% Conf. Interval]	
id: Identity				
sd(_cons)	.8509013	.0278621	.798008	.9073005
sd(Residual)	.1536109	.0018574	.1500132	.1572949

```
LR test vs. linear regression: chibar2(01) =  4576.13 Prob >= chibar2 = 0.0000
```

The estimated coefficients are identical to estimates (not given) from command `xtreg`, `mle`. The standard errors are also the same, aside from a slight difference for those for the coefficients of the regressors $\mathbf{x}_{it}$. Coefficient estimates change very little (less than 0.1%) if the `reml` option of `xtmixed` is used instead. The RE error u_i has an estimated standard deviation that is 5–6 times that for the idiosyncratic error.

9.5.4   Cluster–robust standard errors

A bigger issue is that the reported standard errors require that the errors $\mathbf{u}_i$ and ε_{it} in (9.6) be i.i.d., yet we have already seen from section 8.8.3 that cluster–robust standard errors were 30–40% higher for the RE model (fit using `xtreg`'s `re` option).

One approach is to use a cluster bootstrap. Because there is no `vce()` option for `xtmixed`, we instead need to use the `bootstrap` prefix (explained fully in chapter 13). To do so requires first eliminating the time identifier, since otherwise Stata fails to implement the bootstrap because of "repeated time values within panel". We have

```
. * Cluster robust standard errors after xtmixed using bootstrap
. xtset id
       panel variable:  id (balanced)
. bootstrap, reps(400) seed(10101) cluster(id) nodots:
> xtmixed lwage exp exp2 wks ed || id:, mle
```

```
Mixed-effects ML regression                     Number of obs      =      4165
Group variable: id                              Number of groups   =       595

                                                Obs per group: min =         7
                                                               avg =       7.0
                                                               max =         7

                                                Wald chi2(4)       =   2092.79
Log likelihood =  293.69563                     Prob > chi2        =    0.0000
                                (Replications based on 595 clusters in id)
```

	Observed	Bootstrap			Normal-based
lwage	Coef.	Std. Err.	z	P>|z|	[95% Conf. Interval]
exp	.1079955	.0041447	26.06	0.000	.0998721 .1161189
exp2	-.0005202	.0000831	-6.26	0.000	-.0006831 -.0003573
wks	.0008365	.0008458	0.99	0.323	-.0008212 .0024943
ed	.1378558	.0099856	13.81	0.000	.1182844 .1574273
_cons	2.989858	.1510383	19.80	0.000	2.693829 3.285888

Random-effects Parameters	Observed Estimate	Bootstrap Std. Err.	Normal-based [95% Conf. Interval]
id: Identity			
sd(_cons)	.8509013	.0259641	.8015044 .9033426
sd(Residual)	.1536109	.00824	.1382808 .1706406

```
LR test vs. linear regression: chibar2(01) =  4576.13 Prob >= chibar2 = 0.0000
```

The cluster bootstrap leads to an increase in standard errors for the slope coefficients of time-varying regressors of 20–40%, whereas the standard error of the time-invariant regressor ed has decreased.

Although the regression parameters $\boldsymbol{\beta}$ in (9.6) are consistently estimated if the idiosyncratic errors ε_{it} are serially correlated, the estimates of the variance parameters $\boldsymbol{\Sigma}_u$ and σ_u (reported here as sd(_cons) and sd(Residual)) are inconsistently estimated. This provides motivation for using a random-slopes model.

9.5.5 Random-slopes model

An alternative approach is to use a richer model for the RE portion of the model. If this model is well specified so that the errors $\mathbf{u}_i$ and ε_{it} in (9.6) are i.i.d., then this will lead to more-efficient estimation of $\boldsymbol{\beta}$, correct standard errors for $\widehat{\boldsymbol{\beta}}$, and consistent estimation of $\boldsymbol{\Sigma}_u$ and σ_u.

For our application, we let the random effects depend on exp and wks, and we let $\boldsymbol{\Sigma_u}$ be unstructured. We obtain

```
. * Random-slopes model estimated using xtmixed
. xtmixed lwage exp exp2 wks ed || id: exp wks, covar(unstructured) mle

Performing EM optimization:

Performing gradient-based optimization:

Iteration 0:   log likelihood =  397.61127  (not concave)
Iteration 1:   log likelihood =   482.6225
Iteration 2:   log likelihood =  487.41804
Iteration 3:   log likelihood =  505.61464
Iteration 4:   log likelihood =  508.95847
Iteration 5:   log likelihood =  509.00189
Iteration 6:   log likelihood =  509.00191

Computing standard errors:
```

Mixed-effects ML regression		Number of obs	=	4165
Group variable: id		Number of groups	=	595

```
                                    Obs per group: min =          7
                                                   avg =        7.0
                                                   max =          7

                                    Wald chi2(4)        =    2097.06
Log likelihood =  509.00191         Prob > chi2         =     0.0000
```

lwage	Coef.	Std. Err.	z	P>|z|	[95% Conf. Interval]	
exp	.0527159	.0032966	15.99	0.000	.0462546	.0591772
exp2	.0009476	.0000713	13.28	0.000	.0008078	.0010874
wks	.0006887	.0008267	0.83	0.405	-.0009316	.0023091
ed	.0868604	.0098652	8.80	0.000	.067525	.1061958
_cons	4.317674	.1420957	30.39	0.000	4.039172	4.596177

Random-effects Parameters	Estimate	Std. Err.	[95% Conf. Interval]	
id: Unstructured				
sd(exp)	.043679	.0022801	.0394311	.0483846
sd(wks)	.0081818	.0008403	.00669	.0100061
sd(_cons)	.6042977	.0511419	.5119335	.7133265
corr(exp,wks)	-.2976598	.1000253	-.4792842	-.0915879
corr(exp,_cons)	.0036854	.0859701	-.1633388	.1705042
corr(wks,_cons)	-.4890483	.0835946	-.6352414	-.3090207
sd(Residual)	.1319489	.0017964	.1284745	.1355172

```
LR test vs. linear regression:        chi2(6) =   5006.75   Prob > chi2 = 0.0000
Note: LR test is conservative and provided only for reference.
```

From the first set of output, there is considerable change in the regressor coefficients compared with those from the random-intercept model. The reported standard errors are now similar to those from the cluster bootstrap of the random-intercept model.

From the second set of output, all but one of the RE parameters (corr(exp,_cons)) is statistically significantly different from 0 at 5%. The joint test strongly rejects the null hypothesis that they are all zero. The $\chi^2(6)$ distribution is used to compute p-values, because there are six restrictions. However, these are not six independent restrictions because, for example, if a variance is zero then all corresponding covariances must be zero. The joint-test statistic has a nonstandard and complicated distribution. Using the $\chi^2(6)$ distribution is conservative because it can be shown to overstate the true p-value.

9.5.6 Random-coefficients model

The model that econometricians call the random-coefficients model lets $\mathbf{z}_{it} = \mathbf{x}_{it}$ in (9.6). This can be fit in the same way as the preceding random-slopes model, with the RE portion of the model changed to || id: exp exp2 wks ed.

A similar model can be fit by using the xtrc command. This has exactly the same setup as (9.6) with $\mathbf{z}_{it} = \mathbf{x}_{it}$ and $\boldsymbol{\Sigma}_u$ unstructured. The one difference is that the idiosyncratic error ε_{it} is permitted to be heteroskedastic over i, so that ε_{it} i.i.d. $(0, \sigma_i^2)$. By contrast, the mixed linear model imposes homoskedasticity with $\sigma_i^2 = \sigma^2$. Estimation is by FGLS rather than ML.

In practice, these models can encounter numerical problems, especially if there are many regressors and no structure is placed on $\boldsymbol{\Sigma}_u$. Both xtmixed and xtrc failed numerically for our data, so we apply xtrc to a reduced model, with just exp and wks as regressors. To run xtrc, we must set matsize to exceed the number of groups, here the number of individuals. We obtain

```
. * Random-coefficients model estimated using xtrc
. quietly set matsize 600

. xtrc lwage exp wks, i(id)
Random-coefficients regression              Number of obs     =       4165
Group variable: id                          Number of groups  =        595

                                            Obs per group: min =          7
                                                           avg =        7.0
                                                           max =          7

                                            Wald chi2(2)      =    1692.37
                                            Prob > chi2       =     0.0000
```

lwage	Coef.	Std. Err.	z	P>|z|	[95% Conf. Interval]
exp	.0926579	.0022586	41.02	0.000	.0882312 .0970847
wks	.0006559	.0027445	0.24	0.811	-.0047232 .0060349
_cons	4.915057	.1444991	34.01	0.000	4.631844 5.19827

```
Test of parameter constancy:    chi2(1782) =   5.2e+05      Prob > chi2 = 0.0000
```

These estimates differ considerably from estimates (not given) obtained by using the xtmixed command with the regressors wks and exp.

The matrix $\boldsymbol{\Sigma}_u$ is not output but is stored in e(Sigma). We have

```
. matrix list e(Sigma)
symmetric e(Sigma)[3,3]
                 exp           wks         _cons
    exp     .00263517
    wks    -.00031391     .00355505
  _cons    -.01246813    -.17387686     9.9705349
```

The `xtmixed` command is the more commonly used command. It is much more flexible, because $\mathbf{z}_{it}$ need not equal $\mathbf{x}_{it}$, restrictions can be placed on $\mathbf{\Sigma}$, and it gives estimates of the precision of the variance components. It does impose homoskedasticity of ε_{it}, but this may not be too restrictive because the combined error $\mathbf{z}_{it}\mathbf{u}_i + \varepsilon_{it}$ is clearly heteroskedastic and, depending on the application, the variability in ε_{it} may be much smaller than that of $\mathbf{z}_{it}\mathbf{u}_i$.

9.5.7   Two-way random-effects model

The `xtmixed` command is intended for multilevel models where the levels are nested. The two-way random-effects model (see section 8.2.2) has the error $\alpha_i + \gamma_t + \varepsilon_{it}$, where all three errors are i.i.d. Then the covariance structure is nonnested, because i is not nested in t and t is not nested in i.

Rabe-Hesketh and Skrondal (2008, 476) explain how to nonetheless use `xtmixed` to estimate the parameters of the two-way random-effects model, using a result of Goldstein (1987) that shows how to rewrite covariance models with nonnested structure as nested models. Two levels of random effects are specified as `|| _all: R.t || id:`. We explain each level in turn. At the first level, the RE equation describes the covariance structure due to γ_t. The obvious `t:` cannot be used because it does not nest `id`.

Instead, we use `_all:` because this defines each observation (i, t) to be a separate group (thereby nesting `id`). We then add `R.t` because this ensures the desired correlation pattern due to γ_t by defining a factor structure in `t` with independent factors with identical variance (see [XT] **xtmixed**). At the second level, the RE equation defines the covariance structure for α_i by simply using `id:`. For data where $N < T$ computation is faster if the roles of i and t are reversed, the random effects are specified as `|| _all: R.id || t:`.

Application of the command yields

```
. * Two-way random-effects model estimated using xtmixed
. xtmixed lwage exp exp2 wks ed || _all: R.t || id: , mle

Performing EM optimization:

Performing gradient-based optimization:

Iteration 0:   log likelihood =  891.09366
Iteration 1:   log likelihood =  891.09366

Computing standard errors:

Mixed-effects ML regression                     Number of obs     =       4165
```

Group Variable	No. of Groups	Observations per Group Minimum	Average	Maximum
_all	1	4165	4165.0	4165
id	595	7	7.0	7

		Wald chi2(4)	=	329.99
Log likelihood = 891.09366		Prob > chi2	=	0.0000

lwage	Coef.	Std. Err.	z	P>|z|	[95% Conf.	Interval]
exp	.0297249	.0025537	11.64	0.000	.0247198	.0347301
exp2	-.0004425	.0000501	-8.83	0.000	-.0005407	-.0003443
wks	.0009207	.0005924	1.55	0.120	-.0002404	.0020818
ed	.0736737	.0049275	14.95	0.000	.064016	.0833314
_cons	5.324364	.1036266	51.38	0.000	5.121259	5.527468

Random-effects Parameters	Estimate	Std. Err.	[95% Conf.	Interval]
_all: Identity				
sd(R.t)	.170487	.0457032	.1008106	.2883211
id: Identity				
sd(_cons)	.3216482	.0096375	.3033029	.3411031
sd(Residual)	.1515621	.0017955	.1480836	.1551224

LR test vs. linear regression: chi2(2) = 5770.93   Prob > chi2 = 0.0000
Note: LR test is conservative and provided only for reference.

The random time effects are statistically significant because sd(R.t) is significantly different from zero at a level of 0.05.

9.6 Clustered data

Short-panel data can be viewed as a special case of clustered data, where there is within-individual clustering so that errors are correlated over time for a given individual. Therefore, the xt commands that we have applied to data from short panels can also be applied to clustered data. In particular, xtreg and especially xtmixed are often used.

9.6.1 Clustered dataset

We consider data on use of medical services by individuals, where individuals are clustered within household and, additionally, households are clustered in villages or communes. The data, from Vietnam, are the same as those used in Cameron and Trivedi (2005, 852).

The dependent variable is the number of direct pharmacy visits (pharvis). The independent variables are the logarithm of household medical expenditures (lnhhexp) and the number of illnesses (illness). The data cover 12 months. We have

```
. * Read in Vietnam clustered data and summarize
. use mus09vietnam_ex2.dta, clear

. summarize pharvis lnhhexp illness commune
```

Variable	Obs	Mean	Std. Dev.	Min	Max
pharvis	27765	.5117594	1.313427	0	30
lnhhexp	27765	2.60261	.6244145	.0467014	5.405502
illness	27765	.6219701	.8995068	0	9
commune	27765	101.5266	56.28334	1	194

The commune variable identifies the 194 separate villages. For these data, the lnhhexp variable takes on a different value for each household and can serve as a household identifier.

The pharvis variable is a count that is best modeled by using count regression commands such as poisson and xtpoisson. For illustrative purposes, we use the linear regression model here.

9.6.2   Clustered data using nonpanel commands

One complication of clustering is that the error is correlated within cluster. If that is the only complication, then valid inference simply uses standard cross-section estimators along with cluster–robust standard errors.

Here we contrast no correction for clustering, clustering on household, and clustering on village. We have

```
. * OLS estimation with cluster-robust standard errors
. quietly regress pharvis lnhhexp illness

. estimates store OLS_iid

. quietly regress pharvis lnhhexp illness, vce(robust)

. estimates store OLS_het

. quietly regress pharvis lnhhexp illness, vce(cluster lnhhexp)

. estimates store OLS_hh

. quietly regress pharvis lnhhexp illness, vce(cluster commune)

. estimates store OLS_vill

. estimates table OLS_iid OLS_het OLS_hh OLS_vill, b(%10.4f) se stats(r2 N)
```

Variable	OLS_iid	OLS_het	OLS_hh	OLS_vill
lnhhexp	0.0248	0.0248	0.0248	0.0248
	0.0115	0.0109	0.0140	0.0211
illness	0.6242	0.6242	0.6242	0.6242
	0.0080	0.0141	0.0183	0.0342
_cons	0.0591	0.0591	0.0591	0.0591
	0.0316	0.0292	0.0367	0.0556
r2	0.1818	0.1818	0.1818	0.1818
N	27765	27765	27765	27765

```
legend: b/se
```

The effect of correction for heteroskedasticity is unknown a priori. Here there is little effect on the standard errors for the intercept and `lnhhexp`, though there is an increase for `illness`.

More importantly, controlling for clustering is expected to increase reported standard errors, especially for regressors highly correlated within the cluster. Here standard errors increase by around 30% as we move to clustering on household and by another approximately 50% as we move to clustering on village. In total, clustering on village leads to a doubling of standard errors compared with assuming no heteroskedasticity.

In practice, controlling for clustering can have a bigger effect; see section 3.3.5. Here there are on average 140 people per village, but the within-village correlation of the regressors and of the model errors is fairly low.

9.6.3 Clustered data using panel commands

The Stata `xt` commands enable additional analysis, specifically, more detailed data summary, more-efficient estimation than OLS, and estimation with cluster-specific fixed effects.

In our example, person i is in household or village j, and a cluster-effects model is

$$y_{ij} = \mathbf{x}'_{ij}\boldsymbol{\beta} + \alpha_j + \varepsilon_{ij}$$

It is very important to note an important conceptual change from the panel-data case. For panels, there were multiple observations per individual, so clustering is on the individual (i). Here, instead, there are multiple observations per household or per village, so clustering is on the household or village (j).

It follows that when we use the `xtset` command, the "individual identifier" is really the cluster identifier, so the individual identifier is the household or commune. For clustering on the household, the minimum command is `xtset hh`. We can also declare the analog of time to be a cluster member, here an individual within a household or commune. In doing so, we should keep in mind that, unlike time, there is no natural ordering of individuals in a village.

We consider clustering on the household. We first convert the household identifier `lnhhexp`, unique for each household, to the `hh` variable, which takes on integer values 1, 2, ... by using `egen`'s `group()` function. We then randomly assign the integers 1, 2, ... to each `person` in a household by using the `by hh: generate person = _n` command.

```
. * Generate integer-valued household and person identifiers and xtset
. quietly egen hh = group(lnhhexp)

. sort hh

. by hh: generate person = _n

. xtset hh person
        panel variable:  hh (unbalanced)
         time variable:  person, 1 to 19
                 delta:  1 unit
```

Now that the data are set up, we can use `xt` commands to investigate the data. For example,

```
. xtdescribe
         hh:  1, 2, ..., 5740                                 n =        5740
     person:  1, 2, ..., 19                                   T =          19
              Delta(person) = 1 unit
              Span(person)  = 19 periods
              (hh*person uniquely identifies each observation)
    Distribution of T_i:   min      5%     25%     50%     75%     95%      max
                             1       2       4       5       6       8       19

        Freq.   Percent    Cum. |  Pattern
        1376     23.97    23.97 |  1111..............
        1285     22.39    46.36 |  11111.............
         853     14.86    61.22 |  111111............
         706     12.30    73.52 |  111...............
         471      8.21    81.72 |  1111111...........
         441      7.68    89.41 |  11................
         249      4.34    93.75 |  11111111..........
         126      2.20    95.94 |  1.................
         125      2.18    98.12 |  111111111.........
         108      1.88   100.00 |  (other patterns)

        5740    100.00          |  XXXXXXXXXXXXXXXXXXX
```

There are 5,740 households with 1–19 members, the median household has five members, and the most common household size is four members.

We can estimate the within-cluster correlation of a variable by obtaining the correlation for members of a household. The best way to do so is to fit an intercept-only RE model, because from section 8.7.1 the output includes rho, the estimate of the intraclass correlation parameter. Another way to do so is to use the time-series component of `xt` commands, treat each person in the household like a time period, and find the correlation for adjoining household members by lagging once. We have

```
. * Within-cluster correlation of pharvis
. quietly xtreg pharvis, mle
. display "Intra-class correlation for household: " e(rho)
Intra-class correlation for household: .22283723
. quietly correlate pharvis L1.pharvis
. display "Correlation for adjoining household:   " r(rho)
Correlation for adjoining household:   .1834621
```

The usual `xtreg` commands can be applied. In particular, the RE model assumption of equicorrelated errors within cluster is quite reasonable here because there is no natural ordering of household members. We compare in order OLS, FE, and RE estimates with clustering on household followed by OLS, FE, and RE estimates with clustering on village. We expect RE and within estimators to be more efficient than the OLS estimators. We have

```
. * OLS, RE, and FE estimation with clustering on household and on village
. quietly regress pharvis lnhhexp illness, vce(cluster hh)
. estimates store OLS_hh
. quietly xtreg pharvis lnhhexp illness, re
. estimates store RE_hh
. quietly xtreg pharvis lnhhexp illness, fe
. estimates store FE_hh
. quietly xtset commune
. quietly regress pharvis lnhhexp illness, vce(cluster commune)
. estimates store OLS_vill
. quietly xtreg pharvis lnhhexp illness, re
. estimates store RE_vill
. quietly xtreg pharvis lnhhexp illness, fe
. estimates store FE_vill
. estimates table OLS_hh RE_hh FE_hh OLS_vill RE_vill FE_vill, b(%7.4f) se
```

Variable	OLS_hh	RE_hh	FE_hh	OLS_vill	RE_vill	FE_vill
lnhhexp	0.0248	0.0184	(omitted)	0.0248	-0.0449	-0.0657
	0.0140	0.0168		0.0211	0.0149	0.0158
illness	0.6242	0.6171	0.6097	0.6242	0.6155	0.6141
	0.0183	0.0083	0.0096	0.0342	0.0081	0.0082
_cons	0.0591	0.0855	0.1325	0.0591	0.2431	0.3008
	0.0367	0.0448	0.0087	0.0556	0.0441	0.0426

legend: b/se

The coefficient of `illness` is relatively invariant across models and is fit much more precisely by the RE and within estimators. The coefficient of `lnhhexp` fluctuates considerably, including sign reversal, and is more efficiently estimated by RE and FE when clustering is on the village. Because `lnhhexp` is invariant within household, there is no within estimate for its coefficient when clustering is at the household level, but there is when clustering is at the village level.

9.6.4 Hierarchical linear models

Hierarchical models or mixed models are designed for clustered data such as these, especially when there is clustering at more than one level.

A simple example is to suppose that person i is in household j, which is in village k, and that the model is a variance-components model with

$$y_{ijk} = \mathbf{x}'_{ijk}\boldsymbol{\beta} + u_j + v_k + \varepsilon_{ijk}$$

where u_j, v_k, and ε_{ijk} are i.i.d. errors.

The model can be fit using the `xtmixed` command, detailed in section 9.5. The first level is `commune` and the second is `hh` because households are nested in villages. The `difficult` option was added to ensure convergence of the iterative process. We obtain

```
. * Hierarchical linear model with household and village variance components
. xtmixed pharvis lnhhexp illness || commune: || hh:, mle difficult

Performing EM optimization:

Performing gradient-based optimization:

Iteration 0:   log likelihood = -43224.836
Iteration 1:   log likelihood = -43224.635
Iteration 2:   log likelihood = -43224.635

Computing standard errors:
```

| Mixed-effects ML regression | | | | Number of obs | = | 27765 |

Group Variable	No. of Groups	Observations per Group		
		Minimum	Average	Maximum
commune	194	51	143.1	206
hh	5741	1	4.8	19

```
                                         Wald chi2(2)      =    5570.25
Log likelihood = -43224.635              Prob > chi2       =     0.0000
```

pharvis	Coef.	Std. Err.	z	P>|z|	[95% Conf. Interval]	
lnhhexp	-.0408948	.0184302	-2.22	0.026	-.0770173	-.0047722
illness	.6141189	.0082837	74.14	0.000	.5978831	.6303546
_cons	.235717	.0523176	4.51	0.000	.1331763	.3382576

Random-effects Parameters	Estimate	Std. Err.	[95% Conf. Interval]	
commune: Identity				
sd(_cons)	.2575279	.0162584	.2275546	.2914493
hh: Identity				
sd(_cons)	.4532979	.0103451	.4334687	.4740342
sd(Residual)	1.071804	.0051435	1.06177	1.081932

```
LR test vs. linear regression:       chi2(2) =  1910.44   Prob > chi2 = 0.0000
Note: LR test is conservative and provided only for reference.
```

The estimates are similar to the previously obtained RE estimates using the xtreg, re command, given in the RE_vill column in the table in section 9.6.3. Both variance components are statistically significant. The xtmixed command allows the variance components to additionally depend on regressors, as demonstrated in section 9.5.

9.7   Stata resources

The key Stata reference is [XT] *Longitudinal/Panel-Data Reference Manual*, especially [XT] **xtivreg**, [XT] **xthtaylor**, [XT] **xtabond**, and [XT] **xtmixed**.

Many of the topics in this chapter appear in more specialized books on panel data, notably, Arellano (2003), Baltagi (2008), Hsiao (2003), and Lee (2002). Cameron and Trivedi (2005) present most of the methods in this chapter, including hierarchical models that are generally not presented in econometrics texts.

9.8 Exercises

1. For the model and data of section 9.2, obtain the panel IV estimator in the FE model by applying the `ivregress` command to the mean-differenced model with a mean-differenced instrument. Hint: For example, for variable x, type `by id: egen avex = mean(`x`)` followed by `summarize` x and then `generate mdx = `x` - avex + r(mean)`. Verify that you get the same estimated coefficients as you would with `xtivreg, fe`.

2. For the model and data of section 9.4, use the `xtdpdsys` command given in section 9.4.6, and then perform specification tests using the `estat abond` and `estat sargan` commands. Use `xtdpd` at the end of section 9.4.8, and compare the results with those from `xtdpdsys`. Is this what you expect, given the results from the preceding specification tests?

3. Consider the model and data of section 9.4, except consider the case of just one lagged dependent variable. Throughout, estimate the parameters of the models with the `noconstant` option. Consider estimation of the dynamic model $y_{it} = \alpha_i + \gamma y_{it-1} + \varepsilon_{it}$, when $T = 7$, where ε_{it} are serially uncorrelated. Explain why OLS estimation of the transformed model $\Delta y_{it} = \gamma_1 \Delta y_{it-1} + \Delta \varepsilon_{it}$, $t = 2, \dots, 7$, leads to inconsistent estimation of γ_1. Propose an IV estimator of the preceding model where there is just one instrument. Implement this just-identified IV estimator using the data on `lwage` and the `ivregress` command. Obtain cluster–robust standard errors. Compare with OLS estimates of the differenced model.

4. Continue with the model of the previous question. Consider the Arellano–Bond estimator. For each time period, state what instruments are used by the `estat abond` command. Perform the Arellano–Bond estimator using the data on `lwage`. Obtain the one-step estimator with robust standard errors. Obtain the two-step estimator with robust standard errors. Compare the estimates and their standard errors. Is there an efficiency gain compared with your answer in the previous question? Use the `estat abond` command to test whether the errors ε_{it} are serially uncorrelated. Use the `estat sargan` command to test whether the model is correctly specified.

5. For the model and data of section 9.5, verify that `xtmixed` with the `mle` option gives the same results as `xtreg, mle`. Also compare the results with those from using `xtmixed` with the `reml` option. Fit the two-way RE model assuming random individual and time effects, and compare results with those from when the time effects are allowed to be fixed (in which case time dummies are included as regressors).

10 Nonlinear regression methods

10.1 Introduction

We now turn to nonlinear regression methods. In this chapter, we consider single-equation models fit using cross-section data with all regressors exogenous.

Compared with linear regression, there are two complications. There is no explicit solution for the estimator, so computation of the estimator requires iterative numerical methods. And, unlike the linear model, the marginal effect (ME) of a change in a regressor is no longer simply the associated slope parameter. For standard nonlinear models, the first complication is easily handled. Simply changing the command from `regress y x` to `poisson y x`, for example, leads to nonlinear estimation and regression output that looks essentially the same as the output from `regress`. The second complication can often be dealt with by obtaining MEs by using the `margins` command, although other methods may be better.

In this chapter, we provide an overview of Stata's nonlinear estimation commands and subsequent prediction and computation of MEs. The discussion is applicable for analysis after any Stata estimation command, including the commands listed in table 10.1.

Table 10.1. Available estimation commands for various analyses

Data type	Estimation command
Linear	`regress, cnreg, areg, treatreg, ivregress, qreg, boxcox, frontier,` `mvreg, sureg, reg3, xtreg, xtgls, xtrc, xtpcse, xtregar,` `xtmixed, xtivreg, xthtaylor, xtabond, xtfrontier`
Nonlinear LS	`nl`
Binary	`logit, logistic, probit, cloglog, glogit, slogit, hetprob,` `scobit, ivprobit, heckprob, xtlogit, xtprobit, xtcloglog`
Multinomial	`mlogit, clogit, asclogit, nlogit, ologit, rologit, asroprobit,` `mprobit, asmprobit, oprobit, biprobit`
Censored normal	`tobit, intreg, cnsreg, truncreg, ivtobit, xttobit, xttintreg`
Selection normal	`treatreg, heckman`
Durations	`stcox, stcrreg, streg`
Counts	`poisson, nbreg, gnbreg, zip, zinb, ztp, ztnb, xtpoisson, xtnbreg`

Chapter 11 then presents methods to fit a nonlinear model when no Stata command is available for that model. The discussion of model-specific issues—particularly specification tests that are an integral part of the modeling cycle of estimation, specification testing, and reestimation—is deferred to chapter 12 and the model-specific chapters 14–18.

10.2 Nonlinear example: Doctor visits

As a nonlinear estimation example, we consider Poisson regression to model count data on the number of doctor visits. There is no need to first read chapter 16 on count data because we provide any necessary background here.

Although the outcome is discrete, the only difference this makes is in the choice of log density. The `poisson` command is actually not restricted to counts and can be applied to any variable $y \geq 0$. All the points made with the count-data example could equally well be made with, for example, duration data on completed spells modeled by the exponential distribution and other models.

10.2.1 Data description

We model the number of office-based physician visits (`docvis`) by persons in the United States aged 25–64 years, using data from the 2002 Medical Expenditure Panel Survey (MEPS). The sample is the same as that used by Deb, Munkin, and Trivedi (2006). It excludes those receiving public insurance (Medicare and Medicaid) and is restricted to those working in the private sector but who are not self-employed.

The regressors used here are restricted to health insurance status (`private`), health status (`chronic`), and socioeconomic characteristics (`female` and `income`) to keep Stata output short. We have

```
. * Read in dataset, select one year of data, and describe key variables
. use mus10data.dta, clear

. keep if year02==1
(25712 observations deleted)

. describe docvis private chronic female income

              storage   display    value
variable name   type    format     label      variable label

docvis          int     %8.0g                  number of doctor visits
private         byte    %8.0g                  = 1 if private insurance
chronic         byte    %8.0g                  = 1 if a chronic condition
female          byte    %8.0g                  = 1 if female
income          float   %9.0g                  Income in $ / 1000
```

We then `summarize` the data:

```
. * Summary of key variables
. summarize docvis private chronic female income
    Variable |       Obs        Mean    Std. Dev.       Min        Max
-------------+--------------------------------------------------------
      docvis |      4412    3.957389    7.947601          0        134
     private |      4412    .7853581    .4106202          0          1
     chronic |      4412    .3263826    .4689423          0          1
      female |      4412    .4718948    .4992661          0          1
      income |      4412    34.34018    29.03987    -49.999    280.777
```

The dependent variable is a nonnegative integer count, here ranging from 0 to 134. Thirty-three percent of the sample have a chronic condition, and 47% are female. We use the whole sample, including the three people who have negative income (obtained by using the `tabulate income` command).

The relative frequencies of `docvis`, obtained by using the `tabulate docvis` command, are 36%, 16%, 10%, 7%, and 5% for, respectively, 0, 1, 2, 3, and 4 visits. Twenty-six percent of the sample have 5 or more visits.

10.2.2 Poisson model description

The Poisson regression model specifies the count y to have a conditional mean of the exponential form

$$E(y|\mathbf{x}) = \exp(\mathbf{x}'\boldsymbol{\beta}) \tag{10.1}$$

This ensures that the conditional mean is positive, which should be the case for any random variable that is restricted to be nonnegative. However, the key ME $\partial E(y|\mathbf{x})/\partial x_j = \beta_j \exp(\mathbf{x}'\boldsymbol{\beta})$ now depends on both the parameter estimate β_j and the particular value of $\mathbf{x}$ at which the ME is evaluated; see section 10.6.

The starting point for count analysis is the Poisson distribution, with the probability mass function $f(y|\mathbf{x}) = e^{-\mu}\mu^y/y!$. Substituting in $\mu_i = \exp(\mathbf{x}_i'\boldsymbol{\beta})$ from (10.1) gives the conditional density for the ith observation. This in turn gives the log-likelihood function $Q(\boldsymbol{\beta}) = \sum_{i=1}^{N}\{-\exp(\mathbf{x}_i'\boldsymbol{\beta}) + y_i\mathbf{x}_i'\boldsymbol{\beta} - \ln y_i!\}$, which is maximized by the maximum likelihood estimator (MLE). The Poisson MLE solves the associated first-order conditions that can be shown to be

$$\sum_{i=1}^{N}\{y_i - \exp(\mathbf{x}_i'\boldsymbol{\beta})\}\mathbf{x}_i = \mathbf{0} \tag{10.2}$$

Equation (10.2) has no explicit solution for $\boldsymbol{\beta}$. Instead, $\widehat{\boldsymbol{\beta}}$ is obtained numerically by using methods explained in chapter 11.

What if the Poisson distribution is the wrong distribution for modeling doctor visits? In general, the MLE is inconsistent if the density is misspecified. However, the Poisson MLE requires only the much weaker condition that the conditional mean function given in (10.1) is correctly specified, because then the left-hand side of (10.2) has an expected value of zero. Under this weaker condition, robust standard errors rather than default maximum likelihood (ML) standard errors should be used; see section 10.4.5.

10.3   Nonlinear regression methods

We consider three classes of estimators: ML, nonlinear least squares (NLS), and generalized linear models (GLM). All three are examples of m estimators that maximize (or minimize) an objective function of the form

$$Q(\boldsymbol{\theta}) = \sum_{i=1}^{N} q_i(y_i, \mathbf{x}_i, \boldsymbol{\theta}) \tag{10.3}$$

where y denotes the dependent variable, $\mathbf{x}$ denotes regressors (assumed exogenous), $\boldsymbol{\theta}$ denotes a parameter vector, and $q(\cdot)$ is a specified scalar function that varies with the model and estimator. In the Poisson case, $\boldsymbol{\beta} = \boldsymbol{\theta}$; more generally, $\boldsymbol{\beta}$ is a component of $\boldsymbol{\theta}$.

10.3.1   MLE

MLEs maximize the log-likelihood function. For N independent observations, the MLE $\widehat{\boldsymbol{\theta}}$ maximizes

$$Q(\boldsymbol{\theta}) = \sum_{i=1}^{N} \ln f(y_i | \mathbf{x}_i, \boldsymbol{\theta})$$

where $f(y|\mathbf{x}, \boldsymbol{\theta})$ is the conditional density, for continuous y, or the conditional probability mass function, for discrete y.

If the density $f(y|\mathbf{x}, \boldsymbol{\theta})$ is correctly specified, then the MLE is the best estimator to use. It is consistent for $\boldsymbol{\theta}$, it is asymptotically normally distributed, and it is fully efficient, meaning that no other estimator of $\boldsymbol{\theta}$ has a smaller asymptotic variance–covariance matrix of the estimator (VCE).

Of course, the true density is unknown. If $f(y|\mathbf{x}, \boldsymbol{\theta})$ is incorrectly specified, then in general the MLE is inconsistent. It may then be better to use other methods that, while not as efficient as the MLE, are consistent under weaker assumptions than those necessary for the MLE to be consistent.

The MLE remains consistent even if the density is misspecified, however, provided that 1) the specified density is in the linear exponential family (LEF) and 2) the functional form for the conditional mean $E(y|\mathbf{x})$ is correctly specified. The default estimate of the VCE of the MLE is then no longer correct, so we base the inference on a robust estimate of the VCE. Examples of the LEF are Poisson and negative binomial (with a known dispersion parameter) for count data, Bernoulli for binary data (including logit and probit), one-parameter gamma for duration data (including exponential), normal (with a known variance parameter) for continuous data, and the inverse Gaussian.

The term quasi-MLE, or pseudo-MLE, is used when estimation is by ML, but subsequent inference is done without assuming that the density is correctly specified.

The `ml` command enables ML estimation for user-defined likelihood functions; see sections 11.4–11.6. For commonly used models, this is not necessary, however, because specific Stata commands have been developed for specific models.

10.3.2 The poisson command

For Poisson, the ML estimator is obtained by using the `poisson` command. The syntax of the command is

poisson *depvar* [*indepvars*] [*if*] [*in*] [*weight*] [, *options*]

This syntax is the same as that for `regress`. The only relevant option for our analysis here is the `vce()` option for the type of estimate of the VCE.

The `poisson` command with the `vce(robust)` option yields the following results for the doctor-visits data. As already noted, to restrict Stata output, we use far fewer regressors than should be used to model doctor visits.

```
. * Poisson regression (command poisson)
. poisson docvis private chronic female income, vce(robust)
Iteration 0:   log pseudolikelihood = -18504.413
Iteration 1:   log pseudolikelihood = -18503.549
Iteration 2:   log pseudolikelihood = -18503.549
Poisson regression                             Number of obs   =       4412
                                               Wald chi2(4)    =     594.72
                                               Prob > chi2     =     0.0000
Log pseudolikelihood = -18503.549              Pseudo R2       =     0.1930
```

docvis	Coef.	Robust Std. Err.	z	P>|z|	[95% Conf. Interval]	
private	.7986652	.1090014	7.33	0.000	.5850263	1.012304
chronic	1.091865	.0559951	19.50	0.000	.9821167	1.201614
female	.4925481	.0585365	8.41	0.000	.3778187	.6072774
income	.003557	.0010825	3.29	0.001	.0014354	.0056787
_cons	-.2297262	.1108732	-2.07	0.038	-.4470338	-.0124186

The output begins with an iteration log, because the estimator is obtained numerically by using an iterative procedure presented in sections 11.2 and 11.3. In this case, only two iterations are needed. Each iteration increases the log-likelihood function, as desired, and iterations cease when there was little change in the log-likelihood function. The term pseudolikelihood is used rather than log likelihood because use of `vce(robust)` means that we no longer are maintaining that the data are exactly Poisson distributed. The remaining output from `poisson` is remarkably similar to that for `regress`.

The four regressors are jointly statistically significant at 5%, because the `Wald chi2(4)` test statistic has $p = 0.00 < 0.05$. The pseudo-R^2 is discussed in section 10.7.1. There is no ANOVA table, because this table is appropriate only for linear least squares with spherical errors.

The remaining output indicates that all regressors are individually statistically significant at a level of 0.05, because all p-values are less than 0.05. For each regressor, the output presents in turn:

$$
\begin{aligned}
&\text{Coefficients} && \widehat{\beta}_j \\
&\text{Standard errors} && s_{\widehat{\beta}_j} \\
&z \text{ statistics} && z_j = \widehat{\beta}_j / s_{\widehat{\beta}_j} \\
&p\text{-values} && p_j = \Pr\{\,|z_j| > 0 | z_j \sim N(0,1)\,\} \\
&95\% \text{ confidence intervals} && \widehat{\beta}_j \pm 1.96 \times s_{\widehat{\beta}_j}
\end{aligned}
$$

The z statistics and p-values are computed by using the standard normal distribution, rather than the t distribution with $N - k$ degrees of freedom. The p-values are for a two-sided test of whether $\beta_j = 0$. For a one-sided test of $H_0 \colon \beta_j \leq 0$ against $\beta_j > 0$, the p-value is half of that reported in the table, provided that $z_j > 0$. For a one-sided test of $H_0 \colon \beta_j \geq 0$ against $\beta_j < 0$, the p-value is half of that reported in the table, provided that $z_j < 0$.

A nonlinear model raises a new issue of interpretation of the slope coefficients β_j. For example, what does the value 0.0036 for the coefficient of `income` mean? Given the exponential functional form for the conditional mean in (10.1), it means that a \$1,000 increase in income (a one-unit increase in `income`) leads to a 0.0036 proportionate increase, or a 0.36% increase, in doctor visits. We address this important issue in detail in section 10.6.

Note that test statistics following nonlinear estimation commands such as `poisson` are based on the standard normal distribution and chi-squared distributions, whereas those following linear estimation commands such as `regress`, `ivregress`, and `xtreg` use the t and F distributions. This makes little difference for larger samples, say, $N > 100$.

10.3.3 Postestimation commands

The `ereturn list` command details the estimation results that are stored in `e()`; see section 1.6.2. These include regression coefficients in `e(b)` and the estimated VCE in `e(V)`.

Standard postestimation commands available after most estimators are `predict`, `predictnl`, and `margins` for prediction and MEs (this chapter); `test`, `testnl`, `lincom`, and `nlcom` for Wald tests and confidence intervals; `linktest` for a model-specification test (chapter 12); and `estimates` for storing results (chapter 3).

The `estat vce` command displays the estimate of the VCE, and the `correlation` option displays the correlations for this matrix. The `estat summarize` command summarizes the current estimation sample. The `estat ic` command obtains information criteria (section 10.7.2). More command-specific commands, usually beginning with `estat`, are available for model-specification testing.

To find the specific postestimation commands available after a command, e.g., poisson, see [R] **poisson postestimation** or type help poisson postestimation.

10.3.4 NLS

NLS estimators minimize the sum of squared residuals, so for independent observations, the NLS estimator $\widehat{\boldsymbol{\beta}}$ minimizes

$$Q(\boldsymbol{\beta}) = \sum_{i=1}^{N} \{y_i - m(\mathbf{x}_i, \boldsymbol{\beta})\}^2$$

where $m(\mathbf{x}, \boldsymbol{\beta})$ is the specified functional form for $E(y|\mathbf{x})$, the conditional mean of y given $\mathbf{x}$.

If the conditional mean function is correctly specified, then the NLS estimator is consistent and asymptotically normally distributed. If the data-generating process (DGP) is $y_i = m(\mathbf{x}_i, \boldsymbol{\beta}) + u_i$, where $u_i \sim N(0, \sigma^2)$, then NLS is fully efficient. If $u_i \sim [0, \sigma^2]$, then the NLS default estimate of the VCE is correct; otherwise, a robust estimate should be used.

10.3.5 The nl command

The **nl** command implements NLS regression. The simplest form of the command directly defines the conditional mean rather than calling a program or function. The syntax is

nl (*depvar*=<*sexp*>) [*if*] [*in*] [*weight*] [, *options*]

where <*sexp*> is a substitutable expression. The only relevant option for our analysis here is the vce() option for the type of estimate of the VCE.

The challenge is in defining the expression for the conditional mean $\exp(\mathbf{x}'\boldsymbol{\beta})$; see [R] **nl**. An explicit definition for our example is the command

```
. nl (docvis = exp({private}*private + {chronic}*chronic + {female}*female +
> {income}*income + {intercept}))
(obs = 4412)
Iteration 0:   residual SS =   251743.9
Iteration 1:   residual SS =   242727.6
Iteration 2:   residual SS =   241818.1
Iteration 3:   residual SS =   241815.4
Iteration 4:   residual SS =   241815.4
Iteration 5:   residual SS =   241815.4
Iteration 6:   residual SS =   241815.4
Iteration 7:   residual SS =   241815.4
Iteration 8:   residual SS =   241815.4
```

Source	SS	df	MS
Model	105898.644	5	21179.7289
Residual	241815.356	4407	54.870741
Total	347714	4412	78.8109701

Number of obs =   4412
R-squared   =   0.3046
Adj R-squared =   0.3038
Root MSE   = 7.407479
Res. dev.   = 30185.68

docvis	Coef.	Std. Err.	t	P>|t|	[95% Conf. Interval]	
/private	.7105104	.1170408	6.07	0.000	.4810517	.9399691
/chronic	1.057318	.0610386	17.32	0.000	.9376517	1.176984
/female	.4320224	.0523199	8.26	0.000	.3294491	.5345957
/income	.002558	.0006941	3.69	0.000	.0011972	.0039189
/intercept	-.0405628	.1272218	-0.32	0.750	-.2899814	.2088558

Here the parameter names are given in the braces, {}.

The `nl` coefficient estimates are similar to those from `poisson` (within 15% for all regressors except `income`), and the `nl` robust standard errors are 15% higher for `female` and `income` and are similar for the remaining regressors.

The model diagnostic statistics given include R^2 computed as the model (or explained) sum of squares divided by the total sum of squares, the root mean squared error (MSE) that is the estimate s of the standard deviation σ of the model error, and the residual deviance that is a goodness-of-fit measure used mostly in the GLM literature.

We instead use a shorter equivalent expression for the conditional mean function. Also the `vce(robust)` option is used to allow for heteroskedastic errors, and the `nolog` option is used to suppress the iteration log. We have

```
. * Nonlinear least-squares regression (command nl)
. generate one = 1
. nl (docvis = exp({xb: private chronic female income one})), vce(robust) nolog
(obs = 4412)
Nonlinear regression
```

Number of obs =   4412
R-squared   =   0.3046
Adj R-squared =   0.3038
Root MSE   = 7.407479
Res. dev.   = 30185.68

docvis	Coef.	Robust Std. Err.	t	P>|t|	[95% Conf. Interval]	
/xb_private	.7105104	.1086194	6.54	0.000	.4975618	.923459
/xb_chronic	1.057318	.0558352	18.94	0.000	.947853	1.166783
/xb_female	.4320224	.0694662	6.22	0.000	.2958337	.5682111
/xb_income	.002558	.0012544	2.04	0.041	.0000988	.0050173
/xb_one	-.0405628	.1126216	-0.36	0.719	-.2613576	.180232

The output is the same except for the standard errors, which are now robust to heteroskedasticity.

10.3.6   GLM

The GLM framework is the standard nonlinear model framework in many areas of applied statistics, most notably, biostatistics. For completeness, we present it here, but we do not emphasize its use because it is little used in econometrics.

GLM estimators are a subset of ML estimators that are based on a density in the LEF, introduced in section 10.3.1. They are essentially generalizations of NLS, optimal for a nonlinear regression model with homoskedastic additive errors, but also appropriate for other types of data where not only is there intrinsic heteroskedasticity but there is a natural starting point for modeling the intrinsic heteroskedasticity. For example, for the Poisson the variance equals the mean, and for a binary variable the variance equals the mean times unity minus the mean.

The GLM estimator $\widehat{\boldsymbol{\theta}}$ maximizes the LEF log likelihood

$$Q(\boldsymbol{\theta}) = \sum_{i=1}^{N} [a\{m(\mathbf{x}_i, \boldsymbol{\beta})\} + b(y_i) + c\{m(\mathbf{x}_i, \boldsymbol{\beta})\}y_i]$$

where $m(\mathbf{x}, \boldsymbol{\beta}) = E(y|\mathbf{x})$ is the conditional mean of y, different specified forms of the functions $a(\cdot)$ and $c(\cdot)$ correspond to different members of the LEF, and $b(\cdot)$ is a normalizing constant. For the Poisson, $a(\mu) = -\mu$ and $c(\mu) = \ln \mu$.

Given definitions of $a(\mu)$ and $c(\mu)$, the mean and variance are necessarily $E(y) = \mu = -a'(\mu)/c'(\mu)$ and $\text{Var}(y) = 1/c'(\mu)$. For the Poisson, $a'(\mu) = -1$ and $c'(\mu) = 1/\mu$, so $E(y) = 1/(1/\mu) = \mu$ and $\text{Var}(y) = 1/c'(\mu) = 1/(1/\mu) = \mu$. This is the variance–mean equality property of the Poisson.

GLM estimators have the important property that they are consistent provided only that the conditional mean function is correctly specified. This result arises because the first-order conditions $\partial Q(\boldsymbol{\theta})/\partial \theta = \mathbf{0}$ can be written as $N^{-1} \sum_i c'(\mu_i)(y_i - \mu_i)(\partial \mu_i/\partial \boldsymbol{\beta}) = \mathbf{0}$, where $\mu_i = m(\mathbf{x}_i, \boldsymbol{\beta})$. It follows that estimator consistency requires only that $E(y_i - \mu_i) = 0$, or that $E(y_i|\mathbf{x}_i) = m(\mathbf{x}_i, \boldsymbol{\beta})$. However, unless the variance is correctly specified [i.e., $\text{Var}(y) = 1/c'(\mu)$], we should obtain a robust estimate of the VCE.

10.3.7   The glm command

The GLM estimator can be computed by using the `glm` command, which has the syntax

`glm` *depvar* [*indepvars*] [*if*] [*in*] [*weight*] [, *options*]

Important options are `family()` to define the particular member of the LEF to be considered, and `link()` where the link function is the inverse of the conditional mean function. The `family()` options are `gaussian` (normal), `igaussian` (inverse Gaussian), `binomial` (Bernoulli and binomial), `poisson` (Poisson), `nbinomial` (negative binomial), and `gamma` (gamma).

The Poisson estimator can be obtained by using the options `family(poisson)` and `link(log)`. The link function is the natural logarithm because this is the inverse of the exponential function for the conditional mean. We again use the `vce(robust)` option. We expect the same results as those from `poisson` with the `vce(robust)` option.

```
. * Generalized linear models regression for poisson (command glm)
. glm docvis private chronic female income, family(poisson) link(log)
> vce(robust) nolog
Generalized linear models                        No. of obs      =        4412
Optimization     : ML                            Residual df     =        4407
                                                 Scale parameter =           1
Deviance         =    28131.11439                (1/df) Deviance =     6.38328
Pearson          =    57126.23793                (1/df) Pearson  =    12.96261

Variance function: V(u) = u                      [Poisson]
Link function    : g(u) = ln(u)                  [Log]

                                                 AIC             =    8.390095
Log pseudolikelihood = -18503.54883              BIC             =   -8852.797
```

docvis	Coef.	Robust Std. Err.	z	P>|z|	[95% Conf. Interval]	
private	.7986653	.1090014	7.33	0.000	.5850264	1.012304
chronic	1.091865	.0559951	19.50	0.000	.9821167	1.201614
female	.4925481	.0585365	8.41	0.000	.3778187	.6072774
income	.003557	.0010825	3.29	0.001	.0014354	.0056787
_cons	-.2297263	.1108733	-2.07	0.038	-.4470339	-.0124187

The results are exactly the same as those given in section 10.3.2 for the Poisson quasi-MLE, aside from additional diagnostic statistics (deviance, Pearson) that are used in the GLM literature. Robust standard errors are used because they do not impose the Poisson density restriction of variance–mean equality.

A standard statistics reference is McCullagh and Nelder (1989), Hardin and Hilbe (2007) present Stata for GLM, and an econometrics reference that covers GLM in some detail is Cameron and Trivedi (1998).

10.3.8   The gmm command

Generalized method of moments (GMM) estimators minimize an objective function that is a quadratic form in sums; see section 11.8 for definitions. Optimization is more complicated than the single sum for m-estimators given in (10.3). For some linear models, there are built-in Stata commands, notably, `ivregress gmm` for cross-section data and `xtabond` for dynamic panel data. There are no built-in commands for specific nonlinear models. Instead, we use the `gmm` command, introduced in Stata 11.

The simplest form of the `gmm` command directly defines the conditional mean and has the syntax

gmm ([*eqname1:*]<*mexp_1*>) ([*eqname2:*]<*mexp_2*>) ... [*if*] [*in*] [*weight*]
  [*, options*]

where $<mexp\_j>$ is a substitutable expression for the jth moment equation. Options include `instruments()` to define the instruments; one of the `onestep`, `twostep` (the default), and `igmm` options for, respectively, one-step, two-step, and iterated GMM estimation; and `wmatrix()` to define the weighting matrix if estimation is by two-step or iterated GMM for an overidentified model. For a just-identified model, these different estimation methods lead to the same estimates. For models that are more complicated to specify, we instead use a variant of the `gmm` command that references a separate user-written program that defines the moment conditions.

We apply the `gmm` command to the model and data explained in detail in sections 11.8.2 and 11.8.3. GMM estimation is based on the single moment condition

$$E[\mathbf{z}_i\{y_i - \exp(\mathbf{x}_i'\beta)\}] = \mathbf{0}$$

where $\mathbf{x}_i$ are regressors that may include endogenous variables and $\mathbf{z}_i$ are instruments. The GMM estimator minimizes the quadratic form

$$Q(\beta) = \left[\frac{1}{N}\sum_{i=1}^{N}\mathbf{z}_i\{y_i - \exp(\mathbf{x}_i'\beta)\}\right]' \mathbf{W} \left[\frac{1}{N}\sum_{i=1}^{N}\mathbf{z}_i\{y_i - \exp(\mathbf{x}_i'\beta)\}\right]$$

where different estimation methods and choices of the weighting matrix $\mathbf{W}$ lead to different estimates if the model is overidentified (here $\mathbf{z}_i$ has more entries than $\mathbf{x}_i$).

We use the dependent and independent variables to define a substitutable expression $\{y_i - \exp(\mathbf{x}_i'\beta)\}$, using a syntax that is similar to that for the `nl` command. We use the `instruments()` option to define the variables to be used in the instruments $\mathbf{z}_i$. We continue with the number of doctor visits regression example, except that we now treat the regressor `private` as endogenous, with single instrument `firmsize` (measured in hundreds of employees). We have

```
. * Command gmm for GMM estimation (nonlinear IV) for Poisson model
. gmm (docvis - exp({xb:private chronic female income}+{b0})),
> instruments(firmsize chronic female income) onestep nolog

Final GMM criterion Q(b) =  1.29e-17

GMM estimation

Number of parameters =   5
Number of moments    =   5
Initial weight matrix: Unadjusted                    Number of obs  =    4412
```

	Coef.	Robust Std. Err.	z	P>|z|	[95% Conf.	Interval]
/xb_private	1.340292	1.559015	0.86	0.390	-1.715322	4.395905
/xb_chronic	1.072908	.0762684	14.07	0.000	.9234242	1.222391
/xb_female	.4778178	.0690393	6.92	0.000	.3425032	.6131323
/xb_income	.0027833	.002192	1.27	0.204	-.0015129	.0070795
/b0	-.6832462	1.349606	-0.51	0.613	-3.328424	1.961932

```
Instruments for equation 1: firmsize chronic female income _cons
```

By default, robust standard errors are computed. The biggest change compared with the Poisson regression output given in section 10.3.2 is for the endogenous regressor `private`. This regressor is now much less precisely estimated, with a standard error of 1.559 compared with 0.109 and a coefficient that has increased substantially from 0.798 to 1.340, though it is now statistically insignificant. Similar efficiency loss in estimation of endogenous regressor(s) often occurs with linear IV estimation using cross-section data; see the example in section 6.3.6.

The preceding example was a just-identified model with the same number of instruments as regressors. As a result, the minimized value of the objective function is zero (here 1.29×10^{-17}, reflecting a numerical rounding error). For application of the `gmm` command to an overidentified model, see section 17.5.2. Also, [R] **gmm** provides many more examples.

10.3.9   Other estimators

The preceding part of this chapter covers most of the estimators used in microeconometrics analysis using cross-section data. We now consider some nonlinear estimators that are not covered.

One approach is to specify a linear function for $E\{h(y)|\mathbf{x}\}$, so $E\{h(y)|\mathbf{x}\} = \mathbf{x}'\boldsymbol{\beta}$, where $h(\cdot)$ is a nonlinear function. An example is the Box–Cox transformation in section 3.5. A disadvantage of this alternative approach is the transformation bias that arises if we then wish to predict y or $E(y|\mathbf{x})$.

Nonparametric and semiparametric estimators do not completely specify the functional forms of key model components such as $E(y|\mathbf{x})$. Several methods for nonparametric regression of y on a scalar x, including the `lowess` command, are presented in section 2.6.6.

10.4   Different estimates of the VCE

Given an estimator, there are several different standard methods for computation of standard errors and subsequent test statistics and confidence intervals. The most commonly used methods yield default, robust, and cluster–robust standard errors. This section extends the results in section 3.3 for the OLS estimator to nonlinear estimators.

10.4.1   General framework

We consider inference for the estimator $\widehat{\boldsymbol{\theta}}$ of a $q \times 1$ parameter vector $\boldsymbol{\theta}$ that solves the q equations

$$\sum_{i=1}^{N} \mathbf{g}_i(\widehat{\boldsymbol{\theta}}) = \mathbf{0} \tag{10.4}$$

where $\mathbf{g}_i(\cdot)$ is a $q \times 1$ vector. For m estimators defined in section 10.3, differentiation of objective function (10.3) leads to first-order conditions with $\mathbf{g}_i(\boldsymbol{\theta}) = \partial q_i(y_i, \mathbf{x}_i, \boldsymbol{\theta})/\partial \boldsymbol{\theta}$. It is assumed that

$$E\{\mathbf{g}_i(\boldsymbol{\theta})\} = \mathbf{0}$$

a condition that for standard estimators is necessary and sufficient for consistency of $\widehat{\boldsymbol{\theta}}$. This setup covers most models and estimators, with the notable exception of the overidentified two-stage least-squares and GMM estimators presented in chapter 6.

Under appropriate assumptions, it can be shown that

$$\widehat{\boldsymbol{\theta}} \overset{a}{\sim} N\{\boldsymbol{\theta}, \ \mathrm{Var}(\widehat{\boldsymbol{\theta}})\}$$

where $\mathrm{Var}(\widehat{\boldsymbol{\theta}})$ denotes the (asymptotic) VCE. Furthermore,

$$\mathrm{Var}(\widehat{\boldsymbol{\theta}}) = \left[E\left\{\sum_i \mathbf{H}_i(\boldsymbol{\theta})\right\}\right]^{-1} E\left\{\sum_i \sum_j \mathbf{g}_i(\boldsymbol{\theta})\mathbf{g}_j(\boldsymbol{\theta})'\right\} \left[E\left\{\sum_i \mathbf{H}_i(\boldsymbol{\theta})'\right\}\right]^{-1} \quad (10.5)$$

where $\mathbf{H}_i(\boldsymbol{\theta}) = \partial \mathbf{g}_i / \partial \boldsymbol{\theta}'$. This general expression for $\mathrm{Var}(\widehat{\boldsymbol{\theta}})$ is said to be of "sandwich form" because it can be written as $\mathbf{A}^{-1}\mathbf{B}\mathbf{A}'^{-1}$, with $\mathbf{B}$ sandwiched between $\mathbf{A}^{-1}$ and $\mathbf{A}'^{-1}$. OLS is a special case with $\mathbf{g}_i(\widehat{\boldsymbol{\beta}}) = \mathbf{x}_i'\widehat{u}_i = \mathbf{x}_i'(y_i - \mathbf{x}_i'\widehat{\boldsymbol{\beta}})$ and $\mathbf{H}_i(\boldsymbol{\beta}) = \mathbf{x}_i\mathbf{x}_i'$.

We wish to obtain the estimated asymptotic variance matrix $\widehat{V}(\widehat{\boldsymbol{\theta}})$, and the associated standard errors, which are the square roots of the diagonal entries of $\widehat{V}(\widehat{\boldsymbol{\theta}})$. This obviously entails replacing $\boldsymbol{\theta}$ with $\widehat{\boldsymbol{\theta}}$. The first and third matrices in (10.5) can be estimated using $\widehat{\mathbf{A}} = \sum_i \mathbf{H}_i(\widehat{\boldsymbol{\theta}})$. But estimation of $E\left\{\sum_i \sum_j \mathbf{g}_i(\boldsymbol{\theta})\mathbf{g}_j(\boldsymbol{\theta})'\right\}$ requires additional distributional assumptions, such as independence over i and possibly a functional form for $E\{\mathbf{g}_i(\boldsymbol{\theta})\mathbf{g}_i(\boldsymbol{\theta})'\}$. [Note that the obvious $\sum_i \sum_j \mathbf{g}_i(\widehat{\boldsymbol{\theta}})\mathbf{g}_j(\widehat{\boldsymbol{\theta}})' = \mathbf{0}$ because from (10.4) $\sum_i \mathbf{g}_i(\widehat{\boldsymbol{\theta}}) = \mathbf{0}$.]

10.4.2   The vce() option

Different assumptions lead to different estimates of the VCE. They are obtained by using the vce(*vcetype*) option for the estimation command being used. The specific *vcetype*(s) available varies with the estimation command. Their formulas are detailed in subsequent sections.

For the poisson command, many *vcetypes* are supported.

The vce(oim) and vce(opg) options use the DGP assumptions to evaluate the expectations in (10.5); see section 10.4.4. The vce(oim) option is the default.

The vce(robust) and vce(cluster *clustvar*) options use sandwich estimators that do not use the DGP assumptions to explicitly evaluate the expectations in (10.5). The vce(robust) option assumes independence over i. The vce(cluster *clustvar*) option permits a limited form of correlation over i, within clusters where the clusters are independent and there are many clusters; see section 10.4.6. For commands that already control for clustering, such as xtreg, the vce(robust) option rather than vce(cluster *clustvar*) provides a cluster–robust estimate of the VCE.

The vce(bootstrap) and vce(jackknife) options use resampling schemes that make limited assumptions on the DGP similar to those for the vce(robust) or vce(cluster *clustvar*) options; see section 10.4.8.

The various vce() options need to be used with considerable caution. Estimates of the VCE other than the default estimate are used when some part of the DGP is felt to be misspecified. But then the estimator itself may be inconsistent.

10.4.3   Application of the vce() option

For count data, the natural starting point is the MLE, assuming a Poisson distribution. It can be shown that the default ML standard errors are based on the Poisson distribution restriction of variance–mean inequality. But in practice, count data are often "overdispersed" with $\text{Var}(y|\mathbf{x}) > \exp(\mathbf{x}'\boldsymbol{\beta})$, in which case the default ML standard errors can be shown to be biased downward. At the same time, the Poisson MLE can be shown to be consistent provided only that $E(y|\mathbf{x}) = \exp(\mathbf{x}'\boldsymbol{\beta})$ is the correct specification of the conditional mean.

These considerations make the Poisson MLE a prime candidate for using vce(robust) rather than the default. The vce(cluster *clustvar*) option assumes independence over clusters, however clusters are defined, rather than independence over i. The vce(bootstrap) estimate is asymptotically equivalent to the vce(robust) estimate.

For the Poisson MLE, it can be shown that the default, robust, and cluster–robust estimates of the VCE are given by, respectively,

$$\widehat{V}_{\text{oim}}(\widehat{\boldsymbol{\beta}}) = \left(\sum_i e^{\mathbf{x}_i'\widehat{\boldsymbol{\beta}}}\mathbf{x}_i\mathbf{x}_i'\right)^{-1}$$

$$\widehat{V}_{\text{rob}}(\widehat{\boldsymbol{\beta}}) = \left(\sum_i e^{\mathbf{x}_i'\widehat{\boldsymbol{\beta}}}\mathbf{x}_i\mathbf{x}_i'\right)^{-1} \left(\sum_i (y_i - e^{\mathbf{x}_i'\widehat{\boldsymbol{\beta}}})^2 \mathbf{x}_i\mathbf{x}_i'\right) \left(\sum_i e^{\mathbf{x}_i'\widehat{\boldsymbol{\beta}}}\mathbf{x}_i\mathbf{x}_i'\right)^{-1}$$

$$\widehat{V}_{\text{clu}}(\widehat{\boldsymbol{\beta}}) = \left(\sum_i e^{\mathbf{x}_i'\widehat{\boldsymbol{\beta}}}\mathbf{x}_i\mathbf{x}_i'\right)^{-1} \left(\sum_c \frac{C}{C-1} \sum_c \widehat{\mathbf{g}}_c\widehat{\mathbf{g}}_c'\right) \left(\sum_i e^{\mathbf{x}_i'\widehat{\boldsymbol{\beta}}}\mathbf{x}_i\mathbf{x}_i'\right)^{-1}$$

where $\widehat{\mathbf{g}}_c = \sum_{i:i\in c}(y_i - e^{\mathbf{x}_i'\widehat{\boldsymbol{\beta}}})\mathbf{x}_i$, and $c = 1, \ldots, C$ denotes the clusters.

Implementation is straightforward, except that in this example there is no natural reason for clustering. For illustrative purposes, we cluster on age, in which case we are assuming correlation across individuals of the same age, and independence of individuals of different age. For the bootstrap, we first set the seed, for replicability, and set the number of replications at 400, considerably higher than the Stata default. We obtain

```
. * Different VCE estimates after Poisson regression
. quietly poisson docvis private chronic female income
. estimates store VCE_oim
. quietly poisson docvis private chronic female income, vce(opg)
. estimates store VCE_opg
. quietly poisson docvis private chronic female income, vce(robust)
. estimates store VCE_rob
. quietly poisson docvis private chronic female income, vce(cluster age)
. estimates store VCE_clu
. set seed 10101
```

```
. quietly poisson docvis private chronic female income, vce(boot,reps(400))
. estimates store VCE_boot
. estimates table VCE_oim VCE_opg VCE_rob VCE_clu VCE_boot, b(%8.4f) se
```

Variable	VCE_oim	VCE_opg	VCE_rob	VCE_clu	VCE_boot
private	0.7987	0.7987	0.7987	0.7987	0.7987
	0.0277	0.0072	0.1090	0.1496	0.1100
chronic	1.0919	1.0919	1.0919	1.0919	1.0919
	0.0158	0.0046	0.0560	0.0603	0.0555
female	0.4925	0.4925	0.4925	0.4925	0.4925
	0.0160	0.0046	0.0585	0.0686	0.0588
income	0.0036	0.0036	0.0036	0.0036	0.0036
	0.0002	0.0001	0.0011	0.0012	0.0011
_cons	-0.2297	-0.2297	-0.2297	-0.2297	-0.2297
	0.0287	0.0075	0.1109	0.1454	0.1120

legend: b/se

The first two ML-based standard errors, explained in section 10.4.4, are very different. This indicates a problem with the assumption of a Poisson density. The third column robust standard errors are roughly four times the first column default standard errors. This very large difference often happens when fitting Poisson models. For other estimators, the difference is usually not as great. In particular, for OLS, robust standard errors are often within 20% (higher or lower) of the default. The fourth column cluster–robust standard errors are 8–37% higher than the robust standard errors. In other applications, the difference can be much larger. The fifth column bootstrap standard errors are within 1% of the third column robust standard errors, confirming that they are essentially equivalent.

In this example, it would be misleading to use the default standard errors. We should at least use the robust standard errors. This requires relaxing the assumption of a Poisson distribution so that the model should not be used to predict conditional probabilities. But, at least for the Poisson MLE, $\widehat{\boldsymbol{\beta}}$ is a consistent estimate provided that the conditional mean is indeed the specified $\exp(\mathbf{x}'\boldsymbol{\beta})$.

10.4.4   Default estimate of the VCE

If no option is used, then we obtain "default" standard errors. These make the strongest assumptions, essentially that all relevant parts of the DGP are specified and are specified correctly. This permits considerable simplification, not given here, that leads to the sandwiched form $\mathbf{A}^{-1}\mathbf{B}\mathbf{A}'^{-1}$ simplifying to a multiple of $\mathbf{A}^{-1}$.

For the MLE (with data independent over i), it is assumed that the density is correctly specified. Then the information matrix equality leads to simplification so that

$$\widehat{V}_{\text{def}}(\widehat{\boldsymbol{\theta}}) = -\left[\sum_i E\{\mathbf{H}_i(\boldsymbol{\theta})\}|_{\widehat{\boldsymbol{\theta}}}\right]^{-1}$$

The default `vce(oim)` option, where `oim` is an acronym for the observed information matrix, gives this estimate of the VCE for Stata ML commands. Although it goes under the name `vce(ols)`, this estimator is also the default for `regress`, yielding $\widehat{V}_{\text{def}}(\widehat{\boldsymbol{\beta}}) = s^2 \left(\sum_i \mathbf{x}_i \mathbf{x}_i' \right)^{-1}$ with $s^2 = \sum_i \widehat{u}_i^2 / (N - K)$.

The `vce(opg)` option gives an alternative estimate, called the outer-product of the gradient estimate:

$$\widehat{V}_{\text{opg}}(\widehat{\boldsymbol{\theta}}) = \left\{ \sum_i \mathbf{g}_i(\widehat{\boldsymbol{\theta}}) \mathbf{g}_i(\widehat{\boldsymbol{\theta}})' \right\}^{-1}$$

This is asymptotically equivalent to the default estimate if the density is correctly specified.

10.4.5   Robust estimate of the VCE

The `vce(robust)` option to Stata cross-section estimation commands calculates the sandwich estimate under the assumption of independence. Then $E\{\sum_i \sum_j \mathbf{g}_i(\boldsymbol{\theta}) \mathbf{g}_j(\boldsymbol{\theta})'\} = E\{\sum_i \mathbf{g}_i(\boldsymbol{\theta}) \mathbf{g}_i(\boldsymbol{\theta})'\}$, leading to the VCE robust estimate

$$\widehat{V}_{\text{rob}}(\widehat{\boldsymbol{\theta}}) = \left(\sum_i \widehat{\mathbf{H}}_i \right)^{-1} \left(\frac{N}{N-q} \sum_i \widehat{\mathbf{g}}_i \widehat{\mathbf{g}}_i' \right) \left(\sum_i \widehat{\mathbf{H}}_i' \right)^{-1}$$

where $\widehat{\mathbf{H}}_i = \mathbf{H}_i(\widehat{\boldsymbol{\theta}})$ and $\widehat{\mathbf{g}}_i = \mathbf{g}_i(\widehat{\boldsymbol{\theta}})$. In some special cases, such as NLS, $\widehat{\mathbf{H}}_i$ is replaced by the expected Hessian $E(\mathbf{H}_i)$ evaluated at $\widehat{\boldsymbol{\theta}}$. The factor $N/(N-q)$ in the middle term is an ad hoc degrees of freedom analogous to that for the linear regression model with independent and identically distributed normal errors. This estimator is a generalization of similar results of Huber (1965) for the MLE and the heteroskedasticity-consistent estimate of White (1980) for the OLS estimator. It is often called heteroskedasticity-robust rather than robust.

The `vce(robust)` option should be used with caution. It is robust in the sense that, unlike default standard errors, no assumption is made about the functional form for $E\{\mathbf{g}_i(\boldsymbol{\theta}) \mathbf{g}_i(\boldsymbol{\theta})'\}$. But if $E\{\mathbf{g}_i(\boldsymbol{\theta}) \mathbf{g}_i(\boldsymbol{\theta})'\}$ is misspecified, warranting use of robust standard errors, then it may also be the case that $E\{\mathbf{g}_i(\boldsymbol{\theta})\} \neq \mathbf{0}$. Then we have the much more serious problem of $\widehat{\boldsymbol{\theta}}$ being inconsistent for $\boldsymbol{\theta}$. For example, the tobit MLE and the MLE for any other parametric model with selection or truncation becomes inconsistent as soon as any distributional assumptions are relaxed. The only advantage then of using the robust estimate of the VCE is that it does give a consistent estimate of the VCE. However, it is the VCE of an inconsistent estimator.

There are, however, some commonly used estimators that maintain consistency under relatively weak assumptions. ML and GLM estimators based on the LEF (see section 10.3.1) require only that the conditional mean function be correctly specified. IV estimators are consistent provided only that a valid instrument is used so that the model error u_i and instrument vector $\mathbf{z}_i$ satisfy $E\{u_i | \mathbf{z}_i\} = \mathbf{0}$.

The preceding discussion applies to cross-section estimation commands. For panel data or clustered data, `xt` commands such as `xtreg` with the `vce(robust)` option produce a cluster–robust estimate of the VCE.

10.4.6 Cluster–robust estimate of the VCE

A common alternative to independent observations is that observations fall into clusters, where observations in different clusters are independent, but observations within the same cluster are no longer independent. For example, individuals may be grouped into villages, with correlation within villages but not across villages. Such regional groupings are especially important to control for if the regressor of interest, such as a policy variable, is invariant within the region. Then, for cross-section estimators, the robust estimate of the VCE is incorrect and can be substantially downward biased.

Instead, we use a cluster–robust estimate of the VCE. The first-order conditions can be summed within cluster and reexpressed as

$$\sum_{c=1}^{C} \mathbf{g}_c(\widehat{\boldsymbol{\theta}}) = \mathbf{0}$$

where c denotes the cth cluster, there are C clusters, and $\mathbf{g}_c(\boldsymbol{\theta}) = \sum_{i:i \in c} \mathbf{g}_i(\boldsymbol{\theta})$. The key assumption is that $E\{\mathbf{g}_i(\boldsymbol{\theta})\mathbf{g}_j(\boldsymbol{\theta})'\} = \mathbf{0}$ if i and j are in different clusters. Only minor adaptation of the previous algebra is needed, and we obtain

$$\widehat{V}_{\text{clus}}(\widehat{\boldsymbol{\theta}}) = \left(\sum_c \widehat{\mathbf{H}}_c\right)^{-1} \left(\frac{C}{C-1} \sum_c \widehat{\mathbf{g}}_c \widehat{\mathbf{g}}_c'\right) \left(\sum_c \widehat{\mathbf{H}}_c'\right)^{-1}$$

where $\mathbf{H}_c(\boldsymbol{\theta}) = \partial \mathbf{g}_c(\boldsymbol{\theta})/\partial\boldsymbol{\theta}'$. This estimator was proposed by Liang and Zeger (1986), and the scaling $C/(C-1)$ is a more recent ad hoc degrees of freedom correction. The estimator assumes that the number of clusters $C \to \infty$. When each cluster has only one observation, $\widehat{V}_{\text{clus}}(\widehat{\boldsymbol{\theta}}) = (N - k/N - 1)\widehat{V}_{\text{rob}}(\widehat{\boldsymbol{\theta}})$, the cluster–robust and robust standard errors then differ only by a small degrees-of-freedom correction.

This estimator is obtained by using the vce(cluster *clustvar*) option, where *clustvar* is the name of the variable that defines the cluster, such as a village identifier. For panel data, or clustered data, xt commands such as xtreg already explicitly allow for clustering in estimation, and the cluster–robust estimate of the VCE is obtained by using the vce(robust) option rather than the vce(cluster *clustvar*) option.

The same caveat as in the robust case applies. It is still necessary that $E\{\mathbf{g}_c(\boldsymbol{\theta})\} = \mathbf{0}$ to ensure estimator consistency. Essentially, the joint distribution of the $\mathbf{g}_i(\boldsymbol{\theta})$ within cluster can be misspecified, because of assuming independence when there is in fact dependence, but the marginal distribution of $\mathbf{g}_i(\boldsymbol{\theta})$ must be correctly specified in the sense that $E\{\mathbf{g}_i(\boldsymbol{\theta})\} = \mathbf{0}$ for each component of the cluster.

10.4.7 Heteroskedasticity- and autocorrelation-consistent estimate of the VCE

Heteroskedasticity- and autocorrelation-consistent (HAC) estimates of the VCE, such as the Newey–West (1987) estimator, are a generalization of the robust estimate to time-series data. This permits some correlation of adjacent observations, up to, say, m periods apart.

HAC estimates are implemented in Stata for some linear time-series estimators, such as `newey` and `ivregress`. For nonlinear estimators, HAC estimates are available for `glm` by specifying the `vce(hac kernel)` option, in which case you must `tsset` your data.

In microeconometrics analysis, panel data have a time-series component. For short panels covering few time periods, there is no need to use HAC estimates. For long panels spanning many time periods, there is more reason to use HAC estimates of the VCE. An example using the user-written `xtscc` command is given in section 8.10.6.

10.4.8   Bootstrap standard errors

Stata estimation commands with the `vce(bootstrap)` option provide standard errors using the bootstrap, specifically, a paired bootstrap. The default bootstrap assumes independent observations and is equivalent to computing robust standard errors, provided that the number of bootstraps is large. Similarly, a cluster bootstrap that assumes independence across clusters but not within clusters is equivalent to computing cluster–robust standard errors.

A related option is `vce(jackknife)`. This can be computationally demanding because it involves recomputing the estimator N times, where N is the sample size.

These methods are detailed in chapter 13. In that chapter, we also consider a different use of the bootstrap to implement a more refined asymptotic theory that can lead to t statistics with better size properties, and confidence intervals with better coverage, in finite samples.

10.4.9   Statistical inference

Given a method to estimate the VCE, we can compute standard errors, t statistics, confidence intervals, and Wald hypothesis tests. These are automatically provided by estimation commands such as `poisson`. Some tests—notably, likelihood-ratio tests—are no longer appropriate once DGP assumptions are relaxed to allow, for example, a robust estimate of the VCE.

More complicated statistical inference can be performed by using the `test`, `testnl`, `lincom`, and `nlcom` commands, which are detailed in chapter 12.

10.5   Prediction

In this section, we consider prediction. Most often, the prediction is one of the conditional mean $E(y|\mathbf{x})$. This can be much more precisely predicted than can the actual value of y given $\mathbf{x}$.

10.5.1   The predict and predictnl commands

A new variable that contains the prediction for each observation can be obtained by using the postestimation `predict` command. After single-equation commands, this command has the syntax

`predict` [*type*] *newvar* [*if*] [*in*] [, *options*]

The prediction is stored as the variable *newvar* and is of the data type *type*, the default being single precision. The type of prediction desired is defined with *options*, and several different types of prediction are usually available. The possibilities vary with the preceding estimation command.

After `poisson`, the key option for the `predict` command is the default `n` option. This computes $\exp(\mathbf{x}_i'\widehat{\boldsymbol{\beta}})$, the predicted expected number of events. The `xb` option calculates the linear prediction $\mathbf{x}_i'\widehat{\boldsymbol{\beta}}$, and `stdp` calculates $\{\mathbf{x}_i'\widehat{V}(\widehat{\boldsymbol{\beta}})\mathbf{x}_i\}^{1/2}$, the standard error of $\mathbf{x}_i'\widehat{\boldsymbol{\beta}}$. The `score` option calculates the derivative of the log likelihood with respect to the linear prediction. For the Poisson MLE, this is $y_i - \exp(\mathbf{x}_i'\widehat{\boldsymbol{\beta}})$ and can be viewed as a Poisson residual.

The `predictnl` command enables the user to provide a formula for the prediction. The syntax is

`predictnl` [*type*] *newvar*=*pnl_exp* [*if*] [*in*] [, *options*]

where *pnl_exp* is an expression that is illustrated in the next section. The options provide quantities not provided by `predict` that enable Wald statistical inference on the predictions. In particular, the `se`(*newvar2*) option creates a new variable containing standard errors for the prediction *newvar* for each observation. These standard errors are computed using the delta method detailed in section 12.3.8. Other options include `variance()`, `wald()`, `p()`, and `ci()`.

The `predict` and `predictnl` commands act on the currently defined sample, with the `if` and `in` qualifiers used if desired to predict for a subsample. It is possible to inadvertently predict using a sample different from the estimation sample. The `if e(sample)` qualifier ensures that the estimation sample is used in prediction. At other times, it is desired to deliberately use estimates from one sample to predict using a different sample. This can be done by estimating with one sample, reading a new sample into memory, and then predicting using this new sample.

10.5.2   Application of predict and predictnl

The predicted mean number of doctor visits for each individual in the sample can be computed by using the `predict` command with the default option. We use the `if e(sample)` qualifier to ensure that prediction is for the same sample as the estimation sample. This precaution is not necessary here but is good practice to avoid inadvertent error. We also obtain the same prediction by using `predictnl` with the `se()` option to obtain the standard error of the prediction. We obtain

```
. * Predicted mean number of doctor visits using predict and predictnl
. quietly poisson docvis private chronic female income, vce(robust)
. predict muhat if e(sample), n
. predictnl muhat2 = exp(_b[private]*private + _b[chronic]*chronic
> + _b[female]*female + _b[income]*income + _b[_cons]), se(semuhat2)
. summarize docvis muhat muhat2 semuhat2

    Variable |      Obs        Mean    Std. Dev.        Min        Max
-------------+--------------------------------------------------------
      docvis |     4412    3.957389    7.947601          0        134
       muhat |     4412    3.957389    2.985057    .7947512   15.48004
      muhat2 |     4412    3.957389    2.985057    .7947512   15.48004
    semuhat2 |     4412    .2431483    .1980062    .0881166   3.944615
```

Here the average of the predictions of $E(y|\mathbf{x})$ is 3.957, equal to the average of the y values. This special property holds only for some estimators—OLS, just-identified linear IV, Poisson, logit, and exponential (with exponential conditional mean)—provided that these models include an intercept. The standard deviation of the predictions is 2.985, less than that of y. The predicted values range from 0.8 to 15.5 compared with a sample range of 0–134.

The model quite precisely estimates $E(y|\mathbf{x})$, because from the last row the standard error of $\exp(\mathbf{x}_i'\widehat{\boldsymbol{\beta}})$ as an estimate of $\exp(\mathbf{x}_i'\boldsymbol{\beta})$ is relatively small. This is not surprising because asymptotically $\widehat{\boldsymbol{\beta}} \xrightarrow{p} \boldsymbol{\beta}$, so $\exp(\mathbf{x}_i'\widehat{\boldsymbol{\beta}}) \xrightarrow{p} \exp(\mathbf{x}_i'\boldsymbol{\beta})$. Much more difficult is using $\exp(\mathbf{x}_i'\widehat{\boldsymbol{\beta}})$ to predict $y_i|\mathbf{x}_i$ rather than $E(y_i|\mathbf{x}_i)$, because there is always intrinsic randomness in y_i. In our example, y_i without any regression has a standard deviation of 7.95. Even if Poisson regression explains the data well enough to reduce the standard deviation of $y_i|\mathbf{x}_i$ to, say, 4, then any prediction of $y_i|\mathbf{x}_i$ will have a standard error of prediction of at least 4.

More generally, with microeconometric data and a large sample, we can predict the conditional mean $E(y_i|\mathbf{x}_i)$ well but not $y_i|\mathbf{x}_i$. For example, we may predict well the mean earnings of a white female with 12 years of schooling but will predict relatively poorly the earnings of a randomly chosen white female with 12 years of schooling.

When the goal of prediction is to obtain a sample average predicted value, the sample average prediction should be a weighted average. To obtain a weighted average, specify weights with `summarize` or with `mean`; see section 3.7. This is especially important if one wants to make statements about the population and sampling is not simple random sampling. For average predictions, a simpler method is to use the `margins` command; see section 10.5.7.

10.5.3   Out-of-sample prediction

Out-of-sample prediction is possible. For example, we may want to make predictions for the 2001 sample using parameter estimates from the 2002 sample.

The current estimates are those from the 2002 sample, so we just need to read the 2001 sample into memory and use `predict`. We have

```
. * Out-of-sample prediction for year01 data using year02 estimates
. use mus10data.dta, clear
. quietly poisson docvis private chronic female income if year02==1, vce(robust)
. keep if year01 == 1
(23940 observations deleted)
. predict muhatyear01, n
. summarize docvis muhatyear01
```

Variable	Obs	Mean	Std. Dev.	Min	Max
docvis	6184	3.896345	7.873603	0	152
muhatyear01	6184	4.086984	2.963843	.7947512	15.02366

Note that the average of the predictions of $E(y|\mathbf{x})$, 4.09, no longer equals the average of the y values.

10.5.4 Prediction at a specified value of one of the regressors

Suppose we want to calculate the sample average number of doctor visits if all individuals had private insurance, whereas all other regressors are unchanged.

This can be done by setting `private` $= 1$ and using `predict`. To return to the original data after doing so, we use the commands `preserve` to preserve the current dataset and `restore` to return to the preserved dataset. We have

```
. * Prediction at a particular value of one of the regressors
. use mus10data.dta, clear
. keep if year02 == 1
(25712 observations deleted)
. quietly poisson docvis private chronic female income, vce(robust)
. preserve
. replace private = 1
(947 real changes made)
. predict muhatpeq1, n
. summarize muhatpeq1
```

Variable	Obs	Mean	Std. Dev.	Min	Max
muhatpeq1	4412	4.371656	2.927381	1.766392	15.48004

```
. restore
```

The conditional mean is predicted to be 4.37 visits when all have private insurance, compared with 3.96 in the sample where only 78% had private insurance.

The preceding code calculates and stores the prediction for each individual. Usually, only the average of these predictions is required. Then it is much simpler to use the `margins` command, which also gives a 95% confidence interval for the average prediction; see section 10.5.7.

10.5.5   Prediction at a specified value of all the regressors

We may also want to estimate the conditional mean at a given value of all the regressors. For example, consider the number of doctor visits for a privately insured woman with no chronic conditions and an income of $10,000.

To do so, we can use the lincom and nlcom commands. These commands compute point estimates for linear combinations and associated standard errors, z statistics, p-values, and confidence intervals. They are primarily intended to produce confidence intervals for parameter combinations such as $\beta_3 - \beta_4$ and are presented in detail in chapter 12. They can also be used for prediction, because a prediction is a linear combination of the parameters.

We need to predict the expected number of doctor visits when private = 1, chronic = 0, female = 1, and income = 10. The nlcom command has the form

```
. nlcom exp(_b[_cons]+_b[private]*1+_b[chronic]*0+_b[female]*1+_b[income]*10)

       _nl_1:   exp(_b[_cons]+_b[private]*1+_b[chronic]*0+_b[female]*1+
> _b[income]*10)
```

docvis	Coef.	Std. Err.	z	P>|z|	[95% Conf. Interval]
_nl_1	2.995338	.1837054	16.31	0.000	2.635282 3.355394

A simpler command for our example uses lincom with the eform option to display the exponential. Coefficients are then more simply referred to as private, for example, rather than _b[private]. We have

```
. * Predict at a specified value of all the regressors
. lincom _cons + private*1 + chronic*0 + female*1 + income*10, eform
 ( 1)  [docvis]private + [docvis]female + 10*[docvis]income + [docvis]_cons = 0
```

docvis	exp(b)	Std. Err.	z	P>|z|	[95% Conf. Interval]
(1)	2.995338	.1837054	17.89	0.000	2.656081 3.377929

The predicted conditional mean number of doctor visits is 3.00. The standard error of the prediction is 0.18 and a 95% confidence interval is [2.66, 3.38]. The standard error is computed with the delta method, and the bounds of the confidence interval depend on the standard error; see section 12.3.8. The test against a value of 0 is not relevant here but is relevant when lincom is used to test linear combinations of parameters.

The relatively tight confidence interval is for $\exp(\mathbf{x}'\widehat{\boldsymbol{\beta}})$ as an estimate of $E(y|\mathbf{x}) = \exp(\mathbf{x}'\boldsymbol{\beta})$. If instead we want to predict the actual values of y given $\mathbf{x}$, then the confidence interval will be much, much wider, because we also need to add in variation in y around its conditional mean. There is considerable more noise in the prediction of the actual value than in estimating the conditional mean.

An even simpler method, using the margins command, is presented in section 10.5.7.

10.5.6   Prediction of other quantities

We have focused on prediction of the conditional mean. Options of the `predict` command provide prediction of other quantities of interest, where these quantities vary with the estimation command. Usually, one or more residuals are available. Following `poisson`, the `predict` option `score` computes the residual $y_i - \exp(\mathbf{x}_i'\widehat{\boldsymbol{\beta}})$. An example of more command-specific predictions are those following the survival data command `streg` to produce not only mean survival time but also median survival time, the hazard, and the relative hazard.

For a discrete dependent variable, it can be of interest to obtain the predicted probability of each of the discrete values, i.e., $\Pr(y_i = 0)$, $\Pr(y_i = 1)$, $\Pr(y_i = 2)$, ... For binary logit and probit, the default option of `predict` gives $\Pr(y_i = 1)$. For the use of `predict` in multinomial models, see chapter 15. For count models, `predict` does not have an option to compute predicted probabilities, but the user-written `prcounts` command does; see chapter 17.

10.5.7   The margins command for prediction

The `margins` command, introduced in Stata 11, simplifies prediction at specified values of the regressors.

The syntax of the command is

margins [*marginlist*] [*if*] [*in*] [*weight*] [, *response_options options*]

where *marginlist* is a list of variable names or of factor variables that appear in the current estimation results that are being analyzed, *response_options* specify the particular quantity to be computed, and *options* include particular values of regressors at which computation occurs.

The `margins` command without any options computes the sample average value of the default quantity computed by the `predict` command. Following the `poisson` command, therefore, the `margins` command yields the sample average of the predicted number of events. We have

```
. * Sample average of predicted number of events
. quietly poisson docvis private chronic female income, vce(robust)

. margins
Predictive margins                              Number of obs   =       4412
Model VCE    : Robust

Expression   : Predicted number of events, predict()
```

		Delta-method				
	Margin	Std. Err.	z	P>|z|	[95% Conf.	Interval]
_cons	3.957389	.1115373	35.48	0.000	3.73878	4.175998

A 95% confidence interval for the sample average predicted number of doctor visits is [3.74, 4.18].

To predict at a specified value of the regressors, we use the `at()` option. For example, the average number of doctor visits if all individuals had private insurance, with other regressors equal to sample values, is calculated using the following command:

```
. * Sample average prediction at a particular value of one of the regressors
. margins, at(private=1)
Predictive margins                              Number of obs   =      4412
Model VCE      : Robust

Expression     : Predicted number of events, predict()
at             : private         =          1

                            Delta-method
                  Margin    Std. Err.      z    P>|z|     [95% Conf. Interval]

         _cons   4.371656   .1273427   34.33   0.000     4.122069    4.621243
```

The average equals that given in section 10.5.4, with 95% confidence interval [4.12, 4.62].

The `at()` option can be used to compute the predicted number of doctor visits at a specified value of all the regressors. For example,

```
. * Prediction at a specified value of all regressors
. margins, at(private=1 chronic=0 female=1 income=10)
Adjusted predictions                            Number of obs   =      4412
Model VCE      : Robust

Expression     : Predicted number of events, predict()
at             : private         =          1
                 chronic         =          0
                 female          =          1
                 income          =         10

                            Delta-method
                  Margin    Std. Err.      z    P>|z|     [95% Conf. Interval]

         _cons   2.995338   .1837054   16.31   0.000     2.635282    3.355394
```

The results coincide with those given in section 10.5.5.

The `atmean` option computes the predicted number of doctor visits when regressors are evaluated at the sample mean. From output not given, the `margins, atmean` command yields $\exp(\overline{\mathbf{x}}'\widehat{\beta})$ equal to 3.02968. This differs from $1/N \sum_i \exp(\mathbf{x}_i'\widehat{\beta})$, the average of the predicted number of doctor visits, which from previous output equals 3.957389. The average of a nonlinear function does not equal the nonlinear function evaluated at the average.

When the estimation command variable list includes a factor variable, we can easily obtain average predictions for each value of the factor variable. For example, for private insurance, we obtain

```
. * Sample average prediction at different values of indicator variable private
. quietly poisson docvis i.private chronic female income, vce(robust)

. margins private
Predictive margins                          Number of obs   =      4412
Model VCE     : Robust

Expression    : Predicted number of events, predict()
```

	Margin	Delta-method Std. Err.	z	P>|z|	[95% Conf. Interval]	
private						
0	1.966935	.2054776	9.57	0.000	1.564207	2.369664
1	4.371656	.1273427	34.33	0.000	4.122069	4.621243

The i. operator is used in the estimation command to signal that variable `private` is a categorical variable. If the categorical variable instead took four values, say, then estimation would include three indicator variables (with a base category omitted) and the `margins` command would compute the average prediction for each of the four values of the categorical variable.

10.6 Marginal effects

An ME, or partial effect, most often measures the effect on the conditional mean of y of a change in one of the regressors, say, x_j. In the linear regression model, the ME equals the relevant slope coefficient, greatly simplifying analysis. For nonlinear models, this is no longer the case, leading to remarkably many different methods for calculating MEs. Also other MEs may be desired, such as elasticities and effects on conditional probabilities rather than the conditional mean.

10.6.1 Calculus and finite-difference methods

Calculus methods can be applied for a continuous regressor, and the ME of the jth regressor is then

$$\mathrm{ME}_j = \frac{\partial E(y|\mathbf{x} = \mathbf{x}^*)}{\partial x_j}$$

For the Poisson model with $E(y|\mathbf{x}) = \exp(\mathbf{x}'\boldsymbol{\beta})$, we obtain $\mathrm{ME}_j = \exp(\mathbf{x}^{*\prime}\boldsymbol{\beta})\beta_j$. This ME is not simply the relevant parameter β_j, and it varies with the point of evaluation $\mathbf{x}^*$.

Calculus methods are not always appropriate. In particular, for an indicator variable, say, d, the relevant ME is the change in the conditional mean when d changes from 0 to 1. Let $\mathbf{x} = (\mathbf{z}\ d)$, where $\mathbf{z}$ denotes all regressors other than the jth, which is an indicator variable d. Then the finite-difference method yields the ME

$$\mathrm{ME}_j = E(y|\mathbf{z} = \mathbf{z}^*, d = 1) - E(y|\mathbf{z} = \mathbf{z}^*, d = 0)$$

For the linear regression model, calculus and finite-difference methods give the same result. For nonlinear models, this is no longer the case. Interpretation of coefficients in nonlinear models is clearly not as straightforward as in linear models.

Even for continuous regressors, we may want to consider discrete changes, such as the impact of age increasing from 40 to 60. Letting $\mathbf{x} = (\mathbf{z}\ w)$, the finite-difference method uses

$$\mathrm{ME}_j = E(y|\mathbf{z} = \mathbf{z}^*, w = 60) - E(y|\mathbf{z} = \mathbf{z}^*, w = 40)$$

A common change to consider is an increase of one standard deviation from the sample mean of the regressor of interest.

Finally, as for linear models, interactions in regressors lead to additional complications.

10.6.2 MEs estimates AME, MEM, and MER

For nonlinear models, the ME varies with the point of evaluation. Three common choices of evaluation are 1) at sample values and then average, 2) at the sample mean of the regressors, and 3) at representative values of the regressors. We use the following acronyms, where the first two follow Bartus (2005):

AME	Average marginal effect	Average of ME at each $\mathbf{x} = \mathbf{x}_i$
MEM	Marginal effect at mean	ME at $\mathbf{x} = \overline{\mathbf{x}}$
MER	Marginal effect at a representative value	ME at $\mathbf{x} = \mathbf{x}^*$

These three quantities can be computed using the postestimation `margins` command with the `dydx()` option.

10.6.3 Elasticities and semielasticities

The impact of changes in a regressor on the dependent variable can also be measured by using elasticities and semielasticities.

For simplicity, we consider a scalar regressor x, and the effect of a change in x on $E(y|x)$, which we write more simply as y. Then the ME using the finite-difference method is given by

$$\mathrm{ME} = \frac{\Delta y}{\Delta x}$$

This measures the change in y associated with a one-unit change in x. We present elasticities based on finite differences. If instead calculus methods are used, we replace $\Delta y/\Delta x$ in the equations below with the derivative $\partial y/\partial x$.

An elasticity instead measures the proportionate change in y associated with a given proportionate change in x. More formally, the elasticity ε is given by

$$\varepsilon = \frac{\Delta y/y}{\Delta x/x} = \frac{\Delta y}{\Delta x} \times \frac{x}{y} = \mathrm{ME} \times \frac{x}{y} \tag{10.6}$$

For example, if $y = 1 + 2x$, then $\Delta y / \Delta x = 2$ and the elasticity at $x = 3$ equals $2 \times 3/7 = 6/7 = 0.86$. This can be interpreted as follows: a 1% increase in x is associated with a 0.86% increase in y.

Elasticities can be more useful than MEs, because they are scale-free measures. For example, suppose we estimate that a \$1,000 increase in annual income is associated with 0.1 more doctor visits per year. Whether this is a large or small effect depends on whether these changes in income and doctor visits are large or small. Given knowledge that the sample means of income and doctor visits are, respectively, \$34,000 and 4, the elasticity $\varepsilon = 0.1 \times 34/4 = 0.85$. This is a large effect. For example, a 10% increase in income is associated with an 8.5% increase in doctor visits.

A semielasticity is a hybrid of an ME and an elasticity that measures the proportionate change in y associated with a one-unit change in x. The semielasticity is given by

$$\frac{\Delta y / y}{\Delta x} = \frac{\Delta y}{\Delta x} \times \frac{1}{y} = \text{ME} \times \frac{1}{y} \tag{10.7}$$

For the preceding example, the semielasticity is $0.1/4 = 0.025$, so a \$1,000 increase in income (a one-unit change given that income is measured in thousands of dollars) is associated with a 0.025 proportionate increase, or a 2.5% increase, in doctor visits.

Less used is the unit change in y associated with a proportionate change in x, given by

$$\frac{\Delta y}{\Delta x / x} = \frac{\Delta y}{\Delta x} \times x = \text{ME} \times x \tag{10.8}$$

These four quantities can be computed by using various options of the `margins` command, given in section 10.6.5. An illustration is given in section 10.6.9.

10.6.4 Simple interpretations of coefficients in single-index models

In nonlinear models, coefficients are more difficult to interpret because now $\beta_j \neq \partial E(y|\mathbf{x})/\partial x_j$. Nonetheless, some direct interpretation is possible if the conditional mean is of the single-index form

$$E(y|\mathbf{x}) = m(\mathbf{x}'\boldsymbol{\beta})$$

This single-index form implies that the ME is

$$\text{ME}_j = m'(\mathbf{x}'\boldsymbol{\beta}) \times \beta_j \tag{10.9}$$

where $m'(\mathbf{x}'\boldsymbol{\beta})$ denotes the derivative of $m(\mathbf{x}'\boldsymbol{\beta})$ with respect to $\mathbf{x}'\boldsymbol{\beta}$.

Two important properties follow. First, if $m(\mathbf{x}'\boldsymbol{\beta})$ is monotonically increasing, so $m'(\mathbf{x}'\boldsymbol{\beta}) > 0$ always, then the sign of $\widehat{\beta}_j$ gives the sign of the ME (and if $m(\mathbf{x}'\boldsymbol{\beta})$ is monotonically decreasing the sign of $\widehat{\beta}_j$ is the negative of that of the ME). Second, for any function $m(\cdot)$ and at any value of $\mathbf{x}$, we have

$$\frac{\text{ME}_j}{\text{ME}_k} = \frac{\beta_j}{\beta_k}$$

Therefore, if one coefficient is twice as big as another, then so too is the ME. These two properties apply to most commonly used nonlinear regression models, aside from multinomial models.

For example, from section 10.3.2, the regressor `private` has a coefficient of 0.80 and the regressor `chronic` has a coefficient of 1.09. It follows that having a chronic condition is associated with a bigger change in doctor visits than having private insurance, because $1.09 > 0.80$. The effects for both regressors are positive because the coefficients are positive, and the conditional mean $\exp(\mathbf{x}'\boldsymbol{\beta})$ is a monotonically increasing function.

Additional interpretation of coefficients can be possible for specific single-index models. In particular, for the exponential conditional mean $\exp(\mathbf{x}'\boldsymbol{\beta})$, the $\text{ME}_j = E(y|\mathbf{x}) \times \beta_j$. So $\beta_j = \text{ME}_j / E(y|\mathbf{x})$, and from (10.7) the regression coefficients can be interpreted as semielasticities. From section 10.3.2, the regressor `income` has a coefficient of 0.0036. It follows that a \$1,000 increase in income (a one-unit increase in the rescaled regressor `income`) is associated with a 0.0036 proportionate increase, or a 0.36% increase, in doctor visits.

Using instead the finite-difference method, a one-unit change in x_j implies $\mathbf{x}'\boldsymbol{\beta}$ changes to $\mathbf{x}'\boldsymbol{\beta} + \beta_j$, so $\text{ME}_j = \exp(\mathbf{x}'\boldsymbol{\beta} + \beta_j) - \exp(\mathbf{x}'\boldsymbol{\beta}) = (e^{\beta_j} - 1)\exp(\mathbf{x}'\boldsymbol{\beta})$. This is a proportionate increase of $(e^{\beta_j} - 1)$, or a percentage change of $100 \times (e^{\beta_j} - 1)$.

10.6.5   The margins command for marginal effects

The `dydx()` option of the postestimation `margins` command, introduced in Stata 11, computes the marginal effects. The variant `dydx(*)` computes the marginal effect for all regressors, or a subset of regressors can be explicitly listed in the parentheses.

The default is to compute the AME. Alternatively, the MEM can be computed using the `atmean` option, and the MER can be computed using the `at()` option. The default supersedes the user-written `margeff` command, and the `atmean` and `at()` options supersede the Stata `mfx` command.

All these MEs are, by default, calculated using calculus methods. To instead use the finite-difference method for binary regressors, it is necessary to use the `i.` operator in the preceding estimation command to signal that the regressor is a binary variable. This additional step is strongly preferred because the change of interest is from one value of the binary variable to the other, whereas calculus methods consider an infinitesimally small change whose effect is then scaled up to correspond to a large change. The two estimates differ in a nonlinear model.

MEs are computed for the default prediction for the `predict` command. For many nonlinear models, such as the Poisson count model, the default MEs are therefore computed for $E(y|\mathbf{x})$. For multinomial models, such as multinomial logit, the default MEs are instead computed for $\Pr(y = j|\mathbf{x})$, $j = 1, \ldots, m$, i.e., the probability for each of the m outcomes. For models with censoring or selection, such as the `tobit` or `heckman` commands, the default MEs are computed for the index $\mathbf{x}'\beta$ so that the default ME is simply the estimated regression coefficient. Marginal effects for some other quantities can be computed using the `predict(`*predict_option*`)` option of the `margins` command, where the specific *predict_option* varies with the preceding estimation command.

For a few estimation commands, such as `asmprobit`, the `margins` command is not available. In those cases, one can usually use the postestimation `estat mfx` command.

10.6.6 MEM: Marginal effect at mean

The `margins` command with the `dydx()` and `atmean` options yields the ME evaluated at the mean for the default option of the `predict` command. After `poisson`, the default prediction is one for $E(y|\mathbf{x})$, so the MEs give the change in the expected number of doctor visits when the regressor changes, evaluated at $\mathbf{x} = \bar{\mathbf{x}}$.

The default is to use calculus methods to compute the ME. For the three binary regressors, we instead use the finite-difference method, by using the `i.` operator in the `poisson` estimation command. We have

```
. * Marginal effects at mean (MEM) using margins command and finite differences
. quietly poisson docvis i.private i.chronic i.female income, vce(robust)

. margins, dydx(*) atmean
Conditional marginal effects                      Number of obs   =       4412
Model VCE    : Robust

Expression   : Predicted number of events, predict()
dy/dx w.r.t. : 1.private 1.chronic 1.female income
at           : 0.private      =     .2146419  (mean)
               1.private      =     .7853581  (mean)
               0.chronic      =     .6736174  (mean)
               1.chronic      =     .3263826  (mean)
               0.female       =     .5281052  (mean)
               1.female       =     .4718948  (mean)
               income         =     34.34018  (mean)
```

	dy/dx	Delta-method Std. Err.	z	P>|z|	[95% Conf. Interval]	
1.private	1.978178	.204407	9.68	0.000	1.577547	2.378808
1.chronic	4.200068	.2794096	15.03	0.000	3.652435	4.7477
1.female	1.528406	.1775762	8.61	0.000	1.180363	1.876449
income	.0107766	.0033149	3.25	0.001	.0042796	.0172737

```
Note: dy/dx for factor levels is the discrete change from the base level.
```

The output indicates that evaluation is at regressor values of, respectively, 0.785, 0.326, 0.472, and 34.340. In subsequent examples, we use the `noatlegend` option to suppress this part of the output.

Evaluated at the sample mean of the regressors, the number of doctor visits increases by 1.98 for those with private insurance and by 0.01 with a $1,000 increase in annual income. In this example, the marginal effects are about three times the estimated Poisson coefficients.

Comparison of calculus and finite-difference methods

To show that calculus and finite-difference methods can differ considerably, we repeat the command following the `poisson` command without using the `i.` operator to signal which regressors are binary variables. Then calculus methods are used for all variables. We obtain

```
. * Marginal effects at mean (MEM) using margins command and only calculus method
. quietly poisson docvis private chronic female income, vce(robust)

. margins, dydx(*) atmean
Conditional marginal effects                      Number of obs    =      4412
Model VCE      : Robust

Expression    : Predicted number of events, predict()
dy/dx w.r.t.  : private chronic female income
at            : private     =     .7853581 (mean)
                chronic     =     .3263826 (mean)
                female      =     .4718948 (mean)
                income      =     34.34018 (mean)
```

	dy/dx	Delta-method Std. Err.	z	P>|z|	[95% Conf. Interval]	
private	2.4197	.3057397	7.91	0.000	1.820462	3.018939
chronic	3.308002	.1794551	18.43	0.000	2.956277	3.659728
female	1.492263	.1675949	8.90	0.000	1.163783	1.820743
income	.0107766	.0033149	3.25	0.001	.0042796	.0172737

The MEs for the binary regressors change, respectively, from 1.98 to 2.42, 4.20 to 3.31, and 1.53 to 1.49. For binary regressors, the first method, using finite differences, is conceptually better.

10.6.7   MER: Marginal effect at representative value

The `margins` command with the `dydx()` and `at()` options yields the ME evaluated at a particular value of the regressors.

As an example, we consider computing the MEs for a privately insured woman with no chronic conditions and income of $10,000. We use the `i.` operator in estimation so that the MEs for binary regressors are computed using the finite-difference method. We have

```
. * Marginal effects at representative value (MER) using margins command
. quietly poisson docvis i.private i.chronic i.female income, vce(robust)

. margins, dydx(*) at(private=1 chronic=0 female=1 income=10) noatlegend
Conditional marginal effects               Number of obs   =       4412
Model VCE    : Robust

Expression   : Predicted number of events, predict()
dy/dx w.r.t. : 1.private 1.chronic 1.female income
```

	dy/dx	Delta-method Std. Err.	z	P>|z|	[95% Conf. Interval]	
1.private	1.647648	.2072813	7.95	0.000	1.241385	2.053912
1.chronic	5.930251	.4017655	14.76	0.000	5.142805	6.717697
1.female	1.164985	.1546063	7.54	0.000	.8619621	1.468008
income	.0106545	.0028688	3.71	0.000	.0050317	.0162772

```
Note: dy/dx for factor levels is the discrete change from the base level.
```

For example, having private insurance is estimated to increase the number of doctor visits by 1.65, with 95% confidence interval [1.24, 2.05]. Note that the confidence interval is for the change in the expected number of visits $E(y|\mathbf{x})$. A confidence interval for the change in the actual number of visits $(y|\mathbf{x})$ will be much wider; see the analogous discussion for prediction in section 10.5.2.

Related at() options allow evaluation at a mix of user-provided values and sample means of regressors.

10.6.8   AME: Average marginal effect

The margins command with the dydx() option yields the AMEs. It supersedes the user-written margeff command (Bartus 2005) that provides AMEs for a few selected models.

For the doctor visits data, with the i. operator used in estimation so that the AMEs for binary regressors are computed using the finite-difference method, we obtain the following:

(Continued on next page)

```
. * Average marginal effects (AVE) using margins command and finite differences
. quietly poisson docvis i.private i.chronic i.female income, vce(robust)

. margins, dydx(*)
```

Average marginal effects   Number of obs   =   4412
Model VCE   : Robust

Expression   : Predicted number of events, predict()
dy/dx w.r.t. : 1.private 1.chronic 1.female income

	dy/dx	Delta-method Std. Err.	z	P>|z|	[95% Conf. Interval]	
1.private	2.404721	.2438573	9.86	0.000	1.926769	2.882672
1.chronic	4.599174	.2886176	15.94	0.000	4.033494	5.164854
1.female	1.900212	.2156694	8.81	0.000	1.477508	2.322917
income	.0140765	.0043457	3.24	0.001	.0055591	.0225939

Note: dy/dx for factor levels is the discrete change from the base level.

For example, people with private insurance have on average 2.40 more doctor visits than those without private insurance after controlling for income, gender, and chronic conditions. This value is higher than that of 1.98 obtained by using the atmean option to compute the MEM; see page 347. Similarly, the other MEs increase from 4.20 to 4.60, from 1.53 to 1.90, and from 0.011 to 0.014.

The AMEs for single-index models are in practice quite similar to the coefficients obtained by OLS regression of y on $\mathbf{x}$. This is the case here because OLS regression, not given here, yields coefficients of 1.92, 4.82, 1.89, and 0.016.

For nonlinear models, average behavior of individuals differs from behavior of the average individual. The direction of the difference is generally indeterminate, but for models with an exponential conditional mean, it can be shown that the atmean option (giving the MEM) will produce smaller MEs than the default (giving the AME), a consequence of the exponential function being globally convex.

The three MEs—MEM, MER, and AME—can differ appreciably in nonlinear models. Which ME should be used? It is common to use MEM, using the atmean option of the margins command. However, for policy analysis, one should use either the MER for targeted values of the regressors, by using the at() option of the margins command, or the AME, the default option.

Microeconometric studies often use survey data that are stratified on exogenous regressors, in which case the estimators are consistent but the regressors $\mathbf{x}$ may be unrepresentative of the population. For example, a nonlinear regression of earnings (y) on schooling (x) may use a dataset for which individuals with low levels of schooling are oversampled, so that the sample mean $\bar{x}$ is less than the population mean. Then MER can be used with no modification. But MEM and AME should be evaluated by using sampling weights, introduced in section 3.7. In particular, the AME should be computed as a sample-weighted average of the ME for each individual. The calculations in margins automatically adjust for any weights used during estimation. This can be changed by using the noweights option.

10.6.9   Elasticities and semielasticities

The elasticities and semielasticities defined in section 10.6.3 can be computed by using
`margins`'s options `eyex()`, `eydx()`, and `dyex()`. Evaluation can be at each sample value
and then averaged (the default option), or at the mean value of regressors (the `atmean`
option), or at a representative value of the regressors (the `at()` option).

We continue with the same Poisson regression example, with four regressors, but we
focus on the impact of just the regressor `income`. For illustration, we evaluate elasticities
and semielasticities at the mean value of regressors by using the `atmean` option.

We first obtain the ME with the `dydx` option.

```
. * Usual ME evaluated at mean of regressors
. quietly poisson docvis private chronic female income, vce(robust)

. margins, dydx(income) atmean noatlegend
Conditional marginal effects                       Number of obs   =       4412
Model VCE    : Robust

Expression   : Predicted number of events, predict()
dy/dx w.r.t. : income
```

		Delta-method				
	dy/dx	Std. Err.	z	P>|z|	[95% Conf. Interval]	
income	.0107766	.0033149	3.25	0.001	.0042796	.0172737

The number of doctor visits increases by 0.0108 with a $1,000 increase in annual income.
This repeats the result given in section 10.6.6.

We next compute the elasticity. The `eyex` option yields

```
. * Elasticity evaluated at mean of regressors
. margins, eyex(income) atmean noatlegend
Conditional marginal effects                       Number of obs   =       4412
Model VCE    : Robust

Expression   : Predicted number of events, predict()
ey/ex w.r.t. : income
```

		Delta-method				
	ey/ex	Std. Err.	z	P>|z|	[95% Conf. Interval]	
income	.1221485	.0371729	3.29	0.001	.049291	.195006

The elasticity is 0.122, so a 1% increase in income is associated with a 0.122% increase
in doctor visits, or a 10% increase in income is associated with a 1.22% increase in
doctor visits. The elasticity equals ME $\times$ x/y from (10.6), where evaluation here is
at $x = 34.34$ (the sample mean of `income`) and $y = 3.03$ (the predicted number of
doctor visits for $\mathbf{x} = \overline{\mathbf{x}}$ computed using the `margins, atmean` command). This yields
$0.0108 \times 34.34/3.03 = 0.122$ as given in the above output.

The semielasticity is obtained with the `eydx()` option:

```
. * Semielasticity evaluated at mean of regressors
. margins, eydx(income) atmean noatlegend
Conditional marginal effects                    Number of obs    =       4412
Model VCE     : Robust
Expression    : Predicted number of events, predict()
ey/dx w.r.t.  : income
```

		Delta-method				
	ey/dx	Std. Err.	z	P>|z|	[95% Conf. Interval]	
income	.003557	.0010825	3.29	0.001	.0014354	.0056787

A \$1,000 increase in annual income (a one-unit change in `income`) is associated with a 0.003557 proportionate rise, or a 0.3557% increase in the number of doctor visits. This exactly equals the coefficient of `income` in the original Poisson regression (see section 10.3.2), confirming that if the conditional mean is of exponential form, then the coefficient $\widehat{\beta}_j$ is already a semielasticity, as explained in section 10.6.4.

Finally, the `dyex()` option yields

```
. * Other semielasticity evaluated at mean of regressors
. margins, dyex(income) atmean noatlegend
Conditional marginal effects                    Number of obs    =       4412
Model VCE     : Robust
Expression    : Predicted number of events, predict()
dy/ex w.r.t.  : income
```

		Delta-method				
	dy/ex	Std. Err.	z	P>|z|	[95% Conf. Interval]	
income	.3700708	.1138338	3.25	0.001	.1469607	.5931809

A proportionate increase of one in income (a doubling of income) is associated with 0.37 more doctor visits. Equivalently, a 1% increase in income is associated with 0.0037 more doctor visits.

10.6.10   AME computed manually

The AME can always be computed manually using the following method. Predict at the current sample values for all observations, change one regressor by a small amount, predict at the new values, subtract the two predictions, and divide by the amount of the change. The AME is the average of this quantity. By choosing a very small change, we replicate the calculus method. A finite-difference estimate is obtained by considering a large change such as a one-unit change (whether this is large depends on the scaling of the regressors). In either case, we use the `preserve` and `restore` commands to return to the original data after computing the AME.

We consider the effect of a small change in income on doctor visits. This yields a crude approximation to the derivative. For more precise numerical estimates of the derivative, see section 5.7 of Press et al. (1992). We have

```
. * AME computed manually for a single regressor
. use mus10data.dta, clear
. keep if year02 == 1
(25712 observations deleted)
. quietly poisson docvis private chronic female income, vce(robust)
. preserve
. predict mu0, n
. quietly replace income = income + 0.01
. predict mu1, n
. generate memanual = (mu1-mu0)/0.01
. summarize memanual
```

Variable	Obs	Mean	Std. Dev.	Min	Max
memanual	4412	.0140761	.0106173	.0028253	.055027

```
. restore
```

The AME estimate is 0.0140761, essentially the same as the 0.0140765 obtained by using `margins` in section 10.6.8. This method gives no standard error for the AME estimate. Instead, it computes the standard deviation of the AME for the 4,412 observations.

A better procedure chooses a change that varies with the scaling of each regressor. We use a change equal to the standard deviation of the regressor divided by 1,000. We also use the looping command `foreach` to obtain the AME for each variable. We have the following:

```
. * AME computed manually for all regressors
. global xlist private chronic female income
. preserve
. predict mu0, n
. foreach var of varlist $xlist {
  2.    quietly summarize `var'
  3.    generate delta = r(sd)/1000
  4.    quietly generate orig = `var'
  5.    quietly replace `var' = `var' + delta
  6.    predict mu1, n
  7.    quietly generate me_`var' = (mu1 - mu0)/delta
  8.    quietly replace `var' = orig
  9.    drop mu1 delta orig
 10.    }
. summarize me_*
```

Variable	Obs	Mean	Std. Dev.	Min	Max
me_private	4412	3.16153	2.384785	.6349193	12.36743
me_chronic	4412	4.322181	3.260333	.8679963	16.90794
me_female	4412	1.949399	1.470413	.3915812	7.625329
me_income	4412	.0140772	.0106184	.0028284	.055073

```
. restore
```

The AME estimate for `income` is the average 0.0140772; essentially, the same as the 0.0140765 produced by using the `margins, dydx()` command. The other AMEs differ from those given in section 10.6.8 because here we used calculus methods, whereas in section 10.6.8, the AMEs for binary regressors were computed using the finite-difference method.

The code can clearly be adapted to use finite differences in those cases. In nonstandard models, such as those fit by using the `ml` command, it will be necessary to also provide code to replace the `predict` command.

10.6.11 Polynomial regressors

Regressors may appear as polynomials. Then computing MEs becomes considerably more complicated.

First, consider a linear model that includes a cubic function in regressor z. Then

$$
\begin{aligned}
E(y|\mathbf{x}, z) &= \mathbf{x}'\boldsymbol{\beta} + \alpha_1 z + \alpha_2 z^2 + \alpha_3 z^3 \\
\Rightarrow \qquad \mathrm{ME}_z &= \alpha_1 + 2\alpha_2 z + 3\alpha_3 z^2
\end{aligned}
$$

The AME can be computed by calculating $\widehat{\alpha}_1 + 2\widehat{\alpha}_1 z + 3\widehat{\alpha}_1 z^2$ for each observation and averaging.

We do so for a slightly more difficult example, the Poisson model. Then

$$
\begin{aligned}
E(y|\mathbf{x}, z) &= \exp(\mathbf{x}'\boldsymbol{\beta} + \alpha_1 z + \alpha_2 z^2 + \alpha_3 z^3) \\
\Rightarrow \qquad \mathrm{ME}_z &= \exp(\mathbf{x}'\boldsymbol{\beta} + \alpha_1 z + \alpha_2 z^2 + \alpha_3 z^3) \times (\alpha_1 + 2\alpha_2 z + 3\alpha_3 z^2)
\end{aligned}
$$

A cubic function in income is part of our model for doctor visits. Below we estimate the parameters of the model by ML.

```
. * AME for a polynomial regressor: manual computation
. generate inc2 = income^2
. generate inc3 = income^3
. quietly poisson docvis private chronic female income inc2 inc3, vce(robust)
. predict muhat, n
. generate me_income = muhat*(_b[income]+2*_b[inc2]*income+3*_b[inc3]*inc2)
. summarize me_income
```

Variable	Obs	Mean	Std. Dev.	Min	Max
me_income	4412	.0178233	.0137618	-.0534614	.0483436

The code uses the simplification that $\mathrm{ME}_z = E(y|\mathbf{x}, z) \times (\alpha_1 + 2\alpha_2 z + 3\alpha_3 z^2)$. The AME of a change in income is 0.0178 in the cubic model, compared with 0.0141 when income enters only linearly.

This AME can be more simply computed using factor variables. From section 1.3.4, a polynomial in variable `income` can be specified using the `c.` operator (for a continuous variable) and the # operator for interaction. The marginal effect of `income` can then be computed using the usual `margins` command. We have

```
. * AME for a polynomial regressor: computation using factor variables
. quietly poisson docvis private chronic female c.income c.income#c.income
> c.income#c.income#c.income, vce(robust)

. margins, dydx(income)
Average marginal effects                      Number of obs   =      4412
Model VCE    : Robust

Expression   : Predicted number of events, predict()
dy/dx w.r.t. : income
```

	dy/dx	Delta-method Std. Err.	z	P>|z|	[95% Conf. Interval]
income	.0178233	.0062332	2.86	0.004	.0056064 .0300402

The AME equals that computed manually, but additionally, we have a 95% confidence interval of $[0.0056, 0.0300]$. The MEM and MER can also be computed with the `atmean` and `at()` options.

10.6.12 Interacted regressors

Similar issues arise with regressors interacted with indicator variables. For example,

$$
\begin{aligned}
E(y|\mathbf{x}, z, d) &= \exp(\mathbf{x}'\boldsymbol{\beta} + \alpha_1 z + \alpha_2 d + \alpha_3 d \times z) \\
\Rightarrow \qquad \mathrm{ME}_z &= \exp(\mathbf{x}'\boldsymbol{\beta} + \alpha_1 z + \alpha_2 d + \alpha_3 d \times z) \times (\alpha_1 + \alpha_3 d)
\end{aligned}
$$

Long and Freese (2006) give an extensive discussion of MEs with interactive regressors, performed by using the user-written `prvalue` command. These are oriented toward calculation of the MEM or MER rather than the AME.

Factor variables, introduced in Stata 11, automate the computation of marginal effects in models with interactions. As an example, we modify the doctor visits model to include an interaction between the binary regressor `female` and the continuous regressor `income`.

We first need to specify the model fit with factor variables, using the relevant `c.`, `i.`, and # operators. We have

(Continued on next page)

```
. * Specify model with interacted regressors using factor variables
. poisson docvis private chronic i.female c.income i.female#c.income,
> vce(robust) nolog
```

Poisson regression   Number of obs   =   4412
                                                        Wald chi2(5)   =   606.43
                                                        Prob > chi2   =   0.0000
Log pseudolikelihood = -18475.536   Pseudo R2   =   0.1943

docvis	Coef.	Robust Std. Err.	z	P>|z|	[95% Conf. Interval]	
private	.802035	.1084187	7.40	0.000	.5895383	1.014532
chronic	1.094331	.0563504	19.42	0.000	.9838865	1.204776
1.female	.6328542	.0927712	6.82	0.000	.451026	.8146823
income	.0051734	.0015708	3.29	0.001	.0020946	.0082522
female# c.income 1	-.0035176	.0019089	-1.84	0.065	-.0072589	.0002237
_cons	-.3082426	.1242329	-2.48	0.013	-.5517346	-.0647506

The coefficient of female is labeled 1.female, and the coefficient of female × income is labeled female#c.income1.

We then compute the AME of variables female and income. We have

```
. * AME with interacted regressors given model specified using factor variables
. margins, dydx(female income)
```

Average marginal effects   Number of obs   =   4412
Model VCE   : Robust

Expression   : Predicted number of events, predict()
dy/dx w.r.t.   : 1.female income

	dy/dx	Delta-method Std. Err.	z	P>|z|	[95% Conf. Interval]	
1.female	1.884021	.2159008	8.73	0.000	1.460864	2.307179
income	.0116832	.0037989	3.08	0.002	.0042375	.019129

Note: dy/dx for factor levels is the discrete change from the base level.

The AME of a change in income is 0.0117 compared with 0.00178 in the cubic model and 0.0141 when income enters only linearly. The AME of moving from male (female = 0) to female (female = 1) is 1.884 office visits, compared with 1.900 from section 10.6.8 where there was no interaction. The output includes 95% confidence intervals for the AMEs. The MEMs and MERs can alternatively be computed, using the atmean and at() options of the margins command.

10.6.13   Complex interactions and nonlinearities

MEs in models with interactions can become very difficult to interpret and calculate.

For complex interactions, a simple procedure is to compute the ME by manually changing the relevant variables and interactions, recomputing the predicted conditional mean, and subtracting. We change by an amount Δ, a single variable that, because of interactions and/or polynomials, appears several times as a regressor. Let the original values of $\mathbf{x}_i$ be denoted by $\mathbf{x}_{i0}$, and obtain the prediction $\widehat{\mu}_{i0} = \exp(\mathbf{x}'_{i0}\widehat{\boldsymbol{\beta}})$. Then change the variable by Δ to give new values of $\mathbf{x}_i$ denoted by $\mathbf{x}_{i1}$, and obtain the prediction $\widehat{\mu}_{i1} = \exp(\mathbf{x}'_{i1}\widehat{\boldsymbol{\beta}})$. Then the ME of changing the variable is $(\widehat{\mu}_{i1} - \widehat{\mu}_{i0})/\Delta$.

We illustrate this for the cubic in income example.

```
. * AME computed manually for a complex model
. preserve
. predict mu0, n
. quietly summarize income
. generate delta = r(sd)/100
. quietly replace income = income + delta
. quietly replace inc2 = income^2
. quietly replace inc3 = income^3
. predict mu1, n
. generate me_inc = (mu1 - mu0)/delta
. summarize me_inc
```

Variable	Obs	Mean	Std. Dev.	Min	Max
me_inc	4412	.0116899	.0089482	.0022914	.1083069

```
. restore
```

This reproduces the calculus result because it considers a small change in income, here one-hundredth of the standard deviation of income. If instead we had used `delta=1`, then this program would have given the ME of a one-unit change in the regressor `income` (here a \$1,000 change) computed by using the finite-difference method.

For complex interactions, one can use factor variables, as shown in section 10.6.12; although for complex interactions, care is needed to correctly specify the factor variables and their interactions in the initial model estimation command.

10.7   Model diagnostics

As for the linear model, the modeling process follows a cycle of estimation, diagnostic checks, and model respecification. Here we briefly summarize diagnostics checks with model-specific checks deferred to the models chapters.

10.7.1   Goodness-of-fit measures

The R^2 in the linear model does not extend easily to nonlinear models. When fitting by NLS a nonlinear model with additive errors, $y = m(\mathbf{x}'\boldsymbol{\beta}) + u$, the residual sum of squares (RSS) plus the model sum of squares (MSS) do not sum to the total sum of squares (TSS). So the three measures MSS/TSS, $1 - $ RSS/TSS, and $\widehat{\rho}^2_{y,\widehat{y}}$ [the squared correlation between y

and $m(\mathbf{x}'\widehat{\boldsymbol{\beta}})]$ differ. By contrast, they all coincide for OLS estimation of the linear model with an intercept. Furthermore, many nonlinear models are based on the distribution of y and do not have a natural interpretation as a model with an additive error.

A fairly universal measure in nonlinear models is $\widehat{\rho}^2_{y,\widehat{y}}$, the squared correlation between y and $\widehat{y}$. This has a tangible interpretation and can be used provided that the model yields a fitted value $\widehat{y}$. This is the case for most commonly used models except multinomial models.

For ML estimators, Stata reports a pseudo-R^2 defined as

$$\widetilde{R}^2 = 1 - \ln L_{\text{fit}} / \ln L_0 \tag{10.10}$$

where $\ln L_0$ is the log likelihood of an intercept-only model, and $\ln L_{\text{fit}}$ is the likelihood of the fitted model.

For doctor visits, we have

```
. * Compute pseudo-R-squared after Poisson regression
. quietly poisson docvis private chronic female income, vce(robust)
. display "Pseudo-R^2 = " 1 - e(ll)/e(ll_0)
Pseudo-R^2 = .19303857
```

This equals the statistic `Pseudo R2` that is provided as part of `poisson` output; see section 10.3.2.

For discrete dependent variables, $\widetilde{R}^2$ has the desirable properties that $\widetilde{R}^2 \geq 0$, provided that an intercept is included in the model, and $\widetilde{R}^2$ increases as regressors are added for models fit by ML. For binary and multinomial models, the upper bound for $\widetilde{R}^2$ is 1, whereas for other discrete data models such as Poisson the upper bound for $\widetilde{R}^2$ is less than 1. For continuous data, these desirable properties disappear, and it is possible that $\widetilde{R}^2 > 1$ or $\widetilde{R}^2 < 0$, and $\widetilde{R}^2$ does not increase as regressors are added.

To understand the properties of $\widetilde{R}^2$, let $\ln L_{\max}$ denote the largest possible value of $\ln L(\boldsymbol{\theta})$. Then we can compare the actual gain in the objective function due to inclusion of regressors compared with the maximum possible gain, giving the relative gain measure

$$R^2_{\text{RG}} = \frac{\ln L_{\text{fit}} - \ln L_0}{\ln L_{\max} - \ln L_0} = 1 - \frac{\ln L_{\max} - \ln L_{\text{fit}}}{\ln L_{\max} - \ln L_0}$$

In general, $\ln L_{\max}$ is not known, however, making it difficult to implement this measure (see Cameron and Windmeijer [1997]). For binary and multinomial models, it can be shown that $\ln L_{\max} = 0$, because perfect ability to model the multinomial outcome gives a probability mass function with a value of 1 and a natural logarithm of 0. Then R^2_{RG} simplifies to $\widetilde{R}^2$, given in (10.10). For other discrete models, such as Poisson, $\ln L_{\max} < 0$ because the probability mass function takes a maximum value of less than 1, so the maximum $\widetilde{R}^2 < 1$. For continuous density, the log density can exceed zero, so it is possible that $\ln L_{\text{fit}} > \ln L_0 > 0$ and $\widetilde{R}^2 < 0$. An example is given as an end-of-chapter exercise.

10.7.2   Information criteria for model comparison

For ML models that are nested in each other, we can discriminate between models on the basis of a likelihood-ratio (LR) test of the restrictions that reduce one model to the other; see section 12.4.

For ML models that are nonnested, a standard procedure is to use information criteria. Two standard measures are Akaike's information criterion (AIC) and Schwarz's Bayesian information criterion (BIC). Different references use different scalings of these measures. Stata uses

$$\text{AIC} = -2\ln L + 2k$$
$$\text{BIC} = -2\ln L + k\ln N$$

Smaller AIC and BIC are preferred, because higher log likelihood is preferred. The quantities $2k$ and $k\ln N$ are penalties for model size.

If the models are actually nested, then a LR test statistic [equals $\Delta(2\ln L)$] could be used. Then the larger model is favored at a level of 0.05 if $\Delta(2\ln L)$ increases by $\chi^2_{0.05}(\Delta k)$. By comparison, the AIC favors the larger model if $\Delta(2\ln L)$ increases by $2\Delta k$, which is a smaller amount [e.g., if $\Delta k = 1$ then $2 < \chi^2_{0.05}(1) = 3.84$]. The AIC penalty is too small. The BIC gives a larger model-size penalty and is generally better, especially if smaller models are desired.

These quantities are stored in e() and are easily displayed by using the postestimation estat ic command. For the Poisson regression with the five regressors, including the intercept, we have

```
. * Report information criteria
. estat ic
```

Model	Obs	ll(null)	ll(model)	df	AIC	BIC
.	4412	-22929.9	-18503.55	5	37017.1	37049.06

Note: N=Obs used in calculating BIC; see [R] BIC note

The information criteria and LR test require correct specification of the density, so they should instead be used after the nbreg command for negative binomial estimation because the Poisson density is inappropriate for these data.

It is possible to test one nonnested likelihood-based model against another, using the LR test of Vuong (1989). This test is available as the vuong option for the count model commands zip and zinb. A general discussion of Vuong's test is given in Greene (2003, 751) and Cameron and Trivedi (2005, 280–283).

10.7.3   Residuals

Analysis of residuals can be a useful diagnostic tool, as demonstrated in sections 3.5 and 5.3 for the linear model.

In nonlinear models, several different residuals can be computed. The use of residuals and methods for computing residuals vary with the model and estimator. A natural starting point is the raw residual, $y_i - \widehat{y}_i$. In nonlinear models, this is likely to be heteroskedastic, and a common residual to use is the Pearson residual $(y_i - \widehat{y}_i)/\widehat{\sigma}_i$, where $\widehat{\sigma}_i^2$ is an estimate of $\text{Var}(y_i|\mathbf{x}_i)$.

Residual analysis is well-developed for GLMs, where additional residuals include Anscombe and deviance residuals. These residuals are not presented in econometrics texts but are presented in texts on GLMs. For the Poisson model fit with the glm command, various options of the postestimation predict command yield various residuals and the mu option gives the predicted conditional mean. We have

```
. * Various residuals after command glm
. quietly glm docvis private chronic female income, family(poisson)
. predict mu, mu
. generate uraw = docvis - mu
. predict upearson, pearson
. predict udeviance, deviance
. predict uanscombe, anscombe
. summarize uraw upearson udeviance uanscombe
```

Variable	Obs	Mean	Std. Dev.	Min	Max
uraw	4412	-3.34e-08	7.408631	-13.31806	125.3078
upearson	4412	-.0102445	3.598716	-3.519644	91.16232
udeviance	4412	-.5619514	2.462038	-4.742847	24.78259
uanscombe	4412	-.5995917	2.545318	-5.03055	28.39791

The Pearson residual has a standard deviation much greater than the expected value of 1 because it uses $\widehat{\sigma}_i^2 = \widehat{\mu}_i$, when in fact there is overdispersion and σ_i^2 is several times this. The deviance and Anscombe residuals are quite similar. The various residuals differ mainly in their scaling. For these data, the pairwise correlations between the residuals exceed 0.92.

Other options of predict after glm allow for adjustment of deviance residuals and the standardizing and studentizing of the various residuals. The cooksd and hat options aid in finding outlying and influential observations, as for the linear model. For definitions of all these quantities, see [R] **glm postestimation** or a reference book on GLMs.

Another class of models where residuals are often used as a diagnostic are survival data models. After the streg or stcox commands, the predict options csnell and mgale produce Cox–Snell and martingale-like residuals. Schoenfeld residuals are obtained by specifying the predict option schoenfeld after the stcox or stcrreg commands. For definitions, see [ST] **streg postestimation**, [ST] **stcox postestimation**, and [ST] **stcrreg postestimation**.

10.7.4 Model-specification tests

Most estimation commands include a test of overall significance in the header output above the table of estimated coefficients. This is a test of joint significance of all the regressors. Some estimation commands provide further tests in the output. For example, the `xtmixed` command includes a LR test against the linear regression model; see sections 9.5 and 9.6.

More tests may be requested as postestimation commands. Some commands such as `linktest`, to test model specification, are available after most commands. More model-specific commands begin with `estat`. For example, the `poisson` postestimation command `estat gof` provides a goodness-of-fit test.

Discussion of model-specification tests is given in chapter 12 and in the model-specific chapters.

10.8 Stata resources

A complete listing of estimation commands can be obtained by typing `help estimation commands`. For `poisson`, for example, see the entries [R] **poisson** and [R] **poisson postestimation**, and the corresponding online help. Useful Stata commands include `predict`, `margins`, `lincom`, and `nlincom`.

Graduate econometrics texts give considerable detail on estimation and less on prediction and computation of MEs.

10.9 Exercises

1. Fit the Poisson regression model of section 10.3 by using the `poisson`, `nl`, and `glm` commands. In each case, report default standard errors and robust standard errors. Use the `estimates store` and `estimates table` commands to produce a table with the six sets of output and discuss.

2. In this exercise, we use the medical expenditure data of section 3.4 with the dependent variable $y = $ `totexp/1000` and regressors the same as those in section 3.4. We suppose that $E(y|\mathbf{x}) = \exp(\mathbf{x'\boldsymbol{\beta}})$, which ensures that $E(y|\mathbf{x}) > 0$. The obvious estimator is NLS, but the Poisson MLE is also consistent if $E(y|\mathbf{x}) = \exp(\mathbf{x'\boldsymbol{\beta}})$ and does not require that y be integer values. Repeat the analysis of question 1 with these data.

3. Use the same medical expenditure data as in exercise 2. Compare the different standard errors obtained with `poisson`'s `vce()` option with the *vcetypes* oim, opg, robust, cluster *clustvar*, and bootstrap. For clustered standard errors, cluster on `age`. Comment on your results.

4. Consider Poisson regression of `docvis` on an intercept, `private`, `chronic`, and `income`. This is the model of section 10.5 except `female` is dropped. Find the following: the sample average prediction of `docvis`; the average prediction if we use the Poisson estimates to predict `docvis` for males only; the prediction for

someone who is privately insured, has a chronic condition, and has an income of $20,000 (so `income=20`); and the sample average of the residuals.

5. Continue with the same data and regression model as in exercise 4. Provide a direct interpretation of the estimated Poisson coefficients. Find the ME of changing regressors on the conditional mean in the following ways: the MEM using calculus methods for continuous regressors and finite-difference methods for discrete regressors; the MEM using calculus methods for all regressors; the AME using calculus methods for continuous regressors and finite-difference methods for discrete regressors; the AME using calculus methods for all regressors; and the MER for someone who is privately insured, has a chronic condition, and has an income of $20,000 (so `income=20`).

6. Consider the following simulated data example. Generate 100 observations of $y \sim N(0, 0.001^2)$, first setting the seed to 10101. Let $x = \sqrt{y}$. Regress y on an intercept and x. Calculate the pseudo-R^2 defined in (10.10). Is this necessarily a good measure when data are continuous?

7. Consider the negative binomial regression of `docvis` on an intercept, `private`, `chronic`, `female`, and `income`, using the `nbreg` command (replace `poisson` by `nbreg`). Compare this model with one with `income` excluded. Which model do you prefer on the grounds of 1) AIC, 2) BIC, and 3) a LR test using the `lrtest` command?

11 Nonlinear optimization methods

11.1 Introduction

The previous chapter considered estimation when a built-in Stata command, `poisson`, existed. Here we present methods for nonlinear models where there is no such command and it is instead necessary to provide estimation code. Estimation is more difficult than in the linear case because there is no explicit formula for the estimator. Instead, the estimator is the numerical solution to an optimization problem.

In this chapter, we review optimization methods. The discussion can be relevant even when a built-in command is used. We present and illustrate the `ml` command, which enables maximum likelihood (ML) estimation if, at a minimum, the log-density formula is provided. The command is more generally applicable to other m estimators, such as the nonlinear least-squares (NLS) estimator.

We also present the Mata `optimize()` function for optimization when the objective function is defined using matrix programming language commands.

11.2 Newton–Raphson method

Estimators that maximize an objective function, such as the log likelihood, are obtained by calculating a sequence of estimates $\widehat{\boldsymbol{\theta}}_1$, $\widehat{\boldsymbol{\theta}}_2$, ... that move toward the top of the hill. Gradient methods do so by moving by an amount that is a suitable multiple of the gradient at the current estimate. A standard method is the Newton–Raphson (NR) method, which works especially well when the objective function is globally concave.

11.2.1 NR method

We consider the estimator $\widehat{\boldsymbol{\theta}}$ that is a local maximum to the objective function $Q(\boldsymbol{\theta})$, so $\widehat{\boldsymbol{\theta}}$ solves

$$\mathbf{g}(\widehat{\boldsymbol{\theta}}) = \mathbf{0}$$

where $\mathbf{g}(\boldsymbol{\theta}) = \partial Q(\boldsymbol{\theta})/\partial \boldsymbol{\theta}$ is the gradient vector. Numerical solution methods are needed when these first-order conditions do not yield an explicit solution for $\widehat{\boldsymbol{\theta}}$.

Let $\widehat{\boldsymbol{\theta}}_s$ denote the sth round estimate of $\boldsymbol{\theta}$. Then a second-order Taylor-series expansion around $\widehat{\boldsymbol{\theta}}_s$ approximates the objective function $Q(\boldsymbol{\theta})$ by

$$Q^*(\boldsymbol{\theta}) = Q(\widehat{\boldsymbol{\theta}}_s) + \mathbf{g}(\widehat{\boldsymbol{\theta}}_s)'(\boldsymbol{\theta} - \widehat{\boldsymbol{\theta}}_s) + \frac{1}{2}(\boldsymbol{\theta} - \widehat{\boldsymbol{\theta}}_s)'\mathbf{H}(\widehat{\boldsymbol{\theta}}_s)(\boldsymbol{\theta} - \widehat{\boldsymbol{\theta}}_s)$$

where $\mathbf{H} = \partial g(\boldsymbol{\theta})/\partial\boldsymbol{\theta}' = \partial^2 Q(\boldsymbol{\theta})/\partial\boldsymbol{\theta}\partial\boldsymbol{\theta}'$ is the Hessian matrix. Maximizing this approximating function with respect to $\boldsymbol{\theta}$ leads to

$$\partial Q^*(\boldsymbol{\theta})/\partial\boldsymbol{\theta} = \mathbf{g}(\widehat{\boldsymbol{\theta}}_s) + \mathbf{H}(\widehat{\boldsymbol{\theta}}_s)(\boldsymbol{\theta} - \widehat{\boldsymbol{\theta}}_s) = \mathbf{0}$$

Solving for $\boldsymbol{\theta}$ yields the NR algorithm

$$\widehat{\boldsymbol{\theta}}_{s+1} = \widehat{\boldsymbol{\theta}}_s - \mathbf{H}(\widehat{\boldsymbol{\theta}}_s)^{-1}\mathbf{g}(\widehat{\boldsymbol{\theta}}_s) \tag{11.1}$$

The parameter estimate is changed by a matrix multiple of the gradient vector, where the multiple is minus the inverse of the Hessian matrix.

The final step in deriving (11.1) presumes that the inverse exists. If instead the Hessian is singular, then $\widehat{\boldsymbol{\theta}}_{s+1}$ is not uniquely defined. The Hessian may be singular for some iterations, and optimization methods have methods for still continuing the iterations. However, the Hessian must be nonsingular at the optimum. This complication does not arise if $Q(\boldsymbol{\theta})$ is globally concave because then the Hessian is negative definite at all points of evaluation. In that case, the NR method works well, with few iterations required to obtain the maximum.

11.2.2 NR method for Poisson

The Poisson model is summarized in section 10.2.2. As noted at the start of section 10.2, the fact that we are modeling a discrete random variable places no restriction on the generality of the example. Exactly the same points could be illustrated using, for example, complete spell duration data modeled using the exponential or even a nonlinear model with normally distributed errors.

For the Poisson model, the objective function, gradient, and Hessian are, respectively,

$$\begin{aligned}
Q(\boldsymbol{\beta}) &= \sum_{i=1}^N \{-\exp(\mathbf{x}_i'\boldsymbol{\beta}) + y_i\mathbf{x}_i'\boldsymbol{\beta} - \ln y_i!\} \\
g(\boldsymbol{\beta}) &= \sum_{i=1}^N \{y_i - \exp(\mathbf{x}_i'\boldsymbol{\beta})\}\mathbf{x}_i \\
H(\boldsymbol{\beta}) &= \sum_{i=1}^N -\exp(\mathbf{x}_i'\boldsymbol{\beta})\mathbf{x}_i\mathbf{x}_i'
\end{aligned} \tag{11.2}$$

Note that $H(\boldsymbol{\beta}) = -\mathbf{X}'\mathbf{D}\mathbf{X}$, where $\mathbf{X}$ is the $N \times K$ regressor matrix, and $\mathbf{D} = \text{Diag}\{\exp(\mathbf{x}_i'\boldsymbol{\beta})\}$ is an $N \times N$ diagonal matrix with positive entries. It follows that if $\mathbf{X}$ is of full rank, then $H(\boldsymbol{\beta})$ is negative definite for all $\boldsymbol{\beta}$, and the objective function is globally concave. Combining (11.1) and (11.2), the NR iterations for the Poisson maximum likelihood estimator (MLE) are

$$\widehat{\boldsymbol{\beta}}_{s+1} = \widehat{\boldsymbol{\beta}}_s + \left\{\sum_{i=1}^N \exp(\mathbf{x}_i'\widehat{\boldsymbol{\beta}}_s)\mathbf{x}_i\mathbf{x}_i'\right\}^{-1} \times \sum_{i=1}^N \{y_i - \exp(\mathbf{x}_i'\widehat{\boldsymbol{\beta}}_s)\}\mathbf{x}_i \tag{11.3}$$

11.2.3   Poisson NR example using Mata

To present an iterative method in more detail, we manually code the NR algorithm for
the Poisson model, using Mata functions that are explained in appendix B. The same
example is used in subsequent sections to demonstrate use of the Stata `ml` command
and the Mata `optimize()` function.

Core Mata code for Poisson NR iterations

For expositional purposes, we begin with the Mata code for the core commands to
implement the NR iterative method for the Poisson.

It is assumed that the regressor matrix X and the dependent variable vector y have
already been constructed, as well as a vector b of starting values. Iterations stop when
$\{(\widehat{\boldsymbol{\beta}}_{s+1} - \widehat{\boldsymbol{\beta}}_s)'(\widehat{\boldsymbol{\beta}}_{s+1} - \widehat{\boldsymbol{\beta}}_s)\}/(\widehat{\boldsymbol{\beta}}_s'\widehat{\boldsymbol{\beta}}_s) < 10^{-16}$.

```
* Core Mata code for Poisson MLE NR iterations
mata
  cha = 1                         // initialize stopping criterion
  do
    mu = exp(X*b)
    grad = X'(y-mu)               // k x 1 gradient vector
    hes = cross(X, mu, X)         // negative of the k x k Hessian matrix
    bold = b
    b = bold + cholinv(hes)*grad
    cha = (bold-b)'(bold-b)/(bold'bold)
    iter = iter + 1
  while (cha > 1e-16)             // end of iteration loops
end
```

The $N \times 1$ vector `mu` has the ith entry $\mu_i = \exp(\mathbf{x}_i'\boldsymbol{\beta})$. The $K \times 1$ vector `grad` equals
$\sum_{i=1}^{N}(y_i - \mu_i)\mathbf{x}_i$, and `hes=cross(X, mu, X)` equals $\sum_i \mu_i \mathbf{x}_i \mathbf{x}_i'$. The quickest function
for taking a matrix inverse is `cholinv()` for a positive-definite matrix. For this reason,
we set `hes` to $-H(\widehat{\boldsymbol{\beta}})$, which is positive definite, but then the NR update has a plus sign
rather than the minus sign in (11.1). In some cases, `hes` may not be symmetric because
of a rounding error. Then we would add the `hes = makesymmetric(hes)` command
before calling `cholinv()`.

Complete Stata and Mata code for Poisson NR iterations

The complete program has the following sequence: 1) in Stata, obtain the data and
define any macros used subsequently; 2) in Mata, calculate the parameter estimates
and the estimated variance–covariance matrix of the estimator (VCE), and pass these
back to Stata; and 3) in Stata, output nicely formatted results.

We begin by reading in the data and defining the local macro y for the dependent
variable and the local macro `xlist` for the regressors.

```
. * Set up data and local macros for dependent variable and regressors
. use mus10data.dta
. keep if year02 == 1
(25712 observations deleted)
. generate cons = 1
. local y docvis
. local xlist private chronic female income cons
```

The subsequent Mata program reads in the relevant data and obtains the parameter estimates and the estimate of the VCE. The program first associates vector y and matrix X with the relevant Stata variables by using the st_view() function. The tokens("") function is added to convert `xlist´ to a comma-separated list with each entry in double quotes, the necessary format for st_view(). The starting values are simply zero for slope parameters and one for the intercept. The robust estimate of the VCE is obtained, and this and the parameter estimates are passed back to Stata by using the st_matrix() function. We have

```
. * Complete Mata code for Poisson MLE NR iterations
. mata:
                                    ──────── mata (type end to exit) ────────
:   st_view(y=., ., "`y´")              // read in stata data to y and X
:   st_view(X=., ., tokens("`xlist´"))
:   b = J(cols(X),1,0)                  // compute starting values
:   n = rows(X)
:   iter = 1                            // initialize number of iterations
:   cha = 1                             // initialize stopping criterion
:   do {
>       mu = exp(X*b)
>       grad = X´(y-mu)                 // k x 1 gradient vector
>       hes = cross(X, mu, X)           // negative of the k x k Hessian matrix
>       bold = b
>       b = bold + cholinv(hes)*grad
>       cha = (bold-b)´(bold-b)/(bold´bold)
>       iter = iter + 1
>   } while (cha > 1e-16)               // end of iteration loops
:   mu = exp(X*b)
:   hes = cross(X, mu, X)
:   vgrad = cross(X, (y-mu):^2, X)
:   vb = cholinv(hes)*vgrad*cholinv(hes)*n/(n-cols(X))
:   iter                               // number of iterations
    13
:   cha                                // stopping criterion
    1.11465e-24
:   st_matrix("b",b´)                  // pass results from Mata to Stata
:   st_matrix("V",vb)                  // pass results from Mata to Stata
: end
```

Once back in Stata, we use the ereturn command to display the results, first assigning names to the columns and rows of b and V. We have

```
. * Present results, nicely formatted using Stata command ereturn
. matrix colnames b = `xlist´
. matrix colnames V = `xlist´
. matrix rownames V = `xlist´
. ereturn post b V
. ereturn display
```

	Coef.	Std. Err.	z	P>|z|	[95% Conf.	Interval]
private	.7986654	.1090509	7.32	0.000	.5849295	1.012401
chronic	1.091865	.0560205	19.49	0.000	.9820669	1.201663
female	.4925481	.058563	8.41	0.000	.3777666	.6073295
income	.003557	.001083	3.28	0.001	.0014344	.0056796
cons	-.2297263	.1109236	-2.07	0.038	-.4471325	-.0123202

The coefficients are the same as those from the `poisson` command (see section 10.3.2), and the standard errors are the same to at least the first three significant digits. Thirteen iterations were required because of poor starting values and a tighter convergence criterion than used by `poisson`.

The preceding NR algorithm can be adapted to use Stata `matrix` commands, but it is better to use Mata functions because these can be simpler. Also Mata functions read more like algebraic matrix expressions, and Mata does not have the restrictions on matrix size that are present in Stata.

11.3 Gradient methods

In this section, we consider various gradient methods, stopping criteria, multiple optimums, and numerical derivatives. The discussion is relevant for built-in estimation commands, as well as for user-written commands.

11.3.1 Maximization options

Stata ML estimation commands, such as `poisson`, and the general-purpose `ml` command, presented in the next section, have various maximization options that are detailed in [R] **maximize**.

The default is to provide an iteration log that gives the value of the objective function at each step plus information on the iterative method being used. This can be suppressed using the `nolog` command. Additional information at each iteration can be given by using the `trace` (current parameter values), `gradient` (current gradient vector), `hessian` (current Hessian), and `showstep` (report steps within each iteration) options.

The `technique()` option allows several maximization techniques other than NR. The `nr`, `bhhh`, `dfp`, `bfgs`, and `nm` options are discussed in section 11.3.2.

Four stopping criteria—the `tolerance(#)`, `ltolerance(#)`, `gtolerance(#)`, and `nrtolerance(#)` options—are discussed in section 11.3.4. The default is the option `nrtolerance(1e-5)`.

The `difficult` option uses an alternative method to determine steps when the estimates are in a region where the objective function is nonconcave.

The `from(`*init_specs*`)` option allows starting values to be set.

The maximum number of iterations can be set by using the `iterate(#)` option or by the separate command `set maxiter #`. The default is 16,000, but this can be changed.

11.3.2   Gradient methods

Stata maximization commands use the iterative algorithm

$$\widehat{\boldsymbol{\theta}}_{s+1} = \widehat{\boldsymbol{\theta}}_s + a_s \mathbf{W}_s \mathbf{g}_s, \qquad s = 1, \dots, S \tag{11.4}$$

where a_s is a scalar step-size adjustment and $\mathbf{W}_s$ is a $q \times q$ weighting matrix. A special case is the NR method given in (11.1), which uses $-\mathbf{H}_s^{-1}$ in place of $a_s \mathbf{W}_s$.

If the matrix multiplier $\mathbf{W}_s$ is too small, we will take a long time to reach the maximum, whereas if a multiple is too large, we can overshoot the maximum. The step-size adjustment a_s is used to evaluate $Q(\widehat{\boldsymbol{\theta}}_{s+1})$ at $\widehat{\boldsymbol{\theta}}_{s+1} = \widehat{\boldsymbol{\theta}}_s + a_s \mathbf{W}_s \mathbf{g}_s$ over a range of values of a_s (such as 0.5, 1, and 2), and the value of a_s that leads to the largest value for $Q(\widehat{\boldsymbol{\theta}}_{s+1})$ is chosen. This speeds up computation because calculation of $\mathbf{W}_s \mathbf{g}_s$ takes much more time than several subsequent evaluations of $Q(\widehat{\boldsymbol{\theta}}_{s+1})$. Stata uses a sophisticated method to choose a_s so that convergence occurs quickly, even for difficult problems.

Different weighting matrices $\mathbf{W}_s$ correspond to different gradient methods. Ideally, the NR method can be used, with $\mathbf{W}_s = -\mathbf{H}_s^{-1}$. If $\mathbf{H}_s$ is nonnegative definite, noninvertible, or both, then $\mathbf{H}_s$ is adjusted so that it is invertible. Stata also uses $\mathbf{W}_s = -\{\mathbf{H}_s + c\mathrm{Diag}(\mathbf{H}_s)\}^{-1}$. If this fails, then Stata uses NR for the orthogonal subspace corresponding to nonproblematic eigenvalues of $\mathbf{H}_s$ and steepest ascent ($\mathbf{W}_s = \mathbf{I}_s$) for the orthogonal subspace corresponding to problematic (negative or small positive) eigenvalues of $\mathbf{H}_s$.

Other optimization methods can also be used. These methods calculate alternatives to $\mathbf{H}_s^{-1}$ that can be computationally faster and can be possible even in regions where $\mathbf{H}_s$ is nonnegative definite, noninvertible, or both. The alternative methods available for the `ml` command are the Berndt–Hall–Hall–Hausman (BHHH), Davidon–Fletcher–Powell (DFP), Boyden–Fletcher–Goldfarb–Shannon (BFGS), and Nelder–Mead algorithms. These methods can be selected by using the `technique()` option, with argument, respectively, `nr`, `bhhh`, `dfp`, `bfgs`, and `nm`. The methods are explained in, for example, Greene (2008) and Gould, Pitblado, and Sribney (2006).

Some of these algorithms can converge even if $\mathbf{H}_s$ is still nonnegative definite. Then it is possible to obtain parameter estimates but not standard errors because the latter require inversion of the Hessian. The lack of standard errors is a clear signal of problems.

11.3.3   Messages during iterations

The iteration log can include comments on each iteration.

The message (backed up) is given when the original step size a_s in (11.4) resulted in a *lower* $Q(\widehat{\boldsymbol{\theta}}_{s+1})$. The message (not concave) means that $-\mathbf{H}_s$ was not invertible. In both cases, the ultimate results are fine, provided that these messages are not being given at the last iteration.

11.3.4   Stopping criteria

The iterative process continues until it is felt that $\mathbf{g}(\widehat{\boldsymbol{\theta}}) \simeq \mathbf{0}$ and that $Q(\widehat{\boldsymbol{\theta}})$ is close to a maximum.

Stata has four stopping criteria: small change in the coefficient vector (tolerance()); small change in the objective function (ltolerance()); small gradient relative to the Hessian (nrtolerance()); and small gradient relative to the coefficients (gtolerance()). The Stata default values for these criteria can be changed; see help maximize.

The default and preferred stopping criterion is nrtolerance(), which is based on $\mathbf{g}(\widehat{\boldsymbol{\theta}})'\mathbf{H}(\widehat{\boldsymbol{\theta}})^{-1}\mathbf{g}(\widehat{\boldsymbol{\theta}})$. The default is to stop when nrtolerance() $< 10^{-5}$.

In addition, the user should be aware that even if the iterative method has not converged, estimation will stop after maxiter iterations. If the maximum is reached without convergence, regression results including parameters and standard errors are still provided, along with a warning message that convergence is not achieved.

11.3.5   Multiple maximums

Complicated objective functions can have multiple optimums. The following provides an example:

```
. * Objective function with multiple optima
. graph twoway function
> y=100-0.0000001*(x-10)*(x-30)*(x-50)*(x-50)*(x-70)*(x-80),
> range (5 90) plotregion(style(none))
> title("Objective function Q(theta) as theta varies")
> xtitle("Theta", size(medlarge)) xscale(titlegap(*5))
> ytitle("Q(theta)", size(medlarge)) yscale(titlegap(*5))
```

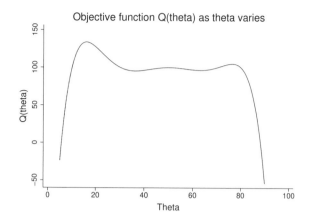

Figure 11.1. Objective function with multiple optimums

From figure 11.1, there are three local maximums—at $\theta \simeq 15$, at $\theta \simeq 50$, and at $\theta \simeq 75$—and two local minimums—at $\theta \simeq 35$ and at $\theta \simeq 65$. Most econometrics estimators are defined as a local maximum, because the asymptotic theory applies to an estimator that sets the gradient to zero. The asymptotic theory usually applies to the largest of the local maximums, which is $\theta \simeq 15$.

What problems might a gradient method encounter? If we start at $\theta < 30$, we will eventually move to the desired optimum at $\theta \simeq 15$. If instead we start at $\theta > 30$, then we will move to smaller local maximums at $\theta = 50$ or $\theta \simeq 75$. Furthermore, the objective function is relatively flat for $30 < \theta < 80$, so it may take quite some time to move to a local maximum.

Even if one obtains parameter estimates, they need not provide the largest local maximum. One method to check for multiple optimums is to use a range of starting values. This problem is more likely with user-written estimators, because most built-in Stata commands apply to models where multiple optimums do not arise.

11.3.6   Numerical derivatives

All gradient methods require first derivatives of the objective function, and most require second derivatives. For the $q \times 1$ vector $\boldsymbol{\theta}$, there are q first derivatives and $q(q + 1)/2$ unique second derivatives that need to be calculated for each observation at each round of the iterative process, so a key component is fast computation of derivatives.

The derivatives can be computed analytically or numerically. Numerical derivatives have the attraction of simplicity but can lead to increased computation time compared with analytical derivatives. For the Poisson example in section 10.2.2, it was easy to obtain and provide analytical derivatives. We now consider numerical derivatives.

The scalar derivative $df(x)/dx = \lim_{h \to 0}[\{f(x+h) - f(x-h)\}/2h]$, so one can approximate the derivative by $\{f(x+h) - f(x-h)\}/2h$ for suitable small choice of h. Applying this to the optimization of $Q(\boldsymbol{\theta})$, where differentiation is now with respect to a vector, for the first derivative of $Q(\widehat{\boldsymbol{\theta}}_s)$ with respect to the jth component of the vector $\boldsymbol{\theta}$, the numerical derivative is

$$\left.\frac{\Delta Q(\boldsymbol{\theta})}{\Delta \theta_j}\right|_{\widehat{\boldsymbol{\theta}}_s} = \frac{Q(\widehat{\boldsymbol{\theta}}_s + h\mathbf{e}_j) - Q(\widehat{\boldsymbol{\theta}}_s - h\mathbf{e}_j)}{2h}, \quad j = 1, \ldots, q$$

where h is small and $\mathbf{e}_j = (0 \ldots 0 \; 1 \; 0 \ldots 0)'$ is a column vector with unity in the jth row and 0s elsewhere. Numerical second derivatives are calculated as the numerical first derivative of the numerical or analytical first derivative. In theory, h should be very small, because formally $\partial Q(\boldsymbol{\theta})/\partial \theta_j$ equals the limit of $\Delta Q(\boldsymbol{\theta})/\Delta \theta_j$ as $h \to 0$. But in practice, too small a value of h leads to inaccuracy due to rounding error. Stata chooses $2h$ so that $f(x + h)$ and $f(x - h)$ differ in about half their digits, or roughly 8 out of 16 digits, because computations are in double precision. This computation of h each time a derivative is taken increases accuracy at the expense of considerable increase in computation time.

The number of derivatives is greatly reduced if the objective function is an index model with few indexes. In the simplest case of a single-index model, $Q(\boldsymbol{\theta}) = N^{-1}\sum_i q(y_i, \mathbf{x}_i'\boldsymbol{\theta})$ so that $\boldsymbol{\theta}$ only appears via $\mathbf{x}_i'\boldsymbol{\theta}$. Then, by the chain rule, the gradient vector is

$$\frac{\partial Q(\boldsymbol{\theta})}{\partial \boldsymbol{\theta}} = \frac{1}{N}\sum_{i=1}^{N} \frac{\partial q(y_i, \mathbf{x}_i'\boldsymbol{\theta})}{\partial \mathbf{x}_i'\boldsymbol{\theta}} \times \mathbf{x}_i$$

The q scalar derivatives $\partial q(y_i, \mathbf{x}_i'\boldsymbol{\theta})/\partial \theta_j$ are simply the same scalar derivative $\partial q(y_i, \mathbf{x}_i'\boldsymbol{\theta})/\partial \mathbf{x}_i'\boldsymbol{\theta}$ times $\mathbf{x}_i$. Similarly, the $q(q+1)/2$ unique second derivatives are simple multiples of a scalar second derivative.

For a multi-index model with J indexes (often $J \leq 2$), there are J first derivatives to be calculated. Because few derivatives need to be computed, computation is slowed down little when numerical derivatives are used rather than analytical derivatives if J is small.

11.4   The ml command: lf method

The Stata optimization command `ml` is deliberately set up for multi-index models to speed up the computation of derivatives. The name `ml` is somewhat misleading because the command can be applied to any m estimator (see the NLS example in section 11.4.5), but in non-ML cases, one should always use a robust estimate of the VCE.

The `lf` method is the simplest method. It requires the formula for a single observation's contribution to the objective function. For ML estimation, this is the log density. The more advanced methods `d0–d2` and `lf0–lf2` are deferred to section 11.6.

11.4.1   The ml command

The key commands for the `ml` method are the `ml model` command to define the model to be fit and the `ml maximize` command to perform the maximization.

The syntax for `ml model` is

`ml model` *method progname eq1* [*eq2 ...*] [*if*] [*in*] [*weight*] [, *options*]

For example, `ml model lf lfpois (y=x1 x2)` will use the `lfpois` program to estimate the parameters of a single-index model with the dependent variable `y`, the regressors `x1` and `x2`, and an intercept.

The `lf`, `d0`, and `lf0` methods use only numerical derivatives, the `d1` and `lf1` methods use analytical first derivatives and numerical second derivatives, and the `d2` and `lf2` methods use only analytical derivatives. The user must provide the formulas for any analytical derivatives. Methods `lf0`, `lf1`, and `lf2` are extensions of method `lf` that allow you to specify analytic derivatives; see section 11.6.4. Finally, for methods `d2` and `lf2` that provide expressions for second derivatives, the `negh` option needs to be used in Stata 11 if, following earlier versions of Stata, the negative Hessian rather than the Hessian is specified.

The syntax for `ml maximize` is

`ml maximize` [, *options*]

where many of the options are the maximization options covered in section 11.3.1.

There are several other `ml` commands. These include `ml check` to check that the objective function is valid; `ml search` to find better starting values; `ml trace` to trace maximization execution; and `ml init` to provide starting values.

11.4.2   The lf method

The simplest method is `lf`. This is intended for the special case where the objective function is an m estimator, simply a sum or average over the N observations of a subfunction $q_i(\boldsymbol{\theta})$, with parameters that enter as a single-index or a multi-index form. Then

$$Q(\boldsymbol{\theta}) = \sum_{i=1}^{N} q(y_i, \mathbf{x}'_{i1}\boldsymbol{\theta}_1, \ldots, \mathbf{x}'_{iJ}\boldsymbol{\theta}_J) \tag{11.5}$$

Usually, $J = 1$, in which case $q_i(\boldsymbol{\theta}) = q(y_i, \mathbf{x}'_{i1}\boldsymbol{\theta}_1)$, or $J = 2$. Most cross-section likelihoods and Stata built-in commands fall into this class, which Stata documentation refers to as meeting the linear-form restrictions.

The `lf` method requires that a program be written to give the formula for the subfunction $q_i(\boldsymbol{\theta})$. This program is subsequently called by the `ml model lf` command.

The Stata documentation refers to the first index $\mathbf{x}'_{i1}\boldsymbol{\theta}_1$ as theta1, the second index $\mathbf{x}'_{i2}\boldsymbol{\theta}_2$ as theta2, and so on. This has the potential to cause confusion with the standard statistical terminology, where $\boldsymbol{\theta}$ is the generic notation for the parameter vector.

11.4.3   Poisson example: Single-index model

For the Poisson MLE, $Q(\boldsymbol{\beta}) = \sum_i q_i(\boldsymbol{\beta})$, where the log density

$$q_i(\boldsymbol{\beta}) = -\exp(\mathbf{x}'_i\boldsymbol{\beta}) + y_i\mathbf{x}'_i\boldsymbol{\beta} - \ln y_i! \tag{11.6}$$

is of single-index form.

We first write the program, referenced in ml model, that evaluates $q_i(\boldsymbol{\beta})$. This program has two arguments: lnf, for the evaluated log density, and theta1, for the single index $\mathbf{x}'_i\boldsymbol{\beta}$. The dependent variable y_i is assigned by Stata to the global macro \$ML_y1.

To improve program readability, we use the local macro y to substitute for \$ML_y1, we define the temporary variable mu equal to exp(theta1), and we define the temporary variable lnyfact equal to $\ln y!$. The program argument lnf stores the result $q_i(\boldsymbol{\beta})$:

```
. * Poisson ML program lfpois to be called by command ml method lf
. program lfpois
  1.    version 11
  2.    args lnf theta1              // theta1=x´b, lnf=ln(y)
  3.    tempvar lnyfact mu
  4.    local y "$ML_y1"             // Define y so program more readable
  5.    generate double `lnyfact´ = lnfactorial(`y´)
  6.    generate double `mu´       = exp(`theta1´)
  7.    quietly replace `lnf´      = -`mu´ + `y´*`theta1´ - `lnyfact´
  8. end
```

We could have more simply defined lnf as

```
`lnf´ = -exp(`theta1´) + $ML_y1*exp(`theta1´) - lnfactorial($ML_y1)
```

The preceding code instead breaks this into pieces, which can be advantageous when lnf is complex. Stata computes lnf using double precision, so the intermediate variables should also be calculated in double precision. The lnfactorial() function is used rather than first computing $y!$ and then taking the natural logarithm, because the latter method is not possible if y is at all large.

The essential commands are ml model and ml maximize. It is good practice to additionally use the ml check and ml search commands before ml maximize.

For Poisson regression of docvis on several regressors, with a robust estimate of the VCE, we have

```
. * Command ml model including defining y and x, plus ml check
. ml model lf lfpois (docvis = private chronic female income), vce(robust)
. ml check
```

Test 1: Calling lfpois to check if it computes log pseudolikelihood and
does not alter coefficient vector...
Passed.

Test 2: Calling lfpois again to check if the same log pseudolikelihood value
is returned...
Passed.

   (*output omitted*)

We then type **ml search** to try to obtain better starting values.

```
. * Search for better starting values
. ml search
initial:       log pseudolikelihood = -22985.527
rescale:       log pseudolikelihood = -22985.527
```

ML estimation then occurs by typing

```
. * Compute the estimator
. ml maximize
initial:       log pseudolikelihood = -22985.527
rescale:       log pseudolikelihood = -22985.527
Iteration 0:   log pseudolikelihood = -22985.527
Iteration 1:   log pseudolikelihood = -19116.796
Iteration 2:   log pseudolikelihood = -18507.763
Iteration 3:   log pseudolikelihood = -18503.554
Iteration 4:   log pseudolikelihood = -18503.549
Iteration 5:   log pseudolikelihood = -18503.549
```

```
                                    Number of obs   =       4412
                                    Wald chi2(4)    =     594.72
Log pseudolikelihood = -18503.549   Prob > chi2     =     0.0000
```

docvis	Coef.	Robust Std. Err.	z	P>|z|	[95% Conf. Interval]	
private	.7986654	.1090015	7.33	0.000	.5850265	1.012304
chronic	1.091865	.0559951	19.50	0.000	.9821167	1.201614
female	.4925481	.0585365	8.41	0.000	.3778187	.6072775
income	.003557	.0010825	3.29	0.001	.0014354	.0056787
_cons	-.2297263	.1108733	-2.07	0.038	-.4470339	-.0124188

Note that an intercept was automatically added. The results are the same as those given
in section 10.3.2 using **poisson** and as those obtained in section 11.2.3 using Mata.

11.4.4 Negative binomial example: Two-index model

A richer model for counts is the negative binomial. The log density, which is explained in chapter 17, is then

$$q_i(\boldsymbol{\beta}, \alpha) = \ln \Gamma(y_i + \alpha^{-1}) - \ln \Gamma(y_i + \alpha^{-1}) - \ln y_i!$$
$$- (y_i + \alpha^{-1}) \ln\{1 + \alpha^{-1} \exp(\mathbf{x}_i'\boldsymbol{\beta})\} + y_i \ln \alpha + y_i \mathbf{x}_i'\boldsymbol{\beta}$$

This introduces an additional parameter, α, so that the model is now a two-index model, with indexes $\mathbf{x}_i'\boldsymbol{\beta}$ and α.

The following program computes $q_i(\boldsymbol{\beta}, \alpha)$ for the negative binomial, where the two indexes are referred to as `theta1` (equals $\mathbf{x}_i'\boldsymbol{\beta}$) and `a` (equals α).

```
. * Negbin ML program lfnb to be called by command ml method lf
. program lfnb
  1.    version 11
  2.    args lnf theta1 a              // theta1=x`b, a=alpha, lnf=ln(y)
  3.    tempvar mu
  4.    local y $ML_y1                 // Define y so program more readable
  5.    generate double `mu` = exp(`theta1`)
  6.    quietly replace `lnf` = lngamma(`y`+(1/`a`)) - lngamma((1/`a`))
  >            - lnfactorial(`y`) - (`y`+(1/`a`))*ln(1+`a`*`mu`)
  >            + `y`*ln(`a`) + `y`*ln(`mu`)
  7. end
```

The program has an additional argument, `a`, so the call to the program using `ml maximize` includes an additional argument `()` indicating that `a` is a constant that does not depend on regressors, unlike `theta1`. We have

```
. * Command lf implemented for negative binomial MLE
. ml model lf lfnb (docvis = private chronic female income) ()

. ml maximize, nolog
initial:       log likelihood =     -<inf>  (could not be evaluated)
feasible:      log likelihood = -14722.779
rescale:       log likelihood = -10743.548
rescale eq:    log likelihood = -10570.445
```

				Number of obs	=	4412
				Wald chi2(4)	=	1159.02
Log likelihood = -9855.1389				Prob > chi2	=	0.0000

docvis	Coef.	Std. Err.	z	P>|z|	[95% Conf. Interval]	
eq1						
private	.8876559	.0594232	14.94	0.000	.7711886	1.004123
chronic	1.143545	.0456778	25.04	0.000	1.054018	1.233071
female	.5613027	.0448022	12.53	0.000	.473492	.6491135
income	.0045785	.000805	5.69	0.000	.0030007	.0061563
_cons	-.4062135	.0611377	-6.64	0.000	-.5260411	-.2863858
eq2						
_cons	1.726868	.05003	34.52	0.000	1.628811	1.824925

The standard errors are based on the default estimate of the VCE, because `vce(robust)` was not used in `ml maximize`. The estimates and standard errors are exactly the same as those obtained by using the `nbreg` command; see section 12.4.1.

11.4.5   NLS example: Nonlikelihood model

The preceding examples were likelihood-based, but other m estimators can be considered.

In particular, consider NLS estimation with an exponential conditional mean. Then $Q_N(\boldsymbol{\beta}) = 1/N \sum_{i=1}^{N} \{y_i - \exp(\mathbf{x}_i'\boldsymbol{\beta})\}^2$. This is easily estimated by typing

```
. * NLS program lfnls to be called by command ml method lf
. program lfnls
  1.    version 11
  2.    args lnf theta1                      // theta1=x´b, lnf=squared residual
  3.    local y "$ML_y1"                      // Define y so program more readable
  4.    quietly replace `lnf´ = -(`y´-exp(`theta1´))^2
  5. end
```

Note the minus sign in the definition of `lnf`, because the program is designed to maximize, rather than minimize, the objective function.

Running this program, we obtain

```
. * Command lf implemented for NLS estimator
. ml model lf lfnls (docvis = private chronic female income), vce(robust)
. ml maximize
  (output omitted )
```

The results, omitted here, give the same coefficient estimates as those obtained from the `nl` command, which are given in section 10.3.5. The corresponding robust standard errors differ, however, by as much as 5%. The reason is that `nl` uses the expected Hessian in forming the robust estimate of the VCE (see section 10.4.5), exploiting additional information about the NLS estimator. The `ml` method instead uses the empirical Hessian. For NLS, these two differ, whereas for Poisson MLE, they do not.

For this example, the default estimate of the VCE, the inverse of the negative Hessian matrix, will always be wrong. To see this, consider ordinary least squares (OLS) in the linear model. Then $Q_N(\boldsymbol{\beta}) = (\mathbf{y} - \mathbf{X}\boldsymbol{\beta})'(\mathbf{y} - \mathbf{X}\boldsymbol{\beta})$ has a Hessian of $-2 \times \mathbf{X}'\mathbf{X}$. Even if the errors are homoskedastic, `ml` would give an estimate of $(1/2)(\mathbf{X}'\mathbf{X})^{-1}$ rather than $s^2(\mathbf{X}'\mathbf{X})^{-1}$. Whenever `ml` is used to optimize models that are not likelihood-based, a robust estimate of the VCE must be used.

11.5   Checking the program

The initial challenge is to debug a program and get it to successfully run, meaning that iterations converge and plausible regression output is obtained. There is a great art to this, and there is no replacement for experience. There are many ways to make errors, especially given the complexity of program syntax.

The next challenge is to ensure that computations are done correctly, to verify that plausible output is indeed correct output. This is feasible if it is possible to generate simulated data that satisfy model assumptions.

We focus on user-written programs for `ml`, but many of the following points apply to evaluating any estimator.

11.5.1   Program debugging using ml check and ml trace

The `ml check` command provides a check of the code to ensure that it is possible to evaluate `lnf`, though this does not ensure that the evaluation is correct. This command is most useful for checking program syntax, because it provides much more detailed information than if we instead proceed directly to `ml maximize`.

For example, suppose in the `lfpois` program we typed the line

```
. generate double `mu´ = exp(`heta1´)
```

The mistake is that `` `heta1´ `` was typed rather than `` `theta1´ ``. The `ml maximize` command leads to failure and the following error message:

```
invalid syntax
r(198);
```

This message is not particularly helpful. If instead we type

```
. ml search
```

before `ml maximize`, the program again fails, but the output now includes

```
- generate double `mu´ = exp(`heta1´)
= generate double __000006 = exp(
invalid syntax
```

which indicates that the error is due to a problem with `` `heta1´ ``.

More complete information is given by the `ml trace` command. If we type

```
. ml trace on
```

before `ml maximize`, the program fails, and we get essentially the same output as when using `ml search`. Once the program is corrected and runs successfully, `ml trace` provides extensive details on the execution of the program. In this example, 980 lines of detail are given.

The trace facility can also be used for commands other than `ml` by typing

```
. set trace on
```

A drawback of using `trace` is that it can produce copious output.

A more manual-targeted method to determine where problems may arise in a program is to include messages in the program. For example, suppose we place in the `lfpois` program the line

```
display "I made it to here"
```

If the program fails after this line is displayed, then we know that the problem arose beyond the line where the `display` statement was given.

11.5.2   Getting the program to run

The `ml check` command essentially checks program syntax. This does not protect against other coding errors such as misspecification of the log density. Suppose, for example, that in the `lfpois` program we typed

```
quietly replace `lnf´ = `mu´ + `y´*ln(`mu´) - `lnyfact´
```

The error here is that we have `mu´ rather than -`mu´. Then we obtain

```
. ml maximize
initial:       log pseudolikelihood = -25075.609
alternative:   log pseudolikelihood = -13483.451
(4412 missing values generated)
rescale:       log pseudolikelihood =  1.01e+226
Iteration 0:   log pseudolikelihood =  1.01e+226  (not concave)
(1 missing value generated)
Iteration 1:   log pseudolikelihood =  1.76e+266  (not concave)
(762 missing values generated)
Hessian has become unstable or asymmetric (NC)
r(504);
```

Here the error has occurred quite early. One possibility is that poor starting values were given. But using `ml search` leads to far worse starting values in this example. In this case, the most likely explanation is an error in the objective function.

Poor starting values can lead to problems if the objective function is not globally concave. For index models, a good approach is to set all parameters to zero aside from the constant, which is set to that value appropriate for an intercept-only model. For example, for the Poisson intercept-only model with the parameter α, we have $\widehat{\alpha} = \ln \overline{y}$, because then $\exp(\widehat{\alpha}) = \overline{y}$. Thus the initial value is $(0, \ldots, \ln \overline{y})$. It can be useful to try the model with few regressors, such as with an intercept and a single regressor.

The `ml` methods `d1`, `d2`, `lf1`, and `lf2`, presented in section 11.6, require analytical expressions for the first and second derivatives. The `ml model d1debug` and `ml model d2debug` commands check these by comparing these expressions with numerical first and second derivatives. Any substantial difference indicates error in coding the derivatives or, possibly, the original objective function. There are similar methods `lf1debug` and `lf2debug`.

11.5.3 Checking the data

A common reason for program failure is that the program is fine, but the data passed to the program are not.

For example, the `lfpois` program includes the `lnfactorial('y')` function, which requires that `` `y' `` be a nonnegative integer. Consider the impact of the following:

```
. replace docvis = 0.5 if docvis == 1
. ml model lf lfpois (docvis = private chronic female income), vce(robust)
. ml maximize
```

The resulting output after `maximize` includes many lines with

```
(700 missing values generated)
```

followed by

```
could not find feasible values
 'r(491);
```

One should always use `summarize` to obtain summary statistics of the dependent variable and regressors ahead of estimation as a check. Indications of problems include an unexpected range, zero standard deviation, and missing values. In this particular example, `summarize` will not detect the problem, but `tabulate docvis` will.

11.5.4 Multicollinearity and near collinearity

If variables are perfectly collinear, then Stata estimation commands, including `ml`, detect multicollinearity and drop the regressor(s) as needed.

If variables are close to perfectly collinear, then numerical instability may cause problems. We illustrate this by adding the additional regressor `extra`, which equals `income` plus au, where u is a draw from the uniform distribution and $a = 0.001$ or 0.01.

First, add the regressor `extra` that equals `income` plus $0.001u$. We have

```
. * Example with high collinearity interpreted as perfect collinearity
. generate extra = income + 0.001*runiform()
. ml model lf lfpois (docvis = private chronic female income extra), vce(robust)
note: extra omitted because of collinearity
```

Here `ml maximize` interprets this as perfect collinearity and drops `income` before maximization.

Next instead add the regressor `extra2` that equals `income` plus the larger amount $0.01u$. We have

```
. * Example with high collinearity not interpreted as perfect collinearity
. generate extra2 = income + 0.01*runiform()
. ml model lf lfpois (docvis = private chronic female income extra2), vce(robust)
```

```
. ml maximize, nolog
```

	Number of obs	=	4412
	Wald chi2(5)	=	604.80
Log pseudolikelihood = -18501.171	Prob > chi2	=	0.0000

docvis	Coef.	Robust Std. Err.	z	P>|z|	[95% Conf.	Interval]
private	.7989052	.1090178	7.33	0.000	.5852342	1.012576
chronic	1.092046	.0559464	19.52	0.000	.9823929	1.201699
female	.4922495	.0585066	8.41	0.000	.3775787	.6069203
income	-5.646463	9.522279	-0.59	0.553	-24.30979	13.01686
extra2	5.650021	9.522036	0.59	0.553	-13.01283	24.31287
_cons	-.2581047	.1157039	-2.23	0.026	-.4848803	-.0313291

Now this is no longer interpreted as perfect collinearity, so estimation proceeds. The coefficients of income and extra2 are very imprecisely estimated, while the remaining coefficients and standard errors are close to those given in section 10.3.2.

Pairwise collinearity can be detected by using the correlate command, and multi-collinearity can be detected with the _rmcoll command. For example,

```
. * Detect multicollinearity using _rmcoll
. _rmcoll income extra
note: extra omitted because of collinearity
. _rmcoll income extra2
```

Another simple data check is to see whether the parameters of the model can be estimated by using a closely related Stata estimation command. For the lfpois program, the obvious test is regression using poisson. If a command to compute Poisson regression was not available, we could at least try OLS regression, using regress.

11.5.5   Multiple optimums

Even when iterations converge and regression output is obtained, there is a possibility that a local rather than a global maximum has been obtained.

A clear signal of lack of global concavity is warning messages of nonconcavity in some of the intermediate iterations. This may or may not be a serious problem, but at the least one should try a range of different starting values in this situation.

Parameter estimates should not be used if iterations fail to converge. They also should not be used if the final iteration has a warning that the objective function is nonconcave or that the Hessian is not negative definite, because this indicates that the model is not identified. Missing standard errors also indicates a problem.

11.5.6 Checking parameter estimation

Once a program runs, we need to check that it is correct. The following approach is applicable to estimation with any method, not just with the `ml` command.

To check parameter estimation, we generate data from the same data-generating process (DGP) as that justifying the estimator, for a large sample size N. Because the desired estimator is consistent as $N \rightarrow \infty$, we expect the estimated parameters to be very close to those of the DGP. A similar exercise was done in chapter 4.

To check a Poisson model estimation program, we generate data from the following DGP:

$$y_i = \text{Poisson}(\beta_1 + \beta_2 x); \quad (\beta_1, \beta_2) = (2, 1); \quad x_i \sim N(0, 0.5); \quad i = 1, \dots, 10,000$$

The following code generates data from this DGP:

```
. * Generate dataset from Poisson DGP for large N
. clear
. set obs 10000
obs was 0, now 10000
. set seed 10101
. generate x = rnormal(0,0.5)
. generate mu = exp(2 + x)
. generate y = rpoisson(mu)
. summarize mu x y
```

Variable	Obs	Mean	Std. Dev.	Min	Max
mu	10000	8.386844	4.509056	.9930276	47.72342
x	10000	.0020266	.4978215	-2.006997	1.865422
y	10000	8.3613	5.277475	0	52

The normal regressor has the desired mean and variance, and the count outcome has a mean of 8.36 and ranges from 0 to 52.

We then run the previously defined `lfpois` program.

```
. * Consistency check: Run program lfpois and compare beta to DGP value
. ml model lf lfpois (y = x)
. ml maximize, nolog
```

		Number of obs	=	10000
		Wald chi2(1)	=	20486.09
Log likelihood = -24027.863		Prob > chi2	=	0.0000

y	Coef.	Std. Err.	z	P>|z|	[95% Conf. Interval]
x	.9833643	.0068704	143.13	0.000	.9698985 .9968301
_cons	2.001111	.0038554	519.05	0.000	1.993555 2.008667

The estimates are $\widehat{\beta}_2 = 0.983$ and $\widehat{\beta}_1 = 2.001$, quite close to the DGP values of 1 and 2. The standard errors for the slope coefficient are 0.007 and 0.004, so in 95% of such simulations, we expect $\widehat{\beta}_2$ to lie within $[0.970, 0.997]$. The DGP value falls just outside this interval, most likely because of randomness. The DGP value of the intercept lies within the similar interval $[1.994, 2.009]$.

If $N = 1,000,000$, for example, estimation is more precise, and we expect estimates to be very close to the DGP values.

This DGP is quite simple. More challenging tests would consider a DGP with additional regressors from other distributions.

11.5.7   Checking standard-error estimation

To check that standard errors for an estimator $\widehat{\beta}$ are computed correctly, we can perform, say, $S = 2000$ simulations that yield S estimates $\widehat{\beta}$ and S computed standard errors $s_{\widehat{\beta}}$. If the standard errors are correctly estimated, then the average of the S computed standard errors, $\overline{s_{\widehat{\beta}}} = S^{-1} \sum_{s=1}^{S} s_{\widehat{\beta}}$, should equal the standard deviation of the S estimates $\widehat{\beta}$, which is $(S-1)^{-1} \sum_{s=1}^{S} (\widehat{\beta} - \overline{\widehat{\beta}})^2$, where $\overline{\widehat{\beta}} = S^{-1} \sum_{s=1}^{S} \widehat{\beta}$. The sample size needs to be large enough that we believe that asymptotic theory provides a good guide for computing the standard errors. We set $N = 500$.

We first write the `secheck` program, which draws one sample from the same DGP as was used in the previous section.

```
. * Program to generate dataset, obtain estimate, and return beta and SEs
. program secheck, rclass
  1.    version 11
  2.    drop _all
  3.    set obs 500
  4.    generate x = rnormal(0,0.5)
  5.    generate mu = exp(2 + x)
  6.    generate y = rpoisson(mu)
  7.    ml model lf lfpois (y = x)
  8.    ml maximize
  9.    return scalar b1 =_b[_cons]
 10.    return scalar se1 = _se[_cons]
 11.    return scalar b2 =_b[x]
 12.    return scalar se2 = _se[x]
 13. end
```

We then run this program 2,000 times, using the `simulate` command. (The `postfile` command could alternatively be used.) We have

```
. * Standard errors check: run program secheck
. set seed 10101

. simulate "secheck" bcons=r(b1) se_bcons=r(se1) bx=r(b2) se_bx=r(se2),
> reps(2000)
command:      secheck
statistics:   bcons       = r(b1)
              se_bcons    = r(se1)
              bx          = r(b2)
              se_bx       = r(se2)
```

```
. summarize
    Variable |       Obs        Mean    Std. Dev.         Min         Max
-------------+--------------------------------------------------------------
       bcons |      2000    1.999649    .0173978    1.944746     2.07103
    se_bcons |      2000    .0172925    .0002185    .0164586    .0181821
          bx |      2000    1.000492    .0308205    .8811092    1.084845
       se_bx |      2000    .0311313    .0014977    .0262835    .0362388
```

The column `Obs` in the summary statistics here refers to the number of simulations ($S = 2000$). The actual sample size, set inside the `secheck` program, is $N = 500$.

For the intercept, we have $\overline{s_{\widehat{\beta}_1}} = 0.0173$ compared with 0.0174 for the standard deviation of the 2,000 estimates for $\widehat{\beta}_1$ (`bcons`). For the slope, we have $\overline{s_{\widehat{\beta}_2}} = 0.0311$ compared with 0.0308 for the standard deviation of the 2,000 estimates for $\widehat{\beta}_2$ (`bx`). The standard errors are correctly estimated.

11.6 The ml command: d0, d1, d2, lf0, lf1, and lf2 methods

The `lf` method is fast and simple when the objective function is of the form $Q(\boldsymbol{\theta}) = \sum_i q(\mathbf{y}_i, \mathbf{x}_i, \boldsymbol{\theta})$ with independence over observation i, a form that Stata manuals refer to as the linear form; when parameters appear as a single index or as just a few indexes; and when you do not wish to specify derivatives.

The `d0`, `d1`, and `d2` methods are more general than `lf`. They can accommodate situations where there are multiple observations or equations for each individual in the sample. This can arise with panel data, where data on an individual are available in several time periods, with conditional logit models, where regressor values vary over each of the potential outcomes, in systems of equations, and in the Cox proportional hazards model, where a risk set is formed at each failure.

Method `d1` allows one to provide analytical expressions for the gradient, and method `d2` additionally allows an analytical expression for the Hessian. This can speed up computation compared with the `d0` method, which relies on numerical derivatives.

Methods `lf0`, `lf1`, and `lf2` are like method `lf` in that they are used when the objective function meets the linear-form restrictions. These methods require you to write evaluator programs that are more complicated than those for method `lf`, in exchange for the ability to specify first and second derivatives with methods `lf1` and `lf2`. You can use method `lf0` to check that you have correctly programmed the likelihood function before programming the derivatives. If you do not plan to provide derivatives, there is no point in using method `lf0`, because method `lf` provides the same functionality and is easier to use.

11.6.1 Evaluator functions

The objective function is one with multiple indexes, say, J, and multiple dependent variables, say, G, for a given observation. So

$$Q(\boldsymbol{\theta}) = \sum_{i=1}^{N} Q(\mathbf{x}'_{1i}\boldsymbol{\theta}_1, \mathbf{x}'_{2i}\boldsymbol{\theta}_2, \dots, \mathbf{x}'_{Ji}\boldsymbol{\theta}_J; y_{1i}, \dots, y_{Gi})$$

where $y_{1i}, \dots, y_{Gi}$ are possibly correlated with each other for a given i though they are independent over i.

For the **d0–d2** and **lf0–lf2** methods, the syntax is

ml model *method progname eq1* $\big[$ *eq2* ...$\big]$ $\big[$ *if* $\big]$ $\big[$ *in* $\big]$ $\big[$ *weight* $\big]$ $\big[$, *options* $\big]$

where *method* is **d0**, **d1**, **d2**, **lf0**, **lf1**, or **lf2**; *progname* is the name of an evaluator program; *eq1* defines the dependent variables and the regressors involved in the first index; *eq2* defines the regressors involved in the second index; and so on.

The evaluator program, *progname*, has five arguments for **d0**, **d1**, and **d2** evaluators: **todo**, **b**, **lnf**, **g**, and **H**. The **ml** command uses the **todo** argument to request no derivative, the gradient, or the gradient and the Hessian. The **b** argument is the row vector of parameters $\boldsymbol{\theta}$. The **lnf** argument is the scalar objective function $Q(\boldsymbol{\theta})$. The **g** argument is a row vector for the gradient $\partial Q(\boldsymbol{\theta})/\partial\boldsymbol{\theta}'$, which needs to be provided only for the **d1** and **d2** methods. The **H** argument is the Hessian matrix $\partial^2 Q(\boldsymbol{\theta})/\partial\boldsymbol{\theta}\partial\boldsymbol{\theta}'$, which needs to be provided for the **d2** method.

The evaluators for the **d0–d2** methods first need to link the parameters $\boldsymbol{\theta}$ to the indexes $\mathbf{x}'_{1i}\boldsymbol{\theta}_1, \dots$. This is done with the **mleval** command, which has the syntax

mleval *newvar* = *vecname* $\big[$, **eq(**#**)** $\big]$

For example, **mleval 'theta1'='b'**, **eq(1)** labels the first index $\mathbf{x}'_{1i}\boldsymbol{\theta}_1$ as **theta1**. The variables in $\mathbf{x}_{1i}$ will be listed in *eq1* in **ml model**.

Next the evaluator needs to compute the objective function $Q(\boldsymbol{\theta})$, unlike the **lf** method, where the ith entry $q_i(\boldsymbol{\theta})$ in the objective function is computed. The **mlsum** command sums $q_i(\boldsymbol{\theta})$ to yield $Q(\boldsymbol{\theta})$. The syntax is

mlsum *scalarname_lnf* = *exp* $\big[$ *if* $\big]$

For example, **mlsum 'lnf'=('y'-'theta1')^2** computes the sum of squared residuals $\sum_i (y_i - \mathbf{x}'_{1i}\boldsymbol{\theta}_1)^2$.

The **d1** and **d2** methods require specification of the gradient of the overall log likelihood $Q(\boldsymbol{\theta})$. For linear-form models, this computation can be simplified with the **mlvecsum** command, which has the syntax

mlvecsum *scalarname_lnf rowvecname* = *exp* $\big[$ *if* $\big]$ $\big[$, **eq(**#**)** $\big]$

For example, **mlvecsum `lnf´ `d1´=`y´-`theta1´** computes the gradient for the subset of parameters that appear in the first index as the row vector $\sum_i (y_i - \mathbf{x}'_{1i}\boldsymbol{\theta}_1)\mathbf{x}_{1i}$. Note that **mlvecsum** automatically multiplies `y´-`theta1´ by the regressors $\mathbf{x}_{1i}$ in the index **theta1** because equation one is the default when **eq()** is not specified.

The `lf0`–`lf2` methods, like the `d0`–`d2` methods, require you to first use `mleval` to obtain the indexes $\mathbf{x}_{1i}\theta_1, \ldots$. Method `lf0` evaluators receive three arguments: `todo`, `b`, and `lnfj`. The variable `lnfj` is to be filled in with the observation-level likelihood $q_i(\theta)$. The variable is named `lnfj`, rather than `lnfi`, because official Stata documentation for the `ml` command uses j, rather than i, to denote the typical observation. The key difference between methods `lf` and `lf0` is that the former passes to *progname* the indexes as `theta1`, `theta2`, etc., while the latter passes to your program the parameter vector `b` from which you must obtain the indexes yourself.

The `lf1` and `lf2` methods require specification of the observation-level scores associated with the log-likelihood function. That is, the derivatives of the log-likelihood function with respect to the indexes $\mathbf{x}_{1i}\theta_1$, $\mathbf{x}_{2i}\theta_2$, ... Whereas you specify the derivatives $\partial Q(\theta)/\partial\theta_j$ with methods `d1` and `d2`, methods `lf1` and `lf2` require you to fill in variables containing $\partial q_i(\theta)/\partial\mathbf{x}_{ji}\theta_j$. The predominant advantage of these methods is speed: evaluating analytic derivatives is much faster than computing them numerically. Moreover, with observation-level first derivatives, `ml` can compute the robust estimate of the VCE, which is not possible with methods `d0`–`d2`.

With method `lf1`, *progname* receives $3 + J$ arguments, where J is the number of indexes. For a single-index model, *progname* will receive four arguments: `todo`, `b`, `lnfj`, and `g1`. When `todo==1`, in addition to filling in `lnfj` with the observation-level log likelihood, you are to fill in the variable `g1` with $\partial q_i(\theta)/\partial\mathbf{x}_{ji}\theta_j$. Method `lf2` receives $4 + J$ arguments; the final argument, `H`, is a matrix to be filled in with the Hessian of the overall log-likelihood function when `todo==2`.

The `d2` and `lf2` methods require specification of the Hessian matrix, the default, or the negative Hessian matrix, in which case the `negh` option is added to the `ml model` command. The `negh` option is included for compatibility with older versions of Stata. Stata 11 users should specify the Hessian matrix and ignore this option. For linear-form models, this computation can be simplified with the `mlmatsum` command, which has the syntax

`mlmatsum` *scalarname_lnf matrixname* = *exp* $\left[\,if\,\right]$ $\left[\,,\ \texttt{eq(\#}\left[\,,\#\,\right]\texttt{)}\,\right]$

For example, `mlmatsum `lnf´ `d1´=`theta1´` computes the negative Hessian matrix for the subset of parameters that appear in the first index as $\sum_i \mathbf{x}'_{1i}\boldsymbol{\theta}_1$. The `mlmatsum` command automatically multiplies `theta1´` by $\mathbf{x}_{1i}\mathbf{x}'_{1i}$, the outer product of the regressors $\mathbf{x}_{1i}$ in the index `theta1`.

11.6.2   The d0 method

We consider the cross-section Poisson model, a single-index model. For multi-index models—such as the Weibull—and panel data, Cox proportional hazards, and conditional logit models, see the Weibull example in [R] **ml** and Gould, Pitblado, and Sribney (2006). Gould, Pitblado, and Sribney (2006) also consider complications such as how to make ado-files and how to incorporate sample weights.

A d0 method evaluator program for the Poisson MLE is the following:

```
. * Method d0: Program d0pois to be called by command ml method d0
. program d0pois
  1.   version 11
  2.   args todo b lnf                  // todo is not used, b=b, lnf=lnL
  3.   tempvar theta1                   // theta1=x`b given in eq(1)
  4.   mleval `theta1' = `b', eq(1)
  5.   local y $ML_y1                   // Define y so program more readable
  6.   mlsum `lnf' = -exp(`theta1') + `y'*`theta1' - lnfactorial(`y')
  7. end
```

The code is similar to that given earlier for the lf method. The mleval command forms the single index $\mathbf{x}_i'\boldsymbol{\beta}$. The mlsum command forms the objective function as the sum of log densities for each observation.

Here there is one dependent variable, docvis, and only one index with the regressors private, chronic, female, and income, plus an intercept.

```
. * Method d0: implement Poisson MLE
. ml model d0 d0pois (docvis = private chronic female income)

. ml maximize
initial:       log likelihood = -33899.609
alternative:   log likelihood = -28031.767
rescale:       log likelihood = -24020.669
Iteration 0:   log likelihood = -24020.669
Iteration 1:   log likelihood = -18845.464
Iteration 2:   log likelihood = -18510.287
Iteration 3:   log likelihood = -18503.552
Iteration 4:   log likelihood = -18503.549
Iteration 5:   log likelihood = -18503.549
```

Number of obs =	4412
Wald chi2(4) =	8052.34
Log likelihood = -18503.549 Prob > chi2 =	0.0000

docvis	Coef.	Std. Err.	z	P>|z|	[95% Conf. Interval]
private	.7986653	.027719	28.81	0.000	.7443371 .8529936
chronic	1.091865	.0157985	69.11	0.000	1.060901 1.12283
female	.4925481	.0160073	30.77	0.000	.4611744 .5239218
income	.003557	.0002412	14.75	0.000	.0030844 .0040297
_cons	-.2297263	.0287022	-8.00	0.000	-.2859815 -.173471

The resulting coefficient estimates are the same as those from the poisson command and those using the lf method given in section 11.4.3. For practice, check the nonrobust standard errors.

11.6.3 The d1 method

The d1 method evaluator program must also provide an analytical expression for the gradient.

```
.  * Method d1: Program d1pois to be called by command ml method d1
. program d1pois
  1.    version 11
  2.    args todo b lnf g                // gradient g added to the arguments list
  3.    tempvar theta1                   // theta1 = x´b given in eq(1)
  4.    mleval `theta1´ = `b´, eq(1)
  5.    local y $ML_y1                   // Define y so program more readable
  6.    mlsum `lnf´ = -exp(`theta1´) + `y´*`theta1´ - lnfactorial(`y´)
  7.    if (`todo´==0 | `lnf´>=.) exit   // Extra code from here on
  8.    tempname d1
  9.    mlvecsum `lnf´ `d1´ = `y´ - exp(`theta1´)
 10.    matrix `g´ = (`d1´)
 11. end
```

The `mlvecsum` command forms the gradient row vector $\sum_i \{y_i - \exp(\mathbf{x}_i'\widehat{\boldsymbol{\beta}})\}\mathbf{x}_i'$, where $\mathbf{x}_i$ are the first-equation regressors.

The model is run in the same way, with `d0` replaced by `d1` and the evaluator function `d0pois` replaced by `d1pois`. The `ml model d1 d1pois` command with the dependent variable `docvis` and the regressors `private`, `chronic`, `female`, and `income` yields the same coefficient estimates and (nonrobust) standard errors as those given in section 11.6.2. These results are not shown; for practice, confirm this.

11.6.4   The lf1 method with the robust estimate of the VCE

Unlike the `lf` method, computation of robust standard errors requires some additional coding in the evaluator program. This code provides variables containing the contributions of each individual observation to the gradient, whereas methods `d1` and `d2` provide only the sum of these (the gradient). In versions of Stata before version 11, once the additional code was provided, estimation proceeded using the `d1` or `d2` methods. In Stata 11, the new `lf1` or `lf2` methods must be used instead.

The Poisson model is a single-index model, so our evaluator program will receive four arguments. We have

```
.  * Method lf1: Program lf1poisrob is variation of program d1pois for robust se´s
. program lf1poisrob
  1.    version 11
  2.    args todo b lnfj g1
  3.    tempvar theta1                       // theta1 = x´b where x given in eq(1)
  4.    mleval `theta1´ = `b´, eq(1)
  5.    local y $ML_y1                       // define y so program more readable
  7.    quietly replace `lnfj´ = -exp(`theta1´) + `y´*`theta1´ - lnfactorial(`y´)
  8.    if (`todo´==0) exit
  9.    quietly replace `g1´ = `y´ - exp(`theta1´)  // extra code for robust
 10. end
```

The robust estimate of the VCE is that given in section 10.4.5, with $\widehat{\mathbf{H}}_i$ computed by using numerical derivatives.

We obtain

```
. * Method lf1: implement Poisson MLE with robust standard errors
. ml model lf1 lf1poisrob (docvis = private chronic female income), vce(robust)
. ml maximize, nolog
                                            Number of obs    =        4412
                                            Wald chi2(4)     =      594.72
      Log pseudolikelihood = -18503.549     Prob > chi2      =      0.0000
```

docvis	Coef.	Robust Std. Err.	z	P>|z|	[95% Conf. Interval]	
private	.7986654	.1090015	7.33	0.000	.5850265	1.012304
chronic	1.091865	.0559951	19.50	0.000	.9821167	1.201614
female	.4925481	.0585365	8.41	0.000	.3778187	.6072775
income	.003557	.0010825	3.29	0.001	.0014354	.0056787
_cons	-.2297263	.1108733	-2.07	0.038	-.4470339	-.0124187

We obtain the same coefficient estimates and robust standard errors as `poisson` with robust standard errors (see section 11.4.3).

11.6.5   The d2 and lf2 methods

The d2 method evaluator program, based on program d1pois, must also provide an analytical expression for the Hessian.

```
. * Method d2: Program d2pois to be called by command ml method d2
. program d2pois
  1.    version 11
  2.    args todo b lnf g H          // Add g and H to the arguments list
  3.    tempvar theta1               // theta1 = x´b where x given in eq(1)
  4.    mleval `theta1´ = `b´, eq(1)
  5.    local y $ML_y1               // Define y so program more readable
  6.    mlsum `lnf´ = -exp(`theta1´) + `y´*`theta1´ - lnfactorial(`y´)
  7.    if (`todo´==0 | `lnf´>=.) exit   // d1 extra code from here
  8.    tempname d1
  9.    mlvecsum `lnf´ `d1´ = `y´ - exp(`theta1´)
 10.    matrix `g´ = (`d1´)
 11.    if (`todo´==1 | `lnf´>=.) exit   // d2 extra code from here
 12.    tempname d11
 13.    mlmatsum `lnf´ `d11´ = exp(`theta1´)
 14.    matrix `H´ = -`d11´
 15. end
```

The `mlmatsum` command forms minus the Hessian matrix $\sum_i \exp(\mathbf{x}_i'\widehat{\boldsymbol{\beta}})\mathbf{x}_i\mathbf{x}_i'$, where $\mathbf{x}_i$ are the first-equation regressors.

```
. * Method d2: Poisson MLE with first and second derivatives provided
. ml model d2 d2pois (docvis = private chronic female income)
. ml maximize
  (output omitted )
```

We obtain the same coefficient estimates and (nonrobust) standard errors as those given in section 11.6.2.

With more than one index, it will be necessary to compute cross-derivatives such as `d12`. The `mlmatbysum` command is an extension that can be applied when the log likelihood for the ith observation involves a grouped sum, such as for panel data. See [R] **ml** for a two-index example, the Weibull MLE.

To obtain a robust estimate of the VCE requires use of the `ml` method `lf2`. We begin with evaluator program `lfpois1` and define the Hessian by adding code similar to that in the evaluator program `d2pois`, using `mlmatsum`.

11.7   The Mata optimize() function

It can be more convenient, even necessary for complicated models, to express an objective function using matrix programming functions. The Mata `optimize()` function, introduced in Stata 10, uses the same optimizer as the Stata `ml` command, though with different syntax.

The Mata `moptimize()` function, introduced in Stata 11, is an intermediate optimization function that we do not cover.

11.7.1   Type d and gf evaluators

Because y and x are used to denote dependent variables and regressors, the Mata documentation uses the generic notation that we want to compute real row vector $\mathbf{p}$ that maximizes the scalar function $f(\mathbf{p})$. Note that $\mathbf{p}$ is a row vector, whereas in this book we usually define vectors (such as $\boldsymbol{\beta}$) to be column vectors.

An evaluator function calculates the value of the objective function at values of the parameter vector. It may optionally calculate the gradient and the Hessian.

There are two distinct types of evaluator functions used by Mata.

A type d evaluator returns the value of the objective as the scalar $v = f(\mathbf{p})$. The minimal syntax is

void evaluator(todo, p, v, g, H)

where *todo* is a scalar, p is the row vector of parameters, v is the scalar function value, g is the gradient row vector $\partial f(\mathbf{p})/\partial \mathbf{p}$, and H is the Hessian matrix $\partial f(\mathbf{p})/\partial \mathbf{p}\partial \mathbf{p}'$. If *todo* equals zero, then numerical derivatives are used (method `d0`), and g and H need not be provided. If *todo* equals one, then g must be provided (method `d1`), and if *todo* equals two, then both g and H must be provided (method `d2`).

A type gf evaluator is more suited to m-estimation problems, where we maximize $Q(\boldsymbol{\theta}) = \sum_{i=1}^{N} q_i(\boldsymbol{\theta})$. Then it may be more convenient to provide an $N \times 1$ vector with the ith entry $q_i(\boldsymbol{\theta})$ rather than the scalar $Q(\boldsymbol{\theta})$. A type gf evaluator returns the column vector $\mathbf{v}$, and $f(\mathbf{p})$ equals the sum of the entries in $\mathbf{v}$. The minimal syntax for a type gf evaluator is *void evaluator*(*todo*, *p*, *v*, *g*, *H*)

where *todo* is a scalar, *p* is a row vector of parameters, *v* is a column vector, *g* is now the gradient matrix $\partial\mathbf{v}/\partial\mathbf{p}$, and *H* is the Hessian matrix. If *todo* equals zero, then numerical derivatives are used (method gf0) and *g* and *H* need not be provided. If *todo* equals one, then *g* must be provided (method gf1), and if *todo* equals two, then both *g* and *H* must be provided (method gf2).

Up to nine additional arguments can be provided in these evaluators, appearing after *p* and before *v*. In that case, these arguments and their relative positions need to be declared by using the optimize_init_arguments() function, illustrated below. For regression with data in **y** and **X**, the arguments will include *y* and *X*.

11.7.2   Optimize functions

The optimize functions fall into four broad categories. First, functions that define the optimization problem, such as the name of the evaluator and the iterative technique to be used, begin with optimize_init. Second, functions that lead to optimization are optimize() or optimize_evaluate(). Third, functions that return results begin with optimize_result. Fourth, the optimize_query() function lists optimization settings and results.

A complete listing of these functions and their syntaxes is given in [M-5] **optimize( )**. The following example essentially uses the minimal set of optimize() functions to perform a (nonlinear) regression of *y* on **x** and to obtain coefficient estimates and an estimate of the associated VCE.

11.7.3   Poisson example

We implement the Poisson MLE, using the Mata optimize() function method gf2.

Evaluator program for Poisson MLE

The key ingredient is the evaluator program, named poissonmle. Because the gf2 method is used, the evaluator program needs to evaluate a vector of log densities, named lndensity, an associated gradient matrix, named g, and the Hessian, named H. We name the parameter vector b. The dependent variable and the regressor matrix, named y and X, are two additional program arguments that will need to be declared by using the optimize_init_argument() function.

For the Poisson MLE, from section 11.2.2, the column vector of log densities has the *i*th entry $\ln f(y_i|\mathbf{x}_i) = -\exp(\mathbf{x}_i'\boldsymbol{\beta}) + \mathbf{x}_i'\boldsymbol{\beta}y_i - \ln y_i!$; the associated gradient matrix has the *i*th row $\{y_i - \exp(\mathbf{x}_i'\boldsymbol{\beta})\}\mathbf{x}_i$; and the Hessian is the matrix $\sum_i -\exp(\mathbf{x}_i'\boldsymbol{\beta})\mathbf{x}_i\mathbf{x}_i'$. A listing of the evaluator program follows:

```
. * Evaluator function for Poisson MLE using optimize gf2 evaluator
. mata
─────────────────────────────────────── mata (type end to exit) ───────
:   void poissonmle(todo, b, y, X, lndensity, g, H)
>   {
>       Xb = X*b´
>       mu = exp(Xb)
>       lndensity = -mu + y:*Xb - lnfactorial(y)
>       if (todo == 0) return
>       g = (y-mu):*X
>       if (todo == 1) return
>       H = - cross(X, mu, X)
>   }
: end
```

A better version of this evaluator function that declares the types of all program arguments and other variables used in the program is given in appendix B.3.1.

The optimize() function for Poisson MLE

The complete Mata code has four components. First, define the evaluator, a repeat of the preceding code listing. Second, associate matrices **y** and **X** with Stata variables by using the st_view() function. In the code below, the names of the dependent variable and the regressors are in the local macros y and xlist, defined in section 11.2.3. Third, optimize, which at a minimum requires the seven optimize() functions, given below. Fourth, construct and list the key results.

```
. * Mata code to obtain Poisson MLE using command optimize
. mata
─────────────────────────────────────── mata (type end to exit) ───────
:   void poissonmle(todo, b, y, X, lndensity, g, H)
>   {
>       Xb = X*b´
>       mu = exp(Xb)
>       lndensity = -mu + y:*Xb - lnfactorial(y)
>       if (todo == 0) return
>       g = (y-mu):*X
>       if (todo == 1) return
>       H = - cross(X, mu, X)
>   }
:   st_view(y=., ., "`y´")
:   st_view(X=., ., tokens("`xlist´"))
:   S = optimize_init()
:   optimize_init_evaluator(S, &poissonmle())
:   optimize_init_evaluatortype(S, "gf2")
:   optimize_init_argument(S, 1, y)
:   optimize_init_argument(S, 2, X)
:   optimize_init_params(S, J(1,cols(X),0))
```

```
:   b = optimize(S)
Iteration 0:    f(p) = -33899.609
Iteration 1:    f(p) = -19668.697
Iteration 2:    f(p) = -18585.609
Iteration 3:    f(p) = -18503.779
Iteration 4:    f(p) = -18503.549
Iteration 5:    f(p) = -18503.549
:   Vbrob = optimize_result_V_robust(S)
:   serob = (sqrt(diagonal(Vbrob)))´
:   b \ serob
```

	1	2	3	4	5
1	.7986653788	1.091865108	.4925480693	.0035570127	-.2297263376
2	.1090014507	.0559951312	.0585364746	.0010824894	.1108732568

```
: end
```

The S = optimize_init() function initiates the optimization, and because S is used, the remaining functions have the first argument S. The next two optimize() functions state that the evaluator is named poissonmle and that optimize() method gf2 is being used. The subsequent two optimize() functions indicate that the first additional argument after b in program poissonmle is y, and the second is X. The next function provides starting values and is necessary. The b = optimize(S) function initiates the optimization. The remaining functions compute robust standard errors and print the results.

The parameter estimates and standard errors are the same as those from the Stata poisson command with the vce(robust) option (see section 11.4.3). Nicely displayed results can be obtained by using the st_matrix() function to pass b´ and Vbrob from Mata to Stata and then by using the ereturn display command in Stata, exactly as in the section 11.2.3 example.

11.8 Generalized method of moments

As an example of GMM estimation, we consider the estimation of a Poisson model with endogenous regressors. Estimation using the gmm command, introduced in Stata 11, was presented in section 10.3.8. Here we present further explanation of the GMM technique and obtain the GMM estimates by using the Mata optimize() function.

The two-stage least-squares interpretation of linear instrumental variables (IV) does not extend to nonlinear models, so we cannot simply do Poisson regression with the endogenous regressor replaced by fitted values from a first-stage regression. And the objective function is not of a form well-suited for the Stata ml command, because it is a quadratic form in sums rather than a simple sum. Instead, we use Mata to code the GMM objective function, and then we use the Mata optimize() function.

11.8.1   Definition

The GMM begins with the population moment conditions

$$E\{\mathbf{h}(\mathbf{w}_i, \boldsymbol{\theta})\} = \mathbf{0} \tag{11.7}$$

where $\boldsymbol{\theta}$ is a $q \times 1$ vector, $\mathbf{h}(\cdot)$ is an $r \times 1$ vector function with $r \geq q$, and the vector $\mathbf{w}_i$ represents all observables including the dependent variable, regressors and, where relevant, IV. A leading example is linear IV (see section 6.2), where $\mathbf{h}(\mathbf{w}_i, \boldsymbol{\theta}) = \mathbf{z}_i(y_i - \mathbf{x}_i'\boldsymbol{\beta})$.

If $r = q$, then the method-of-moments (MM) estimator $\widehat{\boldsymbol{\theta}}_{\text{MM}}$ solves the corresponding sample moment condition $N^{-1}\sum_i \mathbf{h}(\mathbf{w}_i, \boldsymbol{\theta}) = \mathbf{0}$. This is not possible if $r > q$, such as for an overidentified linear IV model, because there are then more equations than parameters.

The GMM estimator $\widehat{\boldsymbol{\theta}}_{\text{GMM}}$ minimizes a quadratic form in $\sum_i \mathbf{h}(\mathbf{w}_i, \boldsymbol{\theta})$, with the objective function

$$Q(\boldsymbol{\theta}) = \left\{ \sum_{i=1}^{N} \mathbf{h}(\mathbf{w}_i, \boldsymbol{\theta}) \right\}' \mathbf{W} \left\{ \sum_{i=1}^{N} \mathbf{h}(\mathbf{w}_i, \boldsymbol{\theta}) \right\} \tag{11.8}$$

where the $r \times r$ weighting matrix $\mathbf{W}$ is positive-definite symmetric, possibly stochastic with a finite probability limit, and does not depend on $\boldsymbol{\theta}$. The MM estimator, the special case $r = q$, can be obtained most simply by letting $\mathbf{W} = \mathbf{I}$, or any other value, and then $Q(\boldsymbol{\theta}) = 0$ at the optimum.

Provided that condition (11.7) holds, the GMM estimator is consistent for $\boldsymbol{\theta}$ and is asymptotically normal with the robust estimate of the VCE

$$\widehat{V}(\widehat{\boldsymbol{\theta}}_{\text{GMM}}) = \left(\widehat{\mathbf{G}}'\mathbf{W}\widehat{\mathbf{G}}\right)^{-1}\widehat{\mathbf{G}}'\mathbf{W}\widehat{\mathbf{S}}\mathbf{W}\widehat{\mathbf{G}}\left(\widehat{\mathbf{G}}'\mathbf{W}\widehat{\mathbf{G}}\right)^{-1}$$

where, assuming independence over i,

$$\widehat{\mathbf{G}} = \sum_{i=1}^{N} \left.\frac{\partial \mathbf{h}_i}{\partial \boldsymbol{\theta}'}\right|_{\widehat{\boldsymbol{\theta}}} \tag{11.9}$$

$$\widehat{\mathbf{S}} = \sum_{i=1}^{N} \mathbf{h}_i(\widehat{\boldsymbol{\theta}})\mathbf{h}_i(\widehat{\boldsymbol{\theta}})'$$

For MM, the variance simplifies to $\left(\widehat{\mathbf{G}}'\widehat{\mathbf{S}}^{-1}\widehat{\mathbf{G}}\right)^{-1}$ regardless of the choice of $\mathbf{W}$. For GMM, different choices of $\mathbf{W}$ lead to different estimators. The best choice of $\mathbf{W}$ is $\mathbf{W} = \widehat{\mathbf{S}}^{-1}$, in which case again the variance simplifies to $(1/N)\left(\widehat{\mathbf{G}}'\widehat{\mathbf{S}}^{-1}\widehat{\mathbf{G}}\right)^{-1}$. For linear IV, an explicit formula for the estimator can be obtained; see section 6.2.

11.8.2   Nonlinear IV example

We consider the Poisson model with an endogenous regressor. There are several possible methods to control for endogeneity; see chapter 17. We consider use of the nonlinear IV (NLIV) estimator.

The Poisson regression model specifies that $E\{y - \exp(\mathbf{x}'\boldsymbol{\beta})|\mathbf{x}\} = 0$, because $E(y|\mathbf{x}) = \exp(\mathbf{x}'\boldsymbol{\beta})$. Suppose instead that $E\{y - \exp(\mathbf{x}'\boldsymbol{\beta})|\mathbf{x}\} \neq 0$, because of endogeneity of one or more regressors, but there are instruments $\mathbf{z}$ such that

$$E[\mathbf{z}_i\{y_i - \exp(\mathbf{x}'\boldsymbol{\beta})\}] = 0$$

Then the GMM estimator minimizes

$$Q(\boldsymbol{\beta}) = \mathbf{h}(\boldsymbol{\beta})'\mathbf{W}\mathbf{h}(\boldsymbol{\beta}) \tag{11.10}$$

where the $r \times 1$ vector $\mathbf{h}(\boldsymbol{\beta}) = \sum_i \mathbf{z}_i\{y_i - \exp(\mathbf{x}_i'\boldsymbol{\beta})\}$. This is a special case of (11.8) with $\mathbf{h}(\mathbf{w}_i, \boldsymbol{\theta}) = \mathbf{z}_i\{y_i - \exp(\mathbf{x}_i'\boldsymbol{\beta})\}$.

Define the $r \times K$ matrix $\mathbf{G}(\boldsymbol{\beta}) = -\sum_i \exp(\mathbf{x}_i'\boldsymbol{\beta})\mathbf{z}_i\mathbf{x}_i'$. Then the $K \times 1$ gradient vector

$$\mathbf{g}(\boldsymbol{\beta}) = \mathbf{G}(\boldsymbol{\beta})'\mathbf{W}\mathbf{h}(\boldsymbol{\beta}) \tag{11.11}$$

and the $K \times K$ expected Hessian is

$$\mathbf{H}(\boldsymbol{\beta}) = \mathbf{G}(\boldsymbol{\beta})'\mathbf{W}\mathbf{G}(\boldsymbol{\beta})'$$

where simplification has occurred by using $E\{\mathbf{h}(\boldsymbol{\beta})\} = \mathbf{0}$.

The estimate of the VCE is that in (11.9) with $\widehat{\mathbf{G}} = \mathbf{G}(\widehat{\boldsymbol{\beta}})$ and $\widehat{\mathbf{S}} = \sum_i\{y_i - \exp(\mathbf{x}_i'\widehat{\boldsymbol{\beta}})\}^2\mathbf{z}_i\mathbf{z}_i'$.

11.8.3 GMM using the Mata optimize() function

The first-order conditions $\mathbf{g}(\boldsymbol{\beta}) = \mathbf{0}$, where $\mathbf{g}(\boldsymbol{\beta})$ is given in (11.11), have no solution for $\boldsymbol{\beta}$, so we need to use an iterative method. The `ml` command is not well suited to this optimization because $Q(\boldsymbol{\beta})$ given in (11.10) is a quadratic form. Instead, we use the Mata `optimize()` function.

We let $\mathbf{W} = (\sum_i \mathbf{z}_i\mathbf{z}_i')^{-1}$ as for linear two-stage least squares. The following Mata expressions form the desired quantities, where we express the parameter vector b and gradient vector g as row vectors because the `optimize()` function requires row vectors. We have

```
: Xb = X*b´                        // b for optimize is 1 x k row vector
: mu = exp(Xb)
: h = Z´(y-mu)                     // h is r x 1 column row vector
: W = cholinv(Z´Z)                 // W is r x r wmatrix
: G = -(mu:*Z)´X                   // G is r x k matrix
: S = ((y-mu):*Z)´((y-mu):*Z)      // S is r x r matrix
: Qb = h´W*h                       // Q(b) is scalar
: g = G´W*h                        // gradient for optimize is 1 x k row vector
: H = G´W*G                        // Hessian for optimize is k x k matrix
: V = luinv(G´W*G)*G´W*S*W*G*luinv(G´W*G)
```

We fit a model for `docvis`, where `private` is endogenous and `firmsize` is used as an instrument, so the model is just-identified. We use `optimize()` method d2, where the objective function is given as a scalar and both a gradient vector and Hessian matrix are provided. The `optimize_result_V_robust(S)` command does not apply to d evaluators, so we compute the robust estimate of the VCE after optimization.

The structure of the Mata code is similar to that for the Poisson example explained in section 11.7.3. We have

```
. * Mata code to obtain GMM estimator for Poisson using command optimize
. mata
                                             ─────── mata (type end to exit) ───────
:     void pgmm(todo, b, y, X, Z, Qb, g, H)
>     {
>       Xb = X*b´
>       mu = exp(Xb)
>       h = Z´(y-mu)
>       W = cholinv(cross(Z,Z))
>       Qb = h´W*h
>       if (todo == 0) return
>       G = -(mu:*Z)´X
>       g = (G´W*h)´
>       if (todo == 1) return
>       H = G´W*G
>       _makesymmetric(H)
>     }
:     st_view(y=., ., "`y´")
:     st_view(X=., ., tokens("`xlist´"))
:     st_view(Z=., ., tokens("`zlist´"))
:     S = optimize_init()
:     optimize_init_which(S,"min")
:     optimize_init_evaluator(S, &pgmm())
:     optimize_init_evaluatortype(S, "d2")
:     optimize_init_argument(S, 1, y)
:     optimize_init_argument(S, 2, X)
:     optimize_init_argument(S, 3, Z)
:     optimize_init_params(S, J(1,cols(X),0))
:     optimize_init_technique(S,"nr")
:     b = optimize(S)
Iteration 0:    f(p) =  71995.212
Iteration 1:    f(p) =  9259.0408
Iteration 2:    f(p) =  1186.8103
Iteration 3:    f(p) =  3.4395408
Iteration 4:    f(p) =  .00006905
Iteration 5:    f(p) =  5.672e-14
Iteration 6:    f(p) =  1.861e-27
:     // Compute robust estimate of VCE and SEs
:     Xb = X*b´
:     mu = exp(Xb)
:     h = Z´(y-mu)
:     W = cholinv(cross(Z,Z))
:     G = -(mu:*Z)´X
:     Shat = ((y-mu):*Z)´((y-mu):*Z)*rows(X)/(rows(X)-cols(X))
:     Vb = luinv(G´W*G)*G´W*Shat*W*G*luinv(G´W*G)
```

```
  :  seb = (sqrt(diagonal(Vb)))´
  :  b \ seb
                        1              2             3             4              5

  1    1.340291853    1.072907529    .477817773    .0027832801   -.6832461817
  2    1.559899278    .0763116698    .0690784466   .0021932119    1.350370916

  :  end
```

The results are the same as those from the **gmm** command given in the section 10.3.8 example.

More generally, we could include additional instruments, which requires changing only the local macro for `zlist`. The model becomes overidentified and GMM estimates vary with choice of weighting matrix $\mathbf{W}$. The one-step GMM estimator is $\widehat{\boldsymbol{\beta}}$, given above. The two-step (or optimal) GMM estimator recalculates $\widehat{\boldsymbol{\beta}}$ by using the weighting matrix $\mathbf{W} = \widehat{\mathbf{S}}^{-1}$. This is illustrated in section 17.5.2 with the **gmm** command.

The Mata code is easily adapted to other cases where $E\{y - m(\mathbf{x}'\boldsymbol{\beta})|\mathbf{z}\} = 0$ for the specified function $m(\cdot)$, so it can be used, for example, for logit and probit models.

11.9 Stata resources

The key references are [R] **ml** and [R] **maximize**. Gould, Pitblado, and Sribney (2006) provides a succinct yet quite comprehensive overview of the **ml** method.

Nonlinear optimization is covered in Cameron and Trivedi (2005, chap. 10), Greene (2003, app. E.6), and Wooldridge (2002, chap. 12.7). GMM is covered in Cameron and Trivedi (2005, chap. 5), Greene (2003, chap. 18), and Wooldridge (2002, chap. 14).

11.10 Exercises

1. Consider estimation of the logit model covered in chapter 14. Then $Q(\boldsymbol{\beta}) = \sum_i \{y_i \ln \Lambda_i + (1 - y_i)\Lambda_i\}$, where $\Lambda_i = \Lambda(\mathbf{x}_i'\boldsymbol{\beta}) = \exp(\mathbf{x}_i'\boldsymbol{\beta})/\{1 + (\mathbf{x}_i'\boldsymbol{\beta})\}$. Show that $g(\boldsymbol{\beta}) = \sum_i (y_i - \Lambda_i)\mathbf{x}_i$ and $H(\boldsymbol{\beta}) = \sum_i -\Lambda_i(1 - \Lambda_i)\mathbf{x}_i\mathbf{x}_i'$. Hint: $\partial\Lambda(z)/\partial z = \Lambda(z)\{1 - \Lambda(z)\}$. Use the data on **docvis** to generate the binary variable d_dv for whether any doctor visits. Using just 2002 data, as in this chapter, use **logit** to perform logistic regression of the binary variable d_dv on **private**, **chronic**, **female**, **income**, and an intercept. Obtain robust estimates of the standard errors. You should find that the coefficient of **private**, for example, equals 1.27266, with a robust standard error of 0.0896928.

2. Adapt the code of section 11.2.3 to fit the logit model of exercise 1 using NR iterations coded in Mata. Hint: in defining an $n \times 1$ column vector with entries Λ_i, it may be helpful to use the fact that J(n,1,1) creates an $n \times 1$ vector of ones.

3. Adapt the code of section 11.4.3 to fit the logit model of exercise 1 using the `ml` command method `lf`.

4. Generate 100,000 observations from the following logit model DGP:

$$y_i = 1 \text{ if } \beta_1 + \beta_2 x_i + u_i > 0 \text{ and } y_i = 0 \text{ otherwise}; \ (\beta_1, \beta_2) = (0, 1); \ x_i \sim N(0, 1)$$

where u_i is logistically distributed. Using the inverse transformation method, a draw u from the logistic distribution can be computed as $u = -\ln\{(1-r)/r\}$, where r is a draw from the uniform distribution. Use data from this DGP to check the consistency of your estimation method in exercise 3 or, more simply, of the `logit` command.

5. Consider the NLS example in section 11.4.5 with an exponential conditional mean. Fit the model using the `ml` command and the `lfnls` program. Also fit the model using the `nl` command, given in section 10.3.5. Verify that these two methods give the same parameter estimates but, as noted in the text, the robust standard errors differ.

6. Continue the preceding exercise. Fit the model using the `ml` command and the `lfnls` program with default standard errors. These implicitly assume that the NLS model error has a variance of $\sigma^2 = 1$. Obtain an estimate of $s^2 = (1/N - K)\sum_i\{y_i - \exp(\mathbf{x}_i'\widehat{\boldsymbol{\beta}})\}^2$, using the postestimation `predictnl` command to obtain $\exp(\mathbf{x}_i'\widehat{\boldsymbol{\beta}})$. Then obtain an estimate of the VCE by multiplying the stored result `e(V)` by s^2. Obtain the standard error of $\widehat{\beta}_{\text{private}}$, and compare this with the standard error obtained when the NLS model is fit using the `nl` command with a default estimate of the VCE.

7. Consider a Poisson regression of `docvis` on the regressors `private`, `chronic`, `female`, and `income` and the programs given in section 11.6. Run the `ml model d0 d0pois` command, and confirm that you get the same output as produced by the code in section 11.6.2. Confirm that the nonrobust standard errors are the same as those obtained using `poisson` with default standard errors. Run `ml model d1 d1pois`, and confirm that you get the same output as produced by the code in section 11.6.2. Run `ml model d2 d2pois`, and confirm that you get the same output as that given in section 11.6.2.

8. Adapt the code of section 11.6.2 to fit the logit model of exercise 1 by using `ml` command method `d0`.

9. Adapt the code of section 11.6.4 to fit the logit model of exercise 1 by using `ml` command method `lf1` with robust standard errors reported.

10. Adapt the code of section 11.6.5 to fit the logit model of exercise 1 by using `ml` command method `d2`.

11. Consider the negative binomial example given in section 11.4.4. Fit this same model by using the `ml` command method `d0`. Hint: see the Weibull example in [R] ml.

12 Testing methods

12.1 Introduction

Econometric modeling is composed of a cycle of initial model specification, estimation, diagnostic checks, and model respecification. The diagnostic checks are often based on hypothesis tests for the statistical significance of key variables and model-specification tests. This chapter presents additional details on hypothesis tests, associated confidence intervals, and model-specification tests that are used widely throughout the book.

The emphasis is on Wald hypothesis tests and confidence intervals. These produce the standard regression output and can also be obtained by using the `test`, `testnl`, `lincom`, and `nlcom` commands. We also present the other two classical testing methods, likelihood-ratio and Lagrange multiplier tests, and Monte Carlo methods for obtaining test size and power. Finally, we discuss model-specification tests, including information matrix tests, goodness-of-fit tests, Hausman tests, and tests of overidentifying restrictions that are applied in various chapters. Model-selection tests for nonnested nonlinear models are not covered, though some of the methods given in chapter 3 for linear models can be extended to nonlinear models, and a brief discussion for likelihood-based nonlinear models is given in section 10.7.2.

12.2 Critical values and p-values

Before discussing Stata estimation and testing commands and associated output, we discuss how critical values and p-values are computed.

Introductory econometrics courses often emphasize use of the $t(n)$ and $F(h, n)$ distributions for hypothesis testing, where n is the degrees of freedom and h is the number of restrictions. For cross-section analysis, often $n = N - K$, where N is the sample size and K is the number of regressors. For clustered data, Stata sets $n = C - 1$, where C is the number of clusters.

These distributions hold exactly only in the very special case of tests of linear restrictions for the ordinary least-squares (OLS) estimator in the linear regression model with independent normal homoskedastic errors. Instead, virtually all inference in microeconometrics is based on asymptotic theory. This is the case not only for nonlinear estimators but also for linear estimators, such as OLS and instrumental variables (IV), with robust standard errors. Then test statistics are asymptotically standard normal (Z) distributed rather than $t(n)$, and chi-squared [$\chi^2(h)$] distributed rather than $F(h, n)$.

12.2.1   Standard normal compared with Student's t

The change from $t(n)$ to standard normal distributions is relatively minor, unless n is small, say, less than 30. The two distributions are identical for $n \to \infty$. The t distribution has fatter tails, leading to larger p-values and critical values than the standard normal at conventional levels of significance such as 0.05. In this book, we rely on asymptotic approximations that require samples much larger than 30 and for which the difference between the two distributions is negligible.

12.2.2   Chi-squared compared with F

Many tests of joint hypotheses use the χ^2 distribution. A $\chi^2(h)$ random variable has a mean of h, a variance of $2h$, and for $h > 7$, the 5% critical value lies between h and $2h$.

The $\chi^2(h)$ distribution is scaled quite differently from the F. As the denominator degrees of freedom of the F go to infinity, we have

$$F(h, n) \to \frac{\chi^2(h)}{h} \text{ as } n \to \infty \tag{12.1}$$

Thus, if asymptotic theory leads to a test statistic that is $\chi^2(h)$ distributed, then division of this statistic by h leads to a statistic that is approximately $F(h, n)$ distributed if n is large. In finite samples, the $F(h, n)$ distribution has fatter tails than $\chi^2(h)/h$, leading to larger p-values and critical values for the F compared with the χ^2.

12.2.3   Plotting densities

We compare the density of a $\chi^2(5)$ random variable with a random variable that is five times a $F(5, 30)$ random variable. From (12.1), the two are the same for large n but will differ for $n = 30$. In practice, $n = 30$ is not large enough for asymptotic theory to approximate the finite distribution well.

One way to compare is to evaluate the formulas for the respective densities at a range of points, say, 0.1, 0.2, ..., 20.0, and graph the density values against the evaluation points. The `graph twoway function` command automates this method. This is left as an exercise.

This approach requires providing density formulas that can be quite complicated and may even be unknown to the user if the density is that of a mixture distribution, for example. A simpler way is to make many draws from the respective distributions, using the methods of section 4.2, and then use the `kdensity` command to compute and graph the kernel density estimate.

We take this latter approach. We begin by taking 10,000 draws from each distribution. We use the `rchi2()` function introduced in an update to Stata 10; see section 4.2.

```
. * Create many draws from chi(5) and 5*F(5,30) distributions
. set seed 10101
. quietly set obs 10000
. generate chi5 = rchi2(5)            // result xc ~ chisquared(10)
. generate xfn = rchi2(5)/5           // result numerator of F(5,30)
. generate xfd = rchi2(30)/30         // result denominator of F(5,30)
. generate f5_30 = xfn/xfd            // result xf ~ F(5,30)
. generate five_x_f5_30 = 5*f5_30
. summarize chi5 five_x_f5_30
```

Variable	Obs	Mean	Std. Dev.	Min	Max
chi5	10000	5.057557	3.203747	.0449371	26.48803
five_x_f5_30	10000	5.314399	3.777524	.1302123	30.91342

For chi5, the average of 5.06 is close to the theoretical mean of 5, and the sample variance of $3.204^2 = 10.27$ is close to the theoretical variance of 10. For five_x_f5_30, the sample variance of $3.778^2 = 14.27$ is much larger than that of chi5, reflecting the previously mentioned fatter tails.

We then plot the kernel density estimates based on these draws. To improve graph readability the kernel density estimates are plotted only for draws less than 25, using a method already explained in section 3.2.7. To produce smoother plots, we increase the default bandwidth to 1.0, an alternative being to increase the number of draws. We have

```
. * Plot the densities for these two distributions using kdensity
. label var chi5 "chi(5)"
. label var five_x_f5_30 "5*F(5,30)"
. kdensity chi5, bw(1.0) generate (kx1 kd1) n(500)
. kdensity five_x_f5_30, bw(1.0) generate (kx2 kd2) n(500)
. quietly drop if (chi5 > 25  |  five_x_f5_30 > 25)
. graph twoway (line kd1 kx1) if kx1 < 25, name(chi)
. graph twoway (line kd2 kx2) if kx2 < 25, name(F)
. graph twoway (line kd1 kx1) (line kd2 kx2, clstyle(p3)) if kx1 < 25,
> scale(1.2) plotregion(style(none))
> title("Chisquared(5) and 5*F(5,30) Densities")
> xtitle("y", size(medlarge)) xscale(titlegap(*5))
> ytitle("Density f(y)", size(medlarge)) yscale(titlegap(*5))
> legend(pos(1) ring(0) col(1)) legend(size(small))
> legend(label(1 "Chi(5)") label(2 "5*F(5,30)"))
```

In figure 12.1, the two densities appear similar, though the density of $5 \times F(5,30)$ has a longer tail than that of $\chi^2(5)$, and it is the tails that are used for tests at a 0.05 level and for 95% confidence intervals. The difference disappears as the denominator degrees of freedom (here 30) goes to infinity.

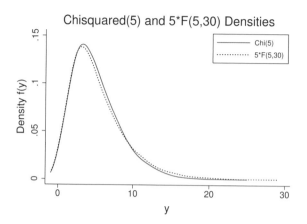

Figure 12.1. $\chi^2(5)$ density compared with 5 times $F(5, 30)$ density

12.2.4   Computing p-values and critical values

Stata output automatically provides p-values but not critical values. The p-values can be obtained manually from the relevant cumulative distribution function (c.d.f.), whereas critical values can be obtained by using the inverse c.d.f. The precise Stata functions vary with the distribution. For details, see [D] **functions** or type `help density functions`.

We compute p-values for the test of a single restriction ($h = 1$). We suppose the test statistic is equal to 2 by using the $t(30)$ or Z distributions. In that case, it is equivalently equal to $2^2 = 4$ by using the $F(1, 30)$ or $\chi^2(1)$ distributions. We have

```
. * p-values for t(30), F(1,30), Z, and chi(1) at y = 2
. scalar y = 2
. scalar p_t30 = 2*ttail(30,y)
. scalar p_f1and30 = Ftail(1,30,y^2)
. scalar p_z = 2*(1-normal(y))
. scalar p_chi1 = chi2tail(1,y^2)
. display "p-values" "  t(30) =" %7.4f p_t30 "  F(1,30)=" %7.4f
> p_f1and30 "  z =" %7.4f p_z "  chi(1)=" %7.4f p_chi1
p-values  t(30) = 0.0546  F(1,30)= 0.0546  z = 0.0455  chi(1)= 0.0455
```

The general properties that $Z^2 = \chi^2(1)$ and $t(n)^2 = F(1, n)$ are confirmed for this example. Also $t(n) \to Z$ and $F(1, n)/1 \to \chi^2(1)$ as $n \to \infty$, but there is still a difference for $n = 30$, with a p-value of 0.0455 compared with 0.0546.

We next compute critical values for these distributions for a two-sided test of a single restriction at a level of 0.05. We have

```
. * Critical values for t(30), F(1,30), Z, and chi(1) at level 0.05
. scalar alpha = 0.05
. scalar c_t30 = invttail(30,alpha/2)
. scalar c_f1and30 = invFtail(1,30,alpha)
. scalar c_z = -invnormal(alpha/2)
. scalar c_chi1 = invchi2(1,1-alpha)
. display "critical values" "  t(30) =" %7.3f c_t30 "  F(1,30)=" %7.3f
> c_f1and30 "  z =" %7.3f c_z "  chi(1)=" %7.3f c_chi1
critical values  t(30) =  2.042  F(1,30)=  4.171  z =  1.960  chi(1)=  3.841
```

Again $t(30)^2 = F(1, 30)$ and $Z^2 = \chi^2(1)$, whereas $t(30) \simeq Z$ and $F(1, 30)/1 \simeq \chi^2(1)$.

12.2.5 Which distributions does Stata use?

In practice, the t and F distributions may continue to be used as an ad hoc finite-sample correction, even when only asymptotic results supporting the Z and χ^2 distributions are available. This leads to more conservative inference, with less likelihood of rejecting the null hypothesis because p-values are larger and with wider confidence intervals because critical values are larger.

Stata uses the t and F distributions for variants of least squares, such as robust standard errors, nonlinear least squares, linear IV, and quantile regression. Stata uses the Z and χ^2 distributions in most other cases, notably, maximum likelihood (ML) and quasi-ML nonlinear estimators, such as logit, tobit, and Poisson.

12.3 Wald tests and confidence intervals

A quite universal method for hypothesis testing and obtaining confidence intervals is the Wald method, based on the estimated variance–covariance matrix of the estimator (VCE) presented in sections 3.3 and 10.4. This method produces the test statistics and p-values for a test of the significance of individual coefficients, the confidence intervals for individual coefficients, and the tests of overall significance that are given in Stata regression output.

Here we provide background on the Wald test, extension to tests of more complicated hypotheses that require the use of the `test` and `testnl` commands, and extension to confidence intervals on combinations of parameters using the `lincom` and `nlcom` commands.

12.3.1 Wald test of linear hypotheses

By a linear hypothesis, we mean one that can be expressed as a linear combination of parameters. Single hypothesis examples include $H_0: \beta_2 = 0$ and $H_0: \beta_2 - \beta_3 - 5 = 0$. A joint hypothesis example tests the two preceding hypotheses simultaneously.

The Wald test method is intuitively appealing. The test is based on how well the corresponding parameter estimates satisfy the null hypothesis. For example, to test $H_0: \beta_2 - \beta_3 - 5 = 0$, we ask whether $\widehat{\beta}_2 - \widehat{\beta}_3 - 5 \simeq 0$. To implement the test, we need to know the distribution of $\widehat{\beta}_2 - \widehat{\beta}_3 - 5$. But this is easy because the estimators used in this book are asymptotically normal, and a linear combination of normals is normal.

We do need to find the variance of this normal distribution. In this example, $\text{Var}(\widehat{\beta}_2 - \widehat{\beta}_3 - 5) = \text{Var}(\widehat{\beta}_2) + \text{Var}(\widehat{\beta}_3) - 2\text{Cov}(\widehat{\beta}_2, \widehat{\beta}_3)$, because for the random variables X and Y, $\text{Var}(X - Y) = \text{Var}(X) + \text{Var}(Y) - 2\text{Cov}(X, Y)$. More generally, it is helpful to use matrix notation, which we now introduce.

Let $\boldsymbol{\beta}$ denote the $K \times 1$ parameter vector, where the results also apply if instead we use the more general notation $\boldsymbol{\theta}$ that includes $\boldsymbol{\beta}$ and any auxiliary parameters. Then, for example, $H_0: \beta_2 = 0$ and $\beta_2 - \beta_3 - 5 = 0$ can be written as

$$
\begin{bmatrix} \beta_2 \\ \beta_2 - \beta_3 - 5 \end{bmatrix} = \begin{bmatrix} 0 & 1 & 0 & 0 & \cdots & 0 \\ 0 & 1 & -1 & 0 & \cdots & 0 \end{bmatrix} \begin{bmatrix} \beta_1 \\ \beta_2 \\ \beta_3 \\ \vdots \\ \beta_k \end{bmatrix} - \begin{bmatrix} 0 \\ 5 \end{bmatrix} = \begin{bmatrix} 0 \\ 0 \end{bmatrix}
$$

This linear combination can be written as $\mathbf{R}\beta - \mathbf{r} = \mathbf{0}$.

For a two-sided test of h linear hypotheses under H_0, we therefore test

$$H_0 : \mathbf{R}\boldsymbol{\beta} - \mathbf{r} = \mathbf{0}$$
$$H_a : \mathbf{R}\boldsymbol{\beta} - \mathbf{r} \neq \mathbf{0}$$

where $\mathbf{R}$ is an $h \times K$ matrix and $\mathbf{r}$ is an $h \times 1$ vector, $h \leq K$. Standard examples include tests of individual exclusions restrictions, $\beta_j = 0$, and tests of joint statistical significance, $\beta_2 = 0, \ldots, \beta_q = 0$ (with β_1 as an intercept coefficient).

The Wald test uses the quite intuitive approach of rejecting $H_0 : \mathbf{R}\boldsymbol{\beta} - \mathbf{r} = \mathbf{0}$ if $\mathbf{R}\widehat{\boldsymbol{\beta}} - \mathbf{r}$ is considerably different from zero. Now

$$
\begin{aligned}
\widehat{\boldsymbol{\beta}} &\stackrel{a}{\sim} N\{\boldsymbol{\beta}, \text{Var}(\widehat{\boldsymbol{\beta}})\} \\
\implies \quad \mathbf{R}\widehat{\boldsymbol{\beta}} - \mathbf{r} &\stackrel{a}{\sim} N\{\mathbf{R}\boldsymbol{\beta} - \mathbf{r}, \mathbf{R}V(\widehat{\boldsymbol{\beta}})\mathbf{R}'\} \\
\implies \quad \mathbf{R}\widehat{\boldsymbol{\beta}} - \mathbf{r} &\stackrel{a}{\sim} N\{\mathbf{0}, \mathbf{R}\text{Var}(\widehat{\boldsymbol{\beta}})\mathbf{R}'\} \qquad \text{under } H_0
\end{aligned}
\qquad (12.2)
$$

For a single hypothesis, $\mathbf{R}\widehat{\boldsymbol{\beta}} - \mathbf{r}$ is a scalar that is univariate normally distributed, so we can transform to a standard normal variate and use standard normal tables.

More generally, there are multiple hypotheses. To avoid using the multivariate normal distribution, we transform to a chi-squared distribution. If the $h \times 1$ vector $\mathbf{y} \sim N(\boldsymbol{\mu}, \boldsymbol{\Sigma})$, then $(\mathbf{y} - \mu)'\boldsymbol{\Sigma}^{-1}(\mathbf{y} - \mu) \sim \chi^2(h)$. Applying this result to (12.2), we obtain the Wald statistic for the test of $H_0: \mathbf{R}\boldsymbol{\beta} - \mathbf{r} = \mathbf{0}$:

$$W = (\mathbf{R}\widehat{\boldsymbol{\beta}} - \mathbf{r})' \left\{ \mathbf{R}\widehat{V}(\widehat{\boldsymbol{\beta}})\mathbf{R}' \right\}^{-1} (\mathbf{R}\widehat{\boldsymbol{\beta}} - \mathbf{r}) \stackrel{a}{\sim} \chi^2(h) \text{ under } H_0 \qquad (12.3)$$

Large values of W lead to rejection of H_0. At a level of 0.05, for example, we reject H_0 if the p-value $p = \Pr\{\chi^2(h) > W\} < 0.05$, or if W exceeds the critical value $c = \chi_{0.05}^2(h)$, where by $\chi_{0.05}^2(h)$ we mean the area in the right tail is 0.05.

In going from (12.2) to (12.3), we also replaced $\mathrm{Var}(\widehat{\boldsymbol{\beta}})$ by an estimate, $\widehat{V}(\widehat{\boldsymbol{\beta}})$. For the test to be valid, the estimate $\widehat{V}(\widehat{\boldsymbol{\beta}})$ must be consistent for $\mathrm{Var}(\widehat{\boldsymbol{\beta}})$, i.e., we need to use a correct estimator of the VCE.

An alternative test statistic is the F statistic, which is the Wald statistic divided by the number of restrictions. Then

$$F = \frac{W}{h} \stackrel{a}{\sim} F(h, N - K) \text{ under } H_0 \tag{12.4}$$

where K denotes the number of parameters in the regression model. Large values of F lead to rejection of H_0. At a level of 0.05, for example, we reject H_0 if the p-value $p = \Pr\{F(h, N - k) > F\} < 0.05$, or if F exceeds the critical value $c = F_{0.05}(h, N - K)$.

12.3.2   The test command

The Wald test can be performed by using the `test` command. Usually, the W in (12.3) is used, though the F in (12.4) is frequently used after fitting linear models. As the equations state, W has a large-sample chi-squared distribution, and F has a large-sample F distribution.

The `test` command has several different syntaxes. The simplest two are

`test` *coeflist* $\big[$ `,` *options* $\big]$

`test` *exp* `=` *exp* $\big[$ `=` `...` $\big]$ $\big[$ `,` *options* $\big]$

The syntax is best explained with examples. More complicated syntax enables testing across equations in multiequation models. A multiequation example following the `sureg` command was given in section 5.4. An example following `nbreg` is given as an end-of-chapter exercise. It can be difficult to know the Stata convention for naming coefficients in some cases; the output from `estat vce` may give the appropriate complete names.

The options are usually not needed. They include `mtest` to test each hypothesis separately if several hypotheses are given and `accumulate` to test hypotheses jointly with previously tested hypotheses.

(Continued on next page)

We illustrate Wald tests using the same model and data as in chapter 10. We have

```
. * Estimate Poisson model of chapter 10
. use mus10data.dta, clear
. quietly keep if year02==1
. poisson docvis private chronic female income, vce(robust) nolog
Poisson regression                                Number of obs   =       4412
                                                  Wald chi2(4)    =     594.72
                                                  Prob > chi2     =     0.0000
Log pseudolikelihood = -18503.549                 Pseudo R2       =     0.1930
```

docvis	Coef.	Robust Std. Err.	z	P>|z|	[95% Conf. Interval]	
private	.7986652	.1090014	7.33	0.000	.5850263	1.012304
chronic	1.091865	.0559951	19.50	0.000	.9821167	1.201614
female	.4925481	.0585365	8.41	0.000	.3778187	.6072774
income	.003557	.0010825	3.29	0.001	.0014354	.0056787
_cons	-.2297262	.1108732	-2.07	0.038	-.4470338	-.0124186

Test single coefficient

To test whether a single coefficient equals zero, we just need to specify the regressor name. For example, to test $H_0: \beta_{\texttt{female}} = 0$, we have

```
. * Test a single coefficient equal 0
. test female
 ( 1)  [docvis]female = 0
            chi2(  1) =     70.80
          Prob > chi2 =     0.0000
```

We reject H_0 because $p < 0.05$ and conclude that female is statistically significant at the level of 0.05. The test statistic is the square of the z statistic given in the regression output $(8.414^2 = 70.80)$, and the p-values are the same.

Test several hypotheses

As an example of testing more than one hypothesis, we test $H_0 : \beta_{\texttt{female}} = 0$ and $\beta_{\texttt{private}} + \beta_{\texttt{chronic}} = 1$. Then

```
. * Test two hypotheses jointly using test
. test (female) (private + chronic = 1)
 ( 1)  [docvis]female = 0
 ( 2)  [docvis]private + [docvis]chronic = 1
            chi2(  2) =    122.29
          Prob > chi2 =     0.0000
```

We reject H_0 because $p < 0.05$.

The `mtest` option additionally tests each hypothesis in isolation. We have

```
. * Test each hypothesis in isolation as well as jointly
. test (female) (private + chronic = 1), mtest
( 1)  [docvis]female = 0
( 2)  [docvis]private + [docvis]chronic = 1
```

	chi2	df	p
(1)	70.80	1	0.0000 #
(2)	56.53	1	0.0000 #
all	122.29	2	0.0000

```
                   # unadjusted p-values
```

As expected, the hypothesis test value of 70.80 for `female` equals that given earlier when the hypothesis was tested in isolation.

The preceding test makes no adjustment to *p*-values to account for multiple testing. Options to `mtest` include several that implement Bonferroni's method and variations.

Test of overall significance

The `test` command can be used to test overall significance. We have

```
. * Wald test of overall significance
. test private chronic female income
( 1)  [docvis]private = 0
( 2)  [docvis]chronic = 0
( 3)  [docvis]female = 0
( 4)  [docvis]income = 0

        chi2(  4) =  594.72
      Prob > chi2 =    0.0000
```

The Wald test statistic value of 594.72 is the same as that given in the `poisson` output.

Test calculated from retrieved coefficients and VCE

For pedagogical purposes, we compute this overall test manually even though we use `test` in practice. The computation requires retrieving $\widehat{\boldsymbol{\beta}}$ and $\widehat{V}(\widehat{\boldsymbol{\beta}})$, defining the appropriate matrices $\mathbf{R}$ and $\mathbf{r}$, and calculating W defined in (12.3). In doing so, we note that Stata stores regression coefficients as a row vector, so we need to transpose to get the $K \times 1$ column vector $\widehat{\boldsymbol{\beta}}$. Because we use Stata estimates of $\widehat{\boldsymbol{\beta}}$ and $\widehat{V}(\widehat{\boldsymbol{\beta}})$, in defining $\mathbf{R}$ and $\mathbf{r}$ we need to follow the Stata convention of placing the intercept coefficient as the last coefficient. We have

```
. * Manually compute overall test of significance using the formula for W
. quietly poisson docvis private chronic female income, vce(robust)
. matrix b = e(b)´
. matrix V = e(V)
```

```
. matrix R = (1,0,0,0,0 \ 0,1,0,0,0 \ 0,0,1,0,0 \ 0,0,0,1,0 )
. matrix r = (0 \ 0 \ 0 \ 0)
. matrix W = (R*b-r)´*invsym(R*V*R´)*(R*b-r)
. scalar Wald = W[1,1]
. scalar h = rowsof(R)
. display "Wald test statistic: " Wald "  with p-value: " chi2tail(h,Wald)
Wald test statistic: 594.72457  with p-value: 2.35e-131
```

The value of 594.72 is the same as that from the `test` command.

12.3.3 One-sided Wald tests

The preceding tests are two-sided tests, such as $\beta_j = 0$ against $\beta_j \neq 0$. We now consider one-sided tests of a single hypothesis, such as a test of whether $\beta_j > 0$.

The first step in conducting a one-sided test is determining which side is H_0 and which side is H_a. The convention is that the claim made is set as the alternative hypothesis. For example, if the claim is made that the jth regressor has a positive marginal effect and this means that $\beta_j > 0$, then we test $H_0 : \beta_j \leq 0$ against $H_a : \beta_j > 0$.

The second step is to obtain a test statistic. For tests on a single regressor, we use the z statistic

$$z = \frac{\widehat{\beta}_j}{s_{\widehat{\beta}_j}} \overset{a}{\sim} N(0,1) \text{ under } H_0$$

where $z^2 = W$ given in (12.3). In some cases, the $t(n-K)$ distribution is used, in which case the z statistic is called a t statistic. Regression output gives this statistic, along with p-values for two-sided tests. For a one-sided test, these p-values should be halved, with the important condition that it is necessary to check that $\widehat{\beta}_j$ has the correct sign. For example, if testing $H_0 : \beta_j \leq 0$ against $H_a : \beta_j > 0$, then we reject H_0 at the level of 0.05 if $\widehat{\beta}_j > 0$ and the reported two-sided p-value is less than 0.10. If instead $\widehat{\beta}_j < 0$, the p-value for a one-sided test must be at least 0.50, because we are on the wrong side of zero, leading to certain rejection at conventional statistical significance levels.

As an example, consider a test of the claim that doctor visits increase with income, even after controlling for chronic conditions, gender, and income. The appropriate test of this claim is one of $H_0 : \beta_{\text{income}} \leq 0$ against $H_a : \beta_{\text{income}} > 0$. The `poisson` output includes $\widehat{\beta}_{\text{income}} = 0.0036$ with $p = 0.001$ for a two-sided test. Because $\widehat{\beta}_{\text{income}} > 0$, we simply halve the two-sided test p-value to get $p = 0.001/2 = 0.0005 < 0.05$. So we reject $H_0 : \beta_{\text{income}} \leq 0$ at the 0.05 level.

More generally, suppose we want to test the single hypothesis $H_0 : \mathbf{R}\beta - r \leq 0$ against $H_a : \mathbf{R}\beta - r > 0$, where here $\mathbf{R}\beta - r$ is a scalar. Then we use

$$z = \frac{\mathbf{R}\widehat{\beta} - r}{s_{\mathbf{R}\widehat{\beta}-r}} \overset{a}{\sim} N(0,1) \text{ under } H_0$$

When squared, this statistic again equals the corresponding Wald test, i.e., $z^2 = W$. The `test` command gives W, but z could be either $\sqrt{W}$ or $-\sqrt{W}$, and the sign of z is needed to be able to perform the one-sided test. To obtain the sign, we can also compute $\mathbf{R}\widehat{\boldsymbol{\beta}} - r$ by using the `lincom` command; see section 12.3.7. If $\mathbf{R}\widehat{\boldsymbol{\beta}} - r$ has a sign that differs from that of $\mathbf{R}\boldsymbol{\beta} - r$ under H_0, then the p-value is one half of the two-sided p-value given by `test` (or by `lincom`); we reject H_0 at the α level if this adjusted p-value is less than α and do not reject otherwise. If instead $\mathbf{R}\widehat{\boldsymbol{\beta}} - r$ has the same sign as that of $\mathbf{R}\boldsymbol{\beta} - r$ under H_0, then we always do not reject H_0.

12.3.4   Wald test of nonlinear hypotheses (delta method)

Not all hypotheses are linear combinations of parameters. A nonlinear hypothesis example is a test of $H_0 : \beta_2/\beta_3 = 1$ against $H_a : \beta_2/\beta_3 \neq 1$. This can be expressed as a test of $g(\boldsymbol{\beta}) = 0$, where $g(\boldsymbol{\beta}) = \beta_2/\beta_3 - 1$. More generally, there can be h hypotheses combined into the $h \times 1$ vector $\mathbf{g}(\boldsymbol{\beta}) = \mathbf{0}$, with each separate hypothesis being a separate row in $\mathbf{g}(\boldsymbol{\beta})$. Linear hypotheses are the special case of $\mathbf{g}(\boldsymbol{\beta}) = \mathbf{R}\boldsymbol{\beta} - \mathbf{r}$.

The Wald test method is now based on the closeness of $\mathbf{g}(\widehat{\boldsymbol{\beta}})$ to $\mathbf{0}$. Because $\widehat{\boldsymbol{\beta}}$ is asymptotically normal, so too is $\mathbf{g}(\widehat{\boldsymbol{\beta}})$. Some algebra that includes linearization of $\mathbf{g}(\widehat{\boldsymbol{\beta}})$ using a Taylor-series expansion yields the Wald test statistic for the nonlinear hypotheses $H_0 : \mathbf{g}(\boldsymbol{\beta}) = \mathbf{0}$:

$$W = \mathbf{g}(\widehat{\boldsymbol{\beta}})' \left\{ \widehat{\mathbf{R}} \widehat{V}(\widehat{\boldsymbol{\beta}}) \widehat{\mathbf{R}}' \right\}^{-1} \mathbf{g}(\widehat{\boldsymbol{\beta}}) \stackrel{a}{\sim} \chi^2(h) \text{ under } H_0, \text{ where } \widehat{\mathbf{R}} = \left. \frac{\partial \mathbf{g}(\boldsymbol{\beta})'}{\partial \boldsymbol{\beta}} \right|_{\widehat{\boldsymbol{\beta}}} \quad (12.5)$$

This is the same test statistic as W in (12.3) upon replacement of $\mathbf{R}\widehat{\boldsymbol{\beta}} - r$ by $\mathbf{g}(\widehat{\boldsymbol{\beta}})$ and replacement of $\mathbf{R}$ by $\widehat{\mathbf{R}}$. Again large values of W lead to rejection of H_0, and $p = \Pr\{\chi^2(h) > W\}$.

The test statistic is often called one based on the delta method because of the derivative used to form $\widehat{\mathbf{R}}$.

12.3.5   The testnl command

The Wald test for nonlinear hypotheses is performed using the `testnl` command. The basic syntax is

```
testnl exp = exp [= ...] [ , options ]
```

The main option is `mtest` to separately test each hypothesis in a joint test.

As an example, we consider a test of $H_0 : \beta_{\text{female}}/\beta_{\text{private}} - 1 = 0$ against $H_a : \beta_{\text{female}}/\beta_{\text{private}} - 1 \neq 0$. Then

```
. * Test a nonlinear hypothesis using testnl
. testnl _b[female]/_b[private] = 1

 (1)  _b[female]/_b[private] = 1

              chi2(1) =        13.51
           Prob > chi2 =        0.0002
```

We reject H_0 at the 0.05 level because $p < 0.05$.

The hypothesis in the preceding example can be equivalently expressed as $\beta_{\text{female}} = \beta_{\text{private}}$. So a simpler test is

```
. * Wald test is not invariant
. test female = private

 ( 1)  - [docvis]private + [docvis]female = 0

            chi2( 1) =      6.85
         Prob > chi2 =      0.0088
```

Surprisingly, we get different values for the test statistic and p-value, even though both methods are valid and are asymptotically equivalent. This illustrates a weakness of Wald tests: in finite samples, they are not invariant to nonlinear transformations of the null hypothesis. With one representation of the null hypothesis, we might reject H_0 at the α level, whereas with a different representation we might not. Likelihood-ratio and Lagrange multiplier tests do not have this weakness.

12.3.6   Wald confidence intervals

Stata output provides Wald confidence intervals for individual regression parameters β_j of the form $\widehat{\beta}_j \pm z_{\alpha/2} \times s_{\widehat{\beta}_j}$, where $z_{\alpha/2}$ is a standard normal critical value. For some linear-model commands, the critical value is from the t distribution rather than the standard normal. The default is a 95% confidence interval, which is $\widehat{\beta}_j \pm 1.96 \times s_{\widehat{\beta}_j}$ if standard normal critical values (with $\alpha = 0.05$) are used. This default can be changed in Stata estimation commands by using the `level()` option, or it can be changed globally by using the `set level` command.

Now consider any scalar, say, γ, that is a function $g(\boldsymbol{\beta})$ of $\boldsymbol{\beta}$. Examples include $\gamma = \beta_2$, $\gamma = \beta_2 + \beta_3$, and $\gamma = \beta_2/\beta_3$. A Wald $100(1-\alpha)\%$ confidence interval for γ is

$$\widehat{\gamma} \pm z_{\alpha/2} \times s_{\widehat{\gamma}} \qquad (12.6)$$

where $\widehat{\gamma} = g(\widehat{\beta})$, and $s_{\widehat{\gamma}}$ is the standard error of $\widehat{\gamma}$. For the nonlinear estimator $\widehat{\boldsymbol{\beta}}$, the critical value $z_{\alpha/2}$ is usually used, and for the linear estimator, the critical value $t_{\alpha/2}$ is usually used. Implementation requires computation of $\widehat{\gamma}$ and $s_{\widehat{\gamma}}$, using (12.7) and (12.8) given below.

12.3.7   The lincom command

The `lincom` command calculates the confidence interval for a scalar linear combination of the parameters $\mathbf{R}\boldsymbol{\beta} - r$. The syntax is

```
lincom exp [ , options ]
```

The `eform` reports exponentiated coefficients, standard errors, and confidence intervals. This is explained in section 12.3.9.

The confidence interval is computed by using (12.6), with $\widehat{\gamma} = \mathbf{R}\widehat{\boldsymbol{\beta}} - r$ and the squared standard error

$$s_{\widehat{\gamma}}^2 = \mathbf{R}\widehat{V}(\widehat{\boldsymbol{\beta}})\mathbf{R}' \tag{12.7}$$

We consider a confidence interval for $\beta_{\texttt{private}} + \beta_{\texttt{chronic}} - 1$. We have

```
. * Confidence interval for linear combinations using lincom
. use mus10data.dta, clear
. quietly keep if year02==1
. quietly poisson docvis private chronic female income if year02==1, vce(robust)
. lincom private + chronic - 1
 ( 1)  [docvis]private + [docvis]chronic = 1
```

docvis	Coef.	Std. Err.	z	P>|z|	[95% Conf. Interval]	
(1)	.8905303	.1184395	7.52	0.000	.6583932	1.122668

The 95% confidence interval is $[0.66, 1.12]$ and is based on standard normal critical values because we used the `lincom` command after `poisson`. If instead it had followed `regress`, then $t(N - K)$ critical values would have been used.

The `lincom` command also provides a test statistic and p-value for the two-sided test of $H_0 : \beta_{\texttt{private}} + \beta_{\texttt{chronic}} - 1 = 0$. Then $z^2 = 7.52^2 \approx (0.8905303/0.1184395)^2 = 56.53$, which equals the W obtained in section 12.3.2 in the example using `test, mtest`. The `lincom` command enables a one-sided test because, unlike using W, we know the sign of z.

12.3.8   The nlcom command (delta method)

The `nlcom` command calculates the confidence intervals in (12.6) for a scalar nonlinear function $g(\boldsymbol{\beta})$ of the parameters. The syntax is

```
nlcom [ name: ] exp [ , options ]
```

The confidence interval is computed by using (12.6), with $\widehat{\gamma} = g(\widehat{\boldsymbol{\beta}})$ and the squared standard error

$$s_{\widehat{\gamma}}^2 = \partial\gamma/\partial\boldsymbol{\theta}|_{\widehat{\boldsymbol{\theta}}}\, \widehat{V}(\widehat{\boldsymbol{\theta}})\, \partial\gamma/\partial\boldsymbol{\theta}'|_{\widehat{\boldsymbol{\theta}}} \tag{12.8}$$

The standard error $s_{\widehat{\gamma}}$ and the resulting confidence interval are said to be computed by the delta method because of the derivative $\partial\gamma/\partial\boldsymbol{\theta}$.

As an example, consider confidence intervals for $\gamma = \beta_{\text{female}}/\beta_{\text{private}} - 1$. We have

```
. * Confidence interval for nonlinear function of parameters using nlcom
. nlcom _b[female] / _b[private] - 1
       _nl_1:  _b[female] / _b[private] - 1
```

docvis	Coef.	Std. Err.	z	P>|z|	[95% Conf. Interval]	
_nl_1	-.383286	.1042734	-3.68	0.000	-.587658	-.1789139

Note that $z^2 = (-3.68)^2 \approx (-0.383286/0.1042734)^2 = 13.51$. This equals the W for the test of $H_0\colon \beta_{\text{female}}/\beta_{\text{private}} - 1$ obtained by using the `testnl` command in section 12.3.5.

12.3.9   Asymmetric confidence intervals

For several nonlinear models, such as those for binary outcomes and durations, interest often lies in exponentiated coefficients that are given names such as hazard ratio or odds ratio depending on the application. In these cases, we need a confidence interval for $e^{\widehat{\beta}}$ rather than $\widehat{\beta}$. This can be done by using either the `lincom` command with the `eform` option, or the `nlcom` command. These methods lead to different confidence intervals, with the former preferred.

We can directly obtain a 95% confidence interval for $\exp(\beta_{\text{private}})$, using the `lincom, eform` command. We have

```
. * CI for exp(b) using lincom option eform
. lincom private, eform
 ( 1)  [docvis]private = 0
```

docvis	exp(b)	Std. Err.	z	P>|z|	[95% Conf. Interval]	
(1)	2.222572	.2422636	7.33	0.000	1.795038	2.751935

This confidence interval is computed by first obtaining the usual 95% confidence interval for β_{private} and then exponentiating the lower and upper bounds of the interval. We have

```
. * CI for exp(b) using lincom followed by exponentiate
. lincom private
 ( 1)  [docvis]private = 0
```

docvis	Coef.	Std. Err.	z	P>|z|	[95% Conf. Interval]	
(1)	.7986652	.1090014	7.33	0.000	.5850263	1.012304

Because $\beta_{\text{private}} \in [0.5850, 1.0123]$, it follows that $\exp(\beta_{\text{private}}) \in [e^{0.5850}, e^{1.0123}]$, so $\exp(\beta_{\text{private}}) \in [1.795, 2.752]$, which is the interval given by `lincom, eform`.

If instead we use `nlcom`, we obtain

```
. * CI for exp(b) using nlcom
. nlcom exp(_b[private])
        _nl_1:  exp(_b[private])
```

docvis	Coef.	Std. Err.	z	P>|z|	[95% Conf. Interval]	
_nl_1	2.222572	.2422636	9.17	0.000	1.747744	2.6974

The interval is instead $\exp(\beta_{\text{private}}) \in [1.748, 2.697]$. This differs from the $[1.795, 2.752]$ interval obtained with `lincom`, and the difference between the two methods can be much larger in other applications.

Which interval should we use? The two are asymptotically equivalent but can differ considerably in small samples. The interval obtained by using `nlcom` is symmetric about $\exp(\widehat{\beta}_{\text{private}})$ and could include negative values (if $\widehat{\beta}$ is small relative to $s_{\widehat{\beta}}$). The interval obtained by using `lincom, eform` is asymmetric and, necessarily, is always positive because of exponentiation. This is preferred.

12.4 Likelihood-ratio tests

An alternative to the Wald test is the likelihood-ratio (LR) test. This is applicable to only ML estimation, under the assumption that the density is correctly specified.

12.4.1 Likelihood-ratio tests

Let $L(\boldsymbol{\theta}) = f(\mathbf{y}|\mathbf{X}, \boldsymbol{\theta})$ denote the likelihood function, and consider testing the h hypotheses $H_0 : \mathbf{g}(\boldsymbol{\theta}) = \mathbf{0}$. Distinguish between the usual unrestricted maximum likelihood estimator (MLE) $\widehat{\boldsymbol{\theta}}_u$ and the restricted MLE $\widetilde{\boldsymbol{\theta}}_r$ that maximizes the log likelihood subject to the restriction $\mathbf{g}(\boldsymbol{\theta}) = \mathbf{0}$.

The motivation for the likelihood-ratio test is that if H_0 is valid, then imposing the restrictions in estimation of the parameters should make little difference to the maximized value of the likelihood function. The LR test statistic is

$$\text{LR} = -2\{\ln L(\widetilde{\boldsymbol{\theta}}_r) - \ln L(\widehat{\boldsymbol{\theta}}_u)\} \overset{a}{\sim} \chi^2(h) \text{ under } H_0$$

At the 0.05 level, for example, we reject if $p = \Pr\{\chi^2(h) > \text{LR}\} < 0.05$, or equivalently if $\text{LR} > \chi^2_{0.05}(h)$. It is unusual to use an F variant of this test.

The LR and Wald tests, under the conditions in which `vce(oim)` is specified, are asymptotically equivalent under H_0 and local alternatives, so there is no a priori reason to prefer one over the other.

Nonetheless, the LR test is preferred in fully parametric settings, in part because the LR test is invariant under nonlinear transformations, whereas the Wald test is not, as was demonstrated in section 12.3.5.

Microeconometricians use Wald tests more often than LR tests because wherever possible fully parametric models are not used. For example, consider a linear regression with cross-section data. Assuming normal homoskedastic errors permits the use of a LR test. But the preference is to relax this assumption, obtain a robust estimate of the VCE, and use this as the basis for Wald tests.

The LR test requires fitting two models, whereas the Wald test requires fitting only the unrestricted model. And restricted ML estimation is not always possible. The Stata ML commands can generally be used with the `constraint()` option, but this supports only linear restrictions on the parameters.

Stata output for ML estimation commands uses LR tests in two situations: first, to perform tests on a key auxiliary parameter, and second, in the test for joint significance of regressors automatically provided as part of Stata output, if the default `vce(oim)` option is used.

We demonstrate this for negative binomial regression for doctor visits by using default ML standard errors. We have

```
. * LR tests output if estimate by ML with default estimate of VCE
. nbreg docvis private chronic female income,  nolog
Negative binomial regression                     Number of obs   =      4412
                                                 LR chi2(4)      =   1067.55
Dispersion     = mean                            Prob > chi2     =    0.0000
Log likelihood = -9855.1389                      Pseudo R2       =    0.0514
```

docvis	Coef.	Std. Err.	z	P>|z|	[95% Conf.	Interval]
private	.8876559	.0594232	14.94	0.000	.7711886	1.004123
chronic	1.143545	.0456778	25.04	0.000	1.054018	1.233071
female	.5613027	.0448022	12.53	0.000	.473492	.6491135
income	.0045785	.000805	5.69	0.000	.0030007	.0061563
_cons	-.4062135	.0611377	-6.64	0.000	-.5260411	-.2863858
/lnalpha	.5463093	.0289716			.4895261	.6030925
alpha	1.726868	.05003			1.631543	1.827762

```
Likelihood-ratio test of alpha=0:  chibar2(01) = 1.7e+04 Prob>=chibar2 = 0.000
```

Here the overall test for joint significance of the four coefficients, given as LR chi2(4) = 1067.55, is a LR test.

The last line of output provides a LR test of $H_0: \alpha = 0$ against $H_a: \alpha > 0$. Rejection of H_0 favors the more general negative binomial model, because the Poisson is the special case $\alpha = 0$. This LR test is nonstandard because the null hypothesis is on the boundary of the parameter space (the negative binomial model restricts $\alpha \geq 0$). In this case, the LR statistic has a distribution that has a probability mass of $1/2$ at zero and a half-$\chi^2(1)$ distribution above zero. This distribution is known as the chibar-0-1 distribution and is used to calculate the reported p-value of 0.000, which strongly rejects the Poisson in favor of the negative binomial model.

12.4.2   The lrtest command

The `lrtest` command calculates a LR test of one model that is nested in another when both are fit by using the same ML command. The syntax is

lrtest *modelspec1* [*modelspec2*] [, *options*]

where ML results from the two models have been saved previously by using `estimates store` with the names *modelspec1* and *modelspec2*. The order of the two models does not matter. The variation `lrtest` *modelspec1* requires applying `estimates store` only to the model other than the most recently fitted model.

We perform a LR test of $H_0 : \beta_{private} = 0, \beta_{chronic} = 0$ by fitting the unrestricted model with all regressors and then fitting the restricted model with `private` and `chronic` excluded. We fit a negative binomial model because this is a reasonable parametric model for these overdispersed count data, whereas the Poisson was strongly rejected in the test of $H_0 : \alpha = 0$ in the previous section. We have

```
. * LR test using command lrtest
. quietly nbreg docvis private chronic female income

. estimates store unrestrict

. quietly nbreg docvis female income

. estimates store restrict

. lrtest unrestrict restrict
Likelihood-ratio test                                LR chi2(2)  =     808.74
(Assumption: restrict nested in unrestrict)          Prob > chi2 =     0.0000
```

The null hypothesis is strongly rejected because $p = 0.000$. We conclude that `private` and `chronic` should be included in the model.

The same test can be performed with a Wald test. Then

```
. * Wald test of the same hypothesis
. quietly nbreg docvis private chronic female income

. test chronic private
 ( 1)  [docvis]chronic = 0
 ( 2)  [docvis]private = 0
         chi2(  2) =   852.26
       Prob > chi2 =     0.0000
```

The results differ somewhat, with test statistics of 809 and 852. The differences can be considerably larger in other applications, especially those with few observations.

12.4.3   Direct computation of LR tests

The default is for the `lrtest` command to compute the LR test statistic only in situations where it is clear that the LR test is appropriate. The command will produce an error when, for example, the `vce(robust)` option is used or when different estimation commands are used. The `force` option causes the LR test statistic to be computed in such settings, with the onus on the user to verify that the test is still appropriate.

As an example, we return to the LR test of Poisson against the negative binomial model, automatically given after the `nbreg` command, as discussed in section 12.4.1. To perform this test using the `lrtest` command, the `force` option is needed because two different estimation commands, `poisson` and `nbreg`, are used. We have

```
. * LR test using option force
. quietly nbreg docvis private chronic female income

. estimates store nb

. quietly poisson docvis private chronic female income

. estimates store poiss

. lrtest nb poiss, force
Likelihood-ratio test                                       LR chi2(1)  =  17296.82
(Assumption: poiss nested in nb)                            Prob > chi2 =    0.0000
. display "Corrected p-value for LR-test = " r(p)/2
Corrected p-value for LR-test = 0
```

As expected, the LR statistic is the same as `chibar2(01)` = 1.7e+04, reported in the last line of output from `nbreg` in section 12.4.1. The `lrtest` command automatically computes p-values using $\chi^2(h)$, where h is the difference in the number of parameters in the two fitted models, here $\chi^2(1)$. As explained in section 12.4.1, however, the half-$\chi^2(1)$ should be used in this particular example, providing a cautionary note for the use of the `force` option.

12.5   Lagrange multiplier test (or score test)

The third major hypothesis testing method is a test method usually referred to as the score test by statisticians and as the Lagrange multiplier (LM) test by econometricians. This test is less often used, aside from some leading model-specification tests in situations where the null hypothesis model is easy to fit but the alternative hypothesis model is not.

12.5.1   LM tests

The unrestricted MLE $\widehat{\boldsymbol{\theta}}_u$ sets $\mathbf{s}(\widehat{\boldsymbol{\theta}}_u) = \mathbf{0}$, where $\mathbf{s}(\boldsymbol{\theta}) = \partial \ln L(\boldsymbol{\theta})/\partial \boldsymbol{\theta}$ is called the score function. An LM test, or score test, is based on closeness of $\mathbf{s}(\widetilde{\boldsymbol{\theta}}_r)$ to zero, where evaluation is now at $\widetilde{\boldsymbol{\theta}}_r$, the alternative restricted MLE that maximizes $\ln L(\boldsymbol{\theta})$ subject to the h restrictions $\mathbf{g}(\boldsymbol{\theta}) = \mathbf{0}$. The motivation is that if the restrictions are supported by the data, then $\widetilde{\boldsymbol{\theta}}_r \simeq \widehat{\boldsymbol{\theta}}_u$, so $\mathbf{s}(\widetilde{\boldsymbol{\theta}}_r) \simeq \mathbf{s}(\widehat{\boldsymbol{\theta}}_u) = \mathbf{0}$.

Because $\mathbf{s}(\widetilde{\boldsymbol{\theta}}_r) \stackrel{a}{\sim} N\{\mathbf{0},\ \mathrm{Var}(\widetilde{\boldsymbol{\theta}}_r)\}$, we form a quadratic form that is a chi-squared statistic, similar to the method in section 12.3.1. This yields the LM test statistic, or score test statistic, for $H_0 : \mathbf{g}(\boldsymbol{\theta}) = \mathbf{0}$:

$$\mathrm{LM} = \mathbf{s}(\widetilde{\boldsymbol{\theta}}_r)' \left[\widehat{V}\{\mathbf{s}(\widetilde{\boldsymbol{\theta}}_r)\} \right]^{-1} \mathbf{s}(\widetilde{\boldsymbol{\theta}}_r) \stackrel{a}{\sim} \chi^2(h) \text{ under } H_0$$

At the 0.05 level, for example, we reject if $p = \Pr\{\chi^2(h) > \text{LM}\} < 0.05$, or equivalently if $\text{LM} > \chi^2_{0.05}(h)$. It is not customary to use an F variant of this test.

The preceding motivation explains the term "score test". The test is also called the LM test for the following reason: Let $\ln L(\boldsymbol{\theta})$ be the log-likelihood function in the unrestricted model. The restricted MLE $\widetilde{\boldsymbol{\theta}}_r$ maximizes $\ln L(\boldsymbol{\theta})$ subject to $\mathbf{g}(\boldsymbol{\theta}) = \mathbf{0}$, so $\widetilde{\boldsymbol{\theta}}_r$ maximizes $\ln L(\boldsymbol{\theta}) - \boldsymbol{\lambda}'\mathbf{g}(\boldsymbol{\theta})$. An LM test is based on whether the associated Lagrange multipliers $\widetilde{\boldsymbol{\lambda}}_r$ of this restricted optimization are close to zero, because $\boldsymbol{\lambda} = \mathbf{0}$ if the restrictions are valid. It can be shown that $\widetilde{\boldsymbol{\lambda}}_r$ is a full-rank matrix multiple of $\mathbf{s}(\widetilde{\boldsymbol{\theta}}_r)$, so the LM and score tests are equivalent.

Under the conditions in which vce(oim) is specified, the LM test, LR test, and Wald test are asymptotically equivalent for H_0 and local alternatives, so there is no a priori reason to prefer one over the others. The attraction of the LM test is that, unlike Wald and LR tests, it requires fitting only the restricted model. This is an advantage if the restricted model is easier to fit, such as a homoskedastic model rather than a heteroskedastic model. Furthermore, an asymptotically equivalent version of the LM test can often be computed by the use of an auxiliary regression. On the other hand, there is generally no universal way to implement an LM test, unlike Wald and LR tests. If the LM test rejects the restrictions, we then still need to fit the unrestricted model.

12.5.2   The estat command

Because LM tests are estimator specific and model specific, there is no lmtest command. Instead, LM tests usually appear as postestimation estat commands to test misspecifications.

A leading example is the estat hettest command to test for heteroskedasticity after regress. This LM test is implemented by auxiliary regression, which is detailed in section 3.5.4. The default version of the test requires that under the null hypothesis, the independent homoskedastic errors must be normally distributed, whereas the iid option relaxes the normality assumption to one of independent and identically distributed errors.

Another example is the xttest0 command to implement an LM test for random effects after xtreg. Yet another example is the LM test for overdispersion in the Poisson model, given in an end-of-chapter exercise.

12.5.3   LM test by auxiliary regression

For ML estimation with a correctly specified density, an asymptotically equivalent version of the LM statistic can always be obtained from the following auxiliary procedure. First, obtain the restricted MLE $\widetilde{\boldsymbol{\theta}}_r$. Second, form the scores for each observation of the unrestricted model, $\mathbf{s}_i(\boldsymbol{\theta}) = \partial \ln f(y_i|\mathbf{x}_i, \boldsymbol{\theta})/\partial \theta$, and evaluate them at $\widetilde{\boldsymbol{\theta}}_r$ to give $\mathbf{s}_i(\widetilde{\boldsymbol{\theta}}_r)$. Third, compute N times the uncentered R^2 (or, equivalently, the model sum of squares) from the auxiliary regression of 1 on $\mathbf{s}_i(\widetilde{\boldsymbol{\theta}}_r)$.

It is easy to obtain restricted model scores evaluated at the restricted MLE or unrestricted model scores evaluated at the unrestricted MLE. However, this auxiliary regression requires computation of the unrestricted model scores evaluated at the restricted MLE. If the parameter restrictions are linear, then these scores can be obtained by using the `constraint` command to define the restrictions before estimation of the unrestricted model.

We illustrate this method for the LM test of whether $H_0 : \beta_{\text{private}} = 0$, $\beta_{\text{chronic}} = 0$ in a negative binomial model for `docvis` that, when unrestricted, includes as regressors an intercept, `female`, `income`, `private`, and `chronic`. The restricted MLE $\widetilde{\beta}_r$ is then obtained by negative binomial regression of `docvis` on all these regressors, subject to the constraint that $\beta_{\text{private}} = 0$ and $\beta_{\text{chronic}} = 0$. The two constraints are defined by using the `constraint` command, and the restricted estimates of the unrestricted model are obtained using the `nbreg` command with the `constraints()` option. Scores can be obtained by using the `predict` command with the `scores` option. However, these scores are derivatives of the log density with respect to model indices (such as $\mathbf{x}_i'\boldsymbol{\beta}$) rather than with respect to each parameter. Thus following `nbreg` only two "scores" are given, $\partial \ln f(y_i)/\partial \mathbf{x}_i'\boldsymbol{\beta}$ and $\partial \ln f(y_i)/\partial \alpha$. These two scores are then expanded to $K + 1$ scores $\partial \ln f(y_i)/\partial \beta_j = \{\partial \ln f(y_i)/\partial \mathbf{x}_i'\boldsymbol{\beta}\} \times x_{ij}$, $j = 1, \dots, K$, where K is the number of regressors in the unrestricted model, and $\partial \ln f(y_i)/\partial \alpha$, where α is the scalar overdispersion parameter. Then 1 is regressed on these $K + 1$ scores.

We have

```
. * Perform LM test that b_private=0, b_chronic=0 using auxiliary regression
. use mus10data.dta, clear
. quietly keep if year02==1
. generate one = 1
. constraint define 1 private = 0
. constraint define 2 chronic = 0
. quietly nbreg docvis female income private chronic, constraints(1 2)
. predict eqscore ascore, scores
. generate s1restb = eqscore*one
. generate s2restb = eqscore*female
. generate s3restb = eqscore*income
. generate s4restb = eqscore*private
. generate s5restb = eqscore*chronic
. generate salpha = ascore*one
. quietly regress one s1restb s2restb s3restb s4restb s5restb salpha, noconstant
. scalar lm = e(N)*e(r2)
. display "LM = N x uncentered Rsq = " lm " and p = " chi2tail(2,lm)
LM = N x uncentered Rsq = 424.17616 and p = 7.786e-93
```

The null hypothesis is strongly rejected with LM = 424. By comparison, in section 12.4.2, the asymptotically equivalent LR and Wald statistics for the same hypothesis were, respectively, 809 and 852.

The divergence of these purportedly asymptotically equivalent tests is surprising given the large sample size of 4,412 observations. One explanation, always a possibility with real data, is that the unknown data-generating process (DGP) for these data is not the fitted negative binomial model—the asymptotic equivalence only holds under H_0, which includes correct model specification. A second explanation is that this LM test has poor size properties even in relatively large samples. This explanation could be pursued by adapting the simulation exercise in section 12.6 to one for the LM test with data generated from a negative binomial model.

Often more than one auxiliary regression is available to implement a specific LM test. The easiest way to implement an LM test is to find a reference that defines the auxiliary regression for the example at hand and then implement the regression. For example, to test for heteroskedasticity in the linear regression model that depends on variables z_i, we calculate N times the uncentered explained sum of squares from the regression of squared OLS residuals $\widehat{u}_i^2$ on an intercept and z_i; all that is needed is the computation of $\widehat{u}_i$. In this case, `estat hettest` implements this anyway.

The auxiliary regression versions of the LM test are known to have poor size properties, though in principle these can be overcome by using the bootstrap with asymptotic refinement.

12.6 Test size and power

We consider computation of the test size and power of a Wald test by Monte Carlo simulation. The goal is to determine whether tests that are intended to reject at, say, a 0.05 level really do reject at a 0.05 level, and to determine the power of tests against meaningful parameter values under the alternative hypothesis. This extends the analysis of section 4.6, which focused on the use of simulation to check the properties of estimators of parameters and estimators of standard errors. Here we instead focus on inference.

12.6.1 Simulation DGP: OLS with chi-squared errors

The DGP is the same as that in section 4.6, with data generated from a linear model with skewed errors, specifically,

$$y = \beta_1 + \beta_2 x + u; \quad u \sim \chi^2(1) - 1; \quad x \sim \chi^2(1)$$

where $\beta_1 = 1$, $\beta_2 = 2$, and the sample size $N = 150$. The $[\chi^2(1) - 1]$ errors have a mean of 0, a variance of 2, and are skewed.

In each simulation, both y and x are redrawn, corresponding to random sampling of individuals. We investigate the size and power of t tests on $H_0 \colon \beta_2 = 2$, the DGP value after OLS regression.

12.6.2 Test size

In testing H_0, we can make the error of rejecting H_0 when H_0 is true. This is called a type I error. The test size is the probability of making this error. Thus

$$\text{Size} = \Pr(\text{Reject } H_0 | H_0 \text{ true})$$

The reported p-value of a test is the estimated size of the test. Most commonly, we reject H_0 if the size is less than 0.05.

The most serious error is one of incorrect test size, even asymptotically, because of, for example, the use of inconsistent estimates of standard errors if a Wald test is used. Even if this threshold is passed, a test is said to have poor finite-sample size properties or, more simply, poor finite-sample properties, if the reported p-value is a poor estimate of the true size. Often the problem is that the reported p-value is much lower than the true size, so we reject H_0 more often than we should.

For our example with DGP value of $\beta_2 = 2$, we want to use simulations to estimate the size of an α-level test of $H_0 \colon \beta_2 = 2$ against $H_a \colon \beta_2 \neq 2$. In section 4.6.2, we did so when $\alpha = 0.05$ by counting the proportion of simulations that led to rejection of H_0 at a level of $\alpha = 0.05$. The estimated size was 0.046 because 46 of the 1,000 simulations led to rejection of H_0.

A computationally more-efficient procedure is to compute the p-value for the test of $H_0 \colon \beta_2 = 2$ against $H_a \colon \beta_2 \neq 2$ in each of the 1,000 simulations, because the 1,000 p-values can then be used to estimate the test size for a range of values of α, as we demonstrate below. The p-values were computed in the **chi2data** program defined in section 4.6.1 and returned as the scalar **p2**, but these p-values were not used in any subsequent analysis of test size. We now do so here.

The simulations in section 4.6 were performed by using the **simulate** command. Here we instead use **postfile** and a **forvalues** loop; the code is for the 1,000 simulations:

```
. * Do 1000 simulations where each gets p-value of test of b2=2
. set seed 10101

. postfile sim pvalues using pvalues, replace

. forvalues i = 1/1000 {
  2.    drop _all
  3.    quietly set obs 150
  4.    quietly generate double x = rchi2(1)
  5.    quietly generate y = 1 + 2*x + rchi2(1)-1
  6.    quietly regress y x
  7.    quietly test x = 2
  8.    scalar p = r(p)          // p-value for test this simulation
  9.    post sim (p)
 10. }

. postclose sim
```

The simulations produce 1,000 p-values that range from 0 to 1.

```
 . * Summarize the p-value from each of the 1000 tests
 . use pvalues, clear
 . summarize pvalues
```

Variable	Obs	Mean	Std. Dev.	Min	Max
pvalues	1000	.5175818	.2890325	.0000108	.9997772

These should actually have a uniform distribution, and the `histogram` command reveals that this is the case.

Given the 1,000 values of `pvalues`, we can find the actual size of the test for any choice of α. For a test at the $\alpha = 0.05$ level, we obtain

```
 . * Determine size of test at level 0.05
 . count if pvalues < .05
    46
 . display "Test size from 1000 simulations = " r(N)/1000
Test size from 1000 simulations = .046
```

The actual test size of 0.046 is reasonably close to the nominal size of 0.05. Furthermore, it is exactly the same value as that obtained in section 4.6.1 because the same seed and sequence of commands was used there.

As noted in section 4.6.2, the size is not estimated exactly because of simulation error. If the true size equals the nominal size of α, then the proportion of times H_0 is rejected in S simulations is a random variable with a mean of α and a standard deviation of $\sqrt{\alpha(1-\alpha)/S} \simeq 0.007$ when $S = 1000$ and $\alpha = 0.05$. Using a normal approximation, the 95% simulation interval for this simulation is $[0.036, 0.064]$, and 0.046 is within this interval. More precisely, the `cii` command yields an exact binomial confidence interval.

```
 . * 95% simulation interval using exact binomial at level 0.05 with S=1000
 . cii 1000 50
```

Variable	Obs	Mean	Std. Err.	— Binomial Exact — [95% Conf. Interval]	
	1000	.05	.006892	.0373354	.0653905

With $S = 1000$, the 95% simulation interval is $[0.037, 0.065]$. With $S = 10,000$ simulations, this interval narrows to $[0.046, 0.054]$.

In general, tests rely on asymptotic theory, and we do not expect the true size to exactly equal the nominal size unless the sample size N is very large and the number of simulations S is very large. In this example, with 150 observations and only one regressor, the asymptotic theory performs well even though the model error is skewed.

12.6.3   Test power

A second error in testing, called a type II error, is to fail to reject H_0 when we should reject H_0. The power of a test is one minus the probability of making this error. Thus

$$\text{Power} = \Pr(\text{Reject } H_0 | H_0 \text{ false})$$

Ideally, test size is minimized and test power is maximized, but there is a trade-off with smaller size leading to lower power. The standard procedure is to set the size at a level such as 0.05 and then use the test procedure that is known from theory or simulations to have the highest power.

The power of a test is not reported because it needs to be evaluated at a specific H_a value, and the alternative hypothesis H_a defines a range of values for β rather than one single value.

We compute the power of our test of $\beta_2 = \beta_2^{Ha}$ against $H_a : \beta_2 = \beta_2^{Ha}$, where β_2^{Ha} takes on a range of values. We do so by first writing a program that determines the power for a given value β_2^{Ha} and then calling this program many times to evaluate at the many values of β_2^{Ha}.

The program is essentially the same as that used to determine test size, except that the command generating y becomes `generate y = 1 + b2Ha*x + rchi2(1)-1`. We allow more flexibility by allowing the user to pass the number of simulations, sample size, H_0 value of β_2, H_a value of β_2, and nominal test size (α) as the arguments, respectively, `numsims`, `numobs`, `b2H0`, `b2Ha`, and `nominalsize`. The r-class program returns the computed power of the test as the scalar `p`. We have

```
* Program to compute power of test given specified H0 and Ha values of b2
program power, rclass
    version 11
    args numsims numobs b2H0 b2Ha nominalsize
                                        // Setup before simulation loops
    drop _all
    set seed 10101
    postfile sim pvalues using power, replace
                                        // Simulation loop
    forvalues i = 1/`numsims' {
      drop _all
      quietly set obs `numobs'
      quietly generate double x = rchi2(1)
      quietly generate y = 1 + `b2Ha'*x + rchi2(1)-1
      quietly regress y x
      quietly test x = `b2H0'
      scalar p = r(p)
      post sim (p)
    }
    postclose sim
    use power, clear
                                        // Determine the size or power
    quietly count if pvalues < `nominalsize'
    return scalar power=r(N)/`numsims'
  end
```

This program can also be used to find the size of the test of $H_0 : \beta_2 = 2$ by setting $\beta_2^{Ha} = 2$. The following command obtains the size using 1,000 simulations and a sample size of 150, for a test of the nominal size 0.05.

```
. * Size = power of test of b2H0=2 when b2Ha=2, S=1000, N=150, alpha=0.05
. power 1000 150 2.00 2.00 0.05
. display r(power) " is the test power"
.046 is the test power
```

The program **power** uses exactly the same coding as that given earlier for size computation, we have the same number of simulations and same sample size, and we get the same size result of 0.046.

To find the test power, we set $\beta_2 = \beta_2^{Ha}$, where β_2^{Ha} differs from the null hypothesis value. Here we set $\beta_2^{Ha} = 2.2$, which is approximately 2.4 standard errors away from the H_0 value of 2.0 because, from section 4.6.1, the standard error of the slope coefficient is 0.084. We obtain

```
. * Power of test of b2H0=2 when b2Ha=2.2, S=1000, N=150, alpha=0.05
. power 1000 150 2.00 2.20 0.05
. display r(power) " is the test power"
.657 is the test power
```

Ideally, the probability of rejecting $H_0 : \beta_2 = 2.0$ when $\beta_2 = 2.2$ is one. In fact, it is only 0.657.

We next evaluate the power for a range of values of β_2^{Ha}, here from 1.60 to 2.40 in increments of 0.025. We use the **postfile** command, which was presented in chapter 4:

```
. * Power of test of H0:b2=2 against Ha:b2=1.6,1.625, ..., 2.4
. postfile simofsims b2Ha power using simresults, replace
. forvalues i = 0/33 {
  2.    drop _all
  3.    scalar b2Ha = 1.6 + 0.025*`i´
  4.    power 1000 150 2.00 b2Ha 0.05
  5.    post simofsims (b2Ha) (r(power))
  6. }
. postclose simofsims
. use simresults, clear
. summarize
```

Variable	Obs	Mean	Std. Dev.	Min	Max
b2Ha	34	2.0125	.2489562	1.6	2.425
power	34	.6103235	.3531139	.046	.997

The simplest way to see the relationship between power and β_2^{Ha} is to plot the power curve.

(*Continued on next page*)

```
. * Plot the power curve
. twoway (connected power b2Ha), scale(1.2) plotregion(style(none))
```

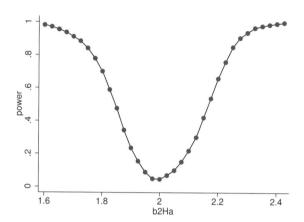

Figure 12.2. Power curve for the test of $H_0: \beta_2 = 2$ against $H_a: \beta_2 \neq 2$ when β_2 takes on the values $\beta_2^{Ha} = 1.6, \ldots, 2.4$ under H_a and $N = 150$ and $S = 1000$

As you can see in figure 12.2, power is minimized at $\beta_2^{Ha} = \beta_2^{H0} = 2$, and then the power size equals 0.05, as desired. As $|\beta_2^{Ha} - \beta_2^{H0}|$ increases, the power goes to one, but power does not exceed 0.9 until $|\beta_2^{Ha} - \beta_2^{H0}| > 0.3$.

The power curve can be made smoother by increasing the number of simulations or by smoothing the curve by, for example, using predictions from a regression of `power` on a quartic in `b2Ha`.

12.6.4 Asymptotic test power

The asymptotic power of the Wald test can be obtained without resorting to simulation. We do this now, for the square of the t test.

We consider $W = \{(\widehat{\beta}_2 - 2)/s_{\widehat{\beta}_2}\}^2$. Then $W \stackrel{a}{\sim} \chi^2(1)$ under $H_0: \beta_2 = 2$. It can be shown that under $H_a: \beta_2 = \beta_2^{Ha}$, the test statistic $W \stackrel{a}{\sim}$ noncentral $\chi^2(h; \lambda)$, where the noncentrality parameter $\lambda = (\beta_2^{Ha} - 2)^2/\sigma_{\widehat{\beta}_2}^2$. [If $\mathbf{y} \sim N(\boldsymbol{\delta}, \mathbf{I})$, then $(\mathbf{y} - \boldsymbol{\delta})'(\mathbf{y} - \boldsymbol{\delta}) \sim \chi^2(1)$ and $\mathbf{y}'\mathbf{y} \sim$ noncentral $\chi^2(h; \boldsymbol{\delta}'\boldsymbol{\delta})$.]

We again consider the power of a test of $\beta_2 = 2$ against $\beta_2^{Ha} = 2.2$. Then $\lambda = (\beta_2^{Ha} - \beta_2^{H0})^2/\sigma_{\widehat{\beta}_2}^2 = (0.2/0.084)^2 = 5.67$, where we recall the earlier discussion that the DGP is such that $\sigma_{\widehat{\beta}_2} = 0.084$. A $\chi^2(1)$ test rejects at a level of $\alpha = 0.05$ if $W > 1.96^2 = 3.84$. So the asymptotic test power equals $\Pr\{W > 3.84 | W \stackrel{a}{\sim}$ noncentral $\chi^2(1; 5.67)\}$. The `nchi2()` function gives the relevant c.d.f., and we use `1-nchi2()` to get the right tail. We have

```
. * Power of chi(1) test when noncentrality parameter lambda = 5.67
. display 1-nchi2(1,5.67,3.84)
.6633429
```

The asymptotic power of 0.663 is similar to the estimated power of 0.657 from the Monte Carlo example at the top of page 423. This closeness is due to the relatively large sample size with just one regressor.

12.7   Specification tests

The preceding Wald, LR, and LM tests are often used for specification testing, particularly of inclusion or exclusion of regressors. In this section, we consider other specification testing methods that differ in that they do not work by directly testing restrictions on parameters. Instead, they test whether moment restrictions implied by the model, or other model properties, are satisfied.

12.7.1   Moment-based tests

A moment-based test, or m test, is one of moment conditions imposed by a model but not used in estimation. Specifically,

$$H_0 \colon E\{\mathbf{m}(y_i, \mathbf{x}_i, \boldsymbol{\theta})\} = \mathbf{0} \tag{12.9}$$

where $\mathbf{m}(\cdot)$ is an $h \times 1$ vector. Several examples follow below. The test statistic is based on whether the sample analogue of this condition is satisfied, i.e., whether $\widehat{\mathbf{m}}(\widehat{\boldsymbol{\theta}}) = \sum_{i=1}^{N} \mathbf{m}(y_i, \mathbf{x}_i, \widehat{\boldsymbol{\theta}}) = \mathbf{0}$. This statistic is asymptotically normal because $\widehat{\boldsymbol{\theta}}$ is, and taking the quadratic form we obtain a chi-squared statistic. The m test statistic is then

$$M = \widehat{\mathbf{m}}(\widehat{\boldsymbol{\theta}})' \left[\widehat{V}\{\widehat{\mathbf{m}}(\widehat{\boldsymbol{\theta}})\} \right]^{-1} \widehat{\mathbf{m}}(\widehat{\boldsymbol{\theta}}) \overset{a}{\sim} \chi^2(h) \text{ under } H_0$$

As usual, we reject at the α level if $p = \Pr\{\chi^2(h) > W\} < \alpha$.

Obtaining $\widehat{V}\{\widehat{\mathbf{m}}(\widehat{\boldsymbol{\theta}})\}$ can be difficult. Often this test is used after ML estimation, because likelihood-based models impose many conditions that can be used as the basis for an m test. Then the auxiliary regression for the LM test (see section 12.5.3) can be generalized. We compute M as N times the uncentered R^2 from the auxiliary regression of 1 on $\mathbf{m}(y_i, \mathbf{x}_i, \widehat{\boldsymbol{\theta}})$ and $\mathbf{s}_i(\widehat{\boldsymbol{\theta}})$, where $\mathbf{s}_i(\boldsymbol{\theta}) = \partial \ln f(y_i | \mathbf{x}_i, \boldsymbol{\theta}) / \partial \boldsymbol{\theta}$. In finite samples, the test statistic has a size that can differ significantly from the nominal size, but this can be rectified by using a bootstrap with asymptotic refinement.

An example of this auxiliary regression, used to test moment conditions implied by the tobit model, is given in section 16.4.

12.7.2   Information matrix test

For a fully parametric model, the expected value of the outer product of the first derivatives of $\ln L(\boldsymbol{\theta})$ equals the negative expected value of the second derivatives. This property, called the information matrix (IM) equality, enables the variance matrix of the MLE to simplify from the general sandwich form $\mathbf{A}^{-1}\mathbf{B}\mathbf{A}^{-1}$ to the simpler form $-\mathbf{A}^{-1}$; see section 10.4.4.

The IM test is a test of whether the IM equality holds. It is a special case of (12.9) with $\mathbf{m}(y_i, \mathbf{x}_i, \boldsymbol{\theta})$ equal to the unique elements in $\mathbf{s}_i(\boldsymbol{\theta})\mathbf{s}_i(\boldsymbol{\theta})' + \partial\mathbf{s}_i(\boldsymbol{\theta})/\partial\boldsymbol{\theta}$. For the linear model under normality, the IM test is performed by using the `estat imtest` command after `regress`; see section 3.5.4 for an example.

12.7.3   Chi-squared goodness-of-fit test

A simple test of goodness of fit is the following. Discrete variable y takes on the values 1, 2, 3, 4, and 5, and we compare the fraction of sample values y that take on each value with the corresponding predicted probability from a fitted parametric regression model. The idea extends easily to partitioning on the basis of regressors as well as y and to continuous regressor y, where we replace a discrete value with a range of values.

Stata implements a goodness-of-fit test by using the `estat gof` command following `logit`, `logistic`, `probit`, and `poisson`. An example of `estat gof` following logit regression is given in section 14.6. A weakness of this command is that it treats estimated coefficients as known, ignoring estimation error. The goodness-of-fit test can instead be set up as an m test that provides additional control for estimation error; see Andrews (1988) and Cameron and Trivedi (2005, 266–271).

12.7.4   Overidentifying restrictions test

In the generalized methods of moments (GMM) estimation framework of section 11.8, moment conditions $E\{\mathbf{h}(y_i, \mathbf{x}_i, \boldsymbol{\theta})\} = \mathbf{0}$ are used as the basis for estimation. In a just-identified model, the GMM estimator solves the sample analog $\sum_{i=1}^{N} \mathbf{h}(y_i, \mathbf{x}_i, \widehat{\boldsymbol{\theta}}) = \mathbf{0}$. In an overidentified model, these conditions no longer hold exactly, and an overidentifying restrictions (OIR) test is based on the closeness of $\sum_{i=1}^{N} \mathbf{h}(y_i, \mathbf{x}_i, \widehat{\boldsymbol{\theta}})$ to $\mathbf{0}$, where $\widehat{\boldsymbol{\theta}}$ is the optimal GMM estimator. The test is chi-squared distributed with degrees of freedom equal to the number of overidentifying restrictions.

This test is most often used in overidentified IV models, though it can be applied to any overidentified model. It is performed in Stata with the `estat overid` command after `ivregress gmm`; see section 6.3.7 for an example.

12.7.5   Hausman test

The Hausman test compares two estimators where one is consistent under both H_0 and H_a while the other is consistent under H_0 only. If the two estimators are dissimilar, then H_0 is rejected. An example is to test whether a single regressor is endogenous by comparing two-stage least-squares and OLS estimates.

We want to test $H_0 \colon \operatorname{plim}(\widehat{\boldsymbol{\theta}} - \widetilde{\boldsymbol{\theta}}) = \mathbf{0}$. Under standard assumptions, each estimator is asymptotically normal and so is their difference. Taking the usual quadratic form,

$$H = (\widehat{\boldsymbol{\theta}} - \widetilde{\boldsymbol{\theta}})' \left\{ \widehat{V}(\widehat{\boldsymbol{\theta}} - \widetilde{\boldsymbol{\theta}}) \right\}^{-1} (\widehat{\boldsymbol{\theta}} - \widetilde{\boldsymbol{\theta}}) \overset{a}{\sim} \chi^2(h) \text{ under } H_0$$

The `hausman` command, available after many estimation commands, implements this test under the strong assumption that $\widehat{\boldsymbol{\theta}}$ is a fully efficient estimator. Then it can be shown that $\widehat{V}(\widehat{\boldsymbol{\theta}} - \widetilde{\boldsymbol{\theta}}) = \widehat{V}(\widetilde{\boldsymbol{\theta}}) - \widehat{V}(\widehat{\boldsymbol{\theta}})$. In some common settings, the Hausman test can be more simply performed with a test of the significance of a subset of variables in an auxiliary regression. Both variants are demonstrated in section 8.8.5.

The standard microeconometrics approach of using robust estimates of the VCE implicitly presumes that estimators are not efficient. Then the preceding test is incorrect. One solution is to use a bootstrapped version of the Hausman test; see section 13.4.6. A second approach is to test the statistical significance in the appropriate auxiliary regression by using robust standard errors; see sections 6.3.6 and 8.8.5 for examples.

12.7.6   Other tests

The preceding discussion only scratches the surface of specification testing. Many model-specific tests are given in model-specific reference books such as Baltagi (2008) for panel data, Hosmer and Lemeshow (2000) for binary data, and Cameron and Trivedi (1998) for count data. Some of these tests are given in estimation command output or through postestimation commands, usually as an `estat` command, but many are not.

12.8   Stata resources

The Stata documentation [D] **functions**, `help functions`, or `help density functions` describe the functions to compute p-values and critical values for various distributions. For testing, see the relevant entries for the commands discussed in this chapter: [R] **test**, [R] **testnl**, [R] **lincom**, [R] **nlcom**, [R] **lrtest**, [R] **hausman**, [R] **regress postestimation** (for `estat imtest`), and [R] **estat**.

Much of the material in this chapter is covered in Cameron and Trivedi (2005, chap. 7 and 8) and various chapters of Greene (2008) and Wooldridge (2002).

12.9   Exercises

1. The density of a $\chi^2(h)$ random variable is $f(y) = \{y^{(h/2)-1} \exp(-y/2)\}/\{2^{h/2} \Gamma(h/2)\}$, where $\Gamma(\cdot)$ is the gamma function and $\Gamma(h/2)$ can be obtained in Stata as `exp(lngamma(h/2))`. Plot this density for $h = 5$ and $y \le 25$.

2. Use Stata commands to find the appropriate p-values for $t(100)$, $F(1, 100)$, Z, and $\chi^2(1)$ distributions at $y = 2.5$. For the same distributions, find the critical values for tests at the 0.01 level.

3. Consider the Poisson example in section 12.3, with a robust estimate of the VCE. Use the `test` or `testnl` commands to test the following hypotheses: 1) $H_0 : \beta_{\text{female}} - 100 \times \beta_{\text{income}} = 0.5$; 2) $H_0 : \beta_{\text{female}} = 0$; 3) test the previous two hypotheses jointly with the `mtest` option; 4) $H_0 : \beta_{\text{female}}^2 = 0$; and 5) $H_0 : \beta_{\text{female}}^{\beta_{\text{income}}} = 1$. Are you surprised that the second and fourth tests lead to different Wald test statistics?

4. Consider the test of $H_0 : \beta_{\text{female}} / \beta_{\text{private}} - 1 = 0$, given in section 12.3.5. It can be shown that, given that $\boldsymbol{\beta}$ has the entries β_{private}, β_{chronic}, β_{female}, β_{income}, and $\beta_{\text{\_cons}}$, then $\widehat{\mathbf{R}}$ defined in (12.5) is given by

$$\widehat{\mathbf{R}} = \left[\; -\frac{\widehat{\beta}_{\text{female}}}{(\widehat{\beta}_{\text{private}})^2} \quad 0 \quad \frac{1}{\widehat{\beta}_{\text{private}}} \quad 0 \quad 0 \; \right]$$

Manually calculate the Wald test statistic defined in (12.5) by adapting the code at the end of section 12.3.2.

5. The claim is made that the effect of private insurance on doctor visits is less than that of having a chronic condition, i.e., that $\beta_{\text{private}} - \beta_{\text{chronic}} < 0$. Test this claim at the 0.05 level. Obtain 95% and 99% confidence intervals for $\beta_{\text{private}} - \beta_{\text{chronic}}$.

6. Consider the negative binomial example in section 12.4.1, where we test $H_0 : \alpha = 0$. Use the output from the `nbreg` command to compute the Wald test statistic, and compare this with the LR test statistic given in the output. Next calculate this Wald test statistic using the `testnl` command as follows: Fit the model by using `nbreg`, and then type `estat vce`. You will see that the estimate `lnalpha` is saved in the equation `lnalpha` with the name `_cons`. We want to test $\alpha = 0$, in which case $\exp(\ln \alpha) = 0$. Give the command `testnl exp([lnalpha]_cons)=0`. Compare the results with your earlier results.

7. Consider the Poisson example of section 12.3.2. The parametric Poisson model imposes the restriction that $\text{Var}(y|\mathbf{x}) = \exp(\mathbf{x}'\boldsymbol{\beta})$. An alternative model is that $\text{Var}(y|\mathbf{x}) = \exp(\mathbf{x}'\boldsymbol{\beta}) + \alpha \times \exp(\mathbf{x}'\boldsymbol{\beta})^2$. A test of overdispersion is a test of $H_0 : \alpha \leq 0$ against $H_a : \alpha > 0$. The LM test statistic can be computed as the t test that $\alpha = 0$ in the auxiliary OLS regression of $\{(y_i - \widehat{\mu}_i)^2 - \widehat{\mu}_i\}/\widehat{\mu}_i$ on $\widehat{\mu}_i$ (no intercept) where $\widehat{\mu}_i = \exp(\mathbf{x}'\widehat{\boldsymbol{\beta}})$. Perform this test.

8. Consider the DGP $y = 0 + \beta_2 x + \varepsilon$, $x \sim N(0, 1)$, $N = 36$. For this DGP, what do you expect $\text{Var}(\widehat{\beta}_2)$ to equal? Consider a test of $H_0 : \beta_2 = 0$ at the 0.05 level. By simulation, find the size of the test and the power when $\beta_2 = 0.25$.

9. Consider the same DGP as in the previous question, but adapt it to a probit model by defining $y^* = 0 + \beta_2 x + \varepsilon$, and define $y = 1$ if $y^* > 0$ and $y = 0$ otherwise. Consider a test of $H_0 : \beta_2 = 0$ at the 0.05 level. By simulation, find the size of the test and the power when $\beta_2 = 0.25$.

13 Bootstrap methods

13.1 Introduction

The chapter considers bootstrap methods. The most common use of the bootstrap is to provide standard-error estimates when analytical expressions are quite complicated. These standard errors are then used to form confidence intervals and test statistics.

Additionally, a more complicated bootstrap with asymptotic refinement can provide tests with an actual size closer to the nominal size and confidence intervals with an actual coverage rate closer to the nominal coverage rate, compared with the standard inferential methods presented in the preceding chapter.

13.2 Bootstrap methods

A bootstrap provides a way to perform statistical inference by resampling from the sample. The statistic being studied is usually a standard error, a confidence interval, or a test statistic.

13.2.1 Bootstrap estimate of standard error

As a leading example, consider calculating the standard error of an estimator $\widehat{\theta}$ when this is difficult to do using conventional methods. Suppose 400 random samples from the population were available. Then we could obtain 400 different estimates of $\widehat{\theta}$ and let the standard error of $\widehat{\theta}$ be the standard deviation of these 400 estimates.

In practice, however, only one sample from the population is available. The bootstrap generates multiple samples by resampling from the current sample. Essentially, the observed sample is viewed as the population, and the bootstrap is a method to obtain multiple samples from this population. Given 400 bootstrap resamples, we obtain 400 estimates and then estimate the standard error of $\widehat{\theta}$ by the standard deviation of these 400 estimates.

Let $\widehat{\theta}_1^*, \ldots, \widehat{\theta}_B^*$ denote the estimates, where here $B = 400$. Then the bootstrap estimate of the variance of $\widehat{\theta}$ is

$$\widehat{\mathrm{Var}}_{\mathrm{Boot}}(\widehat{\theta}) = \frac{1}{B-1} \sum_{b=1}^{B} (\widehat{\theta}_b^* - \overline{\widehat{\theta}^*})^2 \tag{13.1}$$

where $\overline{\widehat{\theta}^*} = 1/B \sum_{b=1}^{B} \widehat{\theta}_b^*$ is the average of the B bootstrap estimates.

The square root of $\widehat{\mathrm{Var}}_{\mathrm{Boot}}(\widehat{\theta})$, denoted by $\mathrm{se}_{\mathrm{Boot}}(\widehat{\theta})$, is called the bootstrap estimate of the standard error of $\widehat{\theta}$. Some authors more correctly call this the bootstrap standard error of $\widehat{\theta}$, or the bootstrap estimate of the standard deviation of $\widehat{\theta}$, because the term "standard error" means estimated standard deviation.

13.2.2 Bootstrap methods

The plural form, bootstrap methods, is used because there is no one single bootstrap. As already noted, the bootstrap can be used to obtain the distribution of many different statistics. There are several different ways to obtain bootstrap resamples. Even for given statistic and bootstrap resampling methods, there are different ways to proceed.

The simplest bootstraps implement standard asymptotic methods. The most common use of the bootstrap in microeconometrics is standard-error estimation. More complicated bootstraps implement more refined asymptotics.

13.2.3 Asymptotic refinement

Consider a statistic such as a Wald test statistic of a single restriction. Asymptotic methods are used to obtain an approximation to the cumulative distribution function of this statistic. For a statistic with a limiting normal distribution based on conventional first-order root-N asymptotics, the approximation error behaves in the limit as a multiple of $N^{-1/2}$ (so the error disappears as $N \to \infty$). For example, a one-sided test with the nominal size of 0.05 will have the true size $0.05 + O(N^{-1/2})$, where $O(N^{-1/2})$ behaves as a constant divided by $\sqrt{N}$.

Asymptotic methods with refinement have an approximation error that disappears at a faster rate. In particular, bootstraps with asymptotic refinement can implement second-order asymptotics that yield an approximation error that behaves as a multiple of N^{-1}. So a one-sided test with a nominal size of 0.05 now has a true size of $0.05 + O(N^{-1})$. This improvement is only asymptotic and is not guaranteed to exist in small samples. But simulation studies usually find that the improvement carries over to small samples.

We present confidence intervals with asymptotic refinement in sections 13.3.6–13.3.8, and both hypothesis tests and confidence intervals with asymptotic refinement in section 13.5.

13.2.4 Use the bootstrap with caution

Caution is needed in applying the bootstrap because it is easy to misapply. For example, it is always possible to compute $\mathrm{se}(\widehat{\theta})$ by using the formula in (13.1). But this estimate is inconsistent if, for example, the bootstrap resampling scheme assumes independent observations when observations are in fact correlated. And in some cases, $\mathrm{Var}(\widehat{\theta})$ does not exist, even asymptotically. Then $\mathrm{se}(\widehat{\theta})$ is estimating a nonexistent standard deviation.

The bootstraps presented in this chapter assume independence of observations or of clusters of observations. This does permit dependence via clustering, provided that observations are combined into clusters that are independent and the bootstrap is over the clusters. Then the bootstrap commands given in this chapter should include the `cluster(`*varlist*`)` option, where *varlist* denotes the clustering variables. The `idcluster(`*newvar*`)` option may additionally be needed; see section 13.3.5.

The bootstraps also assume that the estimator is a smooth estimator that is root-N consistent and asymptotically normal distributed. Some variations of the bootstrap can be applied to more complicated cases than this, but one should first read relevant journal articles. In particular, care is needed for estimators with a nonparametric component, for nonsmooth estimators, and for dependent data.

The Stata defaults for the number of bootstrap replications are set very low to speed up computation time. These values may be adequate for exploratory data analysis but should be greatly increased for published results; see section 13.3.4. And for published results, the seed should be set, using `set seed`, rather than determined by the computer clock, to enable replication.

13.3   Bootstrap pairs using the vce(bootstrap) option

The most common use of the bootstrap is to obtain a consistent estimate of the standard errors of an estimator, with no asymptotic refinement. With standard Stata estimation commands, this can be easily done by using the `vce(bootstrap)` option.

13.3.1   Bootstrap-pairs method to estimate VCE

Let $\mathbf{w}_i$ denote all the data for the *i*th observation. Most often, $\mathbf{w}_i = (y_i, \mathbf{x}_i)$, where y is a scalar dependent variable and $\mathbf{x}$ is a regressor vector. More generally, $\mathbf{w}_i = (\mathbf{y}_i, \mathbf{x}_i, \mathbf{z}_i)$, where now there may be several dependent variables and $\mathbf{z}$ denotes instruments. We assume $\mathbf{w}_i$ is independent over i.

Stata uses the following bootstrap-pairs algorithm:

1. Repeat steps a) and b) B independent times:

   a) Draw a bootstrap sample of size N by sampling with replacement from the original data $\mathbf{w}_1, \ldots, \mathbf{w}_N$. Denote the bootstrap sample by $\mathbf{w}_1^*, \ldots, \mathbf{w}_N^*$.

   b) Calculate an estimate, $\widehat{\boldsymbol{\theta}}^*$, of $\boldsymbol{\theta}$ based on $\mathbf{w}_1^*, \ldots, \mathbf{w}_N^*$.

2. Given the B bootstrap estimates, denoted by $\widehat{\boldsymbol{\theta}}_1^*, \ldots, \widehat{\boldsymbol{\theta}}_B^*$, the bootstrap estimate of the variance–covariance matrix of the estimator (VCE) is

$$\widehat{V}_{\text{Boot}}(\widehat{\boldsymbol{\theta}}) = \frac{1}{B-1} \sum_{b=1}^{B} \left(\widehat{\boldsymbol{\theta}}_b^* - \overline{\widehat{\boldsymbol{\theta}}}^* \right) \left(\widehat{\boldsymbol{\theta}}_b^* - \overline{\widehat{\boldsymbol{\theta}}}^* \right)'$$

where $\overline{\widehat{\boldsymbol{\theta}}}^* = B^{-1} \sum_{b=1}^{B} \widehat{\boldsymbol{\theta}}_b^*$.

The corresponding standard-error estimate of the jth component of $\widehat{\boldsymbol{\theta}}$ is then

$$\mathrm{se}_{\mathrm{Boot}}(\widehat{\theta}_j) = \left\{ \widehat{V}_{\mathrm{Boot},jj}(\boldsymbol{\theta}) \right\}^{1/2}$$

The bootstrap resamples differ in the number of occurrences of each observation. For example, the first observation may appear twice in the first bootstrap sample, zero times in the second sample, once in the third sample, once in the fourth sample, and so on.

The method is called bootstrap pairs or paired bootstrap because in the simplest case $\mathbf{w}_i = (y_i, \mathbf{x}_i)$ and the pair $(y_i, \mathbf{x}_i)$ is being resampled. It is also called a case bootstrap because all the data for the ith case is resampled. It is called a nonparametric bootstrap because no information about the conditional distribution of y_i given $\mathbf{x}_i$ is used. For cross-section estimation commands, this bootstrap gives the same standard errors as those obtained by using the vce(robust) option if $B \to \infty$, aside from possible differences due to degrees-of-freedom correction that disappear for large N.

This bootstrap method is easily adapted to cluster bootstraps. Then $\mathbf{w}_i$ becomes $\mathbf{w}_c$, where $c = 1, \ldots, C$ denotes each of the C clusters, data are independent over c, resampling is over clusters, and the bootstrap resample is of size C clusters.

13.3.2   The vce(bootstrap) option

The bootstrap-pairs method to estimate the VCE can be obtained for most Stata cross-section estimation commands by using the estimator command option

vce(bootstrap [, *bootstrap_options*])

We list many of the options in section 13.4.1 and illustrate some of the options in the following example.

The vce(bootstrap) option is also available for some panel-data estimation commands. The bootstrap is actually a cluster bootstrap over individuals i, rather than one over the individual observations (i, t).

13.3.3   Bootstrap standard-errors example

We demonstrate the bootstrap using the same data on doctor visits (docvis) as that in chapter 10, except that we use one regressor (chronic) and just the first 50 observations. This keeps output short, reduces computation time, and restricts attention to a small sample where the gains from asymptotic refinement may be greater.

```
. * Sample is only the first 50 observations of chapter 10 data
. use mus10data.dta
. quietly keep if year02 == 1
. quietly drop if _n > 50
```

```
. quietly keep docvis chronic age
. quietly save bootdata.dta, replace
```

The analysis sample, saved as **bootdata.dta**, is used a number of times in this chapter.

For standard-error computation, we set the number of bootstrap replications to 400. We have

```
. * Option vce(bootstrap) to compute bootstrap standard errors
. poisson docvis chronic, vce(boot, reps(400) seed(10101) nodots)
Poisson regression                          Number of obs   =          50
                                            Replications    =         400
                                            Wald chi2(1)    =        3.50
                                            Prob > chi2     =      0.0612
Log likelihood = -238.75384                 Pseudo R2       =      0.0917
```

docvis	Observed Coef.	Bootstrap Std. Err.	z	P>|z|	Normal-based [95% Conf. Interval]	
chronic	.9833014	.5253149	1.87	0.061	-.0462968	2.0129
_cons	1.031602	.3497212	2.95	0.003	.3461607	1.717042

The output is qualitatively the same as that obtained by using any other method of standard-error estimation. Quantitatively, however, the standard errors change, leading to different test statistics, z-values, and p-values. For **chronic**, the standard error of 0.525 is similar to the robust estimate of 0.515 given in the last column of the results from **estimates table** in the next section. Both standard errors control for Poisson overdispersion and are much larger than the default standard errors from **poisson**.

13.3.4   How many bootstraps?

The Stata default is to perform 50 bootstrap replications, to minimize computation time. This value may be useful during the modeling cycle, but for final results given in a paper, this value is too low.

Efron and Tibshirani (1993, 52) state that for standard-error estimation "$B = 50$ is often enough to give a good estimate" and "very seldom are more than $B = 200$ replications needed". Some other studies suggest more bootstraps than this. Andrews and Buchinsky (2000) show that the bootstrap estimate of the standard error of $\widehat{\theta}$ with $B = 384$ is within 10% of that with $B = \infty$ with a probability of 0.95, in the special case that $\widehat{\theta}$ has no excess kurtosis. We choose to use $B = 400$ when the bootstrap is used to estimate standard errors. The user-written **bssize** command (Poi 2004) performs the calculations needed to implement the methods of Andrews and Buchinsky (2000).

For uses of the bootstrap other than for standard-error estimation, B generally needs to be even higher. For tests at the α level or at $100(1 - \alpha)\%$ confidence intervals, there are reasons for choosing B so that $\alpha(B + 1)$ is an integer. In subsequent analysis, we use $B = 999$ for confidence intervals and hypothesis tests when $\alpha = 0.05$.

To see the effects of the number of bootstraps on standard-error estimation, the following compares results with very few bootstraps, $B = 50$, using two different seeds, and with a very large number of bootstraps, $B = 2000$. We also present the robust standard error obtained by using the vce(robust) option. We have

```
. * Bootstrap standard errors for different reps and seeds
. quietly poisson docvis chronic, vce(boot, reps(50) seed(10101))
. estimates store boot50
. quietly poisson docvis chronic, vce(boot, reps(50) seed(20202))
. estimates store boot50diff
. quietly poisson docvis chronic, vce(boot, reps(2000) seed(10101))
. estimates store boot2000
. quietly poisson docvis chronic, vce(robust)
. estimates store robust
. estimates table boot50 boot50diff boot2000 robust, b(%8.5f) se(%8.5f)
```

Variable	boot50	boot50~f	boot2000	robust
chronic	0.98330	0.98330	0.98330	0.98330
	0.47010	0.50673	0.53479	0.51549
_cons	1.03160	1.03160	1.03160	1.03160
	0.39545	0.32575	0.34885	0.34467

legend: b/se

Comparing the two replications with $B = 50$ but different seed, the standard error of chronic differs by 5% (0.470 versus 0.507). For $B = 2000$, the bootstrap standard errors still differ from the robust standard errors (0.535 versus 0.515) due in part to the use of $N/(N - K)$ with $N = 50$ in calculating robust standard errors.

13.3.5 Clustered bootstraps

For cross-section estimation commands, the vce(bootstrap) option performs a paired bootstrap that assumes independence over i. The bootstrap resamples are obtained by sampling from the individual observations with replacement.

The data may instead be clustered, with observations correlated within cluster and independent across clusters. The vce(bootstrap, cluster(*varlist*)) option performs a cluster bootstrap that samples the clusters with replacement. If there are C clusters, then the bootstrap resample has C clusters. This may mean that the number of observations $N = \sum_{c=1}^{C} N_c$ may vary across bootstrap resamples, but this poses no problem.

As an example,

```
. * Option vce(boot, cluster) to compute cluster-bootstrap standard errors
. poisson docvis chronic, vce(boot, cluster(age)) reps(400) seed(10101) nodots)
Poisson regression                          Number of obs    =         50
                                            Replications     =        400
                                            Wald chi2(1)     =       4.12
                                            Prob > chi2      =     0.0423
Log likelihood = -238.75384                 Pseudo R2        =     0.0917

                           (Replications based on 26 clusters in age)
```

	Observed Coef.	Bootstrap Std. Err.	z	P>|z|	Normal-based [95% Conf. Interval]	
chronic	.9833014	.484145	2.03	0.042	.0343947	1.932208
_cons	1.031602	.303356	3.40	0.001	.4370348	1.626168

The cluster-pairs bootstrap estimate of the standard error of β_{chronic} is 0.484, similar to the 0.525 using a bootstrap without clustering. If we instead obtain the usual (nonbootstrap) cluster–robust standard errors, using the vce(cluster age) option, the cluster estimate of the standard error is 0.449. In practice, and unlike this example, cluster–robust standard errors can be much larger than those that do not control for clustering.

Some applications use cluster identifiers in computing estimators. For example, suppose cluster-specific indicator variables are included as regressors. This can be done, for example, by using the xi prefix and the regressors i.*id*, where *id* is the cluster identifier. If the first cluster in the original sample appears twice in a cluster-bootstrap resample, then its cluster dummy will be nonzero twice in the resample, rather than once, and the cluster dummies will no longer be unique to each observation in the resample. For the bootstrap resample, we should instead define a new set of C unique cluster dummies that will each be nonzero exactly once. The idcluster(*newvar*) option does this, creating a new variable containing a unique identifier for each observation in the resampled cluster. This is particularly relevant for estimation with fixed effects, including fixed-effects panel-data estimators.

For some xt commands, the vce(bootstrap) option actually performs a cluster bootstrap, because clustering is assumed in a panel setting and xt commands require specification of the cluster identifier.

13.3.6 Bootstrap confidence intervals

The output after a command with the vce(bootstrap) option includes a "normal-based" 95% confidence interval for θ that equals

$$[\widehat{\theta} - 1.96 \times \text{se}_{\text{Boot}}(\widehat{\theta}), \widehat{\theta} + 1.96 \times \text{se}_{\text{Boot}}(\widehat{\theta})]$$

and is a standard Wald asymptotic confidence interval, except that the bootstrap is used to compute the standard error.

Additional confidence intervals can be obtained by using the postestimation `estat bootstrap` command, defined in the next section.

The percentile method uses the relevant percentiles of the empirical distribution of the B bootstrap estimates $\widehat{\theta}_1^*, \ldots, \widehat{\theta}_B^*$. In particular, a percentile 95% confidence interval for θ is

$$(\widehat{\theta}_{0.025}^*, \widehat{\theta}_{0.975}^*)$$

ranging from the 2.5th percentile to the 97.5th percentile of $\widehat{\theta}_1^*, \ldots, \widehat{\theta}_B^*$. This confidence interval has the advantage of being asymmetric around $\widehat{\theta}$ and being invariant to monotonic transformation of θ. Like the normal-based confidence interval, it does not provide an asymptotic refinement, but there are still theoretical reasons to believe it provides a better approximation than the normal-based confidence interval.

The bias-corrected (BC) method is a modification of the percentile method that incorporates a bootstrap estimate of the finite-sample bias in $\widehat{\theta}$. For example, if the estimator is upward biased, as measured by estimated median bias, then the confidence interval is moved to the left. So if 40%, rather than 50%, of $\widehat{\theta}_1^*, \ldots, \widehat{\theta}_B^*$ are less than $\widehat{\theta}$, then a BC 95% confidence interval might use $[\widehat{\theta}_{0.007}^*, \widehat{\theta}_{0.927}^*]$, say, rather than $[\widehat{\theta}_{0.025}^*, \widehat{\theta}_{0.975}^*]$.

The BC accelerated (BCa) confidence interval is an adjustment to the BC method that adds an "acceleration" component that permits the asymptotic variance of $\widehat{\theta}$ to vary with θ. This requires the use of a jackknife that can add considerable computational time and is not possible for all estimators. The formulas for BC and BCa confidence intervals are given in [R] **bootstrap** and in books such as Efron and Tibshirani (1993, 185) and Davison and Hinkley (1997, 204).

The BCa confidence interval has the theoretical advantage over the other confidence intervals that it does offer an asymptotic refinement. So a BCa 95% confidence interval has a coverage rate of $0.95 + O(N^{-1})$, compared with $0.95 + O(N^{-1/2})$ for the other methods.

The percentile-t method also provides the same asymptotic refinement. The `estat bootstrap` command does not provide percentile-t confidence intervals, but these can be obtained by using the `bootstrap` command, as we demonstrate in section 13.5.3. Because it is based on percentiles, the BCa confidence interval is invariant to monotonic transformation of θ, whereas the percentile-t confidence interval is not. Otherwise, there is no strong theoretical reason to prefer one method over the other.

13.3.7   The postestimation estat bootstrap command

The `estat bootstrap` command can be issued after an estimation command that has the `vce(bootstrap)` option, or after the `bootstrap` command. The syntax for `estat bootstrap` is

`estat bootstrap` $\big[$, *options* $\big]$

where the options include `normal` for normal-based confidence intervals, `percentile` for percentile-based confidence intervals, `bc` for BC confidence intervals, and option `bca` for BCa confidence intervals. To use the `bca` option, the preceding bootstrap must be done with `bca` to perform the necessary additional jackknife computation. The `all` option provides all available confidence intervals.

13.3.8   Bootstrap confidence-intervals example

We obtain these various confidence intervals for the Poisson example. To obtain the BCa interval, the original bootstrap needs to include the `bca` option. To speed up bootstraps, we should include only necessary variables in the dataset. For bootstrap precision, we set $B = 999$. We have

```
. * Bootstrap confidence intervals: normal-based, percentile, BC, and BCa
. quietly poisson docvis chronic, vce(boot, reps(999) seed(10101) bca)

. estat bootstrap, all
```

| Poisson regression | | | | Number of obs | = | 50 | |
| | | | | Replications | = | 999 | |

docvis	Observed Coef.	Bias	Bootstrap Std. Err.	[95% Conf. Interval]		
chronic	.98330144	-.0244473	.54040762	-.075878	2.042481	(N)
				-.1316499	2.076792	(P)
				-.0820317	2.100361	(BC)
				-.0215526	2.181476	(BCa)
_cons	1.0316016	-.0503223	.35257252	.3405721	1.722631	(N)
				.2177235	1.598568	(P)
				.2578293	1.649789	(BC)
				.3794897	1.781907	(BCa)

```
(N)    normal confidence interval
(P)    percentile confidence interval
(BC)   bias-corrected confidence interval
(BCa)  bias-corrected and accelerated confidence interval
```

The confidence intervals for β_{chronic} are, respectively, $[-0.08, 2.04]$, $[-0.13, 2.08]$, $[-0.08, 2.10]$, and $[-0.02, 2.18]$. The differences here are not great. Only the normal-based confidence interval is symmetric about $\widehat{\beta}_{\text{chronic}}$.

13.3.9   Bootstrap estimate of bias

Suppose that the estimator $\widehat{\theta}$ is biased for θ. Let $\overline{\widehat{\theta}}^*$ be the average of the B bootstraps and $\widehat{\theta}$ be the estimate from the original model. Note that $\overline{\widehat{\theta}}^*$ is not an unbiased estimate of θ. Instead, the difference $\overline{\widehat{\theta}}^* - \widehat{\theta}$ provides a bootstrap estimate of the bias of the estimate $\widehat{\theta}$. The bootstrap views the data-generating process (DGP) value as $\widehat{\theta}$, and $\overline{\widehat{\theta}}^*$ is viewed as the mean of the estimator given this DGP value.

Below we list e(b_bs), which contains the average of the bootstrap estimates.

```
. matrix list e(b_bs)
e(b_bs)[1,2]
         docvis:    docvis:
        chronic      _cons
y1    .95885413   .9812793
```

The above output indicates that $\overline{\widehat{\theta}}^* = 0.95885413$, and the output from estat bootstrap, all indicates that $\widehat{\theta} = 0.98330144$. Thus the bootstrap estimate of bias is -0.02444731, which is reported in the estat bootstrap, all output. Because $\widehat{\theta} = 0.98330144$ is downward biased by 0.02444731, we must add back this bias to get a BC estimate of θ that equals $0.98330144 + 0.02444731 = 1.0077$. Such BC estimates are not used, however, because the bootstrap estimate of mean bias is a very noisy estimate; see Efron and Tibshirani (1993, 138).

13.4   Bootstrap pairs using the bootstrap command

The bootstrap command can be applied to a wide range of Stata commands such as nonestimation commands, user-written commands, two-step estimators, and Stata estimators without the vce(bootstrap) option. Before doing so, the user should verify that the estimator is one for which it is appropriate to apply the bootstrap; see section 13.2.4.

13.4.1   The bootstrap command

The syntax for bootstrap is

bootstrap *explist* [, *options eform_option*] :   *command*

The command being bootstrapped can be an estimation command, other commands such as summarize, or user-written commands. The argument *explist* provides the quantity or quantities to be bootstrapped. These can be one or more expressions, possibly given names [so *newvarname = (exp)*].

For estimation commands, not setting *explist* or setting *explist* to _b leads to a bootstrap of the parameter estimates. Setting *explist* instead to _se leads to a bootstrap of the standard errors of the parameter estimates. Thus bootstrap: poisson y x bootstraps parameter estimates, as does bootstrap _b: poisson y x. The bootstrap _se: poisson y x command instead bootstraps the standard errors. The bootstrap _b[x]: poisson y x command bootstraps just the coefficient of x and not that of the intercept. The bootstrap bx=_b[x]: poisson y x command does the same, with the results of each bootstrap stored in a variable named bx rather than a variable given the default name of _bs_1.

The options include reps(#) to set the number of bootstrap replications; seed(#) to set the random-number generator seed value to enable reproducibility; nodots to suppress dots produced for each bootstrap replication; cluster(*varlist*) if the bootstrap is over clusters; `idcluster(`*newvar*`)`, which is needed for some cluster bootstraps (see section 13.3.5); `group(`*varname*`)`, which may be needed along with `idcluster()`; `strata(`*varlist*`)` for bootstrap over strata; `size(#)` to draw samples of size #; `bca` to compute the acceleration for a BCa confidence interval; and `saving()` to save results from each bootstrap iteration in a file. The *eform_option* option enables bootstraps for e^θ rather than θ.

If `bootstrap` is applied to commands other than Stata estimation commands, it produces a warning message. For example, the user-written `poissrobust` command defined below, leads to the warning

```
Warning:  Since poissrobust is not an estimation command or does not set
e(sample), bootstrap has no way to determine which observations are
used in calculating the statistics and so assumes that all
observations are used.  This means no observations will be excluded
from the resampling because of missing values or other reasons.

If the assumption is not true, press Break, save the data, and drop
the observations that are to be excluded.  Be sure that the dataset
in memory contains only the relevant data.
```

Because we know that this is not a problem in the examples below and we want to minimize output, we use the `nowarn` option to suppress this warning.

The output from `bootstrap` includes the bootstrap estimate of the standard error of the statistic of interest and the associated normal-based 95% confidence interval. The `estat bootstrap` command after `bootstrap` computes other, better confidence intervals. For brevity, we do not obtain these alternative and better confidence intervals in the examples below.

13.4.2   Bootstrap parameter estimate from a Stata estimation command

The `bootstrap` command is easily applied to an existing Stata estimation command. It gives exactly the same result as given by directly using the Stata estimation command with the `vce(bootstrap)` option, if this option is available and the same values of B and the seed are used.

We illustrate this for doctor visits. Because we are bootstrapping parameter estimates from an estimation command, there is no need to provide *explist*.

```
. * Bootstrap command applied to Stata estimation command
. bootstrap, reps(400) seed(10101) nodots noheader: poisson docvis chronic
```

docvis	Observed Coef.	Bootstrap Std. Err.	z	P>|z|	Normal-based [95% Conf. Interval]	
chronic	.9833014	.5253149	1.87	0.061	-.0462968	2.0129
_cons	1.031602	.3497212	2.95	0.003	.3461607	1.717042

The results are exactly the same as those obtained in section 13.3.3 by using `poisson` with the `vce(bootstrap)` option.

13.4.3 Bootstrap standard error from a Stata estimation command

Not only is $\widehat{\theta}$ not an exact estimate of θ, but $\mathrm{se}(\widehat{\theta})$ is not an exact estimate of the standard deviation of the estimator $\widehat{\theta}$. We consider a bootstrap of the standard error, $\mathrm{se}(\widehat{\theta})$, to obtain an estimate of the standard error of $\mathrm{se}(\widehat{\theta})$.

We bootstrap both the coefficients and their standard errors. We have

```
. * Bootstrap standard-error estimate of the standard error of a coeff estimate
. bootstrap _b _se, reps(400) seed(10101) nodots: poisson docvis chronic
Bootstrap results                               Number of obs    =        50
                                                Replications     =       400
```

	Observed Coef.	Bootstrap Std. Err.	z	P>|z|	Normal-based [95% Conf. Interval]	
docvis						
chronic	.9833014	.5253149	1.87	0.061	-.0462968	2.0129
_cons	1.031602	.3497212	2.95	0.003	.3461607	1.717042
docvis_se						
chronic	.1393729	.0231223	6.03	0.000	.094054	.1846917
_cons	.0995037	.0201451	4.94	0.000	.06002	.1389875

The bootstrap reveals that there is considerable noise in $\mathrm{se}(\widehat{\beta}_{\mathrm{chronic}})$, with an estimated standard error of 0.023 and the 95% confidence interval [0.09, 0.18].

Ideally, the bootstrap standard error of $\widehat{\beta}$, here 0.525, should be close to the mean of the bootstraps of $\mathrm{se}(\widehat{\beta})$, here 0.139. The fact that they are so different is a clear sign of problems in the method used to obtain $\mathrm{se}(\widehat{\beta}_{\mathrm{chronic}})$. The problem is that the default Poisson standard errors were used in `poisson` above, and given the large overdispersion, these standard errors are very poor. If we repeated the exercise with `poisson` and the `vce(robust)` option, this difference should disappear.

13.4.4 Bootstrap standard error from a user-written estimation command

Continuing the previous example, we would like an estimate of the robust standard errors after Poisson regression. This can be obtained by using `poisson` with the `vce(robust)` option. We instead use an alternative approach that can be applied in a wide range of settings.

We write a program named `poissrobust` that returns the Poisson maximum likelihood estimator (MLE) estimates in `b` and the robust estimate of the VCE of the Poisson MLE in `V`. Then we apply the `bootstrap` command to `poissrobust` rather than to `poisson, vce(robust)`.

Because we want to return e and V, the program must be `eclass`. The program is

```
* Program to return b and robust estimate V of the VCE
program poissrobust, eclass
version 11
tempname b V
poisson docvis chronic, vce(robust)
matrix `b´ = e(b)
matrix `V´ = e(V)
ereturn post `b´ `V´
end
```

Next it is good practice to check the program, typing the commands

```
. * Check preceding program by running once
. poissrobust
. ereturn display
```

The omitted output is the same as that from `poisson, vce(robust)`.

We then bootstrap 400 times. The bootstrap estimate of the standard error of $se(\widehat{\theta})$ is the standard deviation of the B values of $se(\widehat{\theta})$. We have

```
. * Bootstrap standard-error estimate of robust standard errors
. bootstrap _b _se, reps(400) seed(10101) nodots nowarn: poissrobust
Bootstrap results                              Number of obs     =         50
                                               Replications      =        400
```

	Observed Coef.	Bootstrap Std. Err.	z	P>|z|	Normal-based [95% Conf. Interval]	
docvis						
chronic	.9833014	.5253149	1.87	0.061	-.0462968	2.0129
_cons	1.031602	.3497212	2.95	0.003	.3461607	1.717042
docvis_se						
chronic	.5154894	.0784361	6.57	0.000	.3617575	.6692213
_cons	.3446734	.0613856	5.61	0.000	.2243598	.464987

There is considerable noise in the robust standard error, with the standard error of $se(\widehat{\beta}_{\text{chronic}})$ equal to 0.078 and a 95% confidence interval of $[0.36, 0.67]$. The upper limit is about twice the lower limit, as was the case for the default standard error. In other examples, robust standard errors can be much less precise than default standard errors.

13.4.5 Bootstrap two-step estimator

The preceding method of applying the `bootstrap` command to a user-defined estimation command can also be applied to a two-step estimator.

A sequential two-step estimator of, say, $\widehat{\beta}$ is one that depends in part on a consistent first-stage estimator, say, $\widehat{\alpha}$. In some examples—notably, feasible generalized least squares (FGLS), where $\widehat{\alpha}$ denotes error variance parameters—one can do regular inference ignoring any estimation error in $\widehat{\alpha}$. More generally, however, the asymptotic distribution of $\widehat{\boldsymbol{\beta}}$ will depend on that of $\widehat{\boldsymbol{\alpha}}$. Asymptotic results do exist that confirm the asymptotic normality of leading examples of two-step estimators, and provide a general formula for Var($\widehat{\boldsymbol{\beta}}$). But this formula is usually complicated, both analytically and in implementation. A much simpler method is to use the bootstrap, which is valid if indeed the two-step estimator is known to be asymptotically normal.

A leading example is Heckman's two-step estimator in the selection model; see section 16.6.4. We use the same example here as in chapter 16. We first read in the data and form the dependent variable dy and the regressor list given in xlist.

```
. * Set up the selection model two-step estimator data of chapter 16
. use mus16data.dta, clear
. generate y = ambexp
. generate dy = y > 0
. generate lny = ln(y)
(526 missing values generated)
. global xlist age female educ blhisp totchr ins
```

The following program produces the Heckman two-step estimator:

```
* Program to return b for Heckman 2-step estimator of selection model
program hecktwostep, eclass
version 11
tempname b V
tempvar xb
capture drop invmills
probit dy $xlist
predict `xb', xb
generate invmills = normalden(`xb')/normal(`xb')
regress lny $xlist invmills
matrix `b' = e(b)
ereturn post `b'
end
```

This program can be checked by typing hecktwostep in isolation. This leads to the same parameter estimates as in section 16.6.4. Here β denotes the second-stage regression coefficients of regressors and the inverse of the Mills' ratio. The inverse of the Mills' ratio depends on the first-stage probit parameter estimates $\widehat{\boldsymbol{\alpha}}$. Because the above code fails to control for the randomness in $\widehat{\boldsymbol{\alpha}}$, the standard errors following hecktwostep differ from the correct standard errors given in section 16.6.4.

To obtain correct standard errors that control for the two-step estimation, we bootstrap.

```
. * Bootstrap for Heckman two-step estimator using chapter 16 example
. bootstrap _b, reps(400) seed(10101) nodots nowarn: hecktwostep
Bootstrap results                      Number of obs    =    3328
                                       Replications     =     400
```

	Observed Coef.	Bootstrap Std. Err.	z	P>|z|	Normal-based [95% Conf. Interval]	
age	.202124	.0233969	8.64	0.000	.1562671	.247981
female	.2891575	.0704133	4.11	0.000	.1511501	.4271649
educ	.0119928	.0114104	1.05	0.293	-.0103711	.0343567
blhisp	-.1810582	.0654464	-2.77	0.006	-.3093308	-.0527856
totchr	.4983315	.0432639	11.52	0.000	.4135358	.5831272
ins	-.0474019	.050382	-0.94	0.347	-.1461488	.051345
invmills	-.4801696	.291585	-1.65	0.100	-1.051666	.0913265
_cons	5.302572	.2890579	18.34	0.000	4.736029	5.869115

The standard errors are generally within 5% of those given in chapter 16, which are based on analytical results.

13.4.6   Bootstrap Hausman test

The Hausman test statistic, presented in section 12.7.5, is

$$ H = (\widehat{\boldsymbol{\theta}} - \widetilde{\boldsymbol{\theta}})' \left\{ \widehat{V}(\widehat{\boldsymbol{\theta}} - \widetilde{\boldsymbol{\theta}}) \right\}^{-1} (\widehat{\boldsymbol{\theta}} - \widetilde{\boldsymbol{\theta}}) \stackrel{a}{\sim} \chi^2(h) \text{ under } H_0 $$

where $\widehat{\boldsymbol{\theta}}$ and $\widetilde{\boldsymbol{\theta}}$ are different estimators of $\boldsymbol{\theta}$.

Standard implementations of the Hausman test, including the `hausman` command presented in section 12.7.5, require that one of the estimators be fully efficient under H_0. Great simplification occurs because $\text{Var}(\widehat{\boldsymbol{\theta}} - \widetilde{\boldsymbol{\theta}}) = \text{Var}(\widetilde{\boldsymbol{\theta}}) - \text{Var}(\widehat{\boldsymbol{\theta}})$ if $\widehat{\boldsymbol{\theta}}$ is fully efficient under H_0. For some likelihood-based estimators, correct model specification is necessary for consistency and in that case the estimator is also fully efficient. But often it is possible and standard to not require that the estimator be efficient. In particular, if there is reason to use robust standard errors, then the estimator is not fully efficient.

The bootstrap can be used to estimate $\text{Var}(\widehat{\boldsymbol{\theta}} - \widetilde{\boldsymbol{\theta}})$, without the need to assume that one of the estimators is fully efficient under H_0. The B replications yield B estimates of $\widehat{\boldsymbol{\theta}}$ and $\widetilde{\boldsymbol{\theta}}$, and hence of $\widehat{\boldsymbol{\theta}} - \widetilde{\boldsymbol{\theta}}$. We estimate $\text{Var}(\widehat{\boldsymbol{\theta}} - \widetilde{\boldsymbol{\theta}})$ with $(1/B - 1) \sum_b (\widehat{\theta}_b - \widetilde{\theta}_b - \overline{\theta}^*_{\text{diff}})(\widehat{\theta}_b - \widetilde{\theta}_b - \overline{\theta}^*_{\text{diff}})'$, where $\overline{\theta}^*_{\text{diff}} = 1/B \sum_b (\widehat{\theta}_b - \widetilde{\theta}_b)$.

As an example, we consider a Hausman test for endogeneity of a regressor based on comparing instrumental-variables and ordinary least-squares (OLS) estimates. Large values of H lead to rejection of the null hypothesis that all regressors are exogenous.

The following program is written for the two-stage least-squares example presented in section 6.3.6.

```
* Program to return (b1-b2) for Hausman test of endogeneity
program hausmantest, eclass
version 11
tempname b bols biv
regress ldrugexp hi_empunion totchr age female blhisp linc, vce(robust)
matrix `bols´ = e(b)
ivregress 2sls ldrugexp (hi_empunion = ssiratio) totchr age female blhisp ///
    linc, vce(robust)
matrix `biv´ = e(b)
matrix `b´ = `bols´ - `biv´
ereturn post `b´
end
```

This program can be checked by typing `hausmantest` in isolation.

We then run the bootstrap.

```
. * Bootstrap estimates for Hausman test using chapter 6 example
. use mus06data.dta, clear

. bootstrap _b, reps(400) seed(10101) nodots nowarn: hausmantest
Bootstrap results                          Number of obs    =      10391
                                           Replications     =        400
```

	Observed Coef.	Bootstrap Std. Err.	z	P>|z|	Normal-based [95% Conf. Interval]	
hi_empunion	.9714701	.2396239	4.05	0.000	.5018158	1.441124
totchr	-.0098848	.00463	-2.13	0.033	-.0189594	-.0008102
age	.0096881	.002437	3.98	0.000	.0049117	.0144645
female	.0782115	.0221073	3.54	0.000	.0348819	.1215411
blhisp	.0661176	.0208438	3.17	0.002	.0252646	.1069706
linc	-.0765202	.0201043	-3.81	0.000	-.1159239	-.0371165
_cons	-.9260396	.2320957	-3.99	0.000	-1.380939	-.4711404

For the single potentially endogenous regressor, we can use the t statistic given above, or we can use the `test` command. The latter yields

```
. * Perform Hausman test on the potentially endogenous regressor
. test hi_empunion

 ( 1)  hi_empunion = 0

        chi2( 1) =   16.44
      Prob > chi2 =    0.0001
```

The null hypothesis of regressor exogeneity is strongly rejected. The `test` command can also be used to perform a Hausman test based on all regressors.

The preceding example has wide applicability for robust Hausman tests.

13.4.7 Bootstrap standard error of the coefficient of variation

The bootstrap need not be restricted to regression models. A simple example is to obtain a bootstrap estimate of the standard error of the sample mean of `docvis`. This can be obtained by using the `bootstrap _se: mean docvis` command.

A slightly more difficult example is to obtain the bootstrap estimate of the standard error of the coefficient of variation ($= s_x/\overline{x}$) of doctor visits. The results stored in r() after `summarize` allow the coefficient of variation to be computed as `r(sd)/r(mean)`, so we bootstrap this quantity.

To do this, we use `bootstrap` with the expression `coeffvar=(r(sd)/r(mean))`. This bootstraps the quantity `r(sd)/r(mean)` and gives it the name `coeffvar`. We have

```
. * Bootstrap estimate of the standard error of the coefficient of variation
. use bootdata.dta, clear

. bootstrap coeffvar=(r(sd)/r(mean)), reps(400) seed(10101) nodots nowarn
> saving(coeffofvar, replace): summarize docvis
Bootstrap results                              Number of obs    =         50
                                               Replications     =        400

      command:  summarize docvis
     coeffvar:  r(sd)/r(mean)
```

	Observed Coef.	Bootstrap Std. Err.	z	P>|z|	Normal-based [95% Conf. Interval]	
coeffvar	1.898316	.2718811	6.98	0.000	1.365438	2.431193

The normal-based bootstrap 95% confidence interval for the coefficient of variation is [1.37, 2.43].

13.5 Bootstraps with asymptotic refinement

Some bootstraps can yield asymptotic refinement, defined in section 13.2.3. The postestimation `estat bootstrap` command automatically provides BCa confidence intervals; see sections 13.3.6–13.3.8. In this section, we focus on an alternative method that provides asymptotic refinement, the percentile-t method. The percentile-t method has general applicability to hypothesis testing and confidence intervals.

13.5.1 Percentile-t method

A general way to obtain asymptotic refinement is to bootstrap a quantity that is asymptotically pivotal, meaning that its asymptotic distribution does not depend on unknown parameters. The estimate $\widehat{\theta}$ is not asymptotically pivotal, because its variance depends on unknown parameters. Percentile methods therefore do not provide an asymptotic refinement unless an adjustment is made, notably, that by the BCa percentile method. The t statistic is asymptotically pivotal, however, and percentile-t methods or bootstrap-t methods bootstrap the t statistic.

We therefore bootstrap the t statistic:

$$t = (\widehat{\theta} - \theta)/\text{se}(\widehat{\theta}) \tag{13.2}$$

The bootstrap views the original sample as the DGP, so the bootstrap sets the DGP value of θ to be $\widehat{\theta}$. So in each bootstrap resample, we compute a t statistic centered on $\widehat{\theta}$:

$$t_b^* = (\widehat{\theta}_b^* - \widehat{\theta})/\text{se}(\widehat{\theta}_b^*) \qquad (13.3)$$

where $\widehat{\theta}_b^*$ is the parameter estimate in the bth bootstrap, and $\text{se}(\widehat{\theta}_b^*)$ is a consistent estimate of the standard error of $\widehat{\theta}_b^*$, often a robust or cluster–robust standard error.

The B bootstraps yield the t-values $t_1^*, \ldots, t_B^*$, whose empirical distribution is used as the estimate of the distribution of the t statistic. For a two-sided test of $H_0 : \theta = 0$, the p-value of the original test statistic $t = \widehat{\theta}/\text{se}(\widehat{\theta})$ is

$$\frac{1}{B} \sum_{b=1}^{B} \mathbf{1}\left(|t| < |t_b^*|\right)$$

which is the fraction of times in B replications that $|t| < |t^*|$. The percentile-t critical values for a nonsymmetric two-sided test at the 0.05 level are $t_{0.025}^*$ and $t_{0.975}^*$. And a percentile-t 95% confidence interval is

$$[\widehat{\theta} + t_{0.025}^* \times \text{se}(\widehat{\theta}), \widehat{\theta} + t_{0.975}^* \times \text{se}(\widehat{\theta})] \qquad (13.4)$$

The formula for the lower bound has a plus sign because $t_{0.025}^* < 0$.

13.5.2   Percentile-t Wald test

Stata does not automatically produce the percentile-t method. Instead, the `bootstrap` command can be used to bootstrap the t statistic, saving the B bootstrap values $t_1^*, \ldots, t_B^*$ in a file. This file can be accessed to obtain the percentile-t p-values and critical values.

We continue with a count regression of `docvis` on `chronic`. A complication is that the standard error given in either (13.2) or (13.3) needs to be a consistent estimate of the standard deviation of the estimator. So we use `bootstrap` to perform a bootstrap of `poisson`, where the VCE is estimated with the `vce(robust)` option, rather than using the default Poisson standard-error estimates that are greatly downward biased.

We store the sample parameter estimate and standard error as local macros before bootstrapping the t statistic given in (13.3).

```
. * Percentile-t for a single coefficient: Bootstrap the t statistic
. use bootdata, clear
. quietly poisson docvis chronic, vce(robust)
. local theta = _b[chronic]
. local setheta = _se[chronic]
```

```
. bootstrap tstar=((_b[chronic]-`theta´)/_se[chronic]), seed(10101) reps(999)
> nodots saving(percentilet, replace): poisson docvis chronic, vce(robust)
Bootstrap results                              Number of obs     =        50
                                               Replications      =       999
        command:  poisson docvis chronic, vce(robust)
          tstar:  (_b[chronic]-.9833014421442415)/_se[chronic]
```

	Observed Coef.	Bootstrap Std. Err.	z	P>|z|	Normal-based [95% Conf. Interval]	
tstar	0	1.3004	0.00	1.000	-2.548736	2.548736

The output indicates that the distribution of t^* is considerably more dispersed than a standard normal, with a standard deviation of 1.30 rather than 1.0 for the standard normal.

To obtain the test p-value, we need to access the 999 values of t^* saved in the percentilet.dta file.

```
. * Percentile-t p-value for symmetric two-sided Wald test of H0: theta = 0
. use percentilet, clear
(bootstrap: poisson)
. quietly count if abs(`theta´/`setheta´) < abs(tstar)
. display "p-value = " r(N)/_N
p-value = .14514515
```

We do not reject $H_0 : \beta_{\mathrm{chronic}} = 0$ against $H_0 : \beta_{\mathrm{chronic}} \neq 0$ at the 0.05 level because $p = 0.145 > 0.05$. By comparison, if we use the usual standard normal critical values, $p = 0.056$, which is considerably smaller.

The above code can be adapted to apply to several or all parameters by using the bootstrap command to obtain _b and _se, saving these in a file, using this file, and computing for each parameter of interest the values t^* given $\widehat{\theta}^*$, $\widehat{\theta}$, and $\mathrm{se}(\widehat{\theta}^*)$.

13.5.3   Percentile-t Wald confidence interval

The percentile-t 95% confidence interval for the coefficient of chronic is obtained by using (13.4), where $t_1^*, \ldots, t_B^*$ were obtained in the previous section. We have

```
. * Percentile-t critical values and confidence interval
. _pctile tstar, p(2.5,97.5)
. scalar lb = `theta´ + r(r1)*`setheta´
. scalar ub = `theta´ + r(r2)*`setheta´
. display "2.5 and 97.5 percentiles of t* distn: " r(r1) ", " r(r2) _n
> "95 percent percentile-t confidence interval is  (" lb "," ub ")"
2.5 and 97.5 percentiles of t* distn: -2.7561963, 2.5686913
95 percent percentile-t confidence interval is  (-.43748842,2.3074345)
```

The confidence interval is $[-0.44, 2.31]$, compared with the $[-0.03, 1.99]$, which could be obtained by using the robust estimate of the VCE after `poisson`. The wider confidence interval is due to the bootstrap-t critical values of -2.76 and 2.57, much larger than the standard normal critical values of -1.96 and 1.96. The confidence interval is also wider than the other bootstrap confidence intervals given in section 13.3.8.

Percentile-t 95% confidence intervals, like BCa confidence intervals, have the advantage of having a coverage rate of $0.95 + O(N^{-1})$ rather than $0.95 + O(N^{-1/2})$. Efron and Tibshirani (1993, 184, 188, 326) favor the BCa method for confidence intervals. But they state that "generally speaking, the bootstrap-t works well for location parameters", and regression coefficients are location parameters.

13.6   Bootstrap pairs using bsample and simulate

The `bootstrap` command can be used only if it is possible to provide a single expression for the quantity being bootstrapped. If this is not possible, one can use the `bsample` command to obtain one bootstrap sample and compute the statistic of interest for this resample, and then use the `simulate` or `postfile` command to execute this command a number of times.

13.6.1   The bsample command

The `bsample` command draws random samples with replacement from the current data in memory. The command syntax is

bsample $\big[\,exp\,\big]$ $\big[\,if\,\big]$ $\big[\,in\,\big]$ $\big[\,,\ options\,\big]$

where *exp* specifies the size of the bootstrap sample, which must be at most the size of the selected sample. The `strata(`*varlist*`)`, `cluster(`*varlist*`)`, `idcluster(`*newvar*`)`, and `weight(`*varname*`)` options allow stratification, clustering, and weighting. The `idcluster()` option is discussed in section 13.3.5.

13.6.2   The bsample command with simulate

An example where `bootstrap` is insufficient is testing $H_0 : h(\beta) = 0$, where $h(\cdot)$ is a scalar nonlinear function of β, using the percentile-t method to get asymptotic refinement. The bootstraps will include computation of $\mathrm{se}\{h(\widehat{\theta}_b^*)\}$, and there is no simple expression for this.

In such situations, we can follow the following procedure. First, write a program that draws one bootstrap resample of size N with replacement, using the `bsample` command, and compute the statistic of interest for this resample. Second, use the `simulate` or `postfile` command to execute the program B times and save the resulting B bootstrap statistics.

We illustrate this method for a Poisson regression of `docvis` on `chronic`, using the same example as in section 13.5.2. We first define the program for one bootstrap replication. The `bsample` command without argument produces one resample of all variables with a replacement of size N from the original sample of size N.

The program returns a scalar, `tstar`, that equals t^* in (13.3). Because we are not returning parameter estimates, we use an r-class program. We have

```
* Program to do one bootstrap replication
program onebootrep, rclass
version 11
drop _all
use bootdata.dta
bsample
poisson docvis chronic, vce(robust)
return scalar tstar = (_b[chronic]-$theta)/_se[chronic]
end
```

Note that robust standard errors are obtained here. The referenced global macro, `theta`, constructed below, is the estimated coefficient of `chronic` in the original sample. We could alternatively pass this as a program argument rather than use a global macro. The program returns `tstar`.

We next obtain the original sample parameter estimate and use the `simulate` command to run the `onebootrep` program B times. We have

```
. * Now do 999 bootstrap replications
. use bootdata.dta, clear
. quietly poisson docvis chronic, vce(robust)
. global theta = _b[chronic]
. global setheta = _se[chronic]
. simulate tstar=r(tstar), seed(10101) reps(999) nodots
> saving(percentilet2, replace): onebootrep
        command:  onebootrep
          tstar:  r(tstar)
```

The `percentilet2` file has the 999 bootstrap values $t_1^*, \dots, t_{999}^*$ that can then be used to calculate the bootstrap p-value.

```
. * Analyze the results to get the p-value
. use percentilet2, clear
(simulate: onebootrep)
. quietly count if abs($theta/$setheta) < abs(tstar)
. display "p-value = " r(N)/_N
p-value = .14514515
```

The p-value is 0.145, leading to nonrejection of H_0: $\beta_{\text{chronic}} = 0$ at the 0.05 level. This result is exactly the same as that in section 13.5.2.

13.6.3   Bootstrap Monte Carlo exercise

One way to verify that the bootstrap offers an asymptotic refinement or improvement in finite samples is to perform a simulation exercise. This is essentially a nested simulation, with a bootstrap simulation in the inner loop and a Monte Carlo simulation in the outer loop.

We first define a program that does a complete bootstrap of B replications, by calling the onebootrep program B times.

```
* Program to do one bootstrap of B replications
program mybootstrap, rclass
use bootdata.dta, clear
quietly poisson docvis chronic, vce(robust)
global theta = _b[chronic]
global setheta = _se[chronic]
simulate tstar=r(tstar), reps(999) nodots   ///
     saving(percentilet2, replace): onebootrep
use percentilet2, clear
quietly count if abs($theta/$setheta) < abs(tstar)
return scalar pvalue =  r(N)/_N
end
```

We next check the program by running it once:

```
. set seed 10101

. mybootstrap

     command:  onebootrep
       tstar:  r(tstar)

(simulate: onebootrep)

. display r(pvalue)
.14514515
```

The p-value is the same as that obtained in the previous section.

To use the mybootstrap program for a simulation exercise, we use data from a known DGP and run the program S times. We draw one sample of chronic, held constant throughout the exercise in the tempx file. Then, S times, generate a sample of size N of the count docvis from a Poisson distribution—or, better, a negative binomial distribution—run the mybootstrap command, and obtain the returned p-value. This yields S p-values, and analysis proceeds similar to the test size calculation example in section 12.6.2. Simulations such as this take a long time because regressions are run $S \times B$ times.

13.7   Alternative resampling schemes

There are many ways to resample other than the nonparametric pairs and cluster-pairs bootstraps methods used by the Stata bootstrap commands. These other methods can be performed by using a similar approach to the one in section 13.6.2, with a program written to obtain one bootstrap resample and calculate the statistic(s) of interest, and this program then called B times. We do so for several methods, bootstrapping regression model parameter estimates.

The programs are easily adapted to bootstrapping other quantities, such as the t statistic to obtain asymptotic refinement. For asymptotic refinement, there is particular benefit in using methods that exploit more information about the DGP than is used by bootstrap pairs. This additional information includes holding $\mathbf{x}$ fixed through the bootstraps, called a design-based or model-based bootstrap; imposing conditions such as $E(u|\mathbf{x}) = 0$ in the bootstrap; and for hypothesis tests, imposing the null hypothesis on the bootstrap resamples. See, for example, Horowitz (2001), MacKinnon (2002), and the application by Cameron, Gelbach, and Miller (2008).

13.7.1 Bootstrap pairs

We begin with bootstrap pairs, repeating code similar to that in section 13.6.2. The following program obtains one bootstrap resample by resampling from the original data with replacement.

```
* Program to resample using bootstrap pairs
program bootpairs
version 11
drop _all
use bootdata.dta
bsample
poisson docvis chronic
end
```

To check the program, we run it once.

```
. * Check the program by running once
. bootpairs
```

We then run the program 400 times. We have

```
. * Bootstrap pairs for the parameters
. simulate _b, seed(10101) reps(400) nodots: bootpairs

      command:  bootpairs

. summarize
```

Variable	Obs	Mean	Std. Dev.	Min	Max
docvis_b_c~c	400	.9741139	.5253149	-.6184664	2.69578
docvis_b_c~s	400	.9855123	.3497212	-.3053816	1.781907

The bootstrap estimate of the standard error of β_{chronic} equals 0.525, as in section 13.3.3.

13.7.2 Parametric bootstrap

A parametric bootstrap is essentially a Monte Carlo simulation. Typically, we hold $\mathbf{x}_i$ fixed at the sample values; replace y_i by a random draw, y_i^*, from the density $f(y_i|\mathbf{x}_i, \boldsymbol{\theta})$

with $\boldsymbol{\theta}$ evaluated at the original sample estimate, $\widehat{\boldsymbol{\theta}}$; and regress y_i^* on $\mathbf{x}_i$. A parametric bootstrap requires much stronger assumptions, correct specification of the conditional density of y given $\mathbf{x}$, than the paired or nonparametric bootstrap.

To implement a parametric bootstrap, the preceding `bootpairs` program is adapted to replace the `bsample` command with code to randomly draw y from $f(y|\mathbf{x}, \widehat{\boldsymbol{\theta}})$.

For doctor visits, which are overdispersed count data, we use the negative binomial distribution rather than the Poisson. We first obtain the negative binomial parameter estimates, $\widehat{\boldsymbol{\theta}}$, using the original sample. In this case, it is sufficient and simpler to obtain the fitted mean, $\widehat{\mu}_i = \exp(\mathbf{x}_i'\widehat{\boldsymbol{\beta}})$, and the dispersion parameter, $\widehat{\alpha}$. We have

```
. * Estimate the model with original actual data and save estimates
. use bootdata.dta
. quietly nbreg docvis chronic
. predict muhat
. global alpha = e(alpha)
```

We use these estimates to obtain draws of y from the negative binomial distribution given $\widehat{\alpha}$ and $\widehat{\mu}_i$, using a Poisson–gamma mixture (explained in section 17.2.2). The `rgamma(1/a,a)` function draws a gamma variable, ν, named `nu` with a mean of 1 and a variance of a, and the `rpoisson(nu*mu)` function then generates negative binomial draws with a mean of μ and a variance of $\mu + a\mu^2$. We have

```
* Program for parametric bootstrap generating from negative binomial
program bootparametric, eclass
  version 11
  capture drop nu dvhat
  generate nu = rgamma(1/$alpha,$alpha)
  generate dvhat = rpoisson(muhat*nu)
  nbreg dvhat chronic
end
```

We check the program by using the `bootparametric` command and then bootstrap 400 times.

```
. * Parametric bootstrap for the parameters
. simulate _b, seed(10101) reps(400) nodots: bootparametric
      command:  bootparametric
```

```
. summarize
```

Variable	Obs	Mean	Std. Dev.	Min	Max
dvhat_b_ch~c	400	.9643141	.4639161	-.4608808	2.293015
dvhat_b_cons	400	.9758856	.2604589	.1053605	1.679171
lnalpha_b_~s	400	.486886	.2769207	-.4448161	1.292826

Because we generate data from a negative binomial model and we fit a negative binomial model, the average of the 400 bootstrap coefficient estimates should be close to the DGP values. This is the case here. Also the bootstrap standard errors are within 10% of those from the negative binomial estimation of the original model, not given here, suggesting that the negative binomial model may be a reasonable one for these data.

13.7.3   **Residual bootstrap**

For linear OLS regression, under the strong assumption that errors are independent and identically distributed, an alternative to bootstrap pairs is a residual bootstrap. This holds $\mathbf{x}_i$ fixed at the sample values and replaces y_i with $y_i^* = \mathbf{x}_i'\widehat{\boldsymbol{\beta}} + \widehat{u}_i^*$, where $\widehat{u}_i^*$ are bootstrap draws from the original sample residuals $\widehat{u}_1, \ldots, \widehat{u}_N$. This bootstrap, sometimes called a design bootstrap, can lead to better performance of the bootstrap by holding regressors fixed.

The `bootpairs` program is adapted by replacing `bsample` with code to randomly draw $\widehat{u}_i^*$ from $\widehat{u}_1, \ldots, \widehat{u}_N$ and then form $y_i^* = \mathbf{x}_i'\widehat{\boldsymbol{\beta}} + \widehat{u}_i^*$. This is not straightforward because the `bsample` command is intended to bootstrap the entire dataset in memory, whereas here we wish to bootstrap the residuals but not the regressors.

As illustration, we continue to use the `docvis` example, even though Poisson regression is more appropriate than OLS regression. The following code performs the residual bootstrap:

```
* Residual bootstrap for OLS with iid errors
use bootdata.dta, clear
quietly regress docvis chronic
predict uhat, resid
keep uhat
save residuals, replace
program bootresidual
version 11
drop _all
use residuals
bsample
merge using bootdata.dta
regress docvis chronic
predict xb
generate ystar =  xb + uhat
regress ystar chronic
end
```

We check the program by using the `bootresidual` command and bootstrap 400 times.

```
. * Residual bootstrap for the parameters
. simulate _b, seed(10101) reps(400) nodots: bootresidual
      command:  bootresidual

. summarize
    Variable |        Obs        Mean    Std. Dev.       Min        Max
-------------+--------------------------------------------------------
   _b_chronic |        400     4.73843    2.184259   -1.135362    11.5334
      _b_cons |        400    2.853534    1.206185    .1126543   7.101852
```

The output reports the average of the 400 slope coefficient estimates (4.738), close to the original sample OLS slope coefficient estimate, not reported, of 4.694. The bootstrap estimate of the standard error (2.18) is close to the original sample OLS default estimate, not given, of 2.39. This is expected because the residual bootstrap assumes that errors are independent and identically distributed.

13.7.4 Wild bootstrap

For linear regression, a wild bootstrap accommodates the more realistic assumption that errors are independent but not identically distributed, permitting heteroskedasticity. This holds $\mathbf{x}_i$ fixed at the sample values and replaces y_i with $y_i^* = \mathbf{x}_i'\widehat{\boldsymbol{\beta}} + \widehat{u}_i^*$, where $\widehat{u}_i^* = a_i\widehat{\mathbf{u}}_i$, and $a_i = (1 - \sqrt{5})/2 \simeq -0.618034$ with the probability $(1 + \sqrt{5})/2\sqrt{5} \simeq 0.723607$ and $a_i = 1 - (1 - \sqrt{5})/2$ with the probability $1 - (1 + \sqrt{5})/2\sqrt{5}$. For each observation, $\widehat{u}_i^*$ takes only two possible values, but across all N observations there are 2^N possible resamples if the N values of $\widehat{\mathbf{u}}_i$ are distinct. See Horowitz (2001, 3215–3217), Davison and Hinkley (1997, 272), or Cameron and Trivedi (2005, 376) for discussion.

The preceding `bootresidual` program is adapted by replacing `bsample` with code to randomly draw $\widehat{u}_i^*$ from $\widehat{u}_i$ and then form $y_i^* = \mathbf{x}_i'\widehat{\boldsymbol{\beta}} + \widehat{u}_i^*$.

The Stata code is the same as that in section 13.7.1, except that the `bsample` command in the `bootpairs` program needs to be replaced with code to randomly draw $\widehat{u}_i^*$ from $\widehat{u}_i$ and then form $y_i^* = \mathbf{x}_i'\widehat{\boldsymbol{\beta}} + \widehat{u}_i^*$.

```
* Wild bootstrap for OLS with iid errors
use bootdata.dta, clear
program bootwild
version 11
drop _all
use bootdata.dta
regress docvis chronic
predict xb
predict u, resid
gen ustar = -0.618034*u
replace ustar = 1.618034*u if runiform() > 0.723607
gen ystar =  xb + ustar
regress ystar chronic
end
```

We check the program by issuing the `bootwild` command and bootstrap 400 times.

```
. * Wild bootstrap for the parameters
. simulate _b, seed(10101) reps(400) nodots: bootwild

      command:  bootwild

. summarize
```

Variable	Obs	Mean	Std. Dev.	Min	Max
_b_chronic	400	4.469173	2.904647	-2.280451	12.38536
_b_cons	400	2.891871	.9687433	1.049138	5.386696

The wild bootstrap permits heteroskedastic errors and yields bootstrap estimates of the standard errors (2.90) that are close to the original sample OLS heteroskedasticity-robust estimates, not given, of 3.06. These standard errors are considerably higher than those obtained by using the residual bootstrap, which is clearly inappropriate in this example because of the inherent heteroskedasticity of count data.

The percentile-t method with the wild bootstrap provides asymptotic refinement to Wald tests and confidence intervals in the linear model with heteroskedastic errors.

13.7.5   Subsampling

The bootstrap fails in some settings, such as a nonsmooth estimator. Then a more robust resampling method is subsampling, which draws a resample that is considerably smaller than the original sample.

The `bsample 20` command, for example, draws a sample of size 20. To perform subsampling where the resamples have one-third as many observations as the original sample, replace the `bsample` command in the bootstrap pairs with `bsample int(_N/3)`, where the `int()` function truncates to an integer toward zero.

Subsampling is more complicated than the bootstrap and is currently a topic of econometric research. See Politis, Romano, and Wolf (1999) for an introduction to this method.

13.8   The jackknife

The delete-one jackknife is a resampling scheme that forms N resamples of size $(N-1)$ by sequentially deleting each observation and then estimating $\boldsymbol{\theta}$ in each resample.

13.8.1   Jackknife method

Let $\widehat{\boldsymbol{\theta}}_i$ denote the parameter estimate from the sample with the ith observation deleted, $i = 1, \ldots, N$, let $\widehat{\boldsymbol{\theta}}$ be the original sample estimate of $\boldsymbol{\theta}$, and let $\overline{\overline{\boldsymbol{\theta}}} = N^{-1} \sum_{i=1}^{N} \widehat{\boldsymbol{\theta}}_i$ denote the average of the N jackknife estimates.

The jackknife has several uses. The BC jackknife estimate of $\boldsymbol{\theta}$ equals $N\widehat{\boldsymbol{\theta}} - (N-1)\overline{\overline{\boldsymbol{\theta}}} = (1/N) \sum_{i=1}^{N} \{N\widehat{\boldsymbol{\theta}} - (N-1)\widehat{\boldsymbol{\theta}}_i\}$. The variance of the N pseudovalues $\widehat{\boldsymbol{\theta}}_i^* = N\widehat{\boldsymbol{\theta}} - (N-1)\widehat{\boldsymbol{\theta}}_i$ can be used to estimate $\mathrm{Var}(\widehat{\boldsymbol{\theta}})$. The BCa method for a bootstrap with asymptotic refinement also uses the jackknife.

There are two variants of the jackknife estimate of the VCE. The Stata default is

$$\widehat{V}_{\mathrm{Jack}}(\widehat{\boldsymbol{\theta}}) = \left\{ \frac{1}{N(N-1)} \sum_{i=1}^{N} (\widehat{\boldsymbol{\theta}}_i^* - \overline{\overline{\boldsymbol{\theta}}})(\widehat{\boldsymbol{\theta}}_i^* - \overline{\overline{\boldsymbol{\theta}}})' \right\}$$

and the `mse` option gives the variation

$$\widehat{V}_{\mathrm{Jack}}(\widehat{\boldsymbol{\theta}}) = \left\{ \frac{N-1}{N} \sum_{i=1}^{N} (\widehat{\boldsymbol{\theta}}_i - \widehat{\boldsymbol{\theta}})(\widehat{\boldsymbol{\theta}}_i - \widehat{\boldsymbol{\theta}})' \right\}$$

The use of the jackknife for estimation of the VCE has been largely superseded by the bootstrap. The method entails N resamples, which requires much more computation than the bootstrap if N is large. The resamples are not random draws, so there is no seed to set.

13.8.2   The vce(jackknife) option and the jackknife command

For many estimation commands, the vce(jackknife) option can be used to obtain the jackknife estimate of the VCE. For example,

```
. * Jackknife estimate of standard errors
. use bootdata.dta, replace
. poisson docvis chronic, vce(jackknife, mse nodots)
Poisson regression                              Number of obs   =         50
                                                Replications    =         50
                                                F(   1,     49) =       2.50
                                                Prob > F        =     0.1205
Log likelihood = -238.75384                     Pseudo R2       =     0.0917
```

docvis	Coef.	Jknife * Std. Err.	t	P>|t|	[95% Conf. Interval]	
chronic	.9833014	.6222999	1.58	0.121	-.2672571	2.23386
_cons	1.031602	.3921051	2.63	0.011	.2436369	1.819566

The jackknife estimate of the standard error of the coefficient of **chronic** is 0.62, larger than the value 0.53 obtained by using the vce(boot, reps(2000)) option and the value 0.52 obtained by using the vce(robust) option; see the **poisson** example in section 13.3.4.

The **jackknife** command operates similarly to **bootstrap**.

13.9   Stata resources

For many purposes, the vce(bootstrap) option of an estimation command suffices (see [R] *vce_option*) possibly followed by **estat bootstrap**. For more-advanced analysis, the **bootstrap** and **bsample** commands can be used.

For applications that use more elaborate methods than those implemented with the vce(bootstrap) option, care is needed, and a good understanding of the bootstrap is recommended. References include Efron and Tibshirani (1993), Davison and Hinkley (1997), Horowitz (2001), Davidson and MacKinnon (2004, chap. 4), and Cameron and Trivedi (2005, chap. 9). Cameron, Gelbach, and Miller (2008) survey a range of bootstraps, including some with asymptotic refinement, for the linear regression model with clustered errors.

13.10   Exercises

1. Use the same data as that created in section 13.3.3, except keep the first 100 observations and keep the variables **educ** and **age**. After a Poisson regression of **docvis** on an intercept and **educ**, give default standard errors, robust standard errors, and bootstrap standard errors based on 1,000 bootstraps and a seed of 10101.

2. For the Poisson regression in exercise 1, obtain the following 95% confidence intervals: normal-based, percentile, BC, and BCa. Compare these. Which, if any, is best?

3. Obtain a bootstrap estimate of the standard deviation of the estimated standard deviation of docvis.

4. Continuing with the regression in exercise 1, obtain a bootstrap estimate of the standard deviation of the robust standard error of $\widehat{\beta}_{educ}$.

5. Continuing with the regression in exercise 1, use the percentile-t method to perform a Wald test with asymptotic refinement of $H_0 : \beta = 0$ against $H_a : \beta \neq 0$ at the 0.05 level, and obtain a percentile-t 95% confidence interval.

6. Use the data of section 13.3.3 with 50 observations. Give the command given at the end of this exercise. Use the data in the percentile.dta file to obtain for the coefficient of the chronic variable: 1) bootstrap standard error; 2) bootstrap estimate of bias; 3) normal-based 95% confidence interval; and 4) percentile-t 95% confidence interval. For the last, you can use the centile command. Compare your results with those obtained from estat bootstrap, all after a Poisson regression with the vce(bootstrap) option.

```
bootstrap bstar=_b[chronic], reps(999) seed(10101) nodots ///
    saving(percentile, replace): poisson docvis chronic
use percentile, clear
```

7. Continuing from the previous exercise, does the bootstrap estimate of the distribution of the coefficient of chronic appear to be normal? Use the summarize and kdensity commands.

8. Repeat the percentile-t bootstrap at the start of section 13.5.2. Use kdensity to plot the bootstrap Wald statistics. Repeat for an estimation by poisson with default standard errors, rather than nbreg. Comment on any differences.

14 Binary outcome models

14.1 Introduction

Regression analysis of a qualitative binary or dichotomous variable is a commonplace problem in applied statistics. Models for mutually exclusive binary outcomes focus on the determinants of the probability p of the occurrence of one outcome rather than an alternative outcome that occurs with a probability of $1 - p$. An example where the binary variable is of direct interest is modeling whether an individual has insurance. In regression analysis, we want to measure how the probability p varies across individuals as a function of regressors. A different type of example is predicting the propensity score p, the conditional probability of participation (rather than nonparticipation) of an individual in a treatment program. In the treatment-effects literature, this prediction given observable variables is an important intermediate step, even though ultimate interest lies in outcomes of that treatment.

The two standard binary outcome models are the logit model and the probit model. These specify different functional forms for p as a function of regressors, and the models are fit by maximum likelihood (ML). A linear probability model (LPM), fit by ordinary least squares (OLS), is also used at times.

This chapter deals with the estimation and interpretation of cross-section binary outcome models using a set of standard commands that are similar to those for linear regression. Several extensions are also considered.

14.2 Some parametric models

Different binary outcome models have a common structure. The dependent variable, y_i, takes only two values, so its distribution is unambiguously the Bernoulli, or binomial with one tail, with a probability of p_i. Logit and probit models correspond to different regression models for p_i.

14.2.1 Basic model

Suppose the outcome variable, y, takes one of two values:

$$y = \begin{cases} 1 & \text{with probability } p \\ 0 & \text{with probability } 1 - p \end{cases}$$

Given our interest in modeling p as a function of regressors $\mathbf{x}$, there is no loss of generality in setting the outcome values to 1 and 0. The probability mass function for the observed outcome, y, is $p^y(1-p)^{1-y}$, with $E(y) = p$ and $\text{Var}(y) = p(1-p)$.

A regression model is formed by parameterizing p to depend on an index function $\mathbf{x}'\boldsymbol{\beta}$, where $\mathbf{x}$ is a $K \times 1$ regressor vector and $\boldsymbol{\beta}$ is a vector of unknown parameters. In standard binary outcome models, the conditional probability has the form

$$p_i \equiv \Pr(y_i = 1|\mathbf{x}) = F(\mathbf{x}'_i\boldsymbol{\beta}) \tag{14.1}$$

where $F(\cdot)$ is a specified parametric function of $\mathbf{x}'\boldsymbol{\beta}$, usually a cumulative distribution function (c.d.f.) on $(-\infty, \infty)$ because this ensures that the bounds $0 \leq p \leq 1$ are satisfied.

14.2.2   Logit, probit, linear probability, and clog-log models

Models differ in the choice of function, $F(\cdot)$. Four commonly used functional forms for $F(\mathbf{x}'\boldsymbol{\beta})$, shown in table 14.1, are the logit, probit, linear probability, and complementary log-log (clog-log) forms.

Table 14.1. Four commonly used binary outcome models

Model	Probability $p = \Pr(y = 1|\mathbf{x})$	Marginal effect $\partial p/\partial x_j$
Logit	$\Lambda(\mathbf{x}'\boldsymbol{\beta}) = e^{\mathbf{x}'\boldsymbol{\beta}}/(1 + e^{\mathbf{x}'\boldsymbol{\beta}})$	$\Lambda(\mathbf{x}'\boldsymbol{\beta})\{1 - \Lambda(\mathbf{x}'\boldsymbol{\beta})\}\beta_j$
Probit	$\Phi(\mathbf{x}'\boldsymbol{\beta}) = \int_{-\infty}^{\mathbf{x}'\boldsymbol{\beta}} \phi(z)dz$	$\phi(\mathbf{x}'\boldsymbol{\beta})\beta_j$
Linear probability	$F(\mathbf{x}'\boldsymbol{\beta}) = \mathbf{x}'\boldsymbol{\beta}$	β_j
Complementary log-log	$C(\mathbf{x}'\boldsymbol{\beta}) = 1 - \exp\{-\exp(\mathbf{x}'\boldsymbol{\beta})\}$	$\exp\{-\exp(\mathbf{x}'\boldsymbol{\beta})\}\exp(\mathbf{x}'\boldsymbol{\beta})\beta_j$

The logit model specifies that $F(\cdot) = \Lambda(\cdot)$, the c.d.f. of the logistic distribution. The probit model specifies that $F(\cdot) = \Phi(\cdot)$, the standard normal c.d.f. Logit and probit functions are symmetric around zero and are widely used in microeconometrics. The LPM corresponds to linear regression and does not impose the restriction that $0 \leq p \leq 1$. The complementary log-log model is asymmetric around zero. Its use is sometimes recommended when the distribution of y is skewed such that there is a high proportion of either zeros or ones in the dataset. The last column in the table gives expressions for the corresponding marginal effects, used in section 14.7, where $\phi(\cdot)$ denotes the standard normal density.

14.3   Estimation

For parametric models with exogenous covariates, the maximum likelihood estimator (MLE) is the natural estimator, because the density is unambiguously the Bernoulli. Stata provides ML procedures for logit, probit, and clog-log models, and for several variants of these models. For models with endogenous covariates, instrumental-variables (IV) methods can instead be used; see section 14.8.

14.3.1   Latent-variable interpretation and identification

Binary outcome models can be given a latent-variable interpretation. This provides a link with the linear regression model, explains more deeply the difference between logit and probit models, and provides the basis for extension to some multinomial models given in chapter 15.

We distinguish between the observed binary outcome, y, and an underlying continuous unobservable (or latent) variable, y^*, that satisfies the single-index model

$$y^* = \mathbf{x}'\boldsymbol{\beta} + u \tag{14.2}$$

Although y^* is not observed, we do observe

$$y = \begin{cases} 1 & \text{if } y^* > 0 \\ 0 & \text{if } y^* \le 0 \end{cases} \tag{14.3}$$

where the zero threshold is a normalization that is of no consequence if $\mathbf{x}$ includes an intercept.

Given the latent-variable models (14.2) and (14.3), we have

$$\begin{aligned} \Pr(y = 1) &= \Pr(\mathbf{x}'\boldsymbol{\beta} + u > 0) \\ &= \Pr(-u < \mathbf{x}'\boldsymbol{\beta}) \\ &= F(\mathbf{x}'\boldsymbol{\beta}) \end{aligned}$$

where $F(\cdot)$ is the c.d.f. of $-u$. This yields the probit model if u is standard normally distributed and the logit model if u is logistically distributed.

Identification of the latent-variable model requires that we fix its scale by placing a restriction on the variance of u, because the single-index model can only identify $\boldsymbol{\beta}$ up to scale. An explanation for this is that we observe only whether $y^* = \mathbf{x}'\boldsymbol{\beta} + u > 0$. But this is not distinguishable from the outcome $\mathbf{x}'\boldsymbol{\beta}^+ + u^+ > 0$, where $\boldsymbol{\beta}^+ = a\boldsymbol{\beta}$ and $u^+ = au$ for any $a > 0$. We can only identify $\boldsymbol{\beta}/\sigma$, where σ is the standard deviation (scale parameter) of u.

To uniquely define the scale of $\boldsymbol{\beta}$, the convention is to set $\sigma = 1$ in the probit model and $\pi/\sqrt{3}$ in the logit model. As a consequence, $\boldsymbol{\beta}$ is scaled differently in the two models; see section 14.4.3.

14.3.2   ML estimation

For binary models other than the LPM, estimation is by ML. This ML estimation is straightforward. The density for a single observation can be compactly written as $p_i^{y_i}(1 - p_i)^{1-y_i}$, where $p_i = F(\mathbf{x}_i'\boldsymbol{\beta})$. For a sample of N independent observations, the MLE, $\widehat{\boldsymbol{\beta}}$, maximizes the associated log-likelihood function

$$Q(\boldsymbol{\beta}) = \sum_{i=1}^{N} [y_i \ln F(\mathbf{x}_i'\boldsymbol{\beta}) + (1 - y_i) \ln\{1 - F(\mathbf{x}_i'\boldsymbol{\beta})\}]$$

The MLE is obtained by iterative methods and is asymptotically normally distributed.

Consistent estimates are obtained if $F(\cdot)$ is correctly specified. When instead the functional form $F(\cdot)$ is misspecified, pseudolikelihood theory applies.

14.3.3   The logit and probit commands

The syntax for the `logit` command is

`logit` *depvar* [*indepvars*] [*if*] [*in*] [*weight*] [, *options*]

The syntax for the `probit` and `cloglog` commands is similar.

Like the `regress` command, available options include `vce(cluster` *clustvar*`)` and `vce(robust)` for variance estimation. The constant is included by default but can be suppressed by using the `noconstant` option.

The `or` option of `logit` presents exponentiated coefficients. The rationale is that for the logit model, the log of the odds ratio $\ln\{p/(1-p)\}$ can be shown to be linear in $\mathbf{x}$ and $\boldsymbol{\beta}$. It follows that the odds ratio $p/(1-p) = \exp(\mathbf{x}'\boldsymbol{\beta})$, so that e^{β_j} measures the multiplicative effect of a unit change in regressor x_j on the odds ratio. For this reason, many researchers prefer logit coefficients to be reported after exponentiation, i.e., as e^{β} rather than β. Alternatively, the `logistic` command estimates the parameters of the logit model and directly reports the exponentiated coefficients.

14.3.4   Robust estimate of the VCE

Binary outcome models are unusual in that there is no advantage in using the robust sandwich form for the variance–covariance matrix of the estimator (VCE) of the MLE if data are independent over i and $F(\mathbf{x}'\boldsymbol{\beta})$ is correctly specified. The reason is that the ML default standard errors are obtained by imposing the restriction $\text{Var}(y|\mathbf{x}) = F(\mathbf{x}'\boldsymbol{\beta})\{1 - F(\mathbf{x}'\boldsymbol{\beta})\}$, and this must necessarily hold because the variance of a binary variable is always $p(1-p)$; see Cameron and Trivedi (2005) for further explanation. If $F(\mathbf{x}'\boldsymbol{\beta})$ is correctly specified, the `vce(robust)` option is not required. Hence, we may infer a misspecified functional form $F(\mathbf{x}'\boldsymbol{\beta})$ if the use of the `vce(robust)` option produces substantially different variances from the default.

At the same time, dependence between observations may arise because of cluster sampling. In that case, the appropriate option is to use `vce(cluster` *clustvar*`)`.

14.3.5   OLS estimation of LPM

If $F(\cdot)$ is assumed to be linear, i.e., $p = \mathbf{x}'\boldsymbol{\beta}$, then the linear conditional mean function defines the LPM. The LPM can be consistently estimated by OLS regression of y on $\mathbf{x}$ using `regress`. A major limitation of the method, however, is that the fitted values $\mathbf{x}'\widehat{\boldsymbol{\beta}}$ will not necessarily be in the $[0, 1]$ interval. And, because $\text{Var}(y|\mathbf{x}) = (\mathbf{x}'\boldsymbol{\beta})(1 - \mathbf{x}'\boldsymbol{\beta})$ for the LPM, the regression is inherently heteroskedastic, so a robust estimate of the VCE should be used.

14.4   Example

We analyze data on supplementary health insurance coverage. Initial analysis estimates the parameters of the models of section 14.2.

14.4.1   Data description

The data come from wave 5 (2002) of the Health and Retirement Study (HRS), a panel survey sponsored by the National Institute of Aging. The sample is restricted to Medicare beneficiaries. The HRS contains information on a variety of medical service uses. The elderly can obtain supplementary insurance coverage either by purchasing it themselves or by joining employer-sponsored plans. We use the data to analyze the purchase of private insurance (`ins`) from any source, including private markets or associations. The insurance coverage broadly measures both individually purchased and employer-sponsored private supplementary insurance, and includes Medigap plans and other policies.

Explanatory variables include health status, socioeconomic characteristics, and spouse-related information. Self-assessed health-status information is used to generate a dummy variable (`hstatusg`) that measures whether health status is good, very good, or excellent. Other measures of health status are the number of limitations (up to five) on activities of daily living (`adl`) and the total number of chronic conditions (`chronic`). Socioeconomic variables used are age, gender, race, ethnicity, marital status, years of education, and retirement status (respectively, `age`, `female`, `white`, `hisp`, `married`, `educyear`, `retire`); household income (`hhincome`); and log household income if positive (`linc`). Spouse retirement status (`sretire`) is an indicator variable equal to 1 if a retired spouse is present.

For conciseness, we use global macros to create variable lists, presenting the variables used in sections 14.4–14.7 followed by additional variables used in section 14.8. We have

```
. * Load data
. use mus14data.dta

. * Interaction variables
. drop age2 agefem agechr agewhi

. * Summary statistics of variables
. global xlist age hstatusg hhincome educyear married hisp

. generate linc = ln(hhinc)
(9 missing values generated)

. global extralist linc female white chronic adl sretire
```

(Continued on next page)

```
. summarize ins retire $xlist $extralist
    Variable |      Obs        Mean    Std. Dev.        Min         Max
-------------+--------------------------------------------------------
         ins |     3206    .3870867    .4871597          0           1
      retire |     3206    .6247661    .4842588          0           1
         age |     3206    66.91391    3.675794         52          86
     hstatusg |    3206    .7046163    .4562862          0           1
     hhincome |    3206    45.26391    64.33936          0    1312.124
-------------+--------------------------------------------------------
     educyear |    3206    11.89863    3.304611          0          17
     married |     3206    .7330006     .442461          0           1
        hisp |     3206    .0726762    .2596448          0           1
        linc |     3197    3.383047    .9393629   -2.292635    7.179402
      female |     3206     .477854    .4995872          0           1
-------------+--------------------------------------------------------
       white |     3206    .8206488     .383706          0           1
     chronic |     3206    2.063319    1.416434          0           8
         adl |     3206     .301622    .8253646          0           5
     sretire |     3206    .3883344    .4874473          0           1
```

14.4.2 Logit regression

We begin with ML estimation of the logit model.

```
. * Logit regression
. logit ins retire $xlist
Iteration 0:   log likelihood = -2139.7712
Iteration 1:   log likelihood = -1996.7434
Iteration 2:   log likelihood = -1994.8864
Iteration 3:   log likelihood = -1994.8784
Iteration 4:   log likelihood = -1994.8784

Logistic regression                             Number of obs   =       3206
                                                LR chi2(7)      =     289.79
                                                Prob > chi2     =     0.0000
Log likelihood = -1994.8784                     Pseudo R2       =     0.0677

-------------+--------------------------------------------------------------
         ins |     Coef.   Std. Err.      z    P>|z|     [95% Conf. Interval]
-------------+--------------------------------------------------------------
      retire |  .1969297   .0842067     2.34   0.019     .0318875    .3619718
         age | -.0145955   .0112871    -1.29   0.196    -.0367178    .0075267
    hstatusg |  .3122654   .0916739     3.41   0.001     .1325878     .491943
    hhincome |  .0023036    .000762     3.02   0.003       .00081    .0037972
     educyear |  .1142626   .0142012     8.05   0.000     .0864288    .1420963
     married |   .578636   .0933198     6.20   0.000     .3957327    .7615394
        hisp | -.8103059   .1957522    -4.14   0.000    -1.193973   -.4266387
       _cons | -1.715578   .7486219    -2.29   0.022     -3.18285   -.2483064
-------------+--------------------------------------------------------------
```

All regressors other than age are statistically significantly different from zero at the
0.05 level. For the logit model, the sign of the coefficient is also the sign of the marginal
effect. Further discussion of these results is deferred to the next section, where we
compare logit parameter estimates with those from other models.

The iteration log shows fast convergence in four iterations. Later output suppresses the iteration log to save space. In actual empirical work, it is best to keep the log. For example, a large number of iterations may signal a high degree of multicollinearity.

14.4.3 Comparison of binary models and parameter estimates

It is well known that logit and probit models have similar shapes for central values of $F(\cdot)$ but differ in the tails as $F(\cdot)$ approaches 0 or 1. At the same time, the corresponding coefficient estimates from the two models are scaled quite differently. It is an elementary mistake to suppose that the different models have different implications simply because the estimated coefficients across models are different. However, this difference is mainly a consequence of different functional forms for the probabilities. The marginal effects and predicted probabilities, presented in sections 14.6 and 14.7, are much more similar across models.

Coefficients can be compared across models, using the following rough conversion factors (Amemiya 1981, 1,488):

$$\widehat{\boldsymbol{\beta}}_{\text{Logit}} \simeq 4\widehat{\boldsymbol{\beta}}_{\text{OLS}}$$
$$\widehat{\boldsymbol{\beta}}_{\text{Probit}} \simeq 2.5\widehat{\boldsymbol{\beta}}_{\text{OLS}}$$
$$\widehat{\boldsymbol{\beta}}_{\text{Logit}} \simeq 1.6\widehat{\boldsymbol{\beta}}_{\text{Probit}}$$

The motivation is that it is better to compare the marginal effect, $\partial p / \partial x_j$, across models, and it can be shown that $\partial p / \partial x_j \leq 0.25\widehat{\beta}_j$ for logit, $\partial p / \partial x_j \leq 0.4\widehat{\beta}_j$ for probit, and $\partial p / \partial x_j = \widehat{\beta}_j$ for OLS. The greatest departures across the models occur in the tails.

We estimate the parameters of the logit and probit models by ML and the LPM by OLS, computing standard errors and z statistics based on both default and robust estimates of the VCE. The following code saves results for each model with the **estimates store** command.

```
. * Estimation of several models
. quietly logit ins retire $xlist
. estimates store blogit
. quietly probit ins retire $xlist
. estimates store bprobit
. quietly regress ins retire $xlist
. estimates store bols
. quietly logit ins retire $xlist, vce(robust)
. estimates store blogitr
. quietly probit ins retire $xlist, vce(robust)
. estimates store bprobitr
. quietly regress ins retire $xlist, vce(robust)
. estimates store bolsr
```

This leads to the following output table of parameter estimates across the models:

```
. * Table for comparing models
. estimates table blogit blogitr bprobit bprobitr bols bolsr, t stats(N ll)
> b(%7.3f) stfmt(%8.2f)
```

Variable	blogit	blogitr	bprobit	bprobitr	bols	bolsr
ins						
retire	0.197	0.197	0.118	0.118	0.041	0.041
	2.34	2.32	2.31	2.30	2.24	2.24
age	-0.015	-0.015	-0.009	-0.009	-0.003	-0.003
	-1.29	-1.32	-1.29	-1.32	-1.20	-1.25
hstatusg	0.312	0.312	0.198	0.198	0.066	0.066
	3.41	3.40	3.56	3.57	3.37	3.45
hhincome	0.002	0.002	0.001	0.001	0.000	0.000
	3.02	2.01	3.19	2.21	3.58	2.63
educyear	0.114	0.114	0.071	0.071	0.023	0.023
	8.05	7.96	8.34	8.33	8.15	8.63
married	0.579	0.579	0.362	0.362	0.123	0.123
	6.20	6.15	6.47	6.46	6.38	6.62
hisp	-0.810	-0.810	-0.473	-0.473	-0.121	-0.121
	-4.14	-4.18	-4.28	-4.36	-3.59	-4.49
_cons	-1.716	-1.716	-1.069	-1.069	0.127	0.127
	-2.29	-2.36	-2.33	-2.40	0.79	0.83
Statistics						
N	3206	3206	3206	3206	3206	3206
ll	-1994.88	-1994.88	-1993.62	-1993.62	-2104.75	-2104.75

legend: b/t

The coefficients across the models tell a qualitatively similar story about the impact of a regressor on $\Pr(\texttt{ins} = 1)$. The rough rules for parameter conversion also stand up reasonably well, because the logit estimates are roughly five times the OLS estimates, and the probit estimates are roughly three times the OLS coefficients. The standard errors are similarly rescaled, so that the reported z statistics for the coefficients are similar across the three models. For the logit and probit coefficients, the robust and default z statistics are quite similar, aside from those for the hhincome variable. For OLS, there is a bigger difference.

In section 14.6, we will see that the fitted probabilities are similar for the logit and probit specifications. The linear functional form does not constrain the fitted values to the $[0, 1]$ interval, however, and we find differences in the fitted-tail values between the LPM and the logit and probit models.

14.5 Hypothesis and specification tests

We next consider several tests of the maintained specification against other alternatives. Some of these tests repeat and demonstrate many of the methods presented in more detail in chapter 12, using commands for the nonlinear logit model that are similar to those presented in chapter 3 for the linear regression model.

14.5.1   Wald tests

Tests on coefficients of variables are most easily performed by using the `test` command, which implements a Wald test. For example, we may test for the presence of interaction effects with age. Four interaction variables (`age2`, `agefem`, `agechr`, and `agewhi`) are created, for example, `agefem` equals `age` times `female`, and then they are included in the logit regression. The null hypothesis is that the coefficients of these four regressors are all zero, because then there are no interaction effects. We obtain

```
. * Wald test for zero interactions
. generate age2 = age*age
. generate agefem = age*female
. generate agechr = age*chronic
. generate agewhi = age*white
. global intlist age2 agefem agechr agewhi
. quietly logit ins retire $xlist $intlist
. test $intlist
 ( 1)  [ins]age2 = 0
 ( 2)  [ins]agefem = 0
 ( 3)  [ins]agechr = 0
 ( 4)  [ins]agewhi = 0
           chi2(  4) =     7.45
         Prob > chi2 =     0.1141
```

The p-value is 0.114, so the null hypothesis is not rejected at the 0.05 level or even the 0.10 level.

14.5.2   Likelihood-ratio tests

A likelihood-ratio (LR) test (see section 12.4) provides an alternative method for testing hypotheses. It is asymptotically equivalent to the Wald test if the model is correctly specified. To implement the LR test of the preceding hypothesis, we estimate parameters of both the general and the restricted models and then use the `lrtest` command. We obtain

```
. * Likelihood-ratio test
. quietly logit ins retire $xlist $intlist
. estimates store B
. quietly logit ins retire $xlist
. lrtest B
Likelihood-ratio test                          LR chi2(4)  =      7.57
(Assumption: . nested in B)                     Prob > chi2 =      0.1088
```

This test has a p-value of 0.109, quite similar to that for the Wald test.

In some situations, the main focus is on the predicted probability of the model and the sign and size of the coefficients are not the focus of the inquiry. An example is the estimation of propensity scores, in which case a recommendation is often made to saturate the model and then to choose the best model by using the Bayesian information criterion (BIC). The Akaike information criterion (AIC) or the BIC are also useful for comparing models that are nonnested and have different numbers of parameters; see section 10.7.2.

14.5.3 Additional model-specification tests

For specific models, there are often specific tests of misspecification. Here we consider two variants of the logit and probit models.

Lagrange multiplier test of generalized logit

Stukel (1988) considered, as an alternative to the logit model, the generalized h-family logit model

$$\Lambda_\alpha(\mathbf{x}'\boldsymbol{\beta}) = \frac{e^{h_\alpha(\mathbf{x}'\boldsymbol{\beta})}}{1 + e^{h_\alpha(\mathbf{x}'\boldsymbol{\beta})}} \tag{14.4}$$

where $h_\alpha(\mathbf{x}'\boldsymbol{\beta})$ is a strictly increasing nonlinear function of $\mathbf{x}'\boldsymbol{\beta}$ indexed by the shape parameters α_1 and α_2 that govern, respectively, the heaviness of the tails and the symmetry of the $\Lambda(\cdot)$ function.

Stukel proposed testing whether (14.4) is a better model by using a Lagrange multiplier (LM), or score, test; see section 12.5. This test has the advantage that it requires estimation only of the null hypothesis logit model rather than of the more complicated model (14.4). Furthermore, the LM test can be implemented by supplementing the logit model regressors with generated regressors that are functions of $\mathbf{x}'\boldsymbol{\beta}$ and by testing the significance of these augmented regressors.

For example, to test for departure from the logit in the direction of an asymmetric h-family, we add the generated regressor $(\mathbf{x}_i'\widehat{\boldsymbol{\beta}})^2$ to the list of regressors, reestimate the logit model, and test whether the added variable significantly improves the fit of the model. We have

```
. * Stukel score or LM test for asymmetric h-family logit
. quietly logit ins retire $xlist
. predict xbhat, xb
. generate xbhatsq = xbhat^2
. quietly logit ins retire $xlist xbhatsq
. test xbhatsq
 ( 1)  [ins]xbhatsq = 0

          chi2(  1) =    37.91
        Prob > chi2 =    0.0000
```

The null hypothesis of correct model specification is strongly rejected because the Wald test of zero coefficient for the added regressor $(\mathbf{x}_i'\widehat{\boldsymbol{\beta}})^2$ yields a $\chi^2(1)$ statistic of 38 with $p = 0.000$.

This test is easy to apply and so are several other score tests suggested by Stukel that use the variable-augmentation approach. At the same time, recall from section 3.5.5 that tests have power in more than one rejection. Thus rejection in the previous example may be for reasons other than the need for an asymmetric h-family logit model. For example, perhaps it is enough to use a logit model with additional inclusion of polynomials in the continuous regressors or inclusion of additional variables as regressors.

Heteroskedastic probit regression

The standard probit and logit models assume homoskedasticity of the errors, u, in the latent-variable model (14.2). This restriction can be tested. One strategy is to have as the null-hypothesis model

$$\Pr(y_i = 1|\mathbf{x}) = \boldsymbol{\Phi}\left(\mathbf{x}_i'\boldsymbol{\beta}/\sigma\right)$$

with the normalization $\sigma^2 = 1$, and as the alternative hypothesis

$$\Pr(y_i = 1|\mathbf{x}) = \boldsymbol{\Phi}\left(\mathbf{x}_i'\boldsymbol{\beta}/\sigma_i\right) \tag{14.5}$$

where now u_i in (14.2) is heteroskedastic with a variance of

$$\sigma_i^2 = \exp(\mathbf{z}_i'\boldsymbol{\delta}) \tag{14.6}$$

where the exogenous variables $(z_1, \ldots, z_m)$ do not contain a constant, because the restriction $\boldsymbol{\delta} = \mathbf{0}$ yields $\sigma_i^2 = 1$ as in the null model. Including a constant in $\mathbf{z}$ would make the model unidentified.

ML estimation can be based on (14.5) and (14.6). The parameters of the probit model with heteroskedasticity can be estimated with ML by using Stata's `hetprob` command. The syntax for `hetprob` is

`hetprob` *depvar* [*indepvars*] [*if*] [*in*] [*weight*], `het`(*varlist*) [*options*]

The two models can be compared by using a LR test of $\boldsymbol{\delta} = \mathbf{0}$ that is automatically implemented when the command is used. Alternatively, a Wald test could be used.

As an illustration, we reconsider the probit model used in the preceding analysis. In specifying the variables in $\mathbf{z}$, it seems desirable to exclude the variables already included in $\mathbf{x}$, because in a binomial model, a variable that affects $\Pr(y = 1)$ must necessarily affect the variance of y. To enter a variable in the specification of both, the mean and the variance cause problems of interpretation. In our application, we choose the single variable `chronic` as our $\mathbf{z}$, where `chronic` denotes the number of chronic conditions experienced by an individual. We obtain

(Continued on next page)

```
. * Heteroskedastic probit model
. hetprob ins retire $xlist, het(chronic) nolog // Heteroskedastic Probit
Heteroskedastic probit model                    Number of obs    =       3206
                                                Zero outcomes    =       1965
                                                Nonzero outcomes =       1241

                                                Wald chi2(7)     =      90.34
Log likelihood = -1992.904                      Prob > chi2      =     0.0000
```

ins	Coef.	Std. Err.	z	P>|z|	[95% Conf. Interval]	
ins						
retire	.1075926	.0476757	2.26	0.024	.0141501	.2010352
age	-.0087658	.0062107	-1.41	0.158	-.0209384	.0034069
hstatusg	.1629653	.0564771	2.89	0.004	.0522722	.2736584
hhincome	.0011135	.000364	3.06	0.002	.0004	.001827
educyear	.0642167	.0094184	6.82	0.000	.0457569	.0826765
married	.3341699	.0563861	5.93	0.000	.2236551	.4446847
hisp	-.4344396	.1055044	-4.12	0.000	-.6412244	-.2276548
_cons	-.9089138	.4318121	-2.10	0.035	-1.75525	-.0625776
lnsigma2						
chronic	-.0442144	.0365848	-1.21	0.227	-.1159193	.0274906

```
Likelihood-ratio test of lnsigma2=0: chi2(1) =    1.44    Prob > chi2 = 0.2303
```

The LR test indicates that at the 0.05 level, there is no statistically significant improvement in the model resulting from generalizing the homoskedastic model, because $p = 0.23$.

As a matter of modeling strategy, however, it is better to test first whether the **z** variables are omitted explanatory variables from the conditional mean model because such a misspecification is also consistent with variance depending on **z**. That is, the finding that **z** enters the variance function is also consistent with it having been incorrectly omitted from the conditional mean function. Accordingly, a variable addition test was also applied by adding `chronic` to the regressors in the probit model, and the p-value of the test was found to be 0.23. Thus the evidence is against the inclusion of `chronic` in the probit model.

14.5.4 Model comparison

A question often arises: which model is better, logit or probit? As will be seen in the next section, in many cases the fitted probability is very similar over a large part of the range of $\mathbf{x}'\boldsymbol{\beta}$. Larger differences may be evident in the tails of the distribution, but a large sample is required to reliably differentiate between models on the basis of tail behavior.

Because logit and probit models are nonnested, a penalized likelihood criterion such as AIC or BIC (see section 10.7.2) is appealing for model selection. However, these two models have the same number of parameters, so this reduces to choosing the model with the higher log likelihood. The probit model has a log likelihood of $-1,993.62$ (see the table on page 466), which is 1.26 higher than the $-1,994.88$ for logit. This favors the probit model, but the difference is not great. For example, an LR test of a single restriction rejects at the 0.05 level if the LR statistic exceeds 3.84 or equivalently if the change in log likelihood is $3.84/2 = 1.92$.

14.6   Goodness of fit and prediction

The Stata output for the logit and probit regressions has a similar format. The log likelihood and the LR test of the joint significance of the regressors and its p-value are given. However, some measures of overall goodness of fit are desirable, including those that are specific to the binary outcome model.

Three approaches to evaluating the fit of the model are pseudo-R^2 measures, comparisons of group-average predicted probabilities with sample frequencies, and comparisons based on classification ($\widehat{y}$ equals zero or one). None of these is the most preferred measure a priori. Below we discuss comparisons of model fit using predicted probabilities.

14.6.1   Pseudo-R^2 measure

In linear regression, the total sum of squared deviations from the mean can be decomposed into explained and residual sums of squares, and R^2 measures the ratio of explained sum of squares to total sum of squares, with 0 and 1 as the lower and upper limits, respectively. These properties do not carry over to nonlinear regression. Yet there are some measures of fit that attempt to mimic the R^2 measure of linear regression. There are several R^2 measures, one of which is included in the Stata output.

McFadden's $\widetilde{R}^2$ is computed as $1 - L_N(\widehat{\boldsymbol{\beta}})/L_N(\overline{y})$, where $L_N(\widehat{\boldsymbol{\beta}})$ denotes the maximized or fitted log-likelihood value, and $L_N(\overline{y})$ denotes the value of the log likelihood in the intercept-only model. When applied to models with binary and multinomial outcomes, the lower and upper bounds of the pseudo-R^2 measure are 0 and 1 (see section 10.7.1), though McFadden's $\widetilde{R}^2$ is not a measure of the proportion of variance of the dependent variable explained by the model. For the fitted logit model, $\widetilde{R}^2 = 0.068$.

14.6.2   Comparing predicted probabilities with sample frequencies

In-sample comparison of the average predicted probabilities, $N^{-1}\sum\widehat{p}_i$, with the sample frequency, $\overline{y}$, is not helpful for evaluating the fit of binary outcome models. In particular, the two are necessarily equal for logit models that include an intercept, because the logit MLE first-order conditions can be shown to then impose this condition.

However, this comparison may be a useful thing to do for subgroups of observations. The Hosmer–Lemeshow specification test evaluates the goodness of fit by comparing the sample frequency of the dependent variable with the fitted probability within subgroups of observations, with the number of subgroups being specified by the investigator. The null hypothesis is that the two are equal. The test is similar to the Pearson chi-squared goodness-of-fit test.

Let $\overline{\widehat{p}_g}$ and $\overline{y}_g$ denote, respectively, the average predicted probability and sample frequency in group g. The test statistic is $\sum_{g=1}^{G} (\overline{\widehat{p}_g} - \overline{y}_g)^2 / \overline{y}_g (1 - \overline{y}_g)$, where g is the group subscript. The groups are based on quantiles of the ordered predicted probabilities. For example, if $G = 10$, then each group corresponds to a decile of the ordered $\widehat{p}_i$. Hosmer and Lemeshow established the null distribution by simulation. Under the null of correct specification, the statistic is distributed as $\chi^2(G - 2)$. However, two caveats should be noted: First, the test outcome is sensitive to the number of groups used in the specification. Second, much of what is known about the properties of the test is based on Monte Carlo evidence of the test's performance. See Hosmer and Lemeshow (1980, 2000). Simulation evidence suggests that a fixed sample size specifying a large number of groups in the test causes a divergence between the empirical c.d.f. and the c.d.f. of the $\chi^2(G - 2)$ distribution.

The goodness-of-fit test is performed by using the postestimation `estat gof` command, which has the syntax

`estat gof` [*if*] [*in*] [*weight*] [, *options*]

where the `group(#)` option specifies the number of quantiles to be used to group the data, with 10 being the default.

After estimating the parameters of the logit model, we perform this test, setting the number of groups to four. We obtain

```
. * Hosmer-Lemeshow gof test with 4 groups
. quietly logit ins retire $xlist

. estat gof, group(4)  // Hosmer-Lemeshow gof test
Logistic model for ins, goodness-of-fit test

   (Table collapsed on quantiles of estimated probabilities)
        number of observations =      3206
              number of groups =         4
      Hosmer-Lemeshow chi2(2) =     14.04
                  Prob > chi2 =    0.0009
```

The outcome indicates misspecification, because the p-value is 0.001.

To check if the same outcome occurs if we use a larger number of groups to perform the test, we repeat the test for ten groups.

```
. quietly logit ins retire $xlist

. * Hosmer-Lemeshow gof test with 10 groups
. estat gof, group(10)  // Hosmer-Lemeshow gof test
Logistic model for ins, goodness-of-fit test

   (Table collapsed on quantiles of estimated probabilities)
        number of observations =      3206
              number of groups =        10
      Hosmer-Lemeshow chi2(8) =     31.48
                  Prob > chi2 =    0.0001
```

Again the test rejects the maintained specification, this time with an even smaller p-value.

14.6.3 Comparing predicted outcomes with actual outcomes

The preceding measure is based on the fitted probability of having private insurance. We may instead want to predict the outcome itself, i.e., whether an individual has private insurance ($\widehat{y} = 1$) or does not have insurance ($\widehat{y} = 0$). Strictly speaking, this depends upon a loss function. If we assume a symmetric loss function, then it is natural to set $\widehat{y} = 1$ if $F(\mathbf{x}'\boldsymbol{\beta}) > 0.5$ and $\widehat{y} = 0$ if $F(\mathbf{x}'\boldsymbol{\beta}) \leq 0.5$. One measure of goodness of fit is the percentage of correctly classified observations.

Goodness-of-fit measures based on classification can be obtained by using the postestimation `estat classification` command.

For the fitted logit model, we obtain

```
. * Comparing fitted probability and dichotomous outcome
. quietly logit ins retire $xlist

. estat classification
Logistic model for ins

              -------- True --------
Classified |       D            ~D    |    Total
-----------+--------------------------+----------
     +     |      345           308   |     653
     -     |      896          1657   |    2553
-----------+--------------------------+----------
   Total   |     1241          1965   |    3206

Classified + if predicted Pr(D) >= .5
True D defined as ins != 0
--------------------------------------------------
Sensitivity                     Pr( +| D)   27.80%
Specificity                     Pr( -|~D)   84.33%
Positive predictive value       Pr( D| +)   52.83%
Negative predictive value       Pr(~D| -)   64.90%
--------------------------------------------------
False + rate for true ~D        Pr( +|~D)   15.67%
False - rate for true D         Pr( -| D)   72.20%
False + rate for classified +   Pr(~D| +)   47.17%
False - rate for classified -   Pr( D| -)   35.10%
--------------------------------------------------
Correctly classified                        62.45%
--------------------------------------------------
```

The table compares fitted and actual values. The percentage of correctly specified values in this case is 62.45. In this example, 308 observations are misclassified as 1 when the correct classification is 0, and 896 values are misclassified as 0 when the correct value is 1. The remaining $345 + 1657$ observations are correctly specified.

The `estat classification` command also produces detailed output on classification errors, using terminology that is commonly used in biostatistics and is detailed in [R] **logistic postestimation**. The ratio 345/1241, called the sensitivity measure, gives the fraction of observations with $y = 1$ that are correctly specified. The ratio $1657/1965$, called the specificity measure, gives the fraction of observations with $y = 0$ that are correctly specified. The ratios $308/1965$ and $896/1241$ are referred to as the false positive and false negative classification error rates.

14.6.4 The predict command for fitted probabilities

Fitted probabilities can be computed by using the postestimation `predict` command, defined in section 10.5.1. The difference between logit and probit models may be small, especially over the middle portion of the distribution. On the other hand, the fitted probabilities from the LPM estimated by OLS may be substantially different.

We first summarize the fitted probability from the three models that include only the `hhincome` variable as a regressor.

```
. * Calculate and summarize fitted probabilities
. quietly logit ins hhincome
. predict plogit, pr
. quietly probit ins hhincome
. predict pprobit, pr
. quietly regress ins hhincome
. predict pols, xb
. summarize ins plogit pprobit pols
```

Variable	Obs	Mean	Std. Dev.	Min	Max
ins	3206	.3870867	.4871597	0	1
plogit	3206	.3870867	.0787632	.3176578	.999738
pprobit	3206	.3855051	.061285	.3349603	.9997945
pols	3206	.3870867	.0724975	.3360834	1.814582

The mean and standard deviation are essentially the same in the three cases, but the range of the fitted values from the LPM includes six inadmissible values outside the $[0, 1]$ interval. This fact should be borne in mind in evaluating the graph given below, which compares the fitted probability from the three models. The deviant observations from OLS stand out at the extremes of the range of distribution, but the results for logit and probit cohere well.

For regressions with a single regressor, plotting predicted probabilities against that variable can be informative, especially if that variable takes a range of values. Such a graph illustrates the differences in the fitted values generated by different estimators. The example given below plots the fitted values from logit, probit, and LPM against household income (`hhincome`). For graph readability, the `jitter()` option is used to jitter the observed zero and one values, leading to a band of outcome values that are around 0 and 1 rather than exactly 0 or 1. The divergence between the first two and the LPM (OLS) estimates at high values of income stands out, though this is not necessarily serious because the number of observations in the upper range of income is quite small. The fitted values are close for most of the sample.

```
. * Following gives Figure mus14fig1.eps
. sort hhincome

. graph twoway (scatter ins hhincome, msize(vsmall) jitter(3)) /*
>   */ (line plogit hhincome, clstyle(p1)) /*
>   */ (line pprobit hhincome, clstyle(p2)) /*
>   */ (line pols hhincome, clstyle(p3)), /*
>   */ scale (1.2) plotregion(style(none)) /*
>   */ title("Predicted Probabilities Across Models") /*
>   */ xtitle("HHINCOME (hhincome)", size(medlarge)) xscale(titlegap(*5)) /*
>   */ ytitle("Predicted probability", size(medlarge)) yscale(titlegap(*5)) /*
>   */ legend(pos(1) ring(0) col(1)) legend(size(small)) /*
>   */ legend(label(1 "Actual Data (jittered)") label(2 "Logit") /*
>   */          label(3 "Probit") label(4 "OLS"))
```

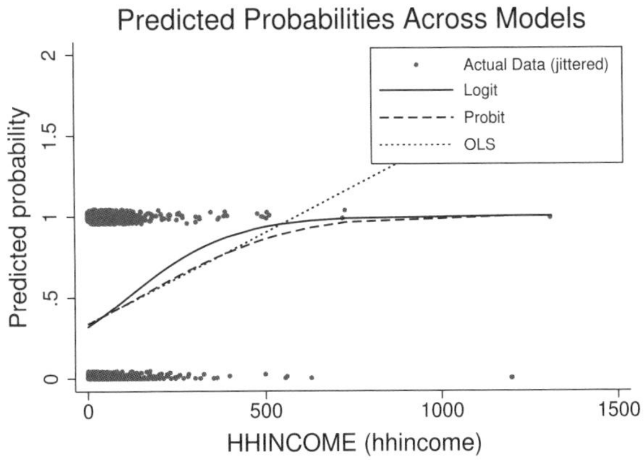

Figure 14.1. Predicted probabilities versus `hhincome`

14.6.5   The prvalue command for fitted probabilities

The `predict` command provides fitted probabilities for each individual, evaluating at $\mathbf{x} = \mathbf{x}_i$. At times, it is useful to instead obtain predicted probabilities at a representative value, $\mathbf{x} = \mathbf{x}^*$. This can be done by using the `nlcom` command, presented in section 10.5.5. It is simpler to instead use the user-written postestimation `prvalue` command (Long and Freese 2006).

(*Continued on next page*)

The syntax of `prvalue` is

```
prvalue [if] [in] [, x(conditions) rest(mean)]
```

where we list two key options. The x(*conditions*) option specifies the conditioning values of the regressors, and the default `rest(mean)` option specifies that the unconditioned variables are to be set at their sample averages. Omitting x(*conditions*) means that the predictions are evaluated at $\mathbf{x} = \overline{\mathbf{x}}$.

The command generates a predicted (fitted) value for each observation, here for a 65-year-old, married, retired non-Hispanic with good health status, 17 years of education, and an income equal to \$50,000 (so the `income` variable equals 50).

```
. * Fitted probabilities for selected baseline
. quietly logit ins retire $xlist
. prvalue, x(age=65 retire=0 hstatusg=1 hhincome=50 educyear=17 married=1 hisp=0)
logit: Predictions for ins
Confidence intervals by delta method
                              95% Conf. Interval
         Pr(y=1|x):    0.5706    [ 0.5226,    0.6186]
         Pr(y=0|x):    0.4294    [ 0.3814,    0.4774]

            retire    age  hstatusg  hhincome  educyear  married    hisp
     x=          0     65         1        50        17        1       0
```

The probability of having private insurance is 0.57 with the 95% confidence interval [0.52, 0.62]. This reasonably tight confidence interval is for the probability that $y = 1$ given $\mathbf{x} = \mathbf{x}^*$. There is much more uncertainty in the outcome that $y = 1$ given $\mathbf{x} = \mathbf{x}^*$. For example, this difficulty in predicting actual values leads to the low $\widetilde{R}^2$ for the logit model. This distinction is similar to that between predicting $E(y|\mathbf{x})$ and $y|\mathbf{x}$ discussed in sections 3.6.1 and 10.5.2.

14.7 Marginal effects

Three variants of marginal effects, previously discussed in section 10.6, are the average marginal effect (AME), marginal effects at a representative value (MER), and marginal effects at the mean (MEM). In a nonlinear model, marginal effects are more informative than coefficients.

The analytical formulas for the marginal effects for the standard binary outcome models were given in table 14.1. For example, for the logit model, the marginal effect with respect to a change in a continuous regressor, x_j, evaluated at $\mathbf{x} = \overline{\mathbf{x}}$, is estimated by $\Lambda(\overline{\mathbf{x}}'\widehat{\boldsymbol{\beta}})\{1 - \Lambda(\overline{\mathbf{x}}'\widehat{\boldsymbol{\beta}})\}\widehat{\beta}_j$. An associated confidence interval can be calculated by using the delta method.

14.7.1 Marginal effect at a representative value (MER)

The postestimation `margins` command with the `dydx()` and `at()` options provides an estimate of the marginal effect at a particular value of $\mathbf{x} = \mathbf{x}^*$. This may be preferred to evaluation at the sample averages of the regressors (the `atmean` option). For example, if the model has several binary regressors, then these are set equal to their sample averages, which is not particularly meaningful. It may be better for the user to create a benchmark value—an index case—for which the MEs are calculated.

We use as a benchmark a 75-year-old, retired, married Hispanic with good health status, 12 years of education, and an income equal to 35. For the four binary regressors, we compute the MEM using the finite-difference method, rather than the calculus method, by using the `i.` operator in the `logit` estimation command; see, for example, section 10.6.6. Then

```
. * Marginal effects (MER) after logit
. quietly logit ins i.retire age i.hstatusg hhincome educyear i.married i.hisp
. margins, dydx(*) at (retire=1 age=75 hstatusg=1 hhincome=35 educyear=12
> married=1 hisp=1) noatlegend   // (MER)
Conditional marginal effects                      Number of obs   =      3206
Model VCE    : OIM

Expression    : Pr(ins), predict()
dy/dx w.r.t. : 1.retire age 1.hstatusg hhincome educyear 1.married 1.hisp
```

	dy/dx	Delta-method Std. Err.	z	P>|z|	[95% Conf. Interval]	
1.retire	.0354151	.0149556	2.37	0.018	.0061026	.0647276
age	-.0027608	.0020524	-1.35	0.179	-.0067834	.0012618
1.hstatusg	.0544316	.0161653	3.37	0.001	.0227482	.086115
hhincome	.0004357	.0001493	2.92	0.004	.0001432	.0007283
educyear	.0216131	.0036845	5.87	0.000	.0143916	.0288346
1.married	.0935092	.0174029	5.37	0.000	.0594001	.1276182
1.hisp	-.1794232	.037961	-4.73	0.000	-.2538254	-.105021

Note: dy/dx for factor levels is the discrete change from the base level.

The MEMs for the regressors are approximately 0.2 times the logit coefficient estimates.

14.7.2   Marginal effect at the mean (MEM)

For comparison, we compute the ME at the mean, using the `margins` command with the `atmean` option. We obtain

(Continued on next page)

```
. * Marginal effects (MEM) after logit
. quietly logit ins i.retire age i.hstatusg hhincome educyear i.married i.hisp
. margins, dydx(*) atmean noatlegend  // (MEM)
Conditional marginal effects                      Number of obs   =     3206
Model VCE    : OIM

Expression   : Pr(ins), predict()
dy/dx w.r.t. : 1.retire age 1.hstatusg hhincome educyear 1.married 1.hisp
```

	dy/dx	Delta-method Std. Err.	z	P>|z|	[95% Conf. Interval]	
1.retire	.0457255	.0193956	2.36	0.018	.0077108	.0837402
age	-.0034129	.0026389	-1.29	0.196	-.008585	.0017592
1.hstatusg	.0716613	.0205694	3.48	0.000	.0313459	.1119766
hhincome	.0005386	.0001785	3.02	0.003	.0001888	.0008885
educyear	.0267179	.0033025	8.09	0.000	.0202452	.0331907
1.married	.1295601	.0197445	6.56	0.000	.0908617	.1682585
1.hisp	-.1677028	.0341779	-4.91	0.000	-.2346902	-.1007154

```
Note: dy/dx for factor levels is the discrete change from the base level.
```

In this particular case, the MEM is 20–30% greater than the MER.

14.7.3 Average marginal effect (AME)

The average marginal effect (AME) is obtained as the default option of the margins, dydx() command. The associated standard errors and confidence interval for the average marginal effect are obtained using the delta method.

For the fitted logit model, we obtain

```
. * Marginal effects (AME) after logit
. quietly logit ins i.retire age i.hstatusg hhincome educyear i.married i.hisp
. margins, dydx(*) noatlegend         // (AME)
Average marginal effects                          Number of obs   =     3206
Model VCE    : OIM

Expression   : Pr(ins), predict()
dy/dx w.r.t. : 1.retire age 1.hstatusg hhincome educyear 1.married 1.hisp
```

	dy/dx	Delta-method Std. Err.	z	P>|z|	[95% Conf. Interval]	
1.retire	.0426943	.0181787	2.35	0.019	.0070647	.0783239
age	-.0031693	.0024486	-1.29	0.196	-.0079686	.00163
1.hstatusg	.0675283	.0196091	3.44	0.001	.0290951	.1059615
hhincome	.0005002	.0001646	3.04	0.002	.0001777	.0008228
educyear	.0248111	.0029705	8.35	0.000	.0189891	.0306332
1.married	.1235562	.0191419	6.45	0.000	.0860388	.1610736
1.hisp	-.1608825	.0339246	-4.74	0.000	-.2273735	-.0943914

```
Note: dy/dx for factor levels is the discrete change from the base level.
```

In this example, the AME is 5–10% less than the MEM. The difference can be larger in other samples.

14.7.4   The prchange command

The marginal change in probability due to a unit change in a specified regressor, conditional on specified values of other regressors, can be calculated by using the user-written `prchange` command (Long and Freese 2006). The syntax is similar to that of `prvalue`, discussed in section 14.6.5:

prchange *varname* $\left[\,if\,\right]$ $\left[\,in\,\right]$ $\left[\,,\; \texttt{x}(conditions)\;\; \texttt{rest(mean)}\,\right]$

where *varname* is the variable that changes. The default for the conditioning variables is the sample mean.

The following gives the marginal effect of a change in income (`hhincome`) evaluated at the mean of regressors evaluated at $\mathbf{x} = \overline{\mathbf{x}}$.

```
. * Computing change in probability after logit
. quietly logit ins retire $xlist

. prchange hhincome

logit: Changes in Probabilities for ins

              min->max       0->1     -+1/2    -+sd/2  MargEfct
hhincome        0.5679     0.0005    0.0005    0.0346    0.0005

                   0         1
Pr(y|x)   0.6272  0.3728

             retire       age  hstatusg  hhincome  educyear   married      hisp
    x=     .624766   66.9139   .704616   45.2639   11.8986   .733001   .072676
 sd_x=     .484259   3.67579   .456286   64.3394   3.30461   .442461   .259645
```

The output supplements the marginal-effect calculation by also reporting changes in probability induced by several types of change in income. The output `min->max` gives the change in probability due to income changing from the minimum to the maximum observed value. The output `0->1` gives the change due to income changing from 0 to 1. The output `-+1/2` gives the impact of income changing from a half unit below to a half unit above the base value. And the output `-+sd/2` gives the impact of income changing from one-half a standard deviation below to one-half a standard deviation above the base value. Adding the `help` option to this command generates explanatory notes for the computer output.

14.8   Endogenous regressors

The probit and logit ML estimators are inconsistent if any regressor is endogenous. Two broad approaches are used to correct for endogeneity.

The structural approach specifies a complete model that explicitly models both nonlinearity and endogeneity. The specific structural model used differs according to whether the endogenous regressor is discrete or continuous. ML estimation is most efficient, but simpler (albeit less efficient) two-step estimators are often used.

The alternative partial model or semiparametric approach defines a residual for the equation of interest and uses the IV estimator based on the orthogonality of instruments and this residual.

As in the linear case, a key requirement is the existence of one or more valid instruments that do not directly explain the binary dependent variable but are correlated with the endogenous regressor. Unlike the linear case, different approaches to controlling for endogeneity can lead to different estimators even in the limit, as the parameters of different models are being estimated.

14.8.1 Example

We again model the binary outcome `ins`, though we use a different set of regressors. The regressors include the continuous variable `linc` (the log of household income) that is potentially endogenous as purchase of supplementary health insurance and household income may be subject to correlated unobserved shocks, even after controlling for a variety of exogenous variables. That is, for the HRS sample under consideration, the choice of supplementary insurance (`ins`), as well as household income (`linc`), may be considered as jointly determined.

Regular probit regression that does not control for this potential endogeneity yields

```
. * Endogenous probit using inconsistent probit MLE
. generate linc = log(hhincome)
(9 missing values generated)
. global xlist2 female age age2 educyear married hisp white chronic adl hstatusg
. probit ins linc $xlist2, vce(robust) nolog
Probit regression                             Number of obs   =       3197
                                              Wald chi2(11)   =     366.94
                                              Prob > chi2     =     0.0000
Log pseudolikelihood = -1933.4275             Pseudo R2       =     0.0946
```

ins	Coef.	Robust Std. Err.	z	P>|z|	[95% Conf. Interval]	
linc	.3466893	.0402173	8.62	0.000	.2678648	.4255137
female	-.0815374	.0508549	-1.60	0.109	-.1812112	.0181364
age	.1162879	.1151924	1.01	0.313	-.109485	.3420608
age2	-.0009395	.0008568	-1.10	0.273	-.0026187	.0007397
educyear	.0464387	.0089917	5.16	0.000	.0288153	.0640622
married	.1044152	.0636879	1.64	0.101	-.0204108	.2292412
hisp	-.3977334	.1080935	-3.68	0.000	-.6095927	-.1858741
white	-.0418296	.0644391	-0.65	0.516	-.168128	.0844687
chronic	.0472903	.0186231	2.54	0.011	.0107897	.0837909
adl	-.0945039	.0353534	-2.67	0.008	-.1637953	-.0252125
hstatusg	.1138708	.0629071	1.81	0.070	-.0094248	.2371664
_cons	-5.744548	3.871615	-1.48	0.138	-13.33277	1.843677

The regressor `linc` has coefficient 0.35 and is quite precisely estimated with a standard error of 0.04. The associated marginal effect at $\mathbf{x} = \bar{\mathbf{x}}$, computed using the `margins`, `dydx(linc)` `atmean` command, is 0.13. This implies that a 10% increase in household income (a change of 0.1 in `linc`) is associated with an increase of 0.013 in the probability of having supplementary health insurance.

14.8.2   Model assumptions

We restrict attention to the case of a single continuous endogenous regressor in a binary outcome model. For a discrete endogenous regressor other methods should be used.

We consider the following linear latent-variable model, in which y_1^* is the dependent variable in the structural equation and y_2 is an endogenous regressor in this equation. These two endogenous variables are modeled as linear in exogenous variables $\mathbf{x}_1$ and $\mathbf{x}_2$. That is,

$$y_{1i}^* = \beta y_{2i} + \mathbf{x}_{1i}'\boldsymbol{\gamma} + u_i \tag{14.7}$$
$$y_{2i} = \mathbf{x}_{1i}'\boldsymbol{\pi}_1 + \mathbf{x}_{2i}'\boldsymbol{\pi}_2 + v_i \tag{14.8}$$

where $i = 1,\ldots,N$; $\mathbf{x}_1$ is a $K_1 \times 1$ vector of exogenous regressors; and $\mathbf{x}_2$ is a $K_2 \times 1$ vector of additional IV that affect y_2 but can be excluded from (14.7) as they do not directly affect y_1. Identification requires that $K_2 \geq 1$.

The variable y_1^* is latent and hence is not directly observed. Instead the binary outcome y_1 is observed, with $y_1 = 1$ if $y_1^* > 0$, and $y_1 = 0$ if $y_1^* \leq 0$.

Equation (14.7) might be referred to as "structural". This structural equation is of main interest and the second equation, called a first-stage equation or reduced-form equation, only serves as a source of identifying instruments. It provides a check on the strength of the instruments and on the goodness of fit of the reduced form.

The reduced-form equation (14.8) explains the variation in the endogenous variable in terms of strictly exogenous variables, including the IV $\mathbf{x}_2$ that are excluded from the structural equation. These excluded instruments, previously discussed in chapter 6 within the context of linear models, are essential for identifying the parameters of the structural equation. Given the specification of the structural and reduced-form equations, estimation can be simultaneous (i.e., joint) or sequential.

14.8.3   Structural-model approach

The structural-model approach completely specifies the distributions of y_1^* and y_2 in (14.7) and (14.8). It is assumed that (u_i, v_i) are jointly normally distributed, i.e., $(u_i, v_i) \sim N(\mathbf{0}, \boldsymbol{\Sigma})$, where $\boldsymbol{\Sigma} = (\sigma_{ij})$. In the binary probit model, the coefficients are identified up to a scale factor only; hence, by scale normalization, $\sigma_{11} = 1$. The assumptions imply that $u_i|v_i = \rho v_i + \varepsilon_i$, where $E(\varepsilon_i|v_i) = 0$. A test of the null hypothesis of exogeneity of y_2 is equivalent to the test of $H_0 : \rho = 0$, because then u_i and v_i are independent.

This approach relies greatly on the distributional assumptions. Consistent estimation requires both normality and homoskedasticity of the errors u_i and v_i.

The ivprobit command

The syntax of `ivprobit` is similar to that of `ivregress`, discussed in chapter 6:

`ivprobit` *depvar* $\begin{bmatrix} varlist1 \end{bmatrix}$ (*varlist2=varlist_iv*) $\begin{bmatrix} if \end{bmatrix}$ $\begin{bmatrix} in \end{bmatrix}$ $\begin{bmatrix} weight \end{bmatrix}$
   $\begin{bmatrix} , mle\_options \end{bmatrix}$

where *varlist2* refers to the endogenous variable y_2 and *varlist_iv* refers to the instruments x_2 that are excluded from the equation for y_1^*. The default version of `ivprobit` delivers ML estimates, and the `twostep` option yields two-step estimates.

Maximum likelihood estimates

For this example, we use as instruments two excluded variables, `retire` and `sretire`. These refer to, respectively, individual retirement status and spouse retirement status. These are likely to be correlated with `linc`, because retirement will lower household income. The key assumption for instrument validity is that retirement status does not directly affect choice of supplementary insurance. This assumption is debatable, and this example is best viewed as merely illustrative. We apply `ivprobit`, obtaining ML estimates:

```
. * Endogenous probit using ivprobit ML estimator
. global ivlist2 retire sretire
. ivprobit ins $xlist2 (linc = $ivlist2), vce(robust) nolog
```

Probit model with endogenous regressors		Number of obs	=	3197
		Wald chi2(11)	=	382.34
Log pseudolikelihood = -5407.7151		Prob > chi2	=	0.0000

	Coef.	Robust Std. Err.	z	P>|z|	[95% Conf. Interval]
linc	-.5338207	.3852263	-1.39	0.166	-1.28885	.221209
female	-.139407	.0494474	-2.82	0.005	-.2363221	-.0424919
age	.2862286	.1280831	2.23	0.025	.0351903	.5372669
age2	-.0021472	.0009318	-2.30	0.021	-.0039735	-.0003209
educyear	.1136878	.0237922	4.78	0.000	.067056	.1603196
married	.7058283	.2377674	2.97	0.003	.2398126	1.171844
hisp	-.5094514	.1049488	-4.85	0.000	-.7151472	-.3037555
white	.1563445	.1035697	1.51	0.131	-.0466484	.3593373
chronic	.0061942	.0275256	0.23	0.822	-.047755	.0601433
adl	-.1347664	.03498	-3.85	0.000	-.2033259	-.0662068
hstatusg	.2341784	.0709763	3.30	0.001	.0950674	.3732895
_cons	-10.00786	4.065785	-2.46	0.014	-17.97665	-2.039065
/athrho	.6745321	.3599835	1.87	0.061	-.0310226	1.380087
/lnsigma	-.331594	.0233799	-14.18	0.000	-.3774178	-.2857703
rho	.5879532	.2355412			-.0310126	.8809707
sigma	.7177787	.0167816			.6856296	.7514352

```
Instrumented:  linc
Instruments:   female age age2 educyear married hisp white chronic adl
               hstatusg retire sretire
```

Wald test of exogeneity (/athrho = 0): chi2(1) =   3.51 Prob > chi2 = 0.0610

The output includes a test of the null hypothesis of exogeneity, i.e., $H_0 : \rho = 0$. The p-value is 0.061, so H_0 is not rejected at the 0.05 level, though it is rejected at the 0.10 level. That the estimated coefficient is positive indicates a positive correlation between u and v. Those unmeasured factors that make it more likely for an individual to have a higher household income also make it more likely that the individual will have supplementary health insurance, conditional on other regressors included in the equation.

Given the large estimated value for ρ ($\widehat{\rho} = 0.59$), we should expect that the coefficients of the estimated `probit` and `ivprobit` models differ. This is indeed the case, for both the endogenous regressor `linc` and for the other regressors. The coefficient of `linc` actually changes signs (from 0.39 to -0.53), so that an increase in household income is estimated to lower the probability of having supplementary insurance. One possible explanation is that richer people are willing to self-insure for medical services not covered by Medicare. At the same time, IV estimation has led to much greater imprecision, with the standard error increasing from 0.04 to 0.39, so that the negative coefficient is not statistically significantly different from zero at the 0.05 level. Taken at face value, however, the result suggests that the `probit` command that neglects endogeneity leads to an overestimate of the effect of household income. The remaining coefficients exhibit the same sign pattern as in the ordinary probit model, and the differences in the point estimates are within the range of estimated standard errors.

Two-step sequential estimates

An alternative estimation procedure for (14.7) and (14.8) with normal errors (Newey 1987) uses a minimum chi-squared estimator. This estimator also assumes multivariate normality and homoskedasticity and is therefore similar to the ML estimator. However, the details of the algorithm are different. The advantage of the two-step sequential estimator over the ML estimator is mainly computational because both methods make the same distributional assumptions.

The estimator is implemented by using `ivprobit` with the `twostep` option.

We do so for our data, using the `first` option, which also provides the least-squares (LPM) estimates of the first stage.

(Continued on next page)

```
. * Endogenous probit using ivprobit 2-step estimator
. ivprobit ins $xlist2 (linc = $ivlist2), twostep first
Checking reduced-form model...
First-stage regression
```

Source	SS	df	MS				
Model	1173.12053	12	97.7600445				
Residual	1647.03826	3184	.517285885				
Total	2820.15879	3196	.882402626				

```
                                    Number of obs =     3197
                                    F( 12,  3184) =   188.99
                                    Prob > F       =   0.0000
                                    R-squared      =   0.4160
                                    Adj R-squared =   0.4138
                                    Root MSE       =   .71923
```

linc	Coef.	Std. Err.	t	P>|t|	[95% Conf. Interval]	
retire	-.0909581	.0288119	-3.16	0.002	-.1474499	-.0344663
sretire	-.0443106	.0317252	-1.40	0.163	-.1065145	.0178932
female	-.0936494	.0297304	-3.15	0.002	-.151942	-.0353569
age	.2669284	.0627794	4.25	0.000	.1438361	.3900206
age2	-.0019065	.0004648	-4.10	0.000	-.0028178	-.0009952
educyear	.094801	.0043535	21.78	0.000	.0862651	.1033369
married	.7918411	.0367275	21.56	0.000	.7198291	.8638531
hisp	-.2372014	.0523874	-4.53	0.000	-.3399179	-.134485
white	.2324672	.0347744	6.69	0.000	.1642847	.3006496
chronic	-.0388345	.0100852	-3.85	0.000	-.0586086	-.0190604
adl	-.0739895	.0173458	-4.27	0.000	-.1079995	-.0399795
hstatusg	.1748137	.0338519	5.16	0.000	.10844	.2411875
_cons	-7.702456	2.118657	-3.64	0.000	-11.85653	-3.548385

```
Two-step probit with endogenous regressors    Number of obs   =      3197
                                               Wald chi2(11)   =    222.51
                                               Prob > chi2     =    0.0000
```

	Coef.	Std. Err.	z	P>|z|	[95% Conf. Interval]	
linc	-.6109088	.5723054	-1.07	0.286	-1.732607	.5107893
female	-.167917	.0773839	-2.17	0.030	-.3195867	-.0162473
age	.3422526	.1915485	1.79	0.074	-.0331756	.7176808
age2	-.0025708	.0014021	-1.83	0.067	-.0053188	.0001773
educyear	.13596	.0543047	2.50	0.012	.0295249	.2423952
married	.8351517	.441743	1.89	0.059	-.0306487	1.700952
hisp	-.6184546	.181427	-3.41	0.001	-.9740451	-.2628642
white	.1818279	.1528281	1.19	0.234	-.1177098	.4813655
chronic	.0095837	.0309618	0.31	0.757	-.0511004	.0702678
adl	-.1630884	.0568288	-2.87	0.004	-.2744709	-.0517059
hstatusg	.2809463	.1228386	2.29	0.022	.0401871	.5217055
_cons	-12.04848	5.928158	-2.03	0.042	-23.66746	-.4295071

```
Instrumented:  linc
Instruments:   female age age2 educyear married hisp white chronic adl
               hstatusg retire sretire

Wald test of exogeneity:     chi2(1) =     3.57          Prob > chi2 = 0.0588
```

The results of the two-step estimator are similar to those from the ivprobit ML estimation. The coefficient estimates are within 20% of each other. The standard errors are
increased by approximately 50%, indicating a loss of precision in two-step estimation
compared with ML estimation. The test statistic for exogeneity of linc has a p-value
of 0.059 compared with 0.061 using ML. The results for the first stage indicate that

one of the two excluded IV has a strong predictive value for `linc`. Because this is a reduced-form equation, we do not attempt an interpretation of the results.

14.8.4 IVs approach

An alternative, less structural approach is to use the IV estimation methods for the linear regression model, presented in chapter 6. This requires fewer distributional assumptions, though if linear IV is used, then the binary nature of the dependent variable y_1 (`ins`) is being ignored.

We have the standard linear formulation for the observed variables (y_1, y_2)

$$y_{1i} = \beta y_{2i} + \mathbf{x}'_{1i}\boldsymbol{\gamma} + u_i$$
$$y_{2i} = x'_{1i}\boldsymbol{\pi}_1 + \mathbf{x}'_{2i}\boldsymbol{\pi}_2 + v_i$$

where y_2 is endogenous and the covariates $\mathbf{x}_2$ are the excluded exogenous regressors (instruments). This is the model (14.7) and (14.8) except that the latent-variable y_1^* is replaced by the binary variable y_1. An important difference is that while (u, v) are zero-mean and jointly dependent they need not be multivariate normal and homoskedastic.

Estimation is by two-stage least-squares (2SLS), using the `ivregress` command. Because y_1 is binary, the error u is heteroskedastic. The 2SLS estimator is then still consistent for $(\beta, \boldsymbol{\gamma})$, but heteroskedasticity-robust standard errors should be used for inference. In chapter 6, we considered several issues, especially that of weak instruments, in applying the IV estimator. These issues remain relevant here also, and the reader is referred back to chapter 6 for a more detailed treatment of the topic.

The `ivregress` command with the `vce(robust)` option yields

```
. * Endogenous probit using ivregress to get 2SLS estimator
. ivregress 2sls ins $xlist2 (linc = $ivlist2), vce(robust) noheader
```

ins	Coef.	Robust Std. Err.	z	P>|z|	[95% Conf. Interval]	
linc	-.167901	.1937801	-0.87	0.386	-.547703	.2119011
female	-.0545806	.0260643	-2.09	0.036	-.1056657	-.0034955
age	.106631	.0624328	1.71	0.088	-.015735	.228997
age2	-.0008054	.0004552	-1.77	0.077	-.0016977	.0000868
educyear	.0416443	.0182207	2.29	0.022	.0059324	.0773562
married	.2511613	.1499264	1.68	0.094	-.042689	.5450116
hisp	-.154928	.0546479	-2.84	0.005	-.2620358	-.0478202
white	.0513327	.0508817	1.01	0.313	-.0483936	.151059
chronic	.0048689	.0103797	0.47	0.639	-.015475	.0252128
adl	-.0450901	.0174479	-2.58	0.010	-.0792874	-.0108928
hstatusg	.0858946	.041327	2.08	0.038	.0048951	.1668941
_cons	-3.303902	1.920872	-1.72	0.085	-7.068743	.4609388

```
Instrumented:  linc
Instruments:   female age age2 educyear married hisp white chronic adl
               hstatusg retire sretire
```

```
. estat overid
  Test of overidentifying restrictions:
  Score chi2(1)           =   .521843  (p = 0.4701)
```

This method yields a coefficient estimate of -0.17 of `linc` that is statistically insignificant at level 0.05, as for `ivprobit`. To compare `ivregress` estimates to `ivprobit` estimates, we need to rescale parameters as in section 14.4.3. Then the rescaled 2SLS parameter estimate is $-0.17 \times 2.5 = -0.42$, comparable to the estimates of -0.53 and -0.61 from the `ivprobit` command.

Advantages of the 2SLS estimator are its computational simplicity and the ability to use tests of validity of overidentifying instruments and diagnostics for weak instruments that were presented in chapter 6. At the same time, the formal tests and inference that require normal homoskedastic errors may be inappropriate due to the intrinsic heteroskedasticity when the dependent variable is binary. Here the single overidentifying restriction is not rejected by the Hansen J test, which yields a $\chi^2(1)$ value of 0.522. Whether the results are sensitive to the choice of instruments can be pursued further by estimating additional specifications, an advisable approach if some instruments are weak.

The linear 2SLS estimator in the current example is based solely on the moment condition $E(u|\mathbf{x}_1, \mathbf{x}_2) = 0$, where $u = y_1 - (\beta y_2 + \mathbf{x}_1'\boldsymbol{\gamma})$; see section 6.2.2. For a binary outcome y_1 modeled using the probit model, it is better to instead use the nonlinear 2SLS estimator based on moment condition $E(u|\mathbf{x}_1, \mathbf{x}_2) = 0$, where the error term, the difference between y_1 and its conditional mean function, is defined as $u = y_1 - \Phi(\beta y_2 + \mathbf{x}_1'\boldsymbol{\gamma})$. This moment condition is not implied by (14.7) and (14.8), so the estimates will differ from those from the `ivprobit` command. There is no Stata command to implement the nonlinear 2SLS estimator, but the nonlinear 2SLS example in section 11.8 can be suitably adapted.

14.9 Grouped data

In some applications, only grouped or aggregate data may be available, yet individual behavior is felt to be best modeled by a binary choice model. For example, we may have a frequency average taken across a sampled population as the dependent variable and averages of explanatory variables for the regressors, which we will assume to be exogenous. We refer to these as grouped data.

Such grouping poses no problem when the grouping is on unique values of the regressors and there are many observations per unique value of the regressors. For example, in the dataset of this chapter, `age` could be the grouping variable. This would generate 33 groups, one for each age between 52 and 86; there are no observations for ages 84 or 85. The number of cases in the 33 groups are as follows:

4	5	2	2	7	8	34	62	72	51	61
67	74	524	470	488	477	286	133	100	91	67
36	29	19	11	8	11	4	6	5	1	1

Observations with no within-group variation will be dropped, and this is likely to occur when the group size is small. In the present sample, there are two groups with two observations each, and two with only one observation. These small groups are dropped, which reduces the sample size to 29.

If the group size is relatively large and the grouping variable is distinct, Berkson's minimum chi-squared estimator is one method of estimating the parameters of the model. As an example, suppose the regressor vector $\mathbf{x}_i$, $i = 1, \ldots, N$, takes only T distinct values, where T is much smaller than N. Then, for each value of the regressors, we have multiple observations on y. This type of grouping involves many observations per cell. Berkson's estimator (see Cameron and Trivedi [2005, 480]) can be computed easily by weighted least squares (WLS).

This method is not suitable for our data because the regressor vector $\mathbf{x}_i$ takes on a large number of values given many regressors, some of which are continuous. We nonetheless group on age to illustrate grouped-data methods.

14.9.1   Estimation with aggregate data

Let $\bar{p}_g$ denote the average frequency in group g ($g = 1, \ldots, G$, $G > K$), and let $\bar{\mathbf{x}}_g$ denote the average of $\mathbf{x}$ across N_g, where the latter is the number of observations in group g. One possible model is OLS regression of $\bar{p}_g$ on $\bar{\mathbf{x}}_g$. Because $0 < \bar{p}_g < 1$, it is common to use the logistic transformation to define the dependent variable that is now unbounded and to estimate the parameters of the model

$$\ln\left(\frac{\bar{p}_g}{1 - \bar{p}_g}\right) = \bar{\mathbf{x}}_g'\boldsymbol{\gamma} + u_g \tag{14.9}$$

where u_g is an error. It is essential to estimate the standard errors of the OLS coefficients in the above model robustly because the average $\bar{p}_g$ is heteroskedastic, since it is given that N_g will vary with g. The logistic transformation may to some extent reduce heteroskedasticity.

The model for aggregated data presented above will potentially yield biased estimates; that is, in general the OLS estimator of $\boldsymbol{\gamma}$ is not a consistent estimator of $\boldsymbol{\beta}$ in a nonlinear model. However, we may interpret the $\boldsymbol{\gamma}$ as an interesting aggregate parameter without any necessary connection with the $\boldsymbol{\beta}$.

14.9.2   Grouped-data application

The full individual dataset of 3,206 observations can be converted to an aggregate dataset by using the following Stata commands that generate group averages and then saving the aggregated data into a separate file.

```
. * Using mus14data.dta to generate grouped data
. sort age
. collapse av_ret=retire av_hhinc=hhincome av_educyear=educyear av_mar=married
> av_adl=adl av_hisp=hisp av_hstatusg=hstatusg av_ins=ins, by(age)
. generate logins = log(av_ins/(1-av_ins))
(4 missing values generated)
. save mus14gdata.dta, replace
file mus14gdata.dta saved
```

Here the `collapse` command is used to form averages by age. For example, `collapse av_hhincome=hhincome, by(age)` creates 29 observations for the `av_hhincome` variable equal to the average of the `hhincome` variable for each of the 29 distinct values taken by the `age` variable. More generally, `collapse` can compute other statistics, such as the median specifying the `median` statistic, and if the `by()` option was not used then just a single observation would be produced. Four observations are lost because the `logins` variable cannot be computed in groups with `av_ins` equal to 0 or 1.

The aggregate regression is estimated as follows:

```
. * Regressions with grouped data
. regress logins av_ret av_hstatusg av_hhinc av_educyear av_mar av_hisp,
> vce(robust)
```

Linear regression

```
                                        Number of obs =        29
                                        F(  6,    22) =      5.26
                                        Prob > F      =    0.0017
                                        R-squared     =    0.4124
                                        Root MSE      =    .44351
```

logins	Coef.	Robust Std. Err.	t	P>|t|	[95% Conf. Interval]	
av_ret	.1460855	.7168061	0.20	0.840	-1.340479	1.63265
av_hstatusg	-.5992984	1.033242	-0.58	0.568	-2.742112	1.543515
av_hhinc	.0016449	.0163948	0.10	0.921	-.0323558	.0356456
av_educyear	.1851466	.1618441	1.14	0.265	-.1504974	.5207906
av_mar	1.514133	1.018225	1.49	0.151	-.5975357	3.625802
av_hisp	-.7119637	.6532035	-1.09	0.288	-2.066625	.6426975
_cons	-3.679837	1.80997	-2.03	0.054	-7.433484	.0738104

The above results are based on 29 grouped observations. Each estimated coefficient reflects the impact of a regressor on the log of the odds ratio. To convert the estimate to reflect the effect on the odds ratio, its coefficient should be exponentiated. The sign pattern of the coefficients in the aggregate regression is similar but not identical to that in the disaggregated logit model in section 14.4.2. Notice that the fit of the model, as measured by R^2, has improved while the standard errors of parameter estimates have deteriorated. The averaged data are less noisy, so the R^2 improves. But the reduction in variance of the regressors and the smaller sample size increase the standard errors.

As was noted above, the parameters in the grouped model cannot be easily related to those in the disaggregated logit model. For example, `hstatusg` had a significant positive coefficient in the logit equation, but `av_hstatusg` has a negative coefficient.

14.10 Stata resources

The main reference for the endogenous regressor case is [R] **ivprobit**. MEs can be estimated by using the `margins, dydx()` command. For grouped or blocked data, Stata provides the `blogit` and `bprobit` commands for ML logit and probit estimation; the variants `glogit` and `gprobit` can be used to perform WLS estimation. For simultaneous-equations estimation, the user-written `cdsimeq` (Keshk 2003) command implements a two-stage estimation method for the case in which one of the endogenous variables is continuous and the other endogenous variable is dichotomous.

14.11 Exercises

1. Consider the example of section 14.4 with dependent variable `ins` and the single regressor `educyear`. Estimate the parameters of logit, probit, and OLS models using both default and robust standard errors. For the regressor `educyear`, compare its coefficient across the models, compare default and robust standard errors of this coefficient, and compare the t statistics based on robust standard errors. For each model, compute the marginal effect of one more year of education for someone with sample mean years of education, as well as the AME. Which model fits the data better—logit or probit?

2. Use the `cloglog` command to estimate the parameters of the binary probability model for `ins` with the same explanatory variables used in the `logit` model in this chapter. Estimate the average marginal effects for the regressors. Calculate the odds ratios of `ins=1` for the following values of the covariates: `age=50`, `retire=0`, `hstatusg=1`, `hhincome=45`, `educyear=12`, `married=1`, and `hisp=0`.

3. Generate a graph of fitted probabilities against years of education (`educyear`) or age (`age`) using as a template the commands used for generating figure 14.1 in this chapter.

4. Estimate the parameters of the logit model of section 14.4.2. Now estimate the parameters of the probit model using the `probit` command. Use the reported log likelihoods to compare the models by the AIC and BIC.

5. Estimate the probit regression of section 14.4.3. Using the conditioning values (`age=65`, `retire=1`, `hstatusg=1`, `hhincome=60`, `educyear=17`, `married=1`, `hisp=0`), estimate and compare the marginal effect of `age` on the $\Pr(\text{ins}=1|x)$, using both the `margins` and `prchange` commands. They should give the same result.

6. Using the `hetprob` command, estimate the parameters of the model of section 14.4, using `hhincome` as the variable determining the variance. Use the LR as a test of the null of homoskedastic probit.

7. Using the example in section 14.9 as a template, estimate a grouped logistic regression using `educyear` as the grouping variable. Comment on what you regard as unsatisfactory features of the grouping variable and the results.

15 Multinomial models

15.1 Introduction

Categorical data are data on a dependent variable that can fall into one of several mutually exclusive categories. Examples include different ways to commute to work (by car, bus, or on foot) and different categories of self-assessed health status (excellent, good, fair, or poor).

The econometrics literature focuses on modeling a single outcome from categories that are mutually exclusive, where the dependent variable outcome must be multinomial distributed, just as binary data must be Bernoulli or binomial distributed. Analysis is not straightforward, however, because there are many different models for the probabilities of the multinomial distribution. These models vary according to whether the categories are ordered or unordered, whether some of the individual-specific regressors vary across the alternative categories, and in some settings, whether the model is consistent with utility maximization. Furthermore, parameter coefficients for any given model can be difficult to directly interpret. The marginal effects (MEs) of interest measure the impact on the probability of observing each of several outcomes rather than the impact on a single conditional mean.

We begin with models for unordered outcomes, notably, multinomial logit, conditional logit, nested logit, and multinomial probit models. We then move to models for ordered outcomes, such as health-status measures, and models for multivariate multinomial outcomes.

15.2 Multinomial models overview

We provide a general discussion of multinomial regression models. Subsequent sections detail the most commonly used multinomial regression models that correspond to particular functional forms for the probabilities of each alternative.

15.2.1 Probabilities and MEs

The outcome, y_i, for individual i is one of m alternatives. We set $y_i = j$ if the outcome is the jth alternative, $j = 1, 2, \ldots, m$. The values $1, 2, \ldots, m$ are arbitrary, and the same regression results are obtained if, for example, we use values $3, 5, 8, \ldots$. The ordering of the values also does not matter, unless an ordered model (presented in section 15.9) is used.

The probability that the outcome for individual i is alternative j, conditional on the regressors $\mathbf{x}_i$, is

$$p_{ij} = \Pr(y_i = j) = F_j(\mathbf{x}_i, \boldsymbol{\theta}), \quad j = 1, \ldots, m, \quad i = 1, \ldots, N \qquad (15.1)$$

where different functional forms, $F_j(\cdot)$, correspond to different multinomial models. Only $m - 1$ of the probabilities can be freely specified because probabilities sum to one. For example, $F_m(\mathbf{x}_i, \boldsymbol{\theta}) = 1 - \sum_{j=1}^{m-1} F_j(\mathbf{x}_i, \boldsymbol{\theta})$. Multinomial models therefore require a normalization. Some Stata multinomial commands, including `asclogit`, permit different individuals to face different choice sets so that, for example, an individual might be choosing only from among alternatives 1, 3, and 4.

The parameters of multinomial models are generally not directly interpretable. In particular, a positive coefficient need not mean that an increase in the regressor leads to an increase in the probability of an outcome being selected. Instead, we compute MEs. For individual i, the ME of a change in the kth regressor on the probability that alternative j is the outcome is

$$\text{ME}_{ijk} = \frac{\partial \Pr(y_i = j)}{\partial x_{ik}} = \frac{\partial F_j(\mathbf{x}_i, \boldsymbol{\theta})}{\partial x_{ik}}$$

For each regressor, there will be m MEs corresponding to the m probabilities, and these m MEs sum to zero because probabilities sum to one. As for other nonlinear models, these marginal effects vary with the evaluation point $\mathbf{x}$.

15.2.2 Maximum likelihood estimation

Estimation is by maximum likelihood (ML). We use a convenient form for the density that generalizes the method used for binary outcome models. The density for the ith individual is written as

$$f(y_i) = p_{i1}^{y_{i1}} \times \cdots \times p_{im}^{y_{im}} = \prod_{j=1}^{m} p_{ij}^{y_{ij}}$$

where $y_{i1}, \ldots, y_{im}$ are m indicator variables with $y_{ij} = 1$ if $y_i = j$ and $y_{ij} = 0$ otherwise. For each individual, exactly one of $y_1, y_2, \ldots, y_m$ will be nonzero. For example, if $y_i = 3$, then $y_{i3} = 1$, the other $y_{ij} = 0$, and upon simplification, $f(y_i) = p_{i3}$, as expected.

The likelihood function for a sample of N independent observations is the product of the N densities, so $L = \prod_{i=1}^{N} \prod_{j=1}^{m} p_{ij}^{y_{ij}}$. The maximum likelihood estimator (MLE), $\widehat{\boldsymbol{\theta}}$, maximizes the log-likelihood function

$$\ln L(\boldsymbol{\theta}) = \sum_{i=1}^{N} \sum_{j=1}^{m} y_{ij} \ln F_j(\mathbf{x}_i, \boldsymbol{\theta}) \qquad (15.2)$$

and as usual $\widehat{\boldsymbol{\theta}} \overset{a}{\sim} N\left(\boldsymbol{\theta}, \ \left[-E\{\partial^2 \ln L(\boldsymbol{\theta})/\partial\boldsymbol{\theta}\partial\boldsymbol{\theta}'\}\right]^{-1}\right)$.

For categorical data, the distribution is necessarily multinomial. There is generally no reason to use standard errors other than the default, unless there is some clustering such as from repeated observations on the same individual, in which case the `vce(cluster` *clustvar*`)` option should be used. Hypothesis tests can be performed by using the `lrtest` command, though it is usually more convenient to perform Wald tests by using the `test` command.

For multinomial models, the pseudo-R^2 has a meaningful interpretation; see section 10.7. Nonnested models can be compared by using the Akaike information criterion (AIC) and related measures.

For multinomial data, the only possible misspecification is that of $F_j(\mathbf{x}_i, \boldsymbol{\theta})$. There is a wide range of models for $F_j(\cdot)$, with the suitability of a particular model depending on the application at hand.

15.2.3 Case-specific and alternative-specific regressors

Some regressors, such as gender, do not vary across alternatives and are called case-specific or alternative-invariant regressors. Other regressors, such as price, may vary across alternatives and are called alternative-specific or case-varying regressors.

The commands used for multinomial model estimation can vary according to the form of the regressors. In the simplest case, all regressors are case specific, and for example, we use the `mlogit` command. In more complicated applications, some or all the regressors are alternative specific, and for example, we use the `asclogit` command. These commands can require data to be organized in different ways; see section 15.5.1.

15.2.4 Additive random-utility model

For unordered multinomial outcomes that arise from individual choice, econometricians favor models that come from utility maximization. This leads to multinomial models that are used much less in other branches of applied statistics.

For individual i and alternative j, we suppose that utility U_{ij} is the sum of a deterministic component, V_{ij}, that depends on regressors and unknown parameters, and an unobserved random component ε_{ij}:

$$U_{ij} = V_{ij} + \varepsilon_{ij} \tag{15.3}$$

This is called an additive random-utility model (ARUM). We observe the outcome $y_i = j$ if alternative j has the highest utility of the alternatives. It follows that

$$
\begin{aligned}
\Pr(y_i = j) &= \Pr(U_{ij} \geq U_{ik}), \quad \text{for all } k \\
&= \Pr(U_{ik} - U_{ij} \leq 0), \quad \text{all } k \\
&= \Pr(\varepsilon_{ik} - \varepsilon_{ij} \leq V_{ij} - V_{ik}), \quad \text{all } k
\end{aligned}
\tag{15.4}
$$

Standard multinomial models specify that $V_{ij} = \mathbf{x}'_{ij}\boldsymbol{\beta} + \mathbf{z}'_i\boldsymbol{\gamma}_j$, where $\mathbf{x}_i$ are alternative-specific regressors and $\mathbf{z}_i$ are case-specific regressors. Different assumptions about the joint distribution of $\varepsilon_{i1}, \ldots, \varepsilon_{im}$ lead to different multinomial models with different specifications for $F_j(\mathbf{x}_i, \boldsymbol{\theta})$ in (15.1). Because the outcome probabilities depend on the difference in errors, only $m - 1$ of the errors are free to vary, and similarly, only $m - 1$ of the γ_j are free to vary.

15.2.5 Stata multinomial model commands

Table 15.1 summarizes Stata commands for the estimation of multinomial models.

Table 15.1. Stata commands for the estimation of multinomial models

Model	Command
Multinomial logit	`mlogit`
Conditional logit	`clogit, asclogit`
Nested logit	`nlogit`
Multinomial probit	`mprobit, asmprobit`
Rank ordered	`rologit, asroprobit`
Ordered	`ologit, oprobit`
Stereotype logit	`slogit`
Bivariate probit	`biprobit`

The `clogit`, `asclogit`, `nlogit`, `asmprobit`, `rologit`, and `asroprobit` commands expect the data to be in long form. The remaining commands expect data to be in wide form. The independent variable lists for all but four of these estimation commands allow factor variables. Hence the postestimation `margins` command can be used to compute marginal effects on the predicted choice probabilities. The four exceptions are the `asclogit`, `asmprobit`, `asroprobit`, and `nlogit` commands. For the first three of these, the postestimation `estat mfx` command can be used to compute the MEMs and MERs, though not the AMEs.

15.3 Multinomial example: Choice of fishing mode

We analyze data on individual choice of whether to fish using one of four possible modes: from the beach, the pier, a private boat, or a charter boat. One explanatory variable is case specific (`income`) and the others [`price` and `crate` (catch rate)] are alternative specific.

15.3.1 Data description

The data from Herriges and Kling (1999) are also analyzed in Cameron and Trivedi (2005). The `mus15data.dta` dataset has the following data:

```
. * Read in dataset and describe dependent variable and regressors
. use mus15data.dta, clear

. describe

Contains data from mus15data.dta
  obs:         1,182
  vars:           16                         12 May 2008 20:46
  size:       80,376 (99.8% of memory free)
```

	storage	display	value	
variable name	type	format	label	variable label
mode	float	%9.0g	modetype	Fishing mode
price	float	%9.0g		price for chosen alternative
crate	float	%9.0g		catch rate for chosen alternative
dbeach	float	%9.0g		1 if beach mode chosen
dpier	float	%9.0g		1 if pier mode chosen
dprivate	float	%9.0g		1 if private boat mode chosen
dcharter	float	%9.0g		1 if charter boat mode chosen
pbeach	float	%9.0g		price for beach mode
ppier	float	%9.0g		price for pier mode
pprivate	float	%9.0g		price for private boat mode
pcharter	float	%9.0g		price for charter boat mode
qbeach	float	%9.0g		catch rate for beach mode
qpier	float	%9.0g		catch rate for pier mode
qprivate	float	%9.0g		catch rate for private boat mode
qcharter	float	%9.0g		catch rate for charter boat mode
income	float	%9.0g		monthly income in thousands $

```
Sorted by:
```

There are 1,182 observations, one per individual. The first three variables are for the chosen fishing mode with the variables `mode`, `price`, and `crate` being, respectively, the chosen fishing mode and the price and catch rate for that mode. The next four variables are mutually exclusive dummy variables for the chosen mode, taking on a value of 1 if that alternative is chosen and a value of 0 otherwise. The next eight variables are alternative-specific variables that contain the price and catch rate for each of the four possible fishing modes (the prefix `q` stands for quality; a higher catch rate implies a higher quality of fishing). These variables are constructed from individual surveys that ask not only about attributes of the chosen fishing mode but also about attributes of alternative fishing modes such as location that allow for determination of price and catch rate. The final variable, `income`, is a case-specific variable: The summary statistics follow:

(Continued on next page)

```
. * Summarize dependent variable and regressors
. summarize, separator(0)
    Variable |       Obs        Mean    Std. Dev.       Min         Max
-------------+---------------------------------------------------------
        mode |      1182    3.005076    .9936162         1           4
       price |      1182    52.08197    53.82997       1.29      666.11
       crate |      1182    .3893684    .5605964       .0002     2.3101
      dbeach |      1182    .1133672    .3171753         0           1
       dpier |      1182    .1505922    .3578023         0           1
    dprivate |      1182    .3536379    .4783008         0           1
    dcharter |      1182    .3824027    .4861799         0           1
      pbeach |      1182     103.422     103.641       1.29     843.186
       ppier |      1182     103.422     103.641       1.29     843.186
    pprivate |      1182    55.25657    62.71344       2.29      666.11
    pcharter |      1182    84.37924    63.54465      27.29      691.11
      qbeach |      1182    .2410113    .1907524       .0678      .5333
       qpier |      1182    .1622237    .1603898       .0014      .4522
    qprivate |      1182    .1712146    .2097885       .0002      .7369
    qcharter |      1182    .6293679    .7061142       .0021     2.3101
      income |      1182    4.099337    2.461964    .4166667      12.5
```

The variable mode takes on the values ranging from 1 to 4. On average, private and charter boat fishing are less expensive than beach and pier fishing. Beach and pier fishing, both close to shore with similar costs, have identical prices. The catch rate for charter boat fishing is substantially higher than for the other modes.

The tabulate command gives the various values and frequencies of the mode variable. We have

```
. * Tabulate the dependent variable
. tabulate mode
    Fishing |
       mode |      Freq.     Percent        Cum.
------------+-----------------------------------
      beach |        134       11.34       11.34
       pier |        178       15.06       26.40
    private |        418       35.36       61.76
    charter |        452       38.24      100.00
------------+-----------------------------------
      Total |      1,182      100.00
```

The shares are roughly one-third fish from the shore (either beach or pier), one-third fish from a private boat, and one-third fish from a charter boat. These shares are the same as the means of dbeach, ..., dcharter given in the summarize table. The mode variable takes on a value from 1 to 4 (see the summary statistics), but the output of describe has a label, modetype, that labels 1 as beach, ..., 4 as charter. This labeling can be verified by using the label list command. There is no obvious ordering of the fishing modes, so unordered multinomial models should be used to explain fishing-mode choice.

15.3.2 Case-specific regressors

Before formal modeling, it is useful to summarize the relationship between the dependent variable and the regressors. This is more difficult when the dependent variable is an unordered dependent variable.

For the case-specific income variable, we could use the bysort mode: summarize income command. More compact output is obtained by instead using the table command. We obtain

```
. * Table of income by fishing mode
. table mode, contents(N income mean income sd income)
```

Fishing mode	N(income)	mean(income)	sd(income)
beach	134	4.051617	2.50542
pier	178	3.387172	2.340324
private	418	4.654107	2.777898
charter	452	3.880899	2.050029

On average, those fishing from the pier have the lowest income and those fishing from a private boat have the highest.

15.3.3 Alternative-specific regressors

The relationship between the chosen fishing mode and the alternative-specific regressor price is best summarized as follows:

```
. * Table of fishing price by fishing mode
. table mode, contents(mean pbeach mean ppier mean pprivate mean pcharter) form
> at(%6.0f)
```

Fishing mode	mean(pbeach)	mean(ppier)	mean(pprivate)	mean(pcharter)
beach	36	36	98	125
pier	31	31	82	110
private	138	138	42	71
charter	121	121	45	75

On average, individuals tend to choose the fishing mode that is the cheapest or second cheapest alternative available for them. For example, for those choosing private, on average the price of private boat fishing is 42, compared with 71 for charter boat fishing and 138 for beach or pier fishing.

Similarly, for the catch rate, we have

```
. * Table of fishing catch rate by fishing mode
. table mode, contents(mean qbeach mean qpier mean qprivate mean qcharter) form
> at(%6.2f)
```

Fishing mode	mean(qbeach)	mean(qpier)	mean(qprivate)	mean(qcharter)
beach	0.28	0.22	0.16	0.52
pier	0.26	0.20	0.15	0.50
private	0.21	0.13	0.18	0.65
charter	0.25	0.16	0.18	0.69

The chosen fishing mode is not on average that with the highest catch rate. In particular, the catch rate is always highest on average for charter fishing, regardless of the chosen mode. Regression analysis can measure the effect of the catch rate after controlling for the price of the fishing mode.

15.4   Multinomial logit model

Many multinomial studies are based on datasets that have only case-specific variables, because explanatory variables are typically observed only for the chosen alternative and not for the other alternatives. The simplest model is the multinomial logit model because computation is simple and parameter estimates are easier to interpret than in some other multinomial models.

15.4.1   The mlogit command

The multinomial logit (MNL) model can be used when all the regressors are case specific. The MNL model specifies that

$$p_{ij} = \frac{\exp(\mathbf{x}_i'\boldsymbol{\beta}_j)}{\sum_{l=1}^{m} \exp(\mathbf{x}_i'\boldsymbol{\beta}_l)}, \qquad j = 1, \ldots, m \tag{15.5}$$

where $\mathbf{x}_i$ are case-specific regressors, here an intercept and income. Clearly, this model ensures that $0 < p_{ij} < 1$ and $\sum_{j=1}^{m} p_{ij} = 1$. To ensure model identification, $\boldsymbol{\beta}_j$ is set to zero for one of the categories, and coefficients are then interpreted with respect to that category, called the base category.

The mlogit command has the syntax

mlogit *depvar* [*indepvars*] [*if*] [*in*] [*weight*] [, *options*]

where *indepvars* are the case-specific regressors, and the default is to automatically include an intercept. The baseoutcome(#) option specifies the value of *depvar* to be used as the base category, overriding the Stata default of setting the most frequently chosen category as the base category. Other options include **rr** to report exponentiated coefficients ($e^{\widehat{\beta}}$ rather than $\widehat{\beta}$).

The `mlogit` command requires that data be in wide form, with one observation per individual. This is the case here.

15.4.2   Application of the mlogit command

We regress fishing mode on an intercept and income, the only case-specific regressor in our dataset. There is no natural base category. The first category, beach fishing, is arbitrarily set to be the base category. We obtain

```
. * Multinomial logit with base outcome alternative 1
. mlogit mode income, baseoutcome(1) nolog
Multinomial logistic regression               Number of obs   =       1182
                                               LR chi2(3)      =      41.14
                                               Prob > chi2     =     0.0000
Log likelihood = -1477.1506                    Pseudo R2       =     0.0137
```

mode	Coef.	Std. Err.	z	P>|z|	[95% Conf. Interval]	
beach	(base outcome)					
pier						
income	-.1434029	.0532884	-2.69	0.007	-.2478463	-.0389595
_cons	.8141503	.228632	3.56	0.000	.3660399	1.262261
private						
income	.0919064	.0406637	2.26	0.024	.0122069	.1716058
_cons	.7389208	.1967309	3.76	0.000	.3533352	1.124506
charter						
income	-.0316399	.0418463	-0.76	0.450	-.1136571	.0503774
_cons	1.341291	.1945167	6.90	0.000	.9600457	1.722537

The model fit is poor with pseudo-R^2, defined in section 10.7.1, equal to 0.014. Nonetheless, the regressors are jointly statistically significant at the 0.05 level, because LR `chi2(3)=41.14`. Three sets of regression estimates are given, corresponding here to $\widehat{\boldsymbol{\beta}}_2$, $\widehat{\boldsymbol{\beta}}_3$, and $\widehat{\boldsymbol{\beta}}_4$, because we used the normalization $\boldsymbol{\beta}_1 = \mathbf{0}$.

Two of the three coefficient estimates of `income` are statistically significant at the 0.05 level, but the results of such individual testing will vary with the omitted category. Instead, we should perform a joint test. Using a Wald test, we obtain

```
. * Wald test of the joint significance of income
. test income
 ( 1)  [beach]income = 0
 ( 2)  [pier]income = 0
 ( 3)  [private]income = 0
 ( 4)  [charter]income = 0
       Constraint 1 dropped
           chi2(  3) =    37.70
         Prob > chi2 =     0.0000
```

Income is clearly highly statistically significant. An asymptotically equivalent alternative test procedure is to use the `lrtest` command (see section 12.4.2), which requires additionally fitting the null hypothesis model that excludes income as a regressor. In this case, with just one regressor, this coincides with the overall test `LR chi2(3)=41.14` reported in the output header.

15.4.3 Coefficient interpretation

Coefficients in a multinomial model can be interpreted in the same way as binary logit model parameters are interpreted, with comparison being to the base category.

This is a result of the multinomial logit model being equivalent to a series of pairwise logit models. For simplicity, we set the base category to be the first category. Then the MNL model defined in (15.5) implies that

$$\Pr(y_i = j | y_i = j \text{ or } 1) = \frac{\Pr(y_i = j)}{\Pr(y_i = j) + \Pr(y_i = 1)} = \frac{\exp(\mathbf{x}_i'\boldsymbol{\beta}_j)}{1 + \exp(\mathbf{x}_i'\boldsymbol{\beta}_j)}$$

using $\boldsymbol{\beta}_1 = \mathbf{0}$ and cancellation of $\sum_{l=1}^{m} \exp(\mathbf{x}_i'\boldsymbol{\beta}_l)$ in the numerator and denominator.

Thus $\widehat{\boldsymbol{\beta}}_j$ can be viewed as parameters of a binary logit model between alternative j and alternative 1. So a positive coefficient from `mlogit` means that as the regressor increases, we are more likely to choose alternative j than alternative 1. This interpretation will vary with the base category and is clearly most useful when there is a natural base category.

Some researchers find it helpful to transform to odds ratios or relative-risk ratios, as in the binary logit case. The odds ratio or relative-risk ratio of choosing alternative j rather than alternative 1 is given by

$$\frac{\Pr(y_i = j)}{\Pr(y_i = 1)} = \exp(\mathbf{x}_i'\boldsymbol{\beta}_j) \tag{15.6}$$

so $e^{\beta_{jr}}$ gives the proportionate change in the relative risk of choosing alternative j rather than alternative 1 when x_{ir} changes by one unit.

The `rrr` option of `mlogit` provides coefficient estimates transformed to relative-risk ratios. We have

```
. * Relative-risk option reports exp(b) rather than b
. mlogit mode income, rr baseoutcome(1) nolog
Multinomial logistic regression              Number of obs   =        1182
                                             LR chi2(3)      =       41.14
                                             Prob > chi2     =      0.0000
Log likelihood = -1477.1506                  Pseudo R2       =      0.0137
```

mode	RRR	Std. Err.	z	P>|z|	[95% Conf. Interval]	
beach	(base outcome)					
pier						
income	.8664049	.0461693	-2.69	0.007	.7804799	.9617896
private						
income	1.096262	.0445781	2.26	0.024	1.012282	1.18721
charter						
income	.9688554	.040543	-0.76	0.450	.8925639	1.051668

Thus a one-unit increase in `income`, corresponding to a \$1,000 monthly increase, leads to relative odds of choosing to fish from a pier rather than the beach that are 0.866 times what they were before the change; so the relative odds have declined. The original coefficient of `income` for the alternative `pier` was -0.1434 and $e^{-0.1434} = 0.8664$.

15.4.4   Predicted probabilities

After most estimation commands, the `predict` command creates one variable. After `mlogit`, however, m variables are created, where m is the number of alternatives. Predicted probabilities for each alternative are obtained by using the `pr` option of `predict`.

Here we obtain four predicted probabilities because there are four alternatives. We have

```
. * Predict probabilities of choice of each mode and compare to actual freqs
. predict pmlogit1 pmlogit2 pmlogit3 pmlogit4, pr
. summarize pmlogit* dbeach dpier dprivate dcharter, separator(4)
```

Variable	Obs	Mean	Std. Dev.	Min	Max
pmlogit1	1182	.1133672	.0036716	.0947395	.1153659
pmlogit2	1182	.1505922	.0444575	.0356142	.2342903
pmlogit3	1182	.3536379	.0797714	.2396973	.625706
pmlogit4	1182	.3824027	.0346281	.2439403	.4158273
dbeach	1182	.1133672	.3171753	0	1
dpier	1182	.1505922	.3578023	0	1
dprivate	1182	.3536379	.4783008	0	1
dcharter	1182	.3824027	.4861799	0	1

Note that the sample average predicted probabilities equal the observed sample frequencies. This is always the case for MNL models that include an intercept, generalizing the similar result for binary logit models.

The ideal multinomial model will predict perfectly. For example, `p1` ideally would take on a value of 1 for the 134 observations with $y = 1$ and would take on a value of 0 for the remaining observations. Here `p1` ranges only from 0.0947 to 0.1154, so the model with income as the only explanatory variable predicts beach fishing very poorly. There is considerably more variation in predicted probabilities for the other three alternatives.

The `margins` command (see section 10.5.7) can be used to compute the average predicted probability of a given outcome, along with an associated confidence interval. For example, for the third outcome, we have

```
. * Sample average predicted probability of the third outcome
. margins, predict(outcome(3)) noatlegend
Predictive margins                               Number of obs   =      1182
Model VCE    : OIM

Expression   : Pr(mode==private), predict(outcome(3))
```

	Margin	Delta-method Std. Err.	z	P>|z|	[95% Conf. Interval]	
_cons	.3536379	.0137114	25.79	0.000	.326764	.3805118

15.4.5   MEs

For an unordered multinomial model, there is no single conditional mean of the dependent variable, y. Instead there are m alternatives, and we model the probabilities of these alternatives. Interest lies in how these probabilities change as regressors change.

For the MNL model, the MEs can be shown to be

$$\frac{\partial p_{ij}}{\partial \mathbf{x}_i} = p_{ij}(\boldsymbol{\beta}_j - \overline{\boldsymbol{\beta}}_i)$$

where $\overline{\boldsymbol{\beta}}_i = \sum_l p_{il}\boldsymbol{\beta}_l$ is a probability weighted average of the $\boldsymbol{\beta}_l$. The marginal effects vary with the point of evaluation, $\mathbf{x}_i$, because p_{ij} varies with $\mathbf{x}_i$. The signs of the regression coefficients do not give the signs of the MEs. For a variable x, the ME is positive if $\boldsymbol{\beta}_j > \overline{\boldsymbol{\beta}}_i$.

The `margins, dydx()` command calculates the ME at the mean (MEM) and the ME at representative values (MER), with separate computation for each alternative. For example, to obtain the ME on $\Pr(y = 3)$ of a change in income evaluated at the sample mean of regressors, we use the `predict(outcome(3))` and `atmean` options to obtain

```
. * Marginal effect at mean of income change for outcome 3
. margins, dydx(*) predict(outcome(3)) atmean noatlegend
Conditional marginal effects                      Number of obs   =       1182
Model VCE    : OIM

Expression   : Pr(mode==private), predict(outcome(3))
dy/dx w.r.t. : income
```

	dy/dx	Delta-method Std. Err.	z	P>|z|	[95% Conf. Interval]
income	.0325985	.005692	5.73	0.000	.0214424 .0437547

A one-unit change in `income`, equivalent to a \$1,000 increase in monthly income, increases by 0.033 the probability of fishing from a private boat rather than from a beach, pier, or charter boat.

To instead obtain the average marginal effect (AME), we drop the `atmean` option. We have

```
. * Average marginal effect of income change for outcome 3
. margins, dydx(*) predict(outcome(3)) noatlegend
Average marginal effects                          Number of obs   =       1182
Model VCE    : OIM

Expression   : Pr(mode==private), predict(outcome(3))
dy/dx w.r.t. : income
```

	dy/dx	Delta-method Std. Err.	z	P>|z|	[95% Conf. Interval]
income	.0317562	.0052589	6.04	0.000	.021449 .0420633

The AME and MEM are quite similar in this example. Usually, `mlogit` leads to much greater differences.

15.5 Conditional logit model

Some multinomial studies use richer datasets that include alternative-specific variables, such as prices and quality measures for all alternatives, not just the chosen alternative. Then the conditional logit model is used.

15.5.1 Creating long-form data from wide-form data

The parameters of conditional logit models are estimated with commands that require the data to be in long form, with one observation providing the data for just one alternative for an individual.

Some datasets will already be in long form, but that is not the case here. Instead, the `mus15data.dta` dataset is in wide form, with one observation containing data for all four alternatives for an individual. For example,

```
. * Data are in wide form
. list mode price pbeach ppier pprivate pcharter in 1, clean
           mode    price   pbeach    ppier  pprivate  pcharter
 1.     charter   182.93   157.93   157.93    157.93    182.93
```

The first observation has data for the price of all four alternatives. The chosen mode
was `charter`, so `price` was set to equal `pcharter`.

To convert data from wide form to long form, we use the `reshape` command, introduced in section 8.11. Here the long form will have four observations for each individual
according to whether the suffix is `beach`, `pier`, `private`, or `charter`. These suffixes
are strings, rather than the `reshape` command's default of numbers, so we use `reshape`
with the `string` option. For completeness, we actually provide the four suffixes.

```
. * Convert data from wide form to long form
. generate id = _n
. reshape long d p q, i(id) j(fishmode beach pier private charter) string
Data                                wide   ->   long

Number of obs.                      1182   ->   4728
Number of variables                   22   ->     14
j variable (4 values)                      ->   fishmode
xij variables:
                dbeach dpier ... dcharter   ->   d
                pbeach ppier ... pcharter   ->   p
                qbeach qpier ... qcharter   ->   q

. save mus15datalong.dta, replace
file mus15datalong.dta saved
```

There are now four observations for the first individual or case. If we had not provided
the four suffixes, the `reshape` command would have erroneously created a fifth alternative, `rice`, from `price` that like `pbeach`, `ppier`, `pprivate`, and `pcharter` also begins
with the letter `p`.

To view the resulting long-form data for the first individual case, we list the first
four observations.

```
. * List data for the first case after reshape
. list in 1/4, clean noobs
     id   fishmode       mode    price    crate       d        p        q      income
> _est_MNL   pmlogit1   pmlogit2   pmlogit3   pmlogit4
      1      beach    charter   182.93    .5391       0   157.93    .0678   7.083332
>          1   .1125092   .0919656   .4516733   .3438518
      1    charter    charter   182.93    .5391       1   182.93    .5391   7.083332
>          1   .1125092   .0919656   .4516733   .3438518
      1       pier    charter   182.93    .5391       0   157.93    .0503   7.083332
>          1   .1125092   .0919656   .4516733   .3438518
      1    private    charter   182.93    .5391       0   157.93    .2601   7.083332
>          1   .1125092   .0919656   .4516733   .3438518
```

The order is no longer pier, beach, private boat, and then charter boat. Instead, it is now
beach, charter boat, pier, and then private boat, because the observations are sorted in
the alphabetical order of `fishmode`. For this first observation, the outcome variable, `d`,

equals 1 for charter boat fishing, as expected. The four separate observations on the alternative-specific variables, p and q, are the different values for price and quality for the four alternatives.

All case-specific variables appear as a single variable that takes on the same value for the four outcomes. For income, this is no problem. But the mode, price, and crate are misleading here. The mode variable indicates that for case 1 the fishing mode was mode=4, because in original wide form this corresponded to charter boat fishing. But d=1 for the second observation of the first case because this corresponds to charter boat fishing in the reordered long form. It would be best to simply drop the misleading variables by typing drop mode price crate, because these variables are not needed.

15.5.2 The asclogit command

When some or all regressors are alternative specific, the conditional logit (CL) model is used. The CL model specifies that

$$p_{ij} = \frac{\exp(\mathbf{x}_{ij}'\boldsymbol{\beta} + \mathbf{z}_i'\boldsymbol{\gamma}_j)}{\sum_{l=1}^{m} \exp(\mathbf{x}_{il}'\boldsymbol{\beta} + \mathbf{z}_i'\boldsymbol{\gamma}_l)}, \qquad j = 1, \ldots, m \tag{15.7}$$

where $\mathbf{x}_{ij}$ are alternative-specific regressors and $\mathbf{z}_i$ are case-specific regressors. To ensure model identification, one of the $\boldsymbol{\gamma}_j$ is set to zero, as for the MNL model. Some authors call the model above a mixed logit model, with conditional logit used to refer to a more restrictive model that has only alternative-specific regressors.

The asclogit command, an acronym for alternative-specific conditional logit, has the syntax

asclogit *depvar* [*indepvars*] [*if*] [*in*] [*weight*] , case(*varname*)

    alternatives(*varname*) [*options*]

where *indepvars* are the alternative-specific regressors, case(*varname*) provides the identifier for each case or individual, and alternatives(*varname*) provides the possible alternatives.

The casevars(*varlist*) option is used to provide the names of any case-specific regressors. The basealternative() option specifies the alternative that is to be used as the base category, which affects only the coefficients of case-specific regressors. The altwise option deletes only the data for an alternative, rather than the entire observation, if data are missing.

The noconstant option overrides the Stata default of including case-specific intercepts. Attributes of each alternative are then explained solely by alternative-specific regressors if noconstant is used. The case-specific intercepts provided by the default estimator are interpreted as reflecting the desirability of each alternative because of unmeasured attributes of the alternative.

The asclogit command allows the choice set to vary across individuals and more than one alternative to be selected.

15.5.3   The clogit command

The conditional logit model can also be fit by using the `clogit` command, yielding the same results. The `clogit` command is designed for grouped data used in matched case–control group studies and is similar to the `xtlogit` command used for panel data grouped over time for an individual.

The `clogit` command does not have an option for case-specific variables. Instead a case-specific variable is interacted with dummies for $m-1$ alternatives, and the $m-1$ variables are entered as regressors. This is illustrated in section 15.8.3, where the same data transformations are needed for the user-written `mixlogit` command. For applications such as the one studied in this chapter, `asclogit` is easier to use than `clogit`.

15.5.4   Application of the asclogit command

We estimate the parameters of the CL model to explain fishing-mode choice given alternative-specific regressors on price and quality; the case-specific regressor, `income`; and case-specific intercepts. As for the MNL model, beach fishing is set to be the base category. We have

```
. * Conditional logit with alternative-specific and case-specific regressors
. asclogit d p q, case(id) alternatives(fishmode) casevars(income)
> basealternative(beach) nolog

Alternative-specific conditional logit          Number of obs      =        4728
Case variable: id                               Number of cases    =        1182

Alternative variable: fishmode                  Alts per case: min =           4
                                                               avg =         4.0
                                                               max =           4

                                                Wald chi2(5)       =      252.98
Log likelihood = -1215.1376                     Prob > chi2        =      0.0000
```

d	Coef.	Std. Err.	z	P>|z|	[95% Conf. Interval]	
fishmode						
p	-.0251166	.0017317	-14.50	0.000	-.0285106	-.0217225
q	.357782	.1097733	3.26	0.001	.1426302	.5729337
beach	(base alternative)					
charter						
income	-.0332917	.0503409	-0.66	0.508	-.131958	.0653745
_cons	1.694366	.2240506	7.56	0.000	1.255235	2.133497
pier						
income	-.1275771	.0506395	-2.52	0.012	-.2268288	-.0283255
_cons	.7779593	.2204939	3.53	0.000	.3457992	1.210119
private						
income	.0894398	.0500671	1.79	0.074	-.0086898	.1875694
_cons	.5272788	.2227927	2.37	0.018	.0906132	.9639444

The first set of estimates are the coefficients $\widehat{\boldsymbol{\beta}}$ for the alternative-specific regressors price and quality. The next three sets of estimates are for the case-specific intercepts and regressors. The coefficients are, respectively, $\widehat{\boldsymbol{\gamma}}_{\text{charter}}$, $\widehat{\boldsymbol{\gamma}}_{\text{pier}}$, and $\widehat{\boldsymbol{\gamma}}_{\text{private}}$, because we used the normalization $\boldsymbol{\gamma}_{\text{beach}} = \mathbf{0}$.

The output header does not give the pseudo-R^2, but this can be computed by using the formula given in section 10.7.1. Here $\ln L_{\text{fit}} = -1215.1$, and estimation of an intercepts-only model yields $\ln L_0 = -1497.7$, so $\widetilde{R}^2 = 1-(-1215.1)/(-1497.7) = 0.189$, much higher than the 0.014 for the MNL model in section 15.4.2. The regressors p, q, and `income` are highly jointly statistically significant with `Wald chi2(5)=253`. The `test` command can be used for individual Wald tests, or the `lrtest` command can be used for likelihood-ratio (LR) tests.

The CL model in this section reduces to the MNL model in section 15.4.2 if $\beta_{\text{p}} = 0$ and $\beta_{\text{q}} = 0$. Using either a Wald test or a LR test, this hypothesis is strongly rejected, and the CL model is the preferred model.

15.5.5 Relationship to multinomial logit model

The MNL and CL models are essentially equivalent. The `mlogit` command is designed for case-specific regressors and data in wide form. The `asclogit` command is designed for alternative-specific regressors and data in long form.

The parameters of the MNL model can be estimated by using `asclogit` as the special case with no alternative-specific regressors. Thus

```
. * MNL is CL with no alternative-specific regressors
. asclogit d, case(id) alternatives(fishmode) casevars(income)
> basealternative(beach)
  (output omitted)
```

yields the same estimates as the earlier `mlogit` command. When all regressors are case specific, it is easiest to use `mlogit` with data in wide form.

Going the other way, it is possible to estimate the parameters of a CL model using `mlogit`. This is more difficult because it requires transforming alternative-specific regressors to deviations from the base category and then imposing parameter-equality constraints. For CL models, `asclogit` is much easier to use than `mlogit`.

15.5.6 Coefficient interpretation

Coefficients of alternative-specific regressors are easily interpreted. The alternative-specific regressor can be denoted by x_r with the coefficient β_r. The effect of a change in x_{rik}, which is the value of x_r for individual i and alternative k, is

$$\frac{\partial p_{ij}}{\partial x_{rik}} = \begin{cases} p_{ij}(1 - p_{ij})\beta_r & j = k \\ -p_{ij}p_{ik}\beta_r & j \neq k \end{cases} \quad (15.8)$$

If $\beta_r > 0$, then the own-effect is positive because $p_{ij}(1 - p_{ij})\beta_r > 0$, and the cross-effect is negative because $-p_{ij}p_{ik}\beta_r < 0$. So a positive coefficient means that if the regressor increases for one category, then that category is chosen more and other categories are chosen less; vice versa for a negative coefficient. Here the negative price coefficient of -0.025 means that if the price of one mode of fishing increases, then demand for that mode decreases and demand for all other modes increases, as expected. For catch rate, the positive coefficient of 0.36 means a higher catch rate for one mode of fishing increases the demand for that mode and decreases the demand for the other modes.

Coefficients of case-specific regressors are interpreted as parameters of a binary logit model against the base category; see section 15.4.3 for the MNL model. The income coefficients of -0.033, -0.128, and 0.089 mean that, relative to the probability of beach fishing, an increase in income leads to a decrease in the probability of charter boat and pier fishing, and an increase in the probability of private boat fishing.

15.5.7 Predicted probabilities

Predicted probabilities can be obtained using the **predict** command with the **pr** option. This provides a predicted probability for each observation, where an observation is one alternative for one individual because the data are in long form.

To obtain predicted probabilities for each of the four alternatives, we need to summarize by **fishmode**. We use the **table** command because this gives condensed output. Much lengthier output is obtained by instead using the **bysort fishmode: summarize** command. We have

```
. * Predicted probabilities of choice of each mode and compare to actual freqs
. predict pasclogit, pr

. table fishmode, contents(mean d mean pasclogit sd pasclogit) cellwidth(15)
```

fishmode	mean(d)	mean(pasclogit)	sd(pasclogit)
beach	.1133672	.1133672	.1285042
charter	.3824027	.3824027	.1565869
pier	.1505922	.1505922	.1613722
private	.3536379	.3536379	.1664636

As for MNL, the sample average predicted probabilities are equal to the sample probabilities. The standard deviations of the CL model predicted probabilities (all in excess of 0.10) are much larger than those for the MNL model, so the CL model predicts better. A summary is also provided by the **estat alternatives** command.

A quite different predicted probability is that of a new alternative. This is possible for the conditional logit model if the parameters of that model are estimated using only alternative-specific regressors, which requires use of the **noconstant** option so that case-specific intercepts are not included, and the values of these regressors are known for the new category.

For example, we may want to predict the use of a new mode of fishing that has a much higher catch rate than the currently available modes but at the same time has a considerably higher price. The parameters, β, in (15.7) are estimated with m alternatives, and then predicted probabilities are computed by using (15.7) with $m + 1$ alternatives.

15.5.8   MEs

The MEM and MER are computed by using the postestimation `estat mfx` command, rather than the usual `margins` command. Options for this command include `varlist()` to compute the marginal effects for a subset of the regressors.

We compute the MEM for just the regressor `price`. We obtain

```
. * Marginal effect at mean of change in price
. estat mfx, varlist(p)
Pr(choice = beach|1 selected) = .05248806
```

variable	dp/dx	Std. Err.	z	P>|z|	[95% C.I.]		X
p							
beach	-.001249	.000121	-10.29	0.000	-.001487	-.001011	103.42
charter	.000609	.000061	9.97	0.000	.000489	.000729	84.379
pier	.000087	.000016	5.42	0.000	.000055	.000118	103.42
private	.000553	.000056	9.88	0.000	.000443	.000663	55.257

```
Pr(choice = charter|1 selected) = .46206853
```

variable	dp/dx	Std. Err.	z	P>|z|	[95% C.I.]		X
p							
beach	.000609	.000061	9.97	0.000	.000489	.000729	103.42
charter	-.006243	.000441	-14.15	0.000	-.007108	-.005378	84.379
pier	.000764	.000071	10.69	0.000	.000624	.000904	103.42
private	.00487	.000452	10.77	0.000	.003983	.005756	55.257

```
Pr(choice = pier|1 selected) = .06584968
```

variable	dp/dx	Std. Err.	z	P>|z|	[95% C.I.]		X
p							
beach	.000087	.000016	5.42	0.000	.000055	.000118	103.42
charter	.000764	.000071	10.69	0.000	.000624	.000904	84.379
pier	-.001545	.000138	-11.16	0.000	-.001816	-.001274	103.42
private	.000694	.000066	10.58	0.000	.000565	.000822	55.257

```
Pr(choice = private|1 selected) = .41959373
```

variable	dp/dx	Std. Err.	z	P>|z|	[	95% C.I.	]	X
p								
beach	.000553	.000056	9.88	0.000	.000443	.000663		103.42
charter	.00487	.000452	10.77	0.000	.003983	.005756		84.379
pier	.000694	.000066	10.58	0.000	.000565	.000822		103.42
private	-.006117	.000444	-13.77	0.000	-.006987	-.005246		55.257

There are 16 MEs in all, corresponding to probabilities of four alternatives times prices for each of the four alternatives. All own-effects are negative and all cross-effects are positive, as explained in section 15.5.6. The header for the first section of the `estat mfx` output gives $p_{11} = \Pr(\text{choice} = \text{beach}|\text{one choice is selected}) = 0.0525$. Using (15.8) and the estimated coefficient of -0.0251, we can estimate the own-effect as $0.0525 \times 0.9475 \times (-0.0251) = -0.001249$, which is the first ME given in the output. This means that a \$1 increase in the price of beach fishing decreases the probability of beach fishing by 0.001249, for a fictional observation with `p`, `q`, and `income` set to sample mean values. The second value of 0.000609 means that a \$1 increase in the price of charter boat fishing increases beach fishing probability by 0.000609, and so on.

The AME cannot be computed with the `margins` command, or the user-written `margeff` command, because these commands do not apply to `asclogit`. Instead, we can compute AME manually, as in section 10.6.9. We do so for a change of beach price only. We obtain

```
. * Alternative-specific example: AME of beach price change computed manually
. preserve
. quietly summarize p
. generate delta = r(sd)/1000
. quietly replace p = p + delta if fishmode == "beach"
. predict pnew, pr
. generate dpdbeach = (pnew - pasclogit)/delta
. tabulate fishmode, summarize(dpdbeach)
```

fishmode	Summary of dpdbeach Mean	Std. Dev.	Freq.
beach	-.00210891	.00195279	1182
charter	.00064641	.00050529	1182
pier	.00090712	.00154869	1182
private	.00055537	.00047725	1182
Total	-9.295e-10	.00178105	4728

```
. restore
```

Only one variable is generated, but this gives four AMEs corresponding to each of the alternatives, similar to the earlier discussion of predicted probabilities. As expected, increasing the price of beach fishing decreases the probability of beach fishing and increases the probability of using any of the other modes of fishing. The AME values compare with MEM values of, respectively, -0.001249, 0.000609, 0.000087, and 0.000553, so the

ME estimates differ substantially for the probability of beach fishing and the probability of pier fishing.

15.6 Nested logit model

The MNL and CL models are the most commonly used multinomial models, especially in other branches of applied statistics. However, in microeconometrics applications that involve individual choice, the models are viewed as placing restrictions on individual decision-making that are unrealistic, as explained below.

The simplest generalization is a nested logit (NL) model. Two variants of the NL model are used. The preferred variant is one based on the ARUM. This is the model we present and is the default model for Stata 10. A second variant was used by most packages in the past, including Stata 9. Both variants have MNL and CL as special cases, and both ensure that multinomial probabilities lie between 0 and 1 and sum to 1. But the variant based on ARUM is preferred because it is consistent with utility maximization.

15.6.1 Relaxing the independence of irrelevant alternatives assumption

The MNL and CL models impose the restriction that the choice between any two pairs of alternatives is simply a binary logit model; see (15.6). This assumption, called the independence of irrelevant alternatives (IIA) assumption, can be too restrictive, as illustrated by the "red bus/blue bus" problem. Suppose commute-mode alternatives are car, blue bus, or red bus. The IIA assumption is that the probability of commuting by car, given commute by either car or red bus, is independent of whether commuting by blue bus is an option. But the introduction of a blue bus, same as a red bus in every aspect except color, should have little impact on car use and should halve use of red bus, leading to an increase in the conditional probability of car use given commute by car or red bus.

This limitation has led to alternative richer models for unordered choice based on the ARUM introduced in section 15.2.4. The MNL and CL models can be shown to arise from the ARUM if the errors, ε_{ij}, in (15.3) are independent and identically distributed as type I extreme value. Instead, in the red bus/blue bus example, we expect the blue bus error, ε_{i2}, to be highly correlated with the red bus error, ε_{i3}, because if we overpredict the red bus utility given the regressors, then we will also overpredict the blue bus utility.

More general multinomial models, presented in this and subsequent sections, allow for correlated errors. The NL is the most tractable of these models.

15.6.2 NL model

The NL model requires that a nesting structure be specified that splits the alternatives into groups, where errors in the ARUM are correlated within group but are uncorrelated across groups. We specify a two-level NL model, though additional levels of nesting can be accommodated, and assume a fundamental distinction between shore and boat fishing. The tree is

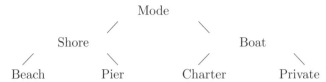

The shore/boat contrast is called level 1 (or a limb), and the next level is called level 2 (or a branch). The tree can be viewed as a decision tree—first decide whether to fish from shore or boat, and then decide between beach and pier (if shore) or between charter and private (if boat). But this interpretation of the tree is not necessary. The key is that the NL model permits correlation of errors within each of the level-2 groupings. Here $(\varepsilon_{i,\text{beach}}, \varepsilon_{i,\text{pier}})$ are a bivariate correlated pair, $(\varepsilon_{i,\text{private}}, \varepsilon_{i,\text{charter}})$ are a bivariate correlated pair, and the two pairs are independent. The CL model is the special case where all errors are independent.

More generally, denote alternatives by subscripts (j, k), where j denotes the limb (level 1) and k denotes the branch (level 2) within the limb, and different limbs can have different numbers of branches, including just one branch. For example, $(2, 3)$ denotes the third alternative in the second limb. The two-level random utility is defined to be

$$U_{jk} + \varepsilon_{jk} = \mathbf{z}_j'\boldsymbol{\alpha} + \mathbf{x}_{jk}'\boldsymbol{\beta}_j + \varepsilon_{jk}, \quad j = 1, \ldots, J, \quad k = 1, \ldots, K_j$$

where $\mathbf{z}_j$ varies over limbs only and $\mathbf{x}_{jk}$ varies over both limbs and branches. For ease of exposition, we have suppressed the individual subscript i, and we consider only alternative-specific regressors. (If all regressors are instead case specific, then we have $\mathbf{z}'\boldsymbol{\alpha}_j + \mathbf{x}'\boldsymbol{\beta}_{jk} + \varepsilon_{jk}$ with one of the $\boldsymbol{\beta}_{jk} = \mathbf{0}$.) The NL model assumes that $(\varepsilon_{j1}, \ldots, \varepsilon_{jK})$ are distributed as Gumbel's multivariate extreme-value distribution. Then the probability that alternative (j, k) is chosen equals

$$p_{jk} = p_j \times p_{k|j} = \frac{\exp\left(\mathbf{z}_j'\boldsymbol{\alpha} + \tau_j I_j\right)}{\sum_{m=1}^{J} \exp\left(\mathbf{z}_m'\boldsymbol{\alpha} + \tau_m I_m\right)} \times \frac{\exp\left(\mathbf{x}_{jk}'\boldsymbol{\beta}_j/\tau_j\right)}{\sum_{l=1}^{K_j} \exp\left(\mathbf{x}_{jl}'\boldsymbol{\beta}_j/\tau_j\right)}$$

where $I_j = \ln\left\{\sum_{l=1}^{K_j} \exp(\mathbf{x}_{jl}'\boldsymbol{\beta}_j/\tau_j)\right\}$ is called the inclusive value or the log sum. The NL probabilities are the product of probabilities p_i and $p_{k|j}$ that are essentially of CL form. The model produces positive probabilities that sum to one for any value of τ_j, called dissimilarity parameters. But the ARUM restricts $0 \leq \tau_j \leq 1$, and values outside this range mean the model, while mathematically correct, is inconsistent with random-utility theory.

15.6.3 The nlogit command

The Stata commands for NL have complicated syntax that we briefly summarize. It is simplest to look at the specific application in this section, and see [R] **nlogit** for further details.

The first step is to specify the tree structure. The `nlogitgen` command has the syntax

`nlogitgen` *newaltvar* = *altvar* (*branchlist*) [, `nolog`]

The *altvar* variable is the original variable defining the possible alternatives, and *newaltvar* is a created variable necessary for `nlogit` to know what nesting structure should be used. Here *branchlist* is

branch, *branch* [, *branch* ...]

and *branch* is

[*label*:] *alternative* [| *alternative* [| *alternative* ...]]

There must be at least two branches, and each branch has one or more alternatives.

The nesting structure can be displayed by using the `nlogittree` command with the syntax

`nlogittree` *altvarlist* [*if*] [*in*] [*weight*] [, *options*]

A useful option is `choice`(*depvar*), which lists sample frequencies for each alternative.

Estimation of model parameters uses the `nlogit` command with the syntax

`nlogit` *depvar* [*indepvars*] [*if*] [*in*] [*weight*] [|| *lev1_equation* [|| *lev2_equation* ...]] || *altvar*: [*byaltvarlist*] `case`(*varname*), [*options*]

where *indepvars* are the alternative-specific regressors and case-specific regressors are introduced in *lev#_equation*. The syntax of *lev#_equation* is

altvar: [*byaltvarlist*] [, `base`(#|*lbl*) `estconst`]

`case`(*varname*) provides the identifier for each case (individual).

The NL commands use data in long form, as did `asclogit`.

15.6.4   Model estimates

We first define the nesting structure by using the `nlogitgen` command. Here we define a variable, `type`, that is called `shore` for the `pier` and `beach` alternatives and is called `boat` for the `private` and `charter` alternatives.

```
. * Define the tree for nested logit
. nlogitgen type = fishmode(shore: pier | beach, boat: private | charter)
new variable type is generated with 2 groups
label list lb_type
lb_type:
            1 shore
            2 boat
```

The tree can be checked by using the **nlogittree** command. We have

```
. * Check the tree
. nlogittree fishmode type, choice(d)
tree structure specified for the nested logit model

type    N       fishmode  N     k

shore  2364  ┬─ beach     1182  134
             └─ pier      1182  178
boat   2364  ┬─ charter   1182  452
             └─ private   1182  418

            total   4728 1182
```
k = number of times alternative is chosen
N = number of observations at each level

The tree is as desired, so we are now ready to estimate with **nlogit**. First, list the dependent variable and the alternative-specific regressors. Then define the level-1 equation for **type**, which here includes no regressors. Finally, define the level-2 equations that here have the regressors **income** and an intercept. We use the **notree** option, which suppresses the tree, because it was already output with the **nlogittree** command. We have

```
. * Nested logit model estimate
. nlogit d p q || type:, base(shore) || fishmode: income, case(id) notree nolog
RUM-consistent nested logit regression      Number of obs      =        4728
Case variable: id                           Number of cases    =        1182

Alternative variable: fishmode              Alts per case: min =           4
                                                           avg =         4.0
                                                           max =           4

                                            Wald chi2(5)       =      212.37
Log likelihood = -1192.4236                 Prob > chi2        =      0.0000
```

d	Coef.	Std. Err.	z	P>|z|	[95% Conf. Interval]	
fishmode						
p	-.0267625	.0018937	-14.13	0.000	-.0304741	-.023051
q	1.340091	.3080519	4.35	0.000	.7363199	1.943861

fishmode equations

	Coef.	Std. Err.	z	P>|z|	[95% Conf. Interval]	
beach						
income	(base)					
_cons	(base)					
charter						
income	-8.40204	78.35484	-0.11	0.915	-161.9747	145.1706
_cons	69.96985	558.8914	0.13	0.900	-1025.437	1165.377
pier						
income	-9.458105	80.30173	-0.12	0.906	-166.8466	147.9304
_cons	58.94372	500.7334	0.12	0.906	-922.4757	1040.363
private						
income	-1.634919	8.588459	-0.19	0.849	-18.46799	15.19815
_cons	37.52542	230.9007	0.16	0.871	-415.0317	490.0825

dissimilarity parameters

	Coef.	Std. Err.	z	P>|z|	[95% Conf. Interval]	
type						
/shore_tau	83.46915	718.5287			-1324.821	1491.76
/boat_tau	52.55972	542.8935			-1011.492	1116.611

```
LR test for IIA (tau = 1):          chi2(2) =     45.43   Prob > chi2 = 0.0000
```

The coefficient of variable p is little changed compared with the CL model, but the other coefficients changed considerably.

The NL model reduces to the CL model if the two dissimilarity parameters are both equal to 1. The bottom of the output includes a LR test statistic of this restriction that leads to strong rejection of CL in favor of NL. However, the dissimilarity parameters are much greater than 1. This is not an unusual finding for NL models; it means that while the model is mathematically correct, with probabilities between 0 and 1 that add up to 1, the fitted model is not consistent with the ARUM.

15.6.5 Predicted probabilities

The `predict` command with the `pr` option provides predicted probabilities for level 1, level 2, and so on. Here there are two levels. The first-level probabilities are for `shore` or `boat`. The second-level probabilities are for each of the four alternatives. We have

```
. * Predict level 1 and level 2 probabilities from NL model
. predict plevel1 plevel2, pr
. tabulate fishmode, summarize(plevel2)
```

fishmode	Summary of Pr(fishmode alternatives)		
	Mean	Std. Dev.	Freq.
beach	.11323521	.1333593	1182
charter	.38070949	.15724226	1182
pier	.15072734	.16982064	1182
private	.35532796	.16444334	1182
Total	.25	.19690015	4728

The average predicted probabilities for NL no longer equal the sample probabilities, but they are quite close. The variation in the predicted probabilities, as measured by the standard deviation, is essentially the same as that for the CL model predictions, given in section 15.5.7.

15.6.6 MEs

The `mfx` and `margins` commands, and the user-written `margeff` command, are not available after `nlogit`.

Instead, we compute the AMEs manually, similar to section 15.5.8 for the CL model. We obtain

```
. * AME of beach price change computed manually
. preserve
. quietly summarize p
. generate delta = r(sd)/1000
. quietly replace p = p + delta if fishmode == "beach"
. predict pnew1 pnew2, pr
. generate dpdbeach = (pnew2 - plevel2)/delta
. tabulate fishmode, summarize(dpdbeach)
```

fishmode	Summary of dpdbeach		
	Mean	Std. Dev.	Freq.
beach	-.00053326	.0004792	1182
charter	.00063591	.00054938	1182
pier	-.00065944	.00057603	1182
private	.00055682	.00051133	1182
Total	8.815e-09	.00079968	4728

```
. restore
```

Compared with the CL model, there is little change in the ME of beach price change on the probability of charter and private boat fishing. But now, surprisingly, the probability of pier fishing falls in addition to the probability of beach fishing.

15.6.7   Comparison of logit models

The following table summarizes key output from fitting the preceding MNL, CL, and NL models. We have

```
. * Summary statistics for the logit models
. estimates table MNL CL NL, keep(p q) stats(N ll aic bic) equation(1) b(%7.3f)
> stfmt(%7.0f)
```

Variable	MNL	CL	NL
p		-0.025	-0.027
q		0.358	1.340
N	1182	4728	4728
ll	-1477	-1215	-1192
aic	2966	2446	2405
bic	2997	2498	2469

The information criteria, AIC and BIC, are presented in section 10.7.2; lower values are preferred. MNL is least preferred, and NL is most preferred.

In this example, the three multinomial models are actually nested, so we can choose between them by using LR tests. From the discussion of the CL and NL models, NL is again preferred to CL, which in turn is preferred to MNL.

All three models use the same amount of data. The CL and NL model entries have an N that is four times that for MNL because they use data in long form, leading to four "observations" per individual.

15.7   Multinomial probit model

The multinomial probit (MNP) model, like the NL model, allows relaxation of the IIA assumption. It has the advantage of allowing a much more flexible pattern of error correlation and does not require the specification of a nesting structure.

15.7.1   MNP

The MNP is obtained from the ARUM of section 15.2.4 by assuming normally distributed errors.

For the ARUM, the utility of alternative j is

$$U_{ij} = \mathbf{x}'_{ij}\boldsymbol{\beta} + \mathbf{z}'_i\boldsymbol{\gamma}_j + \varepsilon_{ij}$$

where the errors are assumed to be normally distributed, with $\boldsymbol{\varepsilon} \sim N(\mathbf{0}, \Sigma)$ where $\boldsymbol{\varepsilon} = (\varepsilon_{i1}, \ldots, \varepsilon_{im})$.

Then from (15.4), the probability that alternative j is chosen equals

$$p_{ij} = \Pr(y_i = j) = \Pr\{\varepsilon_{ik} - \varepsilon_{ij} \leq (\mathbf{x}_{ij} - \mathbf{x}_{ik})'\boldsymbol{\beta} + \mathbf{z}_i'(\boldsymbol{\gamma}_j - \boldsymbol{\gamma}_k)\}, \quad \text{for all } k \quad (15.9)$$

This is an $(m-1)$-dimensional integral for which there is no closed-form solution and computation is difficult. This problem did not arise for the preceding logit models because for those models the distribution of $\boldsymbol{\varepsilon}$ is such that (15.9) has a closed-form solution.

When there are few alternatives, say three or four, or when $\Sigma = \sigma^2 \mathbf{I}$, quadrature methods can be used to numerically compute the integral. Otherwise, maximum simulated likelihood, discussed below, is used.

Regardless of the method used, not all $(m+1)m/2$ distinct entries in the error variance matrix, Σ, are identified. From (15.9), the model is defined for $m-1$ error differences $(\varepsilon_{ik} - \varepsilon_{ij})$ with an $(m-1) \times (m-1)$ variance matrix that has $m(m-1)/2$ unique terms. Because a variance term also needs to be normalized, there are only $\{m(m-1)/2\} - 1$ unique terms in Σ. In practice, further restrictions are often placed on Σ, because otherwise Σ is imprecisely estimated, which can lead to imprecise estimation of $\boldsymbol{\beta}$ and $\boldsymbol{\gamma}$.

15.7.2   The mprobit command

The `mprobit` command is the analogue of `mlogit`. It applies to models with only case-specific regressors and assumes that the alternative errors are independent standard normal so that $\Sigma = \mathbf{I}$. Here the $(m-1)$-dimensional integral in (15.9) can be shown to reduce to a one-dimensional integral that can be approximated by using quadrature methods.

There is little reason to use the `mprobit` command because the model is qualitatively similar to MNL; `mprobit` assumes that alternative-specific errors in the ARUM are uncorrelated, but it is much more computationally burdensome. The syntax for `mprobit` is similar to that for `mlogit`. For a regression with the alternative-invariant regressor `income`, the command is

```
. * Multinomial probit with independent errors and alternative-invariant regressors
. mprobit mode income, baseoutcome(1)
  (output omitted )
```

The output is qualitatively similar to that from `mlogit`, though parameters estimates are scaled differently, as in the binary model case. The fitted log likelihood is $-1,477.8$, very close to the $-1,477.2$ for MNL (see section 15.4.2).

15.7.3   Maximum simulated likelihood

The multinomial log likelihood is given in (15.2), where $p_{ij} = F_j(\mathbf{x}_i, \boldsymbol{\theta})$ and the parameters $\boldsymbol{\theta}$ are $\boldsymbol{\beta}, \boldsymbol{\gamma}_1, \ldots, \boldsymbol{\gamma}_m$ (with one $\boldsymbol{\gamma}$ normalized to zero), and any unspecified entries in $\boldsymbol{\Sigma}$.

Because there is no closed-form solution for $F_j(\mathbf{x}_i, \boldsymbol{\theta})$ in (15.9), the log likelihood is approximated by a simulator, $\widetilde{F}_j(\mathbf{x}_i, \boldsymbol{\theta})$, that is based on S draws. A simple example is a frequency simulator that, given the current estimate $\widehat{\boldsymbol{\theta}}$, takes S draws of $\boldsymbol{\varepsilon}_i \sim N(\mathbf{0}, \widehat{\boldsymbol{\Sigma}})$ and lets $\widetilde{F}_j(\mathbf{x}_i, \boldsymbol{\theta})$ be the proportion of the S draws for which $\varepsilon_{ik} - \varepsilon_{ij} \leq (\mathbf{x}_{ij} - \mathbf{x}_{ik})'\widehat{\boldsymbol{\beta}} + \mathbf{z}_i'(\widehat{\boldsymbol{\gamma}}_j - \widehat{\boldsymbol{\gamma}}_k)$ for all k. This simulator is inadequate, however, because it is very noisy for low-probability events, and for the MNP model, the frequency simulator is nonsmooth in $\boldsymbol{\beta}$ and $\boldsymbol{\gamma}_1, \ldots, \boldsymbol{\gamma}_m$ so that very small changes in these parameters may lead to no change in $\widetilde{F}_j(\mathbf{x}_i, \boldsymbol{\theta})$. Instead, the Geweke–Hajivassiliou–Keane (GHK) simulator—described, for example, in Train (2003)—is used.

The maximum simulated likelihood (MSL) estimator maximizes

$$\ln L(\boldsymbol{\theta}) = \sum_{i=1}^{N} \sum_{j=1}^{m} y_{ij} \ln \widetilde{F}_j(\mathbf{x}_i, \boldsymbol{\theta}) \tag{15.10}$$

The usual ML asymptotic theory applies, provided that both $S \to \infty$ and $N \to \infty$, and $\sqrt{N}/S \to 0$ so that the number of simulations increases at a rate faster than $\sqrt{N}$. Even though default standard errors are fine for a multinomial model, robust standard errors are numerically better when MSL is used.

The MSL estimator can, in principal, be applied to any estimation problem that entails an unknown integral. Some general results are the following: Smooth simulators should be used. Even then, some simulators are much better than others, but this is model specific. When random draws are used, they should be based on the same underlying uniform seed at each iteration, because otherwise the gradient method may fail to converge simply because of different random draws (called chatter). The number of simulations may be greatly reduced for a given level of accuracy by using antithetic draws, rather than independent draws, and by using quasirandom-number sequences such as Halton sequences rather than pseudorandom-uniform draws to generate uniform numbers. The benefits of using Halton and Hammersley rather than uniform draws is exposited in Drukker and Gates (2006). And to reduce the computational burden of gradient methods, it is best to at least use analytical first derivatives. For more explanation, see, for example, Train (2003) or Cameron and Trivedi (2005). The asmprobit command incorporates all these considerations to obtain the MSL estimator for the MNP model.

15.7.4   The asmprobit command

The asmprobit command requires data to be in long form, like the asclogit command, and it has similar syntax:

asmprobit *depvar* [*indepvars*] [*if*] [*in*] [*weight*], case(*varname*)

    alternatives(*varname*) [*options*]

Estimation takes a long time because estimation is by MSL.

Several of the command's options are used to specify the error variance matrix Σ. As already noted, at most $\{m(m-1)/2\} - 1$ unique terms in Σ are identified. The default identification method is to drop the row and column of Σ corresponding to the first alternative (except that Σ_{11} is normalized to 1) and to set $\Sigma_{22} = 1$. These defaults can be changed by using the basealternative() and scalealternative() options. The correlation() and stddev() options are used to place further structure on the remaining off-diagonal and diagonal entries of Σ. The correlation(unstructured) option places no structure, the correlation(exchangeable) option imposes equicorrelation, the correlation(independent) option sets $\Sigma_{jk} = 0$ for all $j \neq k$, and the correlation(pattern) and correlation(fixed) options allow manual specification of the structure. The stddev(homoskedastic) option imposes $\Sigma_{jj} = 1$, the stddev(heteroskedastic) option allows $\Sigma_{jj} \neq 1$, and the stddev(pattern) and stddev(fixed) options allow manual specification of any structure.

Other options allow variations in the MSL computations. The intpoints(S) option sets the number of draws S, where the default of S is $50m$ or $100m$ depending on intmethod(). The intmethod() option specifies whether the uniform numbers are from pseudorandom draws (intmethod(random)), are from a Halton sequence (intmethod(halton)), or are from a Hammersley sequence (intmethod(hammersley)), which is the default. The antithetics option specifies antithetic draws to be used. The intseed() option sets the random-number–generator seed if uniform random draws are used.

15.7.5 Application of the asmprobit command

For simplicity, we restricted attention to a choice between three alternatives: fishing from a pier, private boat, or charter boat. The most general model with unstructured correlation and heteroskedastic errors is used. We use the structural option because then the variance parameter estimates are reported for the $m \times m$ error variance matrix Σ rather than the $(m-1) \times (m-1)$ variance matrix of the difference in errors. We have

```
. * Multinomial probit wuth unstructured errors when charter is dropped
. use mus15datalong.dta, clear

. drop if fishmode=="charter" | mode == 4
(2538 observations deleted)

. asmprobit d p q, case(id) alternatives(fishmode) casevars(income)
> correlation(unstructured) structural vce(robust)
note: variable p has 106 cases that are not alternative-specific: there is no
      within-case variability

Iteration 0:   log simulated-pseudolikelihood =  -493.8207
Iteration 1:   log simulated-pseudolikelihood = -483.41654  (backed up)
Iteration 2:   log simulated-pseudolikelihood = -482.98783  (backed up)
```

```
Iteration 3:   log simulated-pseudolikelihood = -482.9415  (backed up)
Iteration 4:   log simulated-pseudolikelihood = -482.67112
Iteration 5:   log simulated-pseudolikelihood = -482.51402
Iteration 6:   log simulated-pseudolikelihood = -482.44493
Iteration 7:   log simulated-pseudolikelihood = -482.39599
Iteration 8:   log simulated-pseudolikelihood = -482.37574
Iteration 9:   log simulated-pseudolikelihood = -482.35251
Iteration 10:  log simulated-pseudolikelihood = -482.30752
Iteration 11:  log simulated-pseudolikelihood = -482.30473
Iteration 12:  log simulated-pseudolikelihood = -482.30184
Iteration 13:  log simulated-pseudolikelihood = -482.30137
Iteration 14:  log simulated-pseudolikelihood = -482.30128
Iteration 15:  log simulated-pseudolikelihood = -482.30128

Reparameterizing to correlation metric and refining estimates

Iteration 0:   log simulated-pseudolikelihood = -482.30128
Iteration 1:   log simulated-pseudolikelihood = -482.30128
```

Alternative-specific multinomial probit	Number of obs	=	2190
Case variable: id	Number of cases	=	730

Alternative variable: fishmode	Alts per case: min =	3
	avg =	3.0
	max =	3

Integration sequence:	Hammersley		
Integration points:	150	Wald chi2(4) =	12.97
Log simulated-pseudolikelihood = -482.30128		Prob > chi2 =	0.0114

(Std. Err. adjusted for clustering on id)

d	Coef.	Robust Std. Err.	z	P>|z|	[95% Conf. Interval]	
fishmode						
p	-.0233627	.0114346	-2.04	0.041	-.0457741	-.0009513
q	1.399925	.5395423	2.59	0.009	.3424418	2.457409
beach	(base alternative)					
pier						
income	-.097985	.0413117	-2.37	0.018	-.1789543	-.0170156
_cons	.7549123	.2013551	3.75	0.000	.3602636	1.149561
private						
income	.0413866	.0739083	0.56	0.575	-.103471	.1862443
_cons	.6602584	.2766473	2.39	0.017	.1180397	1.202477
/lnsigma3	.4051391	.5009809	0.81	0.419	-.5767654	1.387044
/atanhr3_2	.1757361	.2337267	0.75	0.452	-.2823598	.6338319
sigma1	1	(base alternative)				
sigma2	1	(scale alternative)				
sigma3	1.499511	.7512264			.5617123	4.002998
rho3_2	.173949	.2266545			-.2750878	.5606852

(fishmode=beach is the alternative normalizing location)
(fishmode=pier is the alternative normalizing scale)

As expected, utility is decreasing in price and increasing in quality (catch rate).

The base mode was automatically set to the first alternative, beach, so that the first row and column of Σ are set to 0, except $\Sigma_{11} = 1$. One additional variance restriction is needed, and here that is on the error variance of the second alternative, pier, with $\Sigma_{22} = 1$ (the alternative normalizing scale). With $m = 3$, there are $(3 \times 2)/2 - 1 = 2$ free entries in Σ: one error variance parameter, Σ_{33}, and one correlation, $\rho_{32} = \mathrm{Cor}(\varepsilon_{i3}, \varepsilon_{i2})$. The sigma3 output is $\sqrt{\Sigma_{33}}$, and the rho3_2 output is ρ_{32}.

The estat covariance and estat correlation commands list the complete estimated variance matrix, $\widehat{\Sigma}$, and the associated correlation matrix. We have

```
. * Show correlations and covariance
. estat correlation
```

	beach	pier	private
beach	1.0000		
pier	0.0000	1.0000	
private	0.0000	0.1739	1.0000

```
. estat covariance
```

	beach	pier	private
beach	1		
pier	0	1	
private	0	.2608385	2.248533

If instead the parameters of the model are estimated without the structural option, the same parameter estimates are obtained, aside from estimation error, but the covariances and correlation are given for the variance matrix of the bivariate distribution of $\varepsilon_{i2} - \varepsilon_{i1}$ and $\varepsilon_{i3} - \varepsilon_{i1}$.

15.7.6 Predicted probabilities and MEs

The postestimation predict command with the default pr option predicts p_{ij}, and MEs evaluated at the mean or at a representative value are obtained by using the estat mfx command. The commands are similar to those after asclogit; see sections 15.5.7 and 15.5.8.

15.8 Random-parameters logit

The random-parameters logit, or mixed logit model, relaxes the IIA assumption by allowing parameters in the CL model to be normally distributed or lognormally distributed. Here we estimate the parameters of the models by using individual-level data. Quite different estimation procedures are used if the data are grouped, such as market share data; see Berry (1994).

15.8.1   Random-parameters logit

The random-parameters logit (RPL) model, or mixed logit model, is obtained from the ARUM of section 15.2.4 by assuming that the errors ε_{ij} are type II extreme-value distributed, like for the CL model, and the parameters $\boldsymbol{\beta}$ and $\boldsymbol{\gamma}_j$, $j = 2,\ldots,m$, are normally distributed. Then the utility of alternative j is

$$
\begin{aligned}
U_{ij} &= \mathbf{x}_{ij}'\boldsymbol{\beta}_i + \mathbf{z}_i'\boldsymbol{\gamma}_{ji} + \varepsilon_{ij} \\
&= \mathbf{x}_{ij}'\boldsymbol{\beta} + \mathbf{z}_i'\boldsymbol{\gamma}_j + \mathbf{x}_{ij}'\mathbf{v}_i + \mathbf{z}_i'\mathbf{w}_{ji} + \varepsilon_{ij}
\end{aligned}
$$

where $\boldsymbol{\beta}_i = \boldsymbol{\beta} + \mathbf{v}_i$, $\mathbf{v}_i \sim N(\mathbf{0}, \Sigma_{\boldsymbol{\beta}})$ and $\boldsymbol{\gamma}_{ji} = \boldsymbol{\gamma}_j + \mathbf{w}_{ji}$, $\mathbf{w}_{ji} \sim N(\mathbf{0}, \Sigma_{\boldsymbol{\gamma}j})$. The combined error $(\mathbf{x}_{ij}'\mathbf{v}_i + \mathbf{z}_i'\mathbf{w}_{ji} + \varepsilon_{ij})$ is now correlated across alternatives, whereas the errors ε_{ij} alone were not.

Then conditional on the unobservables $\mathbf{v}_i$ and $\mathbf{w}_{ji}$, we have a CL model with

$$
p_{ij}|\mathbf{v}_i, \mathbf{w}_{ji} = \frac{\exp(\mathbf{x}_{ij}'\boldsymbol{\beta} + \mathbf{z}_i'\boldsymbol{\gamma}_j + \mathbf{x}_{ij}'\boldsymbol{\beta} + \mathbf{z}_i'\mathbf{w}_{ji})}{\sum_{l=1}^m \exp(\mathbf{x}_{il}'\boldsymbol{\beta} + \mathbf{z}_i'\boldsymbol{\gamma}_l + \mathbf{x}_{ij}'\mathbf{v}_i + \mathbf{z}_i'\mathbf{w}_{ji})}, \qquad j = 1,\ldots,m
$$

The MLE is based on p_{ij}, which also requires integrating out $\mathbf{v}_i$ and $\mathbf{w}_{ji}$, a high-dimensional integral.

The MSL estimator instead maximizes (15.10), where $\widetilde{F}_j(\mathbf{x}_i, \boldsymbol{\theta})$ is a simulator for p_{ij}. Here the frequency simulator that makes many draws of $\mathbf{v}_i$ and the $\mathbf{w}_{ji}$ from the normal given current estimates of $\Sigma_{\boldsymbol{\beta}}$ and $\Sigma_{\boldsymbol{\gamma}j}$ is a smooth simulator.

15.8.2   The mixlogit command

The user-written `mixlogit` command (Hole 2007) computes the MSL estimator. The syntax is

`mixlogit` *depvar* [*indepvars*] [*if*] [*in*] [*weight*]`, group(`*varname*`)`
    `rand(`*varlist*`)` [*options*]

which is similar to that for `clogit`, with `group()` used to identify each case or individual. Regressors with random coefficients are listed in `rand()`, and regressors with nonrandom coefficients are listed as *indepvars*.

The `ln(#)` option permits the last `#` variables in `rand()` to be lognormally distributed rather than normally distributed. The `corr` option permits parameters to be correlated; the default is that they are not. The estimator uses the Halton sequence with 50 draws after dropping the first 15 draws. The `nrep(#)` and `burn(#)` options change these defaults, and published results should use many more than 50 draws.

15.8.3 Data preparation for mixlogit

The `mixlogit` command is similar to `clogit`. Unlike `asclogit` and `asmprobit`, there is no option for case-specific regressors.

Instead, we need to manually create regressors for the intercepts and `income`. For case-specific regressors, a normalization is needed. We set $\gamma_{\text{pier}} = \mathbf{0}$ and construct three intercepts and interactions with `income`. We have

```
. * Data set up to include case-invariant regressors
. use mus15datalong.dta, clear
. generate dbeach = fishmode=="pier"
. generate dprivate = fishmode=="private"
. generate dcharter = fishmode=="charter"
. generate ybeach = dbeach*income
. generate yprivate = dprivate*income
. generate ycharter = dcharter*income
```

We next use `mixlogit`. If instead we used `clogit` with the same dependent variables and regressors, then the results would be the same as those from `asclogit` in section 15.5.4.

15.8.4 Application of the mixlogit command

We estimate the same three-choice model as that used in section 15.7.5 for the MNP model, with charter fishing dropped.

The parameters for `p` are specified to be random, using the `rand()` option. All other parameters are specified to be fixed and appear as *indepvars*, though we could, for instance, specify the parameters of the three income variables to also be random. We have

```
. * Mixed logit or random parameters logit estimation
. drop if fishmode=="charter" | mode == 4
(2538 observations deleted)
. mixlogit d q dbeach dprivate ybeach yprivate, group(id) rand(p)
Iteration 0:   log likelihood = -602.33584   (not concave)
Iteration 1:   log likelihood = -447.46013
Iteration 2:   log likelihood = -435.29806
Iteration 3:   log likelihood = -434.56105
Iteration 4:   log likelihood = -434.52856
Iteration 5:   log likelihood = -434.52844
Iteration 6:   log likelihood = -434.52844
```

```
        Mixed logit model                        Number of obs   =      2190
                                                  LR chi2(1)      =     64.57
        Log likelihood = -434.52844              Prob > chi2      =    0.0000
```

d	Coef.	Std. Err.	z	P>|z|	[95% Conf. Interval]	
Mean						
q	.7840088	.9147869	0.86	0.391	-1.008941	2.576958
dbeach	.7742955	.224233	3.45	0.001	.3348069	1.213784
dprivate	.5617395	.3158082	1.78	0.075	-.0572331	1.180712
ybeach	-.1199613	.0492249	-2.44	0.015	-.2164404	-.0234822
yprivate	.0518098	.0721527	0.72	0.473	-.0896068	.1932265
p	-.1069866	.0274475	-3.90	0.000	-.1607827	-.0531904
SD						
p	.0598364	.0191597	3.12	0.002	.022284	.0973888

There is considerable variation across individuals in the effect of price. The random coefficients have a mean of -0.107 and a standard deviation of 0.060, both statistically significant at the 0.05 level. The random-parameters logit model has a log likelihood of -435, substantially higher than the -467 for the CL model. The results of the CL model are not shown but can be obtained by using either the `asclogit` or `clogit` command. The random-parameters model is preferred.

If we want to constrain the effect to be negative, then we should define a variable, `negp`, equal to the negative of variable `p` and use `ln(1)` for lognormal. The subsequent results are for the mean and standard deviation of $\ln \beta_{negp}$ rather than β_{negp} or β_p. These can be converted by using the result that if $\ln \beta \sim N(\mu, \sigma^2)$, then $\beta \sim \{e^{\mu+\sigma^2/2}, e^{2\mu+\sigma^2}(e^{\sigma^2}-1)\}$.

15.9   Ordered outcome models

In some cases, categorical data are naturally ordered. An example is health status that is self-assessed as poor, fair, good, or excellent. The two standard models for such data are the ordered logit and ordered probit models.

15.9.1   Data summary

We use data from the Rand Health Insurance Experiment, described in greater detail in section 18.3. We use one year of this panel, so the data are cross-section data.

The ordered outcome we consider is health status that is, respectively, poor or fair ($y = 1$), good ($y = 2$), or excellent ($y = 3$). This variable needs to be constructed from several binary outcomes for each of the health statuses. The categories poor and fair are combined because only 1.5% of the sample report poor health. The data are constructed as follows:

```
. * Create multinomial ordered outcome variables takes values y = 1, 2, 3
. use mus18data.dta, clear
. quietly keep if year==2
. generate hlthpf = hlthp + hlthf
. generate hlthe = (1 - hlthpf - hlthg)
. quietly generate hlthstat = 1 if hlthpf == 1
. quietly replace hlthstat = 2 if hlthg == 1
. quietly replace hlthstat = 3 if hlthe == 1
. label variable hlthstat "health status"
. label define hsvalue 1 poor_or_fair 2 good 3 excellent
. label values hlthstat hsvalue
. tabulate hlthstat

      health |
      status |      Freq.     Percent        Cum.
-------------+-----------------------------------
poor_or_fair |        523        9.38        9.38
        good |      2,034       36.49       45.87
   excellent |      3,017       54.13      100.00
-------------+-----------------------------------
       Total |      5,574      100.00
```

Health status is poor or fair for roughly 10% of the sample, good for 35%, and excellent for 55%.

The regressors considered are age in years (`age`), log annual family income (`linc`), and number of chronic diseases (`ndisease`). Summary statistics are

```
. * Summarize dependent and explanatory variables
. summarize hlthstat age linc ndisease

    Variable |        Obs        Mean    Std. Dev.        Min        Max
-------------+--------------------------------------------------------
    hlthstat |       5574    2.447435     .659524          1          3
         age |       5574    25.57613    16.73011   .0253251   63.27515
        linc |       5574    8.696929    1.220592          0   10.28324
    ndisease |       5574    11.20526    6.788959          0       58.6
```

The sample is of children and adults but not the elderly.

15.9.2 Ordered outcomes

The ordered outcomes are modeled to arise sequentially as a latent variable, y^*, crosses progressively higher thresholds. In the current example, y^* is an unobserved measure of healthiness. For individual i, we specify

$$y_i^* = \mathbf{x}_i'\boldsymbol{\beta} + u_i$$

where a normalization is that the regressors $\mathbf{x}$ do not include an intercept. For very low y^*, health status is poor; for $y^* > \alpha_1$, health status improves to fair; for $y^* > \alpha_2$, it improves further to good; and so on if there were additional categories.

For an m-alternative ordered model, we define

$$y_i = j \quad \text{if } \alpha_{j-1} < y_i^* \le \alpha_j, \quad j = 1, \dots, m$$

where $\alpha_0 = -\infty$ and $\alpha_m = \infty$. Then

$$
\begin{aligned}
\Pr(y_i = j) &= \Pr(\alpha_{j-1} < y_i^* \le \alpha_j) \\
&= \Pr(\alpha_{j-1} < \mathbf{x}_i'\boldsymbol{\beta} + u_i \le \alpha_j) \\
&= \Pr(\alpha_{j-1} - \mathbf{x}_i'\boldsymbol{\beta} < u_i \le \alpha_j - \mathbf{x}_i'\boldsymbol{\beta}) \\
&= F(\alpha_j - \mathbf{x}_i'\boldsymbol{\beta}) - F(\alpha_{j-1} - \mathbf{x}_i'\boldsymbol{\beta})
\end{aligned}
$$

where F is the cumulative distribution function (c.d.f.) of u_i. The regression parameters, $\boldsymbol{\beta}$, and the $m - 1$ threshold parameters, $\alpha_1, \dots, \alpha_{m-1}$, are obtained by maximizing the log likelihood with $p_{ij} = \Pr(y_i = j)$ as defined above. Stata excludes an intercept from the regressors. If instead an intercept is estimated, then only $m - 2$ threshold parameters are identified.

For the ordered logit model, u is logistically distributed with $F(z) = e^z / (1 + e^z)$. For the ordered probit model, u is standard normally distributed with $F(\cdot) = \Phi(\cdot)$, the standard normal c.d.f.

The sign of the regression parameters, $\boldsymbol{\beta}$, can be immediately interpreted as determining whether the latent variable, y^*, increases with the regressor. If β_j is positive, then an increase in x_{ij} necessarily decreases the probability of being in the lowest category ($y_i = 1$) and increases the probability of being in the highest category ($y_i = m$).

15.9.3 Application of the ologit command

The parameters of the ordered logit model are estimated by using the `ologit` command, which has syntax essentially the same as `mlogit`:

`ologit` *depvar* [*indepvars*] [*if*] [*in*] [*weight*] [*, options*]

Application of this command yields

```
. * Ordered logit estimates
. ologit hlthstat age linc ndisease, nolog
Ordered logistic regression                 Number of obs   =       5574
                                            LR chi2(3)      =     740.39
                                            Prob > chi2     =     0.0000
Log likelihood = -4769.8525                 Pseudo R2       =     0.0720
```

hlthstat	Coef.	Std. Err.	z	P>|z|	[95% Conf. Interval]	
age	-.0292944	.001681	-17.43	0.000	-.0325891	-.0259996
linc	.2836537	.0231098	12.27	0.000	.2383593	.3289481
ndisease	-.0549905	.0040692	-13.51	0.000	-.0629661	-.047015
/cut1	-1.39598	.2061301			-1.799987	-.9919722
/cut2	.9513097	.2054301			.5486741	1.353945

The latent health-status variable is increasing in income and decreasing with age and number of chronic diseases, as expected. The regressors are highly statistically significant. The threshold parameters appear to be statistically significantly different from each other, so the three categories should not be collapsed into two categories.

15.9.4 Predicted probabilities

Predicted probabilities for each of the three outcomes can be obtained by using the `pr` option. For comparison, we also compute the sample frequencies of each outcome.

```
. * Calculate predicted probability that y=1, 2, or 3 for each person
. predict p1ologit p2ologit p3ologit, pr
. summarize hlthpf hlthg hlthe p1ologit p2ologit p3ologit, separator(0)
```

Variable	Obs	Mean	Std. Dev.	Min	Max
hlthpf	5574	.0938285	.2916161	0	1
hlthg	5574	.3649085	.4814477	0	1
hlthe	5574	.541263	.4983392	0	1
p1ologit	5574	.0946903	.0843148	.0233629	.859022
p2ologit	5574	.3651672	.0946158	.1255265	.5276064
p3ologit	5574	.5401425	.1640575	.0154515	.7999009

The average predicted probabilities are within 0.01 of the sample frequencies for each outcome.

15.9.5 MEs

The ME on the probability of choosing alternative j when regressor x_r changes is given by

$$\frac{\partial \Pr(y_i = j)}{\partial x_{ri}} = \{F'(\alpha_{j-1} - \mathbf{x}_i'\boldsymbol{\beta}) - F'(\alpha_j - \mathbf{x}_i'\boldsymbol{\beta})\}\beta_r$$

If one coefficient is twice as big as another, then so too is the size of the ME.

We use the `margins` command with the `atmean` option to obtain the ME evaluated at the mean, for the third outcome (health status excellent). We obtain

```
. * Marginal effect at mean for 3rd outcome (health status excellent)
. margins, dydx(*) predict(outcome(3)) atmean noatlegend
Conditional marginal effects                    Number of obs    =      5574
Model VCE     : OIM

Expression    : Pr(hlthstat==3), predict(outcome(3))
dy/dx w.r.t.  : age linc ndisease
```

	dy/dx	Delta-method Std. Err.	z	P>|z|	[95% Conf. Interval]
age	-.0072824	.0004179	-17.43	0.000	-.0081014 -.0064634
linc	.070515	.0057527	12.26	0.000	.05924 .0817901
ndisease	-.0136704	.0010126	-13.50	0.000	-.015655 -.0116858

The probability of excellent health decreases as people age or have more diseases and increases as income increases.

The AME can be computed by dropping the `atmean` option.

15.9.6   Other ordered models

The parameters of the ordered probit model are estimated by using the `oprobit` command. The command syntax and output are essentially the same as for ordered logit, except that coefficient estimates are scaled differently. Application to the data here yields t statistics and log likelihoods quite close to those from ordered logit.

The user-written `gologit2` command (Williams 2006) estimates a generalization of the ordered logit model that allows the threshold parameters $\alpha_1, \ldots, \alpha_{m-1}$ to depend on regressors.

An alternative model is the MNL model. Although the MNL model has more parameters, the ordered logit model is not nested within the MNL. Estimator efficiency is another way of comparing the two approaches. An ordered estimator makes more assumptions than an MNL estimator. If these additional assumptions are true, the ordered estimator is more efficient than the MNL estimator.

15.10   Multivariate outcomes

We consider the multinomial analog of the seemingly unrelated regression (SUR) model (see section 5.4), where two or more categorical outcomes are being modeled.

In the simplest case, outcomes do not directly depend on each other—there is no simultaneity, but the errors for the outcomes may be correlated. When the errors are correlated, a more-efficient estimator that models the joint distribution of the errors is available.

In more complicated cases, the outcomes depend directly on each other, so there is simultaneity. We do not cover this case, but analysis is much simpler if the simultaneity is in continuous latent variables rather than discrete outcome variables.

15.10.1   Bivariate probit

The bivariate probit model considers two binary outcomes. The outcomes are potentially related after conditioning on regressors. The relatedness occurs via correlation of the errors that appear in the index-function model formulation of the binary outcome model.

Specifically, the two outcomes are determined by two unobserved latent variables,

$$y_1^* = \mathbf{x}_1' \boldsymbol{\beta}_1 + \varepsilon_1$$
$$y_2^* = \mathbf{x}_2' \boldsymbol{\beta}_2 + \varepsilon_2$$

where the errors ε_1 and ε_2 are jointly normally distributed with means of 0, variances of 1, and correlations of ρ, and we observe the two binary outcomes

$$y_1 = \begin{cases} 1 & \text{if } y_1^* > 0 \\ 0 & \text{if } y_1^* \leq 0, \end{cases} \quad \text{and} \quad y_2 = \begin{cases} 1 & \text{if } y_2^* > 0 \\ 0 & \text{if } y_2^* \leq 0 \end{cases}$$

The model collapses to two separate probit models for y_1 and y_2 if $\rho = 0$.

There are four mutually exclusive outcomes that we can denote by y_{10} (when $y_1 = 1$ and $y_2 = 0$), y_{01}, y_{11}, and y_{00}. The log-likelihood function is derived using the expressions for these probabilities and the parameters are estimated by ML. There are two complications. First, there is no analytical expression for the probabilities, because they depend on a one-dimensional integral with no closed-form solution, but this is easily solved with numerical quadrature methods for integration. Second, the resulting expressions for $\Pr(y_1 = 1|\mathbf{x})$ and $\Pr(y_2 = 1|\mathbf{x})$ differ from those for binary probit and probit.

The simplest form of the bivariate command has the syntax

biprobit *depvar1 depvar2* $\big[\,varlist\,\big]$ $\big[\,if\,\big]$ $\big[\,in\,\big]$ $\big[\,weight\,\big]$ $\big[\,,\ options\,\big]$

This version assumes that the same regressors are used for both outcomes. A more general version allows the list of regressors to differ for the two outcomes.

We consider two binary outcomes using the same dataset as that for ordered outcomes models analyzed in section 15.9. The first outcome is the hlthe variable, which takes on a value of 1 if self-assessed health is excellent and 0 otherwise. The second outcome is the dmdu variable, which equals 1 if the individual has visited the doctor in the past year and 0 otherwise. A data summary is

```
. * Two binary dependent variables: hlthe and dmdu
. tabulate hlthe dmdu
```

hlthe	any MD visit = 1 if mdu > 0		Total
	0	1	
0	826	1,731	2,557
1	1,006	2,011	3,017
Total	1,832	3,742	5,574

```
. correlate hlthe dmdu
(obs=5574)
```

	hlthe	dmdu
hlthe	1.0000	
dmdu	-0.0110	1.0000

The outcomes are very weakly negatively correlated, so in this case, there may be little need to model the two jointly.

Bivariate probit model estimation yields the following estimates:

```
. * Bivariate probit estimates
. biprobit hlthe dmdu age linc ndisease, nolog
Bivariate probit regression                    Number of obs   =      5574
                                               Wald chi2(6)    =    770.00
Log likelihood = -6958.0751                    Prob > chi2     =    0.0000
```

	Coef.	Std. Err.	z	P>|z|	[95% Conf. Interval]	
hlthe						
age	-.0178246	.0010827	-16.46	0.000	-.0199466	-.0157025
linc	.132468	.0149632	8.85	0.000	.1031406	.1617953
ndisease	-.0326656	.0027589	-11.84	0.000	-.0380729	-.0272583
_cons	-.2297079	.1334526	-1.72	0.085	-.4912703	.0318545
dmdu						
age	.0020038	.0010927	1.83	0.067	-.0001379	.0041455
linc	.1212519	.0142512	8.51	0.000	.09332	.1491838
ndisease	.0347111	.0028908	12.01	0.000	.0290452	.0403771
_cons	-1.032527	.1290517	-8.00	0.000	-1.285464	-.7795907
/athrho	.0282258	.022827	1.24	0.216	-.0165142	.0729658
rho	.0282183	.0228088			-.0165127	.0728366

```
Likelihood-ratio test of rho=0:     chi2(1) =   1.5295    Prob > chi2 = 0.2162
```

The hypothesis that $\rho = 0$ is not rejected, so in this case, bivariate probit was not necessary. As might be expected, separate probit estimation for each outcome (output not given) yields very similar coefficients to those given above.

Predicted probabilities can be obtained. For example, the marginal probability that $y_1 = 1$ can be obtained with the **pmarg1** option, whereas the joint probability that $(y_1, y_2) = (1, 1)$ is obtained with the **p11** option. We obtain

```
. * Predicted probabilities
. predict biprob1, pmarg1
. predict biprob2, pmarg2
. predict biprob11, p11
. predict biprob10, p10
. predict biprob01, p01
. predict biprob00, p00
. summarize hlthe dmdu biprob1 biprob2 biprob11 biprob10 biprob01 biprob00
```

Variable	Obs	Mean	Std. Dev.	Min	Max
hlthe	5574	.541263	.4983392	0	1
dmdu	5574	.6713312	.4697715	0	1
biprob1	5574	.5414237	.1577588	.0156161	.7853771
biprob2	5574	.6716857	.0976294	.1589158	.9834746
biprob11	5574	.3610553	.0989285	.0090629	.5492701
biprob10	5574	.1803685	.0765047	.0006476	.3680022
biprob01	5574	.3106305	.1434517	.1090853	.9385432
biprob00	5574	.1479458	.064902	.0158778	.6909308

The marginal probabilities that $y_1 = 1$ and $y_2 = 1$ are, respectively, 0.541 and 0.671, very close to the sample frequencies.

15.10.2   Nonlinear SUR

An alternative model is to use the `nlsur` command for nonlinear SUR, where the conditional mean of y_1 is $\Phi(\mathbf{x}_1'\boldsymbol{\beta}_1)$ and of y_2 is $\Phi(\mathbf{x}_2'\boldsymbol{\beta}_2)$. This estimator does not control for the intrinsic heteroskedasticity of binary outcome data, so we use the `vce(robust)` option to obtain standard errors that control for both heteroskedasticity and correlation. We have

```
. * Nonlinear seemingly unrelated regressions estimator
. nlsur (hlthe = normal({a1}*age+{a2}*linc+{a3}*ndisease+{a4}))
> (dmdu = normal({b1}*age+{b2}*linc+{b3}*ndisease+{b4})), vce(robust) nolog
(obs = 5574)
Calculating NLS estimates...
Calculating FGNLS estimates...

FGNLS regression
```

	Equation	Obs	Parms	RMSE	R-sq	Constant
1	hlthe	5574	4	.4727309	0.5871*	(none)
2	dmdu	5574	4	.4595438	0.6854*	(none)

```
* Uncentered R-sq
```

	Coef.	Robust Std. Err.	z	P>|z|	[95% Conf. Interval]
/a1	-.0173125	.0010624	-16.30	0.000	-.0193948	-.0152302
/a2	.1486604	.0184521	8.06	0.000	.1124949	.1848259
/a3	-.0333346	.0028682	-11.62	0.000	-.0389562	-.027713
/a4	-.3790899	.1638203	-2.31	0.021	-.7001719	-.0580079
/b1	.0018343	.0010776	1.70	0.089	-.0002778	.0039464
/b2	.1270039	.0165602	7.67	0.000	.0945465	.1594614
/b3	.0345088	.0030258	11.40	0.000	.0285783	.0404393
/b4	-1.081392	.1496894	-7.22	0.000	-1.374778	-.788006

For this example, the regression coefficients and standard errors are quite similar to those from `biprobit`.

15.11   Stata resources

The key models for initial understanding are the MNL and CL models. In practice, these models are often too restrictive. Stata commands cover most multinomial models, the most notable exception being the random-parameters logit or mixed logit model, which can be estimated with the user-written `mixlogit` command. Train (2003) is an excellent source, especially for models that need to be fit by MSL or Bayesian methods.

15.12   Exercises

1. Consider the health-status multinomial example of section 15.9. Refit this as a multinomial logit model using the `mlogit` command. Comment on the statistical significance of regressors. Obtain the marginal effects of changes in the regressors on the probability of excellent health for the MNL model, and compare these to those given in section 15.9.5 for the ordered logit model. Using BIC, which model do you prefer for these data—multinomial logit or ordered logit?

2. Consider the conditional logit example of section 15.5. Use `mus15datalong.dta`, if necessary to create this file as in section 15.5.1. Drop the charter boat option as in section 15.7.5, using `drop if fishmode=="charter" | mode==4` command, so we have a three choice-model. Estimate the parameters of a conditional logit model with regressors `p` and `q` and income, using the `asclogit` command. What are the MEs on the probability of private boat fishing of a $10 increase in the price of private boat fishing, a one-unit change in the catch rate from private boat fishing, and a $1,000 increase in monthly income? Which model fits these data better—the conditional logit model of this question or the multinomial probit model of section 15.7?

3. Continue the previous question, a three-choice model for fishing mode. Estimate the parameters of the model by nested logit, with errors for the utility of pier and beach fishing correlated with each other and uncorrelated with the error for the utility of private boat fishing. Obtain the ME of a change in the price of private boat fishing, adapting the example of section 15.6.6.

4. Consider the health-status multinomial example of section 15.9. Estimate the parameters of this model as an ordered probit model using the `oprobit` command. Comment on the statistical significance of regressors. Obtain the MEs for the predicted probability of excellent health for the ordered probit model and compare these to those given in section 15.9.5 for the ordered logit model. Which model do you prefer for these data—ordered probit or ordered logit?

16 Tobit and selection models

16.1 Introduction

The tobit model is relevant when the dependent variable of a linear regression is observed only over some interval of its support. Consider the annual household expenditure on a durable item such as a new automobile. A cross-section survey would almost certainly reveal a significant proportion of households with zero expenditure and the rest with a positive level of expenditure. In other words, the sample will be a mixture of observations with zero and positive values. Regression analyses of such data raise new modeling issues that will be considered in this chapter.

Estimating a linear regression in the presence of censoring involves additional computational complications. Ordinary least-squares (OLS) regression will not yield consistent parameter estimates because the censored sample is not representative of the population. For the same reason, statistical inference on the estimated parameters of the model also involves significant extensions of the standard theory.

In this chapter, we will consider two basic approaches to the estimation and inference regarding the tobit model. The first approach is parametric and is based on strong assumptions about the conditional data distribution and functional forms. The second (semiparametric) approach maintains the functional form assumptions but partially relaxes the distributional assumptions.

16.2 Tobit model

Suppose that our data consists of $(y_i, \mathbf{x}_i)$, $i = 1, \ldots, N$. Assume that $\mathbf{x}_i$ is fully observed, but y_i is not always observed. Specifically, some y_i are zero. We first consider the interpretation of zero observed values of y when the corresponding $\mathbf{x}$ is observed.

16.2.1 Regression with censored data

One interpretation is that zero is a censored observation. Suppose the household has a latent (unobserved) demand for goods, denoted by y^*, that is not expressed as a purchase until some known constant threshold, denoted by L, is passed. We observe y^* only when $y^* > L$. Then the zero expenditure can be interpreted as a left-censored variable that equals zero when $y^* \leq L$. Thus the observed sample consists of censored and uncensored observations.

Observations can be left-censored or right-censored. The latter means that the actual value of y^* is not observed when $y^* > U$, where U denotes the upper censoring point. For example, consider the draws y_i, $i = 1, \ldots, N$, from a $N(0,1)$ distribution, that are observed only in the interval $[L, U]$, where L and U are known constants. The distribution has full support over the range $(-\infty, +\infty)$, but we only observe values in the range $[L, U]$. The observations are then said to be censored, and L is the lower (or left) cutoff or censoring point, and U is the upper (or right) cut-off point.

Suppose we are in a regression setting with the observations $(y_i, \mathbf{x}_i)$, $i = 1, \ldots, N$, where $\mathbf{x}_i$ are always completely observed. Censoring is then akin to having missing observations on y. That is, censoring implies a loss of information. In some common cases, $L = 0$, but in other cases, $L = \gamma$, $\gamma > 0$, and furthermore γ may be unknown. For example, the survey may record expenditure on an expense category only when it exceeds, say, \$10. An example of right-censored data occurs when y is top-coded such that one only knows whether $y > U$, but not the precise value itself.

16.2.2 Tobit model setup

The regression of interest is specified as an unobserved latent variable, y^*,

$$y_i^* = \mathbf{x}_i'\boldsymbol{\beta} + \varepsilon_i, \; i = 1, \ldots, N \tag{16.1}$$

where $\varepsilon_i \sim N(0, \sigma^2)$, and $\mathbf{x}_i$ denotes the $(K \times 1)$ vector of exogenous and fully observed regressors. If y^* were observed, we would estimate (β, σ^2) by OLS in the usual way.

The observed variable y_i is related to the latent variable y_i^* through the observation rule

$$y = \begin{cases} y^* & \text{if } y^* > L \\ L & \text{if } y^* \leq L \end{cases}$$

The probability of an observation being censored is $\Pr(y^* \leq L) = \Pr(\mathbf{x}_i'\boldsymbol{\beta} + \varepsilon \leq L) = \Phi\left\{(L - \mathbf{x}_i'\boldsymbol{\beta})/\sigma\right\}$, where $\Phi(\cdot)$ is the standard normal cumulative distribution function.

The truncated mean, or expected value, of y for the noncensored observations can be shown to be

$$E(y_i|\mathbf{x}_i, y_i > L) = \mathbf{x}_i'\boldsymbol{\beta} + \sigma \frac{\phi\{(\mathbf{x}_i'\boldsymbol{\beta} - L)/\sigma\}}{\Phi\left\{(L - \mathbf{x}_i'\boldsymbol{\beta})/\sigma\right\}} \tag{16.2}$$

where $\phi(\cdot)$ is the standard normal density. The conditional mean in (16.2) differs from $\mathbf{x}_i'\boldsymbol{\beta}$ because of the censoring, a difference that leads to OLS being inconsistent. The exact formula in (16.2) relies crucially on the assumption that $\varepsilon \sim N(0, \sigma^2)$.

A sample may instead include right-censored observations. Then we observe that

$$y = \begin{cases} y^* & \text{if } y^* < U \\ U & \text{if } y^* \geq U \end{cases}$$

The leading case of censoring is that in which the data are left-censored only and $L = 0$. A variant of the tobit model, the two-limit tobit, allows for both left and right censoring. Another variant, considered here, is that in which the data are only left-censored but L is unknown.

16.2.3   Unknown censoring point

As Carson and Sun (2007), and others, have pointed out, the censoring point may be unknown. Suppose that the data are left-censored, and there is a constant but unknown threshold, γ. The assumption that the unknown γ can be set to zero as a "normalization" is not innocuous. Instead, $\Pr(y^* < \gamma)$ is $\Phi\{(\gamma - \mathbf{x}'\boldsymbol{\beta})/\sigma\}$, where $(\gamma - \mathbf{x}'\boldsymbol{\beta})/\sigma$ is interpreted as a "threshold". In this case, we can set $\widehat{\gamma} = \min(\text{uncensored } y)$ and proceed as if γ is known. Estimates of the tobit model based on this procedure have been shown to be consistent; see Carson and Sun (2007). In Stata, this only requires that the value of γ should be used in defining ll. So we can again use the tobit command with the ll(#) option. It is simplest to set # equal to $\widehat{\gamma}$. This will treat the observation or observations with $y = \widehat{\gamma}$ as censored, however, and a better alternative is to set # equal to $\widehat{\gamma} - \Delta$ for some small value, Δ, such as 10^{-6}.

16.2.4   Tobit estimation

The foregoing analysis leads to two estimators—maximum likelihood (ML) and two-step regression. We first consider ML estimation under the assumptions that the regression error is homoskedastic and normally distributed.

For the case of left-censored data with the censoring point γ (so $L = \gamma$), the density function has two components that correspond, respectively, to uncensored and censored observations. Let $d = 1$ denote the censoring indicator for the outcome that the observation is not censored, and let $d = 0$ indicate a censored observation. The density can be written as

$$f(y_i) = \left[\frac{1}{\sqrt{2\pi\sigma^2}}\exp\left\{-\frac{1}{2\sigma^2}(y_i - x_i'\boldsymbol{\beta})^2\right\}\right]^{d_i}[\Phi\{(\gamma - \mathbf{x}_i'\boldsymbol{\beta})/\sigma\}]^{1-d_i} \qquad (16.3)$$

The second term in (16.3) reflects the contribution to the likelihood of the censored observation. ML estimates of $(\boldsymbol{\beta}, \sigma^2)$ solve the first-order conditions from maximization of the log likelihood based on (16.3). These equations are nonlinear in parameters, so an iterative algorithm is required.

The tobit ML estimator (MLE) is consistent under the stated assumptions. However, it is inconsistent if the errors are not normally distributed or if they are heteroskedastic. These strong assumptions are likely to be violated in applications, and this makes the tobit MLE a nonrobust estimator. It is desirable to test the assumptions of normality and heteroskedasticity.

Estimation can be based on weaker assumptions than those for the MLE. Equation (16.2) suggests why the OLS regression of y_i on $\mathbf{x}_i$ yields an inconsistent estimate—there will be omitted-variable bias due to the "missing variable" ϕ_i/Φ_i in (16.2). This missing variable can be generated by a probit model that models the probability of the outcome that $y_i^* > 0$. Let $d_i = 1$ denote the outcome that $y_i^* > 0$, and let $d_i = 0$ otherwise. The probit estimator can provide a consistent estimate of $\lambda_i = \phi_i/\Phi_i$. A linear regression of y_i on $\mathbf{x}$ and $\widehat{\phi_i/\Phi_i}$ will provide an estimate of $\boldsymbol{\beta}$.

16.2.5 ML estimation in Stata

The estimation command for tobit regression in Stata has the following basic syntax:

$\texttt{tobit}$ *depvar* $\big[\,indepvars\,\big]$ $\big[\,if\,\big]$ $\big[\,in\,\big]$ $\big[\,weight\,\big]$, $\texttt{ll}\big[\,(\#)\,\big]$ $\texttt{ul}\big[\,(\#)\,\big]$ $\big[\,options\,\big]$

The specifications $\texttt{ll}\big[\,(\#)\,\big]$ and $\texttt{ul}\big[\,(\#)\,\big]$ refer to the lower limit (left-censoring point) and the upper limit (right-censoring point), respectively. If the data are subject to left censoring at zero, for example, then only $\texttt{ll(0)}$ is required. Similarly, only $\texttt{ul(10000)}$ is required for right-censored data at the censoring point 10,000. Both are required if the data are both right-censored and left-censored and if one wants to estimate the parameters of the two-limit tobit model. The postestimation tools for tobit will be discussed later in this chapter.

16.3 Tobit model example

The illustration we consider, ambulatory expenditures, has the very common complication that the data are highly right-skewed. This is best treated by taking the natural logarithm, complicating the analysis of these complications. We present these complications in detail in sections 16.4–16.7, and model diagnostics are deferred to section 16.4.5. In the current section, we instead present the simpler tobit model in levels.

16.3.1 Data summary

The data on the dependent variable for ambulatory expenditure (`ambexp`) and the regressors (`age`, `female`, `educ`, `blhisp`, `totchr`, and `ins`) are taken from the 2001 Medical Expenditure Panel Survey. In this sample of 3,328 observations, there are 526 (15.8%) zero values of `ambexp`.

Descriptive statistics for all the variables follow:

```
. * Raw data summary
. use mus16data.dta, clear

. summarize ambexp age female educ blhisp totchr ins
```

Variable	Obs	Mean	Std. Dev.	Min	Max
ambexp	3328	1386.519	2530.406	0	49960
age	3328	4.056881	1.121212	2.1	6.4
female	3328	.5084135	.5000043	0	1
educ	3328	13.40565	2.574199	0	17
blhisp	3328	.3085938	.4619824	0	1
totchr	3328	.4831731	.7720426	0	5
ins	3328	.3650841	.4815261	0	1

A detailed summary of `ambexp` provides insight into the potential problems in estimating the parameters of the tobit model with a linear conditional mean function.

```
. * Detailed summary to show skewness and kurtosis
. summarize ambexp, detail

                            ambexp

          Percentiles      Smallest
    1%           0             0
    5%           0             0
   10%           0             0       Obs                   3328
   25%         113             0       Sum of Wgt.           3328

   50%       534.5                     Mean              1386.519
                              Largest  Std. Dev.         2530.406
   75%        1618          28269
   90%        3585          30920      Variance           6402953
   95%        5451          34964      Skewness          6.059491
   99%       11985          49960      Kurtosis          72.06738
```

The `ambexp` variable is very heavily skewed and has considerable nonnormal kurtosis. This feature of the dependent variable should alert us to the possibility that the tobit MLE may be a flawed estimator for the model.

To see if these characteristics persist if the zero observations are ignored, we examine the sample distribution of only positive values.

```
. * Summary for positives only
. summarize ambexp if ambexp >0, detail

                            ambexp

          Percentiles      Smallest
    1%          22             1
    5%          67             2
   10%         107             2       Obs                   2802
   25%         275             4       Sum of Wgt.           2802

   50%         779                     Mean                1646.8
                              Largest  Std. Dev.         2678.914
   75%        1913          28269
   90%        3967          30920      Variance           7176579
   95%        6027          34964      Skewness          5.799312
   99%       12467          49960      Kurtosis          65.81969
```

The skewness and nonnormal kurtosis are reduced only a little if the zeros are ignored.

In principle, the skewness and nonnormal kurtosis of `ambexp` could be due to regressors that are skewed. But, from output not listed, an OLS regression of `ambexp` on `age`, `female`, `educ`, `blhisp`, `totchr`, and `ins` explains little of the variation ($R^2 = 0.16$) and the OLS residuals have a skewness statistic of 6.6 and a kurtosis statistic of 92.2. Even after conditioning on regressors, the dependent variable is very nonnormal, and a lognormal model may be more appropriate.

16.3.2   Tobit analysis

As an initial exploratory step, we will run the linear tobit model without any transformation of the dependent variable, even though it appears that the data distribution may be nonnormal.

```
. * Tobit analysis for ambexp using all expenditures
. global xlist age female educ blhisp totchr ins    //Define regressor list $xlist
. tobit ambexp $xlist, ll(0)
```

```
Tobit regression                                  Number of obs   =        3328
                                                  LR chi2(6)      =      694.07
                                                  Prob > chi2     =      0.0000
Log likelihood = -26359.424                       Pseudo R2       =      0.0130
```

ambexp	Coef.	Std. Err.	t	P>|t|	[95% Conf. Interval]	
age	314.1479	42.63358	7.37	0.000	230.5572	397.7387
female	684.9918	92.85445	7.38	0.000	502.9341	867.0495
educ	70.8656	18.57361	3.82	0.000	34.44873	107.2825
blhisp	-530.311	104.2667	-5.09	0.000	-734.7443	-325.8776
totchr	1244.578	60.51364	20.57	0.000	1125.93	1363.226
ins	-167.4714	96.46068	-1.74	0.083	-356.5998	21.65696
_cons	-1882.591	317.4299	-5.93	0.000	-2504.969	-1260.214
/sigma	2575.907	34.79296			2507.689	2644.125

```
Obs. summary:         526  left-censored observations at ambexp<=0
                     2802      uncensored observations
                        0 right-censored observations
```

All regressors aside from `ins` are statistically significant at the 0.05 level. The interpretation of the coefficients is as a partial derivative of the latent variable, y^*, with respect to x. Marginal effects for the observed variable, y, are presented in section 16.3.4.

It is standard to use the default estimate of the variance–covariance matrix of the estimator (VCE) for the tobit MLE, because if the model is misspecified so that a robust estimate of the VCE is needed, it is also likely that the tobit MLE is inconsistent.

16.3.3 Prediction after tobit

The `predict` command is summarized in section 3.6. The command can be used after `tobit` to predict a range of quantities. We begin with the default linear prediction, `xb`, that produces in-sample fitted values of the latent variable, y^*, for all observations.

```
. * Tobit prediction and summary
. predict yhatlin
(option xb assumed; fitted values)
. summarize yhatlin
```

Variable	Obs	Mean	Std. Dev.	Min	Max
yhatlin	3328	1066.683	1257.455	-1564.703	8027.957

A more detailed comparison of the sample statistics for `yhatlin` with those for `ambexp` shows that the tobit model fits especially poorly in the upper tail of the distribution.

This fact notwithstanding, we will use the model to illustrate other prediction options for the observed variable, y, that can be used in combination with the computation of marginal effects.

16.3.4   Marginal effects

In a censored regression, there are a variety of marginal effects (ME) that are of potential interest; see [R] **tobit postestimation**. The ME is the effect on the conditional mean of the dependent variable of changes in the regressors. This effect varies according to whether interest lies in the latent variable mean, or in the truncated or censored means. Omitting derivations given in Cameron and Trivedi (2005, chap. 16), these MEs are as follows:

$$
\begin{aligned}
\text{Latent variable} \quad & \partial E(y^*|\mathbf{x})/\partial \mathbf{x} = \boldsymbol{\beta} \\
\text{Left-truncated (at 0)} \quad & \partial E(y|\mathbf{x}, y > 0)/\partial \mathbf{x} = \{1 - w\lambda(w) - \lambda(w)^2\}\boldsymbol{\beta} \\
\text{Left-censored (at 0)} \quad & \partial E(y|\mathbf{x})/\partial \mathbf{x} = \Phi(w)\boldsymbol{\beta}
\end{aligned}
$$

where $w = \mathbf{x}'\boldsymbol{\beta}/\sigma$ and $\lambda(w) = \phi(w)/\Phi(w)$. The first of these has already been discussed above.

Left-truncated, left-censored, and right-truncated examples

For illustration, we compute MEs for three conditional mean specifications: $E(y|\mathbf{x}, y > 0)$, $E(y|\mathbf{x})$, and $E(y|\mathbf{x}, 0 < y < 535)$, where $b = 535$ is the median value of y. In each case, the estimated conditional mean is followed by the estimated MEs.

The `predict()` option of `margins` is used to obtain MEs with respect to the desired quantity. Here we use the `atmean` option, so the ME at the mean is computed.

We begin with the ME for the left-truncated mean, $E(y|\mathbf{x}, y > 0)$.

```
. * (1) ME on censored expected value E(y|x,y>0)
. quietly tobit ambexp age i.female educ i.blhisp totchr i.ins, ll(0)

. margins, dydx(*) predict(e(0,.)) atmean noatlegend
Conditional marginal effects                    Number of obs   =        3328
Model VCE    : OIM

Expression   : E(ambexp|ambexp>0), predict(e(0,.))
dy/dx w.r.t. : age 1.female educ 1.blhisp totchr 1.ins
```

	dy/dx	Delta-method Std. Err.	z	P>|z|	[95% Conf. Interval]	
age	145.524	19.78076	7.36	0.000	106.7544	184.2935
1.female	317.1037	42.96062	7.38	0.000	232.9024	401.305
educ	32.82734	8.601068	3.82	0.000	15.96956	49.68513
1.blhisp	-240.2953	46.21518	-5.20	0.000	-330.8754	-149.7152
totchr	576.5307	28.50486	20.23	0.000	520.6622	632.3992
1.ins	-77.19554	44.26196	-1.74	0.081	-163.9474	9.556319

```
Note: dy/dx for factor levels is the discrete change from the base level.
```

For these data, the MEs are roughly one-half of the coefficient estimates, $\widehat{\beta}$, given in section 16.3.2.

The MEs for the censored mean, $E(y|\mathbf{x})$, are computed next.

```
. * (2) ME without censoring on E(y|x)
. margins, dydx(*) predict(ystar(0,.)) atmean noatlegend
Conditional marginal effects                    Number of obs   =      3328
Model VCE    : OIM

Expression   : E(ambexp*|ambexp>0), predict(ystar(0,.))
dy/dx w.r.t. : age 1.female educ 1.blhisp totchr 1.ins
```

	dy/dx	Delta-method Std. Err.	z	P>|z|	[95% Conf. Interval]	
age	207.526	28.20535	7.36	0.000	152.2445	262.8074
1.female	451.6399	61.02875	7.40	0.000	332.0257	571.2541
educ	46.81378	12.26549	3.82	0.000	22.77386	70.8537
1.blhisp	-342.4803	65.75642	-5.21	0.000	-471.3606	-213.6001
totchr	822.1678	40.6103	20.25	0.000	742.573	901.7625
1.ins	-110.0883	63.11688	-1.74	0.081	-233.7951	13.61852

Note: dy/dx for factor levels is the discrete change from the base level.

These MEs are larger in absolute value than those for the left-truncated mean and are roughly 70% of the original coefficient estimates.

In the third example, we consider MEs when additionally there is right censoring at the median value of y.

```
. * (3) ME when E(y|0<y<535)
. margins, dydx(*) predict(e(0,535)) atmean noatlegend
Conditional marginal effects                    Number of obs   =      3328
Model VCE    : OIM

Expression   : E(ambexp|0<ambexp<535), predict(e(0,535))
dy/dx w.r.t. : age 1.female educ 1.blhisp totchr 1.ins
```

	dy/dx	Delta-method Std. Err.	z	P>|z|	[95% Conf. Interval]	
age	1.12742	.1554335	7.25	0.000	.8227758	1.432064
1.female	2.458287	.3366104	7.30	0.000	1.798543	3.118032
educ	.2543237	.0668133	3.81	0.000	.123372	.3852754
1.blhisp	-1.903269	.3762307	-5.06	0.000	-2.640668	-1.165871
totchr	4.466565	.2439444	18.31	0.000	3.988442	4.944687
1.ins	-.601031	.3467693	-1.73	0.083	-1.280686	.0786242

Note: dy/dx for factor levels is the discrete change from the base level.

The MEs here are small relative to those in the previous two cases, as expected given the relatively small variation in the range of y being considered.

Left-censored case computed directly

Next we illustrate direct computation of the MEs for the left-censored mean. From the table, this is $\Phi(\overline{\mathbf{x}}'\widehat{\boldsymbol{\beta}}/\widehat{\sigma})\widehat{\beta}_j$ for the jth regressor. This example also illustrates how to retrieve tobit model coefficients.

```
. * Direct computation of marginal effects for E(y|x)
. predict xb1, xb // xb1 is estimate of x´b
. matrix btobit = e(b)
. scalar sigma = btobit[1,11] // sigma is estimate of sigma
. matrix bcoeff = btobit[1,1..9] // bcoeff is betas excl. constant
. quietly summarize xb1
. scalar meanxb = r(mean) // mean of x´b equals (mean of x)´b
. scalar PHI = normal(meanxb/sigma)
. matrix deriv = PHI*bcoeff
. matrix list deriv
deriv[1,9]
            model:     model:     model:     model:     model:     model:
                          0b.         1.                   0b.         1.
              age      female     female       educ      blhisp     blhisp
y1    207.52598            0   452.50523   46.813781           0  -350.32317

            model:     model:     model:
                          0b.         1.
           totchr        ins        ins
y1    822.16778            0   -110.63154
. * The following gives nicer looking results
. ereturn post deriv
. ereturn display
```

	Coef.
model	
age	207.526
1.female	452.5052
educ	46.81378
1.blhisp	-350.3232
totchr	822.1678
1.ins	-110.6315

As expected, the MEs for continuous regressors are identical to those obtained above with `margins`. For binary regressors, there is some difference because `margins` uses the finite-difference method rather than calculus methods; see section 10.6.6.

Marginal impact on probabilities

The impact of a change in a regressor on the probability that y is in a specified interval may be of interest. For illustration, consider the ME on $\Pr(5000 < \texttt{ambexp} < 10000)$.

(Continued on next page)

```
. * Compute margins on Pr(5000<ambexp<10000)
. quietly tobit ambexp age i.female educ i.blhisp totchr i.ins, ll(0) vce(robust)

. margins, dydx(*) predict(pr(5000,10000)) atmean noatlegend
Conditional marginal effects                      Number of obs   =       3328
Model VCE       : Robust

Expression      : Pr(5000<ambexp<10000), predict(pr(5000,10000))
dy/dx w.r.t.    : age 1.female educ 1.blhisp totchr 1.ins
```

	dy/dx	Delta-method Std. Err.	z	P>|z|	[95% Conf.	Interval]
age	.0150449	.0028009	5.37	0.000	.0095551	.0205346
1.female	.0328152	.0068033	4.82	0.000	.0194809	.0461495
educ	.0033938	.0009929	3.42	0.001	.0014478	.0053399
1.blhisp	-.0239676	.0049076	-4.88	0.000	-.0335863	-.0143489
totchr	.0596042	.0090395	6.59	0.000	.0418871	.0773212
1.ins	-.0079162	.0043282	-1.83	0.067	-.0163993	.0005669

```
Note: dy/dx for factor levels is the discrete change from the base level.
```

The effects are very small; only 5% of the sample fall into this range.

16.3.5   The ivtobit command

The preceding analysis applies when the regressors in the tobit model are exogenous. The exogeneity assumption may be thought of as specifying a reduced-form equation. When the equation of interest is "structural", i.e., it involves one or more endogenous variables, so that this exogeneity condition fails, another estimator that controls for the endogeneity of one or more regressors is required. As for binary outcome models (see section 14.8), there are two main estimation approaches available. One is fully parametric and the other is partially parametric (two-step). These are implemented with variants of the ivtobit command.

The theoretical framework underlying the ivtobit procedure is explained in [R] **ivtobit** and in Newey (1987). If there is only one endogenous regressor, then the setup involves two equations—the structural equation of interest and the reduced form for the endogenous regressor. The framework assumes that the endogenous regressor is continuous, so the method should not be used for a discrete endogenous variable. The reduced-form equation for this variable must contain exogenous instrumental variables that affect the outcome variable only through the endogenous regressor. That is, these instrumental variables are excluded from the structural equation. The parametric (MLE) estimation method assumes that the structural and reduced-form equation errors are jointly normally distributed. The semiparametric method drops the joint normality assumption and uses a minimum chi-squared criterion proposed by Newey. ML is the default.

The ivtobit command has similar syntax to ivregress and has options for marginal effects, prediction, and variance estimation similar to those for tobit. We do not provide a data example, but one can be found in [R] **ivtobit**.

16.3.6 Additional commands for censored regression

The censored least absolute-deviations estimator of Powell (1984) provides consistent estimates for left-censored or right-censored data under the weaker assumption that the error, ε, in (16.1) is independent and identically distributed and symmetrically distributed. This is implemented with the user-written `clad` command (Jolliffe, Krushelnytskyy, and Semykina 2000). For these data, the method is best implemented for the data in logs.

The `intreg` command is a generalization of `tobit` for data observed in intervals. For example, expenditures might be observed in the ranges $y \leq 0$, $0 < y \leq 1000$, $1000 < y \leq 10000$, and $y \geq 10000$.

A quite different type of right-censored data is duration data on length of unemployment spell or survival data on time until death. The standard approach for such data is to model the conditional hazard of the spell ending rather than the conditional mean. This approach has the advantage of permitting the use of the Cox proportional hazards model that allows semiparametric estimation without strong distributional assumptions such as an exponential or Weibull distribution for durations. For details, see [ST] *Survival Analysis and Epidemiological Tables Reference Manual*, especially the entries [ST] **stset**, [ST] **sts**, [ST] **stcox**, [ST] **stcrreg**, and [ST] **streg**.

16.4 Tobit for lognormal data

The tobit model relies crucially on normality, but expenditure data are often better modeled as lognormal. A tobit regression model for lognormal data introduces two complications: a nonzero threshold and lognormal y.

Now introduce lognormality by specifying

$$y^* = \exp(\mathbf{x}'\boldsymbol{\beta} + \varepsilon), \quad \varepsilon \sim N(0, \sigma^2)$$

where we observe that

$$y = \begin{cases} y^* & \text{if } \ln y^* > \gamma \\ 0 & \text{if } \ln y^* \leq \gamma \end{cases}$$

Here it is known that $y = 0$ when data are censored, and in general $\gamma \neq 0$. The parameters of this model can be estimated by using the `tobit` command with the `ll(#)` option, where the dependent variable is $\ln y$ rather than y, and the threshold $\#$ equals the minimum uncensored value of $\ln y$. The censored values of $\ln y$ must be set to a value equal to or less than the minimum uncensored value of $\ln y$.

In this model, interest lies in the prediction of expenditures in levels rather than logs. The issues are similar to those considered in chapter 3 for the lognormal model. Some algebra yields the censored mean

$$E(y|\mathbf{x}) = \exp\left(\mathbf{x}'\boldsymbol{\beta} + \frac{\sigma^2}{2}\right)\left\{1 - \Phi\left(\frac{\gamma - \mathbf{x}'\boldsymbol{\beta} - \sigma^2}{\sigma}\right)\right\} \tag{16.4}$$

The truncated mean $E(y|\mathbf{x}, y > 0)$ equals $E(y|\mathbf{x})/[1 - \Phi\{(\gamma - \mathbf{x}'\boldsymbol{\beta})/\sigma\}]$.

16.4.1   Data example

The illustrative application of the tobit model considered here uses the same data as in section 16.3. We remind the reader that in this sample of 3,328 observations, there are 526 (15.8%) zero values of ambexp. A detailed summary of ln(ambexp), denoted by lambexp, follows.

```
. * Summary of log(expenditures) for positives only
. summarize lambexp, detail
                            lambexp

          Percentiles     Smallest
 1%        3.091043             0
 5%        4.204693       .6931472
10%        4.672829       .6931472       Obs               2802
25%        5.616771       1.386294       Sum of Wgt.       2802

50%        6.65801                       Mean          6.555066
                          Largest        Std. Dev.      1.41073
75%        7.556428       10.24952
90%        8.285766       10.33916       Variance      1.990161
95%        8.704004       10.46207       Skewness     -.3421614
99%        9.43084        10.81898       Kurtosis      3.127747
```

The summary shows that ln(ambexp) is almost symmetrically distributed and has negligible nonnormal kurtosis. This is in stark contrast to ambexp even after conditioning on regressors; see section 16.3.1. We anticipate that the tobit model is better suited to modeling lambexp than ambexp.

16.4.2   Setting the censoring point for data in logs

It may be preferred at times to apply a transformation to the dependent variable to make it more suitable for a tobit application. In the present instance, we work with ln(ambexp) as the dependent variable. This variable is originally set to missing if ambexp = 0, but to use the tobit command, it needs to be set to a nonmissing value, the lower limit.

A complication here is that the smallest positive value of ambexp is 1, in which case ln(ambexp) equals 0. Then Stata's ll or ll(0) option mistakenly treats this observation as censored rather than as zero, leading to shrinkage in the sample size for noncensored observations. In our sample, one observation would be thus "lost". To avoid this loss, we "trick" Stata by setting all censored observations of ln y to an amount slightly smaller than the minimum noncensored value of ln y, as follows:

```
. * "Tricking" Stata to handle log transformation
. generate y = ambexp
. generate dy = ambexp > 0
. generate lny = ln(y)                 // Zero values will become missing
(526 missing values generated)
. quietly summarize lny
. scalar gamma = r(min)                // This could be negative
. display "gamma = " gamma
gamma = 0
```

```
. replace lny = gamma - 0.0000001 if lny == .
(526 real changes made)
. tabulate y if y < 0.02            // .02 is arbitrary small value
```

y	Freq.	Percent	Cum.
0	526	100.00	100.00
Total	526	100.00	

```
. tabulate lny if lny < gamma + 0.02
```

lny	Freq.	Percent	Cum.
-1.00e-07	526	99.81	99.81
0	1	0.19	100.00
Total	527	100.00	

```
. * Label the variables
. label variable y "ambexp"
. label variable lny "lnambexp"
. label variable dy "dambexp"
```

Note that the dependent variables have been relabeled. This makes the Stata code given later easier to adapt for other applications. In what follows, y is the `ambexp` variable and $\ln y$ is the `lny` variable.

16.4.3 Results

We first obtain the tobit MLE, where now log expenditures is the dependent variable.

```
. * Now do tobit on lny and calculate threshold and lambda
. tobit lny $xlist, ll
```

Tobit regression				Number of obs	=	3328
				LR chi2(6)	=	831.03
				Prob > chi2	=	0.0000
Log likelihood =	-7494.29			Pseudo R2	=	0.0525

lny	Coef.	Std. Err.	t	P>|t|	[95% Conf. Interval]	
age	.3630699	.0453222	8.01	0.000	.2742077	.4519321
female	1.341809	.0986074	13.61	0.000	1.148471	1.535146
educ	.138446	.0196568	7.04	0.000	.0999054	.1769866
blhisp	-.8731611	.1102504	-7.92	0.000	-1.089327	-.6569955
totchr	1.161268	.0649655	17.88	0.000	1.033891	1.288644
ins	.2612202	.102613	2.55	0.011	.0600292	.4624112
_cons	.9237178	.3350343	2.76	0.006	.2668234	1.580612
/sigma	2.781234	.0392269			2.704323	2.858146

```
Obs. summary:        526  left-censored observations at lny<=-1.000e-07
                    2802       uncensored observations
                       0 right-censored observations
```

All estimated coefficients are statistically significant at the 0.05 level and have the expected signs.

To assess the impact of using the censored regression framework instead of treating the zeros like observations from the same data-generating process as the positives, let us compare the results with those from the OLS regression of lny on the regressors.

```
. * OLS, not tobit
. regress lny $xlist, noheader
```

lny	Coef.	Std. Err.	t	P>|t|	[95% Conf. Interval]
age	.3247317	.038348	8.47	0.000	.2495436	.3999199
female	1.144695	.0833418	13.73	0.000	.9812886	1.308102
educ	.114108	.0165414	6.90	0.000	.0816757	.1465403
blhisp	-.7341754	.0928854	-7.90	0.000	-.9162938	-.5520571
totchr	1.059395	.0553699	19.13	0.000	.9508324	1.167958
ins	.2078343	.0869061	2.39	0.017	.0374394	.3782293
_cons	1.728764	.2812597	6.15	0.000	1.177304	2.280224

All the OLS slope coefficients are in absolute terms smaller than those for the ML tobit, the reduction being 10–15%, but the OLS intercept is larger. The impact of censoring (zeros) on the OLS results depends on the proportion of censored observations, which in our case is around 15%.

16.4.4 Two-limit tobit

In less than 1.5% of the sample (48 observations) ambexp exceeds $10,000. Suppose that we want to exclude these high values that contribute to the nonnormal kurtosis. Or suppose that the data above an upper cutoff point are reported as falling in an interval. Choosing $10,000 as the upper censoring point, we estimate a two-limit tobit version of the tobit model. We see that the impact of dropping the 48 observations is relatively small. This is not too surprising because a small proportion of the sample size is right-censored.

```
. * Now do two-limit tobit
. scalar upper = log(10000)
. display upper
9.2103404
```

```
. tobit lny $xlist, ll ul(9.2103404)
Tobit regression                          Number of obs   =       3328
                                          LR chi2(6)      =     840.33
                                          Prob > chi2     =     0.0000
Log likelihood = -7451.7623               Pseudo R2       =     0.0534
```

lny	Coef.	Std. Err.	t	P>|t|	[95% Conf. Interval]	
age	.3711061	.0459354	8.08	0.000	.2810416	.4611706
female	1.348768	.0999154	13.50	0.000	1.152866	1.54467
educ	.1402643	.0199113	7.04	0.000	.1012246	.1793039
blhisp	-.8759505	.1116504	-7.85	0.000	-1.094861	-.65704
totchr	1.20494	.0664951	18.12	0.000	1.074565	1.335316
ins	.2466838	.1039194	2.37	0.018	.0429313	.4504363
_cons	.8638458	.3394729	2.54	0.011	.1982487	1.529443
/sigma	2.812304	.0401377			2.733607	2.891001

```
Obs. summary:          526   left-censored observations at lny<=-1.000e-07
                      2754         uncensored observations
                        48  right-censored observations at lny>=9.2103404
```

16.4.5 Model diagnostics

To test the validity of the key tobit assumptions of normality and homoskedasticity, we need to apply some diagnostic checks. For the ordinary linear regression model, the `sktest` and `hettest` commands are available to test for normality and homoskedasticity. These tests are based on the OLS residuals. These postestimation tests are invalid for censored data because the fitted values and residuals from a censored model do not share the properties of their ordinary regression counterparts. Generalized residuals for censored regression, as discussed in Cameron and Trivedi (2005, chap. 18.7.2) and in Verbeek (2008, 238–240), provide the key component for generating test statistics for testing the null hypotheses of homoskedasticity and normality.

In linear regression, tests of homoskedasticity typically use squared residuals, and tests of normality use residuals raised to a power of 3 or 4. The first step is then to construct analogous quantities for the censored regression. For uncensored observations, we use $\widehat{\varepsilon}_i = (y_i - \mathbf{x}_i'\widehat{\boldsymbol{\beta}})/\widehat{\sigma}$ raised to the relevant powers, where y_i is a generic notation for the dependent variable, which here is ln(`ambexp`). For observations left-censored at γ, we use the quantities listed in table 16.1, evaluated at $\widehat{\boldsymbol{\beta}}$ and $\widehat{\sigma}$.

(Continued on next page)

Table 16.1. Quantities for observations left-censored at γ

Moments	Expression
$E(\varepsilon_i \mid d_i = 0)$	$-\lambda_i$, where $\lambda_i = \frac{\phi(\mathbf{x}_i'\boldsymbol{\beta}/\sigma)}{1-\Phi(\mathbf{x}_i'\boldsymbol{\beta}/\sigma)}$
$E(\varepsilon_i^2 \mid d_i = 0)$	$1 - z_i\lambda_i$, where $z_i = (\gamma - \mathbf{x}_i'\boldsymbol{\beta})/\sigma$
$E(\varepsilon_i^3 \mid d_i = 0)$	$-(2 + z_i^2)\lambda_i$
$E(\varepsilon_i^4 \mid d_i = 0)$	$3 - (3z_i + z_i^3)\lambda_i$

The components $\phi(\cdot)$ and $\Phi(\cdot)$ can be evaluated by using Stata's `normalden()` and `normal()` functions. Given these and predicted values from the ML regression, the four "generalized" components given in the table can be readily computed.

16.4.6 Tests of normality and homoskedasticity

Lagrange multiplier (LM), or score, tests of heteroskedasticity and nonnormality are appealing because they only require estimation of the models under the hypothesis of normality and homoskedasticity. The test statistics are quadratic forms that can be calculated in several different ways. One way is by using an auxiliary regression; see section 12.5.3 and Cameron and Trivedi (2005, chap. 8).

Conditional moment tests can also be performed by using a similar approach; see section 12.7.1, Newey (1985), and Pagan and Vella (1989). Such regression-based tests have been developed with generalized residuals. Although they are not currently available as a part of the official Stata package, they can be constructed from Stata output, as illustrated below. The key component of the auxiliary regression is the uncentered R^2, denoted by R_u^2, from the auxiliary regression of 1 on generated regressors that are themselves functions of generalized residuals. The specific regressors depend upon the alternative to the null.

Generalized residuals and scores

To implement the test, we first compute and store various components of the test statistic. The inverse of the Mills' ratio, λ_i, and related variables are calculated first, including the generalized residuals.

```
. * Mills´ ratio
. quietly tobit lny $xlist, ll
. predict xb, xb                           // xb is estimate of x´b
. matrix btobit = e(b)
. scalar sigma = btobit[1,e(df_m)+2]       // sigma is estimate of sigma
. generate threshold = (gamma-xb)/sigma    // gamma: lower censoring point
. generate lambda = normalden(threshold)/normal(threshold)
```

Next we calculate generalized residuals and functions of them. For example, `gres3` equals $\{(y_i - \mathbf{x}'_i\widehat{\boldsymbol{\beta}})/\widehat{\sigma}\}^3$ for an uncensored observation and equals $-(2 + \widehat{z}_i^2)\widehat{\lambda}_i$, where z_i and λ_i are defined in the table. The generalized residuals `gres1` and `gres2` can be shown to be the contributions to the score for, respectively, the intercept β_1 and σ, so they must sum to zero over the sample. The generalized residuals `gres3` and `gres4` satisfy the same zero-mean property only if the model is correctly specified.

The generalized residuals are computed as follows:

```
. * Generalized residuals
. * gres1 and gres2 should have mean zero by the first-order conditions
. * gres3 and gres4 have mean zero if model correctly specified
. * Residual (scaled by sigma) for positive values
. quietly generate uifdyeq1 = (lny - xb)/sigma if dy == 1
. * First-sample moment
. quietly generate double gres1 = uifdyeq1
. quietly replace gres1 = -lambda if dy == 0
. summarize gres1
```

Variable	Obs	Mean	Std. Dev.	Min	Max
gres1	3328	4.49e-09	.9877495	-3.129662	2.245604

The zero-mean property of `gres1` is thus verified. The remaining three variables are computed next.

```
. * Second- to fourth-sample moments
. quietly generate double gres2 = uifdyeq1^2 - 1
. quietly replace gres2 = -threshold*lambda if dy == 0
. quietly generate double gres3 = uifdyeq1^3
. replace gres3 = -(2 + threshold^2)*lambda if dy == 0
(526 real changes made)
. quietly generate double gres4 = uifdyeq1^4 - 3
. quietly replace gres4 = -(3*threshold + threshold^3)*lambda if dy == 0
```

Test of normality

To apply the LM test for normality, we need the likelihood scores. The components of the scores with respect to $\boldsymbol{\beta}$ are $\widehat{\lambda}_i$ times the relevant component of $\mathbf{x}$, i.e., $\widehat{\lambda}_i\mathbf{x}_i$. These can be computed by using the `foreach` command:

```
. * Generate the scores to use in the LM test
. foreach var in $xlist {
  2.   generate score`var´ = gres1*`var´
  3. }
. global scores score* gres1 gres2
```

Recall that **gres1** is the score with respect to the intercept β_1, and **gres2** is the score with respect to the intercept σ.

To execute the regression-based test of normality, we regress 1 on **scores** and compute the NR^2 statistic.

```
. * Test of normality in tobit regression
. * NR^2 from the uncentered regression has chi-squared distribution
. generate one = 1
. quietly regress one gres3 gres4 $scores, noconstant
. display "N R^2 = " e(N)*e(r2) " with p-value = " chi2tail(2,e(N)*e(r2))
N R^2 = 1832.1279 with p-value = 0
```

The outcome of the test is a very strong rejection of the normality hypothesis, even though the expenditure variable was transformed to logarithms.

The properties of the conditional moment approach implemented here have been investigated by Skeels and Vella (1999), who found that using the asymptotic distribution of this test produces severe size distortions, even in moderately large samples. This is an important limitation of the test. Drukker (2002) developed a parametric bootstrap to correct the size distortion by using bootstrap critical values. His Monte Carlo results show that the test based on bootstrap critical values has reasonable power for samples larger than 500.

The user-written **tobcm** command (Drukker 2002) implements this better variant of the test. The command only works after **tobit**, and with the left-censoring point 0 and no right censoring. To compare the above outcome of the normality test with that from the improved bootstrap version, the interested reader can perform the **tobcm** command quite easily.

Test of homoskedasticity

For testing homoskedasticity, the alternative hypothesis is that the variance is of the form $\sigma^2 \exp(\mathbf{w}'_i\boldsymbol{\alpha})$. This leads to an auxiliary regression of 1 on $\widehat{\lambda}_i$, $\widehat{\lambda}_i\mathbf{x}_i$, and $z_i\widehat{\lambda}_i\mathbf{w}_i$. The auxiliary regressors, $z_i\widehat{\lambda}_i$, can be generated after specifying $\mathbf{w}$. Often $\mathbf{w}$ is specified to be the same as $\mathbf{x}$. If $\dim(\mathbf{x}_i) = K$ and $\dim(\mathbf{w}_i) = J$, then $\text{NR}^2_u \sim \chi^2(K+J+1)$ under the null hypothesis. The following additional commands that follow on from those for the normality test generate the additional needed components, $z_i\widehat{\lambda}_i\mathbf{w}_i$, for the test of homoskedasticity.

```
. * Test of homoskedasticity in tobit regression
. foreach var in $xlist {
  2.    generate score2`var´ = gres2*`var´
  3. }
. global scores2 score* score2* gres1 gres2
. * summarize $scores2
. quietly regress one gres3 gres4 $scores2, noconstant
. display "N R^2 = " e(N)*e(r2) " with p-value = " chi2tail(2,e(N)*e(r2))
N R^2 = 2585.9089 with p-value = 0
```

The redundant regressors (scores) are dropped from the auxiliary regression. This outcome also leads to a strong rejection of the null hypothesis of homoskedasticity against the alternative that the variance is of the form specified. If an investigator wants to specify different components of $\mathbf{w}$, then the required modifications to the above commands are trivial.

16.4.7 Next step?

Despite the apparently satisfactory estimation results for the tobit model, the diagnostic tests reveal weaknesses. The failure of normality and homoskedasticity assumptions have serious consequences for censored-data regression that do not arise in the case of linear regression. A natural question that arises concerns the direction in which additional modeling effort might be directed to arrive at a more general model.

Two approaches to such generalization will be considered. The two-part model, given in the next section, specifies one model for the censoring mechanism and a second distinct model for the outcome conditional on the outcome being observed. The sample-selection model, presented in the subsequent section, instead specifies a joint distribution for the censoring mechanism and outcome, and then finds the implied distribution conditional on the outcome observed.

16.5 Two-part model in logs

The tobit regression makes a strong assumption that the same probability mechanism generates both the zeros and the positives. It is more flexible to allow for the possibility that the zero and positive values are generated by different mechanisms. Many applications have shown that an alternative model, the two-part model or the hurdle model, can provide a better fit by relaxing the tobit model assumptions.

This model is the natural next step in our modeling strategy. Again we apply it to a model in logs rather than levels.

16.5.1 Model structure

The first part of the two-part model is a binary outcome equation that models $\Pr(\texttt{ambexp} > 0)$, using any of the binary outcome models considered in chapter 11 (usually probit). The second part uses linear regression to model $E(\ln \mathtt{ambexp}|\mathtt{ambexp} > 0)$. The two parts are assumed to be independent and are usually estimated separately.

Let y denote $\mathtt{ambexp}$. Define a binary indicator, d, of positive expenditure such that $d = 1$ if $y > 0$ and $d = 0$ if $y = 0$. When $y = 0$, we observe only $\Pr(d = 0)$. For those with $y > 0$, let $f(y|d = 1)$ be the conditional density of y. The two-part model for y is then given by

$$f(y|\mathbf{x}) = \left\{ \begin{array}{ll} \Pr(d = 0|\mathbf{x}) & \text{if } y = 0 \\ \Pr(d = 1|\mathbf{x})f(y|d = 1, \mathbf{x}) & \text{if } y > 0 \end{array} \right. \tag{16.5}$$

The same regressors often appear in both parts of the model, but this can and should be relaxed if there are obvious exclusion restrictions.

The probit or the logit is an obvious choice for the first part. If a probit model is used, then $\Pr(d = 1|\mathbf{x}) = \Phi(\mathbf{x}_1'\boldsymbol{\beta}_1)$. If a lognormal model for $y|y > 0$ is given, then $(\ln y|d = 1, \mathbf{x}) \sim N(\mathbf{x}_2'\boldsymbol{\beta}_2, \sigma_2^2)$. Combining these, we have for the model in logs

$$E(y|\mathbf{x}_1, \mathbf{x}_2) = \Phi(\mathbf{x}_1'\boldsymbol{\beta}_1)\exp(\mathbf{x}_2'\boldsymbol{\beta}_2 + \sigma_2^2/2)$$

where the second term uses the result that if $\ln y \sim N(\mu, \sigma^2)$ then $E(y) = \exp(\mu + \sigma^2/2)$.

ML estimation of (16.5) is straightforward because it separates the estimation of a discrete choice model using all observations and the estimation of the parameters of the density $f(y|d = 1, \mathbf{x})$ using only the observations with $y > 0$.

16.5.2   Part 1 specification

In the example considered here, $\mathbf{x}_1 = \mathbf{x}_2$, but there is no reason why this should always be so. It is an advantage of the two-part model that it provides the flexibility to have different regressors in the two parts. In this example, the first part is modeled through a probit regression, and again one has the flexibility to change this to logit or cloglog. Comparing the results from the tobit, two-part, and selection models is a little easier if we use the probit form.

```
. * Part 1 of the two-part model
. probit dy $xlist, nolog
Probit regression                               Number of obs   =        3328
                                                LR chi2(6)      =      509.53
                                                Prob > chi2     =      0.0000
Log likelihood = -1197.6644                     Pseudo R2       =      0.1754
```

dy	Coef.	Std. Err.	z	P>|z|	[95% Conf. Interval]	
age	.097315	.0270155	3.60	0.000	.0443656	.1502645
female	.6442089	.0601499	10.71	0.000	.5263172	.7621006
educ	.0701674	.0113435	6.19	0.000	.0479345	.0924003
blhisp	-.3744867	.0617541	-6.06	0.000	-.4955224	-.2534509
totchr	.7935208	.0711156	11.16	0.000	.6541367	.9329048
ins	.1812415	.0625916	2.90	0.004	.0585642	.3039187
_cons	-.7177087	.1924667	-3.73	0.000	-1.094937	-.3404809

```
. scalar llprobit = e(ll)
```

The probit regression indicates that all covariates are statistically significant determinants of the probability of positive expenditure. The standard marginal effects calculations can be done for the first part, as illustrated in chapter 14.

16.5.3 Part 2 of the two-part model

The second part is a linear regression of `lny`, here ln(ambexp), on the regressors in the global macro `xlist`.

```
. * Part 2 of the two-part model
. regress lny $xlist if dy==1
```

Source	SS	df	MS		Number of obs =	2802
					F(6, 2795) =	110.58
Model	1069.37332	6	178.228887		Prob > F =	0.0000
Residual	4505.06629	2795	1.61183051		R-squared =	0.1918
					Adj R-squared =	0.1901
Total	5574.43961	2801	1.99016052		Root MSE =	1.2696

lny	Coef.	Std. Err.	t	P>|t|	[95% Conf. Interval]	
age	.2172327	.0222225	9.78	0.000	.1736585	.2608069
female	.3793756	.0485772	7.81	0.000	.2841247	.4746265
educ	.0222388	.0097615	2.28	0.023	.0030983	.0413793
blhisp	-.2385321	.0551952	-4.32	0.000	-.3467597	-.1303046
totchr	.5618171	.0305078	18.42	0.000	.501997	.6216372
ins	-.020827	.0500062	-0.42	0.677	-.1188797	.0772258
_cons	4.907825	.1681512	29.19	0.000	4.578112	5.237538

```
. scalar lllognormal = e(ll)

. predict rlambexp, residuals
```

The coefficients of regressors in the second part have the same sign as those in the first part, aside from the `ins` variable, which is highly statistically insignificant in the second part.

Given the assumption that the two parts are independent, the joint likelihood for the two parts is the sum of two log likelihoods, i.e., −5,838.8. The computation is shown below.

```
. * Create two-part model log likelihood
. scalar lltwopart = llprobit + lllognormal  //two-part model log likelihood
. display "lltwopart = " lltwopart
lltwopart = -5838.8218
```

By comparison, the log likelihood for the tobit model is −7,494.29. The two-part model fits the data considerably better, even if AIC or BIC is used to penalize the two-part model for its additional parameters.

Does the two-part model eliminate the twin problems of heteroskedasticity and nonnormality? This is easily checked using the `hettest` and `sktest` commands.

```
. * hettest and sktest commands
. quietly regress lny $xlist if dy==1
. hettest
Breusch-Pagan / Cook-Weisberg test for heteroskedasticity
        Ho: Constant variance
        Variables: fitted values of lny
        chi2(1)      =     19.25
        Prob > chi2  =    0.0000
. sktest rlambexp
```

```
                    Skewness/Kurtosis tests for Normality
                                                        ———— joint ————
      Variable │    Obs   Pr(Skewness)   Pr(Kurtosis)  adj chi2(2)    Prob>chi2
    ───────────┼───────────────────────────────────────────────────────────────
      rlambexp │  3.3e+03    0.0000         0.0592          .           0.0000
```

The tests unambiguously reject the homoskedasticity and normality hypotheses. However, unlike the tobit model, neither condition is necessary for consistency of the estimator. The key assumption needed is that $E(\ln y|d = 1, \mathbf{x})$ is linear in $\mathbf{x}$. On the other hand, it is known that the OLS estimate of the residual variance will be biased in the presence of heteroskedasticity. This deficiency will extend to those predictors of y that involve the residual variance. This point is pursued further in section 16.8.

From the viewpoint of interpretation, the two-part model is flexible and attractive because it allows different covariates to have a different impact on the two parts of the model. For example, it allows a variable to make its impact entirely by changing the probability of a positive outcome, with no impact on the size of the outcome conditional on it being positive. In our example, the coefficient of `ins` in the conditional regression has a small and statistically insignificant coefficient but has a positive and significant coefficient in the probit equation.

16.6   Selection model

The two-part model attains some of its flexibility and computational simplicity by assuming that the two parts—the decision to spend and the amount spent—are independent. This is a potential restriction on the model. If it is conceivable that, after controlling for regressors, those with positive expenditure levels are not randomly selected from the population, then the results of the second-stage regression suffer from selection bias. The selection model used in this section considers the possibility of such bias by allowing for possible dependence in the two parts of the model. This new model is an example of a bivariate sample-selection model, also known as the type-2 tobit model.

The application in this section uses expenditures in logs. The same methods can be applied without modification to expenditures in levels.

16.6.1   Model structure and assumptions

Throughout this section, an asterisk will denote a latent variable. Let y_2^* denote the outcome of interest, here expenditure. In the standard tobit model, this outcome is observed if $y_2^* > 0$. A more general model introduces a second latent variable, y_1^*, and the outcome y_2^* is observed if $y_1^* > 0$. In the present case, y_1^* determines whether an individual has any ambulatory expenditure, y_2^* determines the level of expenditure, and $y_1^* \neq y_2^*$.

The two-equation model comprises a selection equation for y_1, where

$$y_1 = \begin{cases} 1 & \text{if } y_1^* > 0 \\ 0 & \text{if } y_1^* \leq 0 \end{cases}$$

and a resultant outcome equation for y_2, where

$$y_2 = \begin{cases} y_2^* & \text{if } y_1^* > 0 \\ - & \text{if } y_1^* \leq 0 \end{cases}$$

Here y_2 is observed only when $y_1^* > 0$, possibly taking a negative value, whereas y_2 need not take on any meaningful value when $y_1^* \leq 0$. The classic version of the model is linear with additive errors, so

$$y_1^* = \mathbf{x}_1' \boldsymbol{\beta}_1 + \varepsilon_1$$
$$y_2^* = \mathbf{x}_2' \boldsymbol{\beta}_2 + \varepsilon_2$$

with ε_1 and ε_2 possibly correlated. The tobit model is a special case where $y_1^* = y_2^*$.

It is assumed that the correlated errors are jointly normally distributed and homoskedastic, i.e.,

$$\begin{bmatrix} \varepsilon_1 \\ \varepsilon_2 \end{bmatrix} \sim N \left(\begin{bmatrix} 0 \\ 0 \end{bmatrix}, \begin{bmatrix} 1 & \sigma_{12} \\ \sigma_{12} & \sigma_2^2 \end{bmatrix} \right)$$

where the normalization $\sigma_1^2 = 1$ is used because only the sign of y_1^* is observed. Estimation by ML is straightforward.

The likelihood function for this model is

$$L = \prod_{i=1}^{n} \{\Pr(y_{1i}^* \leq 0)\}^{1-y_{1i}} \{f(y_{2i} \mid y_{1i}^* > 0) \times \Pr(y_{1i}^* > 0)\}^{y_{1i}}$$

where the first term is the contribution when $y_{1i}^* \leq 0$, because then $y_{1i} = 0$, and the second term is the contribution when $y_{1i}^* > 0$. This likelihood function can be specialized to models other than the linear model considered here. In the case of linear models with jointly normal errors, the bivariate density, $f^*(y_1^*, y_2^*)$, is normal, and hence the conditional density in the second term is univariate normal.

The essential structure of the model and the ML estimation procedure are not affected by the decision to model positive expenditure on the log (rather than the linear) scale, although this does affect the conditional prediction of the level of expenditure. This step is taken here even though tests implemented in the previous two sections show that the normality and homoskedasticity assumptions are both questionable.

16.6.2   ML estimation of the sample-selection model

ML estimation of the bivariate sample-selection model with the `heckman` command is straightforward. The basic syntax for this command is

`heckman` *depvar* [*indepvars*] [*if*] [*in*] [*weight*]`,` `select(`[*depvar_s* `=`] *varlist_s*
    [`, noconstant`]`)` [*options*]

where `select()` is the option for specifying the selection equation. One needs to specify variable lists for both the selection equation and for the outcome equation. In many cases, the investigator might use the same set of regressors in both equations. When this is done, it is often referred to as the case in which model identification is based solely upon the nonlinearity in the functional form. Because the selection equation is nonlinear, it potentially allows the higher powers of regressors to affect the selection variable. In the linear outcome equation, of course, the higher powers do not appear. Therefore, the nonlinearity of the selection regression automatically generates exclusion restrictions. That is, it allows for independent source of variation in the probability of a positive outcome; hence the term "identification through nonlinear functional form".

The specification of the selection equation involves delicate identification issues. For example, if the nonlinearity implied by the probit model is slight, then the identification will be fragile. For this reason, it is common in applied work to look for exclusion restrictions. The investigator seeks a variable(s) that can generate nontrivial variation in the selection variable but does not affect the outcome variable directly. This is exactly the same argument as was encountered in earlier chapters in the context of instrumental variables. A valid exclusion restriction arises if a suitable instrument is available and this may vary from case to case. We will illustrate the practical importance of these ideas in the examples that follow.

16.6.3   Estimation without exclusion restrictions

We first estimate the parameters of the selection model without exclusion restrictions.

```
. * Heckman MLE without exclusion restrictions
. heckman lny $xlist, select(dy = $xlist) nolog
```

Heckman selection model   Number of obs   =   3328
(regression model with sample selection) Censored obs   =   526
                                          Uncensored obs   =   2802

                                          Wald chi2(6)   =   294.42
Log likelihood = -5838.397   Prob > chi2   =   0.0000

	Coef.	Std. Err.	z	P>|z|	[95% Conf. Interval]	
lny						
age	.2122921	.022958	9.25	0.000	.1672952	.257289
female	.349728	.0596734	5.86	0.000	.2327704	.4666856
educ	.0188724	.0105254	1.79	0.073	-.0017569	.0395017
blhisp	-.2196042	.0594788	-3.69	0.000	-.3361804	-.103028
totchr	.5409537	.0390624	13.85	0.000	.4643929	.6175145
ins	-.0295368	.051042	-0.58	0.563	-.1295772	.0705037
_cons	5.037418	.2261901	22.27	0.000	4.594094	5.480743
dy						
age	.0984482	.0269881	3.65	0.000	.0455526	.1513439
female	.6436686	.0601399	10.70	0.000	.5257966	.7615407
educ	.0702483	.0113404	6.19	0.000	.0480216	.092475
blhisp	-.3726284	.0617336	-6.04	0.000	-.4936241	-.2516328
totchr	.7946708	.0710278	11.19	0.000	.6554588	.9338827
ins	.1821233	.0625485	2.91	0.004	.0595305	.3047161
_cons	-.7244413	.192427	-3.76	0.000	-1.101591	-.3472913
/athrho	-.124847	.1466391	-0.85	0.395	-.4122544	.1625604
/lnsigma	.2395983	.0143319	16.72	0.000	.2115084	.2676882
rho	-.1242024	.1443771			-.3903852	.1611435
sigma	1.270739	.018212			1.23554	1.30694
lambda	-.1578287	.1842973			-.5190448	.2033874

LR test of indep. eqns. (rho = 0):   chi2(1) =   0.85   Prob > chi2 = 0.3569

The log likelihood for this model is very slightly higher than that for the two-part model— $-5{,}838.4$ compared with $-5{,}838.8$ (see section 16.5.3). Consistent with this small difference is the finding that $\hat{\rho} = -0.124$ with the 95% confidence interval $[-0.390, 0.161]$. The likelihood-ratio test has a p-value of 0.36. Thus the estimated correlation between the errors is not significantly different from zero, and the hypothesis that the two parts are independent cannot be rejected.

The foregoing conclusion should be treated with caution because the model is based on a bivariate normality assumption that is itself suspect. The two-step estimation, considered next, relies on a univariate normality assumption and is expected to be relatively more robust.

16.6.4   Two-step estimation

The two-step method is based on the conditional expectation

$$E(y_2|\mathbf{x},\ y_1^* > 0) = \mathbf{x}_2'\boldsymbol{\beta}_2 + \sigma_{12}\lambda(\mathbf{x}_1'\boldsymbol{\beta}_1) \qquad (16.6)$$

where $\lambda(\cdot) = \phi(\cdot)/\Phi(\cdot)$. The motivation is that because $y_2^* = \mathbf{x}_2'\boldsymbol{\beta}_2 + \varepsilon_2$, $E(y_2|\mathbf{x}, y_1^* > 0) = \mathbf{x}_2'\boldsymbol{\beta}_2 + E(\varepsilon_2|y_1^* > 0)$ and, given normality of the errors, $E(\varepsilon_2|y_1^* > 0) = \sigma_{12}\lambda(\mathbf{x}_1'\boldsymbol{\beta}_1)$.

The second term in (16.6) can be estimated by $\lambda(\mathbf{x}_1'\widehat{\boldsymbol{\beta}}_1)$, where $\widehat{\boldsymbol{\beta}}_1$ is obtained by probit regression of y_1 on $\mathbf{x}_1$. The OLS regression of y_2 on $\mathbf{x}_2$ and the generated regressor, $\lambda(\mathbf{x}_1'\widehat{\boldsymbol{\beta}}_1)$, called the inverse of the Mills' ratio or the nonselection hazard, yields a semiparametric estimate of $(\boldsymbol{\beta}_2, \sigma_{12})$. The calculation of the standard errors, however, is complicated by the presence in the regression of the generated regressor, $\lambda(\mathbf{x}_1'\widehat{\boldsymbol{\beta}}_1)$.

The addition of the `twostep` option to `heckman` yields the two-step estimator.

```
. * Heckman 2-step without exclusion restrictions
. heckman lny $xlist, select(dy = $xlist) twostep
```

Heckman selection model -- two-step estimates	Number of obs	=	3328
(regression model with sample selection)	Censored obs	=	526
	Uncensored obs	=	2802
	Wald chi2(6)	=	189.46
	Prob > chi2	=	0.0000

	Coef.	Std. Err.	z	P>|z|	[95% Conf. Interval]
lny						
age	.202124	.0242974	8.32	0.000	.1545019	.2497462
female	.2891575	.073694	3.92	0.000	.1447199	.4335951
educ	.0119928	.0116839	1.03	0.305	-.0109072	.0348928
blhisp	-.1810582	.0658522	-2.75	0.006	-.3101261	-.0519904
totchr	.4983315	.0494699	10.07	0.000	.4013724	.5952907
ins	-.0474019	.0531541	-0.89	0.373	-.151582	.0567782
_cons	5.302572	.2941363	18.03	0.000	4.726076	5.879069
dy						
age	.097315	.0270155	3.60	0.000	.0443656	.1502645
female	.6442089	.0601499	10.71	0.000	.5263172	.7621006
educ	.0701674	.0113435	6.19	0.000	.0479345	.0924003
blhisp	-.3744867	.0617541	-6.06	0.000	-.4955224	-.2534509
totchr	.7935208	.0711156	11.16	0.000	.6541367	.9329048
ins	.1812415	.0625916	2.90	0.004	.0585642	.3039187
_cons	-.7177087	.1924667	-3.73	0.000	-1.094936	-.3404809
mills						
lambda	-.4801696	.2906565	-1.65	0.099	-1.049846	.0895067
rho	-0.37130					
sigma	1.2932083					
lambda	-.4801696	.2906565				

The standard errors for the regression coefficients, $\widehat{\boldsymbol{\beta}}_2$, are computed, allowing for the estimation error of $\lambda(\mathbf{x}_1'\widehat{\boldsymbol{\beta}}_1)$; see [R] **heckman**. These standard errors are in general larger than those from the ML estimation. Although no standard error is provided for `rho=lambda/sigma`, the hypothesis of independence of ε_1 and ε_2 can be tested directly by using the coefficient of `lambda`, because from (16.6), this is the error covariance σ_{12}. The coefficient of `lambda` has a larger z statistic, -1.65, than in the ML case, and it is significantly different from zero at any p-value higher than 0.099. Thus the two-step estimator produces somewhat stronger evidence of selection than does the ML estimator.

The standard errors of the two-step estimator are larger than those of the ML estimator in part because the variable $\lambda(\mathbf{x}_1'\widehat{\boldsymbol{\beta}}_1)$ can be collinear with the other regressors in the outcome equation ($\mathbf{x}_2$). This is highly likely if $\mathbf{x}_1 = \mathbf{x}_2$, as would be the case when there are no exclusion restrictions. Having exclusion restrictions, so that $\mathbf{x}_1 \neq \mathbf{x}_2$, may reduce the collinearity problem, especially in small samples.

16.6.5   Estimation with exclusion restrictions

For more robust identification, it is usually recommended, as has been explained above, that exclusion restrictions be imposed. This requires that the selection equation have an exogenous variable that is excluded from the outcome equation. Moreover, the excluded variable should have a substantial (nontrivial) impact on the probability of selection. Because it is often hard to come up with an excluded variable that does not directly affect the outcome and does affect the selection, the investigator should have strong justification for imposing the exclusion restriction.

We repeat the computation of the two-step Heckman model with an additional regressor, `income`, in the selection equation.

(*Continued on next page*)

```
. * Heckman MLE with exclusion restriction
. heckman lny $xlist, select(dy = $xlist income) nolog
```

Heckman selection model
(regression model with sample selection)

				Number of obs	=	3328
				Censored obs	=	526
				Uncensored obs	=	2802
				Wald chi2(6)	=	288.88
Log likelihood = -5836.219				Prob > chi2	=	0.0000

	Coef.	Std. Err.	z	P>|z|	[95% Conf. Interval]	
lny						
age	.2119749	.0230072	9.21	0.000	.1668816	.2570682
female	.3481441	.0601142	5.79	0.000	.2303223	.4659658
educ	.018716	.0105473	1.77	0.076	-.0019563	.0393883
blhisp	-.2185714	.0596687	-3.66	0.000	-.3355199	-.101623
totchr	.53992	.0393324	13.73	0.000	.4628299	.61701
ins	-.0299871	.0510882	-0.59	0.557	-.1301182	.0701439
_cons	5.044056	.2281259	22.11	0.000	4.596938	5.491175
dy						
age	.0879359	.027421	3.21	0.001	.0341917	.14168
female	.6626649	.0609384	10.87	0.000	.5432278	.7821021
educ	.0619485	.0120295	5.15	0.000	.0383711	.0855258
blhisp	-.3639377	.0618734	-5.88	0.000	-.4852073	-.2426682
totchr	.7969518	.0711306	11.20	0.000	.6575383	.9363653
ins	.1701367	.0628711	2.71	0.007	.0469117	.2933618
income	.0027078	.0013168	2.06	0.040	.000127	.0052886
_cons	-.6760546	.1940288	-3.48	0.000	-1.056344	-.2957652
/athrho	-.1313456	.1496292	-0.88	0.380	-.4246134	.1619222
/lnsigma	.2398173	.0144598	16.59	0.000	.2114767	.268158
rho	-.1305955	.1470772			-.4008098	.1605217
sigma	1.271017	.0183786			1.235501	1.307554
lambda	-.1659891	.1878698			-.5342072	.2022291

LR test of indep. eqns. (rho = 0):   chi2(1) =   0.91   Prob > chi2 = 0.3406

The results are only slightly different from those reported above, although `income` appears to have significant additional explanatory power. Furthermore, the use of this exclusion restriction is debatable because there are reasons to expect that `income` should appear in the outcome equation also.

16.7   Prediction from models with outcome in logs

For the models considered in this chapter, conditional prediction is an important application of the estimated parameters of the model. Such an exercise may be of the within-sample type, or it may involve comparison of fitted values under alternative scenarios, as illustrated in section 3.6. Whether a model predicts well within the sample is obviously an important consideration in model comparison and selection.

Calculation and comparison of predicted values is relatively simpler in the levels form of the model because there is no retransformation involved. In the current analysis, the dependent variable is log transformed but one wants predictions in levels, and hence the retransformation problem, first mentioned in chapter 3, must be confronted.

Table 16.2 provides expressions for the conditional and unconditional means for the three models with outcome in logs rather than levels, presented in sections 16.4–16.6. The predictors are functions that depend upon the linear-index function, $\mathbf{x}'\boldsymbol{\beta}$, and variance and covariance parameters, σ^2, σ_2^2, and σ_{12}. These formulas are derived under the twin assumptions of normality and homoskedasticity. The dependence of the predictor on variances estimated under the assumption of homoskedastic errors is potentially problematic for all three models because if that assumption is incorrect, then the usual estimators of variance and covariance parameters will be biased.

Table 16.2. Expressions for conditional and unconditional means

Moment	Model	Prediction function
$E(y\|\mathbf{x}, y > 0)$	Tobit	$\exp(\mathbf{x}'\boldsymbol{\beta} + \sigma^2/2)[1 - \Phi\{(\gamma - \mathbf{x}'\boldsymbol{\beta})/\sigma\}]^{-1}$ $[1 - \Phi\{(\gamma - \mathbf{x}'\boldsymbol{\beta} - \sigma^2)/\sigma\}]$
$E(y\|\mathbf{x})$	Tobit	$\exp(\mathbf{x}'\boldsymbol{\beta} + \sigma^2/2)[1 - \Phi\{(\gamma - \mathbf{x}'\boldsymbol{\beta} - \sigma^2)/\sigma\}]$
$E(y_2\|\mathbf{x}, y_2 > 0)$	Two-part	$\exp(\mathbf{x}_2'\boldsymbol{\beta}_2 + \sigma_2^2/2)$
$E(y_2\|\mathbf{x})$	Two-part	$\exp(\mathbf{x}_2'\boldsymbol{\beta}_2 + \sigma_2^2/2)\Phi(\mathbf{x}_1'\boldsymbol{\beta}_1)$
$E(y_2\|\mathbf{x}, y_2 > 0)$	Selection	$\exp(\mathbf{x}_2'\boldsymbol{\beta}_2 + \sigma_2^2/2)\{1 - \Phi(-\mathbf{x}_1'\boldsymbol{\beta}_1)\}^{-1}$ $\{1 - \Phi(-\mathbf{x}_1'\boldsymbol{\beta}_1 - \sigma_{12}^2)\}$
$E(y_2\|\mathbf{x})$	Selection	$\exp(\mathbf{x}_2'\boldsymbol{\beta}_2 + \sigma_2^2/2)\{1 - \Phi(-\mathbf{x}_1'\boldsymbol{\beta}_1 - \sigma_{12}^2)\}$

16.7.1   Predictions from tobit

We begin by estimating $E(y|\mathbf{x})$ and $E(y|\mathbf{x}, y > 0)$ for the tobit model in logs.

```
. * Prediction from tobit on lny
. generate yhat = exp(xb+0.5*sigma^2)*(1-normal((gamma-xb-sigma^2)/sigma))
. generate ytrunchat = yhat / (1 - normal(threshold)) if dy==1
(526 missing values generated)
. summarize y yhat
```

Variable	Obs	Mean	Std. Dev.	Min	Max
y	3328	1386.519	2530.406	0	49960
yhat	3328	45805.91	273444.6	133.9767	1.09e+07

```
. summarize y yhat ytrunchat if dy==1
    Variable |     Obs       Mean    Std. Dev.       Min        Max
-------------+-----------------------------------------------------
           y |    2802     1646.8     2678.914         1      49960
        yhat |    2802    53271.5     297386.3   283.4537   1.09e+07
    ytrunchat |   2802   53536.84     297376.5   383.6245   1.09e+07
```

The estimates, denoted by **yhat** and **ytrunchat**, confirm that these predictors are very poor. Mean expenditure is overpredicted in both cases and more so in the censored case. The reported results reflect the high sensitivity of the estimator to estimates of σ^2.

16.7.2 Predictions from two-part model

Predictions of $E(y_2|\mathbf{x})$ and $E(y_2|\mathbf{x}, y_2 > 0)$ from the two-part model are considerably better but still biased. We first transform the fitted log values from the conditional part of the two-part model, assuming normality.

```
. * Two-part model predictions
. quietly probit dy $xlist
. predict dyhat, pr
. quietly regress lny $xlist if dy==1
. predict xbpos, xb
. generate yhatpos = exp(xbpos+0.5*e(rmse)^2)
```

Next we generate an estimate of the unconditional values, denoted by **yhat2step**, by multiplying by the fitted probability of the positive expenditure **dyhat** from the probit regression.

```
. * Unconditional prediction from two-part model
. generate yhat2step = dyhat*yhatpos
. summarize yhat2step y
    Variable |     Obs       Mean    Std. Dev.       Min        Max
-------------+-----------------------------------------------------
   yhat2step |    3328   1680.978     2012.084   87.29432   40289.03
           y |    3328   1386.519     2530.406          0      49960
. summarize yhatpos y if dy==1
    Variable |     Obs       Mean    Std. Dev.       Min        Max
-------------+-----------------------------------------------------
     yhatpos |    2802   1995.981     2087.072   430.8354   40289.03
           y |    2802     1646.8     2678.914         1      49960
```

The mean of the predicted values is considerably closer to the sample average than to the corresponding tobit estimator, confirming the greater robustness of the two-part model.

16.7.3   Predictions from selection model

Finally, we predict $E(y_2|\mathbf{x})$ and $E(y_2|\mathbf{x}, y_2 > 0)$ for the selection model.

```
. * Heckman model predictions
. quietly heckman lny $xlist, select(dy = $xlist)
. predict probpos, psel
. predict x1b1, xbsel
. predict x2b2, xb
. scalar sig2sq = e(sigma)^2
. scalar sig12sq = e(rho)*e(sigma)^2
. display "sigma1sq = 1" " sigma12sq = " sig12sq " sigma2sq = " sig2sq
sigma1sq = 1 sigma12sq = -.20055906 sigma2sq = 1.6147766
. generate yhatheck = exp(x2b2 + 0.5*(sig2sq))*(1 - normal(-x1b1-sig12sq))
. generate yhatposheck = yhatheck/probpos
. summarize yhatheck y probpos dy
```

Variable	Obs	Mean	Std. Dev.	Min	Max
yhatheck	3328	1659.802	1937.095	74.32413	37130.18
y	3328	1386.519	2530.406	0	49960
probpos	3328	.8415738	.1411497	.2029135	1
dy	3328	.8419471	.3648454	0	1

```
. summarize yhatposheck probpos dy y if dy==1
```

Variable	Obs	Mean	Std. Dev.	Min	Max
yhatposheck	2802	1970.923	2003.406	389.4755	37130.18
probpos	2802	.8661997	.1237323	.2867923	1
dy	2802	1	0	1	1
y	2802	1646.8	2678.914	1	49960

Qualitatively, the predictions from the selection model, denoted by yhatheck, are closer to those from the two-part model than to the tobit, as expected. The main difference from the two-part model comes from the dependence of the conditional mean on the covariance, which is unrestricted. The larger the covariance, the more likely is a greater difference between the two models. Although its predictions exhibit a positive bias, the selection model avoids the extremely large errors of prediction of the tobit model.

The poor prediction performance of the tobit model confirms the earlier conclusions about its unsuitability for modeling the current dataset.

16.8   Stata resources

For tobit estimation, the relevant entries are [R] **tobit**, [R] **tobit postestimation**, [R] **ivtobit**, and [R] **intreg**. Useful user-written commands are clad and tobcm. Various marginal effects can be computed by using margins with several different predict options. For tobit panel estimation, the relevant command is [XT] **xttobit**, whose application is covered in chapter 18.

16.9   Exercises

1. Consider the "linear version" of the tobit model used in this chapter. Using tests of homoskedasticity and normality, compare the outcome of the tests with those for the log version of the model.

2. Using the linear form of the tobit model in the preceding exercise, compare average predicted expenditure levels for those with insurance and those without insurance (`ins=0`). Compare these results with those from the tobit model for log(`ambexp`).

3. Suppose we want to study the sensitivity of the predicted expenditure from the log form of the tobit model to neglected homoskedasticity. Observe from the table in section 16.7 that the prediction formula involves the variance parameter, σ^2, that will be replaced by its estimate. Using the censoring threshold 0, draw a simulated heteroskedastic sample from a lognormal regression model with a single exogenous variable. Consider two levels of heteroskedasticity, low and high. By considering variations in the estimated σ^2, show how the resulting biases in the estimate of σ^2 from the homoskedastic tobit model lead to biases in the mean prediction.

4. Repeat the simulation exercise using regression errors that are drawn from a $\chi^2(5)$ distribution. Recenter the simulated draws by subtracting the mean so that the recentered errors have a zero mean. Summarize the results of the prediction exercise for this case.

5. A conditional predictor for levels $E(y|\mathbf{x}, y > 0)$ mentioned in section 3.6, given parameters of a model estimated in logs, is $\exp(\mathbf{x}'\widehat{\boldsymbol{\beta}})N^{-1}\sum_i\exp(\widehat{\varepsilon}_i)$. This expression is based on the assumption that ε_i are independent and identically distributed but normality is not assumed. Apply this conditional predictor to both the parameters of the two-part and selection models estimated by the two-step procedure, and obtain estimates of $E(y|\mathbf{x}, y > 0)$ and $E(y|\mathbf{x})$. Compare the results with those given in section 16.7.

6. Repeat the calculations of scores, `gres1`, and `gres2` reported in section 16.4.6. Test that the calculations are done correctly; all the score components should have a zero-mean property.

17 Count-data models

17.1 Introduction

In many contexts, the outcome of interest is a nonnegative integer, or a count, denoted by y, $y \in \mathbb{N}_0 = \{0, 1, 2, \ldots\}$. Examples can be found in demography, economics, ecology, environmental studies, insurance, and finance, to mention just a few of the areas of application.

The objective is to analyze y in a regression setting, given a vector of K covariates, $\mathbf{x}$. Because the response variable is discrete, its distribution places probability mass at nonnegative integer values only. Fully parametric formulations of count models accommodate this property of the distribution. Some semiparametric regression models only accommodate $y \geq 0$ but not discreteness. Count regressions are nonlinear; $E(y|\mathbf{x})$ is usually a nonlinear function, most commonly a single-index function like $\exp(\mathbf{x}'\boldsymbol{\beta})$. Several special features of count regression models are intimately connected to discreteness and nonlinearity.

Some of the standard complications in analyzing count data include the following: presence of unobserved heterogeneity akin to omitted variables; the small-mean property of y as manifested in the presence of many zeros, sometimes an "excess" of zeros; truncation in the observed distribution of y; and endogenous regressors. To deal with these topics, it is necessary to go beyond the basic commands in Stata.

The chapter begins with the basic Poisson and negative binomial models, using the `poisson` and `nbreg` commands, and then details some standard extensions including the hurdle, finite-mixture, and zero-inflated models. The last part of the chapter deals with complications arising from endogenous regressors.

17.2 Features of count data

The natural starting point for analyses of counts is the Poisson distribution and the Poisson model. The univariate Poisson distribution, denoted by Poisson$(y|\mu)$, for the number of occurrences of the event y over a fixed exposure period has the probability mass function

$$\Pr(Y = y) = \frac{e^{-\mu}\mu^y}{y!}, \qquad y = 0, 1, 2, \ldots \tag{17.1}$$

where μ is the intensity or rate parameter. The first two moments are

$$E(Y) = \mu$$
$$\mathrm{Var}(Y) = \mu \tag{17.2}$$

This shows the well-known equality of mean and variance property, also called the equidispersion property of the Poisson distribution.

The standard mean parameterization is $\mu = \exp(\mathbf{x}'\boldsymbol{\beta})$ to ensure that $\mu > 0$. This implies, based on (17.2), that the model is intrinsically heteroskedastic.

17.2.1   Generated Poisson data

To illustrate some features of Poisson-distributed data, we use the `rpoisson()` function, introduced in an update to Stata 10, to make draws from the $\mathrm{Poisson}(y|\mu = 1)$ distribution.

```
. * Poisson (mu=1) generated data
. quietly set obs 10000
. set seed 10101                    // set the seed !
. generate xpois= rpoisson(1) // draw from Poisson(mu=1)
. summarize xpois
    Variable |       Obs        Mean    Std. Dev.       Min        Max
       xpois |     10000       .9933    1.001077          0          6
. tabulate xpois
       xpois |      Freq.     Percent        Cum.
           0 |      3,721       37.21       37.21
           1 |      3,653       36.53       73.74
           2 |      1,834       18.34       92.08
           3 |        607        6.07       98.15
           4 |        142        1.42       99.57
           5 |         35        0.35       99.92
           6 |          8        0.08      100.00
       Total |     10,000      100.00
```

The expected frequency of zeros from (17.1) is $\mathrm{Pr}(Y = 0|\mu = 1) = e^{-1} = 0.368$. The simulated sample has 37.2% zeros. Clearly, the larger is μ, and the smaller will be the proportion of zeros; e.g., for $\mu = 5$, say, the expected proportion of zeros will be just 0.0067%. For data with a small mean, as for example in the case of number of children born in a family (or annual number of accidents or hospitalizations), zero observations are an important feature of the data. Further, when the mean is small, a high proportion of the sample will cluster on a relatively few distinct values. In this example, about 98% of the observations cluster on just four distinct values. The generated data also reflect the equidispersion property, i.e., equality of mean and variance of Y, because the standard deviation and hence variance are close to 1.

17.2.2 Overdispersion and negative binomial data

The equidispersion property is commonly violated in applied work, because overdispersion is common. Then the (conditional) variance exceeds the (conditional) mean. Such additional dispersion can be accounted for in many ways, of which the presence of unobserved heterogeneity is one of the most common.

Unobserved heterogeneity, which generates additional variability in y, can be generated by introducing multiplicative randomness. We replace μ with $\mu\nu$, where ν is a random variable, hence $y \sim \text{Poisson}(y|\mu\nu)$. Suppose we specify ν such that $E(\nu) = 1$ and $\text{Var}(\nu) = \sigma^2$. Then it is straightforward to show that ν preserves the mean but increases dispersion. Specifically, $E(y) = \mu$ and $\text{Var}(y) = \mu(1 + \mu\sigma^2) > E(y) = \mu$. The term "overdispersion" describes the feature $\text{Var}(y) > E(y)$, or more precisely $\text{Var}(y|\mathbf{x}) > E(y|\mathbf{x})$, in a regression model.

In the well-known special case that $\nu \sim \text{Gamma}(1, \alpha)$, where α is the variance parameter of the gamma distribution, the marginal distribution of y is a Poisson–gamma mixture with a closed form—the negative binomial (NB) distribution denoted by $\text{NB}(\mu, \alpha)$—whose probability mass function is

$$\Pr(Y = y|\mu, \alpha) = \frac{\Gamma(\alpha^{-1} + y)}{\Gamma(\alpha^{-1})\Gamma(y + 1)} \left(\frac{\alpha^{-1}}{\alpha^{-1} + \mu}\right)^{\alpha^{-1}} \left(\frac{\mu}{\mu + \alpha^{-1}}\right)^y \qquad (17.3)$$

where $\Gamma(\cdot)$ denotes the gamma integral that specializes to a factorial for an integer argument. The NB model is more general than the Poisson model, because it accommodates overdispersion and it reduces to the Poisson model as $\alpha \to 0$. The moments of the NB2 are $E(y|\mu, \alpha) = \mu$ and $\text{Var}(y|\mu, \alpha) = \mu(1 + \alpha\mu)$. Empirically, the quadratic variance function is a versatile approximation in a wide variety of cases of overdispersed data.

The NB regression model lets $\mu = \exp(\mathbf{x}'\boldsymbol{\beta})$ and leaves α as a constant. The default option for the NB regression in Stata is the version with a quadratic variance (NB2). Another variant of NB in the literature has a linear variance function, $\text{Var}(y|\mu, \alpha) = (1 + \alpha)\mu$, and is called the NB1 model. See Cameron and Trivedi (2005, chap. 20.4).

Using the mixture interpretation of the NB model, we simulate a sample from the $\text{NB}(\mu = 1, \alpha = 1)$ distribution. We first use the `rgamma(1,1)` function to obtain the gamma draw, v, with a mean of $1 \times 1 = 1$ and a variance of $\alpha = 1 \times 1^2 = 1$; see section 4.2.4. We then obtain Poisson draws with $\mu v = 1 \times v = v$, using the `rpoisson()` function with the argument v.

```
. * Negative binomial (mu=1 var=2) generated data
. set seed 10101                    // set the seed !
. generate xg = rgamma(1,1)
. generate xnegbin = rpoisson(xg)   // NB generated as a Poisson-gamma mixture
. summarize xnegbin
```

Variable	Obs	Mean	Std. Dev.	Min	Max
xnegbin	10000	1.0059	1.453092	0	17

```
. tabulate xnegbin
    xnegbin |      Freq.      Percent         Cum.
------------+-----------------------------------
          0 |      5,048        50.48        50.48
          1 |      2,436        24.36        74.84
          2 |      1,264        12.64        87.48
          3 |        607         6.07        93.55
          4 |        324         3.24        96.79
          5 |        151         1.51        98.30
          6 |         78         0.78        99.08
          7 |         46         0.46        99.54
          8 |         19         0.19        99.73
          9 |         11         0.11        99.84
         10 |          9         0.09        99.93
         11 |          3         0.03        99.96
         12 |          1         0.01        99.97
         14 |          1         0.01        99.98
         16 |          1         0.01        99.99
         17 |          1         0.01       100.00
------------+-----------------------------------
      Total |     10,000       100.00
```

As expected, the mean is close to 1 and the variance of $1.45^2 = 2.10$ is close to $(1+1) \times 1 = 2$. Relative to the Poisson(1) pseudorandom draws, this sample has more zeros, a longer right tail, and a variance-to-mean ratio in excess of 1. These features are a consequence of introducing the multiplicative heterogeneity term.

The `rnbinomial()` function can instead be used to make direct draws from the NB distribution, but because it uses an alternative parameterization of the NB distribution, it is easier to use the above Poisson–gamma mixture.

17.2.3 Modeling strategies

Given (17.1), $\mu = \exp(\mathbf{x}'\boldsymbol{\beta})$, and the assumption that the observations $(y_i|\mathbf{x}_i)$ are independent, Poisson maximum likelihood (ML) is often the starting point of a modeling exercise, especially if the entire distribution and not just the conditional mean is the object of interest.

Count data are often overdispersed. One approach is to maintain the conditional mean assumption $E(y|\mathbf{x}) = \exp(\mathbf{x}'\boldsymbol{\beta})$. Then one can continue to use the Poisson maximum likelihood estimator (MLE), which retains its consistency, but relax the equivariance assumption to obtain a robust estimate of the variance–covariance matrix of the estimator (VCE). Alternatively, the NB model, which explicitly models overdispersion, can be used with estimation by ML.

A quite different approach is still parametric but broader, in the sense that both the conditional mean and variance functions are allowed to be more flexible than the Poisson model. In particular, now $E(y|\mathbf{x}) \neq \exp(\mathbf{x}'\boldsymbol{\beta})$. The empirical examples in sections 17.3 and 17.4 illustrate several alternatives to the Poisson model.

17.2.4 Estimation methods

ML is the basic estimation method for a variety of count models that will be covered in the rest of this chapter. Nonlinear estimation based only on the exponential conditional mean moment condition is also feasible, using nonlinear regression or generalized method of moments (GMM), covered in chapters 10 and 11. Although chapters 10 and 11 explicitly covered only the Poisson and NB models, the programs provided there can be extended to other count models. Whereas ML is widely used, the GMM estimators are of special interest when the relationship of interest includes endogenous regressors. Stata also provides inference methods for count models within the generalized linear models (GLM) class. These models may differ in their parameterizations, but the estimation methods used are either ML or nonlinear generalized least squares (GLS).

17.3 Empirical example 1

In this section, we will estimate several parameters of count-data models for the annual number of doctor visits (docvis). The models are Poisson, NB2, and richer extensions of these models. The example in section 17.4 also considers the complication of excess zeros.

17.3.1 Data summary

The data are a cross-section sample from the U.S. Medical Expenditure Panel Survey for 2003. We model the annual number of doctor visits (docvis) using a sample of the Medicare population aged 65 and higher.

The covariates in the regressions are age (age), squared age (age2), years of education (educyr), presence of activity limitation (actlim), number of chronic conditions (totchr), having private insurance that supplements Medicare (private), and having public Medicaid insurance for low-income individuals that supplements Medicare (medicaid).

Summary statistics for the dependent variable and regressors are as follows:

```
. * Summary statistics for doctor visits data
. use mus17data.dta
. global xlist private medicaid age age2 educyr actlim totchr
. summarize docvis $xlist
```

Variable	Obs	Mean	Std. Dev.	Min	Max
docvis	3677	6.822682	7.394937	0	144
private	3677	.4966005	.5000564	0	1
medicaid	3677	.166712	.3727692	0	1
age	3677	74.24476	6.376638	65	90
age2	3677	5552.936	958.9996	4225	8100
educyr	3677	11.18031	3.827676	0	17
actlim	3677	.333152	.4714045	0	1
totchr	3677	1.843351	1.350026	0	8

The sampled individuals are aged 65–90 years, and a considerable portion have an activity limitation or chronic condition. The sample mean of docvis is 6.82, and the sample variance is $7.39^2 = 54.61$, so there is great overdispersion.

For count data, one should always obtain a frequency distribution or histogram. To reduce output, we create a variable, dvrange, with counts of 11–40 recoded as 40 and counts of 41–143 recoded as 143. We have

```
. * Tabulate docvis after recoding values > 10 to ranges 11-40 or 41-143
. generate dvrange = docvis

. recode dvrange (11/40 = 40) (41/143 = 143)
(dvrange: 786 changes made)

. tabulate dvrange
```

dvrange	Freq.	Percent	Cum.
0	401	10.91	10.91
1	314	8.54	19.45
2	358	9.74	29.18
3	334	9.08	38.26
4	339	9.22	47.48
5	266	7.23	54.72
6	231	6.28	61.00
7	202	5.49	66.49
8	179	4.87	71.36
9	154	4.19	75.55
10	108	2.94	78.49
40	774	21.05	99.54
143	16	0.44	99.97
144	1	0.03	100.00
Total	3,677	100.00	

The distribution has a long right tail. Twenty-two percent of observations exceed 10, and the maximum is 144. More than 99% of the values are under 40. The proportion of zeros is 10.9%. This is relatively low for this type of data, partly because the data pertain to the elderly population. Samples of the younger and usually healthier population often have as many as 90% zero observations for some health outcomes.

17.3.2 Poisson model

For the Poisson model, the probability mass function is the Poisson distribution given in (17.1), and the default is the exponential mean parameterization

$$\mu_i = \exp(\mathbf{x}_i'\boldsymbol{\beta}), \quad i = 1, \ldots, N \tag{17.4}$$

where by assumption there are K linearly independent covariates, usually including a constant. This specification restricts the conditional mean to be positive.

The Poisson MLE, denoted by $\widehat{\boldsymbol{\beta}}_P$, is the solution to K nonlinear equations corresponding to the ML first-order conditions

$$\sum_{i=1}^{N} \{y_i - \exp(\mathbf{x}_i'\boldsymbol{\beta})\}\mathbf{x}_i = \mathbf{0} \tag{17.5}$$

If $\mathbf{x}_i$ includes a constant term, then the residuals $y_i - \exp(\mathbf{x}_i'\boldsymbol{\beta})$ sum to zero based on (17.5). Because the log-likelihood function is globally concave, the iterative solution algorithm, usually the Newton–Raphson (see section 11.2), converges fast to a unique global maximum.

By standard ML theory, if the Poisson model is parametrically correctly specified, the estimator $\widehat{\boldsymbol{\beta}}_P$ is consistent for $\boldsymbol{\beta}$, with a covariance matrix estimated by

$$\widehat{V}(\widehat{\boldsymbol{\beta}}_P) = \left(\sum_{i=1}^{N} \widehat{\mu}_i \mathbf{x}_i \mathbf{x}_i' \right)^{-1} \tag{17.6}$$

where $\widehat{\mu}_i = \exp(\mathbf{x}_i'\widehat{\boldsymbol{\beta}}_P)$. We show below that it is usually very misleading to use (17.6) to estimate the VCE of $\widehat{\boldsymbol{\beta}}_P$, and we present better alternative methods.

The Poisson MLE is implemented with the **poisson** command, and the default estimate of the VCE is that in (17.6). The syntax for **poisson**, similar to that for **regress**, is

poisson *depvar* $\big[$ *indepvars* $\big]$ $\big[$ *if* $\big]$ $\big[$ *in* $\big]$ $\big[$ *weight* $\big]$ $\big[$, *options* $\big]$

The vce(robust) option yields a robust estimate of the VCE.

Two commonly used options are **offset()** and **exposure()**. Suppose regressor z is an exposure variable, such as time. Then, as z doubles, we expect the count, y, to double. Then $E(y|z, \mathbf{x}_2) = z \exp(\mathbf{x}_2'\boldsymbol{\beta}) = \exp(\ln z + \mathbf{x}_2'\boldsymbol{\beta})$. If the variable z appears in the regressor list, this constraint is imposed by using the **offset(z)** option. If instead the variable $\texttt{lnz} = \ln(\texttt{z})$ appears in the regressor list, this constraint is imposed by using the **exposure(lnz)** option.

Poisson model results

We first obtain and discuss the results for Poisson ML estimation.

```
. * Poisson with default ML standard errors
. poisson docvis $xlist, nolog
Poisson regression                          Number of obs   =       3677
                                            LR chi2(7)      =    4477.98
                                            Prob > chi2     =     0.0000
Log likelihood =  -15019.64                 Pseudo R2       =     0.1297
```

docvis	Coef.	Std. Err.	z	P>|z|	[95% Conf.	Interval]
private	.1422324	.0143311	9.92	0.000	.114144	.1703208
medicaid	.0970005	.0189307	5.12	0.000	.0598969	.134104
age	.2936722	.0259563	11.31	0.000	.2427988	.3445457
age2	-.0019311	.0001724	-11.20	0.000	-.0022691	-.0015931
educyr	.0295562	.001882	15.70	0.000	.0258676	.0332449
actlim	.1864213	.014566	12.80	0.000	.1578726	.2149701
totchr	.2483898	.0046447	53.48	0.000	.2392864	.2574933
_cons	-10.18221	.9720115	-10.48	0.000	-12.08732	-8.277101

The top part of the output lists sample size, the likelihood-ratio (LR) test for the joint significance of the seven regressors, the p-value associated with the test, and the pseudo-R^2 statistic that is intended to serve as a measure of the goodness of fit of the model (see section 10.7.1). On average, docvis in increasing in age, education, number of chronic conditions, being limited in activity, and having either type of supplementary health insurance. These results are also consistent with a priori expectations.

Another measure of the fit of the model is the squared coefficient of correlation between the fitted and observed values of the dependent variable. This is not provided by poisson but is easily computed as follows:

```
. * Squared correlation between y and yhat
. drop yphat
. predict yphat, n
. quietly correlate docvis yphat
. display "Squared correlation between y and yhat = " r(rho)^2
Squared correlation between y and yhat = .1530784
```

The squared correlation coefficient is low but reasonable for cross-section data.

The variables in the Poisson model appear to be highly statistically significant, but this is partly due to great underestimation of the standard errors, as we explain next.

Robust estimate of VCE for Poisson MLE

As explained in section 10.3.1, the Poisson MLE retains consistency if the count is not actually Poisson distributed, provided that the conditional mean function in (17.4) is correctly specified.

When the count is not Poisson distributed, but the conditional mean function is specified by (17.4), we can use the pseudo-ML or quasi-ML approach, which maximizes the Poisson MLE but uses the robust estimate of the VCE,

$$\widehat{V}_{\text{Rob}}(\widehat{\boldsymbol{\beta}}_P) = \left(\sum\nolimits_{i=1}^{N} \widehat{\mu}_i \mathbf{x}_i \mathbf{x}_i'\right)^{-1} \left\{\sum\nolimits_{i=1}^{N} (y_i - \widehat{\mu}_i)^2 \mathbf{x}_i \mathbf{x}_i'\right\} \left(\sum\nolimits_{i=1}^{N} \widehat{\mu}_i \mathbf{x}_i \mathbf{x}_i'\right)^{-1} \quad (17.7)$$

where $\widehat{\mu}_i = \exp(\mathbf{x}_i'\widehat{\boldsymbol{\beta}}_P)$. That is, we use the Poisson MLE to obtain our point estimates, but we obtain robust estimates of the VCE. With overdispersion, the variances will be larger using (17.7) than (17.6) because (17.7) reduces to (17.6), but with overdispersion, $(y_i - \widehat{\mu}_i)^2 > \widehat{\mu}_i$, on average. In the rare case of underdispersion, this ordering is reversed.

This preferred estimate of the VCE is obtained by using the vce(robust) option of poisson. We obtain

```
. * Poisson with robust standard errors
. poisson docvis $xlist, vce(robust) nolog  // Poisson robust SEs
Poisson regression                              Number of obs   =      3677
                                                Wald chi2(7)    =    720.43
                                                Prob > chi2     =    0.0000
Log pseudolikelihood =  -15019.64               Pseudo R2       =    0.1297
```

docvis	Coef.	Robust Std. Err.	z	P>|z|	[95% Conf. Interval]	
private	.1422324	.036356	3.91	0.000	.070976	.2134889
medicaid	.0970005	.0568264	1.71	0.088	-.0143773	.2083783
age	.2936722	.0629776	4.66	0.000	.1702383	.4171061
age2	-.0019311	.0004166	-4.64	0.000	-.0027475	-.0011147
educyr	.0295562	.0048454	6.10	0.000	.0200594	.039053
actlim	.1864213	.0396569	4.70	0.000	.1086953	.2641474
totchr	.2483898	.0125786	19.75	0.000	.2237361	.2730435
_cons	-10.18221	2.369212	-4.30	0.000	-14.82578	-5.538638

Compared with the Poisson MLE, the robust standard errors are 2–3 times larger. This is a very common feature of results for Poisson regression applied to overdispersed data.

Test of overdispersion

A formal test of the null hypothesis of equidispersion, $\text{Var}(y|\mathbf{x}) = E(y|\mathbf{x})$, against the alternative of overdispersion can be based on the equation

$$\text{Var}(y|\mathbf{x}) = E(y|\mathbf{x}) + \alpha^2 E(y|\mathbf{x})$$

which is the variance function for the NB2 model. We test $H_0 : \alpha = 0$ against $H_1 : \alpha > 0$.

The test can be implemented by an auxiliary regression of the generated dependent variable, $\{(y - \hat{\mu})^2 - y\}/\hat{\mu}$ on $\hat{\mu}$, without an intercept term, and performing a t test of whether the coefficient of $\hat{\mu}$ is zero; see Cameron and Trivedi (2005, 670–671) for details of this and other specifications of overdispersion.

```
. * Overdispersion test against V(y|x) = E(y|x) + a*{E(y|x)^2}
. quietly poisson docvis $xlist, vce(robust)
. predict muhat, n
. quietly generate ystar = ((docvis-muhat)^2 - docvis)/muhat
. regress ystar muhat, noconstant noheader
```

ystar	Coef.	Std. Err.	t	P>|t|	[95% Conf. Interval]	
muhat	.7047319	.1035926	6.80	0.000	.5016273	.9078365

The outcome indicates the presence of significant overdispersion. One way to model this feature of the data is to use the NB model. But this commonly chosen alternative is by no means the only one. For example, we can simply use poisson with the vce(robust) option.

Coefficient interpretation and marginal effects

Section 10.6 discusses coefficient interpretation and marginal effects (MEs) estimation, both in general and for the exponential conditional mean, $\exp(\mathbf{x}'\boldsymbol{\beta})$. From section 10.6.4, for the exponential conditional mean, the coefficients can be interpreted as a semielasticity. Thus the coefficient of educyr of 0.030 can be interpreted as one more year of education being associated with a 3.0% increase in the number of doctor visits. The irr option of poisson produces exponentiated coefficients, $e^{\widehat{\beta}}$, that can be given a multiplicative interpretation. Thus one more year of education is associated with doctor visits increasing by the multiple $e^{0.030} \approx 1.030$.

The ME of a unit change in a continuous regressor, x_j, equals $\partial E(y|\mathbf{x})/\partial x_j = \beta_j \exp(\mathbf{x}'\boldsymbol{\beta})$, which depends on the evaluation point, $\mathbf{x}$. From section 10.6.2, there are three standard ME measures. It can be shown that for the Poisson model with an intercept, the average marginal effect (AME) equals $\widehat{\beta}_j \overline{y}$. For example, one more year of education is associated with $0.02956 \times 6.823 = 0.2017$ additional doctor visits. The same result, along with a confidence interval, can be obtained by using the margins, dydx() command. To ensure that MEs for binary regressors are calculated using the finite-difference method, we use the factor-variable i. operator in defining the model to be fit by the poisson command. We obtain

```
. * Average marginal effect for Poisson
. quietly poisson docvis i.private i.medicaid age age2 educyr i.actlim totchr,
> vce(robust)

. margins, dydx(*)

Average marginal effects                          Number of obs    =       3677
Model VCE      : Robust

Expression     : Predicted number of events, predict()
dy/dx w.r.t.   : 1.private 1.medicaid age age2 educyr 1.actlim totchr
```

	dy/dx	Delta-method Std. Err.	z	P>|z|	[95% Conf. Interval]	
1.private	.9701906	.2473149	3.92	0.000	.4854622	1.454919
1.medicaid	.6830664	.4153252	1.64	0.100	−.130956	1.497089
age	2.003632	.4303207	4.66	0.000	1.160219	2.847045
age2	−.0131753	.0028473	−4.63	0.000	−.0187559	−.0075947
educyr	.2016526	.0337805	5.97	0.000	.1354441	.2678612
1.actlim	1.295942	.2850588	4.55	0.000	.7372367	1.854647
totchr	1.694685	.0908883	18.65	0.000	1.516547	1.872823

Note: dy/dx for factor levels is the discrete change from the base level.

For example, one more year of education is associated with 0.202 additional doctor visits. The output also provides confidence intervals for the ME. The ME at the mean (MEM) is calculated with the atmean option of margins, and the ME at a representative value (MER) is calculated with the at() option.

17.3.3   NB2 model

The NB2 model with a quadratic variance function is consistent with overdispersion gen-erated by a Poisson–gamma mixture (see section 17.2.2), but it can also be considered simply as a more flexible functional form for overdispersed count data.

The NB2 model MLE, denoted by $\widehat{\boldsymbol{\beta}}_{\mathrm{NB2}}$, maximizes the log likelihood based on the probability mass function (17.3), where again $\mu = \exp(\mathbf{x}'\boldsymbol{\beta})$, whereas α is simply a constant parameter. The estimators $\widehat{\boldsymbol{\beta}}_{\mathrm{NB2}}$ and $\widehat{\alpha}_{\mathrm{NB2}}$ are the solution to the $K + 1$ nonlinear equations corresponding to the ML first-order conditions

$$\sum_{i=1}^{N} \frac{y_i - \mu_i}{1 + \alpha\mu_i} \mathbf{x}_i = \mathbf{0}$$
$$\sum_{i=1}^{N} \left[\frac{1}{\alpha^2} \left\{ \ln(1 + \alpha\mu_i) - \sum_{j=0}^{y_i - 1} \frac{1}{(j + \alpha^{-1})} \right\} + \frac{y_i - \mu_i}{\alpha\,(1 + \alpha\mu_i)} \right] = 0 \tag{17.8}$$

The K-element $\boldsymbol{\beta}$ equations, the first line in (17.8), are in general different from (17.5) and are sometimes harder to solve using the iterative algorithms. Very large or small values of α can generate numerical instability, and convergence of the algorithm is not guaranteed.

Unlike the Poisson MLE, the NB2 MLE is not consistent if the variance specification $\mathrm{Var}(y|\mu,\alpha) = \mu(1 + \alpha\mu)$ is incorrect. However, this quadratic specification is often a very good approximation to a more general variance function, a feature that might explain why this model usually works well in practice. The variance function parameter, α, enters the probability equation (17.3). This means that the probability distribution over the counts depends upon α, even though the conditional mean does not. It follows that the fitted probability distribution of the NB can be quite different from that of the Poisson, even though the conditional mean is similarly specified in both. If the data are indeed overdispersed, then the NB model is preferred if the goal is to model the probability distribution and not just the conditional mean.

The NB model is not a panacea. There are other reasons for overdispersion, including misspecification due to restriction to an exponential conditional mean. Alternative models are presented in sections 17.3.5 and 17.3.6.

The partial syntax for the MLE for the NB model is similar to that for the **poisson** command:

nbreg *depvar* $\big[$ *indepvars* $\big]$ $\big[$ *if* $\big]$ $\big[$ *in* $\big]$ $\big[$ *weight* $\big]$ $\big[$, *options* $\big]$

The **vce(robust)** option can be used if the variance specification is suspect, but in practice, the default is used and usually differs little from **vce(robust)**. The default fits an NB2 model, and the **dispersion(constant)** option fits an NB1 model.

NB2 model results

Given the presence of considerable overdispersion in our data, the NB2 model should be considered. We obtain

```
. * Standard negative binomial (NB2) with default SEs
. nbreg docvis $xlist, nolog
Negative binomial regression                    Number of obs   =      3677
                                                LR chi2(7)      =    773.44
Dispersion     = mean                           Prob > chi2     =    0.0000
Log likelihood = -10589.339                     Pseudo R2       =    0.0352
```

docvis	Coef.	Std. Err.	z	P>|z|	[95% Conf. Interval]	
private	.1640928	.0332186	4.94	0.000	.0989856	.2292001
medicaid	.100337	.0454209	2.21	0.027	.0113137	.1893603
age	.2941294	.0601588	4.89	0.000	.1762203	.4120384
age2	-.0019282	.0004004	-4.82	0.000	-.0027129	-.0011434
educyr	.0286947	.0042241	6.79	0.000	.0204157	.0369737
actlim	.1895376	.0347601	5.45	0.000	.121409	.2576662
totchr	.2776441	.0121463	22.86	0.000	.2538378	.3014505
_cons	-10.29749	2.247436	-4.58	0.000	-14.70238	-5.892595
/lnalpha	-.4452773	.0306758			-.5054007	-.3851539
alpha	.6406466	.0196523			.6032638	.6803459

```
Likelihood-ratio test of alpha=0:  chibar2(01) = 8860.60 Prob>=chibar2 = 0.000
```

The parameter estimates are all within 15% of those for the Poisson MLE and are often much closer than this. The standard errors are 5%–20% smaller, indicating efficiency gains due to using a more appropriate parametric model. The parameters and MEs are interpreted in the same way as for a Poisson model, because both models have the same conditional mean.

The NB2 estimate of the overdispersion parameter of 0.64 is similar to the 0.70 from the auxiliary regression used in testing for overdispersion. The computer output also includes a LR test of $H_0: \alpha = 0$, and here it is conclusively rejected. The improvement in log likelihood is $\{-10589.3 - (-15019.6)\} = 4430.3$, at the cost of one additional overdispersion parameter, α. The LR statistic is simply twice this value, leading to a highly significant LR test statistic. Recall that α may be interpreted as a measure of the variance of heterogeneity; it is significantly different from zero—a result that is consistent with large improvement in the fit of the model.

The pseudo-R^2 is 0.035 compared with the 0.130 for the Poisson model. This difference, a seemingly worse fit for the Poisson model, is because the pseudo-R^2 is not directly comparable across classes of models, here NB2 and Poisson.

More directly comparable is the squared correlation between fitted and actual counts. We obtain

```
. * Squared correlation between y and yhat
. predict ynbhat, n

. quietly correlate docvis ynbhat

. display "Squared correlation between y and yhat = " r(rho)^2
Squared correlation between y and yhat = .14979846
```

This is similar to the 0.153 for the Poisson model, so the two models provide a similar fit for the conditional mean. The real advantage of the NB2 model is in fitting probabilities, considered next.

Fitted probabilities for Poisson and NB2 models

To get more insight into the improvement in the fit, we should compare what the parameter estimates from the Poisson and the NB2 models imply for the fitted probability distribution of `docvis`.

Using the fitted models, we can compare actual and fitted cell frequencies of `docvis`. The fitted cell frequencies are calculated by using $\widehat{p}(i, y)$, $i = 1, 2, \ldots, N$ and $y = 0, 1, 2, \ldots$, which denote the fitted probability that individual i experiences y events. These are calculated for each i by plugging in the estimated $\boldsymbol{\beta}$ in (17.1) for the Poisson model, and the estimated $\boldsymbol{\beta}$ and α in (17.3) for the NB2 model. Then the fitted frequency in cell y is calculated as $N\bar{p}(y)$, where

$$\bar{p}(y) = \frac{1}{N} \sum_{i=1}^{N} \widehat{p}(i, y), \qquad y = 0, 1, 2, \ldots \tag{17.9}$$

A large deviation between $\bar{p}(y)$ and the observed sample frequency for a given y indicates a lack of fit.

Alternatively, we can evaluate the probabilities at a particular value, $\mathbf{x} = \mathbf{x}^*$, where often $\mathbf{x}^* = \bar{\mathbf{x}}$, the sample mean. Then we use

$$p(y|\mathbf{x} = \mathbf{x}^*) = \widehat{p}(y|\mathbf{x} = \mathbf{x}^*), \qquad y = 0, 1, 2, \ldots \tag{17.10}$$

where $\bar{\mathbf{x}}$ is the K-element vector of the sample averages of the regressors. The difference between $\bar{p}(y)$ and $p(y|\mathbf{x} = \mathbf{x}^*)$ is that the former averages over N sample values of $\mathbf{x}_i$, whereas the latter is conditional on $\bar{\mathbf{x}}$ and has less variability.

Several user-written postestimation commands following count regression are detailed in Long and Freese (2006). Here we illustrate the `countfit` and `prvalue` commands, which compute the quantities defined, respectively, in (17.9) and (17.10). The `prcounts` command, which also computes the quantity defined in (17.9), is illustrated in section 17.4.3.

The countfit command

The user-written `countfit` command (Long and Freese 2006) computes the average predicted probabilities, $\bar{p}(y)$, defined in (17.9). The `prm` option fits the Poisson model and the `nbreg` option fits the NB2 model. Additional options control the amount of output produced by the command. In particular, the `maxcount(#)` option sets the maximum count for which predicted probabilities are evaluated; the default is `maxcount(9)`. For the Poisson model, we obtain

```
. * Poisson: Sample vs avg predicted probabilities of y = 0, 1, ..., 5
. countfit docvis $xlist, maxcount(5) prm nograph noestimates nofit
Comparison of Mean Observed and Predicted Count
              Maximum       At       Mean
Model        Difference    Value     |Diff|
-----------------------------------------------
PRM           0.102          0       0.045

PRM: Predicted and actual probabilities
Count   Actual    Predicted    |Diff|   Pearson
-----------------------------------------------
0       0.109      0.007       0.102   5168.233
1       0.085      0.030       0.056    387.868
2       0.097      0.063       0.034     69.000
3       0.091      0.095       0.005      0.789
4       0.092      0.116       0.024     17.861
5       0.072      0.121       0.049     72.441
-----------------------------------------------
Sum     0.547      0.432       0.269   5716.192
```

The Poisson model seriously underestimates the probability mass at low counts. In particular, the predicted probabilities at 0 and 1 counts are 0.007 and 0.030 compared with sample frequencies of 0.109 and 0.085. For the NB2 model that allows for overdispersion, we obtain

```
. * NB2: Sample vs average predicted probabilities of y = 0, 1, ..., 5
. countfit docvis $xlist, maxcount(5) nbreg nograph noestimates nofit
Comparison of Mean Observed and Predicted Count
              Maximum       At       Mean
Model        Difference    Value     |Diff|
-----------------------------------------------
NBRM         -0.023          1       0.010

NBRM: Predicted and actual probabilities
Count   Actual    Predicted    |Diff|   Pearson
-----------------------------------------------
0       0.109      0.091       0.018    12.708
1       0.085      0.108       0.023    17.288
2       0.097      0.105       0.008     2.270
3       0.091      0.096       0.005     1.086
4       0.092      0.085       0.007     2.333
5       0.072      0.074       0.001     0.072
-----------------------------------------------
Sum     0.547      0.559       0.062    35.757
```

The fit is now much better. The greatest discrepancy is for $y = 1$, with a predicted probability of 0.108, which exceeds the sample frequency of 0.085. The final column, marked `Pearson`, gives N times $(\texttt{Diff})^2/\texttt{Predicted}$, where `Diff` is the difference between average fitted and empirical frequencies, for each value of `docvis` up to that given by the `maxcount()` option. Although these values are a good rough indicator of goodness of fit, caution should be exercised in using these numbers as the basis of a Pearson chi-squared goodness-of-fit test because the fitted probabilities are functions of estimated coefficients; see Cameron and Trivedi (2005, 266).

The comparison confirms that the NB2 model provides a much better fit of the probabilities than the Poisson model (even though for the conditional mean, the MEs are similar for the two models).

The prvalue command

The user-written `prvalue` command (Long and Freese 2006) predicts probabilities for given values of the regressors, computed using (17.10). As an example, we obtain predicted probabilities for a person with private insurance and access to Medicaid, with other regressors set to their sample means. The `prvalue` command, with options used to minimize the length of output, following the `nbreg` command, yields

```
. * NB2: Predicted NB2 probabilities at x = x* of y = 0, 1, ..., 5
. quietly nbreg docvis $xlist
. prvalue, x(private=1 medicaid=1) max(5) brief

nbreg: Predictions for docvis

                            95% Conf. Interval
    Rate:          7.34    [ 6.6477,   8.0322]
    Pr(y=0|x):     0.0660  [ 0.0580,   0.0741]
    Pr(y=1|x):     0.0850  [ 0.0761,   0.0939]
    Pr(y=2|x):     0.0898  [ 0.0818,   0.0977]
    Pr(y=3|x):     0.0879  [ 0.0816,   0.0942]
    Pr(y=4|x):     0.0826  [ 0.0781,   0.0872]
    Pr(y=5|x):     0.0758  [ 0.0728,   0.0787]
```

These predicted probabilities at a specific value of the regressors are within 30% of the average predicted probabilities for the NB2 model previously computed by using the `countfit` command.

Discussion

The assumption of gamma heterogeneity underlying the mixture interpretation of the NB2 model is very convenient, but there are other alternatives. For example, one could assume that heterogeneity is lognormally distributed. Unfortunately, this specification does not lead to an analytical expression for the mixture distribution and will therefore require an estimation method involving one-dimensional numerical integration, e.g., simulation-based or quadrature-based estimation. The official version of Stata does not currently support this option.

Generalized NB model

The generalized NB model is an extension of the NB2 model that permits additional parameterization of the overdispersion parameter, α, in (17.3), whereas it is simply a positive constant in the NB2 model. The overdispersion parameter can then vary across individuals, and the same variable can affect both the location and the scale parameters of the distribution, complicating the computation of MEs. Alternatively, the model may be specified such that different variables may separately affect the location and scale of the distribution.

Even though in principle flexibility is desirable, such models are currently not widely used. The parameters of the model can be estimated by using the `gnbreg` command that has a syntax similar to that of `nbreg`, with the addition of the `lnalpha()` option to specify the variables in the model for $\ln(\alpha)$.

We parameterize $\ln(\alpha)$ for the dummy variables `female` and `bh` (black/Hispanic).

```
. * Generalized negative binomial with alpha parameterized
. gnbreg docvis $xlist, lnalpha(female bh) nolog
Generalized negative binomial regression          Number of obs   =       3677
                                                  LR chi2(7)      =     759.49
                                                  Prob > chi2     =     0.0000
Log likelihood = -10576.261                       Pseudo R2       =     0.0347
```

docvis	Coef.	Std. Err.	z	P>|z|	[95% Conf. Interval]	
docvis						
private	.1571795	.0329147	4.78	0.000	.0926678	.2216912
medicaid	.0860199	.0462092	1.86	0.063	-.0045486	.1765883
age	.30188	.0598412	5.04	0.000	.1845934	.4191665
age2	-.0019838	.0003981	-4.98	0.000	-.0027641	-.0012036
educyr	.0284782	.0043246	6.59	0.000	.0200021	.0369544
actlim	.1875403	.0346287	5.42	0.000	.1196693	.2554112
totchr	.2761519	.0120868	22.85	0.000	.2524623	.2998415
_cons	-10.54756	2.23684	-4.72	0.000	-14.93169	-6.163434
lnalpha						
female	-.1871933	.0634878	-2.95	0.003	-.311627	-.0627595
bh	.3103148	.0706505	4.39	0.000	.1718423	.4487873
_cons	-.4119142	.0512708	-8.03	0.000	-.512403	-.3114253

There is some improvement in the log likelihood relative to the NB2 model. The dispersion is greater for blacks and Hispanics and smaller for females. However, these two variables could also have been introduced into the conditional mean function. The decision to let a variable affect α rather than μ can be difficult to justify.

17.3.4 Nonlinear least-squares estimation

Suppose one wants to avoid any parametric specification of the conditional variance function. Instead, one may fit the exponential mean model by nonlinear least squares (NLS) and use a robust estimate of the VCE. For count data, this estimator is likely to be less efficient than the Poisson MLE, because the Poisson MLE explicitly models the intrinsic heteroskedasticity of count data, whereas the NLS is based on homoskedastic errors.

The NLS objective function is

$$Q(\boldsymbol{\beta}) = \sum_{i=1}^{N} \{y_i - \exp(\mathbf{x}_i'\boldsymbol{\beta})\}^2$$

Section 10.3.5 provides a NLS application, using the `nl` command, for doctor visits in a related dataset.

A practical complication not mentioned in section 10.3.5 is that if most observations are 0, then the NLS estimator can encounter numerical problems. The NLS estimator can be shown to solve

$$\sum_{i=1}^{N} \{y_i - \exp(\mathbf{x}_i'\boldsymbol{\beta})\} \exp(\mathbf{x}_i'\boldsymbol{\beta})\mathbf{x}_i = \mathbf{0}$$

Compared with (17.5) for the Poisson MLE, there is an extra multiple, $\exp(\mathbf{x}_i'\boldsymbol{\beta})$, which can lead to numerical problems if most counts are 0. NLS estimation using the `nl` command yields

```
. * Nonlinear least squares
. nl (docvis = exp({xb: $xlist one})), vce(robust) nolog
(obs = 3677)
```

Nonlinear regression

```
Number of obs =      3677
R-squared       =    0.5436
Adj R-squared =      0.5426
Root MSE       =    6.804007
Res. dev.      =   24528.25
```

docvis	Coef.	Robust Std. Err.	t	P>|t|	[95% Conf. Interval]	
/xb_private	.1235144	.0395179	3.13	0.002	.0460351	.2009937
/xb_medicaid	.0856747	.0649936	1.32	0.188	-.0417525	.2131018
/xb_age	.2951153	.0720509	4.10	0.000	.1538516	.4363789
/xb_age2	-.0019481	.0004771	-4.08	0.000	-.0028836	-.0010127
/xb_educyr	.0309924	.0051192	6.05	0.000	.0209557	.0410291
/xb_actlim	.1916735	.0413705	4.63	0.000	.110562	.2727851
/xb_totchr	.2191967	.0151021	14.51	0.000	.1895874	.248806
/xb_one	-10.12438	2.713159	-3.73	0.000	-15.44383	-4.804931

The NLS coefficient estimates are within 20% of the Poisson and NB2 ML estimates, with similar differences for the implied MEs. The robust standard errors for the NLS estimates are about 20% higher than those for the Poisson MLE, confirming the expected efficiency loss.

Unless there is good reason to do otherwise, for count data it is better to use Poisson or NB2 MLEs than to use the NLS estimator.

17.3.5   Hurdle model

We now consider the first of two types of mixture models that involve new specifications of both the conditional mean and variance of the distributions.

The hurdle model, or two-part model, relaxes the assumption that the zeros and the positives come from the same data-generating process. The zeros are determined by the density $f_1(\cdot)$, so that $\Pr(y = 0) = f_1(0)$ and $\Pr(y > 0) = 1 - f_1(0)$. The positive counts come from the truncated density $f_2(y|y > 0) = f_2(y)/\{1 - f_2(0)\}$, which is multiplied by $\Pr(y > 0)$ to ensure that probabilities sum to 1. Thus suppressing regressors for notational simplicity,

$$f(y) = \begin{cases} f_1(0) & \text{if } y = 0, \\ \dfrac{1 - f_1(0)}{1 - f_2(0)} f_2(y) & \text{if } y \geq 1 \end{cases}$$

This specializes to the standard model only if $f_1(\cdot) = f_2(\cdot)$. Although the motivation for this model is to handle excess zeros, it is also capable of modeling too few zeros.

A hurdle model has the interpretation that it reflects a two-stage decision-making process, each part being a model of one decision. The two parts of the model are functionally independent. Therefore, ML estimation of the hurdle model can be achieved by separately maximizing the two terms in the likelihood, one corresponding to the zeros and the other to the positives. This is straightforward. The first part uses the full sample, but the second part uses only the positive count observations.

For certain types of activities, such a specification is easy to rationalize. For example, in a model that explains the amount of cigarettes smoked per day, the survey may include both smokers and nonsmokers. One model determines whether one smokes, and a second model determines the number of cigarettes (or packs of cigarettes) smoked given that at least one is smoked.

As an illustration, we obtain draws from a hurdle model as follows. The positives are generated by Poisson(2) truncated at 0. One way to obtain these truncated draws is to draw from Poisson(2) and then replace any zero draw for any observation by a nonzero draw, until all draws are nonzero. This can be shown to be equivalent to the accept–reject method for drawing random variates that is defined in, for example, Cameron and Trivedi (2005, 414). This method is simple but is computationally inefficient if a high fraction of draws are truncated at zero. To then obtain draws from the hurdle model, we randomly replace some of the truncated Poisson draws with zeros. A draw is replaced with a probability of π and kept with a probability $1 - \pi$. We set $\pi = 1 - (1 - e^{-2})/2 \simeq 0.568$ because this can be shown to yield a mean of 1 for the hurdle model draws. The proportion of positives is then 0.432. We have

```
. * Hurdle: Pr(y=0)=pi and Pr(y=k)=(1-pi) x Poisson(2) truncated at 0
. quietly set obs 10000
. set seed 10101            // set the seed !
. scalar pi=1-(1-exp(-2))/2  // Probability y=0
. generate xhurdle = 0
. scalar minx = 0
. while minx == 0 {
  2.   generate xph = rpoisson(2)
  3.   quietly replace xhurdle = xph if xhurdle==0
  4.   drop xph
  5.   quietly summarize xhurdle
  6.   scalar minx = r(min)
  7. }
. replace xhurdle = 0 if runiform() < pi
(5663 real changes made)
```

```
. summarize xhurdle
    Variable |       Obs        Mean    Std. Dev.       Min        Max
-------------+--------------------------------------------------------
     xhurdle |     10000        .999    1.415698          0          9
```

The setup is such that the random variable has a mean of 1. From the summary statistics, this is the case. The model has induced overdispersion because the variance $1.4157^2 = 2.004 > 1$.

The hurdle model changes the conditional mean specification. Under the hurdle model, the conditional mean is

$$E(y|\mathbf{x}) = \Pr(y_1 > 0|\mathbf{x}_1) \times E_{y_2>0}(y_2|y_2 > 0, \mathbf{x}_2) \qquad (17.11)$$

and the two terms on the right are determined by the two respective parts of the model. Because of the form of the conditional mean specification, the calculation of MEs, $\partial E(y|\mathbf{x})/\partial x_j$, is more complicated.

Variants of the hurdle model

Any binary outcome model can be used for modeling the zero-versus-positive outcome. Logit is a popular choice. The second part can use any truncated parametric count density, e.g., Poisson or NB. In application, the covariates in the hurdle part that models the zero/one outcome need not be the same as those that appear in the truncated part, although in practice they are often the same.

The hurdle model is widely used, and the hurdle NB model is quite flexible. The main drawback is that the model is not very parsimonious. A competitor to the hurdle model is the zero-inflated class of models, presented in section 17.4.2.

Two variants of the hurdle count model are provided by the user-written `hplogit` and `hnblogit` commands (Hilbe 2005a,b). They use the logit model for the first part and either the zero-truncated Poisson (ZTP) or the zero-truncated NB (ZTNB) model for the second part. (Zero-inflated models are discussed in section 17.4.2.) The partial syntax is

`hplogit` *depvar* [*indepvars*] [*if*] [*in*] [, *options*]

where *options* include `robust` and `nolog`, as well as many of those for the regression command.

Application of the hurdle model

We implement ML estimation of the hurdle model with two-step estimation using official Stata commands, rather than the user-written commands, because the user-written commands require the same set of regressors in each part.

The first step involves estimating the parameters of a binary outcome model, popular choices being binary logit or probit estimated by using `logit` or `probit`.

The second step estimates the parameters of a ZTP or ZTNB model, using the `ztp` command or the `ztnb` command. The syntax and options for these commands are the same as those for the `poisson` and `nbreg` commands. In particular, the default for `ztnb` is to estimate the parameters of a zero-truncated NB2 model.

We first use `logit`. We do not need to transform `docvis` to a binary variable before running the logit because Stata does this automatically. This is easy to verify by doing the transformation and then running the logit.

```
. * Hurdle logit-nb model manually
. logit docvis $xlist, nolog
Logistic regression                             Number of obs   =       3677
                                                LR chi2(7)      =     453.08
                                                Prob > chi2     =     0.0000
Log likelihood = -1040.3258                     Pseudo R2       =     0.1788
```

docvis	Coef.	Std. Err.	z	P>|z|	[95% Conf. Interval]	
private	.6586978	.1264608	5.21	0.000	.4108392	.9065563
medicaid	.0554225	.1726693	0.32	0.748	-.2830032	.3938482
age	.5428779	.2238845	2.42	0.015	.1040724	.9816834
age2	-.0034989	.0014957	-2.34	0.019	-.0064304	-.0005673
educyr	.047035	.0155706	3.02	0.003	.0165171	.0775529
actlim	.1623927	.1523743	1.07	0.287	-.1362553	.4610408
totchr	1.050562	.0671922	15.64	0.000	.9188676	1.182256
_cons	-20.94163	8.335137	-2.51	0.012	-37.2782	-4.605057

The second-step regression is based only on the sample with positive observations for `docvis`.

```
. * Second step uses positives only
. summarize docvis if docvis > 0
```

Variable	Obs	Mean	Std. Dev.	Min	Max
docvis	3276	7.657814	7.415095	1	144

Dropping zeros from the sample has raised the mean and lowered the standard deviation of `docvis`.

The parameters of the ZTNB model are then estimated next by using `ztnb`.

```
. * Zero-truncated negative binomial
. ztnb docvis $xlist if docvis>0, nolog
Zero-truncated negative binomial regression      Number of obs   =        3276
                                                 LR chi2(7)      =      509.10
Dispersion     = mean                            Prob > chi2     =      0.0000
Log likelihood = -9452.899                       Pseudo R2       =      0.0262
```

docvis	Coef.	Std. Err.	z	P>|z|	[95% Conf. Interval]	
private	.1095567	.0345239	3.17	0.002	.0418911	.1772223
medicaid	.0972309	.0470358	2.07	0.039	.0050425	.1894193
age	.2719032	.0625359	4.35	0.000	.1493352	.3944712
age2	-.0017959	.000416	-4.32	0.000	-.0026113	-.0009805
educyr	.0265974	.0043937	6.05	0.000	.0179859	.035209
actlim	.1955384	.0355161	5.51	0.000	.1259281	.2651488
totchr	.2226969	.0124128	17.94	0.000	.1983683	.2470254
_cons	-9.19017	2.337591	-3.93	0.000	-13.77176	-4.608576
/lnalpha	-.5259629	.0418671			-.6080209	-.443905
alpha	.590986	.0247429			.5444273	.6415264

```
Likelihood-ratio test of alpha=0:  chibar2(01) = 7089.37 Prob>=chibar2 = 0.000
```

A positively signed coefficient in the logit model means that the corresponding regressor increases the probability of a positive observation. In the second part, a positive coefficient means that, conditional on a positive count, the corresponding variable increases the value of the count. The results show that all the variables except `medicaid` and `actlim` have statistically significant coefficients and that they affect both the outcomes in the same direction.

For this example with a common set of regressors in both parts of the model, the user-written `hnblogit` command can instead be used. Then

```
. * Same hurdle model fit using the user-written hnblogit command
. hnblogit docvis $xlist, robust
   (output omitted)
```

yields the same parameter estimates as the separate estimation of the two components of the model.

Computation of MEs for the hurdle model are complicated, because change in a regressor may change both the logit and the truncated count components of the model. A complete analysis specializes the expression for the conditional mean given in (17.11) to one for a logit–truncated Poisson hurdle model or a logit–truncated NB2 hurdle model, and then computes the ME using calculus or finite-difference methods. Here we simply calculate MEs for the two components separately. To ensure that MEs for binary regressors are calculated using the finite-difference method, we need to use the factor-variable `i.` operator in defining the model to be fit. The user-written `hnblogit` command does not support factor variables, however, so we instead estimate each part separately by using the `logit` and `ztnb` commands. We then use the `margins, dydx()` `atmean` command, so evaluation is at the sample mean of the regressors.

The MEs for the first part are obtained by using `margins` after `logit`.

```
. * margins for marginal effects of first part
. quietly logit docvis i.private i.medicaid age age2 educyr i.actlim totchr,
> vce(robust)

. margins, dydx(*) atmean noatlegend
Conditional marginal effects                      Number of obs    =        3677
Model VCE      : Robust

Expression     : Pr(docvis), predict()
dy/dx w.r.t.   : 1.private 1.medicaid age age2 educyr 1.actlim totchr
```

	dy/dx	Delta-method Std. Err.	z	P>|z|	[95% Conf. Interval]	
1.private	.0363352	.0074238	4.89	0.000	.0217848	.0508856
1.medicaid	.0029777	.009205	0.32	0.746	-.0150638	.0210192
age	.0296458	.0126672	2.34	0.019	.0048186	.054473
age2	-.0001911	.0000849	-2.25	0.024	-.0003574	-.0000247
educyr	.0025685	.0008484	3.03	0.002	.0009056	.0042314
1.actlim	.0086643	.0079693	1.09	0.277	-.0069552	.0242838
totchr	.0573697	.0031875	18.00	0.000	.0511224	.0636171

```
Note: dy/dx for factor levels is the discrete change from the base level.
```

The MEs for the second part are also obtained by using `margins` applied to the ZTNB estimates.

```
. * margins for marginal effects of second part
. quietly ztnb docvis i.private i.medicaid age age2 educyr i.actlim totchr
> if docvis>0, vce(robust)

. margins, dydx(*) atmean noatlegend
Conditional marginal effects                      Number of obs    =        3276
Model VCE      : Robust

Expression     : Predicted number of events, predict()
dy/dx w.r.t.   : 1.private 1.medicaid age age2 educyr 1.actlim totchr
```

	dy/dx	Delta-method Std. Err.	z	P>|z|	[95% Conf. Interval]	
1.private	.7370146	.2570038	2.87	0.004	.2332963	1.240733
1.medicaid	.6763898	.424458	1.59	0.111	-.1555326	1.508312
age	1.830569	.4512706	4.06	0.000	.9460952	2.715043
age2	-.0120909	.0029917	-4.04	0.000	-.0179544	-.0062273
educyr	.1790653	.0343792	5.21	0.000	.1116834	.2464473
1.actlim	1.35768	.2932072	4.63	0.000	.7830047	1.932356
totchr	1.499291	.0901593	16.63	0.000	1.322582	1.676

```
Note: dy/dx for factor levels is the discrete change from the base level.
```

In this example, `age` appears quadratically, complicating calculation of the ME. By replacing `age` and `age2` in the regressor list with `c.age` and `c.age#c.age`, the `margins` command will give the correct ME with respect to `age`; see section 10.6.11.

The parameters of the Poisson hurdle model can be estimated by replacing `ztnb` with `ztp`, because the first part of the model is the same. The ZTNB regression gives a much better fit than the ZTP because of the overdispersion in the data. The majority of ZTP coefficients are slightly larger or of the same magnitude as the ZTNB coefficients, but the substantive conclusions from ZTP and ZTNB are similar.

The hurdle model estimates are more fragile because any distributional misspecification leads to inconsistency of the MLE. This should be clear from the conditional mean expression in (17.11). This includes a truncated mean, $E_{y>0}(y|y > 0, \mathbf{x})$, that will differ according to whether we use ZTP or ZTNB.

The discussion of model selection is postponed to later in this chapter.

17.3.6 Finite-mixture models

The NB model is an example of a continuous mixture model, because the heterogeneity variable, or mixing random variable, ν, was assumed to have a continuous distribution (gamma). An alternative approach instead uses a discrete representation of unobserved heterogeneity. This generates a class of models called finite-mixture models (FMMs)— a particular subclass of latent-class models; see Deb (2007) and Cameron and Trivedi (2005, sec. 20.4.3).

FMM specification

An FMM specifies that the density of y is a linear combination of m different densities, where the jth density is $f_j(y|\boldsymbol{\beta}_j)$, $j = 1, 2, \ldots, m$. Thus an m-component finite mixture is

$$f(y|\boldsymbol{\beta}, \boldsymbol{\pi}) = \sum_{j=1}^{m} \pi_j f_j(y|\boldsymbol{\beta}_j), \quad 0 \le \pi_j \le 1, \ \sum_{j=1}^{m} \pi_j = 1$$

A simple example is a two-component ($m = 2$) Poisson mixture of Poisson(μ_1) and Poisson(μ_2). This may reflect the possibility that the sampled population contains two "types" of cases, whose y outcomes are characterized by the distributions $f_1(y|\boldsymbol{\beta}_1)$ and $f_2(y|\boldsymbol{\beta}_2)$, which are assumed to have different moments. The mixing fraction, π_1, is in general an unknown parameter. In a more general formulation, it too can be parameterized for the observed variable(s) z.

Simulated FMM sample with comparisons

As an illustration, we generate a mixture of Poisson(0.5) and Poisson(5.5) in proportions 0.9 and 0.1, respectively.

```
. * Mixture: Poisson(.5) with prob .9 and Poisson(5.5) with prob .1
. set seed 10101        // set the seed !
. generate xp1= rpoisson(.5)
. generate xp2= rpoisson(5.5)
```

```
. summarize xp1 xp2
    Variable |      Obs        Mean    Std. Dev.       Min         Max
-------------+--------------------------------------------------------
         xp1 |    10000       .5064    .7114841         0           5
         xp2 |    10000      5.4958    2.335793         0          16

. rename xp1 xpmix

. quietly replace xpmix = xp2 if runiform() > 0.9

. summarize xpmix
    Variable |      Obs        Mean    Std. Dev.       Min         Max
-------------+--------------------------------------------------------
       xpmix |    10000       .9936    1.761894         0          15
```

The setup yields a random variable with a mean of $0.9 \times 0.5 + 0.1 \times 5.5 = 1$. But the data are overdispersed, with a variance in this sample of $1.762^2 = 3.10$. This dispersion is greater than those for the preceding generated data samples from Poisson, NB2, and hurdle models.

```
. tabulate xpmix
     xpmix |     Freq.     Percent       Cum.
-----------+-----------------------------------
         0 |     5,414       54.14       54.14
         1 |     2,770       27.70       81.84
         2 |       764        7.64       89.48
         3 |       245        2.45       91.93
         4 |       195        1.95       93.88
         5 |       186        1.86       95.74
         6 |       151        1.51       97.25
         7 |       108        1.08       98.33
         8 |        73        0.73       99.06
         9 |        42        0.42       99.48
        10 |        27        0.27       99.75
        11 |        12        0.12       99.87
        12 |         6        0.06       99.93
        13 |         4        0.04       99.97
        14 |         2        0.02       99.99
        15 |         1        0.01      100.00
-----------+-----------------------------------
     Total |    10,000      100.00
```

As for the NB2, the distribution has a long right tail. Although the component means are far apart, the mixture distribution is not bimodal; see the histogram in figure 17.1. This is because only 10% of the observations come from the high-mean distribution.

It is instructive to view graphically the four distributions generated in this chapter—Poisson, NB2, hurdle, and finite mixture. All have the same mean of 1, but they have different dispersion properties. The generated data were used to produce four histograms that we now combine into a single graph.

```
. * Compare the four distributions, all with mean 1
. graph combine mus17xp.gph mus17negbin.gph mus17pmix.gph
> mus17hurdle.gph, title("Four different distributions with mean = 1")
> ycommon xcommon
```

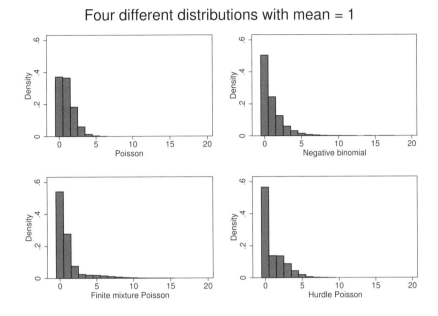

Figure 17.1. Four count distributions

It is helpful for interpretation to supplement this graph with summary statistics for the distributions:

```
. * Compare the four distributions, all with mean 1
. summarize xpois xnegbin xpmix xhurdle
    Variable |       Obs        Mean    Std. Dev.        Min         Max
-------------+--------------------------------------------------------
       xpois |     10000       .9933    1.001077          0           6
     xnegbin |     10000      1.0129    1.442339          0          12
       xpmix |     10000       .9936    1.761894          0          15
     xhurdle |     10000        .999    1.415698          0           9
```

ML estimation of the FMM

The components of the mixture may be assumed, for generality, to differ in all their parameters. This is a more flexible specification because all moments of the distribution depend upon $(\pi_j, \boldsymbol{\beta}_j, j = 1, \ldots, m)$. But such flexibility comes at the expense of parsimonious parameterization. More parsimonious formulations assume that only some parameters differ across the components, e.g., the intercepts, and the remaining parameters are common to the mixture components.

ML estimation of an FMM is computationally challenging because the log-likelihood function may be multimodal and not logconcave and because individual components may be hard to identify empirically. The presence of outliers in the sample may cause further identification problems.

The fmm command

The user-written `fmm` command (Deb 2007) enables ML estimation of finite-mixture count models. The command can be used to estimate mixtures of several continuous and count models. Here only the count models are covered.

The partial syntax for this command is as follows:

fmm *depvar* [*indepvars*] [*if*] [*in*] [*weight*], `components(#)`

  `mixtureof(`*density*`)` [*options*]

where `components(#)` refers to the number of components in the specification and `mixtureof(`*density*`)` refers to the specification of the distribution. For count models, there are three choices: Poisson, NB2 (`negbin2`), and NB1 (`negbin1`). Specific examples are

fmm *depvar* [*varlist1*], `components(2) mixtureof(poisson) vce(robust)`

fmm *depvar* [*varlist1*], `components(3) mixtureof(negbin2) vce(robust)`

fmm *depvar* [*varlist1*], `components(2) mixtureof(negbin1)`

  `probability(`*varlist2*`) vce(robust)`

The algorithm works sequentially with the number of components. If the specification with three components is desired, then one should first run the specification with two components to provide initial values for the algorithm for the three-component model. An important option is `probability(`*varlist2*`)`, which allows the π_j to be parameterized as a function of the variables in *varlist2*.

The default setup assumes constant class probabilities. The command supports the `vce()` option with all the usual types of VCE.

Application: Poisson finite-mixture model

Next we apply the FMM to the doctor-visit data. Both Poisson and NB variants are considered. In a 2-component Poisson mixture, denoted by FMM2-P, each component is a Poisson distribution with a different mean, i.e., Poisson$\{\exp(\mathbf{x}'\boldsymbol{\beta}_j)\}$, $j = 1, 2$, and the proportion π_j of the sample comes from each subpopulation. This model will have $2K + 1$ unknown parameters, where K is the number of exogenous variables in the model. For the 2-component NB mixture, denoted by FMM2-NB, a similar interpretation applies, but now the overdispersion parameters also vary between subpopulations. This model has $2(K + 1) + 1$ unknown parameters.

We first consider the FMM2-P model.

```
. * Finite-mixture model using fmm command with constant probabilities
. use mus17data.dta, clear
. fmm docvis $xlist, vce(robust) components(2) mixtureof(poisson)
Fitting Poisson model:
Iteration 0:   log likelihood = -15019.656
Iteration 1:   log likelihood =  -15019.64
Iteration 2:   log likelihood =  -15019.64
Fitting 2 component Poisson model:
Iteration 0:   log pseudolikelihood = -14985.068  (not concave)
Iteration 1:   log pseudolikelihood = -12233.072  (not concave)
Iteration 2:   log pseudolikelihood = -11752.598
Iteration 3:   log pseudolikelihood =  -11518.01
Iteration 4:   log pseudolikelihood = -11502.758
Iteration 5:   log pseudolikelihood = -11502.686
Iteration 6:   log pseudolikelihood = -11502.686
```

```
2 component Poisson regression            Number of obs   =      3677
                                          Wald chi2(14)   =    576.86
Log pseudolikelihood = -11502.686         Prob > chi2     =    0.0000
```

docvis	Coef.	Robust Std. Err.	z	P>|z|	[95% Conf. Interval]	
component1						
private	.2077415	.0560256	3.71	0.000	.0979333	.3175497
medicaid	.1071618	.0964233	1.11	0.266	-.0818245	.2961481
age	.3798087	.100821	3.77	0.000	.1822032	.5774143
age2	-.0024869	.0006711	-3.71	0.000	-.0038022	-.0011717
educyr	.029099	.0067908	4.29	0.000	.0157893	.0424087
actlim	.1244235	.0558883	2.23	0.026	.0148844	.2339625
totchr	.3191166	.0184744	17.27	0.000	.2829074	.3553259
_cons	-14.25713	3.759845	-3.79	0.000	-21.62629	-6.887972
component2						
private	.138229	.0614901	2.25	0.025	.0177106	.2587474
medicaid	.1269723	.1329626	0.95	0.340	-.1336297	.3875742
age	.2628874	.1140355	2.31	0.021	.0393819	.486393
age2	-.0017418	.0007542	-2.31	0.021	-.00322	-.0002636
educyr	.0241679	.0076208	3.17	0.002	.0092314	.0391045
actlim	.1831598	.0622267	2.94	0.003	.0611977	.3051218
totchr	.1970511	.0263763	7.47	0.000	.1453545	.2487477
_cons	-8.051256	4.28211	-1.88	0.060	-16.44404	.3415266
/imlogitpi1	.877227	.0952018	9.21	0.000	.690635	1.063819
pi1	.7062473	.0197508			.6661082	.7434197
pi2	.2937527	.0197508			.2565803	.3338918

Interpretation

Here the computer output separates the parameter estimates for the two components. If the two latent classes differ a lot in their responses to the changes in the regressors, we would expect the parameters to differ also. In this example, the differentiation does not appear to be very sharp at the level of individual coefficients. But as we see below, this is misleading because the two components have substantially different mean numbers of doctor visits, leading to quite different MEs even though the slope parameters do not seem to be all that different.

The last two lines in the output give $\widehat{\pi}_1$ and $\widehat{\pi}_2 (= 1 - \widehat{\pi}_1)$. The algorithm parameterizes π as a logistic function to constrain it to have a positive value. After the algorithm converges, $\widehat{\pi}_1$ is recovered by transformation. The interpretation of pi1 is that it represents the proportion of observations in class 1. Here about 70% are in class 1 and the remaining 30% come from class 2.

These classes are latent, so it is helpful to give them some interpretation. One natural interpretation is that classes differ in terms of the mean of their respective distributions, i.e., $\exp(\mathbf{x}'\boldsymbol{\beta}_1) \neq \exp(\mathbf{x}'\boldsymbol{\beta}_2)$. To make this comparison, we generate fitted values by using the predict command. For the Poisson model, the predictions are $\widehat{y}_i^j = \exp(\mathbf{x}_i'\widehat{\boldsymbol{\beta}}_j)$, $j = 1, 2$.

The predictions from the two components are stored as the yfit1 and yfit2 variables.

```
. * Predict y for two components
. quietly fmm docvis $xlist, vce(robust) components(2) mixtureof(poisson)
. predict yfit1, equation(component1)
. predict yfit2, equation(component2)
. summarize yfit1 yfit2
```

Variable	Obs	Mean	Std. Dev.	Min	Max
yfit1	3677	3.801692	2.176922	.9815563	27.28715
yfit2	3677	13.95943	5.077463	5.615584	55.13366

The summary statistics make explicit the implication of the mixture model. The first component has a relatively low mean number of doctor visits, around 3.80. The second component has a relatively high mean number of doctor visits, around 13.96. The probability-weighted average of the two classes is $0.7062 \times 3.8017 + 0.2938 \times 13.9594 = 6.79$, which is close to the overall sample average of 6.82.

So the FMM has the interpretation that the data are generated by two classes of individuals, the first of which accounts for about 70% of the population who are relatively low users of doctor visits and the second that accounts for about 30% of the population who are high users of doctor visits.

Comparing marginal effects

The two classes also differ in their response to changes in regressors. To compare these, we use the margins, dydx(*) atmean command, which evaluates the MEs at the same value of the regressors, the sample mean $\overline{\mathbf{x}}$. The user-written fmm command does not handle factor variables, so we cannot use the i. operator to identify the binary regressors in the regression. As a result, the MEs of all regressors, including binary regressors, are computed using calculus methods. We have

```
. * Marginal effects for component 1
. margins, dydx(*) predict(eq(component1)) atmean noatlegend
Warning: cannot perform check for estimable functions.
Conditional marginal effects                       Number of obs   =        3677
Model VCE     : Robust

Expression    : predicted mean: component1, predict(eq(component1))
dy/dx w.r.t.  : private medicaid age age2 educyr actlim totchr
```

	dy/dx	Delta-method Std. Err.	z	P>|z|	[95% Conf. Interval]	
private	.6952778	.1788904	3.89	0.000	.3446591	1.045897
medicaid	.3586536	.3277206	1.09	0.274	-.2836669	1.000974
age	1.27116	.3300602	3.85	0.000	.6242534	1.918066
age2	-.0083233	.0021979	-3.79	0.000	-.0126311	-.0040156
educyr	.0973898	.0235727	4.13	0.000	.0511881	.1435915
actlim	.4164256	.1898125	2.19	0.028	.0444	.7884513
totchr	1.068033	.0641102	16.66	0.000	.9423791	1.193686

```
. * Marginal effects for component 2
. margins, dydx(*) predict(eq(component2)) atmean noatlegend
Warning: cannot perform check for estimable functions.
Conditional marginal effects                       Number of obs   =        3677
Model VCE     : Robust

Expression    : predicted mean: component2, predict(eq(component2))
dy/dx w.r.t.  : private medicaid age age2 educyr actlim totchr
```

	dy/dx	Delta-method Std. Err.	z	P>|z|	[95% Conf. Interval]	
private	1.822005	.795072	2.29	0.022	.2636927	3.380318
medicaid	1.673629	1.784231	0.94	0.348	-1.823399	5.170657
age	3.465136	1.482792	2.34	0.019	.5589159	6.371355
age2	-.0229591	.0098162	-2.34	0.019	-.0421986	-.0037196
educyr	.3185591	.1043185	3.05	0.002	.1140986	.5230197
actlim	2.41424	.8438944	2.86	0.004	.7602377	4.068243
totchr	2.597344	.3458016	7.51	0.000	1.919585	3.275102

The MEs for the high-use group, the second group, are several times those for the low-use group. For the two key insurance status variables, the MEM is roughly 3 and 4 times larger for the high-use group.

The following code produces histograms of the distributions of the fitted means for the two components.

```
. * Create histograms of fitted values
. quietly histogram yfit1, name(_comp_1, replace)
. quietly histogram yfit2, name(_comp_2, replace)
. quietly graph combine _comp_1 _comp_2
```

These histograms are plotted in figure 17.2. Clearly, the second component experiences more doctor visits.

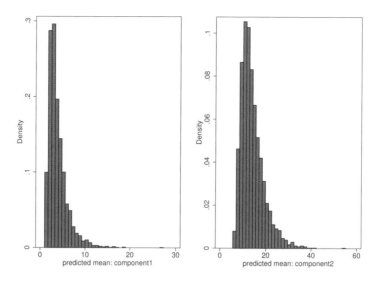

Figure 17.2. Fitted values distribution, FMM2-P

Application: NB finite-mixture model

The fmm command with the mixtureof(negbin1) option can be used to estimate a mixture distribution with NB components. This model involves additional overdispersion parameters that can potentially create problems for convergence of the numerical algorithm. This may happen if an overdispersion parameter is too close to zero. Further, the number of parameters increases linearly with the number of components, and the likelihood function quickly becomes high dimensional when the specification includes many regressors. Typically, the mixtureof(negbin1) or mixtureof(negbin2) option requires many more iterations than the mixtureof(poisson) option.

A 2-component NB1 finite-mixture model example follows.

```
. * 2-component mixture of NB1
. fmm docvis $xlist, vce(robust) components(2) mixtureof(negbin1)
Fitting Negative Binomial-1 model:
Iteration 0:   log likelihood = -15019.656
Iteration 1:   log likelihood =  -15019.64
Iteration 2:   log likelihood =  -15019.64
Iteration 0:   log likelihood = -12739.566
Iteration 1:   log likelihood = -11125.786
Iteration 2:   log likelihood = -10976.314
Iteration 3:   log likelihood = -10976.058
Iteration 4:   log likelihood = -10976.058
Iteration 0:   log likelihood = -10976.058
Iteration 1:   log likelihood = -10566.829
Iteration 2:   log likelihood = -10531.205
Iteration 3:   log likelihood = -10531.054
Iteration 4:   log likelihood = -10531.054
```

```
Fitting 2 component Negative Binomial-1 model:
Iteration 0:   log pseudolikelihood = -10531.611  (not concave)
Iteration 1:   log pseudolikelihood = -10529.012  (not concave)
Iteration 2:   log pseudolikelihood =  -10515.85  (not concave)
Iteration 3:   log pseudolikelihood = -10500.668  (not concave)
Iteration 4:   log pseudolikelihood = -10495.501  (not concave)
Iteration 5:   log pseudolikelihood = -10494.709
Iteration 6:   log pseudolikelihood = -10493.449
Iteration 7:   log pseudolikelihood = -10493.333
Iteration 8:   log pseudolikelihood = -10493.324
Iteration 9:   log pseudolikelihood = -10493.324
```

```
2 component Negative Binomial-1 regression     Number of obs   =      3677
                                               Wald chi2(14)   =    560.31
Log pseudolikelihood = -10493.324              Prob > chi2     =    0.0000
```

docvis	Coef.	Robust Std. Err.	z	P>|z|	[95% Conf. Interval]	
component1						
private	.137827	.0610423	2.26	0.024	.0181863	.2574676
medicaid	.0379753	.0628139	0.60	0.545	-.0851377	.1610883
age	.253357	.0633567	4.00	0.000	.12918	.3775339
age2	-.0016569	.0004261	-3.89	0.000	-.002492	-.0008218
educyr	.0228524	.0055063	4.15	0.000	.0120602	.0336446
actlim	.1060655	.0514198	2.06	0.039	.0052845	.2068464
totchr	.2434641	.0294843	8.26	0.000	.1856759	.3012523
_cons	-8.645394	2.352187	-3.68	0.000	-13.2556	-4.035192
component2						
private	.372013	.5124233	0.73	0.468	-.6323182	1.376344
medicaid	.3344168	.856897	0.39	0.696	-1.34507	2.013904
age	.5260549	.6902627	0.76	0.446	-.8268352	1.878945
age2	-.0034424	.0047508	-0.72	0.469	-.0127539	.005869
educyr	.0457671	.0499026	0.92	0.359	-.0520402	.1435743
actlim	.3599301	.3852059	0.93	0.350	-.3950595	1.11492
totchr	.4150389	.1332826	3.11	0.002	.1538097	.6762681
_cons	-19.3304	25.16197	-0.77	0.442	-68.64696	29.98615
/imlogitpi1	2.382195	2.159316	1.10	0.270	-1.849987	6.614377
/lndelta1	1.210492	.2343343	5.17	0.000	.7512047	1.669778
/lndelta2	2.484476	.7928709	3.13	0.002	.9304772	4.038474
delta1	3.355133	.7862229			2.119552	5.310991
delta2	11.99483	9.510352			2.535719	56.7397
pi1	.9154595	.1671169			.1358744	.9986608
pi2	.0845405	.1671169			.0013392	.8641256

The two classes are very different in probability of occurrence, because 92% of the population falls in the low-use category and only 8% fall in the high-use category. Only one coefficient, that of `totchr`, is significantly different from zero in the second category. The maximized value of the log likelihood is about the same as in the case of hurdle NB, but there are three more parameters in the mixture model. The coefficients in the two classes do not differ much from the corresponding FMM2-P results. As expected, there is evidence of overdispersion in both components; `delta1` and `delta2` are the overdispersion parameters.

A comparison of the mean and variance of the two components is possible by using the fitted values from each component.

```
. * Fitted values for 2-component NB1 mixture
. drop yfit1 yfit2
. predict yfit1, equation(component1)
. predict yfit2, equation(component2)
. summarize yfit1 yfit2
    Variable |      Obs       Mean    Std. Dev.       Min        Max
-------------+--------------------------------------------------------
       yfit1 |     3677   6.366216    2.634751   2.382437   28.68904
       yfit2 |     3677   12.39122    11.20933   1.507496   186.8094
```

The first component has a mean of 6.37, slightly below the sample average; the second component has mean of 12.39. The variance of the second distribution is very high, which indicates a substantial overlap between the two distributions. This means that the distinction between the two components is more blurred than in the case of the FMM2-P model but the fit is significantly better.

Model selection

Choosing the "best" model involves trade-offs between fit, parsimony, and ease of interpretation. Which of the six models that have been estimated best fits the data?

Table 17.1 summarizes three commonly used model-comparison statistics—log likelihood, and Akaike and Bayes information criteria (AIC and BIC)—explained in section 10.7.2.

The log likelihood for the hurdle model is simply the sum of log likelihoods for the two parts of the model, whereas for the other models, it is directly given as command output. All three criteria suggest that the NB2 hurdle model provides the best fitting and the most parsimonious specification. Such an unambiguous outcome is not always realized.

Table 17.1. Goodness-of-fit criteria for six models

Model	Parameters	Log likelihood	AIC	BIC
Poisson	8	−15,019.64	30,055	30,113
NB2	9	−10,589.34	21,197	21,253
Poisson hurdle	16	−14,037.91[a]	28,108	28,207
NB2 hurdle	17	−10,493.23	21,020	21,126
FMM2-P	17	−11,502.69	23,039	23,145
FMM2-NB1	19	−10,493.32	21,025	21,143

[a] The log-likelihood value for the Poisson-hurdle model can be obtained by using hplogit instead of hnblogit in the model fit on page 587.

Most of these models are nonnested, so LR tests are not possible. The LR test can be used to test the Poisson against the NB2 model and leads to strong rejection of the Poisson model.

Cautionary note

It is easy to overparameterize mixture models. When the number of components is small, say, 2, and the means of the component distribution are far apart, clear discrimination between the components will emerge. However, if this is not the case, and a larger value of m is specified, unambiguous identification of all components may be difficult because of the increasing overlap in the distributions. In particular, the presence of outliers may give rise to components that account for a small proportion of the observations. For example, if $m = 3$, $\pi_1 = 0.6$, $\pi_2 = 0.38$, and $\pi_3 = (1 - 0.6 - 0.38) = 0.02$, then this means that the third component accounts for just 2% of the data. If 2% of the sample is a small number, one might regard the result as indicating the presence of extreme observations. The `fmm` command allows the number of components to be between 2 and 9.

There are a number of indications of failure of identification or fragile identification of mixture components. We list several examples. First, the log likelihood may only increase slightly when additional components are added. Second, the log likelihood may "fall" when additional components are added, which could be indicative of a multimodal objective function. Third, one or more mixture components may be small in the sense of accounting for few observations. Fourth, the iteration log may persistently generate the message "`not concave`". Finally, convergence may be very slow, which could indicate a flat log likelihood. Therefore, it is advisable to use contextual knowledge and information when specifying and evaluating an FMM.

17.4   Empirical example 2

We now consider the application of a class of count-data models that permits the mechanism generating the zero observations to differ from the one for positive observations. A subclass of these models is the so-called zero-inflated class of models designed to deal with the "excess zeros" problem. These models are generalizations of several that were considered in the previous section, so it is natural to ask at an appropriate point in the investigation whether they are statistically superior to their restricted versions.

17.4.1   Zero-inflated data

The dataset used in this section overlaps heavily with those used in the last section. The most important difference is that the variable we choose to analyze is different. In place of `docvis` as the dependent variable, we use the `er` variable, defined as the number of emergency room visits by the survey respondent. An emergency room visit is a rare event for the Medicare elderly population who have access to care through their public insurance program and hence do not need to use emergency room facilities as the only available means of getting care. There is a high degree of randomness in this variable, which will become apparent.

The full set of explanatory variables in the model was initially the same as that used in the `docvis` example. However, after some preliminary analysis, this list was reduced to just three health-status variables—age, `actlim`, and `totchr`—that appeared to have some predictive power for `er`. The summary statistics follow, along with a tabulation of the frequency distribution for `er`.

```
. * Summary stats for ER use model
. use mus17data_z.dta
. global xlist1 age actlim totchr
. summarize er $xlist1
```

Variable	Obs	Mean	Std. Dev.	Min	Max
er	3677	.2774001	.6929326	0	10
age	3677	74.24476	6.376638	65	90
actlim	3677	.333152	.4714045	0	1
totchr	3677	1.843351	1.350026	0	8

```
. tabulate er
```

# Emergency Room Visits	Freq.	Percent	Cum.
0	2,967	80.69	80.69
1	515	14.01	94.70
2	128	3.48	98.18
3	40	1.09	99.27
4	15	0.41	99.67
5	8	0.22	99.89
6	2	0.05	99.95
7	1	0.03	99.97
10	1	0.03	100.00
Total	3,677	100.00	

Compared with `docvis`, the `er` variable has a much higher proportion (80.7%) of zeros. The first four values (0, 1, 2, 3) account for over 99% of the probability mass of `er`.

In itself, this does not imply that we have the "excess zero" problem. Given the mean value of 0.2774, a Poisson distribution predicts that $\Pr(Y = 0) = e^{-0.2774} = 0.758$. The observed proportion of 0.807 is higher than this, but the difference could potentially be explained by the regressors in the model. So there is no need to jump to the conclusion that a zero-inflated variant is essential.

17.4.2   Models for zero-inflated data

The zero-inflated model was originally proposed to handle data with excess zeros relative to the Poisson model. Like the hurdle model, it supplements a count density, $f_2(\cdot)$, with a binary process with a density of $f_1(\cdot)$. If the binary process takes on a value of 0, with a probability of $f_1(0)$, then $y = 0$. If the binary process takes on a value of 1, with a probability of $f_1(1)$, then y takes on the count values $0, 1, 2, \ldots$ from the count density

$f_2(\cdot)$. This lets zero counts occur in two ways: as a realization of the binary process and as a realization of the count process when the binary random variable takes on a value of 1.

Suppressing regressors for notational simplicity, the zero-inflated model has a density of

$$f(y) = \begin{cases} f_1(0) + \{1 - f_1(0)\}f_2(0) & \text{if } y = 0, \\ \{1 - f_1(0)\}f_2(y) & \text{if } y \geq 1 \end{cases}$$

As in the case of the hurdle model, the probability $f_1(0)$ may be a constant or may be parameterized through a binomial model like the logit or probit. Once again, the set of variables in the $f_1(\cdot)$ density need not be the same as those in the $f_2(\cdot)$ density.

To estimate the parameters of the zero-inflated Poisson (ZIP) and zero-inflated NB (ZINB) models, the estimation commands `zip` and `zinb`, respectively, are used. The partial syntax for `zip` is

`zip` *depvar* $\big[$ *indepvars* $\big]$ $\big[$ *if* $\big]$ $\big[$ *in* $\big]$ $\big[$ *weight* $\big]$ **,** `inflate(`*varlist*`)` $\big[$ *options* $\big]$

where `inflate(`*varlist*`)` specifies the variables, if any, that determine the probability that the count is logit (the default) or probit (the `probit` option). Other options are essentially the same as for `poisson`.

The partial syntax for `zinb` is essentially the same as that for `zip`. Other options are the same as for `nbreg`. The only NB model estimated is a (truncated) NB2 model.

For the Poisson and NB models, the count process has the conditional mean $\exp(\mathbf{x}_2'\boldsymbol{\beta})$ and the corresponding with-zeros model can be shown to have the conditional mean

$$E(y|\mathbf{x}) = \{1 - f_1(0|\mathbf{x}_1)\} \times \exp(\mathbf{x}_2'\boldsymbol{\beta}_2) \qquad (17.12)$$

where $1 - f_1(0|\mathbf{x}_1)$ is the probability that the binary process variable equals 1. The MEs are complicated by the presence of regressors in both parts of the model, as for the hurdle model. But if the binary process does not depend on regressors, so $f_1(0|\mathbf{x}_1) = f_1(0)$, then the parameters, $\boldsymbol{\beta}_2$, can be directly interpreted as semielasticities, as for the regular Poisson and NB models.

After the `zip` and `zinb` commands, the predicted mean in (17.12) can be obtained by using the postestimation `predict` command, and the `margins` command can be used to obtain the MEM, MER, or AME.

17.4.3   Results for the NB2 model

Our starting point is the NB2 model.

```
. * NB2 for er
. nbreg er $xlist1, nolog
Negative binomial regression                        Number of obs   =        3677
                                                    LR chi2(3)      =      225.15
Dispersion      = mean                              Prob > chi2     =      0.0000
Log likelihood = -2314.4927                         Pseudo R2       =      0.0464
```

er	Coef.	Std. Err.	z	P>|z|	[95% Conf. Interval]
age	.0088528	.0061341	1.44	0.149	-.0031697 .0208754
actlim	.6859572	.0848127	8.09	0.000	.5197274 .8521869
totchr	.2514885	.0292559	8.60	0.000	.1941481 .308829
_cons	-2.799848	.4593974	-6.09	0.000	-3.700251 -1.899446
/lnalpha	.4464685	.1091535			.2325315 .6604055
alpha	1.562783	.1705834			1.26179 1.935577

```
Likelihood-ratio test of alpha=0:  chibar2(01) =  237.98 Prob>=chibar2 = 0.000
```

There is statistically significant overdispersion with $\alpha = 1.56$. The coefficient estimates are similar to those from Poisson model (not given). The regression equation has low but statistically significant explanatory power. For an event that is expected to have a high degree of inherent randomness, low overall explanatory power is to be expected. Having an activity limitation and a high number of chronic conditions is positively associated with er visits.

The prcounts command

One indication of the fit of the model is obtained from the average fitted probabilities of the NB2 model. This can be done by using the user-written countfit command, discussed in section 17.3.3. Instead, we demonstrate the use of the user-written prcounts command (Long and Freese 2006), which computes predicted probabilities and cumulative probabilities for each observation. We use the max(3) option because for these data most counts are at most 3.

```
. * Sample average fitted probabilities of y = 0 to max()
. prcounts erpr, max(3)
. summarize erpr*
```

Variable	Obs	Mean	Std. Dev.	Min	Max
erprrate	3677	.2782362	.1833994	.1081308	1.693112
erprpr0	3677	.8073199	.0855761	.4370237	.9049199
erprpr1	3677	.1387214	.0389334	.0837048	.2136777
erprpr2	3677	.0355246	.0243344	.0099214	.1207627
erprpr3	3677	.0112286	.0122574	.001262	.0771202
erprcu0	3677	.8073199	.0855761	.4370237	.9049199
erprcu1	3677	.9460414	.0485141	.6399685	.9886248
erprcu2	3677	.981566	.0249449	.7607312	.9985461
erprcu3	3677	.9927946	.0130371	.8378514	.9998082
erprprgt	3677	.0072054	.0130371	.0001918	.1621486

The output begins with the `erprrate` variable, which is the fitted mean and has an average value of 0.278, close to the sample mean of 0.277. The `erprpr0`–`erprpr3` variables are predictions of $\Pr(y_i = j)$, $j = 0, 1, 2, 3$, that have averages of 0.807, 0.139, 0.036, and 0.011 compared with sample frequencies of 0.807, 0.140, 0.035, and 0.011, given in the output in section 17.4.1. The fitted frequencies and observed frequencies are very close, an improved fit compared with the Poisson model, which is not given. The `erprcu0`–`erprcu3` variables are the corresponding cumulative predicted probabilities.

17.4.4   Results for ZINB

The parameters of the ZINB model are estimated by using the `zinb` command. We use the same set of regressors in the two parts of the model

```
. * Zero-inflated negative binomial for er
. zinb er $xlist1, inflate($xlist1) vuong nolog
Zero-inflated negative binomial regression      Number of obs    =       3677
                                                Nonzero obs      =        710
                                                Zero obs         =       2967

Inflation model = logit                         LR chi2(3)       =      34.29
Log likelihood  = -2304.868                     Prob > chi2      =     0.0000
```

er	Coef.	Std. Err.	z	P>|z|	[95% Conf. Interval]	
er						
age	.0035485	.0076344	0.46	0.642	-.0114146	.0185116
actlim	.2743106	.1768941	1.55	0.121	-.0723954	.6210165
totchr	.1963408	.0558635	3.51	0.000	.0868504	.3058313
_cons	-1.822978	.6515914	-2.80	0.005	-3.100074	-.5458825
inflate						
age	-.0236763	.0284226	-0.83	0.405	-.0793835	.0320309
actlim	-4.22705	18.91192	-0.22	0.823	-41.29372	32.83962
totchr	-.3471091	.2052892	-1.69	0.091	-.7494686	.0552505
_cons	1.846526	2.071003	0.89	0.373	-2.212565	5.905618
/lnalpha	.1602371	.235185	0.68	0.496	-.3007171	.6211913
alpha	1.173789	.2760576			.7402871	1.861144

```
Vuong test of zinb vs. standard negative binomial: z =    1.99  Pr>z = 0.0233
```

The estimated coefficients differ from those from the NB2 model. The two models have different conditional means—see (17.12)—so the coefficients are not directly comparable.

The `vuong` option of `zinb` implements the LR test of Vuong (1989) to discriminate between the NB and ZINB models. This test corrects for the complication that the ZINB model only reduces to the NB model at the boundary of the parameter space for the logit model [so that $f_1(0) = 0$]. Furthermore, Vuong's test does not require that either of the two models be correctly specified under the null hypothesis. The test statistic is standard normally distributed, with large positive values favoring the ZINB model and large negative values favoring the NB model. Here the test statistic of 1.99 with a one-sided p-value of 0.023 favors the ZINB model at a significance level of 0.05.

17.4.5 Model comparison

Count models, even those nonnested, can be compared on the basis of goodness of fit.

The countfit command

Although we could simply stop at this point and base our substantive conclusions on these estimates, we should examine whether zero-inflated models improve the fit to the data. The user-written `countfit` command (Long and Freese 2006) facilitates the task of multiple model comparisons for the four candidate models: Poisson, NB2, ZIP, and ZINB.

Model comparison using countfit

We apply the `countfit` command, using several options to restrict the output, most notably not reporting model estimates and just comparing the NB2 and ZINB models. We obtain

```
. * Comparison of NB and ZINB using countfit
. countfit er $xlist1, nbreg zinb nograph noestimates
Comparison of Mean Observed and Predicted Count

              Maximum      At      Mean
Model       Difference    Value   |Diff|
-------------------------------------------------
NBRM          0.001         1      0.000
ZINB          0.006         1      0.001

NBRM: Predicted and actual probabilities

Count    Actual    Predicted   |Diff|   Pearson
-------------------------------------------------
0        0.807      0.807       0.000     0.001
1        0.140      0.139       0.001     0.047
2        0.035      0.036       0.001     0.053
3        0.011      0.011       0.000     0.040
4        0.004      0.004       0.000     0.001
5        0.002      0.002       0.001     0.558
6        0.001      0.001       0.000     0.181
7        0.000      0.000       0.000     0.052
8        0.000      0.000       0.000     0.610
9        0.000      0.000       0.000     0.308
-------------------------------------------------
Sum      1.000      1.000       0.004     1.850
```

```
ZINB: Predicted and actual probabilities
Count   Actual    Predicted   |Diff|   Pearson
-------------------------------------------------
0        0.807      0.808      0.001    0.009
1        0.140      0.135      0.006    0.834
2        0.035      0.039      0.004    1.467
3        0.011      0.012      0.001    0.444
4        0.004      0.004      0.000    0.003
5        0.002      0.001      0.001    1.499
6        0.001      0.001      0.000    0.003
7        0.000      0.000      0.000    0.087
8        0.000      0.000      0.000    0.300
9        0.000      0.000      0.000    0.125
-------------------------------------------------
Sum      1.000      1.000      0.013    4.770

Tests and Fit Statistics
----------------------------------------------------------------------
NBRM             BIC=-25517.593  AIC=    1.262  Prefer  Over  Evidence
----------------------------------------------------------------------
    vs ZINB      BIC=-25504.004  dif=  -13.589  NBRM    ZINB  Very strong
                 AIC=     1.259  dif=    0.003  ZINB    NBRM
                 Vuong=   1.991  prob=   0.023  ZINB    NBRM  p=0.023
```

The first set of output gives average predicted probabilities for, respectively, the NB2 model (`nbreg`) and the ZINB model (`zinb`). Both are close to actual frequencies, and the ZINB actually does better.

The second set of output provides the penalized log-likelihood–based statistics AIC (`aic`) and BIC (`bic`), which are alternative scalings to those detailed in section 10.7.2. The BIC, which penalizes model complexity (the number of parameters estimated) more severely than the AIC, favors the NB2 model, whereas AIC favors the ZINB model.

This example indicates that having many zeros in the dataset does not automatically mean that a zero-inflated model is necessary. For these data, the ZINB model is only a slight improvement on the NB2 model and is actually no improvement at all if BIC is used as the model-selection criterion. It is easier to interpret the estimates of the parameters of the NB2 model.

17.5   Models with endogenous regressors

So far, the regressors in the count regression are assumed to be exogenous. We now consider a more general model in which one regressor is endogenous. Specifically, the empirical example used in this chapter has assumed that the regressor `private` is exogenous. But individuals can and do choose whether they want supplementary private insurance and hence potentially this variable is endogenous, i.e., jointly determined with `docvis`. If endogeneity is ignored, the standard single-equation estimator will be inconsistent.

The general issues are similar to those already presented in section 14.8 for endogeneity in the probit model. We present two distinct methods to control for endogeneity—a structural-model approach and a less parametric nonlinear instrumental-variables (IV) approach.

17.5.1 Structural-model approach

The structural-model approach defines explicit models for both the dependent variable of interest (y_1) and the endogenous regressor (y_2).

Model and assumptions

First, the structural equation for the count income is a Poisson model with a mean that depends on an endogenous regressor:

$$y_{1i} \sim \text{Poisson}(\mu_i)$$

$$\mu_i = E(y_{1i}|y_{2i}, \mathbf{x}_{1i}, u_{i1}) = \exp(\beta_1 y_{2i} + \mathbf{x}'_{1i}\boldsymbol{\beta}_2 + u_{1i}) \tag{17.13}$$

where y_2 is endogenous and $\mathbf{x}_1$ is a vector of exogenous variables. The term u_1 is an error term that can be interpreted as unobserved heterogeneity correlated with the endogenous regressor, y_2, but is uncorrelated with the exogenous regressors, $\mathbf{x}_1$. The error term, u_1, is added to allow for endogeneity. Also it induces overdispersion, so that the Poisson model has been generalized to control for overdispersion as would be the case if a NB model was used.

Next, to clarify the nature of interdependence between y_2 and u_1, we specify a linear reduced-form equation for y_2. This is

$$y_{2i} = \mathbf{x}'_{1i}\boldsymbol{\gamma}_1 + \mathbf{x}'_{2i}\boldsymbol{\gamma}_2 + \varepsilon_i \tag{17.14}$$

where $\mathbf{x}_2$ is a vector of exogenous variables that affects y_2 nontrivially but does not directly affect y_1, and hence is an independent source of variation in y_2. It is standard to refer to this as an exclusion restriction and to refer to $\mathbf{x}_2$ as excluded exogenous variables or IV. By convention, a condition for robust identification of (17.13), as in the case of the linear model, is that there is available at least one valid excluded variable (instrument). When only one such variable is present in (17.14), the model is said to be just-identified, and it is said to be overidentified if there are additional excluded variables.

Assume that the errors u_1 and ε are related via

$$u_{1i} = \rho\varepsilon_i + \eta_i \tag{17.15}$$

where $\eta_i \sim [0, \sigma_\eta^2]$ is independent of $\varepsilon_i \sim [0, \sigma_\varepsilon^2]$.

This assumption can be interpreted to mean that ε is a common latent factor that affects both y_1 and y_2 and is the only source of dependence between them, after controlling for the influence of the observable variables $\mathbf{x}_1$ and $\mathbf{x}_2$. If $\rho = 0$, then y_2 can be treated as exogenous. Otherwise, y_2 is endogenous, since it is correlated with u_1 in (17.14) because both y_2 and u_1 depend on ε.

Two-step estimation

ML estimation of this model is computationally challenging. A two-step estimator is much simpler to implement.

Substituting (17.14) for u_1 in (17.13) yields $\mu = \exp(\beta_1 y_2 + \mathbf{x}_1' \boldsymbol{\beta}_2 + \rho \varepsilon) e^{\eta}$. Taking the expectation with respect to η yields $E_{\eta}(\mu) = \exp(\beta_1 y_2 + \mathbf{x}_1' \boldsymbol{\beta}_2 + \rho \varepsilon) \times E(e^{\eta}) = \exp(\beta_1 y_2 + \ln E(e^{\eta}) + \mathbf{x}_1' \boldsymbol{\beta}_2 + \rho \varepsilon)$. The constant term $\ln E(e^{\eta})$ can be absorbed in the coefficient of the intercept, a component of $\mathbf{x}_1$. It follows that

$$\mu_i | \mathbf{x}_{1i}, y_{2i}, \varepsilon_i = \exp(\beta_1 y_{2i} + \mathbf{x}_{1i}' \boldsymbol{\beta}_2 + \rho \varepsilon_i) \tag{17.16}$$

where ε_i is a new additional variable, and the intercept has absorbed $E(e^{\eta_i})$.

If ε were observable, including it as a regressor would control for the endogeneity of y_2. Given that it is unobservable, the estimation strategy is to replace it by a consistent estimate. The following two-step estimation procedure is used: First, estimate (17.14) by OLS, and generate the residuals $\widehat{\varepsilon}_i$. Second, estimate parameters of the Poisson model given in (17.16) after replacing ε_i by $\widehat{\varepsilon}_i$. As discussed below, if $\rho = 0$, then we can use the vce(robust) option, but if $\rho \neq 0$ then the VCE needs to be estimated with the bootstrap method detailed in section 13.4.5 that controls for the estimation of ε_i by $\widehat{\varepsilon}_i$.

Application

We apply this two-step procedure to the Poisson model for the doctor visits data analyzed in section 17.3, with the important change that private is now treated as endogenous. Two excluded variables used as instruments are income and ssiratio. The first is a measure of total household income and the second is the ratio of social security income to total income. Jointly, the two variables reflect the affordability of private insurance. A high value of income makes private insurance more accessible, whereas a high value of ssiratio indicates an income constraint and is expected to be negatively associated with private. For these to be valid instruments, we need to assume that for people aged 65–90 years, doctor visits are not determined by income or ssiratio, after controlling for other regressors that include a quadratic in age, education, health-status measures, and access to Medicaid.

The first step generates residuals from a linear probability regression of private on regressors and instruments.

(*Continued on next page*)

```
. * First-stage linear regression
. use mus17data.dta
. global xlist2 medicaid age age2 educyr actlim totchr
. regress private $xlist2 income ssiratio, vce(robust)
```

Linear regression

```
                                                Number of obs =      3677
                                                F(  8,  3668) =    249.61
                                                Prob > F      =    0.0000
                                                R-squared     =    0.2108
                                                Root MSE      =    .44472
```

private	Coef.	Robust Std. Err.	t	P>|t|	[95% Conf. Interval]	
medicaid	-.3934477	.0173623	-22.66	0.000	-.4274884	-.3594071
age	-.0831201	.0293734	-2.83	0.005	-.1407098	-.0255303
age2	.0005257	.0001959	2.68	0.007	.0001417	.0009098
educyr	.0212523	.0020492	10.37	0.000	.0172345	.02527
actlim	-.0300936	.0176874	-1.70	0.089	-.0647718	.0045845
totchr	.0185063	.005743	3.22	0.001	.0072465	.0297662
income	.0027416	.0004736	5.79	0.000	.0018131	.0036702
ssiratio	-.0647637	.0211178	-3.07	0.002	-.1061675	-.0233599
_cons	3.531058	1.09581	3.22	0.001	1.3826	5.679516

```
. predict lpuhat, residual
```

The two instruments, income and ssiratio, are highly statistically significant with expected signs.

The second step fits a Poisson model on regressors that include the first-step residual.

```
. * Second-stage Poisson with robust SEs
. poisson docvis private $xlist2 lpuhat, vce(robust) nolog
```

Poisson regression

```
                                                Number of obs =      3677
                                                Wald chi2(8)  =    718.87
                                                Prob > chi2   =    0.0000
Log pseudolikelihood = -15010.614               Pseudo R2     =    0.1303
```

docvis	Coef.	Robust Std. Err.	z	P>|z|	[95% Conf. Interval]	
private	.5505541	.2453175	2.24	0.025	.0697407	1.031368
medicaid	.2628822	.1197162	2.20	0.028	.0282428	.4975217
age	.3350604	.0696064	4.81	0.000	.1986344	.4714865
age2	-.0021923	.0004576	-4.79	0.000	-.0030893	-.0012954
educyr	.018606	.0080461	2.31	0.021	.002836	.034376
actlim	.2053417	.0414248	4.96	0.000	.1241505	.286533
totchr	.24147	.0129175	18.69	0.000	.2161523	.2667878
lpuhat	-.4166838	.249347	-1.67	0.095	-.9053949	.0720272
_cons	-11.90647	2.661445	-4.47	0.000	-17.1228	-6.69013

The z statistic for the coefficient of `lpuhat` provides the basis for a robust Wald test of the null hypothesis of exogeneity, $H_0: \rho = 0$. The z statistic has a p-value of 0.095 against $H_1: \rho \neq 0$, leading to nonrejection of H_0 at the 0.05 level. But a one-sided test against $H_1: \rho < 0$ may be appropriate because this was proposed on a priori grounds. Then the p-value is 0.047, leading to rejection of H_0 at the 0.05 level.

If $\rho \neq 0$, then the VCE of the second-step estimator needs to be adjusted for the replacement of ε_i with $\widehat{\varepsilon}_i$ by using the bootstrap method given in section 13.4.5. We have

```
. * Program and bootstrap for Poisson two-step estimator
. program endogtwostep, eclass
  1.     version 11
  2.     tempname b
  3.     capture drop lpuhat2
  4.     regress private $xlist2 income ssiratio
  5.     predict lpuhat2, residual
  6.     poisson docvis private $xlist2 lpuhat2
  7.     matrix `b' = e(b)
  8.     ereturn post `b'
  9. end

. bootstrap _b, reps(400) seed(10101) nodots nowarn: endogtwostep
Bootstrap results                          Number of obs    =        3677
                                           Replications     =         400
```

	Observed Coef.	Bootstrap Std. Err.	z	P>|z|	Normal-based [95% Conf. Interval]	
private	.5505541	.2567815	2.14	0.032	.0472716	1.053837
medicaid	.2628822	.1205813	2.18	0.029	.0265473	.4992172
age	.3350604	.0707275	4.74	0.000	.1964371	.4736838
age2	-.0021923	.0004667	-4.70	0.000	-.0031071	-.0012776
educyr	.018606	.0083042	2.24	0.025	.0023301	.034882
actlim	.2053417	.0412756	4.97	0.000	.124443	.2862405
totchr	.24147	.0134522	17.95	0.000	.2151042	.2678359
lpuhat2	-.4166838	.2617964	-1.59	0.111	-.9297953	.0964276
_cons	-11.90647	2.698704	-4.41	0.000	-17.19583	-6.617104

The standard errors differ little from the previous standard errors obtained by using the option `vce(robust)`. From section 17.3.2, the Poisson ML estimate of the coefficient on `private` was 0.142 with a robust standard error of 0.036. The two-step estimate of the coefficient on `private` is 0.551 with a standard error of 0.256. The precision of estimation is much less, because the standard error is seven times larger. This large increase is very common for cross-section data, where instruments are not very highly correlated with the regressor being instrumented. At the same time, the coefficient is four times larger, and so the regressor retains statistical significance. The effect is now very large, with private insurance leading to a $100(e^{0.551} - 1) = 73\%$ increase in doctor visits.

The negative coefficient of `lpuhat2` can be interpreted to mean that the latent factor, which increases the probability of purchasing private insurance lowers the number of doctor visits—an effect consistent with favorable selection, according to which the relatively healthy individuals self-select into insurance. Controlling for endogeneity has a substantial effect on the ME of an exogenous change in private insurance because the coefficient of `private` and the associated MEs are now much higher.

17.5.2 Nonlinear IV method

An alternative method for controlling for endogeneity is the nonlinear IV (NLIV), or GMM, method presented in section 11.8. In the notation of section 17.5.1, this assumes the existence of the instruments $\mathbf{z}_i = (\mathbf{x}'_{1i}\ \mathbf{x}'_{2i})'$ that satisfy

$$E[\mathbf{z}_i\{y_{1i} - \exp(\beta_1 y_{2i} + \mathbf{x}'_{1i}\boldsymbol{\beta}_2)\}] = \mathbf{0}$$

Equations (17.13)–(17.15) do not imply this moment condition, so this less parametric approach will lead to an estimator that differs from that using the structural approach even in the limit as $N \to \infty$.

To apply nonlinear IV to our example, we use a similar `gmm` command to that given in section 10.3.8. For one-step GMM estimation, we have

```
. * Command gmm for one-step GMM (nonlinear IV) of overidentified Poisson model
. gmm (docvis - exp({xb:private medicaid age age2 educyr actlim totchr}+{b0})),
> instruments(income ssiratio medicaid age age2 educyr actlim totchr) onestep nolog

Final GMM criterion Q(b) = .0495772

GMM estimation

Number of parameters =    8
Number of moments    =    9
Initial weight matrix: Unadjusted                    Number of obs   =    3677
```

	Coef.	Robust Std. Err.	z	P>|z|	[95% Conf.	Interval]
/xb_private	.592014	.3397353	1.74	0.081	-.073855	1.257883
/xb_medicaid	.3186684	.1909955	1.67	0.095	-.0556759	.6930127
/xb_age	.3323178	.0705395	4.71	0.000	.1940628	.4705728
/xb_age2	-.002176	.0004643	-4.69	0.000	-.003086	-.0012659
/xb_educyr	.0190887	.0092216	2.07	0.038	.0010147	.0371627
/xb_actlim	.2084978	.0433758	4.81	0.000	.1234827	.2935129
/xb_totchr	.241843	.0129869	18.62	0.000	.2163892	.2672968
/b0	-11.86322	2.732891	-4.34	0.000	-17.21959	-6.506856

```
Instruments for equation 1: income ssiratio medicaid age age2 educyr actlim
    totchr _cons
```

The results are qualitatively similar to the others given above. The coefficient of `private` is now statistically insignificant at level 0.05 using a two-sided test, because of a larger standard error than that obtained using the two-step estimation method of section 17.5.1. However, it remains statistically significant at the 0.05 level using a one-sided test against the alternative that the coefficient is negative.

Because the model is overidentified, with one more instrument than there are regressors, there is potential efficiency gain to two-step GMM estimation, using the `twostep` option.

```
. * Command gmm for two-step GMM (nonlinear IV) of overidentified Poisson model
. gmm (docvis - exp({xb:private medicaid age age2 educyr actlim totchr}+{b0})),
> instruments(income ssiratio medicaid age age2 educyr actlim totchr) twostep nolog

Final GMM criterion Q(b) =  .0009926

GMM estimation

Number of parameters =   8
Number of moments    =   9
Initial weight matrix: Unadjusted                    Number of obs    =    3677
GMM weight matrix:     Robust
```

	Coef.	Robust Std. Err.	z	P>|z|	[95% Conf. Interval]	
/xb_private	.5863841	.341268	1.72	0.086	-.0824889	1.255257
/xb_medicaid	.2875693	.1910994	1.50	0.132	-.0869785	.6621172
/xb_age	.3495993	.0706441	4.95	0.000	.2111394	.4880593
/xb_age2	-.0022813	.0004653	-4.90	0.000	-.0031932	-.0013694
/xb_educyr	.0174443	.0092449	1.89	0.059	-.0006753	.0355639
/xb_actlim	.1842652	.041791	4.41	0.000	.1023563	.2661741
/xb_totchr	.2461328	.0128855	19.10	0.000	.2208777	.271388
/b0	-12.53958	2.737666	-4.58	0.000	-17.90531	-7.173858

```
Instruments for equation 1: income ssiratio medicaid age age2 educyr actlim
    totchr _cons
```

Compared with one-step GMM, the coefficients have changed by at most 10%, and the standard errors have changed very little.

The overidentifying restrictions test presented for the linear IV model in section 6.3.7 extends to nonlinear GMM models. For this example, following two-step GMM estimation, we have

```
. * Test of overidentifying restriction following two-step GMM
. estat overid

    Test of overidentifying restriction:

    Hansen's J chi2(1) = 3.64969 (p = 0.0561)
```

The test statistic is $\chi^2(1)$ distributed because there is one overidentifying restriction (income and ssiratio are instruments for private and all other regressors are instruments for themselves). Because $p > 0.05$, we do not reject the null hypothesis and conclude that the overidentifying restriction is valid.

17.6   Stata resources

The single-equation Stata commands [R] **poisson** and [R] **nbreg** (for nbreg and gnbreg) cover the basic count regression. See also [R] **poisson postestimation** and [R] **nbreg postestimation** for guidance on testing hypotheses and calculating MEs. For zero-inflated and truncated models, see [R] **zip**, [R] **zinb**, [R] **ztp**, and [R] **ztnb**. For estimating hurdle and finite-mixture models, the user-written hplogit, hnblogit, and fmm commands are relevant. The user-written prvalue, prcount, and countfit commands are useful for model evaluation and comparison. For panel count-data analysis, the basic commands [XT] **xtpoisson** and [XT] **xtnbreg** are covered in chapter 18. Quantile regression for counts is covered in section 7.5. Finally, Deb and Trivedi (2006) provide the `mtreatnb` command for estimating the parameters of a treatment-effects model that can be used to analyze the effects of an endogenous multinomial treatment (when one treatment is chosen from a set of more than two choices) on a nonnegative integer-valued outcome modeled using the NB regression.

17.7   Exercises

1. Consider the Poisson distribution with $\mu = 2$ and a multiplicative mean-preserving lognormal heterogeneity with a variance of 0.25. Using the pseudorandom generators for Poisson and lognormal distributions, and following the approach used for generating a simulated sample from the NB2 distribution, generate a draw from the Poisson–lognormal mixture distribution. Following the approach of section 17.2.2, generate another sample with a mean-preserving gamma distribution with a variance of 0.25. Using the **summarize, detail** command, compare the quantiles of the two samples. Which distribution has a thicker right tail? Repeat this exercise for a count-data regression with the conditional mean function $\mu(x) = \exp(1 + 1x)$, where x is an exogenous variable generated as a draw from the uniform$(0, 1)$ distribution.

2. For each regression sample generated in the previous exercise, estimate the parameters of the NB2 model. Compare the goodness of fit of the NB2 model in the two cases. Which of the two datasets is better explained by the NB2 model? Can you explain the outcome?

3. Suppose it is suggested that the use of the `ztp` command to estimate the parameters of the ZTP model is unnecessary. Instead, simply subtract 1 from all the counts y, replacing them with $y^* = y - 1$, and then apply the regular Poisson model using the new dependent variable y^*; $E(y^*) = E(y) - 1$. Using generated data from Poisson$\{\mu(x) = 1 + x\}$, $x = $ uniform$(0, 1)$, verify whether this method is equivalent to the `ztp`.

4. Using the finite-mixture command (`fmm`), estimate 2- and 3-component NB2 mixture models for the univariate (intercept only) version of the `docvis` model. [The models should be fit sequentially for the two values of m because `fmm` uses results from the $(m-1)$-component mixture to obtain starting values for the m-component mixture model.] Use the BIC to select the "better" model. For the selected model, use the `predict` command to compute the means of the m components. Explain and interpret the estimates of the component means, and the estimates of the mixing fractions. Is the identification of the two and three components robust? Explain your answer.

5. For this exercise, use the data from section 17.4. Estimate the parameters of the Poisson and ZIP models using the same covariates as in section 17.4. Test whether there is statistically significant improvement in the log likelihood. Which model has a better BIC? Contrast this outcome with that for the NB2/ZINB pair and rationalize the outcome.

6. Consider the data application in section 17.5.1. Drop all observations for which the `medicaid` variable equals one, and therefore drop `medicaid` as a covariate in the regression. For this reduced sample, estimate the parameters of the Poisson model treating the `private` variable first as exogenous and then as endogenous. Obtain and compare the two estimates of the ME of `private` on `docvis`. Implement the test for endogeneity given in section 17.5.1.

18 Nonlinear panel models

18.1 Introduction

The general approaches to nonlinear panel models are similar to those for linear models, such as pooled, population-averaged, random effects, and fixed effects.

We focus exclusively on short panels in which consistent estimation of fixed-effects (FE) models is not possible in some standard nonlinear models, such as binary probit. Unlike the linear case, the slope parameters in pooled and random-effects (RE) models lead to different estimators. More generally, results for linear models do not always carry over to nonlinear models, and methods used for one type of nonlinear model may not be applicable to another type.

We begin with a general treatment of nonlinear panel models. We then give a lengthy treatment of the panel methods for the logit model. Other data types are given shorter treatment.

18.2 Nonlinear panel-data overview

We assume familiarity with the material in chapter 8. We use the individual-effects models as the starting point to survey the various panel methods for nonlinear models.

18.2.1 Some basic nonlinear panel models

We consider nonlinear panel models for the scalar dependent variable y_{it} with the regressors $\mathbf{x}_{it}$, where i denotes the individual and t denotes time.

In some cases, a fully parametric model may be specified, with the conditional density

$$f(y_{it}|\alpha_i, \mathbf{x}_{it}) = f(y_{it}, \alpha_i + \mathbf{x}_{it}'\boldsymbol{\beta}, \boldsymbol{\gamma}), \quad t = 1, \ldots, T_i, \, i = 1, \ldots, N \qquad (18.1)$$

where $\boldsymbol{\gamma}$ denotes additional model parameters such as variance parameters, and α_i is an individual effect.

In other cases, a conditional mean model may be specified, with the additive effects

$$E(y_{it}|\alpha_i, \mathbf{x}_{it}) = \alpha_i + g(\mathbf{x}_{it}'\boldsymbol{\beta}) \qquad (18.2)$$

or with the multiplicative effects

$$E(y_{it}|\alpha_i, \mathbf{x}_{it}) = \alpha_i \times g(\mathbf{x}_{it}'\boldsymbol{\beta}) \qquad (18.3)$$

for the specified function $g(\cdot)$. In these models, $\mathbf{x}_{it}$ includes an intercept, so α_i is a deviation from the average centered on zero in (18.1) and (18.2) and centered on unity in (18.3).

FE models

An FE model treats α_i as an unobserved random variable that may be correlated with the regressors $\mathbf{x}_{it}$. In long panels, this poses no problems.

But in short panels, joint estimation of the FE $\alpha_1, \ldots, \alpha_N$ and the other model parameters, $\boldsymbol{\beta}$ and possibly $\boldsymbol{\gamma}$, usually leads to inconsistent estimation of all parameters. The reason is that the N incidental parameters α_i cannot be consistently estimated if T_i is small, because there are only T_i observations for each α_i. This inconsistent estimation of α_i can spill over to inconsistent estimation of $\boldsymbol{\beta}$.

For some models, it is possible to eliminate α_i by appropriate conditioning on a sufficient statistic for $y_{i1}, \ldots, y_{iT_i}$. This is the case for logit (but not probit) models for binary data and for Poisson and negative binomial models for count data. For other models, it is not possible, though recent work has proposed bias-corrected estimators in those cases.

Even when $\boldsymbol{\beta}$ is consistently estimated, it may not be possible to consistently estimate the marginal effects (MEs). It is possible for additive effects, because then $\partial E(y_{it}|\alpha_i, \mathbf{x}_{it})/\partial \mathbf{x}_{it} = \boldsymbol{\beta}$ from (18.2). But for multiplicative effects, (18.3) implies that $\partial E(y_{it}|\alpha_i, \mathbf{x}_{it})/\partial \mathbf{x}_{it} = \alpha_i \boldsymbol{\beta}$, which depends on α_i in addition to $\boldsymbol{\beta}$. For other nonlinear models, the dependence on α_i is even more complicated.

RE models

An RE model treats the individual-specific effect α_i as an unobserved random variable with the specified distribution $g(\alpha_i|\boldsymbol{\gamma})$, often the normal distribution. Then α_i is eliminated by integrating over this distribution. Specifically, the unconditional density for the ith observation is

$$f(y_{it}, \ldots, y_{iT_i}|\mathbf{x}_{i1}, \ldots, \mathbf{x}_{iT_i}, \boldsymbol{\beta}, \boldsymbol{\gamma}, \boldsymbol{\eta}) = \int \left\{ \prod_{t=1}^{T_i} f(y_{it}|\mathbf{x}_{it}, \alpha_i, \boldsymbol{\beta}, \boldsymbol{\gamma}) \right\} g(\alpha_i|\boldsymbol{\eta})d\alpha_i \quad (18.4)$$

In nonlinear models, this integral usually has no analytical solution, but numerical integration works well because only univariate integration is required.

This approach can be generalized to random slope parameters (random coefficients), not just a random intercept, with a greater computational burden because the integral is then of a higher dimension.

Pooled models or population-averaged models

Pooled models set $\alpha_i = \alpha$. For parametric models, it is assumed that the marginal density for a single (i, t) pair,

$$f(y_{it}|\mathbf{x}_{it}) = f(\alpha + \mathbf{x}_{it}'\boldsymbol{\beta}, \boldsymbol{\gamma})$$

is correctly specified, regardless of the (unspecified) form of the joint density $f(y_{it}, \ldots, y_{iT} | \mathbf{x}_{i1}, \ldots, \mathbf{x}_{iT}, \boldsymbol{\beta}, \boldsymbol{\gamma})$. The parameter of the pooled model is easily estimated, using the cross-section command for the appropriate parametric model, which implicitly assumes independence over both t and i. A panel–robust or cluster–robust (with clustering on i) estimate of the variance–covariance matrix of the estimator (VCE) can then be used to correct standard errors for any dependence over time for a given individual. This approach is the analog of pooled ordinary least squares (OLS) for linear models.

Potential efficiency gains can occur if estimation accounts for the dependence over time that is inherent in panel data. This is possible for generalized linear models, defined in section 10.3.7, where you can weight the first-order conditions for the estimator to account for correlation over time for a given individual but still have estimator consistency provided that the conditional mean is correctly specified as $E(y_{it} | \mathbf{x}_{it}) = g(\alpha + \mathbf{x}'_{it}\boldsymbol{\beta})$, for a specified function $g(\cdot)$. This is called the population-averaged (PA) approach, or generalized estimating equations approach, and is the analog of pooled feasible generalized least squares (FGLS) for linear models.

Unlike the linear model, in nonlinear models the PA approach generally leads to inconsistent estimates of the RE model and vice versa (the notable exception is given in section 18.6). This important distinction between RE and PA estimates in nonlinear models needs to be emphasized.

Comparison of models

If the FE model is appropriate, then an FE estimator must be used, if one is available.

The RE model has a different conditional mean than that for pooled and PA models, unless the random individual effects are additive or multiplicative. So, unlike the linear case, pooled estimation in nonlinear models leads to inconsistent parameter estimates if the assumed RE model is appropriate and vice versa.

18.2.2   Dynamic models

Dynamic models with individual effects can be estimated in some cases, most notably conditional mean models with additive or multiplicative effects as in (18.2) and (18.3). The methods are qualitatively similar to those in the linear case. Stata does not currently provide built-in commands to estimate dynamic nonlinear panel models.

18.2.3   Stata nonlinear panel commands

The Stata commands for PA, RE, and FE estimators of nonlinear panel models are the same as for the corresponding cross-section model, with the prefix `xt`. For example, `xtlogit` is the command for panel logit. The `re` option fits an RE model, the `fe` option fits an FE model if this is possible, and the `pa` option fits a PA model. The `xtgee` command with appropriate options is equivalent to the `xtlogit, pa` command, but `xtgee` is available for a wider range of models, including gamma and inverse Gaussian.

Models with random slopes, in addition to a random intercept, can be estimated for logit and Poisson models by using the `xtmelogit` and `xtmepoisson` commands. The user-written `gllamm` command can be applied to a wider range of mixed models than these two models. Table 18.1 lists the Stata commands for pooled, PA, RE, random slopes, and FE estimators of nonlinear panel models.

Table 18.1. Stata nonlinear panel commands

	Binary	Tobit	Counts
Pooled	logit probit	tobit	poisson nbreg
PA	xtlogit, pa xtprobit, pa		xtpoisson, pa xtnbreg, pa
RE	xtlogit, re xtprobit, re	xttobit	xtpoisson, re xtnbreg, re
Random slopes	xtmelogit		xtmepoisson
FE	xtlogit, fe		xtpoisson, fe xtnbreg, fe

The default for all these commands is to report standard errors that are not cluster–robust. Cluster–robust standard errors for pooled estimators can be obtained with the `vce(cluster id)` option, where `id` is the individual identifier. For PA, RE, and FE commands that in principle control for clustering, it can still be necessary to also compute cluster–robust errors. For PA estimators, this can be done by using the `vce(robust)` option. The other `xt` commands for nonlinear models do not have this option, but the `vce(bootstrap)` option is available. For the `xtpoisson, fe` command, the user-written `xtpqml` command calculates cluster–robust standard errors.

18.3   Nonlinear panel-data example

The example dataset we consider is an unbalanced panel from the Rand Health Insurance Experiment. This social experiment randomly assigned different health insurance policies to families that were followed for several years. The goal was to see how the use of health services varied with the coinsurance rate, where a coinsurance rate of 25%, for example, means that the insured pays 25% and the insurer pays 75%. Key results from the experiment were given in Manning et al. (1987). The data extract we use was prepared by Deb and Trivedi (2002).

18.3.1   Data description and summary statistics

Descriptive statistics for the dependent variables and regressors follow.

```
. * Describe dependent variables and regressors
. use mus18data.dta, clear
. describe dmdu med mdu lcoins ndisease female age lfam child id year
```

variable name	storage type	display format	value label	variable label
dmdu	float	%9.0g		any MD visit = 1 if mdu > 0
med	float	%9.0g		medical exp excl outpatient men
mdu	float	%9.0g		number face-to-fact md visits
lcoins	float	%9.0g		log(coinsurance+1)
ndisease	float	%9.0g		count of chronic diseases -- ba
female	float	%9.0g		female
age	float	%9.0g		age that year
lfam	float	%9.0g		log of family size
child	float	%9.0g		child
id	float	%9.0g		person id, leading digit is sit
year	float	%9.0g		study year

The corresponding summary statistics are

```
. * Summarize dependent variables and regressors
. summarize dmdu med mdu lcoins ndisease female age lfam child id year
```

Variable	Obs	Mean	Std. Dev.	Min	Max
dmdu	20186	.6875062	.4635214	0	1
med	20186	171.5892	698.2689	0	39182.02
mdu	20186	2.860696	4.504765	0	77
lcoins	20186	2.383588	2.041713	0	4.564348
ndisease	20186	11.2445	6.741647	0	58.6
female	20186	.5169424	.4997252	0	1
age	20186	25.71844	16.76759	0	64.27515
lfam	20186	1.248404	.5390681	0	2.639057
child	20186	.4014168	.4901972	0	1
id	20186	357971.2	180885.6	125024	632167
year	20186	2.420044	1.217237	1	5

We consider three different dependent variables. The dmdu variable is a binary indicator for whether the individual visited a doctor in the current year (69% did). The med variable measures annual medical expenditures (in dollars), with some observations being zero expenditures (other calculations show that 22% of the observations are zero). The mdu variable is the number of (face-to-face) doctor visits, with a mean of 2.9 visits. The three variables are best modeled by, respectively, logit or probit models, tobit models, and count models.

The regressors are lcoins, the natural logarithm of the coinsurance rate plus one; a health measure, ndisease; and four demographic variables. Children are included in the sample.

(Continued on next page)

18.3.2 Panel-data organization

We declare the individual and time identifiers and use the `xtdescribe` command to describe the panel-data organization.

```
. * Panel description of dataset
. xtset id year
       panel variable:  id (unbalanced)
        time variable:  year, 1 to 5, but with gaps
                delta:  1 unit

. xtdescribe
        id:  125024, 125025, ..., 632167              n =        5908
      year:  1, 2, ..., 5                              T =           5
             Delta(year) = 1 unit
             Span(year)  = 5 periods
             (id*year uniquely identifies each observation)

   Distribution of T_i:   min      5%     25%     50%     75%     95%     max
                            1       2       3       3       5       5       5

       Freq.  Percent    Cum. |  Pattern
    -----------------------------------------
        3710    62.80   62.80 |  111..
        1584    26.81   89.61 |  11111
         156     2.64   92.25 |  1....
         147     2.49   94.74 |  11...
          79     1.34   96.07 |  ..1..
          66     1.12   97.19 |  .11..
          33     0.56   97.75 |  ..111
          33     0.56   98.31 |  .1111
          29     0.49   98.80 |  ...11
          71     1.20  100.00 |  (other patterns)
    -----------------------------------------
        5908   100.00         |  XXXXX
```

The panel is unbalanced. Most individuals (90% of the sample of 5,908 individuals) were in the sample for the first three years or for the first five years, which was the sample design. There was relatively small panel attrition of about 5% over the first two years. There was also some entry, presumably because of family reconfiguration.

18.3.3 Within and between variation

Before analysis, it is useful to quantify the relative importance of within and between variation. For the dependent variables, we defer this until the relevant sections of this chapter.

The regressor variables lcoins, ndisease, and female are time-invariant, so their within variation is zero. We therefore apply the xtsum command to only the other three regressors. We have

```
. * Panel summary of time-varying regressors
. xtsum age lfam child
```

Variable		Mean	Std. Dev.	Min	Max	Observations	
age	overall	25.71844	16.76759	0	64.27515	N =	20186
	between	.	16.97265	0	63.27515	n =	5908
	within		1.086687	23.46844	27.96844	T-bar =	3.41672
lfam	overall	1.248404	.5390681	0	2.639057	N =	20186
	between		.5372082	0	2.639057	n =	5908
	within		.0730824	.3242075	2.44291	T-bar =	3.41672
child	overall	.4014168	.4901972	0	1	N =	20186
	between		.4820984	0	1	n =	5908
	within		.1096116	-.3985832	1.201417	T-bar =	3.41672

For the regressors age, lfam, and child, most of the variation is between variation rather than within variation. We therefore expect that FE estimators will not be very efficient because they rely on within variation. Also the FE parameter estimates may differ considerably from the other estimators if the within and between variation tell different stories.

18.3.4 FE or RE model for these data?

More generally, for these data we expect a priori that there is no need to use FE models. The point of the Rand experiment was to eliminate the endogeneity of health insurance choice, and hence endogeneity of the coinsurance rate, by randomly assigning this to individuals. The most relevant models for these data are RE or PA, which essentially just correct for the panel complication that observations are correlated over time for a given individual.

18.4 Binary outcome models

We fit logit models for whether an individual visited a doctor (dmdu). Similar methods apply for probit and complementary log-log models. The PA and RE estimators can be obtained with the xtprobit and xtcloglog commands, but there is no FE estimator and no mixed models command analogous to xtmelogit.

18.4.1 Panel summary of the dependent variable

The dependent variable dmdu has within variation and between variation of similar magnitude.

```
. * Panel summary of dependent variable
. xtsum dmdu
    Variable   |      Mean   Std. Dev.        Min        Max  |  Observations
    dmdu  overall |  .6875062   .4635214          0          1  |  N =     20186
          between |             .3571059          0          1  |  n =      5908
           within |             .3073307  -.1124938   1.487506  |  T-bar = 3.41672
. * Year-to-year transitions in whether visit doctor
. xttrans dmdu
    any MD |  any MD visit = 1 if
   visit = 1 |       mdu > 0
 if mdu > 0 |        0          1  |   Total
          0 |    58.87      41.13  |  100.00
          1 |    19.73      80.27  |  100.00
      Total |    31.81      68.19  |  100.00
```

There is considerable persistence from year to year: 59% of those who did not visit a doctor one year also did not visit the next, while 80% of those who did visit a doctor one year also visited the next.

```
. * Correlations in the dependent variable
. corr dmdu l.dmdu l2.dmdu
(obs=8626)
                         |              L.        L2.
                 |     dmdu       dmdu       dmdu
            dmdu |
            --.  |   1.0000
            L1.  |   0.3861   1.0000
            L2.  |   0.3601   0.3807   1.0000
```

The correlations in the dependent variable, dmdu, vary little with lag length, unlike the chapter 8 example of log wage where correlations decrease as lag length rises.

18.4.2 Pooled logit estimator

The pooled logit model is the usual cross-section model,

$$\Pr(y_{it} = 1 | \mathbf{x}_{it}) = \Lambda(\mathbf{x}'_{it}\boldsymbol{\beta}) \tag{18.5}$$

where $\Lambda(z) = e^z/(1+e^z)$. A cluster–robust estimate for the VCE is then used to correct for error correlation over time for a given individual.

The `logit` command with the `vce(cluster id)` option yields

```
. * Logit cross-section with panel-robust standard errors
. logit dmdu lcoins ndisease female age lfam child, vce(cluster id) nolog
Logistic regression                        Number of obs    =     20186
                                           Wald chi2(6)     =    488.18
                                           Prob > chi2      =    0.0000
Log pseudolikelihood = -11973.392          Pseudo R2        =    0.0450
                              (Std. Err. adjusted for 5908 clusters in id)
```

dmdu	Coef.	Robust Std. Err.	z	P>|z|	[95% Conf. Interval]	
lcoins	-.1572107	.0109064	-14.41	0.000	-.1785869	-.1358345
ndisease	.050301	.0039657	12.68	0.000	.0425285	.0580735
female	.3091573	.0445772	6.94	0.000	.2217876	.396527
age	.0042689	.0022307	1.91	0.056	-.0001032	.008641
lfam	-.2047573	.0470287	-4.35	0.000	-.2969317	-.1125828
child	.0921709	.0728107	1.27	0.206	-.0505355	.2348773
_cons	.6039411	.1107712	5.45	0.000	.3868335	.8210486

The first four regressors have the expected signs. The negative sign of `lfam` may be due to family economies of scale in health care. The positive coefficient of `child` may reflect a u-shaped pattern of doctor visits with `age`. The estimates imply that a child of age 10, say, is as likely to see the doctor as a young adult of age 31 because $0.092 + 0.0043 \times 10 \simeq 0.0043 \times 31 = 0.1333$.

The estimated coefficients can be converted to MEs by using the `margins, dydx(*)` command, or, approximately, by multiplying by $\overline{y}(1 - \overline{y}) = 0.69 \times 0.31 = 0.21$. For example, the probability of a doctor visit at some stage during the year is 0.07 higher for a woman than for a man, because $0.31 \times 0.21 = 0.07$.

In output not given, the default standard errors are approximately two-thirds those given here, so the use of cluster–robust standard errors is necessary.

18.4.3 The xtlogit command

The pooled logit command assumes independence over i and t, leading to potential efficiency loss, and ignores the possibility of FE that would lead to inconsistent parameter estimates.

These panel complications are accommodated by the `xtlogit` command, which has the syntax

xtlogit *depvar* [*indepvars*] [*if*] [*in*] [*weight*] [, *options*]

The options are for PA (`pa`), RE (`re`), and FE (`fe`) models. Panel–robust standard errors can be calculated by using the `vce(robust)` option with the `pa` option. This is not possible for the other estimators, but the `vce(bootstrap)` option can be used. Model-specific options are discussed below in the relevant model section.

18.4.4   The xtgee command

The `pa` option for the `xtlogit` command is also available for some other nonlinear panel commands, such as `xtpoisson`. It is a special case of the `xtgee` command. This command has the syntax

$$\texttt{xtgee } depvar \; [\,indepvars\,] \; [\,if\,] \; [\,in\,] \; [\,weight\,] \; [\,,\; options\,]$$

The `family()` and `link()` options define the specific model. For example, the linear model is `family(gaussian) link(identity)`, and the logit model is `family(binomial) link(logit)`. Other `family()` options are `poisson`, `nbinomial`, `gamma`, and `igaussian` (inverse Gaussian).

The `corr()` option defines the pattern of time-series correlation assumed for observations on the ith individual. These patterns include `exchangeable` for equicorrelation, `independent` for no correlation, and various time-series models that have been detailed in section 8.4.3.

In the examples below, we obtain the PA estimator by using commands such as `xtlogit` with the `pa` option. If instead the corresponding `xtgee` command is used, then the postestimation `estat wcorrelation` command produces the estimated matrix of the within-group correlations.

18.4.5   PA logit estimator

The PA estimator of the parameters of (18.5) can be obtained by using the `xtlogit` command with the `pa` option. Different arguments for the `corr()` option, presented in section 8.4.3 and in [XT] **xtgee**, correspond to different models for the correlation

$$\rho_{ts} = \text{Cor}[\{y_{it} - \Lambda(\mathbf{x}'_{it}\boldsymbol{\beta})\}\{y_{is} - \Lambda(\mathbf{x}'_{is}\boldsymbol{\beta})\}], \; s \neq t$$

The exchangeable model assumes that correlations are the same regardless of how many years apart the observations are, so $\rho_{ts} = \alpha$. For our data, this model may be adequate because, from section 18.4.1, the correlations of `dmdu` varied little with the lag length. Even with equicorrelation, the covariances can vary across individuals and across year pairs because, given $\text{Var}(y_{it}|\mathbf{x}_{it}) = \Lambda_{it}(1 - \Lambda_{it})$, the implied covariance is $\alpha\sqrt{\Lambda_{it}(1 - \Lambda_{it}) \times \Lambda_{is}(1 - \Lambda_{is})}$.

Estimation with the `xtlogit, pa` command yields

```
. * Pooled logit cross-section with exchangeable errors and panel-robust VCE
. xtlogit dmdu lcoins ndisease female age lfam child, pa corr(exch) vce(robust)
> nolog
GEE population-averaged model          Number of obs      =      20186
Group variable:                   id   Number of groups   =       5908
Link:                          logit   Obs per group: min =          1
Family:                     binomial                  avg =        3.4
Correlation:             exchangeable                  max =          5
                                       Wald chi2(6)       =     521.45
Scale parameter:                   1   Prob > chi2        =     0.0000

                                     (Std. Err. adjusted for clustering on id)
```

dmdu	Coef.	Semirobust Std. Err.	z	P>|z|	[95% Conf. Interval]	
lcoins	-.1603179	.0107779	-14.87	0.000	-.1814422	-.1391935
ndisease	.0515445	.0038528	13.38	0.000	.0439931	.0590958
female	.2977003	.0438316	6.79	0.000	.211792	.3836086
age	.0045675	.0021001	2.17	0.030	.0004514	.0086836
lfam	-.2044045	.0455004	-4.49	0.000	-.2935837	-.1152254
child	.1184697	.0674367	1.76	0.079	-.0137039	.2506432
_cons	.5776986	.106591	5.42	0.000	.368784	.7866132

The pooled logit and PA logit parameter estimates are very similar. The cluster–robust standard errors are slightly lower for the PA estimates, indicating a slight efficiency gain. Typing `matrix list e(R)` shows that $\widehat{\rho}_{ts} = \widehat{\alpha} = 0.34$. The parameter estimates can be interpreted in exactly the same way as those from a cross-section logit model.

18.4.6   RE logit estimator

The logit individual-effects model specifies that

$$\Pr(y_{it} = 1|\mathbf{x}_{it}, \boldsymbol{\beta}, \alpha_i) = \Lambda(\alpha_i + \mathbf{x}'_{it}\boldsymbol{\beta}) \tag{18.6}$$

where α_i may be an FE or an RE.

The logit RE model specifies that $\alpha_i \sim N(0, \sigma_\alpha^2)$. Then the joint density for the ith observation, after integrating out α_i, is

$$f(y_{it}, \ldots, y_{iT}) = \int \left[\prod_{t=1}^{T} \Lambda(\alpha_i + \mathbf{x}'_{it}\boldsymbol{\beta})^{y_{it}} \{1 - \Lambda(\alpha_i + \mathbf{x}'_{it}\boldsymbol{\beta})\}^{1-y_{it}} \right] g(\alpha_i|\sigma^2) d\alpha_i \tag{18.7}$$

where $g(\alpha_i|\sigma^2)$ is the $N(0, \sigma_\alpha^2)$ density. After α_i is integrated out, $\Pr(y_{it} = 1|\mathbf{x}_{it}, \boldsymbol{\beta}) \neq \Lambda(\mathbf{x}'_{it}\boldsymbol{\beta})$, so the RE model parameters are not comparable to those from pooled logit and PA logit.

There is no analytical solution to the univariate integral (18.7), so numerical methods are used. The default method is adaptive 12-point Gauss–Hermite quadrature. The `intmethod()` option allows other quadrature methods to be used, and the `intpoints()`

option allows the use of a different number of quadrature points. The `quadchk` command checks whether a good approximation has been found by using a different number of quadrature points and comparing solutions; see [XT] **xtlogit** and [XT] **quadchk** for details.

The RE estimator is implemented by using the `xtlogit` command with the `re` option. We have

```
. * Logit random-effects estimator
. xtlogit dmdu lcoins ndisease female age lfam child, re nolog
Random-effects logistic regression              Number of obs      =       20186
Group variable: id                              Number of groups   =        5908

Random effects u_i ~ Gaussian                   Obs per group: min =           1
                                                               avg =         3.4
                                                               max =           5

                                                Wald chi2(6)       =      549.76
Log likelihood  = -10878.687                    Prob > chi2        =      0.0000
```

dmdu	Coef.	Std. Err.	z	P>|z|	[95% Conf. Interval]	
lcoins	-.2403864	.0162836	-14.76	0.000	-.2723017	-.208471
ndisease	.078151	.0055456	14.09	0.000	.0672819	.0890201
female	.4631005	.0663209	6.98	0.000	.3331138	.5930871
age	.0073441	.0031508	2.33	0.020	.0011687	.0135194
lfam	-.3021841	.0644721	-4.69	0.000	-.4285471	-.175821
child	.1935357	.1002267	1.93	0.053	-.002905	.3899763
_cons	.8629898	.1568968	5.50	0.000	.5554778	1.170502
/lnsig2u	1.225652	.0490898			1.129438	1.321866
sigma_u	1.84564	.045301			1.758953	1.936599
rho	.5087003	.0122687			.4846525	.532708

```
Likelihood-ratio test of rho=0: chibar2(01) =  2189.41 Prob >= chibar2 = 0.000
```

The coefficient estimates are roughly 50% larger in absolute value than those of the PA model. The standard errors are also roughly 50% larger, so the t statistics are little changed. Clearly, the RE model has a different conditional mean than the PA model, and the parameters are not directly comparable.

The standard deviation of the RE, σ_α, is given in the output as **sigma_u**, so it is estimated that $\alpha_i \sim N(0, 1.846^2)$. The logit RE model can be motivated as coming from a latent-variable model, with $y_{it} = 1$ if $y_{it}^* = \mathbf{x}_{it}'\boldsymbol{\beta} + \alpha_i + \varepsilon_{it} > 0$, where ε_{it} is logistically distributed with a variance of $\sigma_\varepsilon^2 = \pi^2/3$. By a calculation similar to that in section 8.3.10, the intraclass error correlation in the latent-variable model is $\rho = \sigma_\alpha^2/(\sigma_\alpha^2 + \sigma_\varepsilon^2)$. Here $\widehat{\rho} = 1.846^2/(1.846^2 + \pi^2/3) = 0.509$, the quantity reported as rho.

Consistent estimation of $\boldsymbol{\beta}$ does not allow predicting for the individual because, from (18.7), the probability depends on α_i, which is not estimated. Similarly, the associated ME for the RE model

$$\partial \Pr(y_{it} = 1|\mathbf{x}_{it}, \boldsymbol{\beta}, \alpha_i)/\partial x_{ji,t} = \beta_j \Lambda(\alpha_i + \mathbf{x}_{it}'\boldsymbol{\beta})\{1 - \Lambda(\alpha_i + \mathbf{x}_{it}'\boldsymbol{\beta})\}$$

also depends on the unknown α_i. The `margins, dydx() predict(pu0)` command computes this ME at $\alpha_i = 0$, but this can be a nonrepresentative evaluation point and, in this example, understates the MEs. We can still make some statements, using the analysis in section 10.6.4 for single-index models. If one coefficient is twice as large as another, then so too is the ME. The sign of the ME equals that of β_j, because $\Lambda()\{1 - \Lambda()\} > 0$. And the log of the odds-ratio interpretation for logit models, given in chapter 14, is still applicable because $\ln\{p_i/(1 - p_i)\} = \alpha_i + \mathbf{x}'_{it}\boldsymbol{\beta}$ so that $\partial \ln\{p_i/(1 - p_i)\}/\partial x_{ji,t} = \beta_j$. For example, the coefficient of `age` implies that aging one year increases the log of the odds ratio of visiting a doctor by 0.0073 or, equivalently, by 0.73%.

18.4.7   FE logit estimator

In the FE model, the α_i may be correlated with the covariates in the model. Parameter estimation is difficult, and many of the approaches in the linear case fail. In particular, the least-squares dummy-variable estimator of section 8.5.4 yielded a consistent estimate of $\boldsymbol{\beta}$, but a similar dummy-variables estimator for the logit model leads to inconsistent estimation of $\boldsymbol{\beta}$ in the logit model, unless $T \to \infty$.

One method of consistent estimation eliminates the α_i from the estimation equation. This method is the conditional maximum likelihood estimator (MLE), which is based on a log density for the ith individual that conditions on $\sum_{t=1}^{T_i} y_{it}$, the total number of outcomes equal to 1 for a given individual over time.

We demonstrate this in the simplest case of two time periods. Condition on $y_{i1} + y_{i2} = 1$, so that $y_{it} = 1$ in exactly one of the two periods. Then, in general,

$$\Pr(y_{i1} = 0, y_{i2} = 1 | y_{i1} + y_{i2} = 1) = \frac{\Pr(y_{i1} = 0, y_{i2} = 1)}{\Pr(y_{i1} = 0, y_{i2} = 1) + \Pr(y_{i1} = 1, y_{i2} = 0)} \quad (18.8)$$

Now $\Pr(y_{i1} = 0, y_{i2} = 1) = \Pr(y_{i1} = 0) \times \Pr(y_{i2} = 1)$, assuming that y_{1i} and y_{2i} are independent given α_i and $\mathbf{x}_{it}$. For the logit model (18.6), we obtain

$$\Pr(y_{i1} = 0, y_{i2} = 1) = \frac{1}{1 + \exp(\alpha_i + \mathbf{x}'_{i1}\boldsymbol{\beta})} \times \frac{\exp(\alpha_i + \mathbf{x}'_{i2}\boldsymbol{\beta})}{1 + \exp(\alpha_i + \mathbf{x}'_{i2}\boldsymbol{\beta})}$$

Similarly,

$$\Pr(y_{i1} = 1, y_{i2} = 0) = \frac{\exp(\alpha_i + \mathbf{x}'_{i1}\boldsymbol{\beta})}{1 + \exp(\alpha_i + \mathbf{x}'_{i1}\boldsymbol{\beta})} \times \frac{1}{1 + \exp(\alpha_i + \mathbf{x}'_{i2}\boldsymbol{\beta})}$$

Substituting these two expressions into (18.8), the denominators cancel and we obtain

$$\begin{aligned}
\Pr(y_{i1} &= 0, y_{i2} = 1 | y_{i1} + y_{i2} = 1) \\
&= \exp(\alpha_i + \mathbf{x}'_{i2}\boldsymbol{\beta})/\{\exp(\alpha_i + \mathbf{x}'_{i1}\boldsymbol{\beta}) + \exp(\alpha_i + \mathbf{x}'_{i2}\boldsymbol{\beta})\} \\
&= \exp(\mathbf{x}'_{i2}\boldsymbol{\beta})/\{\exp(\mathbf{x}'_{i1}\boldsymbol{\beta}) + \exp(\mathbf{x}'_{i2}\boldsymbol{\beta})\} \\
&= \exp\{(\mathbf{x}_{i2} - \mathbf{x}_{i1})'\boldsymbol{\beta}\}/[1 + \exp\{(\mathbf{x}_{i2} - \mathbf{x}_{i1})'\boldsymbol{\beta}\}]
\end{aligned} \quad (18.9)$$

There are several results. First, conditioning eliminates the problematic FE α_i. Second, the resulting conditional model is a logit model with the regressor $\mathbf{x}_{i2} - \mathbf{x}_{i1}$. Third, coefficients of time-invariant regressors are not identified, because $x_{i2} - x_{i1} = 0$.

More generally, with up to T outcomes, we can eliminate α_i by conditioning on $\sum_{t=1}^{T} y_{it} = 1$ and on $\sum_{t=1}^{T} y_{it} = 2, \ldots, \sum_{t=1}^{T} y_{it} = T - 1$. This leads to the loss of those observations where y_{it} is 0 for all t or y_{it} is 1 for all T. The resulting conditional model is more generally a multinomial logit model. For details, see, for example, Cameron and Trivedi (2005) or [R] **clogit**.

The FE estimator is obtained by using the `xtlogit` command with the `fe` option. We have

```
. * Logit fixed-effects estimator
. xtlogit dmdu lcoins ndisease female age lfam child, fe nolog
note: multiple positive outcomes within groups encountered.
note: 3459 groups (11161 obs) dropped because of all positive or
      all negative outcomes.
note: lcoins omitted because of no within-group variance.
note: ndisease omitted because of no within-group variance.
note: female omitted because of no within-group variance.
```

```
Conditional fixed-effects logistic regression    Number of obs      =      9025
Group variable: id                               Number of groups   =      2449

                                                 Obs per group: min =         2
                                                                avg =       3.7
                                                                max =         5

                                                 LR chi2(3)         =     10.74
Log likelihood  = -3395.5996                     Prob > chi2        =    0.0132
```

dmdu	Coef.	Std. Err.	z	P>|z|	[95% Conf. Interval]	
lcoins	(omitted)					
ndisease	(omitted)					
female	(omitted)					
age	-.0341815	.0183827	-1.86	0.063	-.070211	.001848
lfam	.478755	.2597327	1.84	0.065	-.0303116	.9878217
child	.270458	.1684974	1.61	0.108	-.0597907	.6007068

As expected, coefficients of the time-invariant regressors are not identified and these variables are dropped. The 3,459 individuals with $\sum_{i=1}^{T_i} y_{it} = 0$ (all zeros) or $\sum_{i=1}^{T_i} y_{it} = T_i$ (all ones) are dropped because there is then no variation in y_{it} over t, leading to a loss of 11,161 of the original 20,186 observations. Standard errors are substantially larger for FE estimation because of this loss of observations and because only within variation of the regressors is used.

The coefficients are considerably different from those from the RE logit model, and in two cases, the sign changes. The interpretation of parameters is similar to that given at the end of section 18.4.6 for the RE model. Also one can use an interpretation that conditions on $\sum_{i=1}^{T_i} y_{it}$; see section 18.4.9.

18.4.8   Panel logit estimator comparison

We combine the preceding estimators into a single table that makes comparison easier.
We have

```
. * Panel logit estimator comparison
. global xlist lcoins ndisease female age lfam child
. quietly logit dmdu $xlist, vce(cluster id)
. estimates store POOLED
. quietly xtlogit dmdu $xlist, pa corr(exch) vce(robust)
. estimates store PA
. quietly xtlogit dmdu $xlist, re      // SEs are not cluster-robust
. estimates store RE
. quietly xtlogit dmdu $xlist, fe      // SEs are not cluster-robust
. estimates store FE
. estimates table POOLED PA RE FE, equations(1) se b(%8.4f) stats(N ll)
> stfmt(%8.0f)
```

Variable	POOLED	PA	RE	FE
#1				
lcoins	-0.1572	-0.1603	-0.2404	(omitted)
	0.0109	0.0108	0.0163	
ndisease	0.0503	0.0515	0.0782	(omitted)
	0.0040	0.0039	0.0055	
female	0.3092	0.2977	0.4631	(omitted)
	0.0446	0.0438	0.0663	
age	0.0043	0.0046	0.0073	-0.0342
	0.0022	0.0021	0.0032	0.0184
lfam	-0.2048	-0.2044	-0.3022	0.4788
	0.0470	0.0455	0.0645	0.2597
child	0.0922	0.1185	0.1935	0.2705
	0.0728	0.0674	0.1002	0.1685
_cons	0.6039	0.5777	0.8630	
	0.1108	0.1066	0.1569	
lnsig2u				
_cons			1.2257	
			0.0491	
Statistics				
N	20186	20186	20186	9025
ll	-11973		-10879	-3396

legend: b/se

The pooled logit and PA logit models lead to very similar parameter estimates and
cluster–robust standard errors. The RE logit estimates differ quite substantially from
the PA logit estimates though, as already noted, the associated t statistics are quite
similar. The FE estimates are much less precise, differ considerably from the other
estimates, and are available only for time-varying regressors.

18.4.9 Prediction and marginal effects

The postestimation `predict` command has several options that vary depending on whether the `xtlogit` command was used with the `pa`, `re`, or `fe` option.

After the `xtlogit, pa` command, the default `predict` option is `mu`, which gives the predicted probability given in (18.5).

After the `xtlogit, re` command, the default `predict` option is `xb`, which computes $\mathbf{x}'_{it}\widehat{\boldsymbol{\beta}}$. To predict the probability, one can use `pu0`, which predicts the probability when $\alpha_i = 0$. This is of limited usefulness because (18.6) conditions on α_i, which is not observed or estimated. Interest lies in the unconditional probability

$$\Pr(y_{it} = 1 | \mathbf{x}_{it}, \boldsymbol{\beta}) = \int \Lambda(\alpha_i + \mathbf{x}'_{it}\boldsymbol{\beta}) g(\alpha_i | \sigma^2) d\alpha_i \qquad (18.10)$$

where $g(\alpha_i | \sigma^2)$ is the $N(0, \sigma_\alpha^2 r)$ density, and this does not equal $\Lambda(\mathbf{x}'_{it}\boldsymbol{\beta})$, which is what `pu0` computes. One could, of course, calculate (18.10) by using the simulation methods presented in section 4.5. Or one can estimate the parameters of the RE model by using the `xtmelogit` command, presented in the next section, followed by the postestimation `predict` command with the `reffects` option to calculate posterior modal estimates of the RE; see [XT] **xtmelogit postestimation**.

After the `xtlogit, fe` command, the `predict` options `xb` and `pu0` are available. The default option is `pc1`, which produces the conditional probability that $y_{it} = 1$ given that exactly one of $y_{i1}, \ldots, y_{iT_i}$ equals one. This is used because this conditional probability does not depend on α_i; the formula in the special case $T_i = 2$ is given in (18.9).

The postestimation `margins, dydx()` command with the `predict()` option computes the corresponding ME, evaluated at the mean value of regressors. After `xtlogit, pa`, the MEs are interpreted the same as for cross-section logit models. After `xtlogit, re` and `xtlogit, fe`, interpretation is more difficult because of the presence of α_i. Some discussion has already been given at the end of the relevant sections. For nonlinear panel models, the ease of computing MEs by using PA rather than RE models is emphasized by Drukker (2008).

18.4.10 Mixed-effects logit estimator

The RE logit model specifies only the intercept to be normally distributed. Slope parameters may also be normally distributed. The `xtmelogit` command estimates the parameters of this model, which is the logit extension of `xtmixed` for linear models, presented in section 9.5.2.

For example, the following yields the same estimates as `xtlogit` with the `re` option, aside from minor computational difference:

```
. xtmelogit dmdu lcoins ndisease female age lfam child || id:
  (output omitted )
```

Adding the `lcoins` and `ndisease` variables allows the intercept and slope parameters for `lcoins` and `ndisease` to be jointly normally distributed. Then a trivariate integral is computed with the Gauss–Hermite quadrature, and estimation is very computationally intensive; without restrictions on the variance matrix, the model may not even be estimable.

As in the linear case, the mixed logit model is used more for clustered data than for panel data and is mostly used in areas of applied statistics other than econometrics.

18.5 Tobit model

We fit a panel tobit model for medical expenditures (`med`). Then the only panel estimator available is the `re` option, which introduces a normally distributed RE.

18.5.1 Panel summary of the dependent variable

For simplicity, we model expenditures in levels, though from chapter 16 the key assumption of normality for the tobit model is more reasonable for the natural logarithm of expenditures.

The dependent variable, `med`, has within variation and between variation of similar magnitude, because the `xtsum` command yields

```
. * Panel summary of dependent variable
. xtsum med
```

Variable		Mean	Std. Dev.	Min	Max	Observations
med	overall	171.5892	698.2689	0	39182.02	N = 20186
	between		503.2589	0	19615.14	n = 5908
	within		526.269	-19395.28	20347.2	T-bar = 3.41672

18.5.2 RE tobit model

The RE panel tobit model specifies the latent variable y_{it}^* to depend on regressors, an idiosyncratic error, and an individual-specific error, so

$$y_{it}^* = \mathbf{x}_{it}'\boldsymbol{\beta} + \alpha_i + \varepsilon_{it} \tag{18.11}$$

where $\alpha_i \sim N(0, \sigma_\alpha^2)$ and $\varepsilon_{it} \sim N(0, \sigma_\varepsilon^2)$ and the regressor vector $\mathbf{x}_{it}$ includes an intercept. For left censoring at L, we observe the y_{it} variable, where

$$y_{it} = \begin{cases} y_{it}^* & \text{if } y_{it}^* > L \\ L & \text{if } y_{it}^* \leq L \end{cases} \tag{18.12}$$

The `xttobit` command has a similar syntax to the cross-section `tobit` command. The `ll()` option is used to define the lower limit for left censoring, and the `ul()` option is used to define the upper limit for right censoring. The limit can be a variable, not just a number, so more generally we can have the limit L_i rather than the limit L in

(18.12). Like the RE logit model, estimation requires univariate numerical integration, using Gauss–Hermite quadrature.

For our data, we obtain

```
. * Tobit random-effects estimator
. xttobit med lcoins ndisease female age lfam child, ll(0) nolog
Random-effects tobit regression                 Number of obs     =      20186
Group variable: id                              Number of groups  =       5908

Random effects u_i ~ Gaussian                   Obs per group: min =          1
                                                               avg =        3.4
                                                               max =          5

                                                Wald chi2(6)      =     573.45
Log likelihood  = -130030.45                    Prob > chi2       =     0.0000
```

med	Coef.	Std. Err.	z	P>|z|	[95% Conf. Interval]	
lcoins	-31.10247	3.578498	-8.69	0.000	-38.1162	-24.08875
ndisease	13.49452	1.139156	11.85	0.000	11.26182	15.72722
female	60.10112	14.95966	4.02	0.000	30.78072	89.42152
age	4.075582	.7238253	5.63	0.000	2.656911	5.494254
lfam	-57.75023	14.68422	-3.93	0.000	-86.53077	-28.96968
child	-52.02314	24.21619	-2.15	0.032	-99.48599	-4.560284
_cons	-98.27203	36.05977	-2.73	0.006	-168.9479	-27.59618
/sigma_u	371.3134	8.64634	42.94	0.000	354.3668	388.2599
/sigma_e	715.1779	4.704581	152.02	0.000	705.9571	724.3987
rho	.2123246	.0086583			.1957541	.2296872

```
Observation summary:      4453  left-censored observations
                         15733       uncensored observations
                            0 right-censored observations
```

About 22% of the observations are censored (4,453 of 20,186). All regressor coefficients are statistically significant and have the expected sign. The RE α_i has an estimated standard deviation of 371.3, which is highly statistically significant. The quantity labeled **rho** equals $\sigma_\alpha^2/(\sigma_\alpha^2 + \sigma_\varepsilon^2)$ and measures the fraction of the total variance, $\sigma_\alpha^2 + \sigma_\varepsilon^2$, that is due to the RE α. In an exercise, we compare these estimates with those from the **tobit** command, which treats observations as independent over i and t (so $\alpha_i = 0$). The estimates are similar.

18.5.3 Generalized tobit models

The **xtintreg** command estimates the parameters of interval-data models where continuous data are reported only in ranges. For example, annual medical expenditure data may be reported only as \$0, between \$0 and \$100, between \$100 and \$1,000, and more than \$1,000. The unobserved continuous variable, y_{it}^*, is modeled as (18.11), and the observed variable, y_{it}, arises as y_{it}^* falls into the appropriate range.

Stochastic production frontier models introduce into the production function a strictly negative error term that pushes production below the efficient level. In the simplest panel model, this error term is time invariant and has a truncated normal distribution, so the model has some commonalities with the panel tobit model. The `xtfrontier` command is used to estimate the parameters of these models.

All three commands—`xttobit`, `xtintreg`, and `xtfrontier`—rely heavily on the assumption of homoskedastic normally distributed errors for consistency and, like their cross-section counterparts, are more fragile to distributional misspecification than, for example, linear models and logit models. In particular, in many applications where a tobit model is used, a more general sample-selection model may be warranted. Stata does not provide panel commands in this case, though methods have been proposed.

18.5.4   Parametric nonlinear panel models

More generally, it can be difficult to generalize highly parameterized cross-section nonlinear models to the panel case, even without introducing FE.

The PA, or pooled, approach uses the cross-section estimator, but then bases inference on a panel–robust estimate of the VCE that can be obtained with a panel bootstrap. This approach requires the assumption that the specified marginal distribution for y_{it} is correct even if y_{it} is correlated over t. The RE approach introduces an RE α_i, similar to the `re` option. The user-written `gllamm` command (Rabe-Hesketh, Skrondal, and Pickles 2002) does so for a wide range of generalized linear models.

These two approaches are quite distinct and lead to differently scaled parameters.

18.6   Count-data models

We fit count models for the number of doctor visits (`mdu`). Many of the relevant issues have already been raised for the `xtlogit` command. One difference is that analytical solutions are possible for count RE models, by appropriate choice of (nonnormal) distribution for the RE. A second difference is that Poisson panel estimators have the same robustness properties as Poisson cross-section estimators. They are consistent even if the data are not Poisson distributed, provided the conditional mean is correctly specified. At the same time, count data are often overdispersed, and the need to use heteroskedasticity-robust standard errors in the cross-section case carries over to a need to use panel–robust standard errors in the panel case.

18.6.1   The xtpoisson command

The `xtpoisson` command has the syntax

`xtpoisson` *depvar* [*indepvars*] [*if*] [*in*] [*weight*] [, *options*]

The options include PA (`pa`), RE (`re`), and FE (`fe`) models.

Then PA, RE, and FE estimators are available for both the Poisson model, using `xtpoisson`, and for the negative binomial model, using `xtnbreg`.

18.6.2   Panel summary of the dependent variable

The dependent variable, mdu, is considerably overdispersed because, from section 18.3.1, the sample variance of $4.50^2 = 20.25$ is seven times the sample mean of 2.86. This makes it very likely that default standard errors for both cross-section and panel Poisson estimators will considerably understate the true standard errors.

The mdu variable has a within variation of similar magnitude to the between variation.

```
. * Panel summary of dependent variable
. xtsum mdu
    Variable   |      Mean   Std. Dev.       Min        Max |    Observations

mdu  overall   |  2.860696    4.504765         0         77 | N   =    20186
     between   |             3.785971         0   63.33333 | n   =     5908
     within    |             2.575881  -34.47264    40.0607 | T-bar = 3.41672
```

To provide more detail on the variation in mdu over time, it is useful to look at transition probabilities, after first aggregating all instances of four or more doctor visits into a single category. We have

```
. * Year-to-year transitions in doctor visits
. generate mdushort = mdu

. replace mdushort = 4 if mdu >= 4
(4039 real changes made)

. xttrans mdushort
                                  mdushort
mdushort |        0         1         2         3         4 |    Total

      0  |    58.87     19.61      9.21      4.88      7.42 |   100.00
      1  |    33.16     24.95     17.58     10.14     14.16 |   100.00
      2  |    23.55     24.26     17.90     12.10     22.19 |   100.00
      3  |    17.80     20.74     18.55     12.14     30.77 |   100.00
      4  |     8.79     11.72     12.32     11.93     55.23 |   100.00

   Total |    31.81     19.27     13.73      9.46     25.73 |   100.00
```

There is considerable persistence: over half of people with zero doctor visits one year also have zero visits the next year, and over half of people with four or more visits one year also have four or more visits the next year.

18.6.3   Pooled Poisson estimator

The pooled Poisson estimator assumes that y_{it} is Poisson distributed with a mean of

$$E(y_{it}|\mathbf{x}_{it}) = \exp(\mathbf{x}_{it}'\boldsymbol{\beta}) \tag{18.13}$$

as in the cross-section case. Consistency of this estimator requires that (18.13) is correctly specified but does not require that the data actually be Poisson distributed. If the data are not Poisson distributed, however, then it is essential that robust standard errors be used.

The pooled Poisson estimator can be estimated by using the `poisson` command, with cluster–robust standard errors that take care of both overdispersion and serial correlation. We have

```
. * Pooled Poisson estimator with cluster-robust standard errors
. poisson mdu lcoins ndisease female age lfam child, vce(cluster id)

Iteration 0:   log pseudolikelihood = -62580.248
Iteration 1:   log pseudolikelihood = -62579.401
Iteration 2:   log pseudolikelihood = -62579.401

Poisson regression                              Number of obs   =      20186
                                                Wald chi2(6)    =     476.93
                                                Prob > chi2     =     0.0000
Log pseudolikelihood = -62579.401               Pseudo R2       =     0.0609

                          (Std. Err. adjusted for 5908 clusters in id)
```

		Robust				
mdu	Coef.	Std. Err.	z	P>|z|	[95% Conf.	Interval]
lcoins	-.0808023	.0080013	-10.10	0.000	-.0964846	-.0651199
ndisease	.0339334	.0026024	13.04	0.000	.0288328	.039034
female	.1717862	.0342551	5.01	0.000	.1046473	.2389251
age	.0040585	.0016891	2.40	0.016	.000748	.0073691
lfam	-.1481981	.0323434	-4.58	0.000	-.21159	-.0848062
child	.1030453	.0506901	2.03	0.042	.0036944	.2023961
_cons	.748789	.0785738	9.53	0.000	.5947872	.9027907

The importance of using cluster–robust standard errors cannot be overemphasized. For these data, the correct cluster–robust standard errors are 50% higher than the heteroskedasticity-robust standard errors and 300% higher than the default standard errors; see the end-of-chapter exercises. Here failure to control for autocorrelation and failure to control for overdispersion both lead to considerable understatement of the true standard errors.

18.6.4   PA Poisson estimator

The PA Poisson estimator is a variation of the pooled Poisson estimator that relaxes the assumption of independence of y_{it} to allow different models for the correlation $\rho_{ts} = \text{Cor}[\{y_{it} - \exp(\mathbf{x}'_{it}\boldsymbol{\beta})\}\{y_{is} - \exp(\mathbf{x}'_{is}\boldsymbol{\beta})\}]$.

The PA estimator is obtained by using the `xtpoisson` command with the `pa` option. Different correlation models are specified by using the `corr()` option; see section 8.4.3. Consistency of this estimator requires only that (18.13) be correct. But if the data are non-Poisson and are overdispersed, then the `vce(robust)` option should be used because otherwise default standard errors will understate the true standard errors.

We use the `corr(unstructured)` option so that ρ_{ts} can vary freely over t and s. We obtain

```
. * Poisson PA estimator with unstructured error correlation and robust VCE
. xtpoisson mdu lcoins ndisease female age lfam child, pa corr(unstr) vce(robust)

Iteration 1: tolerance = .01585489
Iteration 2: tolerance = .00034066
Iteration 3: tolerance = 2.334e-06
Iteration 4: tolerance = 1.939e-08

GEE population-averaged model                    Number of obs      =      20186
Group and time vars:                  id year    Number of groups   =       5908
Link:                                     log     Obs per group: min =          1
Family:                               Poisson                      avg =        3.4
Correlation:                     unstructured                      max =          5
                                                 Wald chi2(6)       =     508.61
Scale parameter:                            1     Prob > chi2        =     0.0000

                                         (Std. Err. adjusted for clustering on id)
```

mdu	Coef.	Semirobust Std. Err.	z	P>|z|	[95% Conf. Interval]	
lcoins	-.0804454	.0077782	-10.34	0.000	-.0956904	-.0652004
ndisease	.0346067	.0024238	14.28	0.000	.0298561	.0393573
female	.1585075	.0334407	4.74	0.000	.0929649	.2240502
age	.0030901	.0015356	2.01	0.044	.0000803	.0060999
lfam	-.1406549	.0293672	-4.79	0.000	-.1982135	-.0830962
child	.1013677	.04301	2.36	0.018	.0170696	.1856658
_cons	.7764626	.0717221	10.83	0.000	.6358897	.9170354

The coefficient estimates are quite similar to those from pooled Poisson. The standard errors are as much as 10% lower, reflecting efficiency gain due to better modeling of the correlations. A more detailed comparison of estimators and methods to estimate the VCE (see the end-of-chapter exercises) shows that failure to use the vce(robust) option leads to erroneous standard errors that are one-third of the robust standard errors, and similar estimates are obtained by using the corr(exchangeable) or corr(ar2) options.

18.6.5 RE Poisson estimators

The Poisson individual-effects model assumes that y_{it} is Poisson distributed with a mean of

$$E(y_{it}|\alpha_i, \mathbf{x}_{it}) = \exp(\gamma_i + \mathbf{x}'_{it}\boldsymbol{\beta}) = \alpha_i \exp(\mathbf{x}'_{it}\boldsymbol{\beta}) \qquad (18.14)$$

where $\gamma_i = \ln \alpha_i$, and here $\mathbf{x}_{it}$ includes an intercept. The conditional mean can be viewed either as one with effects that are additive before exponentiation or as one with multiplicative effects.

The standard Poisson RE estimator assumes that α_i is gamma distributed with a mean of 1 and a variance of η. This assumption has the attraction that there is a closed-form expression for the integral (18.4), so the estimator is easy to compute. Furthermore, $E(y_{it}|\mathbf{x}_{it}) = \exp(\mathbf{x}'_{it}\boldsymbol{\beta})$ so that predictions and MEs are easily obtained and interpreted. This is the conditional mean given in (18.13) for the PA and pooled models, so for the special case of the Poisson, the PA and pooled estimators are consistent estimators of the RE model. Finally, the first-order conditions for the Poisson RE estimator, $\widehat{\boldsymbol{\beta}}$, can be shown to be

$$\sum_{i=1}^{N}\sum_{t=1}^{T}\mathbf{x}_{it}\left(y_{it}-\lambda_{it}\frac{\overline{y}_i+\eta/T}{\overline{\lambda}_i+\eta/T}\right)=\mathbf{0} \tag{18.15}$$

where $\overline{\lambda}_i=T^{-1}\sum_t\exp(\mathbf{x}'_{it}\boldsymbol{\beta})$, so the estimator is consistent if $E(y_{it}|\alpha_i,\mathbf{x}_{i1},\dots,\mathbf{x}_{iT})=\alpha_i\exp(\mathbf{x}'_{it}\beta)$ because then the left-hand side of (18.15) has an expected value of 0.

The RE estimator is obtained by using the `xtpoisson` command with the `re` option. There is no option for cluster–robust standard errors, so we use the `vce(bootstrap)` option, which performs a cluster bootstrap. We have

```
. * Poisson random-effects estimator with cluster-robust standard errors
. xtpoisson mdu lcoins ndisease female age lfam child, re
> vce(boot, reps(400) seed(10101) nodots)
Random-effects Poisson regression          Number of obs      =      20186
Group variable: id                         Number of groups   =       5908

Random effects u_i ~ Gamma                 Obs per group: min =          1
                                                          avg =        3.4
                                                          max =          5

                                           Wald chi2(6)       =     534.34
Log likelihood  = -43240.556               Prob > chi2        =     0.0000
                        (Replications based on 5908 clusters in id)
```

mdu	Observed Coef.	Bootstrap Std. Err.	z	P>|z|	Normal-based [95% Conf. Interval]	
lcoins	-.0878258	.0081916	-10.72	0.000	-.103881	-.0717706
ndisease	.0387629	.0024574	15.77	0.000	.0339466	.0435793
female	.1667192	.0376166	4.43	0.000	.0929921	.2404463
age	.0019159	.0016831	1.14	0.255	-.001383	.0052148
lfam	-.1351786	.0338651	-3.99	0.000	-.201553	-.0688042
child	.1082678	.0537636	2.01	0.044	.0028931	.2136426
_cons	.7574177	.0827935	9.15	0.000	.5951454	.91969
/lnalpha	.0251256	.0257423			-.0253283	.0755796
alpha	1.025444	.0263973			.9749897	1.078509

```
Likelihood-ratio test of alpha=0: chibar2(01) =  3.9e+04 Prob>=chibar2 = 0.000
```

Compared with the PA estimates, the RE coefficients are within 10% and the RE cluster–robust standard errors are about 10% higher. The cluster–robust standard errors for the RE estimates are 20–50% higher than the default standard errors, so cluster–robust standard errors are needed. The problem is that the Poisson RE model is not sufficiently flexible because the single additional parameter, η, needs to simultaneously account for both overdispersion and correlation. Cluster–robust standard errors can correct for this, or the richer negative binomial RE model may be used.

For the RE model, $E(y_{it}|\mathbf{x}_{it})=\exp(\mathbf{x}'_{it}\boldsymbol{\beta})$, so the fitted values $\exp(\mathbf{x}'_{it}\widehat{\boldsymbol{\beta}})$, created by using `predict` with the `nu0` option, can be interpreted as estimates of the conditional mean after integrating out the RE. And `margins, dydx()` with the `predict(nu0)` option gives the corresponding MEs. If instead we want to also condition on α_i, then

$E(y_{it}|\alpha_i, \mathbf{x}_{it}) = \alpha_i \exp(\mathbf{x}'_{it}\boldsymbol{\beta})$ implies that $\partial E(y_{it}|\alpha_i, \mathbf{x}_{it})/\partial x_{j,it} = \beta_j \times E(y_{it}|\alpha_i, \mathbf{x}_{it})$, so β_j can still be interpreted as a semielasticity.

An alternative Poisson RE estimator assumes that $\gamma_i = \ln \alpha_i$ is normally distributed with a mean of 0 and a variance of σ_α^2, similar to the `xtlogit` and `xtprobit` commands. Here estimation is much slower because Gauss–Hermite quadrature is used to perform numerical univariate integration. And similarly to the logit RE estimator, prediction and computation of MEs is difficult. This alternative Poisson RE estimator can be computed by using `xtpoisson` with the `re` and `normal` options. Estimates from this method are presented in section 18.6.7.

The RE model permits only the intercept to be random. We can also allow slope coefficients to be random. This is the mixed-effects Poisson estimator implemented with `xtmepoisson`. The method is similar to that for `xtmelogit`, presented in section 18.4.10. The method is computationally intensive.

18.6.6   FE Poisson estimator

The FE model is the Poisson individual-effects model (18.14), where α_i is now possibly correlated with $\mathbf{x}_{it}$, and in short panels, we need to eliminate α_i before estimating $\boldsymbol{\beta}$.

These effects can be eliminated by using the conditional ML estimator based on a log density for the ith individual that conditions on $\sum_{t=1}^{T_i} y_{it}$, similar to the treatment of FE in the logit model. Some algebra leads to the Poisson FE estimator with first-order conditions

$$\sum_{i=1}^{N}\sum_{t=1}^{T} \mathbf{x}_{it}\left(y_{it} - \frac{\lambda_{it}}{\overline{\lambda}_i}\overline{y}_i\right) = \mathbf{0} \tag{18.16}$$

where $\lambda_{it} = \exp(\mathbf{x}'_{it}\boldsymbol{\beta})$ and $\overline{\lambda}_i = T^{-1}\sum_t \exp(\mathbf{x}'_{it}\boldsymbol{\beta})$. The Poisson FE estimator is therefore consistent if $E(y_{it}|\alpha_i, \mathbf{x}_{i1}, \dots, \mathbf{x}_{iT}) = \alpha_i \exp(\mathbf{x}'_{it}\beta)$ because then the left-hand side of (18.16) has the expected value of zero.

The Poisson FE estimator can be obtained by using the `xtpoisson` command with the `fe` option. To obtain cluster–robust standard errors, we can use the `vce(bootstrap)` option. It is quicker, however, to use the user-written `xtpqml` command (Simcoe 2007), which directly calculates cluster–robust standard errors. We have

```
. * Poisson fixed-effects estimator with cluster-robust standard errors
. xtpoisson mdu lcoins ndisease female age lfam child, fe vce(boot, reps(400)
> seed(10101) nodots)
Conditional fixed-effects Poisson regression     Number of obs     =      17791
Group variable: id                               Number of groups  =       4977

                                                 Obs per group: min =          2
                                                               avg =        3.6
                                                               max =          5

                                                 Wald chi2(3)      =       4.39
Log likelihood  = -24173.211                     Prob > chi2       =     0.2221
                                   (Replications based on 4977 clusters in id)
```

mdu	Observed Coef.	Bootstrap Std. Err.	z	P>|z|	Normal-based [95% Conf. Interval]	
age	-.0112009	.0094595	-1.18	0.236	-.0297411	.0073394
lfam	.0877134	.1152712	0.76	0.447	-.138214	.3136407
child	.1059867	.0758987	1.40	0.163	-.0427721	.2547454

Only the coefficients of time-varying regressors are identified, similar to other FE model estimators. The Poisson FE estimator requires that there be at least two periods of data, leading to a loss of 265 observations, and that the count for an individual be nonzero in at least one period ($\sum_{t=1}^{T_i} y_{it} > 0$), leading to a loss of 666 individuals because mdu equals zero in all periods for 666 people. The cluster–robust standard errors are roughly two times those of the default standard errors; see the end-of-chapter exercises. In theory, the individual effects, α_i, could account for overdispersion, but for these data, they do not completely do so. The standard errors are also roughly twice as large as the PA and RE standard errors, reflecting a loss of precision due to using only within variation.

For the FE model, results should be interpreted based on $E(y_{it}|\alpha_i, \mathbf{x}_{it}) = \alpha_i \exp(\mathbf{x}'_{it} \boldsymbol{\beta})$. The `predict` command with the nu0 option gives predictions when $\gamma_i = 0$ so $\alpha_i = 1$, and the `margins, dydx()` command with the `predict(nu0)` option gives the corresponding MEs. If we do not want to consider only the case of $\alpha_i = 1$, then the model implies that $\partial E(y_{it}|\alpha_i, \mathbf{x}_{it})/\partial x_{j,it} = \beta_j \times E(y_{it}|\alpha_i, \mathbf{x}_{it})$, so β_j can still be interpreted as a semielasticity.

Given the estimating equations given by (18.16), the Poisson FE estimator can be applied to any model with multiplicative effects and an exponential conditional mean, essentially whenever the dependent variable has a positive conditional mean. Then the Poisson FE estimator uses the quasi-difference, $y_{it} - (\lambda_{it}/\bar{\lambda}_i)\bar{y}_i$, whereas the linear model uses the mean-difference, $y_{it} - \bar{y}_i$.

In the linear model, one can instead use the first-difference, $y_{it} - y_{i,t-1}$, to eliminate the FE, and this has the additional advantage of enabling estimation of FE dynamic linear models using the Arellano–Bond estimator. Similarly, here one can instead use the alternative quasi-difference, $(\lambda_{i,t-1}/\lambda_{it})y_{it} - y_{i,t-1}$, to eliminate the FE and use this as the basis for estimation of dynamic panel count models.

18.6.7   Panel Poisson estimators comparison

We summarize the results using several panel Poisson estimators. The RE and FE estimators were estimated with the default estimate of the VCE to speed computation, though as emphasized in preceding sections, any reported standard errors should be based on the cluster–robust estimate of the VCE.

```
. * Comparison of Poisson panel estimators
. quietly xtpoisson mdu lcoins ndisease female age lfam child, pa corr(unstr)
> vce(robust)
. estimates store PPA_ROB
. quietly xtpoisson mdu lcoins ndisease female age lfam child, re
. estimates store PRE
. quietly xtpoisson mdu lcoins ndisease female age lfam child, re normal
. estimates store PRE_NORM
. quietly xtpoisson mdu lcoins ndisease female age lfam child, fe
. estimates store PFE
. estimates table PPA_ROB PRE PRE_NORM PFE, equations(1) b(%8.4f) se
> stats(N ll) stfmt(%8.0f)
```

Variable	PPA_ROB	PRE	PRE_NORM	PFE
#1				
lcoins	-0.0804	-0.0878	-0.1145	
	0.0078	0.0069	0.0073	
ndisease	0.0346	0.0388	0.0409	
	0.0024	0.0022	0.0023	
female	0.1585	0.1667	0.2084	
	0.0334	0.0286	0.0305	
age	0.0031	0.0019	0.0027	-0.0112
	0.0015	0.0011	0.0012	0.0039
lfam	-0.1407	-0.1352	-0.1443	0.0877
	0.0294	0.0260	0.0265	0.0555
child	0.1014	0.1083	0.0737	0.1060
	0.0430	0.0341	0.0345	0.0438
_cons	0.7765	0.7574	0.2873	
	0.0717	0.0618	0.0642	
lnalpha				
_cons		0.0251		
		0.0210		
lnsig2u				
_cons			0.0550	
			0.0255	
Statistics				
N	20186	20186	20186	17791
ll		-43241	-43227	-24173

```
                                             legend: b/se
```

The PA and RE parameter estimates are quite similar; the alternative RE estimates based on normally distributed RE are roughly comparable, whereas the FE estimates for the time-varying regressors are quite different.

18.6.8   Negative binomial estimators

The preceding analysis for the Poisson can be replicated for the negative binomial. The negative binomial has the attraction that, unlike Poisson, the estimator is designed to explicitly handle overdispersion, and count data are usually overdispersed. This may lead to improved efficiency in estimation and a default estimate of the VCE that should be much closer to the cluster–robust estimate of the VCE, unlike for Poisson panel commands. At the same time, the Poisson panel estimators rely on weaker distributional assumptions—essentially, correct specification of the mean—and it may be more robust to use the Poisson panel estimators with cluster–robust standard errors.

For the pooled negative binomial, the issues are similar to those for pooled Poisson. For the pooled negative binomial, we use the `nbreg` command with the `vce(cluster id)` option. For the PA negative binomial, we can use the `xtnbreg` command with the `pa` and `vce(robust)` options.

For the panel negative binomial RE and FE models, we use `xtnbreg` with the `re` or `fe` option. The negative binomial RE model introduces two parameters in addition to β that accommodate both overdispersion and within correlation. The negative binomial FE estimator is unusual among FE estimators because it is possible to estimate the coefficients of time-invariant regressors in addition to time-varying regressors. A more complete presentation is given in, for example, Cameron and Trivedi (1998, 2005) and in [XT] **xtnbreg**.

We apply the Poisson PA and negative binomial PA, RE, and FE estimators to the doctor visits data. We have

```
. * Comparison of negative binomial panel estimators
. quietly xtpoisson mdu lcoins ndisease female age lfam child, pa corr(exch)
> vce(robust)
. estimates store PPA_ROB
. quietly xtnbreg mdu lcoins ndisease female age lfam child, pa corr(exch)
> vce(robust)
. estimates store NBPA_ROB
. quietly xtnbreg mdu lcoins ndisease female age lfam child, re
. estimates store NBRE
. quietly xtnbreg mdu lcoins ndisease female age lfam child, fe
. estimates store NBFE
```

(Continued on next page)

```
. estimates table PPA_ROB NBPA_ROB NBRE NBFE, equations(1) b(%8.4f) se
> stats(N ll) stfmt(%8.0f)
```

Variable	PPA_ROB	NBPA_ROB	NBRE	NBFE
#1				
lcoins	-0.0815	-0.0865	-0.1073	-0.0885
	0.0079	0.0078	0.0062	0.0139
ndisease	0.0347	0.0376	0.0334	0.0154
	0.0024	0.0023	0.0020	0.0040
female	0.1609	0.1649	0.2039	0.2460
	0.0338	0.0343	0.0263	0.0586
age	0.0032	0.0026	0.0023	-0.0021
	0.0016	0.0016	0.0012	0.0020
lfam	-0.1487	-0.1633	-0.1434	-0.0008
	0.0299	0.0291	0.0251	0.0477
child	0.1121	0.1154	0.1145	0.2032
	0.0444	0.0452	0.0385	0.0543
_cons	0.7755	0.7809	0.8821	0.9243
	0.0724	0.0730	0.0663	0.1156
ln_r				
_cons			1.1280	
			0.0269	
ln_s				
_cons			0.7259	
			0.0313	
Statistics				
N	20186	20186	20186	17791
ll			-40661	-21627

```
                                                          legend: b/se
```

The Poisson and negative binomial PA estimates and their standard errors are similar. The RE estimates differ more and are closer to the Poisson RE estimates given in section 18.6.4. The FE estimates differ much more, especially for the time-invariant regressors.

18.7 Stata resources

The Stata panel commands cover the most commonly used panel methods, especially for short panels. This topic is exceptionally vast, and there are many other methods that provide less-used alternatives to the methods covered in Stata as well as methods to handle complications not covered in Stata, especially the joint occurrence of several complications such as a dynamic FE logit model. Many of these methods are covered in the panel-data books by Arellano (2003), Baltagi (2008), Hsiao (2003), and Lee (2002); see also Rabe-Hesketh and Skrondal (2008) for the mixed-model approach. Cameron and Trivedi (2005) and Wooldridge (2002) also cover some of these methods.

18.8 Exercises

1. Consider the panel logit estimation of section 18.4. Compare the following three sets of estimated standard errors for the pooled logit estimator: default, heteroskedasticity-robust, and cluster–robust. How important is it to control for heteroskedasticity and clustering? Show that the `pa` option of the `xtlogit` command yields the same estimates as the `xtgee` command with the `family(binomial)`, `link(logit)`, and `corr(exchangeable)` options. Compare the PA estimators with the `corr(exchangeable)`, `corr(AR2)`, and `corr(unstructured)` options, in each case using the `vce(robust)` option.

2. Consider the panel logit estimation of section 18.4. Drop observations with `id > 125200`. Estimate the parameters of the FE logit model by using `xtlogit` as in section 18.4. Then estimate the parameters of the same model by using `logit` with dummy variables for each individual (so use `xi: logit` with regressors including `i.id`). This method is known to give inconsistent parameter estimates. Compare the estimates with those from command `xtlogit`. Are the same parameters identified?

3. For the parameters of the panel logit models in section 18.4, estimate by using `xtlogit` with the `pa`, `re`, and `fe` options. Compute the following predictions: for `pa`, use `predict` with the `mu` option; for `re`, use `predict` with the pu0 option; for `pa`, use `predict` with the pu0 option. For these predictions and for the original dependent variable, `dmdu`, compare the sample average value and the sample correlations. Then use the `margins, dydx()` command with these predict options, and compare the resulting MEs for the `lcoins` variable.

4. For the panel tobit model in section 18.5, compare the results from `xttobit` with those from `tobit`. Which do you prefer? Why?

5. Consider the panel Poisson estimation of section 18.6. Compare the following three sets of estimated standard errors for the pooled Poisson estimator: default, heteroskedasticity-robust, and cluster–robust. How important is it to control for heteroskedasticity and clustering? Compare the PA estimators with the `corr(exchangeable)`, `corr(AR2)`, and `corr(unstructured)` options, in each case using both the default estimate of the VCE and the `vce(robust)` option.

6. Consider the panel count estimation of section 18.6. To reduce computation time, use the `drop if id > 127209` command to use 10% of the original sample. Compare the standard errors obtained by using default standard errors with those obtained by using the `vce(boot)` option for the following estimators: Poisson RE, Poisson FE, negative binomial RE, and negative binomial FE. How important is it to use panel–robust standard errors for these estimators?

A Programming in Stata

In this appendix, we build on the introduction to Stata programming given in chapter 1. We first present Stata matrix commands, introduced in section 1.5. The rest of the appendix focuses on aspects of writing Stata programs, using the `program` command introduced in section 4.3.1. We discuss programs to be included within a Stata do-file, ado-files that are programs intended to be used by other Stata users, and some tips for program debugging that are relevant for even the simplest Stata coding.

A.1 Stata matrix commands

Here we consider Stata matrix commands, initiated with the `matrix` prefix. These provide a limited set of matrix commands sufficient for many uses, especially postestimation manipulation of results, as introduced in section 1.6, and are comparable to matrix commands provided in other econometrics packages.

The separate appendix B presents Mata matrix commands, introduced in Stata 9. Mata is a full-blown matrix programming language, comparable to Gauss and Matlab.

A.1.1 Stata matrix overview

Key considerations are inputting matrices, either directly or by converting data variables into matrices, and performing operations on matrices or on subcomponents of the matrix such as individual elements.

The basics are given in [U] **14 Matrix expressions** and in [P] **matrix**. Useful online help commands include `help matrix`, `help matrix operators`, and `help matrix functions`.

A.1.2 Stata matrix input and output

There are several ways to input matrices in Stata.

Matrix input by hand

Matrix entries can be entered by using the `matrix define` command. For example, consider a 2×3 matrix with the first row entries 1, 2, and 3, and the second row entries 4, 5, and 6. Column entries are separated by commas, and rows are separated by a backslash. We have

```
. * Define a matrix explicitly and list the matrix
. matrix define A = (1,2,3 \ 4,5,6)
. matrix list A
A[2,3]
    c1  c2  c3
r1   1   2   3
r2   4   5   6
```

The word `define` can be omitted from the above command.

The default names for the matrix rows are `r1`, `r2`, ..., and the column defaults are `c1`, `c2`, ... These names can be changed by using the `matrix rownames` and `matrix colnames` commands. For example, to give the names `one` and `two` to the two rows of matrix A, type the command

```
. * Matrix row and column names
. matrix rownames A = one two
. matrix list A
A[2,3]
      c1  c2  c3
one    1   2   3
two    4   5   6
```

An alternative matrix naming command is `matname`.

Matrix input from Stata estimation results

Matrices can be constructed from matrices created by the Stata estimation command results stored in `e()` or `r()`. For example, after ordinary least-squares (OLS) regression, the variance–covariance matrix is stored in `e(V)`. To give it a more obvious name or to save it for later analysis, we define a matrix equal to `e(V)`.

As a data example, we use the same dataset as in chapter 3. We use the first 100 observations and regress medical expenditures (`ltotexp`) on an intercept and chronic problems (`totchr`). We have

```
. * Read in data, summarize and run regression
. use mus03data.dta
. keep if _n <= 100
(2964 observations deleted)
. drop if ltotexp == . | totchr == .
(0 observations deleted)
. summarize ltotexp totchr
```

Variable	Obs	Mean	Std. Dev.	Min	Max
ltotexp	100	4.533688	.8226942	1.098612	5.332719
totchr	100	.48	.717459	0	3

```
. regress ltotexp totchr, noheader
```

ltotexp	Coef.	Std. Err.	t	P>|t|	[95% Conf. Interval]	
totchr	.1353098	.1150227	1.18	0.242	-.0929489	.3635685
_cons	4.468739	.0989462	45.16	0.000	4.272384	4.665095

A command to drop observations with missing values from the dataset in memory is included, because not all matrix operators considered below handle missing values.

We then obtain the variance matrix stored in e(V) and list its contents.

```
. * Create a matrix from estimation results
. matrix vbols = e(V)

. matrix list vbols
symmetric vbols[2,2]
             totchr        _cons
totchr   .01323021
 _cons   -.0063505    .00979036
```

Stata has incorporated the regressor names into the estimate of the variance–covariance matrix of the estimator (VCE) so that vbols has rows and columns named totchr and _cons.

A.1.3   Stata matrix subscripts and combining matrices

Matrix subscripts are represented in square brackets. The entry (i, j) in a matrix is denoted [i,j]. For example, to set the $(1, 1)$ entry in matrix A to equal the $(1, 2)$ entry, type the command

```
. * Change value of an entry in matrix
. matrix A[1,1] = A[1,2]

. matrix list A
A[2,3]
       c1  c2  c3
one     2   2   3
two     4   5   6
```

If the row or column has a name, one can alternatively use this name. For example, because row 1 of A is named one, we could have typed matrix A[1,1] = A["one",2].

For a column vector, the ith entry is denoted by [i,1] rather than simply [i]. Similarly, for a row vector, the jth entry is denoted by [1,j] rather than simply [j].

Matrix subscripts can be given as a range, permitting a submatrix to be extracted from a matrix. For example, to extract all the rows and columns 2–3 from matrix A, type

```
. * Select part of matrix
. matrix B = A[1...,2..3]

. matrix list B
B[2,2]
       c2  c3
one     2   3
two     5   6
```

Here k... selects the kth entry on, and k..l selects the kth–lth entry.

To add or append rows to a matrix, the vertical concatenation operator \ is used. For example, A \ B adds rows of B after the rows of A. Similarly, to add columns to a matrix, the horizontal concatenation operator , is used. For example,

```
. * Add columns to an existing matrix
. matrix C = B, B

. matrix list C
C[2,4]
       c2  c3  c2  c3
one     2   3   2   3
two     5   6   5   6
```

A.1.4   Matrix operators

All the standard matrix operators can be applied, provided that the matrices are conformable. The operators are + to add, − to subtract, * to multiply, and # for the Kronecker product. In addition, the multiplication command can be used for multiplication by a scalar, e.g., 2*A or A*2, and scalar division is possible, e.g., A/2. A single apostrophe, ´, gives the matrix transpose. To compute A´A, we use A´*A. For example,

```
. * Matrix operators
. matrix D = C + 3*C

. matrix list D
D[2,4]
       c2  c3  c2  c3
one     8  12   8  12
two    20  24  20  24
```

A.1.5   Matrix functions

Standard matrix functions are defined by using parentheses, ( ). Some commands lead to a scalar result, for example,

```
. * Matrix functions
. matrix r = rowsof(D)

. matrix list r
symmetric r[1,1]
      c1
r1     2
```

In this example, it is more convenient to store the result as a scalar, rather than in a 1×1 matrix. For example,

```
. * Can use scalar if 1x1 matrix
. scalar ralt = rowsof(D)

. display ralt
2
```

Functions that produce scalars include colsof(A), det(A), rowsof(A), and trace(A).

Other commands produce matrices. For example, matrix B, created earlier, is a nonsymmetric square matrix, with the inverse

```
. * Inverse of nonsymmetric square matrix
. matrix Binv = inv(B)
. matrix list Binv
Binv[2,2]
            one        two
 c2          -2          1
 c3   1.6666667  -.66666667
```

Here are some functions that produce matrices: `cholesky(A)`, `corr(A)`, `diag(A)`, `hadamard(A,B)`, `I(n)`, `inv(A)`, `invsym(A)`, `vec(A)`, and `vecdiag(A)`.

A.1.6   Matrix accumulation commands

Most estimators, such as the OLS estimator $(\mathbf{X}'\mathbf{X})^{-1}\mathbf{X}'\mathbf{y}$, require the computation of matrix cross products. We strongly recommend that you do not put your data into Stata matrices. Stata has accumulation commands that compute cross products from variables and return the results in Stata matrices. If you really want to put your data into a matrix, refer to appendix B on Mata.

Stata's matrix accumulation commands compute the matrix cross products $\mathbf{X}'\mathbf{X}$ and $\mathbf{X}'\mathbf{y}$ without requiring the intermediate step of forming the much larger matrices $\mathbf{X}$ and $\mathbf{y}$.

As an example, the `matrix accum A = v1 v2` command creates a 3×3 matrix $\mathbf{A} = \mathbf{Z}'\mathbf{Z}$, where $\mathbf{Z}$ is an $N \times 3$ matrix with columns of the variables v1 and v2, and a column of ones that `accum` automatically appends unless the `noconstant` option is used. The companion `matrix vecaccum A = w v1 v2` command creates a 1×3 row vector $\mathbf{A} = \mathbf{w}'\mathbf{Z}$, where $\mathbf{w}$ is an $N \times 1$ column vector with a column of the variable w, and $\mathbf{Z}$ is an $N \times 3$ matrix with columns of the variables v1 and v2, and a column of ones that again `accum` automatically adds at the end unless the `noconstant` option is used. Related commands are `matrix glsaccum`, which forms weighted cross products of the form $\mathbf{X}'\mathbf{BX}$, and `matrix opaccum`.

The following code produces the same point estimates as `regress ltotexp totchr`:

```
. * OLS estimator using matrix accumulation operators
. matrix accum XTX = totchr           // Form X'X including constant
(obs=100)
. matrix vecaccum yTX = ltotexp totchr   // Form y'X including constant
. matrix cols = invsym(XTX)*(yTX)'
. matrix list cols
cols[2,1]
            ltotexp
totchr   .13530976
 _cons   4.4687394
```

A.1.7 OLS using Stata matrix commands

The following example runs an OLS regression of `ltotexp` on an intercept and `totchr`, and it also reports the default OLS standard errors and associated t statistics. We use matrix accumulation commands so that large problems can be handled. The challenge in using these commands is to obtain $s^2 = \sum_{i=1}^{N}(y_i - \widehat{y}_i)^2$ without having to form an $N \times 1$ vector of predicted values. One way to do this is to use the result that for OLS $\sum_{i=1}^{N}(y_i - \widehat{y}_i)^2 = \mathbf{y'y} - \widehat{\boldsymbol{\beta}}'\mathbf{X'X}\widehat{\boldsymbol{\beta}}$.

We have

```
. * Illustrate Stata matrix commands: OLS with output
. matrix accum XTX = totchr            // Form X´X including constant
(obs=100)
. matrix vecaccum yTX = ltotexp totchr // Form y´X including constant
. matrix b = invsym(XTX)*(yTX)´
. matrix accum yTy = ltotexp, noconstant
(obs=100)
. scalar k = rowsof(XTX)
. scalar n = _N
. matrix s2 = (yTy - b´*XTX´*b)/(n-k)
. matrix V = s2*invsym(XTX)
. matrix list b
b[2,1]
          ltotexp
totchr  .13530976
 _cons  4.4687394
. matrix list V
symmetric V[2,2]
           totchr       _cons
totchr   .01323021
 _cons  -.0063505   .00979036
```

This yields the same estimates of the coefficients and VCE as listed in section A.1.2.

We now want to obtain output formatted in the usual way with columns of regressor names, coefficient estimates, standard errors, and t statistics. This is not straightforward using Stata matrix commands. We wish to form the column vector `t` with the jth entry $t_j = b_j/s_j = b_j/\sqrt{V_{jj}}$. But Stata provides no facility for element-by-element division and also no easy way to take the element-by-element square root of a matrix. One fix is to first form a column vector, `seinv`, with the jth entry $1/s_j$ by creating a diagonal matrix with the entries s_j^2, taking the inverse of this matrix, taking the square root of this matrix, and forming a column vector with the resulting diagonal entries. Then form the column vector `t` by using the Hadamard product of `b` and `seinv`, where for matrices A and B of the same dimension, `C=hadamard(A,B)` gives the matrix C with the ijth entry $C_{ij} = A_{ij} \times B_{ij}$. We obtain

```
.  * Stata matrix commands to compute SEs and t statistics given b and V
.  matrix se = (vecdiag(cholesky(diag(vecdiag(V)))))´
.  matrix seinv = (vecdiag(cholesky(invsym(diag(vecdiag(V))))))´
.  matrix t = hadamard(b,seinv)
.  matrix results = b, se, t
.  matrix colnames results = coeff sterror tratio
.  matrix list results, format(%7.0g)

results[2,3]
          coeff   sterror   tratio
totchr   .13531   .11502    1.1764
 _cons   4.4687   .09895    45.163
```

It is much easier to instead use Stata's `ereturn` commands to produce this output, based on a row vector of coefficient estimates and an estimated variance matrix. The preceding code led to a column vector of coefficients, so we first need to transpose. We obtain

```
.  * Easier is to use ereturn post and display given b and V
.  matrix brow = b´
.  ereturn post brow V
.  ereturn display
```

	Coef.	Std. Err.	z	P>|z|	[95% Conf. Interval]	
totchr	.1353098	.1150227	1.18	0.239	-.0901305	.36075
_cons	4.468739	.0989462	45.16	0.000	4.274808	4.66267

Similar code for OLS that instead uses Mata functions is provided in section 3.8.

A.2   Programs

Do-files, ado-files, and program files are collections of Stata commands that are useful whenever the same analysis is to be repeated exactly or with relatively minor variation. For many analyses, a do-file that enacts Stata commands (that are themselves often ado-files written in Stata or Mata) is sufficient.

More advanced analysis, however, may require actual programming in Stata. These programs can be defined and executed as a component of a do-file, or they can be converted to an ado-file to enable their being called by other programs. Useful references are [U] **18 Programming Stata** and [P] **program**.

A.2.1   Simple programs (no arguments or access to results)

A program is defined by using the `program define` command followed by the program name. Subsequent lines give the program, which concludes with the line `end`.

The simplest programs do not have any inputs, and the program output is simply displayed. The following program displays the current time and date.

```
. * Program with no arguments
. program time
  1.    display c(current_time) c(current_date)
  2. end
```

The word **define** is optional in the above input—it is sufficient to simply type **program time**.

The program is executed by typing the name of the program. We have

```
. * Run the program
. time
12:36:13 5 Oct 2009
```

Unlike the execution of a do-file, only the program results are listed, here the current date and time; the program commands that were executed are not listed.

A.2.2   Modifying a program

Stata does not allow one to redefine an existing program. So it is necessary to first remove any previous program with the same name, should such a program already exist.

The **program drop time** command will drop the **time** program. If this program does not already exist, however, Stata will stop executing and generate an error message. The **capture** prefix ensures that Stata will continue to run, even if the **time** program does not already exist.

Thus the preferred way to define and then execute the **time** program is

```
. * Drop program if it already exists, write program and run
. capture program drop time

. program time
  1.    display c(current_time) c(current_date)
  2. end

. time
12:36:13 5 Oct 2009
```

The **clear** command does not drop programs, though **clear all** will. To specifically drop all programs, use the **clear programs** command or the **program drop _all** command.

A.2.3   Programs with positional arguments

More complicated programs have inputs called arguments. For example, the Stata **regress** command has as arguments the dependent variable and any regressor variables. Then execution of the command requires that one gives both the command name and the command arguments, e.g., **regress y x1 x2**. These arguments need to be passed into the program and then referred to appropriately within the program.

   Program arguments can be passed as positional arguments. The first argument is referred to by the local macro `` `1´ `` within the program, the second by local macro `` `2´ ``, and so on. For example, for **regress**, the dependent variable, say, y, may be referred to internally as `` `1´ ``. The quotation marks differ from how they appear on this printed page. On most keyboards, the left quote is located in the upper left, and the right quote is located in the middle right. When viewed using a text editor, the single quote appears correctly. But often when viewed in LaTeX documents, they misleadingly appear as '1' rather than the correct `` `1´ ``.

   We present a program to report the median of the difference between two variables, where the two variables need to be passed to the program. Using positional arguments, we have

```
. * Program with two positional arguments
. program meddiff
  1.    tempvar diff
  2.    generate `diff´ = `1´ - `2´
  3.    _pctile `diff´, p(50)
  4.    display "Median difference = " r(r1)
  5. end
```

The program uses a temporary variable, **diff**, explained in the next section, to store the difference between the two variables. Several commands calculate the median. Here we use the **_pctile** command with the **p(50)** option. This command stores the resulting median in **r(r1)**, which we then output by using the **display** command.

   We now run the **meddiff** program, using the same dataset and variables, **ltotchr** and **totchr**, as used in section A.1. We have

```
. * Run the program with two arguments
. meddiff ltotexp totchr
Median difference = 4.2230513
```

A.2.4   Temporary variables

The **meddiff** program requires the computation of the intermediate variable we have named **diff**. To ensure that this name does not conflict with the names of variables elsewhere and that the variable is dropped as soon as the program ends, we use the **tempvar** command to define a temporary variable that is local only to the program and is dropped after the program has executed. This temporary variable is declared by using **tempvar** and is then referred to in the same left and right quotation marks as are used for local macros. Similarly, the **tempname** command can be used to declare temporary scalars and matrices, and the **tempfile** command can be used to declare temporary files.

A.2.5   Programs with named positional arguments

It is much easier to read the program if it gives names to the positional arguments `` `1´ ``, `` `2´ ``, ... To use named positional arguments, we first define the arguments within the program in the order that they will appear in the command. For example,

```
. * Program with two named positional arguments
. capture program drop meddiff

. program meddiff
  1.    args y x
  2.    tempvar diff
  3.    generate `diff´ = `y´ - `x´
  4.    _pctile `diff´, p(50)
  5.    display "Median difference = " r(r1)
  6. end

. meddiff ltotexp totchr
Median difference = 4.2230513
```

As for temporary variables, the arguments are declared without quotes, but Stata stores arguments as local macros, so we need to use quotes to refer to the arguments.

A.2.6   Storing and retrieving program results

The preceding examples simply displayed results. Often we want to store program results for further data analysis. This can be done by storing the results in `r()` and `e()`, introduced in section 1.6, and `s()`. To do this, we need to define the program to be of the relevant class and to return the results to the named entries in `r()`, `e()`, or `s()`.

For our example, we declare the program to be `rclass`, with just one result that will be stored in `r(medylx)`. We have

```
. * Program with results stored in r()
. capture program drop meddiff

. program meddiff, rclass
  1.    args y x
  2.    tempvar diff
  3.    generate `diff´ = `y´ - `x´
  4.    _pctile `diff´, p(50)
  5.    return scalar medylx = r(r1)
  6. end
```

Executing the program produces no output; the results of executing the program are instead stored in `r()`. To list the program results in `r()`, we use the `return list` command, and to display the scalars in `r()`, we use the `display` command.

```
. * Running the program does not immediately display the result
. meddiff ltotexp totchr

. return list

scalars:
            r(medylx) =  4.223051309585571
. display r(medylx)
4.2230513
```

An example of an `eclass` program, returning e and V for subsequent analysis, is given in section 13.4.4.

A.2.7   **Programs with arguments using standard Stata syntax**

Program arguments can be quite lengthy and can include optional arguments, but many commands use arguments in a standard format. In particular, if commands use the standard Stata syntax, then tools exist to parse the command, breaking down the long command into its various arguments.

The full standard Stata syntax is

command $\big[$ *varlist* | *namelist* | *anything* $\big]$ $\big[$ *if* $\big]$ $\big[$ *in* $\big]$ $\big[$ **using** *filename* $\big]$ $\big[$ =*exp* $\big]$
$\big[$ *weight* $\big]$ $\big[$, *options* $\big]$

Square brackets denote optional items. For some commands, some of the items in square brackets will be required to run that specific command. In the syntax for that specific command, these required items will not be surrounded by square brackets.

As an example, consider the command

```
. regress ltotexp totchr if ltotexp < . in 1/100, vce(robust)
```

To enact this command, Stata interprets **regress** as *command*, **ltotexp totchr** as *varlist*, **if ltotexp < .** as *if*, **in 1/100** as *in*, and **vce(robust)** as an *option*. For the **regress** command, Stata needs to further break down *varlist*, with the first variable being the dependent variable and any remaining variables being regressors.

We now show how this is done. We write a program, **myols**, that duplicates **regress**. Specifically, we want to be able to break the command

```
. myols ltotexp totchr if ltotexp < . in 1/100, vce(robust)
```

into its arguments and then execute **regress** with these arguments.

To do so, we use the **syntax** and **gettoken** commands as illustrated in the following program.

```
 . * Program that uses Stata commands syntax and gettoken to parse arguments
 . program myols
 1.    syntax varlist [if] [in] [,vce(string)]
 2.    gettoken y xvars : varlist
 3.    display "varlist contains: "  "`varlist´"
 4.    display "and  if contains: "  "`if´"
 5.    display "and  in contains: "  "`in´"
 6.    display "and vce contains: "  "`vce´"
 7.    display "and   y contains: "  "`y´"
 8.    display "& xvars contains: "  "`xvars´"
 9.    regress `y´ `xvars´ `if´ `in´, `vce´ noheader
10. end
```

The **syntax** command lists required arguments—here a list of variable names (**varlist**) —and optional arguments—here an **if** qualifier (**[if]**), an **in** *range* qualifier (**[in]**), and the **vce()** option with a *string* argument for the specific option to be used (**[, vce(string)]**). The **syntax** command will put in the local `varlist´ macro any list of variable names that appears after **myols** and before the **if** or **in** qualifiers; in the local `if´ macro any if qualifier; in the local `in´ macro any in *range* qualifier; and in the local `vce´ macro any *vce_option*. The names in `varlist´ are space-separated tokens. The specific form of the gettoken command used here puts the first token in `varlist´ into the local `y´ macro and the remaining tokens into the local `x´ macro.

The unnecessary display commands are included to demonstrate that the parsing occurs as desired. Note the use of compound quotes. For example, to display the name in the local `y´ macro, we use the display "`y´" command. If instead we use display `y´, then we would see the value of the variable in the local `y´ macro.

We then execute the myols program for an example.

```
. * Execute program myols for an example
. myols ltotexp totchr if ltotexp < . in 1/100, vce(robust)
varlist contains: ltotexp totchr
and   if contains: if ltotexp < .
and   in contains: in 1/100
and vce contains: robust
and    y contains: ltotexp
& xvars contains:  totchr
```

ltotexp	Coef.	Robust Std. Err.	t	P>|t|	[95% Conf. Interval]
totchr	.1353098	.1089083	1.24	0.217	−.0808151 .3514347
_cons	4.468739	.1089425	41.02	0.000	4.252547 4.684932

The arguments of the myols command have been parsed successfully, leading to the expected output from regress.

A.2.8   Ado-files

Some Stata commands, such as summarize, are built-in commands. But many Stata commands are defined by an ado-file, which is a collection of Stata commands. For example, the file logit.ado defines the logit command for logit regression. Furthermore, Stata users can also define their own Stata commands by using ado-files. We use many such user-written commands throughout this book.

An ado-file is a program file similar to those already presented. But because they are intended for wider use, they are generally more tightly written. Temporary variables, scalars, and matrices are used to avoid potential name conflicts with the program calling the ado-file. Variables may be generated in double precision. Care is given to the output from the program, such as by using the quietly prefix to suppress the unnecessary printing of intermediate results. Comments are provided, such as the current version number and date. And there should be various checks to ensure that the command is being correctly used (e.g., if an input to the program should be positive, then send an error message if this is not the case).

A good example of the development of an ado-file is given in [U] **18.11 Ado-files**. For an estimation command, see Gould, Pitblado, and Sribney (2006, chap. 10).

Here we provide a brief example, converting the `meddiff` program from earlier into an ado-file. Specifically, the `meddiff.ado` file comprises

```
*! version 1.1.0   05oct2009
program meddiff, rclass
    version 11
    args y x
    tempvar diff
    quietly {
            generate double `diff´ = `y´ - `x´
            _pctile `diff´, p(50)
            return scalar medylx = r(r1)
    }
    display "Median of first variable - second variable = " r(r1)
end
```

The program begins with the version and date. The program is written for Stata 11. The `quietly` prefix suppresses output. For example, if `y´ or `x´ has any missing values, then the `generate` statement will lead to a statement that missing values were generated. This statement will be suppressed here. The `diff´ variable is in double precision for increased accuracy.

To execute the commands in `meddiff.ado`, we simply type `meddiff` with the appropriate arguments. For example,

```
. * Execute program meddiff for an example
. meddiff ltotexp totchr
Median of first variable - second variable = 4.2230513
```

The `meddiff.ado` file needs to be in a directory that Stata automatically accesses. For a Microsoft Windows computer, these directories include `C:\ado` and `C:\Program Files\Stata11`, and the current directory. See [U] **17 Ado-files** for further details.

A.3   Program debugging

This section provides advice relevant to even the most basic uses of Stata.

There are two challenges: to get the program to execute without stopping because of an error and to ensure that the program is doing what is intended once it is executing.

We focus here on the first challenge. The simplest way to debug a program is to work with a simplified example and print out intermediate results. Stata also provides error messages and a trace facility to track every step of the execution of a program.

The second challenge is easily ignored, but it should not be skipped. Come up with an example where there is a known result or a way to verify the result. For example, to test an estimation procedure, generate many observations from a known data-generating process, and see whether the estimation procedure yields the known data-generating process parameters; see chapter 4. Printing intermediate results is again very helpful. In particular, always use the `summarize` command to verify that you are working with the intended dataset.

A.3.1 Some simple tips

A simple way to debug Stata code is to display the intermediate output. For example, in the following listing, we can see whether the correct dimension matrices are obtained. If the program failed, we could look at the intermediate results before the failure to see where the failure occurs.

```
. * Display intermediate output to aid debugging
. matrix accum XTX = totchr              // Recall constant is added
(obs=100)
. matrix list XTX                        // Should be 2 x 2
symmetric XTX[2,2]
        totchr     _cons
totchr      74
 _cons      48       100
. matrix vecaccum yTX = ltotexp totchr
. matrix list yTX                        // Should be 1 x 2
yTX[1,2]
             totchr       _cons
ltotexp  224.51242   453.36881
. matrix bOLS = invsym(XTX)*(yTX)´
. matrix list bOLS                       // Should be 2 x 1
bOLS[2,1]
          ltotexp
totchr  .13530976
 _cons  4.4687394
```

Even when there seems to be no problem, if the program is still being debugged, it can be useful to comment out an extraneous command, such as **matrix list**, rather than to delete the command, in case there is reason to use it again later.

Debugging can be quicker and simpler if one works with a simplified program. For example, rather than work with the full dataset and many regressors, one might initially work with a small subset of the data and a single regressor. This may also reduce the chance that problems are arising merely because of data problems, such as multi-collinearity.

To further save time, it can be worthwhile to use **/*** and ***/** to comment out those parts of the program that are not needed during the debugging exercise. This is especially the case for computationally intensive tasks that are not necessary, such as graphs to be used in the final analysis but not needed during the program development stage.

A.3.2 Error messages and return code

Stata produces error messages. The message given can be brief, but a fuller explanation can be obtained from the manual or directly from Stata.

For example, if we regress y on z but one or both of these variables does not exist, we get

```
. regress y x
variable y not found
r(111);
```

For a more detailed explanation of the return code 111, type the command

```
. search rc 111
  (output omitted)
```

If a Stata program is being debugged, then program failure can lead to an error message that is not at all helpful. More useful error messages can be given if the code is not embedded in a program. Thus rather than work with a program in the program environment, it can be helpful at first to work with the commands in a Stata do-file but not within a program. For example, a nonprogram version of the `meddiff` program is

```
. * Debug an initial nonprogram version of a program
. tempvar y x diff
. generate `y´ = ltotexp
. generate `x´ = totchr
. generate double `diff´ = `y´ - `x´
. _pctile `diff´, p(50)
. scalar medylx = r(r1)
. display "Median of first variable - second variable = " medylx
Median of first variable - second variable = 4.2230513
```

A.3.3 Trace

The `trace` command traces the execution of a program. To initiate a trace, type the command

```
. set trace on
  (output omitted)
```

To stop the trace, type the command

```
. set trace off
```

The `trace` facility can generate a large amount of output. For this reason, it can be more useful to manually insert commands that give intermediate results. The default is `set trace off`.

B Mata

Mata, introduced in version 9 of Stata, is a powerful matrix programming language comparable to Gauss and Matlab. Compared to the Stata `matrix` commands, it is computationally faster, supports larger matrices (Mata has no restriction on matrix size so the only restriction is computer specific), has a wider range of matrix commands, and has commands that are closer in syntax to the matrix notation used in mathematics.

Mata is a component of Stata that can be used on its own. Additionally, it is possible to blend Stata and Mata functions.

B.1 How to run Mata

Mata commands are usually run in Mata, which is initiated by first giving the `mata` command in Stata. Single Mata commands can be given in Stata, and single Stata commands can be given in Mata.

B.1.1 Mata commands in Mata

Mata can be initiated by the Stata `mata` command. In Mata, the command prompt is a semicolon (:) rather than a period. Mata commands are separated by line breaks or by semicolons. To exit Mata and return to Stata, use the Mata `end` command.

The following sample Mata session creates a 2×2 identity matrix, I, and then displays the elements of matrix I.

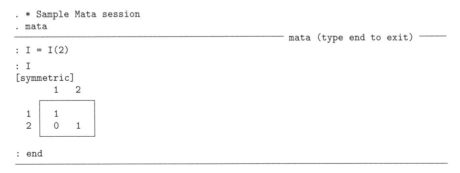

```
. * Sample Mata session
. mata
                                        ──── mata (type end to exit) ────
: I = I(2)
: I
[symmetric]
        1   2

    1 │ 1
    2 │ 0   1

: end
```

For symmetric matrices, such as the identity matrix, only the lower triangle is listed. Here the unlisted $(1, 2)$ element equals the listed $(2, 1)$ element, which is 0.

661

B.1.2   Mata commands in Stata

A single Mata command can be issued in Stata by adding the `mata:` prefix before the Mata command.

For example, to create a 2×2 identity matrix, `I`, and to display the elements of `I`, type the commands

```
. * Mata commands issued from Stata
. mata: I = I(2)

. mata: I
[symmetric]
       1   2

 1 │ 1
 2 │ 0   1
```

B.1.3   Stata commands in Mata

Mata commands are distinct from Stata commands. It is possible to enact a Stata command within a Mata program, however, by using the `stata()` function within Mata.

For example, suppose we are in Mata and want to find the mean of the `ltotexp` variable, which is in the Stata dataset currently in memory. In Stata, we would type the `summarize ltotexp` command. In Mata, we use the `stata()` function with the desired Stata command in double quotes as the argument.

```
. // Stata commands issued from Mata
. mata
───────────────────────────────── mata (type end to exit) ───────
: stata("summarize ltotexp")

    Variable │      Obs        Mean    Std. Dev.        Min         Max
─────────────┼──────────────────────────────────────────────────────────
     ltotexp │      100    4.533688     .8226942    1.098612    5.332719
: end
─────────────────────────────────────────────────────────────────────
```

B.1.4   Interactive versus batch use

There are differences between what is possible in Mata interactive use and what is possible in a Mata program. For example, comments cannot be included in Mata in interactive use.

B.1.5   Mata help

We provide some basic Mata code in this appendix. The two-volume set of Mata manuals is very complete but does not provide as many data-oriented examples as appear in the other Stata manuals.

The `help` command for Mata works at either Stata's dot prompt or Mata's colon prompt.

If you know the name of the matrix command, operator, or function, then type the `help mata` *name* command. For example, if you know that the `det()` function takes the determinant of a matrix, then type the command

```
: help mata det
  (output omitted)
```

In this example, the command was typed in Mata, but exactly the same `help` command can be typed in Stata.

If you do not know the specific name, then it is harder. For example, suppose we want to find help on the category `matrix`. Then no help entry is obtained after `help mata matrix`. However,

```
: help mata m4 matrix
  (output omitted)
```

does work because M-4 is the relevant section of the manuals for Mata. More generally, the command is `help m#` *name*, but this requires knowing the relevant section of the manuals.

Often it is necessary to start with the `help mata` command and then selectively choose from the subsequent entries.

B.2   Mata matrix commands

We present the various basics of creating matrices and matrix operators and functions. Explanatory comments begin with `//` because Mata does not recognize comments beginning with `*`.

B.2.1   Mata matrix input

Matrix input by hand

Matrices can be input by hand. For example, consider a 2×3 matrix `A` with the first row entries 1, 2, and 3, and the second row entries 4, 5, and 6. This can be defined as follows:

```
: // Create a matrix
: A = (1,2,3 \ 4,5,6)
```

Like the `matrix define` command in Stata, a comma is used to separate column entries, and a backslash is used to separate rows.

To see the matrix, simply type the matrix name:

```
: // List a matrix
: A
       1   2   3
    ┌─────────────┐
  1 │ 1   2   3   │
  2 │ 4   5   6   │
    └─────────────┘
```

Identity matrices, unit vectors, and matrices of constants

An $n \times n$ identity matrix is created with $I(n)$. For example,

```
: // Create a 2x2 identity matrix
: I = I(2)
```

A $1 \times n$ row vector with zeros in all entries aside from the ith is created with $e(i,n)$. For example,

```
: // Create a 1x5 unit row vector with 1 in second entry and zeros elsewhere
: e = e(2,5)
: e
       1   2   3   4   5
    ┌─────────────────────┐
  1 │ 0   1   0   0   0   │
    └─────────────────────┘
```

An $r \times c$ matrix of constants equal to the value v is created with $J(r,c,v)$. For example,

```
: // Create a 2x5 matrix with entry 3
: J = J(2,5,3)
: J
       1   2   3   4   5
    ┌─────────────────────┐
  1 │ 3   3   3   3   3   │
  2 │ 3   3   3   3   3   │
    └─────────────────────┘
```

Range operators create vectors with entries that increment by one for each entry by using `a..b` for a row vector and `a::b` for a column vector. For example,

```
: // Create a row vector with entries 8 to 15
: a = 8..15
: a
       1   2    3    4    5    6    7    8
    ┌────────────────────────────────────────┐
  1 │ 8   9   10   11   12   13   14   15   │
    └────────────────────────────────────────┘
```

creates a row vector with the entries 8, 9, ..., 15.

For creation of other standard matrices, type `help m4 standard`.

Matrix input from Stata data

Matrices can be associated with variables in the current Stata dataset in memory by using the Mata st_view() function.

For example, suppose the current Stata dataset includes the variables ltotexp, totchr, and cons. Then

```
: // Create Mata matrices from variables stored in Stata
: st_view(y=., ., "ltotexp")
: st_view(X=., ., ("totchr", "cons"))
```

associates the column vector, y, with the observations on the variable ltotexp, and a matrix, X, with the observations on the variables totchr and cons.

A brief summary of the syntax follows, for the second st_view() function above. The first entry is X=. because this eliminates the need to previously define the vector X. If instead we had first entered simply X, we would have received the error message <istmt>: 3499 X not found. The second entry is a period, meaning that all the observations will be selected. The argument could instead be a list of observations. The third entry is a row vector selecting the particular variables, with variable names given in quotes and commas separating the column entries in the row vector. If totchr and cons were the 31st and 45th entries in the dataset, we could equally well type st_view(X=.,.,(31,45)).

The st_view() function creates a view of the Stata dataset that does not require that the actual data be physically loaded into Mata, saving time and memory. For example, to subsequently form the ordinary least-squares (OLS) estimator $(\mathbf{X'X})^{-1}\mathbf{X'y}$ in Mata, only the $K \times K$ matrix $(\mathbf{X'X})^{-1}$ and the $K \times 1$ matrix $\mathbf{X'y}$ need to be loaded, not the much larger $N \times K$ matrix $\mathbf{X}$.

The related st_data() function does actually load matrices, but this is usually not necessary. As an example,

```
. // Create a Mata matrix from variables stored in Stata
. Xloaded = st_data(., ("totchr", "cons"))
```

creates a matrix, Xloaded, with the ith row the ith observation on the totchr and cons variables.

Matrix input from Stata matrix

Mata matrices can be created from matrices created by Stata commands, using the Mata st_matrix() function. For example,

```
: // Read Stata matrix (created in first line below) into Mata
: stata("matrix define B = I(2)")
: C = st_matrix("B")
```

```
: C
[symmetric]
        1    2

  1 │ 1
  2 │ 0    1
```

The st_matrix() function can also be used to transfer a Mata matrix to Stata; see section B.2.6.

Stata interface functions

Stata interface functions begin with st_ and link matrices and data in Mata with those in Stata. Examples already given are st_view(), st_data(), and st_matrix(). The st_addvar() and st_store() functions are presented in section B.2.6. A summary is given in [M-4] **stata**, and individual st_ functions are given in [M-5] **intro**.

B.2.2   Mata matrix operators

The arithmetic operators for conformable matrices are + to add, - to subtract, * to multiply, and # for the Kronecker product. The multiplication command can also be used for multiplication by a scalar, e.g., 2*A or A*2, and scalar division is possible, e.g., A/2. A scalar can be raised to a scalar power, e.g., a^b. The matrix -A is the negative of A.

A single apostrophe, ´, gives the matrix transpose (or conjugate transpose if the matrix is complex). To compute A´A, we can use A´A or A´*A.

The Kronecker product of two matrices is given by A#B. If A is $m \times n$ and B is $r \times s$, then A#B is $mr \times ns$.

Element-by-element operators

Key arithmetic operators are the colon operators for element-by-element operations. A leading example is element-by-element multiplication of two matrices of the same dimension (the Hadamard product). Then C=A:*B has an ijth element equal to the ijth element of A times the ijth element of B.

Element-by-element multiplication of a column vector and a matrix is possible if they have the same number of rows. Similarly, element-by-element multiplication of a row vector and a matrix is possible if they have the same number of columns. For the column vector case,

```
: // Element-by-element multiplication of matrix by column vector
: b = 2::3
: J = J(2,5,3)
```

```
: b:*J
     1   2   3   4   5

 1   6   6   6   6   6
 2   9   9   9   9   9
```

The column vector **b** has the entries 2 and 3, and the 2×5 matrix J has all entries equal to 3. The first row of matrix J is multiplied by 2 (the first entry in column vector **b**) and the second row of J is multiplied by 3 (the second entry in **b**).

Let **w** be an $N \times 1$ column vector and **X** be an $N \times K$ matrix with ith row $\mathbf{x}'_i$. Then **w:*X** is the $N \times K$ matrix with the ith row $w_i\mathbf{x}'_i$, and (**w:*X**)´**X** is the $K \times K$ matrix equal to $\sum_{i=1}^{N} w_i\mathbf{x}_i\mathbf{x}'_i$.

Other colon operators are available for division (:/), subtraction (:-), power (:^), equality (:==), inequality (:!=), specific inequalities (such as :>=), and (:&), and or (:|) . These operators are a particular advantage of a matrix programming language.

Additional classes of operators are detailed in [M-2] **intro**.

B.2.3 Mata functions

Standard matrix functions have arguments provided in parentheses, ( ).

Scalar and matrix functions

Some matrix commands produce scalars, for example,

```
: // Scalar functions of a matrix
: r = rows(A)

: r
2
```

Commonly used examples include those for matrix determinant (**det()**), rank (**rank()**), and trace (**trace()**). Statistical functions include **mean()**.

Some matrix commands produce matrices by element-by-element transformation. For example,

```
: // Matrix function that returns matrix by element-by-element transformation
: D = sqrt(A)

: D
                  1               2               3

 1                1    1.414213562    1.732050808
 2                2    2.236067977    2.449489743
```

Mathematical functions include absolute value (**abs()**), sign (**sign()**), natural logarithm (**ln()**), exponentiation (**exp()**), log factorial (**lnfactorial()**), modulus (**mod()**), and truncation to integer (`trunc()`). Statistical functions include uniform draws (`runiform()`), standard normal density (`normal()`), and many other densities and cumulative distribution functions.

Some matrix commands produce vectors and matrices by acting on the whole matrix. A leading example is matrix inversion, discussed below. The `mean()` function finds the mean of columns of a matrix, and `corr()` forms a correlation matrix from a variance matrix.

Eigenvalues and eigenvectors of a square matrix can be obtained by using the Mata `eigensystem()` function. For example,

```
: // Calculate eigenvalues and eigenvectors
: E = (1, 2 \ 4, 3)
: lamda = .
: eigvecs = .
: eigensystem(E,eigvecs,lamda)
: lamda
        1    2

1       5   -1

: eigvecs
                    1               2

1       -.447213595     -.707106781
2       -.894427191      .707106781
```

The eigenvalues are in the row vector `lamda`, and the eigenvectors are the corresponding columns of the square matrix `eigvecs`. The command requires that `lamda` and `eigvecs` already exist, so we initialized them as missing values.

Mata has many functions; see [M-4] **intro** for an index and guide to functions.

Matrix inversion

There are several different matrix inversion functions. The `cholinv()` function computes the inverse of a positive-definite symmetric matrix and is the fastest. The `invsym()` function computes the inverse of a real symmetric matrix, `luinv()` computes the inverse of a square matrix, `qrinv()` computes the generalized inverse of a matrix, and `pinv()` computes the Moore–Penrose pseudoinverse.

For the full column rank matrix X, the matrix X´X is positive-definite symmetric, so `cholinv(X´X)` is best. But this function will fail if X´X is not precisely symmetric, because of a rounding error in calculations. The `makesymmetric()` function forms a symmetric matrix by copying elements below the diagonal into the corresponding position above the diagonal. For example,

```
: // Use of makesymmetric() before cholinv()
: F = 0.5*I(2)
```

```
: G = makesymmetric(cholinv(F´F))
: E
[symmetric]
        1     2

1       4
2       0     4
```

B.2.4   Mata cross products

The matrix `cross()` function creates matrix cross products. For example, `cross(X,X)` forms $X´X$, `cross(X,Z)` forms $X´Z$, and `cross(X,w,Z)` forms $X´diag(w)Z$. For the data loaded earlier into X and y, the OLS estimator can be computed as

```
: // Matrix cross product
: beta = (cholinv(cross(X,X)))*(cross(X,y))

: beta
              1

1     .1353097647
2     4.468739434
```

These estimates equal those given in section A.1.2.

The advantages of using `cross()` rather than the arithmetic multiplication operator are faster computation and less memory use. Rows with missing observations are dropped, whereas $X´Z$ will produce missing values everywhere if there are any missing observations. And `cross(X´X)` produces a symmetric result so that there is no longer a need to use the `makesymmetric()` function before `cholinv()` or `invsym()`.

B.2.5   Mata matrix subscripts and combining matrices

The (i, j)th entry in a matrix is denoted by `[i,j]`. For example, to set the $(1, 2)$ entry in matrix A to equal the $(1, 1)$ entry, type the command

```
: // Matrix subscripts
: A[1,2] = A[1,1]

: A
        1   2   3

1       1   1   3
2       4   5   6
```

For a column vector, the ith entry is denoted by `[i,1]` rather than simply `[i]`. Similarly, for a row vector, the jth entry is denoted by `[1,j]` rather than simply `[j]`.

To add columns to a matrix, the horizontal concatenation operator, a comma, is used. Thus A,B adds the columns of B after the columns of A, assuming the two matrices have the same number of rows. For example,

```
: // Combining matrices: add columns
: M = A, A
: M
        1   2   3   4   5   6
    ┌                          ┐
  1 │ 1   1   3   1   1   3    │
  2 │ 4   5   6   4   5   6    │
    └                          ┘
```

To add or append rows to a matrix, the vertical concatenation operator, a backslash, is used. Thus A \ B adds the rows of B after the rows of A, assuming the two matrices have the same number of columns. For example,

```
: // Combining matrices: add rows
: N = A \ A
: N
        1   2   3
    ┌             ┐
  1 │ 1   1   3   │
  2 │ 4   5   6   │
  3 │ 1   1   3   │
  4 │ 4   5   6   │
    └             ┘
```

A submatrix can be extracted from a matrix by using list subscripts that give as a first argument the rows being extracted and as a second argument the columns being extracted. For example, to extract the submatrix formed by rows 1–2 and columns 5–6 of the matrix M, we type

```
: // Form submatrix using list subscripts
: O = M[(1\2), (5::6)]
: O
        1   2
    ┌         ┐
  1 │ 1   3   │
  2 │ 5   6   │
    └         ┘
```

An alternative is to use range subscripts that give the subscripts for the upper-left entry and the lower-right entry of the portion to be extracted. Thus

```
: // Form submatrix using range subscripts
: P = M[|1,5 \ 2,6|]
: P
        1   2
    ┌         ┐
  1 │ 1   3   │
  2 │ 5   6   │
    └         ┘
```

Where both list and range subscripts can be used, range subscripts are preferred because they execute quicker. For more details, see [M-2] **subscripts**.

B.2.6 Transferring Mata data and matrices to Stata

Mata functions beginning with st_ provide an interface with Stata.

Creating Stata matrices from Mata matrices

A Stata matrix can be created from a Mata matrix by using the Mata st_matrix() function.

For example, to create a Stata matrix, Q, from the Mata matrix P and then list the Stata matrix, type

```
: // Output mata matrix to Stata
: st_matrix("Q", P)

: stata("matrix list Q")
Q[2,2]
    c1  c2
r1   1   3
r2   5   6
```

Section 3.8 provides an example where the parameter vector b and the estimate of the variance–covariance matrix of the estimator (VCE) are computed in Stata, passed from Mata to Stata with st_matrix(), and then results are posted and nicely displayed by using the Stata ereturn command.

Creating Stata data from a Mata vector

The st_addvar() function adds a new variable to a Stata dataset, though it creates only the name of the variable and not its values. The st_store() function modifies the values of a variable currently in a Stata dataset. Thus, to create a new variable in Stata and give that new variable values, we type st_addvar() followed by st_store().

Recall that X is a matrix with the variables totchr and cons, and beta is a column vector of OLS coefficients from the regression of ltotexp on totchr and cons. We create the vector of fitted values, yhat, in Mata, pass these to Stata as the ltotexphat variable, and use the summarize command to check the results. We have

```
: // Output Mata matrix to Stata
: yhat = X*beta

: st_addvar("float", "ltotexphat")
  46

: st_store(.,"ltotexphat", yhat)

: stata("summarize ltotexp ltotexphat")
    Variable |      Obs       Mean    Std. Dev.       Min        Max
-------------+--------------------------------------------------------
     ltotexp |      100    4.533688    .8226942   1.098612   5.332719
  ltotexphat |      100    4.533688    .0970792    4.46874   4.874669
```

As expected after OLS regression, the average of the fitted values equals the average of the dependent variable.

B.3   Programming in Mata

Detailed examples using Mata code are presented in section 3.8 and in sections 11.2, 11.7, and 11.8. These examples pass data from Stata to Mata, calculate parameter estimates and estimates of the VCE in Mata, and pass these back to Stata.

Here we present a very introductory treatment of programming in Mata.

B.3.1   Declarations

The examples in sections 11.7 and 11.8 include a Mata program with arguments to be passed to and from the Mata `optimize()` function.

The code in these examples does not declare matrices and scalars ahead of their use. This makes coding easier but makes it more likely that errors may go undetected. For example, if an operation is expected to create a scalar but a matrix is the result, there may be no message to this effect. If instead we had previously declared the expected result to be a scalar, then an error would occur if a matrix was erroneously created.

The following Mata code rewrites the `optimize()` function evaluator `poissonmle` program in section 11.7.3 to declare the types of all program arguments and all other variables used in the program.

```
      :   void poissonmle(real scalar todo,
      >     real rowvector b,
      >     real colvector y,
      >     real matrix X,
      >     real colvector lndensity,
      >     real matrix g,
      >     real matrix H)
      >   {
      >     real colvector Xb
      >     real colvector mu
      >     Xb = X*b´
      >     mu = exp(Xb)
      >     lndensity = -mu + y:*Xb - lnfactorial(y)
      >     if (todo == 0) return
      >     g = (y-mu):*X
      >     if (todo == 1) return
      >     H = - cross(X, mu, X)
      >   }
```

B.3.2   Mata program

As an example, we create a Mata program, `calcsum`, that calculates the column sum of a column vector. This example, based on the example in [M-1] **ado**, is purely illustrative because the Mata `colsum()` function does this anyway.

The column vector x is obtained from a variable named **varname** in the Stata dataset currently in memory by using the `st_view()` function. The **varname** string is a program argument supplied when the program is called. The actual calculation of the column sum is done with the Mata `colsum()` function. The result is put in the real scalar `resultissum`, a second program argument. To apply the program to the `ltotexp` variable, we call the `calcsum` program with the `"ltotexp"` and `sum` arguments. The result is in `sum`, and to see the result, we simply type `sum`. We have

```
. mata:
                                    ─── mata (type end to exit) ───
: void calcsum(varname, resultissum)
>   {
>      st_view(x=., ., varname)
>      resultissum = colsum(x)
>   }
: sum = .
: calcsum("ltotexp", sum)
: sum
  453.3688121
: end
```

The result, 453.3688, is that expected because the sample mean of the 100 observations on `ltotexp` was 4.533688 from output given in section B.2.6.

B.3.3   Mata program with results output to Stata

The preceding Mata program passes the result, `resultissum`, back to Mata. We next consider a variation that passes the result, renamed `sum`, to Stata.

To transfer the result to Stata, we use the Mata `st_numscalar()` function and drop the second argument in the `calcsum` program because the result is no longer passed to Mata. Because the result is now in Stata, we need to use the Stata `display` command to display the result. We have

```
. mata:
                                    ─── mata (type end to exit) ───
: void function calcsum2(varname)
>   {
>      st_view(x=., ., varname)
>      st_numscalar("r(sum)",colsum(x))
>   }
: calcsum2("ltotexp")
: stata("display r(sum)")
453.36881
: end
```

B.3.4   Stata program that calls a Mata program

The preceding two programs call the Mata program from within Mata. We now create a Stata program, `varsum`, that calls the Mata program `calcsum2` from Stata.

The `varsum` program uses standard Stata syntax (see section A.2.7) rather than positional arguments. This syntax recognizes the argument in the call `varsum ltotexp` as a variable name that is placed in `varlist`. The Mata program `calcsum2`, already defined in the preceding section, is called with `varname` being the variable name in `varlist`. We have

```
. program varsum
  1.    version 11
  2.    syntax varname
  3.    mata: calcsum2("`varlist'")
  4.    display r(sum)
  5. end

. varsum ltotexp
453.36881
```

B.3.5 Using Mata in ado-files

The main construct for writing new commands in Stata is a Stata ado-file. When computation in Mata is convenient, the ado-file can include Mata code or call a Mata function.

A Mata function defined in an ado-file requires compilation every time it is called. To save computer time, compiled functions can be reused without the need for recompilation by using the `mata mosave` and `mata mlib` commands. For details, see [M-1] **ado**, which presents the preceding column sum example in much more generality.

Glossary of abbreviations

2SLS — two-stage least squares

3SLS — three-stage least squares

AIC — Akaike information criterion

AME — average marginal effect

ARUM — additive random-utility model

BC — bias-corrected

BCa — bias-corrected accelerated

BIC — Bayesian information criterion

CL — conditional logit

CV — coefficient of variation

DGP — data-generating process

FD — first difference

FE — fixed effects

FGLS — feasible generalized least squares

FMM — finite-mixture models

GLM — generalized linear models

GLS — generalized least squares

GMM — generalized method of moments

HAC — heteroskedasticity- and autocorrelation-consistent

IIA — independence of irrelevant alternatives

IM — information matrix

IV — instrumental variables

JIVE — jackknife instrumental-variables estimator

LEF — linear exponential family

LIML — limited-information maximum likelihood

LM — Lagrange multiplier

LPM — linear probability model

LR — likelihood ratio

LS — least squares

LSDV — least-squares dummy variable

ME — marginal effect

MEM — marginal effect at mean

MEPS — Medical Expenditure Panel Survey

MER — marginal effect at representative value

ML — maximum likelihood

MM — method of moments

MNL — multinomial logit

MNP — multinomial probit

MSE — mean squared error

MSL — maximum simulated likelihood

MSS — model sum of squares

NB — negative binomial

NL — nested logit

NLIV — nonlinear instrumental variables

NLS — nonlinear least squares

NR — Newton–Raphson

OLS — ordinary least squares

PA — population averaged

PFGLS — pooled feasible generalized least squares

PSID — Panel Study of Income Dynamics

PSU — primary sampling unit

QCR — quantile count regression

QR — quantile regression

RE — random effects

RPL — random-parameters logit

RSS — residual sum of squares

SUR — seemingly unrelated regressions

TSS — total sum of squares

VCE — variance–covariance matrix of the estimator

WLS — weighted least squares

ZINB — zero-inflated negative binomial

ZIP — zero-inflated Poisson

ZTNB — zero-truncated negative binomial

ZTP — zero-truncated Poisson

.

References

Amemiya, T. 1981. Qualitative response models: A survey. *Journal of Economic Literature* 19: 1483–1536.

Anderson, T. W., and C. Hsiao. 1981. Estimation of dynamic models with error components. *Journal of the American Statistical Association* 76: 598–606.

Andrews, D. W. K. 1988. Chi-square diagnostic tests for econometric models: Introduction and applications. *Journal of Econometrics* 37: 135–156.

Andrews, D. W. K., and M. Y. Buchinsky. 2000. On the number of bootstrap repetitions for BC_a confidence intervals. Cowles Foundation Discussion Papers 1250, Cowles Foundation, Yale University.

Andrews, D. W. K., M. J. Moreira, and J. H. Stock. 2007. Performance of conditional Wald tests in IV regression with weak instruments. *Journal of Econometrics* 139: 116–132.

Angrist, J. D., G. W. Imbens, and A. B. Krueger. 1999. Jackknife instrumental variables estimation. *Journal of Applied Econometrics* 14: 57–67.

Arellano, M. 2003. *Panel Data Econometrics*. New York: Oxford University Press.

Arellano, M., and S. Bond. 1991. Some tests of specification for panel data: Monte Carlo evidence and an application to employment equations. *Review of Economic Studies* 58: 277–297.

Arellano, M., and O. Bover. 1995. Another look at the instrumental variable estimation of error-components models. *Journal of Econometrics* 68: 29–51.

Azevedo, J. P. 2004. grqreg: Stata module to graph the coefficients of a quantile regression. Statistical Software Components S437001, Boston College Department of Economics. Downloadable from http://ideas.repec.org/c/boc/bocode/s437001.html.

Baltagi, B. H. 2008. *Econometric Analysis of Panel Data*. 4th ed. Chichester, UK: Wiley.

Baltagi, B. H., J. M. Griffin, and W. Xiong. 2000. To pool or not to pool: Homogeneous versus heterogeneous estimators applied to cigarette demand. *Review of Economics and Statistics* 82: 117–126.

Baltagi, B. H., and S. Khanti-Akom. 1990. On efficient estimation with panel data: An empirical comparison of instrumental variables. *Journal of Applied Econometrics* 5: 401–406.

Bartus, T. 2005. Estimation of marginal effects using margeff. *Stata Journal* 5: 309–329.

Baum, C. F., M. E. Schaffer, and S. Stillman. 2007. Enhanced routines for instrumental variables/generalized method of moments estimation and testing. *Stata Journal* 7: 465–506.

Beck, N., and J. N. Katz. 1995. What to do (and not to do) with time-series cross-section data. *American Political Science Review* 89: 634–647.

Berry, S. T. 1994. Estimating discrete-choice models of product differentiation. *Rand Journal of Economics* 25: 242–262.

Bhattacharya, D. 2005. Asymptotic inference from multi-stage samples. *Journal of Econometrics* 126: 145–171.

Blackburne, E. F., and M. W. Frank. 2007. Estimation of nonstationary heterogeneous panels. *Stata Journal* 7: 197–208.

Blomquist, S., and M. Dahlberg. 1999. Small sample properties of LIML and jackknife IV estimators: Experiments with weak instruments. *Journal of Applied Econometrics* 14: 69–88.

Blundell, R., and S. Bond. 1998. Initial conditions and moment restrictions in dynamic panel data models. *Journal of Econometrics* 87: 115–143.

Bornhorst, F., and C. F. Baum. 2006. levinlin: Stata module to perform Levin–Lin–Chu panel unit root test. Statistical Software Components S419702, Boston College Department of Economics. Downloadable from http://ideas.repec.org/c/boc/bocode/s419702.html.

———. 2007. ipshin: Stata module to perform Im–Pesaran–Shin panel unit root test. Statistical Software Components S419704, Boston College Department of Economics. Downloadable from http://ideas.repec.org/c/boc/bocode/s419704.html.

Brady, T. 2002. reformat: Stata module to reformat regression output. Statistical Software Components S426304, Boston College Department of Economics. Downloadable from http://ideas.repec.org/c/boc/bocode/s426304.html.

Breitung, J., and M. H. Pesaran. 2005. Unit roots and cointegration in panels. Manuscript. Downloadable from http://ideas.repec.org/p/ces/ceswps/_1565.html.

Cameron, A. C., J. B. Gelbach, and D. L. Miller. 2008. Bootstrap-based improvements for inference with clustered errors. *Review of Economics and Statistics* 90: 414–427.

Cameron, A. C., and P. K. Trivedi. 1998. *Regression Analysis of Count Data*. Cambridge: Cambridge University Press.

————. 2005. *Microeconometrics: Methods and Applications*. Cambridge: Cambridge University Press.

Cameron, A. C., and F. A. G. Windmeijer. 1997. An R-squared measure of goodness of fit for some common nonlinear regression models. *Journal of Econometrics* 77: 329–342.

Carson, R. T., and Y. Sun. 2007. The Tobit model with a non-zero threshold. *Econometrics Journal* 10: 488–502.

Cornwell, C., and P. Rupert. 1988. Efficient estimation with panel data: An empirical comparison of instrumental variables estimators. *Journal of Applied Econometrics* 3: 149–155.

Cox, N. J. 2005. Speaking Stata: The protean quantile plot. *Stata Journal* 5: 442–460.

Cragg, J. G., and S. G. Donald. 1993. Testing identifiability and specification in instrumental variable models. *Econometric Theory* 9: 222–240.

Davidson, J. 2000. *Econometric Theory*. Oxford: Blackwell.

Davidson, R., and J. G. MacKinnon. 2004. *Econometric Theory and Methods*. New York: Oxford University Press.

————. 2006. The case against JIVE. *Journal of Applied Econometrics* 21: 827–833.

Davison, A. C., and D. V. Hinkley. 1997. *Bootstrap Methods and Their Application*. Cambridge: Cambridge University Press.

Deb, P. 2007. fmm: Stata module to estimate finite mixture models. Statistical Software Components S456895, Boston College Department of Economics. Downloadable from http://ideas.repec.org/c/boc/bocode/s456895.html.

Deb, P., M. K. Munkin, and P. K. Trivedi. 2006. Private insurance, selection, and health care use: A bayesian analysis of a Roy-type model. *Journal of Business and Economic Statistics* 24: 403–415.

Deb, P., and P. K. Trivedi. 2002. The structure of demand for medical care: latent class versus two-part models. *Journal of Health Economics* 21: 601–625.

————. 2006. Maximum simulated likelihood estimation of a negative binomial regression model with multinomial endogenous treatment. *Stata Journal* 6: 246–255.

Driscoll, J. C., and A. C. Kraay. 1998. Consistent covariance matrix estimation with spatially dependent panel data. *Review of Economics and Statistics* 80: 549–560.

Drukker, D. M. 2002. Bootstrapping a conditional moments test for normality after tobit estimation. *Stata Journal* 2: 125–139.

————. 2008. Treatment effects highlight use of population-averaged estimates. Unpublished manuscript.

Drukker, D. M., and R. Gates. 2006. Generating Halton sequences using Mata. *Stata Journal* 6: 214–228.

Duan, N. 1983. Smearing estimate: A nonparametric retransformation method. *Journal of the American Statistical Association* 78: 605–610.

Efron, B., and R. J. Tibshirani. 1993. *An Introduction to the Bootstrap*. New York: Chapman & Hall.

Goldstein, H. 1987. Multilevel covariance component models. *Biometrika* 74: 430–431.

Gould, W., J. Pitblado, and W. Sribney. 2006. *Maximum Likelihood Estimation with Stata*. 3rd ed. College Station, TX: Stata Press.

Greene, W. H. 2003. *Econometric Analysis*. 5th ed. Upper Saddle River, NJ: Prentice Hall.

———. 2008. *Econometric Analysis*. 6th ed. Upper Saddle River, NJ: Prentice Hall.

Hahn, J., and J. Hausman. 2002. A new specification test for the validity of instrumental variables. *Econometrica* 70: 163–189.

Hall, A. 1987. The information matrix test for the linear model. *Review of Economic Studies* 54: 257–263.

Hardin, J. W., and J. M. Hilbe. 2007. *Generalized Linear Models and Extensions*. 2nd ed. College Station, TX: Stata Press.

Herriges, J. A., and C. L. Kling. 1999. Nonlinear income effects in random utility models. *Review of Economics and Statistics* 81: 62–72.

Hilbe, J. 2005a. hnblogit: Stata module to estimate negative binomial-logit hurdle regression. Statistical Software Components S456401, Boston College Department of Economics. Downloadable from http://ideas.repec.org/c/boc/bocode/s456401.html.

———. 2005b. hplogit: Stata module to estimate Poisson-logit hurdle regression. Statistical Software Components S456405, Boston College Department of Economics. Downloadable from http://ideas.repec.org/c/boc/bocode/s456405.html.

Hoechle, D. 2007. Robust standard errors for panel regressions with cross-sectional dependence. *Stata Journal* 7: 281–312.

Hole, A. R. 2007. Fitting mixed logit models by using maximum simulated likelihood. *Stata Journal* 7: 388–401.

Holtz-Eakin, D., W. Newey, and H. S. Rosen. 1988. Estimating vector autoregressions with panel data. *Econometrica* 56: 1371–1395.

Horowitz, J. L. 2001. The bootstrap. In *Handbook of Econometrics*, ed. J. J. Heckman and E. Leamer, vol. 5, 3159–3228. Amsterdam: Elsevier.

Hosmer, D. W., Jr., and S. Lemeshow. 1980. Goodness-of-fit tests for the multiple logistic regression model. *Communications in Statistics: Theory and Methods* 9: 1043–1069.

———. 2000. *Applied Logistic Regression*. 2nd ed. New York: Wiley.

Hsiao, C. 2003. *Analysis of Panel Data*. 2nd ed. Cambridge: Cambridge University Press.

Huber, P. J. 1965. The behavior of maximum likelihood estimates under nonstandard conditions. In *Proceedings of the Fifth Berkeley Symposium on Mathematical Statistics and Probability*, vol. 1, 221–233. Berkeley, CA: University of California Press.

Im, K. S., M. H. Pesaran, and Y. Shin. 2003. Testing for unit roots in heterogeneous panels. *Journal of Econometrics* 115: 53–74.

Jann, B. 2005. Making regression tables from stored estimates. *Stata Journal* 5: 288–308.

———. 2007. Making regression tables simplified. *Stata Journal* 7: 227–244.

Jolliffe, D., B. Krushelnytskyy, and A. Semykina. 2000. sg153: Censored least absolute deviations estimator: CLAD. *Stata Technical Bulletin* 58: 13–16. Reprinted in *Stata Technical Bulletin Reprints*, vol. 10, pp. 240–244. College Station, TX: Stata Press.

Keshk, O. M. G. 2003. CDSIMEQ: A program to implement two-stage probit least squares. *Stata Journal* 3: 157–167.

Koenker, R. 2005. *Quantile Regression*. Cambridge: Cambridge University Press.

Kreuter, F., and R. Valliant. 2007. A survey on survey statistics: What is done and can be done in Stata. *Stata Journal* 7: 1–21.

Lee, M. 2002. *Panel Data Econometrics: Methods-of-Moments and Limited Dependent Variables*. San Diego, CA: Academic Press.

Levin, A., C.-F. Lin, and C.-S. J. Chu. 2002. Unit root tests in panel data: Asymptotic and finite-sample properties. *Journal of Econometrics* 108: 1–24.

Liang, K.-Y., and S. L. Zeger. 1986. Longitudinal data analysis using generalized linear models. *Biometrika* 73: 13–22.

Long, J. S., and J. Freese. 2006. *Regression Models for Categorical Dependent Variables Using Stata*. 2nd ed. College Station, TX: Stata Press.

Machado, J. A. F., and J. M. C. Santos Silva. 2005. Quantiles for counts. *Journal of the American Statistical Association* 100: 1226–1237.

MacKinnon, J. G. 2002. Bootstrap inference in econometrics. *Canadian Journal of Economics* 35: 615–645.

Manning, W. G., J. P. Newhouse, N. Duan, E. B. Keeler, and A. Leibowitz. 1987. Health insurance and the demand for medical care: Evidence from a randomized experiment. *American Economic Review* 77: 251–277.

McCullagh, P., and J. A. Nelder. 1989. *Generalized Linear Models*. 2nd ed. London: Chapman & Hall.

Mikusheva, A., and B. P. Poi. 2006. Tests and confidence sets with correct size when instruments are potentially weak. *Stata Journal* 6: 335–347.

Miller, G. E. 1991. Asymptotic test statistics for coefficients of variation. *Communications in Statistics: Theory and Methods* 20: 3351–3363.

Miranda, A. 2007. qcount: Stata program to fit quantile regression models for count data. Statistical Software Components S456714, Boston College Department of Economics. Downloadable from http://ideas.repec.org/c/boc/bocode/s456714.html.

Mitchell, M. N. 2008. *A Visual Guide to Stata Graphics*. 2nd ed. College Station, TX: Stata Press.

Newey, W. K. 1985. Maximum likelihood specification testing and conditional moment tests. *Econometrica* 53: 1047–1070.

———. 1987. Efficient estimation of limited dependent variable models with endogenous explanatory variables. *Journal of Econometrics* 36: 231–250.

Newey, W. K., and K. D. West. 1987. A simple, positive semi-definite, heteroskedasticity and autocorrelation consistent covariance matrix. *Econometrica* 55: 703–708.

Pagan, A., and F. Vella. 1989. Diagnostic tests for models based on individual data: A survey. Special issue, *Journal of Applied Econometrics* 4: S229–S259.

Papps, K. L. 2006. outsum: Stata module to write formatted descriptive statistics to a text file. Statistical Software Components S456780, Boston College Department of Economics. Downloadable from http://ideas.repec.org/c/boc/bocode/s456780.html.

Pesaran, M. H., Y. Shin, and R. P. Smith. 1999. Pooled mean group estimation of dynamic heterogeneous panels. *Journal of the American Statistical Association* 94: 621–634.

Pesaran, M. H., and R. Smith. 1995. Estimating long-run relationships from dynamic heterogeneous panels. *Journal of Econometrics* 68: 79–113.

Poi, B. P. 2004. From the help desk: Some bootstrapping techniques. *Stata Journal* 4: 312–328.

———. 2006. Jackknife instrumental variables estimation in Stata. *Stata Journal* 6: 364–376.

Politis, D. N., J. P. Romano, and M. Wolf. 1999. *Subsampling*. New York: Springer.

Powell, J. L. 1984. Least absolute deviations estimation for the censored regression model. *Journal of Econometrics* 25: 303–325.

Press, W. H., S. A. Teukolsky, W. T. Vetterling, and B. P. Flannery. 1992. *Numerical Recipes in C: The Art of Scientific Computing.* 2nd ed. Cambridge: Cambridge University Press.

Rabe-Hesketh, S., and A. Skrondal. 2008. *Multilevel and Longitudinal Modeling Using Stata.* 2nd ed. College Station, TX: Stata Press.

Rabe-Hesketh, S., A. Skrondal, and A. Pickles. 2002. Reliable estimation of generalized linear mixed models using adaptive quadrature. *Stata Journal* 2: 1–21.

Salgado-Ugarte, I. H., M. Shimizu, and T. Taniuchi. 1996. snp10: Nonparametric regression: Kernel, WARP, and k-NN estimators. *Stata Technical Bulletin* 30: 15–30. Reprinted in *Stata Technical Bulletin Reprints*, vol. 5, pp. 197–218. College Station, TX: Stata Press.

Schaffer, M. E. 2007. xtivreg2: Stata module to perform extended IV/2SLS, GMM and AC/HAC, LIML, and k-class regression for panel data models. Statistical Software Components S456501, Boston College Department of Economics. Downloadable from http://ideas.repec.org/c/boc/bocode/s456501.html.

Simcoe, T. 2007. xtpqml: Stata module to estimate fixed-effects Poisson (quasi-ML) regression with robust standard errors. Statistical Software Components S456821, Boston College Department of Economics. Downloadable from http://ideas.repec.org/c/boc/bocode/s456821.html.

Skeels, C. L., and F. Vella. 1999. A Monte Carlo investigation of the sampling behavior of conditional moment tests in tobit and probit models. *Journal of Econometrics* 92: 275–294.

Staiger, D., and J. H. Stock. 1997. Instrumental variables regression with weak instruments. *Econometrica* 65: 557–586.

Stock, J. H., and M. Yogo. 2005. Testing for weak instruments in linear IV regression. In *Identification and Inference for Econometric Models: Essays in Honor of Thomas Rothenberg*, ed. D. W. K. Andrews and J. H. Stock, 80–108. Cambridge: Cambridge University Press.

Stukel, T. A. 1988. Generalized logistic models. *Journal of the American Statistical Association* 83: 426–431.

Train, K. 2003. *Discrete Choice Methods with Simulation.* Cambridge: Cambridge University Press.

Verbeek, M. 2008. *A Guide to Modern Econometrics.* 3rd ed. Chichester, UK: Wiley.

Vuong, Q. H. 1989. Likelihood ratio tests for model selection and nonnested hypotheses. *Econometrica* 57: 307–333.

White, H. 1980. A heteroskedasticity-consistent covariance matrix estimator and a direct test for heteroskedasticity. *Econometrica* 48: 817–838.

Williams, R. 2006. Generalized ordered logit/partial proportional odds models for ordinal dependent variables. *Stata Journal* 6: 58–82.

Windmeijer, F. 2005. A finite sample correction for the variance of linear efficient two-step GMM estimators. *Journal of Econometrics* 126: 25–51.

Wolfe, F. 2002. fsum: Stata module to generate and format summary statistics. Statistical Software Components S426501, Boston College Department of Economics. Downloadable from http://ideas.repec.org/c/boc/bocode/s426501.html.

Wooldridge, J. M. 2002. *Econometric Analysis of Cross Section and Panel Data.* Cambridge, MA: MIT Press.

Author index

Subject index